Lecture Notes in Mathematics

Edited by A. Dold, F. Takens and B. Teissier

Editorial Policy
for the publication of monographs

1. Lecture Notes aim to report new developments in all areas of mathematics – quickly, informally and at a high level. Monograph manuscripts should be reasonably self-contained and rounded off. Thus they may, and often will, present not only results of the author but also related work by other people. They may be based on specialized lecture courses. Furthermore, the manuscripts should provide sufficient motivation, examples and applications. This clearly distinguishes Lecture Notes from journal articles or technical reports which normally are very concise. Articles intended for a journal but too long to be accepted by most journals, usually do not have this "lecture notes" character. For similar reasons it is unusual for doctoral theses to be accepted for the Lecture Notes series.

2. Manuscripts should be submitted (preferably in duplicate) either to one of the series editors or to Springer-Verlag, Heidelberg. In general, manuscripts will be sent out to 2 external referees for evaluation. If a decision cannot yet be reached on the basis of the first 2 reports, further referees may be contacted: the author will be informed of this. A final decision to publish can be made only on the basis of the complete manuscript, however a refereeing process leading to a preliminary decision can be based on a pre-final or incomplete manuscript. The strict minimum amount of material that will be considered should include a detailed outline describing the planned contents of each chapter, a bibliography and several sample chapters.
Authors should be aware that incomplete or insufficiently close to final manuscripts almost always result in longer refereeing times and nevertheless unclear referees' recommendations, making further refereeing of a final draft necessary.
Authors should also be aware that parallel submission of their manuscript to another publisher while under consideration for LNM will in general lead to immediate rejection.

3. Manuscripts should in general be submitted in English.
Final manuscripts should contain at least 100 pages of mathematical text and should include

- a table of contents;
- an informative introduction, with adequate motivation and perhaps some historical remarks: it should be accessible to a reader not intimately familiar with the topic treated;
- a subject index: as a rule this is genuinely helpful for the reader.

Continued on back inside cover

Lecture Notes in Mathematics 1741

Springer
Berlin
Heidelberg
New York
Barcelona
Hong Kong
London
Milan
Paris
Singapore
Tokyo

Eric Lombardi

Oscillatory Integrals and Phenomena Beyond all Algebraic Orders

with Applications to Homoclinic Orbits in Reversible Systems

Springer

Author

Eric Lombardi
Institut Non Linéaire de Nice
1361 route des Lucioles
Sophia Antipolis
F-06560 Valbonne
France

E-mail: lombardi@inln.cnrs.fr

Cataloging-in-Publication Data applied for

Die Deutsche Bibliothek - CIP-Einheitsaufnahme

Lombardi, Eric:
Oscillatory integrals and phenomena beyond all algebraic orders : with applications to homoclinic orbits in reversible systems / Eric Lombardi. - Berlin ; Heidelberg ; New York ; Barcelona ; Hong Kong ; London ; Milan ; Paris ; Singapore ; Tokyo : Springer, 2000
(Lecture notes in mathematics ; 1741)
ISBN 3-540-67785-2

Mathematics Subject Classification (2000): 34C23, 34C37, 37G25, 37G40, 76B25

ISSN 0075-8434
ISBN 3-540-67785-2 Springer-Verlag Berlin Heidelberg New York

Typesetting: Camera-ready TEX output by the author
Printed on acid-free paper SPIN: 10724282 41/3142/du 543210

A mes vieux,
qui m'ont appris que la valeur d'un homme
ne tient ni à ses titres, ni à ses richesses,
mais à la passion qu'il met dans son travail.

Preface

In many physical problems one can construct "solutions" in terms of power series of a small parameter ε which reads

$$Y(t,\varepsilon) = \sum_{n=0}^{k} \varepsilon^n Y_n(t) + o(\varepsilon^k).$$

In some cases, this power series is defined at any order but it diverges. This divergence may express that the system has a solution for which such an expansion misses some exponentially small term like $e^{-1/\varepsilon^2} Z(t)$. *Such a term is said to lie beyond any algebraic orders.* This kind of phenomena typically occurs for physical problems governed by reversible vector fields near resonances because of the degeneracy induced by the symmetry and the resonance. Such problems, in which these very small terms have great practical interest, are known in many branches of science including dendritic crystals growth, quantum tunneling, KAM theory, theory of water-waves (which was our original motivation for this work) and others. A collection of these apparently unrelated problems can be found in [STL91].

Many works dealing with this subject use the Matched Asymptotic Expansions (M.A.E.) method for catching the exponentially small terms. Although the "beyond all order asymptotics" are *convincing heuristics* which can be used for a very large class of problems, they provide *no rigorous proof.* In these notes we present *rigorous mathematical methods* partially inspired by M.A.E. approach, which enable a systematic study of these nonlinear problems in finite or infinite dimensions. For instance, treating vector fields as perturbation of their normal forms by higher order terms, we prove that near a $0^{2+}i\omega$ resonant fixed point (the definition and the nomenclature of the resonances are given in chapter 3), any reversible analytic vector field in finite or infinite dimensions, admits reversible solutions homoclinic to periodic orbits of *exponentially small amplitude.* Generically in such a case, there is no reversible homoclinic connections to the fixed points, whereas the normal form systems at any order do admit such connections. Similar results are proved for the $(i\omega_0)^2\ i\omega_1$ resonance. Applications of these results to water waves are given in the third part of the book. The crucial point of the analysis is the description of the analytic continuation of the solutions which enable to catch exponentially small terms which are "hidden beyond all orders on the real axis". The main difficulty is to determine the complex singularities of solutions of non linear differential equation or at least to describe very precisely their behavior near the singularities.

I am grateful to Klaus Kirchgässner for the interest he showed in inviting me to write this book. All his comments and suggestions helped me to improve it.

I am indebted to Alan Champneys, who read the first draft of the manuscript, corrected errors and suggested improvements concerning contents and style.

Finally, I should like to extend very special thanks to my former advisor Gérard Iooss who played a mayor role in my education in mathematics and their applications and who encouraged me whenever needed. In any circumstances he found time to listen to me and he was the first reader of all the chapters.

Eric Lombardi
Nice, April 2000

List of Figures

Table of Contents

Part II. Toolbox for reversible systems studied near resonances

Part III. Applications to homoclinic orbits near resonances in reversible systems

1. Introduction

1.1 A little toy model : from phenomena beyond any algebraic order to oscillatory integrals

This work is devoted to "phenomena beyond any algebraic order" in dynamical systems. To illustrate how these phenomena occur in dynamical system we begin with a toy model in $\mathbb{R}^3$ which reads

$$\begin{cases} \dfrac{dA}{dx} &= -\omega B, \\ \dfrac{dB}{dx} &= \omega A + \rho(\varepsilon^2 - \alpha^2), \\ \dfrac{d\alpha}{dx} &= \varepsilon^2 - \alpha^2. \end{cases} \tag{1.1}$$

where $(A, B, \alpha) \in \mathbb{R}^3$ and ω is a positive fixed number. This system has two parameters, $\varepsilon > 0$ which is small, and ρ which can be any real number.

This system develops typical "phenomena beyond any algebraic order" which can be seen without any difficulty because all the bounded solutions can be computed explicitly. Hence, this system enables to understand what kind of mathematical difficulties are hidden behind these phenomena, and what kind of mathematical tools have to be developed for dealing with any system where such phenomena are involved.

To study this system it is more convenient to set

$$\alpha = \varepsilon\beta, \qquad Z = A + \mathrm{i}B, \qquad x = t/\varepsilon,$$

and to rewrite (1.1) with these new coordinates. We obtain

$$\begin{cases} \dfrac{dZ}{dt} &= \dfrac{\mathrm{i}\omega}{\varepsilon} Z + \mathrm{i}\rho\varepsilon(1 - \beta^2), \\ \dfrac{d\beta}{dt} &= 1 - \beta^2. \end{cases} \tag{1.2}$$

This system is *reversible*, which means that if $(Z(t), \beta(t))$ is a solution, then $S(Z(-t), \beta(-t))$ is another solution where S is the reflection given by $S(Z, \beta) = (\overline{Z}, -\beta)$. We call a *reversible solution* of (1.2) a solution which satisfies $S(Z(-t), \beta(-t)) = (Z(t), \beta(t))$.

For $\rho = 0$, the system is uncoupled and the phase portrait is fully symmetric (See Fig. 1.1) : the truncated system admits two families of periodic orbits *of arbitrary size* given by

$$P^{\pm}_{k,\varphi}(t) = (k e^{i(\omega t/\varepsilon)+i\varphi}, \pm 1),$$

($k = 0$ corresponds to the two fixed points) and a family of heteroclinic orbits connecting $P^{\pm}_{k,\varphi}$ which read

$$H_{k,\varphi,t_0}(t) = (k e^{i(\omega t/\varepsilon)+i\varphi}, \tanh(t + t_0)).$$

Among these solutions there is a one parameter family of reversible ones given by $H_{k,0,0}$ with $k \in \mathbb{R}$ ($H_{-k,0,0}(t) = H_{k,\pi,0}(t)$) and there is a unique (up to a phase shift) front connecting $(0, \pm 1)$ which reads $h(t) = H_{0,0,0}(t) = (0, \tanh t)$. At last, observe that the periodic orbits are exchanged by symmetry, i.e. $SP^{+}_{k,\varphi}(t) = P^{-}_{k,-\varphi}(-t)$. The question is then to determine how this

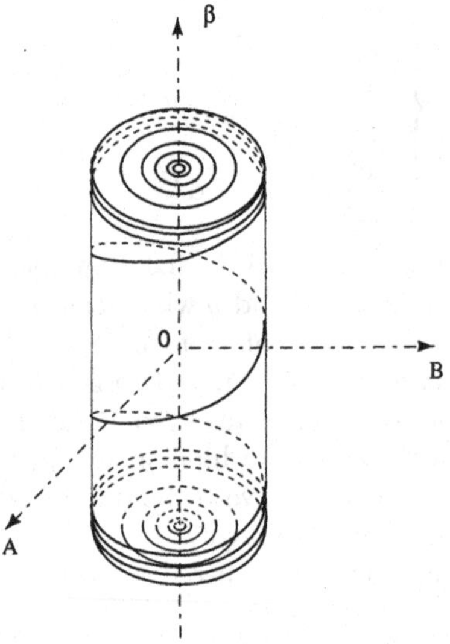

Fig. 1.1. Phase portrait for $\rho = 0$

phase portrait is deformed by the higher order terms ($\rho \neq 0$). In other words, do the previously found orbits persist for the perturbed system?

For periodic solutions the answer is yes. Moreover, because of the particular form of the perturbation, these orbits persist without any deformations. Indeed $P^{\pm}_{k,\varphi}$ is still a solution of the perturbed system. For more general perturbation, the periodic orbits are deformed, and the proof of their persistence can be given by the Lyapunov Schmidt method.

For the heteroclinic connections the situation is not so clear. A first question is the persistence of the *reversible* connections.

A preliminary way to make up one's mind is to look for a reversible front connecting the two fixed points $(0,-1)$ and $(0,1)$ using a classical asymptotic approach: one can compute a formal reversible solution of (1.2) connecting $(0,-1)$ to $(0,1)$ in the form

$$Y(t) = \left(\sum_{n\geq 0} \varepsilon^n Z_n(t), \tanh t\right)$$

where

$$Z_0 = Z_1 = 0 \qquad \text{and} \qquad Z_n(t) = -\frac{\mathrm{i}}{(\mathrm{i}\omega)^{n-1}} \frac{d^{n-2}}{dt^{n-2}}\left(\frac{1}{\cosh^2(t)}\right) \text{ for } n \geq 2.$$

So this first approach predicts the persistence of the reversible front (provided that the power series converges.)

A second approach for studying the persistence of the reversible connection is to formulate the problem in terms of stable manifolds and to illustrate it geometrically. We denote by $\mathcal{W}^{+,*}_{\mathrm{s},k}$ the stable manifold of the periodic solution P_k^+ of the truncated system. The manifold $\mathcal{W}^{+,*}_{\mathrm{s},k}$ is a two dimensional cylinder in $\mathbb{R}^3$ (one dimension for time and one dimension for the phase shift).The radius of this cylinder is the size of the periodic orbit, i.e. k. A fundamental remark is the following :

Since P_k^+ and P_k^- are exchanged by symmetry, there exists a reversible heteroclinic orbit connecting P_k^- and P_k^+ if and only if the stable manifold of P_k^+ intersects the symmetry space $\mathcal{E}^+ = \{Y/SY = Y\}$.

In this case, $\mathcal{E}^+$ is a line in $\mathbb{R}^3$. For $k > 0$, the intersection consists of two points which leads to the two reversible heteroclinic orbits connecting the periodic orbits P_k^- and P_k^+. For $k = 0$, there is a unique point of intersection, which leads to the unique reversible front of the truncated system. Figure 1.2 represents the intersection between $\mathcal{W}^{+,*}_{\mathrm{s},k}$ and $\mathcal{E}^+$ in two case : $k > 0$ and $k = 0$ for a fixed value of ε.

Let us denote by $\mathcal{W}^{+}_{\mathrm{s},k}$ the stable manifold of the periodic solution P_k^+ of the perturbed system. The stable manifold $\mathcal{W}^{+}_{\mathrm{s},k}$ is obtained by perturbation of the stable manifold $\mathcal{W}^{+,*}_{\mathrm{s},k}$ of the periodic solution of the truncated system. On figure 1.2 we can observe that for $k = 0$ the situation is not robust: unless there is a miracle, there should not exist any reversible front for the perturbed system, whereas for k large enough, the two points of intersection between $\mathcal{W}^{+,*}_{\mathrm{s},k}$ and $\mathcal{E}^+$ should persist, i.e. there should exist two reversible heteroclinic orbits connecting P_k^- and P_k^+. The natural question is then to determine, for a fixed value of the bifurcation parameter ε, the smallest size $k_c(\varepsilon)$ of the periodic solutions P_k^- and P_k^+ which admit a reversible heteroclinic orbit

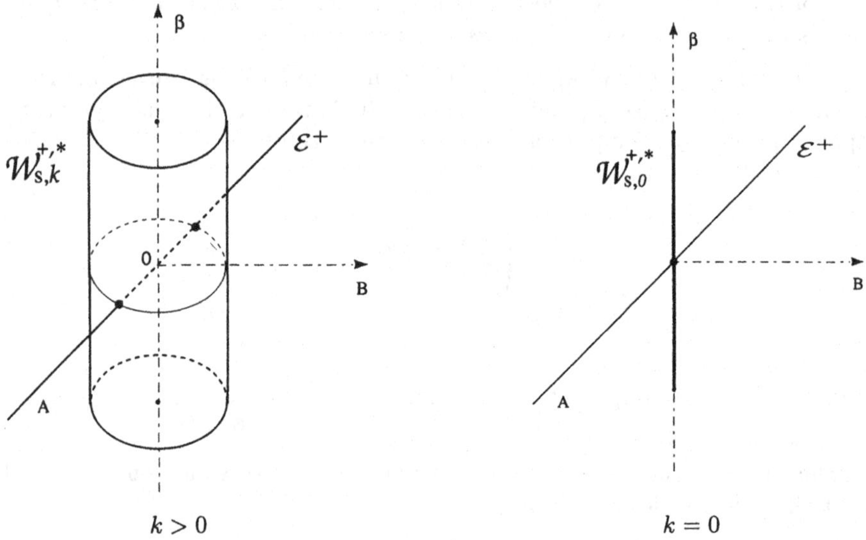

Fig. 1.2. Stable manifold of the periodic solution P_k^+ of the truncated system

connecting themselves : is it 0 or not? And if it is not 0 we would like to compute k_c with respect to ε.

When $k_c > 0$, a subsidiary question is then to determine the behavior when t tends to $-\infty$ of the one dimensional stable manifold $\mathcal{W}^+_{s,0}$ of the fixed point $(0, 1)$.

This second approach predicts the non persistence of the reversible front. So, we have two heuristic arguments (an asymptotic one and a geometrical one) which lead to two opposite predictions.

Usually the analytical implementation of our former geometric arguments is performed with a Melnikov approach [Me63] (for a good introduction to this theory see [GH83]). For this system, this strategy fails, because the Melnikov function depends on ε and is exponentially small.

However, here we can overcome this difficulty since all the bounded solutions of the perturbed system can be computed explicitly. For a more general perturbation, explicit computations cannot be done, and this exponential smallness of the Melnikov function leads to serious difficulties. One aim of this work is to give mathematical tools to study such general systems.

Now, let us perform explicit computations to check the validity of our heuristic arguments.

Lemma 1.1.1. *The bounded solutions of the perturbed system (1.2) are*

$$P^{\pm}_{k,\varphi}(t) = (k\mathrm{e}^{\mathrm{i}(\omega t/\varepsilon)+\mathrm{i}\varphi}, \pm 1),$$

and

$$Y_{t_0,z_0}(t) = \left(z_0 \mathrm{e}^{\mathrm{i}\omega t/\varepsilon} + \mathrm{i}\rho\varepsilon \int_0^t \mathrm{e}^{\mathrm{i}\omega(t-s)/\varepsilon} \frac{1}{\cosh^2(s+t_0)} ds,\ \tanh(t+t_0) \right)$$

where $k, \varphi, t_0 \in \mathbb{R}$ and $z_0 \in \mathbb{C}$.

The proof of this lemma is made in two steps. We first observe that the bounded solutions of (1.2b) are given by $\beta(t) = \tanh(t+t_0)$. Then it remains to solve (1.2a) which is a *linear oscillator of high frequency ω/ε forced by an explicitly known, analytic, exponentially decaying second term.*

Using the explicit formula giving Y_{t_0,z_0} we easily compute the reversible solutions.

Lemma 1.1.2.

(a) *The perturbed system, admits a one parameter family of reversible heteroclinic orbits H_λ explicitly given by*

$$H_\lambda = \left(\lambda \mathrm{e}^{\mathrm{i}\omega t/\varepsilon} + \mathrm{i}\rho\varepsilon \int_0^t \mathrm{e}^{\mathrm{i}\omega(t-s)/\varepsilon} \frac{1}{\cosh^2(s)} ds, \tanh t \right)$$

with $\lambda \in \mathbb{R}$.

(b) *The solution H_λ connects $P^-_{k(\lambda,\varepsilon),-\varphi(\lambda,\varepsilon)}$ and $P^+_{k(\lambda,\varepsilon),\varphi(\lambda,\varepsilon)}$ where*

$$k(\lambda,\varepsilon)\mathrm{e}^{\mathrm{i}\varphi(\lambda,\varepsilon)} = \lambda + \rho\varepsilon \int_0^{+\infty} \frac{\sin(\omega s/\varepsilon)}{\cosh^2(s)} ds + \mathrm{i}\rho\varepsilon \int_0^{+\infty} \frac{\cos(\omega s/\varepsilon)}{\cosh^2(s)} ds.$$

(c) *When λ varies from $-\infty$ to $+\infty$, $k(\lambda,\varepsilon)$ takes twice all the values in $]k_c(\varepsilon), +\infty[$ and once the value $k_c(\varepsilon)$ which reads*

$$k_c(\varepsilon) = \rho\varepsilon \int_0^{+\infty} \frac{\cos(\omega s/\varepsilon)}{\cosh^2(s)} ds = \frac{\pi\omega\rho}{2\sinh(\omega\pi/2\varepsilon)} \underset{\varepsilon\to 0}{\sim} \pi\rho\omega \mathrm{e}^{-\omega\pi/2\varepsilon}$$

This lemma confirms our former geometric arguments: for the perturbed system, and for each $\varepsilon > 0$ fixed, there exists a critical size $k_c(\varepsilon)$ (exponentially small) such that for every k, $0 \le k < k_c(\varepsilon)$ there is no reversible heteroclinic orbit which connects $P^-_{k,-\varphi}$ and $P^+_{k,\varphi}$, whereas for the truncated system, there exists heteroclinic orbits which connect two periodic orbits of the same arbitrary small size. The "surprise" is that this critical size is exponentially small. It lies beyond any algebraic order of ε. This phenomenon cannot be detected using a classical expansion of the solutions in powers of ε.

The next question is to determine the behavior of the one-dimensional stable manifold $\mathcal{W}^+_{s,0}$ of the fixed point $(0,1)$, when t tends to $-\infty$. Here again, using the explicit formulas giving the solutions we obtain

Lemma 1.1.3. *A parameterization of the stable manifold $\mathcal{W}^+_{s,0}$ of the fixed point (0,1) is given by*

$$Y_s(t) = \left(-i\rho\varepsilon \int_t^{+\infty} e^{i\omega(t-s)/\varepsilon} \frac{1}{\cosh^2(s)} ds \ , \tanh(t) \right).$$

Moreover,

$$Y_s(t) - P^-_{K(\varepsilon),-\pi/2} \underset{t\to-\infty}{=} \mathcal{O}\left(e^{-2|t|}\right)$$

with

$$K(\varepsilon) = \rho\varepsilon \int_{-\infty}^{+\infty} e^{-i\omega s/\varepsilon} \frac{1}{\cosh^2(s)} ds = \frac{\pi\omega\rho}{\sinh(\omega\pi/2\varepsilon)} \underset{\varepsilon\to 0}{\sim} 2\pi\rho\omega e^{-\omega\pi/2\varepsilon}.$$

So the front does not persist. The stable manifold of (0,1) does not connect (0,-1) to (0,1), but it connects an exponentially small periodic orbit $P^-_{K(\varepsilon),\pi/2}$ to (0,1). The stable manifold of $(0,1)$ for the perturbed system develops exponentially small oscillations at $-\infty$ which cannot be detected with a classical asymptotic expansion of the solution in powers of ε.

A third phenomenon beyond any algebraic order can be found when computing the distance $d(\varepsilon)$ between the stable manifold $\mathcal{W}^+_{s,0}$ of $(0,1)$ and the symmetry line $\mathcal{E}^+$. Once again, this distance is given by an oscillatory integral and it is exponentially small.

In all cases, the exponential smallness of $k_c(\varepsilon)$, $K(\varepsilon)$ and $d(\varepsilon)$ is due to the fact that theses quantities are given by oscillatory integrals of the form

$$\kappa(\varepsilon) = \int_{-\infty}^{+\infty} e^{i\omega t/\varepsilon} f(X(t))dt$$

where f is an analytic function and $X(t)$ is a particular solution of the system.

The toy model has been designed, so that all the solutions can be explicitly computed. Hence, in this particular case the integrand of the oscillatory integral is explicitly known whereas in general this is not true. So, in general we have to face the following problem:

Problem 1.1.4. *Assume we study a nonlinear differential equation in finite (O.D.E. case) or infinite (P.D.E case) dimensions of the form*

$$\frac{dY}{dt} = F(Y,t,\varepsilon) \tag{1.3}$$

and that we want to compute

$$I(\varepsilon) = \int_{-\infty}^{+\infty} e^{\frac{i\omega t}{\varepsilon}} g(Y_0(t,\varepsilon))dt$$

where ω is positive; g is a given function and Y_0 is a particular solution of (1.3) characterized by its initial value or more frequently by its behavior at infinity (for instance, it tends to a fixed point, a periodic orbit...).
The problem is to determine what kind of information on Y_0 we need to know for being able to compute or at least to bound the oscillatory integral $I(\varepsilon)$.

The first part of this book is devoted to the study of . We explain in chapter 2 how to obtain *exponentially small equivalents* for ε close to 0 of such oscillatory integrals involving solutions of nonlinear differential equations.

Remark 1.1.5. The method of stationary phase is a classical tool for evaluating the asymptotic behavior of an integral of the form

$$J(\lambda) = \int e^{i\lambda\varphi(t)} g(t)\, dt$$

where φ, g are smooth and where $\lambda \to +\infty$ (see for instance [Er56] and [Ho9094]–I). This method is one of main analytical tool of the Theory of Fourier Integral Operators developed by Hörmander and Duistermaat [Ho71], [DH72], [Du96], [Ho9094]. However the method of stationary phase was developed to deal with integrals $J(\lambda)$ for which the phase function φ has *stationary points*, i.e. critical point of φ ($d\varphi = 0$) which leads to polynomial equivalents of $J(\lambda)$. In this context, exponentially small terms are seen as irrelevant and the case corresponding to a phase function φ with no stationary points (which is precisely the case for Problem 1.1.4) was not investigated because it was seen as "degenerated" since integrations by parts ensure that $J(\lambda) \underset{\lambda\to+\infty}{=} \mathcal{O}(\lambda^{-p})$ for every p. So, for solving Problem 1.1.4 we had to develop other tools which enable us to obtain exponentially small equivalents of oscillatory integrals with non stationary phase.

1.2 Examples of oscillatory integrals hidden in dynamical systems

Problem 1.1.4 is the heart of many problems governed by a system of equations where coexist "a rapid oscillatory part and a slow hyperbolic one".

- A first example is a model of crystal growth governed by the equation

$$\varepsilon^2\theta''' + \theta' = \cos\theta. \tag{1.4}$$

For $\varepsilon = 0$ this equation admits a front connecting the two fixed points $\pm\frac{\pi}{2}$. The question is then to determine whether this front persists for the perturbed equation $\varepsilon \neq 0$. It appears that the stable manifold $\theta_s(t, \varepsilon)$ of $\frac{\pi}{2}$ connects $-\frac{\pi}{2}$ to $\frac{\pi}{2}$ if and only if $\theta_s''(0) = 0$, and that $\theta_s''(0)$ is given by an oscillatory integral.

This problem studied by Dashen et al. in [DKLS86], by Hammersley and Mazzarino in [HM89b] and by Amick and McLeod in [AL90]. Their arguments are rigorous but unfortunately very specific to this equation, and thus, they cannot be used for larger class of equations.

In [KS91] Kruskal and Segur give a very nice formal argument which ensures that $\theta_s''(0)$ is exponentially small but does not vanish. So, this gives a formal proof of the non persistence of the front. Their approach is based on Matched Asymptotics Expansions which enable to capture exponentially small terms and this kind of approach is often named "Asymptotics beyond all orders". The paper of Kruskal and Segur [KS91] which appeared as a preprint in 1985 inspired a lot of others works where relevant exponentially small terms are involved. It is not possible to list all of them, however we can mention several papers of Hakim and co-authors: [CHDPP88] for the study of Saffman-Taylor fingers, [Hk91] for the crystal growth model and [HM93] for computing the splitting of separatrices in rapidly forced system. In [KS91] the constant in front the exponentially small term is computed numerically whereas in the works of Hakim it is computed using Borel summation.

We can also mention several works of Grimshaw [Gr92], Joshi [GJ95] and Yang and Akylas [YA96] in hydrodynamics which where inspired by the approach of Kruskal and Segur in [KS91].

Finally, we should mention the work of Tovbis [To94] who gives a proof of non existence of symmetric, heteroclinic or homoclinic connections based on the study of formal power series solutions. His approach works for several model equations among which the upper geometrical model of crystal growth and the perturbed KdV equation below.

- A second example is a perturbed KdV equation

$$\varepsilon'^2 \frac{\partial^5 u}{\partial x^5} + \frac{\partial^3 u}{\partial x^3} + 6u\frac{\partial u}{\partial x} + \frac{\partial u}{\partial t} = 0.$$

This equation was formally derived from Euler equation by Hunter and Scheurle [HS88] for studying water waves in the presence of small surface tension (Bond Number$< \frac{1}{3}$). For $\varepsilon' = 0$, the KdV equation admits a one parameter family of solitary waves explicitly given by

$$u(x,t) = \frac{c}{2\cosh^2\left(\frac{(x-ct)\sqrt{c}}{2}\right)}.$$

Then one wants to know whether these solitary waves persist for the perturbed problem ($\varepsilon' \neq 0$): in other words, does the perturbed KdV equation admit a solution of the form

$$u(x,t) = \frac{c}{3} Y\left((x-ct)\sqrt{c}\right)$$

where $Y(\xi) \underset{\xi \to \pm\infty}{\longrightarrow} 0$. This amounts to looking for a homoclinic connection to 0 of the fourth order equation ($\varepsilon = \varepsilon' \sqrt{c}$)

$$\varepsilon^2 \frac{d^4 Y}{d\xi^4} + \frac{d^2 Y}{d\xi^2} - Y + Y^2 = 0, \tag{1.5}$$

This model equation was also derived directly from the Euler equation by Amick and Kirchgässner in [AK89] for studying the existence of water solitary waves for a certain range of the parameters values. The linearized part of this equation has eigenvalues $\pm 1 + \mathcal{O}(\varepsilon)$ and $\pm \frac{i}{\varepsilon} + \mathcal{O}(1)$. Several different authors have studied this equation. All their approaches are based on the same remark: for a half orbit $Y_-(x)$ vanishing as $x \longrightarrow -\infty$, it is possible to choose the time origin $x = 0$ such that $Y_-'(0) = 0$. Then, one can show that for a smooth continuation of $Y_-(x)$ into a homoclinic connection to 0, one needs $Y_-'''(0) = 0$. $Y_-'''(0)$ is given by an oscillatory integral and one wants to know whether $Y_-'''(0)$ vanishes or not.

Amick and McLeod prove in [AL92] that $Y_-'''(0) > 0$ holds by extending the solution in the complex plane. However no order of magnitude is given and their proof heavily relies on the exact form of (1.5): a fourth order equation with a quadratic nonlinearity. Their proof cannot be generalized to the study of a general nonlinearity.

In [HM89a], Hammersley and Mazzarino study this problem, but they restricted their analysis to strictly decaying half orbit ($Y_-'(x) < 0$ for $x > 0$). In this context, they obtain an exponential estimate of $Y_-'''(0)$ which does not vanish for small ε.

Notice that W. Eckhaus in [Ec92] gives a formal argument of non-existence of homoclinic connections for (1.5). Other formal arguments are given by Sun in [Su98], and a very elegant one is given by Pomeau et al. in [PYG88].

In their pioneering paper [HS88], Hunter and Scheurle did not solve the problem of the existence of homoclinic connections to 0 for (1.5), but they prove that for every $\rho > 0$, there exits ε_0 sufficiently small *depending on* ρ such that for every $\varepsilon \in]0, \varepsilon_0]$, (1.5) admits homoclinic reversible connection to periodic orbits of size ρ. Moreover, Amick and Toland proved in [AT92] the existence of reversible connections homoclinic to periodic orbits of size $\rho \leq C(n)\varepsilon^n$ for any $n \in \mathbb{N}$ and every sufficiently small ε. However, they did not compute *for a fixed small value of* ε, the smallest size of a periodic orbit which admits a reversible homoclinic connection to itself. As we shall see below, this smallest size is given by an oscillatory integral which is exponentially small with respect to ε but which does not vanish.

- For the true P.D.E. water wave problem (Euler equations + gravity + surface tension for a Froude number close to 1 and a Bond number less than $\frac{1}{3}$), the existence of generalized solitary waves with exponentially small oscillations at infinity is proved by Sun in [SS93] and by Lombardi in [Lo97]. The non existence of true solitary waves for Froude number close to 1 and for Bond number less than $\frac{1}{3}$ and close to $\frac{1}{3}$ was proved by Sun in [Su99]. This is an example of infinite dimensional vector field where problem 1.1.4 is hidden. Indeed, the size of oscillations at infinity is given by an oscillatory integral.

• A fourth example is a chain of nonlinear oscillators coupled to their nearest neighbors, governed by

$$\ddot{X}_n + V'(X_n) = \gamma(X_{n+1} - 2X_n + X_{n-1}), \qquad n \in \mathbb{Z}, \tag{1.6}$$

where X_n is a function of $t \in \mathbb{R}$; γ is assumed to be positive and where V is smooth function such that $V'(0) = 0$ and $V''(0) = 1$. The existence of breathers, i.e of solutions which are periodic in time and localized in space is proved by R.S. Mac Kay and Aubry in [MKA94]. A second problem is the existence of traveling waves, i.e. solutions of (1.6) of the form

$$X_n(t) = x(t - n\tau)$$

which amounts to find solutions $x(t)$ of the advance and delay differential equation

$$\ddot{x}(t) + V'[x(t)] = \gamma[x(t-\tau) - 2x(t) + x(t+\tau)].$$

G.Iooss and K.Kirchässgner proved in [IK99] that for not too large coupling constants γ the solutions constitute a one parameter family of periodic orbits. They also proved in [IK99] that there exist in the (τ, γ) parameter plane curves in the neighborhood of which there always exist *nanopterons*, i.e homoclinic connections to exponentially small periodic orbits, and there is generically no homoclinic connection to 0. The existence of *nanopterons* was suggested by the arguments given by Aubry in [Au97]. Here again, the size of the oscillation at infinity is given by an oscillatory integral.

Now if we want to determine the common features between all these examples, we need to understand what causes the appearance of exponentially small oscillations and the disappearance of hetero or homoclinic connections:

we first observe that all these problems can be reformulated as dynamical systems

$$\frac{dX}{d\xi} = V(X, \mu), \qquad \mu \in \mathbb{R}$$

in finite dimensions (O.D.E. case, $X \in \mathbb{R}^n$) or infinite dimensions (P.D.E. case $X \in H$, H Hilbert or Banach space), studied near a fixed point placed at the origin (this is also true for the advance and delay differential equation, which can be reformulated as a P.D.E. problem [IK99]). Moreover, the parameter can be chosen such that the phenomena of appearance of exponentially small oscillations and disappearance of hetero or homoclinic connections occurs for

$\mu = 0$. Finally the time ξ of the dynamical system is not necessarily the physical time . For instance for the traveling waves "$\xi = x - ct$".

A second observation is that all these systems are reversible, i.e. that the vector field V anticommutes with some symmetry S.

The last observation is that for $\mu = 0$ when the phenomena of appearance or disappearance occurs, the spectrum of the differential $D_X V(0,0)$ presents a very specific configuration. A first example is the $0^2 i\omega$ resonance which occurs for the perturbed KdV equation and for the true water water wave problem. In this case one part of the spectrum of $D_X V(0,\mu)$ is bounded away from the imaginary axis and a second part admits the bifurcation described in Figure 1.3. A second example is the $(i\omega_0)^2 i\omega_1$ resonance, which occurs for the chain of coupled nonlinear oscillators. The bifurcation of the spectrum corresponding to this resonance is described in Figure 1.3.

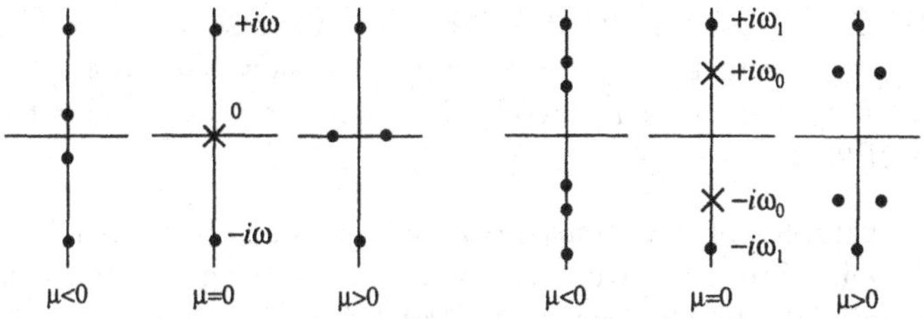

Fig. 1.3. $0^2 i\omega$ reversible resonance and $(i\omega_0)^2 i\omega_1$ reversible resonance

We can observe that for these two resonances, even after bifurcation, for $\mu > 0$ remains a pair of purely imaginary eigenvalues which coexists with a set of hyperbolic eigenvalues with *small* real parts. *Hence, a "rapid oscillatory part and a slow hyperbolic part" coexist in such vector fields* admitting a $0^2 i\omega$ or a $(i\omega_0) i\omega_1$ resonance. When studying the existence of homoclinic connections this coexistence leads to problem 1.1.4, i.e. the problem of computing oscillatory integrals involving solutions of nonlinear differential equations.

Surprisingly, in none of the previously mentioned papers, is the word "oscillatory integral" mentioned although in each case, the crucial point of the analysis is to determine either the size of a term given by an oscillatory integral or whether this term vanishes or not.

The aim of this book is to provide a unified and systematic treatment of these problems via the computation of the size of oscillatory integrals.

So in Part I, we introduce "exponential tools" which enable us to compute equivalents and upper bounds of oscillatory integrals involving solutions of nonlinear differential equations.

In Part III we use these tools for studying the existence of homoclinic connections for one parameter family of vector fields admitting a $0^2 i\omega$ resonance (Chapters 7 and 8) or a $(i\omega_0)^2 i\omega_1$ resonance (Chapter 9).

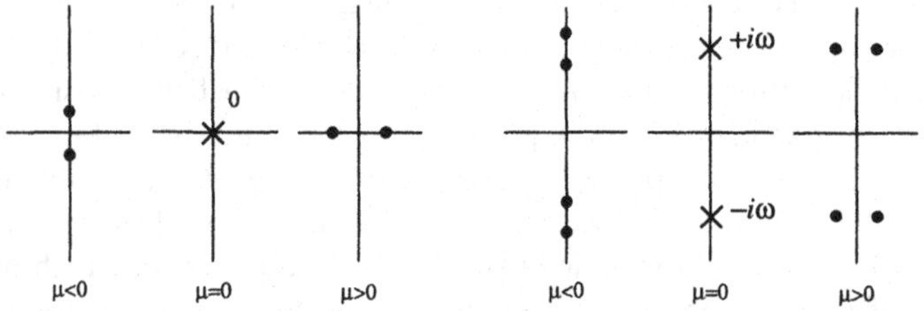

Fig. 1.4. 0^2 reversible resonance and $(i\omega)^2$ reversible resonance

Observe from Figure 1.4 that for the two first resonances, i.e. for the 0^2 and the $(i\omega)^2$ resonances such a coexistence of "slow hyperbolic part with a rapid oscillatory one" does not exist. Thus they can be studied with classical tools (see [IP93]).

Other examples of equations involving Problem 1.1.4 can be found in the Hamiltonian literature when studying the splitting of separatrices. This study was initiated by Poincaré in [Po1893]. Later on, a regular method for studying separatrices splitting was proposed by Melnikov [Me63] and Arnold gave this method an elegant form [Ar64]. This theory can also be found in the book of Guckenheimer and Holmes [GH83]. Sanders has pointed out [Sa82] that the direct application of the Melnikov theory fails when the Melnikov function depends on a small parameter and is given by an *oscillatory integral.*

Upper bounds for the exponentially small splitting of separatrices in the case of Hamiltonian systems with two degrees of freedom were given by Neishtadt [Ne84]. Fontich [Fo93] obtained also upper bounds for ordinary differential equations of the form

$$\ddot{x} = f(x) + \varepsilon^p g(\frac{t}{\varepsilon}),$$

for $p > -2$, and zero-mean function g. This result is improved in [Fo95].

Starting from the work of Holmes, Marsden and Scheurle [HMS88] and Scheurle [Sc89] the rapidly forced pendulum governed by the equation

$$\ddot{x} = \sin x + \mu\varepsilon^p \sin\frac{t}{\varepsilon} \tag{1.7}$$

became one of the most popular models involving this difficulty. Holmes, Marsden and Scheurle gave exponentially small estimates of the splitting for $p > 8$. This result was improved by several authors up to $p = 0$ (see [DS92] for the case $p = 0$ and for a good historic of this subject). More recently Gelfreich gave a proof of the asymptotic formula of the splitting up to $p = -2$ [Ge97]. Here again the difficulty comes from the fact that the splitting is given by an oscillatory integral involving solutions of nonlinear differential equations.

The case of fast quasiperiodic forcing was initiated by Simó [Si94], and studied by Delsham and al. [DGJS97] and by Sauzin [Sz99]. This problem leads to delicate analysis due to the presence of small denominators in the integrals which give the size of the splitting. A similar difficulty occurs for reversible systems when studying homoclinic connections to tori near a $0^2 i\omega_0 i\omega_1$ resonant fixed point (see Fig. 1.5).

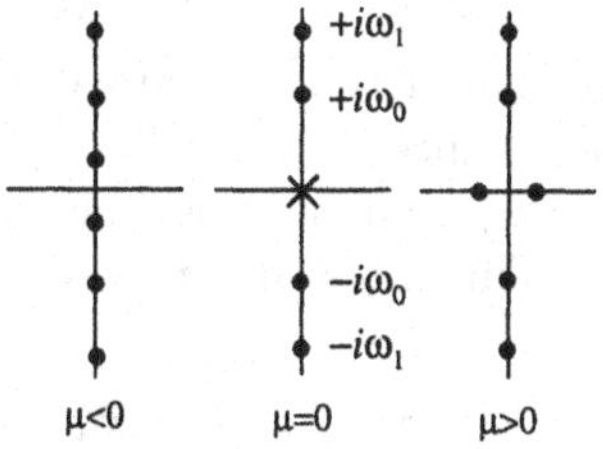

Fig. 1.5. $0^2 i\omega_0 i\omega_1$ reversible resonance

The phenomena of exponentially small splitting of separatrices also occurs for maps. However, in this case the exponentially small quantities are not given by oscillatory integrals but by infinite "oscillatory" sums. The study of the splitting of separatrices for maps was initiated by R.W. Easton [Ea84] and Jean Marc Gambaudo [Ga83], [Ga85]. In 1984, V.F Lazutkin [La84] formally obtained exponentially small formula for the separatrices splitting of the standard map

$$SM : (x, y) \mapsto (x + y + \varepsilon \sin x, y + \varepsilon \sin x).$$

This formula was refined by Gelfreich, Lazutkin and Svanidze [GLS94] and finally rigorously established by Gelfreich [Ge99] with a proof inspired by the original ideas of Lazutkin [La84].

Exponentially small upper bounds of the splitting of separatrices for families of area preserving diffeomorphism close to identity, having homoclinic points were obtained by Fontich and Simó [FS90].

1.3 From mono-frequency oscillatory integrals to singularities of solutions of complex differential equations

To solve problem 1.1.4 a preliminary idea is to integrate by parts. It requires only the knowledge of the regularity of Y_0 and upper bounds on its derivatives which can be easily obtained. However, with such a strategy, we can only obtain polynomial upper bounds of the form $|I(\varepsilon)| \leq M\varepsilon^n$.

A basic example of oscillatory integrals which arises in our toy model is the following

$$\kappa(\varepsilon) = \int_{-\infty}^{+\infty} e^{i\omega t/\varepsilon} \frac{1}{\cosh^2(t)} dt.$$

This example is the prototype of the oscillatory integral which occurs in a system when there is a coexistence of an oscillatory part induced by a pair of imaginary eigenvalues of order 1, and a hyperbolic part coming from a *pair of real eigenvalues of order* ε. This integral can be computed using the method of residues. We obtain

$$\kappa(\varepsilon) = \frac{\pi\omega}{\varepsilon \sinh(\frac{\omega\pi}{2\varepsilon})} \underset{\varepsilon\to 0}{\sim} \frac{2\pi\omega}{\varepsilon} e^{-\frac{\omega\pi}{2\varepsilon}}.$$

Observe that the coefficient $-\frac{\pi}{2}$ in the exponential comes from the position of the singularities of $\cosh^{-2}(t)$ and the coefficient $1/\varepsilon$ comes from the fact that $\cosh^{-2}(t)$ has a pole of order 2 at $i\frac{\pi}{2}$.

So, if we want to solve problem 1.1.4 using residues when F and g are analytic functions, we have to face the following new problem

Problem 1.3.1. *Assume we study a nonlinear differential equation in finite (O.D.E. case) or infinite dimensions (P.D.E case) of the form*

$$\frac{dY}{dt} = F(Y, t, \varepsilon) \tag{1.8}$$

where F is an analytic function.
For a particular solution Y_0 of (1.8) characterized by its initial value or by its behavior at infinity (for instance, it tends towards a fixed point, a periodic orbit...) the problem is to determine

(a) *whether Y_0 admits an analytic continuation in $\mathbb{C}$,*
(b) *the nature and position of the singularities of Y_0 in $\mathbb{C}$.*

Remark 1.3.2. Restriction to equations with analytic vector fields does not limit the application to problems of physical interest, since almost all problems coming from physics are governed by analytic equations.

Unfortunately, there is no general theory for computing the nature and the position of singularities of solutions of complex differential equations. The only available theory is the one of Fuchs for linear time dependent equations and there are some works of Painlevé in dimension 2. Using the theory of Ecalle, Sauzin in [Sz95] has succeeded in determining the singularities of the solutions necessary to compute the exponential splitting of the seperatrix of a "modified rapidly forced pendulum" governed by the equation

$$\ddot{q} = \sin q + \rho \sin q \; \mathrm{e}^{\mathrm{i}\frac{t}{\varepsilon}} .$$

However, his proof is specific to this equation and does not work for the real equation

$$\ddot{q} = \sin q + \rho \sin q \; \sin \tfrac{t}{\varepsilon} .$$

All the difficulties come from the fact that the nature and the position of the singularities of analytic functions are not stable under addition, multiplication, composition, or integration. For instance, accumulation of poles may lead to essential singularities and logarithmic singularities may appear after integration... Moreover, we study O.D.E's or P.D.E's with a small parameter ε. So the nature and the position of the singularities of the solution may depend on ε. Finally, knowing that the solution admits an essential singularity at $\mathrm{i}\frac{\pi}{2}$ is not sufficient to compute the size of $I(\varepsilon)$.The following example illustrates this fact:

$$\begin{aligned} I_p(\varepsilon) &= \int_{-\infty}^{+\infty} \mathrm{e}^{\frac{i\tau}{\varepsilon}} \left[\exp\left(\frac{\mathrm{i}\varepsilon^p}{\tau - \frac{i\pi}{2}} \right) - 1 - \frac{\mathrm{i}\varepsilon^p}{\tau - \frac{i\pi}{2}} \right] d\tau \\ &= 2\pi \mathrm{e}^{\frac{-\pi}{2\varepsilon}} \sum_{n \geq 2} \frac{(-1)^n}{n! \; (n-1)!} \varepsilon^{(p-1)n+1} \underset{\varepsilon \to 0}{\sim} \pi \varepsilon^{2p-1} \mathrm{e}^{\frac{-\pi}{2\varepsilon}} . \end{aligned}$$

We must know very precisely how the small parameter is involved in the singularity. So, at the present time, determining the nature and the position of singularities of solutions of general nonlinear O.D.E (or P.D.E) seems beyond possibilities. Fortunately, in most of the cases where problem 1.1.4 is involved, the exact value of the oscillatory integral $I(\varepsilon)$ is not necessary. An equivalent for $\varepsilon \to 0$ or an exponentially small upper bound on $I(\varepsilon)$ is most often sufficient.

In the first section of Chapter 2 we prove that for obtaining an exponential estimate of $I(\varepsilon)$, it is sufficient to prove that Y_0 is holomorphic in a complex strip of the form $\mathcal{B}_\ell = \{z \in \mathbb{C}, |\mathcal{I}m\,(z)| < \ell\}$. The holomorphy of solutions can be obtained with the contraction mapping theorem in appropriate spaces of holomorphic functions defined on $\mathcal{B}_\ell$.

Moreover, we also prove that it is possible to obtain an equivalent for $\varepsilon \to 0$ of the oscillatory integral $I(\varepsilon)$ without computing the exact position of singularities of the solutions. For that purpose, we prove in the first section of Chapter 2, that it is sufficient to delimit very small regions (the size of which are of order ε) where the singularities may be located, and to describe very precisely how the solution blows up near these regions. Following the idea developed by Kruskal and Segur in [KS91], the division of $\mathbb{C}$ into several subdomains necessary to describe the "explosion" of the solution is inspired by the formal method of Matched Asymptotic Expansions (M.A.E.). The M.A.E. method gives a *formal description* of the solutions on each subdomain in terms of asymptotic expansions which "match" on the boundary of each subdomain following some heuristic rules. The method explained in the first section of Chapter 2 gives an *exact description* of the solution on each subdomain in terms of sums of holomorphic functions which coincide on the boundary of the subdomains. The estimates of the size of the different holomorphic functions indicate how the solution explodes near the singularities. It is worth mentioning that the main technique used by Lazutkin [La84] (later joined by colleagues [GLS94][Ge99]) to study the standard map is in fact parallel to that of Kruskal and Segur [KS91] and was developed at approximately the same time. Finally, the method of the "reference system" introduced by Gelfreich in [Ge97] for studying the exponential splitting of the separatrices of the rapidly forced pendulum is based on the same kind of ideas.

1.4 From bi-frequency oscillatory integrals to singularities of solutions of complex partial differential equations

A second basic example of oscillatory integral which occurs for instance in the $(i\omega_0)^2 i\omega_1$ reversible resonance is the following

$$J(\varepsilon) = \int_{-\infty}^{+\infty} \mathrm{e}^{\frac{i\omega_1 t}{\varepsilon}}\, g\left(\frac{\varepsilon\cos(\omega_0 t/\varepsilon)}{\cosh(t)}\right)\, dt$$

which leads to the problem of computing for $p \in \mathbb{Z}$ and $n \in \mathbb{N}$

$$\begin{aligned} J_{p,n} &= \int_{-\infty}^{+\infty} \mathrm{e}^{\frac{i\omega_1 t}{\varepsilon}}\, \mathrm{e}^{\frac{ip\omega_0 t}{\varepsilon}} \left(\frac{\varepsilon}{\cosh(t)}\right)^n dt \\ &\underset{\varepsilon\to 0}{\sim} \frac{2\pi\varepsilon|\omega_1 + p\omega_0|^{n-1}}{(n-1)!}\, \mathrm{e}^{-\frac{|\omega_1+p\omega_0|\pi}{2\varepsilon}}. \end{aligned}$$

The integral J is the prototype for the oscillatory integrals which occur in a system where there coexist an oscillatory part, induced by a pair of imaginary eigenvalues of order 1, and a hyperbolic part coming from a set of eigenvalues $\pm\varepsilon \pm i\omega_0$. The resonance between the two frequencies modifies the size of the oscillatory integral. Indeed, for a holomorphic function g, using the residues we get

$$K(\varepsilon) = \int_{-\infty}^{+\infty} \mathrm{e}^{\frac{i\omega_1 t}{\varepsilon}}\, g\left(\frac{\varepsilon}{\cosh(t)}\right) dt \underset{\varepsilon\to 0}{\sim} C(g)\mathrm{e}^{\frac{-\pi\omega_1}{2\varepsilon}}$$

whereas

$$J(\varepsilon) = \int_{-\infty}^{+\infty} \mathrm{e}^{\frac{\mathrm{i}\omega_1 t}{\varepsilon}} g\left(\frac{\varepsilon \cos(\omega_0 t/\varepsilon)}{\cosh(t)}\right) dt \underset{\varepsilon\to 0}{\sim} C'(g)\mathrm{e}^{\frac{-\pi\omega_\star}{2\varepsilon}}$$

where $\omega_\star = \min_{p\in\mathbb{Z}} |\omega_1 + p\omega_0|$. The coefficient $\frac{\pi}{2}$ in the exponential comes from the position of the singularity of $(\cosh z)^{-1}$ at $\mathrm{i}\frac{\pi}{2}$. The methods developed for computing the size of mono frequency oscillatory integrals are based on a complexification of time which determines the holomorphic continuation of solutions and says how this continuation "explodes near singularities". In the first case, the function $H_1 : (z,\varepsilon) \mapsto \varepsilon(\cosh z)^{-1}$ is bounded near the real axis and explodes near its singularity at $\mathrm{i}\frac{\pi}{2}$. In the second case, $H_2 : (z,\varepsilon) \mapsto \varepsilon(\cosh z)^{-1} \cos(\omega_0 z/\varepsilon)$ is no longer bounded near the real axis, since for every arbitrary small $\ell > 0$, $\sup_{\varepsilon\in]0,\varepsilon_0]} |H_2(\mathrm{i}\ell,\varepsilon)| = +\infty$ because of the oscillation $\left(\cos(\mathrm{i}\ell\omega_0/\varepsilon) \underset{\varepsilon\to 0}{\longrightarrow} +\infty\right)$. We are interested in the explosion induced by the singularity and not by the one coming from oscillations. To overcome this difficulty, a *partial complexification* of time is necessary, i.e we look for solutions of the system

$$\frac{dY}{dt} = F(Y,\varepsilon) \tag{1.9}$$

of the form $Y_0(t) = y_0(t,\omega_0 t/\varepsilon)$ where $y_0(z,s)$ is holomorphic with respect to $z \in \mathbb{C}$ and 2π-periodic with respect to s. For our example it amounts to writing $H_2(t,\varepsilon) = h_2(t,\omega_0 t/\varepsilon,\varepsilon)$ where $h_2(z,s,\varepsilon) = \varepsilon \mathrm{e}^{\mathrm{i}s}(\cosh z)^{-1}$. Now the function h_2 is bounded with respect to (z,s,ε) for z near the real axis and it explodes at $z = i\frac{\pi}{2}$. In order to find solutions of (1.9) in the form $y(t,\omega_0 t/\varepsilon)$, we look for solutions of the P.D.E.

$$\frac{\partial y}{\partial z} + \frac{\omega_0}{\varepsilon}\frac{\partial y}{\partial s} = F(Y,\varepsilon). \tag{1.10}$$

The position of the singularities of the solutions of this P.D.E. gives the size of the bi-frequency oscillatory integral. In section 2.2 we give a "P.D.E. version of the tools" described in section 2.1 for mono frequency oscillatory integrals.

1.5 On the contents

Part I proposes a complete set of "exponential tools" for evaluating oscillatory integrals.

2. In Chapter 2 we explain what kind of information it is necessary to know on a function f to be able to compute an upper bound or an equivalent of the oscillatory integral induced by f. Moreover we explain how to obtain this kind of information when f involves the solution of nonlinear differential equations. Section 2.1 is devoted to mono-frequency oscillatory integrals and Section 2.2 deals with bi-frequency oscillatory integrals.

Part II gives all the necessary prerequisites for studying reversible vector fields near resonances.

3. In section 3.1 of Chapter 3, we recall basic properties of reversible vector fields and define resonances. We also give a theorem of classification of reversible matrices up to simultaneous conjugacy. This theorem enables us to introduce a convenient way of naming resonances (see Remark 3.1.16 and 3.1.12). Vector fields are seen as perturbation of their normal forms by higher order terms. Elphick and al. introduced in [ETBCI87] a very efficient way to compute normal forms. In section 3.2, we recall how to compute them and we give some examples. Complete proofs and many examples can be found in the book of Iooss and Adelmeyer [IA92]. There then remains the question of persistence for the full vector field of solutions obtained to the normal form.

4. When considering a perturbed system, a first task before studying the persistence of connections heteroclinic or homoclinic to periodic orbits, is to prove that periodic orbits persist. A very efficient tool for that purpose is the Lyapunov Schmidt method. For analytic vector fields an adaptation of this method described in chapter 4, gives a description of the periodic orbit as a power series with trigonometric polynomial coefficients. This description appears to be very useful when the time is complexified.

5. When linearizing around a periodic orbit, one is often disappointed to observe that Floquet theory is not constructive. For periodic orbits close to a fixed point, we give, in Chapter 5, a constructive version of Floquet theory using the implicit function theorem. This constructive version appears to be invaluable when complexifying time.

6. A key step of the study of a perturbed equation near a given solution of the unperturbed system, is the inversion of affine equations obtained by linearization around this solution.
   Chapter 6 is devoted to affine equations obtained by linearization around homoclinic or heteroclinic orbits. We first recall the different method commonly used for computing a basis of the homogeneous linear equation. Then, we recall the main result of the theory of Fuchs, which gives a basis of solutions of the homogeneous linear equation characterized by their singularities in the complex field. The inversion of the affine equations lead to compatibility conditions, "the principal part" of which are nothing else than a generalized Melnikov function for systems of dimension greater than 2. The tools given in chapter 2, were developed for the study of these compatibility conditions in the hyper-degenerated case when the principal part is given by an oscillatory integral which is exponentially small.
   The complexification of time necessary for using the tools given in Section 2.1 does not induce new difficulties for solving affine equations whereas the partial complexification of time necessary for the use of tools given in Section 2.2 leads to the study of linear partial differential equations of the

form

$$\frac{\partial v}{\partial z} + \frac{\omega_0}{\varepsilon}\frac{\partial v}{\partial s} + DF(h).v = f,$$

the inversion of which is far more intricate. This inversion is done in section 6.5

Part III is devoted to the problem of the existence of homoclinic connections near resonances for reversible systems.

7. Chapter 7 is devoted to the study of the $0^2 i\omega$ reversible resonance in $\mathbb{R}^4$ for analytic vector fields. We prove the existence of periodic solutions of arbitrary small size, the existence of reversible connections homoclinic to exponentially small periodic orbits and the generic non existence of homoclinic connections to 0.
8. In chapter 8, we extend the result of the previous section to infinite dimensional analytic vector fields. This extension cannot be done via a center manifold reduction since we loose the analyticity with the reduction.
As already mentioned, one example of such a vector field in infinite dimensions occurs when describing the irrotational flow of an inviscid fluid layer of finite depth under the influence of gravity and small surface tension (Bond number $b < 1/3$), for a Froude number F close to 1. In fact, most of the present work was undertaken to solve this problem. In this context a homoclinic solution to a periodic connection is called a generalized solitary wave. In section 8.5 we explain how the previously obtained results ensure that there exist generalized solitary waves with exponentially small oscillations at infinity.
9. Chapter 9 is devoted to the $(i\omega_0)^2\omega_1$ reversible resonance. At first sight, this resonance seems very similar to the previous one. However the interaction between the two frequency ω_0 and ω_1 makes the study of this resonance far more intricate and leads to a partial complexification of time. We prove the existence of periodic solutions of arbitrary small size, the existence of reversible connections homoclinic to exponentially small periodic orbits and the generic non existence of reversible homoclinic connections to 0. Such a resonance occurs in infinite dimensions for the chain of coupled oscillators (1.6) (see [IK99]). Following a pioneering work of M.Haragus-Courcelle and A. Illichev [HI98], M.Groves and A.Mielke prove in [GM99] that it also occurs when studying 3 dimensional steady water wave problem in which the waves are uniformly translating in one horizontal direction and even and periodic in the other.

Part I

Toolbox for oscillatory integrals

2. "Exponential tools" for evaluating oscillatory integrals

This chapter is devoted to the study of problem 1.1.4, i.e. to the computation of upper bounds and equivalents of oscillatory integrals involving solutions of nonlinear analytic differential equations.

2.1 Mono-frequency oscillatory integrals: complexification of time

In this section we define spaces of functions which enable us to compute simultaneously equivalents of the mono-frequency oscillatory integrals

$$I^{\pm}(f,\varepsilon) = \int_{-\infty}^{+\infty} f(t,\varepsilon) e^{\pm i\omega t/\varepsilon}\, dt.$$

Moreover we explain how to prove that solutions of nonlinear differential equations belong to these spaces.

All these spaces are composed of holomorphic functions defined in domains symmetric with respect to the real axis. If one is only interested in the computation of an equivalent of I^+ (resp. I^-), it is sufficient to work with functions defined on a half domain, i.e. holomorphic for $\mathcal{I}m\,(\xi) \geq 0$ (resp. holomorphic for $\mathcal{I}m\,(\xi) \leq 0$). However, in most of the cases, the functions are real on the real axis. Thus, their holomorphic continuation is symmetric with respect to the real axis and hence, automatically defined on domains symmetric with respect to the real axis.

2.1.1 Rough exponential upper bounds

We begin with a very simple lemma which gives exponential upper bounds for mono-frequency oscillatory integrals.

Lemma 2.1.1 (First Mono-Frequency Exponential Lemma).
Let ω, ℓ, λ be three real positive numbers. Let $\mathcal{B}_\ell$ be the strip in the complex field: $\mathcal{B}_\ell = \{\xi \in \mathbb{C} / |\mathcal{I}m\,(\xi)| < \ell\}$.
Let H_ℓ^λ be the set of functions f satisfying

(a) $f : \mathcal{B}_\ell \times]0,1] \longrightarrow \mathbb{C}$,
(b) $\xi \mapsto f(\xi,\varepsilon)$ *is holomorphic in* $\mathcal{B}_\ell$,
(c) $\|f\|_{H_\ell^\lambda} := \sup_{\varepsilon\in]0,1],\xi\in\mathcal{B}_\ell} (|f(\xi,\varepsilon)|e^{\lambda|\mathcal{R}e(\xi)|}) < +\infty$.

Then for every $f \in H_\ell^\lambda$ *and* $\varepsilon \in]0,1]$, $I^\pm(f,\varepsilon) = \int_{-\infty}^{+\infty} f(t,\varepsilon)\ e^{\pm i\omega t/\varepsilon}\ dt$ *satisfies*

$$|I^\pm(f,\varepsilon)| \leq \frac{2\,\|f\|_{H_\ell^\lambda}}{\lambda} e^{-\omega\ell/\varepsilon}.$$

Proof. We only do the proof for I^+. For I^-, perform the change of time $t' = -t$ in the integral and observe that $f^- : \xi \mapsto f(-\xi,\varepsilon)$ also belongs to H_ℓ^λ and that $\|f^-\|_{H_\ell^\lambda} = \|f\|_{H_\ell^\lambda}$.

Let $f \in H_\ell^\lambda$, ε, $\ell' < \ell$ be fixed. Since f is holomorphic in $\mathcal{B}_\ell$, the integral of $fe^{i\omega t/\varepsilon}$ along the path Γ_1 given in Fig. 2.1, is equal to zero.

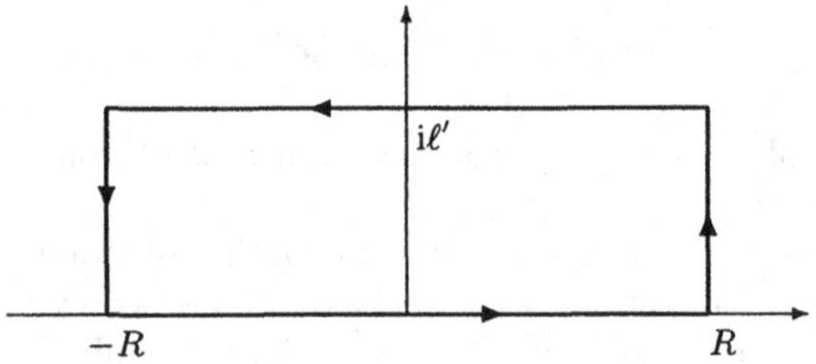

Fig. 2.1. Path Γ_1.

Pushing R to $+\infty$, we get

$$I^+(f,\varepsilon) = \int_{-\infty}^{+\infty} f(i\ell' + t,\varepsilon)\ e^{i\omega(i\ell'+t)/\varepsilon}\ dt.$$

The estimate then follows, where the exponential comes from the oscillating term computed on the line $\mathcal{I}m\,(t) = \ell'$. □

Lemma 2.1.1 gives a very efficient way to obtain exponential upper bounds because the membership to H_ℓ^λ is stable by addition, multiplication, and "composition", which can be summed up as follows

Lemma 2.1.2.

(a) H_ℓ^λ *is an algebra.*
(b) *If* $f \in H_\ell^\lambda$ *and* g *is holomorphic in a domain containing the range of* f *and satisfies* $g(0) = 0$, *then* $g \circ f \in H_\ell^\lambda$.

Remark 2.1.3. Lemma 2.1.1 is called mono-frequency because functions belonging to H_ℓ^λ cannot have a second "high frequency" like $f_0(\xi, \varepsilon) = \cos(\omega_0 \xi/\varepsilon) \cosh^{-1} \xi$. Indeed, for every $\ell > 0$, $\|f_0\|_{H_\ell^\lambda} = +\infty$ holds because of the oscillation term. Oscillatory integrals involving such functions are studied in Section 2.2.

Remark 2.1.4. In Lemma 2.1.1 the exponential decay of the functions at infinity is not fundamental. Indeed, one can obtain a similar exponential upper bound of the oscillatory integral with another set of holomorphic functions in $\mathcal{B}_\ell$ which only decay polynomialy at infinity, i.e. which satisfies

$$\|f\| := \sup_{\varepsilon \in]0,1], \xi \in \mathcal{B}_\ell} \left(|f(\xi, \varepsilon)| (1 + |\xi|^2) \right) < +\infty.$$

However, analyticity is fundamental. For instance take $f(t) = \mathrm{e}^{-|t|}$; then

$$I(f, \varepsilon) = \int_{-\infty}^{+\infty} \mathrm{e}^{-|t|} \mathrm{e}^{\mathrm{i}\omega t/\varepsilon} dt = \frac{2\varepsilon^2}{\omega^2 + \varepsilon^2}.$$

Remark 2.1.5. How to use Lemma 2.1.1 for solving problem 1.1.4 Our aim is to compute the size, or at least an upper bound of the size, of an integral of the form

$$I(\varepsilon) = \int_{-\infty}^{+\infty} g(Y_0(t), t, \varepsilon) \mathrm{e}^{\mathrm{i}\omega t/\varepsilon} dt,$$

where g is an analytic function and Y_0 is a particular solution of a differential equation of the form

$$\frac{dY}{dt} = F(Y, t, \varepsilon)$$

with F analytic with respect to (Y, t). To apply Lemma 2.1.1 we have to proceed via three steps:

- We must find the singularity which has the smallest imaginary part of $\xi \mapsto F(\cdot, \xi, \cdot)$. This determines the width of the complex strip $\mathcal{B}_\ell$: ℓ must be chosen such that F has no singularity in $\mathcal{B}_\ell$.

– Studying the holomorphic equation

$$\frac{dY}{d\xi} = F(Y, \xi, \varepsilon), \qquad \xi \in \mathcal{B}_\ell,$$

we must extend the solutions of our first real equation into $\mathcal{B}_\ell$.
This step is often performed using the Contraction Mapping Theorem in an appropriate space of functions of the type H_ℓ^λ.
– Then, we can apply Lemma 2.1.1, to obtain an exponential upper bound of $I(\varepsilon)$.

This strategy works either in finite or infinite dimensions.

Example 2.1.6. We study here a second toy model

$$\begin{cases} \dfrac{dA}{dx} &= -\omega B, \\ \dfrac{dB}{dx} &= \omega A + \rho(\varepsilon^2 - \alpha^2), \\ \dfrac{d\alpha}{dx} &= \varepsilon^2 - \alpha^2 + \rho(\varepsilon^2 - \alpha^2)\alpha^2 \end{cases} \tag{2.1}$$

obtained by perturbation of the same initial system ($\rho = 0$) as the one studied in the introduction (see Eq. (1.1)). Here again it is more convenient to set

$$\alpha = \varepsilon\beta, \qquad Z = A + iB, \qquad x = t/\varepsilon,$$

and to rewrite (2.1) with these new coordinates. We obtain

$$\begin{cases} \dfrac{dZ}{dt} &= \dfrac{i\omega}{\varepsilon} Z + i\rho\varepsilon(1 - \beta^2), \\ \dfrac{d\beta}{dt} &= 1 - \beta^2 + \rho\varepsilon^2(1 - \beta^2)\beta^2. \end{cases} \tag{2.2}$$

System (2.2) is again reversible with the same symmetry as (1.2). The phase portrait of the unperturbed system ($\rho = 0$) is described in the beginning of the introduction (see p. 2).

As for our first toy model (1.1), the question of the persistence for $\rho \neq 0$ of the front and of the reversible heteroclinic connections reduces to the computation of an oscillatory integral (see Question 2.1.7 below). Indeed, for $|\rho\varepsilon^2| < 1$, (2.2b) has a unique (up to time shift) bounded solution which is a front connecting -1 to $+1$. We denote by $\beta^\star$ the parameterization of this orbit such that $\beta^\star(0) = 0$ (observe that $\beta^\star$ is odd). Then, because System (2.2) is only partially coupled, all its bounded solutions can be expressed as a function of $\beta^\star$. These bounded solutions are the periodic orbits

$$P_{k,\varphi}^{\pm}(t) = (k e^{i(\omega t/\varepsilon) + i\varphi}, \pm 1),$$

and the two parameters family of solutions

$$Y_{t_0,z_0}(t) = \left(z_0 e^{i\omega t/\varepsilon} + i\rho\varepsilon \int_0^t e^{i\omega(t-s)/\varepsilon} \left(1 - (\beta^\star)^2 (s+t_0)\right) ds,\ \beta^\star(t+t_0) \right)$$

where $k, \varphi, t_0 \in \mathbb{R}$ and $z_0 \in \mathbb{C}$. This last formula enables us to check that

(a) The stable manifold Y_s of (0,1) connects (0,1) to the periodic orbit $P^-_{I(\varepsilon,\rho),-\frac{\pi}{2}}$ of size $I(\varepsilon, \rho)$ with

$$I(\varepsilon, \rho) = \rho\varepsilon \int_{-\infty}^{+\infty} e^{i\omega t/\varepsilon} \left(1 - (\beta^\star)^2 (t, \varepsilon, \rho)\right) dt.$$

So unless $I(\varepsilon, \rho)$ vanishes, the front does not persist since it develops for $\rho \neq 0$ oscillations at $-\infty$ of size $I(\varepsilon, \rho)$.

(b) There exists a smallest size $k_c(\varepsilon, \rho)$ such that for every $k > k_c$ there exists two *reversible heteroclinic orbits* connecting the periodic orbits $P^\pm_{k,\varphi}$ of size k; for $k = k_c$ there is a *unique reversible heteroclinic orbit* connecting $P^-_{k,\varphi}$ to$P^+_{k,\varphi}$, and for $k \in [0, k_c[$ there is *no reversible heteroclinic* orbit connecting $P^\pm_{k,\varphi}$. This smallest size is given by

$$k_c(\varepsilon, \rho) = \frac{1}{2} I(\varepsilon, \rho).$$

So the question of the persistence of the front and of the reversible heteroclinic connections of (2.1) and for ε small and $\rho \neq 0$ reduces to the following question :

Question 2.1.7. *Compute an equivalent for ε small of the oscillatory integral*

$$I(\varepsilon, \rho) = \rho\varepsilon \int_{-\infty}^{+\infty} e^{i\omega t/\varepsilon} \left(1 - (\beta^\star)^2\right) dt$$

where $\beta^\star$ is the solution of

$$\frac{d\beta}{dt} = 1 - \beta^2 + \rho\varepsilon^2 \left(1 - \beta^2\right) \beta^2 \tag{2.3}$$

which vanishes for $t = 0$. A subsidiary question is the following: is the equivalent given by the Melnikov function $M_e(\varepsilon, \rho)$

$$M_e(\varepsilon, \rho) = \rho\varepsilon \int_{-\infty}^{+\infty} e^{i\omega t/\varepsilon} \left(1 - (\beta^h)^2\right) dt. \tag{2.4}$$

obtained by computation of the integral along the solution $\beta^h(t) = \tanh(t)$ of the unperturbed system ($\rho\varepsilon = 0$)?

Remark 2.1.8. A non zero equivalent ensures that the front of (2.1) for $\rho = 0$ does not persists for $\rho \neq 0$ because it develops oscillations at ∞ of size $I(\varepsilon, \rho)$. Moreover, since the solution of the unperturbed system is explicitly known, the Melnikov function $M_e(\varepsilon, \rho)$ can be computed explicitly using residues

$$M_e(\varepsilon, \rho) = \frac{\pi\omega\rho}{\sinh(\pi\omega/2\varepsilon)} \underset{\varepsilon \to 0}{\sim} 2\pi\omega\rho e^{-\pi\omega/2\varepsilon}.$$

To give a first answer to Question 2.1.7, we use the strategy proposed in Remark 2.1.5. As explained we proceed in three steps:

Step 1. We look for a holomorphic continuation of $\beta^\star$ in the form $\beta^\star(\xi, \varepsilon) = \beta^h(\xi) + \gamma^\star(\xi, \varepsilon)$, where $\beta^h(\xi) = \tanh \xi$. To work with holomorphic functions we restrict ξ to $\mathcal{B}_\ell$ with $\ell < \frac{\pi}{2}$.

Step 2. Our aim is now to prove that for ε small, $\gamma^\star(\cdot, \varepsilon)$ is well defined on $\mathcal{B}_\ell$ and that it belongs to $\mathcal{H}_\ell^\lambda$ for $\ell < \frac{\pi}{2}$ and $\lambda < 2$ where

$$\mathcal{H}_\ell^\lambda := \{f : \mathcal{B}_\ell \to \mathbb{C}, \text{ holomorphic in } \mathcal{B}_\ell, \ |f|_{\mathcal{H}_\ell^\lambda} < +\infty\}$$

with $|f|_{\mathcal{H}_\ell^\lambda} := \sup_{\xi \in \mathcal{B}_\ell} |f(\xi)| e^{\lambda |\mathcal{R}e(\xi)|}$. Observe that at this step, we see $\gamma^\star$ as a function of ξ with parameter ε. So we have introduced $\mathcal{H}_\ell^\lambda$ which is a space of functions of the unique variable ξ. For ε sufficiently small, $\gamma^\star(\cdot, \varepsilon)$ will be built using the contraction mapping theorem in $\mathcal{H}_\ell^\lambda$ and we will get that $|\gamma(\cdot, \varepsilon)|_{\mathcal{H}_\ell^\lambda} \leq M\varepsilon^2$. This estimate will ensure that $\varepsilon^{-2}\gamma^\star$ seen as a function of (ξ, ε) belongs to H_ℓ^λ.

$\gamma^\star$ is characterized by the fact that it is the unique solution of

$$\frac{d\gamma}{d\xi} + 2\beta^h\gamma = -\gamma^2 + \rho\varepsilon^2 \left(1 - (\beta^h + \gamma)^2\right) (\beta^h + \gamma)^2 := G(\gamma)$$

vanishing for ξ=0. Uniqueness for the Cauchy problem and the variation of constants formula ensures that $\gamma^\star$ is the unique holomorphic function defined in a neighborhood of 0 which satisfies $\gamma^\star = \mathcal{F}(\gamma^\star)$ where

$$\mathcal{F}(\gamma)(\xi) = \cosh^{-2}(\xi) \int_{[0,\xi]} \cosh^2(\zeta)\ G(\gamma(\zeta))\ d\zeta.$$

The estimates

$$|\mathcal{F}(\gamma)|_{\mathcal{H}_\ell^\lambda} \leq M(\ell, \lambda) \left(|\gamma|^2_{\mathcal{H}_\ell^\lambda} + \rho\varepsilon^2\right)$$

$$|\mathcal{F}(\gamma) - \mathcal{F}(\gamma')|_{\mathcal{H}_\ell^\lambda} \leq M(\ell, \lambda) \left(|\gamma|_{\mathcal{H}_\ell^\lambda} + |\gamma'|_{\mathcal{H}_\ell^\lambda} + \rho\varepsilon^2\right) |\gamma - \gamma'|_{\mathcal{H}_\ell^\lambda}$$

for every $\gamma, \gamma' \in \mathcal{H}_\ell^\lambda$, $|\gamma|_{\mathcal{H}_\ell^\lambda} \leq 1$, $|\gamma'|_{\mathcal{H}_\ell^\lambda} \leq 1$ ensure that for $\rho\varepsilon^2$ sufficiently small, $\mathcal{F}$ is a contraction mapping from $\mathcal{BH}_\ell^\lambda(2M(\ell, \lambda)\rho\varepsilon^2)$ to itself where

$\mathcal{BH}_\ell^\lambda(d) = \{f \in \mathcal{H}_\ell^\lambda / \ |f|_{\mathcal{H}_\ell^\lambda} \leq d\}$. We can then conclude that $\beta^\star$ has a holomorphic continuation of the form

$$\beta^\star = \beta^h + \gamma^\star \qquad \text{where } \gamma^\star(\cdot,\varepsilon) \in \mathcal{H}_\ell^\lambda, \text{ and } |\gamma^\star|_{\mathcal{H}_\ell^\lambda} \leq 2M(\ell,\lambda)\rho\varepsilon^2.$$

Hence $(\xi,\varepsilon) \mapsto \varepsilon^{-2}\gamma^\star \in H_\ell^\lambda$ and $\left\|\varepsilon^{-2}\gamma^\star\right\|_{H_\ell^\lambda} \leq 2M(\ell,\lambda)\rho$.

Step 3. For estimating $I(\varepsilon,\rho)$ we split it into two parts: the first one coming from β^h, is called the Melnikov function which can be explicitly computed whereas the second one involving $\gamma^\star$ can only be bounded using the exponential Lemma 2.1.1. We get for $\ell < \frac{\pi}{2}$,

$$\begin{aligned} I(\varepsilon,\rho) &= \rho\varepsilon\int_{-\infty}^{+\infty} \mathrm{e}^{\mathrm{i}\omega s/\varepsilon}\left(1-(\beta^h)^2\right)ds + \rho\varepsilon\int_{-\infty}^{+\infty} \mathrm{e}^{\mathrm{i}\omega s/\varepsilon}\left(-2\beta^h\gamma^\star - (\gamma^\star)^2\right)ds \\ &= \frac{\pi\omega\rho}{\sinh(\pi\omega/2\varepsilon)} + \underset{\varepsilon\to 0}{\mathcal{O}}\left((\rho^2\varepsilon^3+\rho^3\varepsilon^5)\mathrm{e}^{-\omega\ell/\varepsilon}\right). \end{aligned}$$

The explicitly known part

$$\frac{\pi\omega\rho}{\sinh(\pi\omega/2\varepsilon)} = 2\pi\omega\rho\mathrm{e}^{-\pi\omega/2\varepsilon}(1+\mathcal{O}(\varepsilon))$$

being smaller than the estimated one (because $\ell < \frac{\pi}{2}$) we cannot determine whether $I(\varepsilon,\rho)$ vanishes or not. We can only finally obtain the following exponentially small upper bounds for $I(\varepsilon,\rho)$,

$$|I(\varepsilon,\rho)| \leq M(\ell)\ (\rho+\rho^2\varepsilon^3+\rho^3\varepsilon^5)\mathrm{e}^{-\omega\ell/\varepsilon} \qquad \text{for } \ell < \frac{\pi}{2}.$$

The Exponential Lemma 2.1.1 does not give an equivalent of the oscillatory integral. To obtain an equivalent we would need to set $\ell = \frac{\pi}{2}$. In fact $\ell = \frac{\pi}{2} - \varepsilon$ is sufficient to balance the exponentially small terms. However, either for $\ell = \frac{\pi}{2}$ or $\ell = \frac{\pi}{2} - \varepsilon$, $\mathcal{F}$ is not a contraction mapping in $\mathcal{H}_\ell^\lambda$. So we cannot obtain the holomorphic continuation of $\beta^\star$ in this way. To build the holomorphic continuation of $\beta^\star$ in $\mathcal{B}_\ell$ with $\ell = \frac{\pi}{2} - \varepsilon$, we must describe more carefully what happens near the singularities. This is the subject of the two next sections.

2.1.2 Sharp exponential upper bounds

In order to determine the subtlety of Lemma 2.1.1 one can apply it to the basic example $J(\varepsilon) = \displaystyle\int_{-\infty}^{+\infty} \mathrm{e}^{\mathrm{i}\omega t/\varepsilon}\cosh^{-2}(t)\ dt$ which can be explicitly computed using residues,

$$J(\varepsilon) = \frac{\pi\omega}{\varepsilon\ \sinh(\pi\omega/2\varepsilon)} \underset{\varepsilon\to 0}{\sim} \frac{2\pi\omega}{\varepsilon}\ \mathrm{e}^{-\pi\omega/2\varepsilon}.$$

The function $\xi \mapsto \cosh^{-2}\xi$ has a double pole at $\mathrm{i}\frac{\pi}{2}$. This pole limits the size of the complex strip to $\ell < \frac{\pi}{2}$. Applying Lemma 2.1.1, one then find

$$|J(\varepsilon)| \leq \left\| \frac{1}{\cosh^2(\xi)} \right\|_{H^2_\ell} \mathrm{e}^{-\omega\ell/\varepsilon} = \frac{1}{\cos^2 \ell}\, \mathrm{e}^{-\omega\ell/\varepsilon} \quad \forall \varepsilon > 0,\ \forall \ell < \frac{\pi}{2}.$$

To obtain the right coefficient in the exponential we can set $\ell = \frac{\pi}{2} - \delta\varepsilon$. We get

$$|J(\varepsilon)| \leq \frac{\mathrm{e}^{\delta\omega}}{\sin^2 \delta\varepsilon} \mathrm{e}^{-\omega\pi/2\varepsilon} \underset{\varepsilon\to 0}{\sim} \frac{\mathrm{e}^{\delta\omega}}{\delta^2\varepsilon^2} \mathrm{e}^{-\omega\pi/2\varepsilon}.$$

With this "trick" we obtain the right coefficient in the exponential, but the polynomial degeneracy is not correct. This is due to the fact that $\|\cdot\|_{H^\lambda_\ell}$ is uniform in the strip $\mathcal{B}_\ell$ and does not describe what happens near the singularities. So we give in this section a refined version of the Exponential Lemma 2.1.1 which gives the right coefficient in the exponential and the right polynomial degeneracy.

Lemma 2.1.9 (Second Mono-Frequency Exponential Lemma). *Let ω, γ, σ, λ be four positive numbers. Let δ be in $[1, +\infty[$. Let $H^{\gamma,\lambda}_{\sigma,\delta}$ be the Banach set of functions satisfying*

$$f: \bigcup_{0<\varepsilon<\sigma/\delta} \mathcal{B}_{\sigma-\delta\varepsilon} \times \{\varepsilon\} \longrightarrow \mathbb{C},$$

$$f(\cdot,\varepsilon) \text{ is holomorphic in } \mathcal{B}_{\sigma-\delta\varepsilon},$$

$$\|f\|_{H^{\gamma,\lambda}_{\sigma,\delta}} < +\infty$$

where

$$\|f\|_{H^{\gamma,\lambda}_{\sigma,\delta}} := \sup_{\substack{0<\varepsilon<\sigma/\delta \\ \xi\in\mathcal{B}_{\sigma-\delta\varepsilon}}} \left[(|f(\xi,\varepsilon)| \left(|\sigma^2 + \xi^2|^\gamma \chi_1(\xi) + (1-\chi_1(\xi))\mathrm{e}^{\lambda|\mathcal{R}e(\xi)|} \right) \right]$$

with $\chi_1(\xi) = \begin{cases} 1 & \text{for } |\mathcal{R}e\,(\xi)\,| < 1, \\ 0 & \text{for } |\mathcal{R}e\,(\xi)\,| \geq 1. \end{cases}$

Then for every $\gamma > 1$, there exits $M(\gamma, \lambda, \delta, \omega)$ such that for every $f \in H^{\gamma,\lambda}_{\sigma,\delta}$ and $\varepsilon \in]0, \sigma/\delta[$, $I^\pm(f,\varepsilon) = \int_{-\infty}^{+\infty} f(t,\varepsilon)\, \mathrm{e}^{\pm\mathrm{i}\omega t/\varepsilon}\, dt$ satisfies

$$|I^\pm(f,\varepsilon)| \leq M \frac{1}{\varepsilon^{\gamma-1}} \mathrm{e}^{-\sigma\omega/\varepsilon} \|f\|_{H^{\gamma,\lambda}_{\sigma,\delta}}.$$

Remark 2.1.10. Observe that $|\sigma^2+\xi^2|^\gamma = |(\xi-\mathrm{i}\sigma)(\xi+\mathrm{i}\sigma)|^\gamma$. Hence, roughly speaking, $H^{\gamma,\lambda}_{\sigma,\delta}$ is the space of the holomorphic functions in $\mathcal{B}_{\sigma-\delta\varepsilon}$ which decay exponentially at infinity and which explode at most as $|\xi \pm \mathrm{i}\sigma|^{-\gamma}$ for ξ near $\pm\mathrm{i}\sigma$.

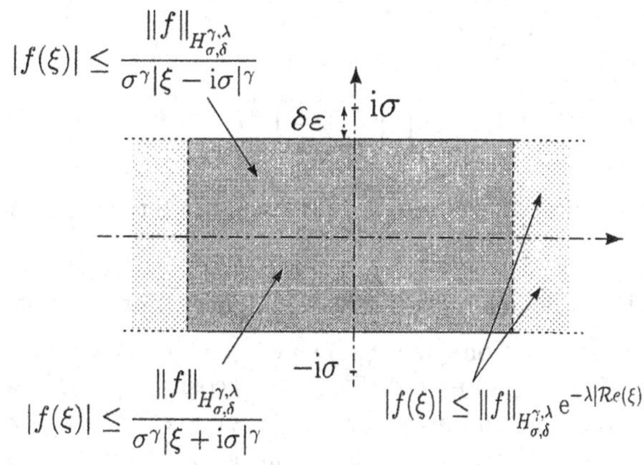

Fig. 2.2. Description of $f \in H^{\gamma,\lambda}_{\sigma,\delta}$

Proof of the Second Mono-Frequency Exponential Lemma 2.1.9. Here again, we only give the proof for I^+. For I^-, perform the change of time $t' = -t$ in the integral and observe that $f^- : \xi \mapsto f(-\xi, \varepsilon)$ also belongs to $H^{\gamma,\lambda}_{\sigma,\delta}$ and that $\|f^-\|_{H^{\gamma,\lambda}_{\sigma,\delta}} = \|f\|_{H^{\gamma,\lambda}_{\sigma,\delta}}$.

The function being holomorphic in $\mathcal{B}_{\sigma-\delta\varepsilon}$ and exponentially decaying at infinity, for every $\eta > \delta$ we get,

$$I^+(f,\varepsilon) = \int_{\mathbb{R}+i(\sigma-\eta\varepsilon)} e^{i\omega\xi/\varepsilon} f(\xi,\varepsilon)d\xi = e^{-\sigma\omega/\varepsilon} e^{\eta\omega} \int_{-\infty}^{+\infty} e^{i\omega t/\varepsilon} f(i(\sigma-\eta\varepsilon)+t,\varepsilon)dt.$$

Hence,

$$\begin{aligned}|I^+(f,\varepsilon)| &\le e^{\eta\omega-\sigma\omega/\varepsilon} \|f\|_{H^{\gamma,\lambda}_{\sigma,\delta}}\left(\left(\int_{-\infty}^{-1} + \int_1^{+\infty}\right) e^{-\lambda|t|}dt + \int_{-1}^{1} \frac{dt}{\sigma^\gamma(\eta^2\varepsilon^2+t^2)^{\gamma/2}}\right) \\ &\le e^{\eta\omega} e^{-\sigma\omega/\varepsilon} \|f\|_{H^{\gamma,\lambda}_{\sigma,\delta}} \left(2\lambda^{-1}e^{-\lambda} + \frac{\varepsilon^{1-\gamma}}{\sigma^\gamma}\int_{-\infty}^{+\infty} \frac{du}{(1+u^2)^{\gamma/2}}\right).\end{aligned}$$

Pushing η to δ, the result follows. □

Remark 2.1.11. We check for our basic example $J(\varepsilon)$ that this lemma gives the right polynomial degeneracy,

$$|J(\varepsilon)| \le M \left\|\frac{1}{\cosh^2(\xi)}\right\|_{H^{2,2}_{\frac{\pi}{2},1}} \frac{1}{\varepsilon} e^{-\pi\omega/2\varepsilon} = M \sup_{[0,\frac{\pi}{2}]} \left(\frac{x^2}{\sin^2 x}\right) \frac{1}{\varepsilon} e^{-\pi\omega/2\varepsilon}.$$

Observe for this basic example, that Lemma 2.1.9 works not only for $\gamma = 2$ but for every $\gamma \in]1,2]$. Indeed, for every $\gamma \in]1,2]$, $\varepsilon^{2-\gamma}\cosh^{-2}(\xi)$ lies in $H^{\gamma,2}_{\frac{\pi}{2},1}$. Then Lemma 2.1.9 ensures that

$$\begin{aligned} |J(\varepsilon)| &\leq \frac{1}{\varepsilon^{2-\gamma}} M(\gamma, 2, 1, \omega) \left\| \varepsilon^{2-\gamma} \cosh^{-2}(\xi) \right\|_{H^{\gamma,2}_{\frac{\pi}{2},1}} \frac{1}{\varepsilon^{\gamma-1}} \mathrm{e}^{-\pi\omega/2\varepsilon}, \\ &\leq M(\gamma, 2, 1, \omega) \sup_{[0,\frac{\pi}{2}]} \left(\frac{x^2}{\sin^2 x} \right) \frac{1}{\varepsilon} \mathrm{e}^{-\pi\omega/2\varepsilon}. \end{aligned}$$

This degree of freedom on γ is very useful when the integrand is not explicitly known.

Remark 2.1.12. The membership to $H^{\gamma,\lambda}_{\sigma,\delta}$ *does not mean that functions have a pole at* $\pm \mathrm{i}\sigma$. Functions like $(\xi, \varepsilon) \mapsto \mathrm{e}^{-\xi^2}$ which is analytic everywhere; $(\xi, \varepsilon) \mapsto \cosh^{-2}(\xi/2)$, $(\xi, \varepsilon) \mapsto \cosh^{-1}(\xi + \mathrm{i}\varepsilon^2)\cosh^{-1}(\xi - \mathrm{i}\varepsilon^2)$ which respectively have poles at $\pm \mathrm{i}\pi$ and at $\pm \mathrm{i}\frac{\pi}{2} \pm \mathrm{i}\varepsilon^2$; $(\xi, \varepsilon) \mapsto \mathrm{e}^{\mathrm{i}\varepsilon \tanh(\xi)} \cosh^{-2}(\xi)$ which has an essential singularity at $\mathrm{i}\frac{\pi}{2}$ all belong to $H^{2,2}_{\frac{\pi}{2},1}$. This comes from the fact that we work in $\mathcal{B}_{\frac{\pi}{2}-\varepsilon}$ and not in $\mathcal{B}_{\frac{\pi}{2}}$. This lemma simply describes how the function grows near $\mathrm{i}(\sigma - \delta\varepsilon)$.

Remark 2.1.13. Unlike H^{λ}_{ℓ}, $H^{\gamma,\lambda}_{\sigma,\delta}$ *is not an algebra.* However the following lemma enables us to use it in nonlinear differential equations.

Lemma 2.1.14.

(a) *For every* $f \in H^{\gamma,\lambda}_{\sigma,\delta}$ *and every* $g \in H^{\gamma',\lambda'}_{\sigma,\delta}$, fg *lies in* $H^{\gamma+\gamma',\lambda+\lambda'}_{\sigma,\delta}$.
(b) *For every* $f \in H^{\gamma,\lambda}_{\sigma,\delta}$ *and every* $p \in]0, \gamma[$, $\varepsilon^p f$ *lies in* $H^{\gamma-p,\lambda}_{\sigma,\delta}$
(c) *For every* $f, g \in H^{\gamma,\lambda}_{\sigma,\delta}$, $\varepsilon^\gamma fg$ *lies in* $H^{\gamma,\lambda}_{\sigma,\delta}$.

So, Lemma 2.1.9 can be used like the Exponential Lemma 2.1.9 to solve Problem 1.1.4 (see Remark 2.1.5) with holomorphic continuations of solutions in $H^{\gamma,\lambda}_{\sigma,\delta}$ instead of H^{λ}_{ℓ}. Here again, these continuations can be obtained by the Contraction Mapping Theorem.

Example 2.1.15. We come back to our second toy model (Example 2.1.6) and more precisely to Question 2.1.7. We give a second answer to this question using the strategy explained in Remark 2.1.5 with $H^{\gamma,\lambda}_{\sigma,\delta}$ instead of H^{λ}_{ℓ}. As we did for Example 2.1.6, we proceed in three steps.

Step 1. We look for a continuation of $\beta^\star$ in the form $\beta^\star(\xi, \varepsilon) = \beta^h(\xi) + \gamma^\star(\xi, \varepsilon)$ with $\gamma^\star(\cdot, \varepsilon) \in \mathcal{H}^{2,\lambda}_{\varepsilon}$ where

$$\mathcal{H}^{2,\lambda}_{\varepsilon} := \{ f : \mathcal{B}_{\frac{\pi}{2}-\varepsilon} \to \mathbb{C},\ \text{holomorphic in } \mathcal{B}_{\frac{\pi}{2}-\varepsilon},\ |f|_{\mathcal{H}^{2,\lambda}_{\varepsilon}} < +\infty \}$$

with $|f|_{\mathcal{H}^{2,\lambda}_{\varepsilon}} := \sup_{\xi \in \mathcal{B}_{\frac{\pi}{2}-\varepsilon}} \left[|f(\xi)| \left(\left| \xi^2 + \frac{\pi^2}{4} \right|^2 \chi_1(\xi) + (1 - \chi_1(\xi)) \mathrm{e}^{\lambda |\mathcal{R}e(\xi)|} \right) \right]$ and $\lambda \in]0, 2[$.

Here again, we first see $\gamma^\star$ as a function of ξ with parameter ε and $\mathcal{H}_\varepsilon^{2,\lambda}$ as a space of functions of ξ, which depends on ε. We expect $\gamma^\star$ to lie in $H^{1,\lambda}_{\frac{\pi}{2},1}$ and more precisely to behave like the function $\widetilde{\gamma} := \varepsilon^2 \cosh^{-3} \xi$. Observe that

$$\sup_{\xi \in \mathcal{B}_{\frac{\pi}{2}-\varepsilon}} \left[|\widetilde{\gamma}(\xi,\varepsilon)| \left(\left|\xi^2 + \tfrac{\pi^2}{4}\right|^1 \chi_1(\xi) + (1-\chi_1(\xi)) e^{\lambda|\mathcal{R}e(\xi)|} \right) \right] \leq M$$

holds for ε small, whereas

$$\sup_{\xi \in \mathcal{B}_{\frac{\pi}{2}-\varepsilon}} \left[|\widetilde{\gamma}(\xi,\varepsilon)| \left(\left|\xi^2 + \tfrac{\pi^2}{4}\right|^2 \chi_1(\xi) + (1-\chi_1(\xi)) e^{\lambda|\mathcal{R}e(\xi)|} \right) \right] \leq M\varepsilon.$$

These two estimates ensure that $\widetilde{\gamma} \in H^{1,\lambda}_{\frac{\pi}{2},1}$ but only the later one can be obtained for $\gamma^\star$ using the Contraction Mapping Theorem because we need small norms. This explains the apparently strange choice of the space $\mathcal{H}_\varepsilon^{2,\lambda}$ for finally obtaining that $\gamma^\star$ belongs to $H^{1,\lambda}_{\frac{\pi}{2},1}$.

Step 2. In Example 2.1.6, we proved that $\gamma^\star$ is the unique holomorphic function defined in a neighborhood of 0 which satisfies $\gamma^\star = \mathcal{F}(\gamma^\star)$ where

$$\mathcal{F}(\gamma)(\xi) = \cosh^{-2}(\xi) \int_{[0,\xi]} \cosh^2(\zeta)\, G(\gamma(\zeta))\, d\zeta.$$

The estimates

$$\begin{aligned} |\mathcal{F}(\gamma)|_{\mathcal{H}_\varepsilon^{2,\lambda}} &\leq M \left(\frac{|\gamma|^2_{\mathcal{H}_\varepsilon^{2,\lambda}}}{\varepsilon} + \rho\varepsilon \left(1 + \frac{|\gamma|_{\mathcal{H}_\varepsilon^{2,\lambda}}}{\varepsilon}\right)^4 \right) \\ |\mathcal{F}(\gamma) - \mathcal{F}(\gamma')|_{\mathcal{H}_\varepsilon^{2,\lambda}} &\leq M \\ & \left(\frac{|\gamma|_{\mathcal{H}_\varepsilon^{2,\lambda}} + |\gamma'|_{\mathcal{H}_\varepsilon^{2,\lambda}}}{\varepsilon} + \rho\varepsilon \left(1 + \frac{|\gamma|_{\mathcal{H}_\varepsilon^{2,\lambda}}}{\varepsilon}\right)^3 \right) |\gamma - \gamma'|_{\mathcal{H}_\varepsilon^{2,\lambda}} \end{aligned}$$

which hold for every $\gamma, \gamma' \in \mathcal{H}_\varepsilon^{2,\lambda}$, $|\gamma|_{\mathcal{H}_\varepsilon^{2,\lambda}} \leq 1$, $|\gamma'|_{\mathcal{H}_\varepsilon^{2,\lambda}} \leq 1$ ensure that for $|\rho|$ sufficiently small, $\mathcal{F}$ is a contraction mapping from $\mathcal{BH}_\varepsilon^{2,\lambda}(2M\rho\varepsilon)$ to itself where $\mathcal{BH}_\varepsilon^{2,\lambda}(d) = \{f \in \mathcal{H}_\varepsilon^{2,\lambda} /\ |f|_{\mathcal{H}_\varepsilon^{2,\lambda}} \leq d\}$. We can then conclude that for $|\rho|$ sufficiently small, $\beta^\star$ has a holomorphic continuation of the form

$$\beta^\star = \beta^h + \gamma^\star \qquad \text{where } \gamma^\star \in \mathcal{H}_\varepsilon^{2,\lambda} \text{ and } |\gamma^\star|_{\mathcal{H}_\varepsilon^{2,\lambda}} \leq 2M\rho\varepsilon.$$

Hence, $\varepsilon^{-1}\gamma^\star$ lies in $H^{2,\lambda}_{\frac{\pi}{2},1}$. Thus, $\gamma^\star \in H^{1,\lambda}_{\frac{\pi}{2},1}$ and $\|\gamma^\star\|_{H^{1,\lambda}_{\frac{\pi}{2},1}} \leq 2M\rho$.

Step 3. For estimating $I(\varepsilon,\rho)$ we split it once more, into two parts: on one side the Melnikov function which can be explicitly computed

$$M_e(\varepsilon,\rho)=\rho\varepsilon\int_{-\infty}^{+\infty} e^{i\omega s/\varepsilon}\left(1-(\beta^h)^2\right)ds=\frac{\pi\omega\rho}{\sinh(\pi\omega/2\varepsilon)}.$$

and on the other side a second part involving $\gamma^\star$ which reads

$$I_2(\varepsilon,\rho)=\rho\varepsilon\int_{-\infty}^{+\infty} e^{i\omega s/\varepsilon}\left(-2\beta^h\gamma^\star-(\gamma^\star)^2\right)ds.$$

Since $\left(-2\beta^h\gamma^\star-(\gamma^\star)^2\right)\in H^{2,\lambda}_{\frac{\pi}{2},1}$, I_2 can be bounded using Lemma 2.1.9. We get for $|\rho|$ sufficiently small,

$$\begin{aligned} I(\varepsilon,\rho) &= \frac{\pi\omega\rho}{\sinh(\pi\omega/2\varepsilon)}+e^{-\pi\omega/2\varepsilon}\underset{\rho\to 0}{\mathcal{O}}\left(\rho^2\right), \\ &= \rho e^{-\pi\omega/2\varepsilon}\left(2\pi\omega+\underset{\varepsilon\to 0}{\mathcal{O}}\left(e^{-\pi\omega/\varepsilon}\right)+\underset{\rho\to 0}{\mathcal{O}}(\rho)\right). \end{aligned} \tag{2.5}$$

Lemma 2.1.9 gives an equivalent for $I(\varepsilon,\rho)$ for ε and ρ small (but not connected). This equivalent is given by the Melnikov function. It ensures that for $\rho\neq 0$ small and for ε small, $I(\varepsilon,\rho)$ does not vanish. So the front of (2.1) does not persists for $\rho\neq 0$ small and for ε small. For $\rho=1$, we need a more refined method given in next section. The description of the holomorphic continuation of $\beta^\star$ obtained previously is not sharp enough to obtain an equivalent for $\rho=1$.

2.1.3 Exponential equivalent : general theory

In this section, our aim is to determine what kind of informations on a function f we must know to compute an equivalent of the oscillatory integral

$$I(f,\varepsilon)=\int_{-\infty}^{+\infty} f(t,\varepsilon)\, e^{i\omega t/\varepsilon}\, dt.$$

Moreover, we must be able to obtain this kind of information when f is not explicitly known and involves solutions of O.D.Es. or P.D.Es. The membership to $H^{\gamma,\lambda}_{\sigma,\delta}$ only gives upper bounds of $I(f,\varepsilon)$.

A preliminary idea is to believe that the equivalent of the oscillatory integral $I(f,\varepsilon)$, is given by the leading part of f on $\mathbb{R}$. The following example ensures that this is not always true.

Example 2.1.16. Let f be given by $f(t,\varepsilon)=f_1(t,\varepsilon)+f_2(t,\varepsilon)$ with

$$f_1(t,\varepsilon)=\frac{1}{\cosh^2(t)}\qquad\text{and}\qquad f_2(t,\varepsilon)=\frac{\varepsilon^\nu}{\cosh^4(t)}.$$

The leading part of f on $\mathbb{R}$ is f_1 ($f\underset{\varepsilon\to 0}{\sim} f_1$). The oscillatory integrals can be computed using residues

$$I_1 := I(f_1,\varepsilon) = \frac{\pi\omega}{\varepsilon\ \sinh(\pi\omega/2\varepsilon)},$$
$$I_2 := I(f_2,\varepsilon) = \frac{\pi\omega}{\varepsilon\ \sinh(\pi\omega/2\varepsilon)} \left(\frac{\omega^2}{6\varepsilon^2}+\frac{1}{3}\right)\varepsilon^{\nu}.$$

Hence,

$$I(f,\varepsilon) \underset{\varepsilon\to 0}{\sim} I_1 \underset{\varepsilon\to 0}{\sim} \frac{2\pi\omega}{\varepsilon}\mathrm{e}^{-\pi\omega/2\varepsilon} \qquad \text{for } \nu > 2,$$
$$I(f,\varepsilon) \underset{\varepsilon\to 0}{\sim} I_1 + I_2 \underset{\varepsilon\to 0}{\sim} \frac{2\pi\omega}{\varepsilon}\left(1+\frac{\omega^2}{6}\right)\mathrm{e}^{-\pi\omega/2\varepsilon} \qquad \text{for } \nu = 2,$$
$$I(f,\varepsilon) \underset{\varepsilon\to 0}{\sim} I_2 \underset{\varepsilon\to 0}{\sim} \frac{\pi\omega^3}{\varepsilon^{3-\nu}}\mathrm{e}^{-\pi\omega/2\varepsilon} \qquad \text{for } 0 < \nu < 2.$$

A second idea, comes from the previous explicit computations. We check that the coefficient in the exponential comes from the position of the singularity of f and that the polynomial coefficient ε^{-p^*} comes from the order of the pole. So, for computing oscillatory integrals involving solutions of analytic differential equations we have to face Problem 1.3.1, i.e. we must be able to determine whether a solution of an analytic differential equation admits an analytic continuation. If it is the case, we must then determine the position and the nature of its singularities in $\mathbb{C}$. Unfortunately, there is no general theory for computing the nature and the position of singularities of solutions of complex differential equations. The only available theory is the theory of Fuchs for linear time dependent equations. All the difficulties come from the fact that the nature and the position of singularities of analytic functions are not stable by addition, multiplication, composition, integration. For instance, accumulation of poles may lead to essential singularities and logarithmic singularities may appear after integration... Moreover, we study O.D.Es or P.D.Es with a small parameter ε. So the nature and the position of the singularities of the solution may depend on ε. At last, we observe in Example 2.1.16, that the value of the equivalent depends in a non trivial way of the coefficient ε^{ν} in front of the pole of order 4 at $\pm\mathrm{i}\frac{\pi}{2}$. So we must know very precisely how the small parameter is involved in the singularity. At the present time, such an approach seems beyond possibilities.

A third idea, comes once more, from meromorphic functions for which explicit computations are available : let $f(.,\varepsilon)$ be a holomorphic function in $\mathcal{B}_\ell \setminus \{\pm\mathrm{i}\sigma\}$ with $\ell > \sigma$. Assume that $f(.,\varepsilon)$ has poles of order p at $\pm\mathrm{i}\sigma$ and that $|f(\xi,\varepsilon)|$ decreases exponentially to 0 when $\mathcal{R}e\,(\xi) \to \pm\infty$. Hence f reads

$$f(\xi,\varepsilon) = \sum_{k=1}^{p} \frac{a_k(\varepsilon)\varepsilon^{n_k}}{(\xi - \mathrm{i}\sigma)^k} + g(\xi,\varepsilon)$$

where $g(.,\varepsilon)$ is holomorphic in $\mathcal{B}_\ell \setminus \{-\mathrm{i}\sigma\}$ and $a_k(\varepsilon)$ is a smooth function of ε satisfying $a_k(0) \neq 0$.

Denote by

$$p^* = \max_{0\le k\le p} (k - n_k) \qquad \text{and} \qquad \mathcal{K} = \{k \in [1..p]/k - n_k = p^*\}.$$

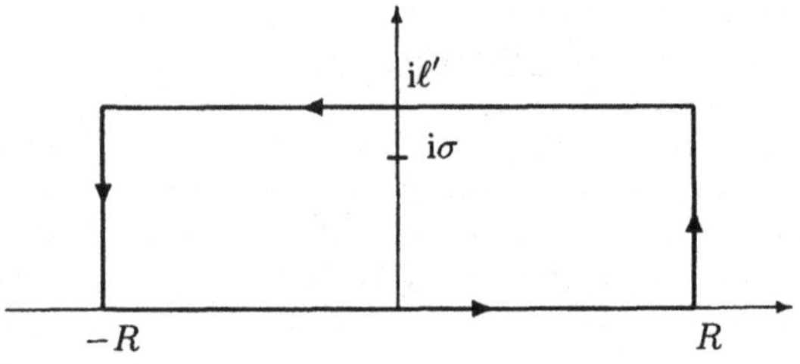

Fig. 2.3. Path $\Gamma_{\ell',R}$.

Using the residues with the path $\Gamma_{\ell',R}$ (see Fig. 2.3) for $\sigma < \ell' < \ell$ and pushing R to $+\infty$, we get

$$\begin{aligned} I(f,\varepsilon) &\underset{\varepsilon\to 0}{\sim} \frac{\mathrm{e}^{-\omega\sigma/\varepsilon}}{\varepsilon^{p^*-1}} 2\pi\mathrm{i} \sum_{k\in\mathcal{K}} a_k(0)\,\frac{(\mathrm{i}\omega)^{k-1}}{(k-1)!}, \\ &\underset{\varepsilon\to 0}{\sim} \int_{\mathbb{R}} f_0(t)\mathrm{e}^{\mathrm{i}\omega t/\varepsilon} dt \end{aligned}$$

where

$$f_0(\xi) = \sum_{k\in\mathcal{K}} \frac{a_k(0)\varepsilon^{n_k}}{(\xi - \mathrm{i}\sigma)^k}.$$

We clearly see on this example that not all the terms in the Laurent expansion of f contribute to the equivalent. The only relevant ones are those which belong to $\mathcal{K}$. Hence the relevant part of f is not the leading part of f on $\mathbb{R}$, but it is f_0. *The fundamental remark is that* for $p^* > 0$, f_0 is given by

$$f_0(\xi) = \frac{1}{\varepsilon^{p^*}} \widehat{f_0}\left(\frac{\xi - \mathrm{i}\sigma}{\varepsilon}\right) \qquad \text{with } \widehat{f_0}(z) = \lim_{\varepsilon\to 0} \varepsilon^{p^*} f(\mathrm{i}\sigma + \varepsilon z, \varepsilon). \tag{2.6}$$

and that

$$\begin{aligned} I(f,\varepsilon) &\underset{\varepsilon\to 0}{\sim} \int_{\mathbb{R}} f_0(t)\mathrm{e}^{\mathrm{i}\omega t/\varepsilon} dt \\ &\underset{\varepsilon\to 0}{\sim} \frac{\mathrm{e}^{-\omega\sigma/\varepsilon}}{\varepsilon^{p^*-1}} \int_{-\mathrm{i}\eta+\mathbb{R}} \widehat{f_0}(z)\mathrm{e}^{\mathrm{i}\omega z} dz \end{aligned} \tag{2.7}$$

for every $\eta > 0$.

Observe that the last integral is a fixed number which does not depend on η *and* ε.

Thus, formulas (2.6) and (2.7) give a process to compute an equivalent of oscillatory integrals involving meromorphic functions with $p^* > 0$. The next lemmas show that this process can be used for a class of functions far more general than the meromorphic ones and that it can be used with nonlinear differential equations.

Definition 2.1.17. *Let $\gamma, \lambda > 0$, $\sigma > 0$, $\delta \geq 1$, $m_0, m_1 > 1$, $0 \leq r < 1$ and $\nu > 0$. Denote by $E^{\gamma,\lambda}_{\sigma,\delta}(m_0, m_1, r, \nu)$ the set of functions f such that*

(a) $f \in H^{\gamma,\lambda}_{\sigma,\delta}$,

(b) *for every $\varepsilon \in]0, \varepsilon_0]$ and every $z \in \widehat{\Sigma}_{\delta,r,\varepsilon} :=]-\varepsilon^{r-1}, \varepsilon^{r-1}[\times]-\varepsilon^{r-1}, -\delta[$ f reads*

$$f(\mathrm{i}\sigma + \varepsilon z, \varepsilon) = \frac{1}{\varepsilon^\gamma}\widehat{f_0^+}(z) + \frac{\varepsilon^\nu}{\varepsilon^\gamma}\widehat{f_1^+}(z, \varepsilon)$$

where $\widehat{f_0^+} \in \widehat{H}^{m_0}_\delta$ and $\widehat{f_1^+} \in \widehat{H}^{m_1}_{\delta,r}$ with

$$\widehat{H}^{m_0}_\delta = \Big\{ \widehat{f} : \widehat{\Omega}_\delta \to \mathbb{C},\ \widehat{f} \text{ holomorphic in } \widehat{\Omega}_\delta := \mathbb{R}\times] - \infty, -\delta[,$$
$$\left\|\widehat{f}\right\|_{\widehat{H}^{m_0}_\delta} := \sup_{z\in\widehat{\Omega}_\delta} \left(|\widehat{f}(z)||z|^{m_0}\right) < +\infty \Big\}$$

and

$$\widehat{H}^{m_1}_{\delta,r} = \Big\{ \widehat{f} : \bigcup_{\varepsilon\in]0,\varepsilon_0]} \left(\widehat{\Sigma}_{\delta,r,\varepsilon} \times \{\varepsilon\}\right) \to \mathbb{C},\ \widehat{f}(\cdot, \varepsilon) \text{ holomorphic in } \widehat{\Sigma}_{\delta,r,\varepsilon},$$
$$\left\|\widehat{f}\right\|_{\widehat{H}^{m_1}_{\delta,r}} := \sup_{\substack{\varepsilon\in]0,\varepsilon_0] \\ z\in\widehat{\Sigma}_{\delta,r,\varepsilon}}} \left(|\widehat{f}(z)||z|^{m_1}\right) < +\infty \Big\}$$

(c) *for every $\varepsilon \in]0, \varepsilon_0]$ and every $z \in \widehat{\Sigma}_{\delta,r,\varepsilon}$ f reads*

$$f(-\mathrm{i}\sigma - \varepsilon z, \varepsilon) = \frac{1}{\varepsilon^\gamma}\widehat{f_0^-}(z) + \frac{\varepsilon^\nu}{\varepsilon^\gamma}\widehat{f_1^-}(z, \varepsilon)$$

where $\widehat{f_0^-} \in \widehat{H}^{m_0}_\delta$ and $\widehat{f_1^-} \in \widehat{H}^{m_1}_{\delta,r}$

Remark 2.1.18. The four functions $\widehat{f_0^\pm}$ and $\widehat{f_1^\pm}$ are uniquely defined.
Indeed, they are given by

$$\widehat{f_0^\pm}(z) := \lim_{\varepsilon\to 0} \varepsilon^\gamma f(\pm(\mathrm{i}\sigma + \varepsilon z), \varepsilon)$$

and

$$\widehat{f_1^\pm}(z) = \lim_{\varepsilon\to 0} \frac{\varepsilon^\gamma f(\pm(\mathrm{i}\sigma + \varepsilon z)) - \widehat{f_0^\pm}(z)}{\varepsilon^\nu}.$$

Hence, we have

Lemma 2.1.19.
Let us denote by $\mathrm{P}_{\mathrm{r}}^{\pm}$ *the operators*

$$\mathrm{P}_{\mathrm{r}}^{\pm} : \begin{array}{ccc} E_{\sigma,\delta}^{\gamma,\lambda}(m_0, m_1, r, \nu) & \rightarrow & \widehat{H}_{\delta}^{m_0} \\ f & \longmapsto & \widehat{f}_0^{\pm} \end{array}.$$

Then, the operators $\mathrm{P}_{\mathrm{r}}^{\pm}$ *satisfy*

(a) *When* f *is even ,* $\mathrm{P}_{\mathrm{r}}^{-}(f)(z) = \mathrm{P}_{\mathrm{r}}^{+}(f)(z)$ *holds for* $z \in \widehat{\Omega}_{\delta}$,
(b) *When* f *is odd ,* $\mathrm{P}_{\mathrm{r}}^{-}(f)(z) = -\mathrm{P}_{\mathrm{r}}^{+}(f)(z)$ *holds for* $z \in \widehat{\Omega}_{\delta}$,
(c) *When* f *is real valued on the real axis,* $\mathrm{P}_{\mathrm{r}}^{-}(f)(z) = \overline{\mathrm{P}_{\mathrm{r}}^{+}(f)(-\overline{z})}$ *holds for* $z \in \widehat{\Omega}_{\delta}$.

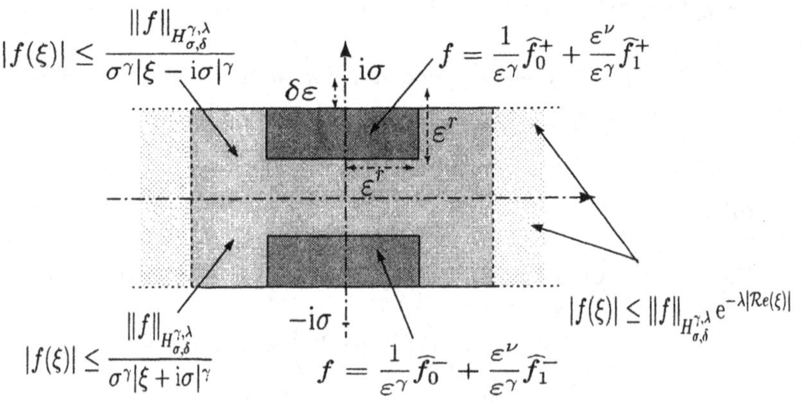

Fig. 2.4. Description of $f \in E_{\sigma,\delta}^{\gamma,\lambda}$

Remark 2.1.20. The membership to $H_{\sigma,\delta}^{\gamma,\lambda}$ simply ensures that $f(\cdot,\varepsilon)$ decays exponentially for $|\mathcal{R}e\,(\xi)\,| \rightarrow \infty$ and explodes at most as $|\xi \mp \mathrm{i}\sigma|^{\gamma}$ for ξ close to $\pm\mathrm{i}(\sigma - \delta\varepsilon)$. Membership of $E_{\sigma,\delta}^{\gamma,\lambda}$ specifies the behavior of the function $f(\cdot,\varepsilon)$ near $\pm\mathrm{i}(\sigma - \delta\varepsilon)$. In the domain $]-\varepsilon^r, \varepsilon^r[\times]\sigma - \varepsilon^r, \sigma - \delta\varepsilon[$, (see Fig. 2.4) $f(\xi,\varepsilon)$ is given by

$$f(\xi,\varepsilon) = \frac{1}{\varepsilon^{\gamma}}\widehat{f}_0^{+}\left(\frac{\xi - \mathrm{i}\sigma}{\varepsilon}\right) + \frac{\varepsilon^{\nu}}{\varepsilon^{\gamma}} f_1^{+}\left(\frac{\xi - \mathrm{i}\sigma}{\varepsilon}\right)$$

and for ξ close to $\mathrm{i}(\sigma - \delta\varepsilon)$, $f(\xi,\varepsilon)$ explodes like $\frac{1}{\varepsilon^{\gamma}}\widehat{f}_0^{+}(\mathrm{i}\delta) = \frac{1}{\varepsilon^{\gamma}}\mathrm{P}_{\mathrm{r}}^{+}(f)(\mathrm{i}\delta)$. So, $\mathrm{P}_{\mathrm{r}}^{+}(f)$ is the "leading part of f near $\mathrm{i}(\sigma - \delta\varepsilon)$". We call it the principal part of f. Similarly, $\mathrm{P}_{\mathrm{r}}^{-}(f)$ is the principal part of f near $-\mathrm{i}(\sigma - \delta\varepsilon)$

Moreover, the membership to $E_{\sigma,\delta}^{\gamma,\lambda}(m_0, m_1, r, \nu)$ *does not mean that functions have poles at* $\pm\mathrm{i}\sigma$. All the following functions belong to the space

$E^{2,2}_{\frac{\pi}{2},1}\left(2,2,\frac{1}{2},\frac{1}{2}\right)$, but only first two have poles at $\pm i\frac{\pi}{2}$. This comes from the fact that we work in $\mathcal{B}_{\frac{\pi}{2}-\varepsilon}$ and not in $\mathcal{B}_{\frac{\pi}{2}}$.

- The function $f : (\xi,\varepsilon) \mapsto \cosh^{-2}(\xi)$ has poles of order 2 at $\pm\frac{i\pi}{2}$ and $\mathrm{P}^{\pm}_{\Gamma}(f) = -z^{-2}$;
- The function $f : (\xi,\varepsilon) \mapsto \varepsilon^2\cosh^{-4}(\xi)$ has poles of order 4 at $\pm\frac{i\pi}{2}$ and $\mathrm{P}^{\pm}_{\Gamma}(f) = z^{-4}$;
- The function $f : (\xi,\varepsilon) \mapsto e^{-\xi^2}$ is entire and $\mathrm{P}^{\pm}_{\Gamma}(f) = 0$;
- The function $f : (\xi,\varepsilon) \mapsto \cosh^{-2}(\xi/2)$ has poles at $\pm i\pi$ and $\mathrm{P}^{\pm}_{\Gamma}(f) = 0$;
- The function $f : (\xi,\varepsilon) \mapsto \cosh^{-1}(\xi + i\varepsilon^2)\cosh^{-1}(\xi - i\varepsilon^2)$ has poles at $\pm i\frac{\pi}{2} \pm i\varepsilon^2$ and $\mathrm{P}^{\pm}_{\Gamma}(f) = -z^{-2}$;
- The function $f : (\xi,\varepsilon) \mapsto e^{i\varepsilon\tanh(\xi)}\cosh^{-2}(\xi)$ has an essential singularity at $i\frac{\pi}{2}$ and $\mathrm{P}^{\pm}_{\Gamma}(f) = -z^{-2}e^{\pm i/z}$.

The membership to $E^{\gamma,\lambda}_{\sigma,\delta}$ gives enough information on f to compute an equivalent of the oscillatory integral involving it.

Lemma 2.1.21 (Third Mono-Frequency Exponential Lemma). *Let $\gamma > 1$, $\delta \geq 1$, $\sigma,\lambda > 0$, $m_0, m_1 > 1$, $0 < r < 1$, $\nu,\omega > 0$. For every function $f \in E^{\gamma,\lambda}_{\sigma,\delta}(m_0,m_1,r,\nu)$ the oscillatory integral*

$$I^{\pm}(f,\varepsilon) := \int_{-\infty}^{+\infty} e^{\pm i\omega t/\varepsilon} f(t,\varepsilon)\, dt$$

satisfy

$$I^{\pm}(f,\varepsilon) = \frac{e^{-\sigma\omega/\varepsilon}}{\varepsilon^{\gamma-1}}\left(\Lambda^{\pm} + \underset{\varepsilon\to 0}{\mathcal{O}}\left(\varepsilon^{\nu^*}\right)\right)$$

where $\nu^* := \min\big(\nu, (1-r)(m_0-1), (1-r)(\gamma-1)\big)$ and

$$\Lambda^{\pm} := \int_{-i\eta+\mathbb{R}} \mathrm{P}^{\pm}_{\Gamma}(f)(z)\, e^{i\omega z} dz,$$

for $\eta > \delta$ ($\Lambda^{\pm}$ does not depend on η).

This lemma ensures that the relevant parts of $f \in E^{\gamma,\lambda}_{\sigma,\delta}$ for computing an equivalent of the oscillatory integrals $I^{\pm}(f,\varepsilon)$ are $\mathrm{P}^{\pm}_{\Gamma}(f)$. Moreover, it gives a very efficient way to compute equivalents of oscillatory integrals involving solutions of nonlinear differential equations since *the principal part $\mathrm{P}^{\pm}_{\Gamma}$ of a sum (resp. of a product) is the sum (resp. the product) of the principal parts.*

Lemma 2.1.22.

(a) $E^{\gamma,\lambda}_{\sigma,\delta}(m_0, m_1, r, \nu)$ *is a linear space and* $\mathrm{Pr}^\pm : E^{\gamma,\lambda}_{\sigma,\delta}(m_0, m_1, r, \nu) \mapsto \widehat{H}^{m_0}_\delta$ *are linear operators.*

(b) *The two Banach spaces* $\widehat{H}^{m_0}_\delta$ *and* $\widehat{H}^{m_1}_{\delta,r}$ *are algebras.*

(c) *For every* $f \in E^{\gamma,\lambda}_{\sigma,\delta}$ *and every* $g \in E^{\gamma',\lambda'}_{\sigma,\delta}$, fg *lies in* $E^{\gamma+\gamma',\lambda+\lambda'}_{\sigma,\delta}$ *and*

$$\mathrm{Pr}^\pm_{E^{\gamma+\gamma',\lambda+\lambda'}_{\sigma,\delta}}(fg) = \mathrm{Pr}^\pm_{E^{\gamma,\lambda}_{\sigma,\delta}}(f)\mathrm{Pr}^\pm_{E^{\gamma',\lambda'}_{\sigma,\delta}}(g).$$

(d) *For every* $f \in E^{\gamma,\lambda}_{\sigma,\delta}$ *and every* $p \in [0, \gamma[$, $\varepsilon^p f$ *lies in* $E^{\gamma-p,\lambda}_{\sigma,\delta}$ *and*

$$\mathrm{Pr}^\pm_{E^{\gamma-p,\lambda}_{\sigma,\delta}}(\varepsilon^p f) = \mathrm{Pr}^\pm_{E^{\gamma,\lambda}_{\sigma,\delta}}(f).$$

(e) *For every* $f \in E^{\gamma,\lambda}_{\sigma,\delta}$ *and every* $g \in E^{\gamma,\lambda}_{\sigma,\delta}$, $\varepsilon^\gamma fg$ *lies in* $E^{\gamma,\lambda}_{\sigma,\delta}$ *and*

$$\mathrm{Pr}^\pm(\varepsilon^\gamma fg) = \mathrm{Pr}^\pm(f)\mathrm{Pr}^\pm(g).$$

Proof of the Third Mono-Frequency Exponential Lemma 2.1.21. Here again, we only do the proof for I^+. For I^-, perform the change of time $t' = -t$ in the integral and observe that $f^- : \xi \mapsto f(-\xi)$ also belongs to $E^{\gamma,\lambda}_{\sigma,\delta}$, $\mathrm{Pr}^+(f^-) = \mathrm{Pr}^-(f)$ and that $\mathrm{Pr}^-(f^-) = \mathrm{Pr}^+(f)$.

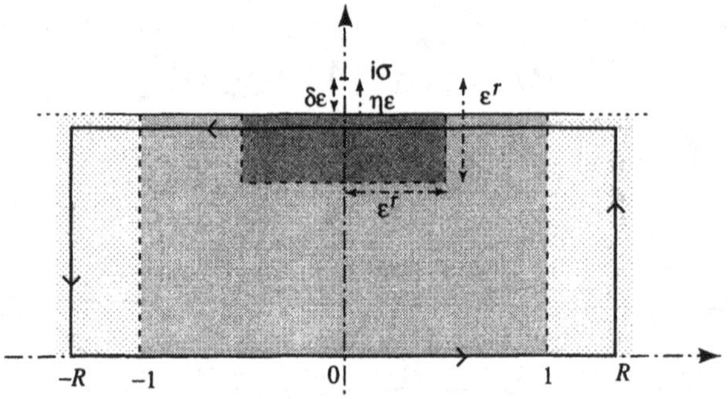

Fig. 2.5. Path $\Gamma_{\eta,R}$

Let η be fixed in $]\delta, 1/\varepsilon^{1-r}[$. The function $\xi \mapsto f(\xi,\varepsilon)\mathrm{e}^{\mathrm{i}\omega\xi/\varepsilon}$ decreases exponentially for $|\mathcal{R}e(\xi)| \to \infty$ and is holomorphic in $\mathcal{B}_{\sigma-\delta\varepsilon}$. Hence, using the path $\Gamma_{\eta,R}$ (see Fig. 2.5) and pushing R to ∞ we get

$$I^+(f,\varepsilon) = \int_{\mathrm{i}(\sigma-\eta\varepsilon)+\mathbb{R}} \mathrm{e}^{\mathrm{i}\omega\xi/\varepsilon} f(\xi,\varepsilon)\, d\xi.$$

For estimating $I^+(f,\varepsilon)$ we split it in three parts corresponding to the three subdomains crossed by the line $\mathrm{i}(\sigma-\eta\varepsilon)+\mathbb{R}$.

$$\begin{aligned}
I_1 &= \left(\int_{-\infty}^{-1}+\int_{1}^{+\infty}\right) f(\mathrm{i}(\sigma-\eta\varepsilon)+t)\ \mathrm{e}^{-\omega\sigma/\varepsilon}\mathrm{e}^{\eta\omega}\mathrm{e}^{\mathrm{i}\omega t/\varepsilon}dt,\\
I_2 &= \left(\int_{-1}^{-\varepsilon^r}+\int_{\varepsilon^r}^{1}\right) f(\mathrm{i}(\sigma-\eta\varepsilon)+t)\ \mathrm{e}^{-\omega\sigma/\varepsilon}\mathrm{e}^{\eta\omega}\mathrm{e}^{\mathrm{i}\omega t/\varepsilon}dt,\\
I_3 &= \int_{-\varepsilon^r}^{\varepsilon^r} f(\mathrm{i}(\sigma-\eta\varepsilon)+t)\ \mathrm{e}^{-\omega\sigma/\varepsilon}\mathrm{e}^{\eta\omega}\mathrm{e}^{\mathrm{i}\omega t/\varepsilon}dt.
\end{aligned}$$

Hence,

$$\begin{aligned}
|I_1| &\le 2\lambda^{-1}\mathrm{e}^{-\lambda}\,\|f\|_{H^{\gamma,\lambda}_{\sigma,\delta}}\ \mathrm{e}^{\eta\omega}\ \mathrm{e}^{-\omega\sigma/\varepsilon},\\
&\le M(\lambda)\,\|f\|_{H^{\gamma,\lambda}_{\sigma,\delta}}\,\mathrm{e}^{\eta\omega}\ \frac{\mathrm{e}^{-\omega\sigma/\varepsilon}}{\varepsilon^{\gamma-1}}\ \varepsilon^{\gamma-1},
\end{aligned}\tag{2.8}$$

and

$$\begin{aligned}
|I_2| &\le 2\int_{\varepsilon^r}^{1}\frac{\mathrm{e}^{\eta\omega}\mathrm{e}^{-\omega\sigma/\varepsilon}}{\sigma^\gamma\,|\eta^2\varepsilon^2+t^2|^{\gamma/2}}\ \|f\|_{H^{\gamma,\lambda}_{\sigma,\delta}}\ dt,\\
&\le \frac{2\mathrm{e}^{\eta\omega}}{\sigma^\gamma}\ \frac{\mathrm{e}^{-\omega\sigma/\varepsilon}}{\varepsilon^{\gamma-1}}\,\|f\|_{H^{\gamma,\lambda}_{\sigma,\delta}}\int_{\varepsilon^{r-1}}^{\varepsilon^{-1}}\frac{ds}{|\eta^2+s^2|^{\gamma/2}},\\
&\le \frac{2\mathrm{e}^{\eta\omega}}{\sigma^\gamma}\ \frac{\mathrm{e}^{-\omega\sigma/\varepsilon}}{\varepsilon^{\gamma-1}}\,\|f\|_{H^{\gamma,\lambda}_{\sigma,\delta}}\int_{\varepsilon^{r-1}}^{\varepsilon^{-1}}\frac{ds}{s^\gamma},\\
&\le M(\sigma,\gamma)\,\|f\|_{H^{\gamma,\lambda}_{\sigma,\delta}}\ \mathrm{e}^{\eta\omega}\ \frac{\mathrm{e}^{-\omega\sigma/\varepsilon}}{\varepsilon^{\gamma-1}}\ \varepsilon^{(1-r)(\gamma-1)}.
\end{aligned}\tag{2.9}$$

Setting $t=\varepsilon s$, we get

$$\begin{aligned}
I_3 &= \varepsilon\int_{-1/\varepsilon^{1-r}}^{1/\varepsilon^{1-r}}\left(\frac{1}{\varepsilon^\gamma}\widehat{f}_0(-\mathrm{i}\eta+s)+\frac{\varepsilon^\nu}{\varepsilon^\gamma}\widehat{f}_1(-\mathrm{i}\eta+s)\right)\ \mathrm{e}^{-\omega\sigma/\varepsilon}\mathrm{e}^{\eta\omega}\mathrm{e}^{\mathrm{i}\omega s}ds,\\
&= \frac{\mathrm{e}^{-\omega\sigma/\varepsilon}}{\varepsilon^{\gamma-1}}\left(\Lambda^+(\eta)+\Lambda_0(\eta,\varepsilon)+\Lambda_1(\eta,\varepsilon)\right)
\end{aligned}$$

with

$$\begin{aligned}
\Lambda^+(\eta) &= \int_{-\mathrm{i}\eta+\mathbb{R}}\widehat{f}_0(z)\ \mathrm{e}^{\mathrm{i}\omega z}dz,\\
\Lambda_0(\eta,\varepsilon) &= -\left(\int_{-\infty}^{-1/\varepsilon^{1-r}}+\int_{1/\varepsilon^{1-r}}^{+\infty}\right)\widehat{f}_0(-\mathrm{i}\eta+s)\mathrm{e}^{\mathrm{i}\omega(-\mathrm{i}\eta+s)}\ ds,\\
\Lambda_1(\eta,\varepsilon) &= \varepsilon^\nu\int_{-1/\varepsilon^{1-r}}^{1/\varepsilon^{1-r}}\widehat{f}_1(-\mathrm{i}\eta+s)\mathrm{e}^{\mathrm{i}\omega(-\mathrm{i}\eta+s)}\ ds.
\end{aligned}$$

The functions $\widehat{f}_0$, $\widehat{f}_1$ belong respectively to $\widehat{H}_\delta^{m_0}$ and $\widehat{H}_{\delta,r}^{m_1}$. So, Λ_0 and Λ_1 can be easily bounded

$$\begin{aligned}|\Lambda_0(\eta,\varepsilon)| &\le 2\mathrm{e}^{\eta\omega}\left\|\widehat{f}_0\right\|_{\widehat{H}_\delta^{m_0}}\int_{1/\varepsilon^{1-r}}^{+\infty}\frac{ds}{(\eta^2+s^2)^{m_0/2}},\\ &\le 2\mathrm{e}^{\eta\omega}\left\|\widehat{f}_0\right\|_{\widehat{H}_\delta^{m_0}}\int_{1/\varepsilon^{1-r}}^{+\infty}\frac{1}{s^{m_0}}ds,\\ &\le M(m_0)\left\|\widehat{f}_0\right\|_{\widehat{H}_\delta^{m_0}}\mathrm{e}^{\eta\omega}\varepsilon^{(1-r)(m_0-1)},\end{aligned} \tag{2.10}$$

and

$$\begin{aligned}|\Lambda_1(\eta,\varepsilon)| \le\ &\varepsilon^\nu\mathrm{e}^{\eta\omega}\left\|\widehat{f}_1\right\|_{\widehat{H}_{\delta,r}^{m_1}}\int_{-1/\varepsilon^{1-r}}^{1/\varepsilon^{1-r}}\frac{ds}{(\eta^2+s^2)^{m_1/2}},\\ &\le \varepsilon^\nu\mathrm{e}^{\eta\omega}\left\|\widehat{f}_1\right\|_{\widehat{H}_{\delta,r}^{m_1}}\int_{-\infty}^{-\infty}\frac{ds}{(1+s^2)^{m_1/2}},\\ &\le M(m_1)\left\|\widehat{f}_1\right\|_{\widehat{H}_{\delta,r}^{m_1}}\mathrm{e}^{\eta\omega}\varepsilon^\nu.\end{aligned} \tag{2.11}$$

Collecting (2.8), (2.9), (2.10), (2.11) we finally get

$$I^+(f,\varepsilon) = \frac{\mathrm{e}^{-\omega\sigma/\varepsilon}}{\varepsilon^{\gamma-1}}\left(\Lambda^+(\eta)+J(\varepsilon,\eta)\right) \tag{2.12}$$

where

$$|J(\varepsilon,\eta)| \le M(\sigma,\lambda,\gamma,m_0,m_1)\mathrm{e}^{\omega\eta}\varepsilon^{\nu^*}$$

with $\nu^* := \min\big(\nu,(1-r)(m_0-1),(1-r)(\gamma-1)\big)$.

The function $z \mapsto \widehat{f}_0(z)\mathrm{e}^{\mathrm{i}\omega z}$ is holomorphic on $\widehat{\Omega}_\delta$ and decays polynomialy at ∞. Then using the path $\Gamma_{\eta,\eta',R}$ (see Fig. 2.6) and pushing R to ∞ we obtain that $\Lambda^+(\eta) = \Lambda^+(\eta')$ for every $\eta > \delta$, $\eta' > \delta$. So, Λ^+ *does not depend on* η. Thus, J *does not depend on* η too. Finally, pushing $\eta \to \delta$ we get

$$|J(\varepsilon)| \le M(\sigma,\lambda,\gamma,m_0,m_1)\mathrm{e}^{\omega\delta}\varepsilon^{\nu^*}. \qquad \square$$

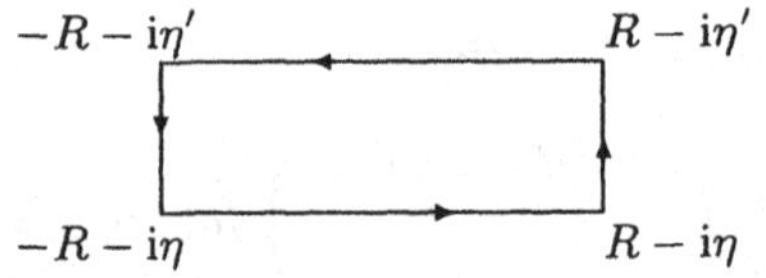

Fig. 2.6. Path $\Gamma_{\eta,\eta',R}$.

2.1.4 Exponential equivalent: strategy for nonlinear differential equations

To use the Third Mono-Frequency Exponential Lemma 2.1.21 for solving Problem 1.1.4, i.e. for computing equivalents of oscillatory integrals

$$I^{\pm}(f,\varepsilon)=\int_{-\infty}^{+\infty} f(t,\varepsilon)\ \mathrm{e}^{\pm\mathrm{i}\omega t/\varepsilon}\ dt$$

when f involves solutions of real nonlinear differential equations, we must be able to prove that a given solution of a differential equation admits a holomorphic continuation which belongs to $E^{\gamma,\lambda}_{\sigma,\delta}$. The set $E^{\gamma,\lambda}_{\sigma,\delta}$ can be normed with

$$\|f\|_{E^{\gamma,\lambda}_{\sigma,\delta}}=\|f\|_{H^{\gamma,\lambda}_{\sigma,\delta}}+\left\|\widehat{f_0}\right\|_{\widehat{H}^{m_0}_{\delta}}+\left\|\widehat{f_1}\right\|_{\widehat{H}^{m_1}_{\delta,r}}$$

to obtain a Banach space. So, a first idea is to build the holomorphic continuation using the Contraction Mapping Theorem in $E^{\gamma,\lambda}_{\sigma,\delta}$. This strategy happens to be quite intractable because the norm $\|\cdot\|_{E^{\gamma,\lambda}_{\sigma,\delta}}$ involves $\widehat{f_0}$ and $\widehat{f_1}$ which are indirectly linked to f. We propose here a strategy which enables us to prove step by step that a given solution $\mathcal{Y}$ of a *real* nonlinear analytic differential equation

$$\frac{dY}{dt}=F(Y,t,\varepsilon) \tag{2.13}$$

admits a holomorphic continuation which belongs to $E^{\gamma,\lambda}_{\sigma,\delta}$. In particular, this strategy gives a simple way for determining $\mathrm{P}^{+}_{\mathrm{r}}(\mathcal{Y})$ which is the relevant part of $\mathcal{Y}$ used for computing oscillatory integrals involving it. Since $\mathcal{Y}$ is real valued on the real axis $\mathrm{P}^{-}_{\mathrm{r}}(\mathcal{Y})(z)=\overline{\mathrm{P}^{+}_{\mathrm{r}}(\mathcal{Y})(-\overline{z})}$ holds.

The strategy proposed in this section is described for simplicity in the one-dimensional case. It can be readily extended either to finite or infinite dimensions with spaces of type $\left(E^{\gamma,\lambda}_{\sigma,\delta}\right)^n$ or more generally of type $\prod_{k=1}^{n} E^{\gamma_k,\lambda_k}_{\sigma,\delta}$.

Step 1. Continuation far away from singularities. For proving the existence of a holomorphic continuation of $\mathcal{Y}$ in

$$\mathcal{D}_{\sigma,r,\varepsilon}=\mathcal{B}_\sigma\setminus\{\xi\in\mathbb{C}/|\mathcal{R}e\,(\xi)|\le\frac{1}{2}\varepsilon^r,\ |\mathcal{I}m\,(\xi\pm\mathrm{i}\sigma)|\le\frac{1}{2}\varepsilon^r\},$$

(See Fig. 2.7), one can use the Contraction Mapping Theorem in the space of holomorphic functions defined on $\mathcal{D}_{\sigma,r,\varepsilon}$ such that

$$\sup_{\substack{0<\varepsilon<\varepsilon_0\\ \xi\in\mathcal{D}_{\sigma,r,\varepsilon}}}\left[|f(\xi,\varepsilon)|\left(|\sigma^2+\xi^2|^\gamma\chi_1(\xi)+(1-\chi_1(\xi))\mathrm{e}^{\lambda|\mathcal{R}e(\xi)|}\right)\right]<+\infty.$$

where χ_1 is defined in Lemma 2.1.9.

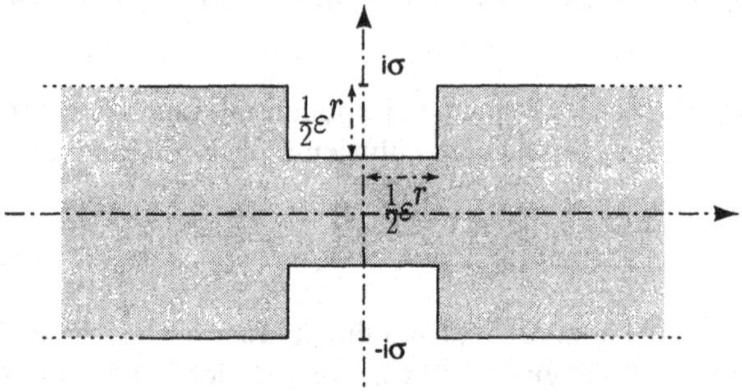

Fig. 2.7. Step 1 : continuation in domain $\mathcal{D}_{\sigma,r,\varepsilon}$

Step 2. Continuation near iσ. For proving the existence of a holomorphic continuation of $\mathcal{Y}(\xi)$ in $]-\varepsilon^r,\varepsilon^r[\times]\sigma-\varepsilon^r,\sigma-\delta\varepsilon[$ which satisfies

$$\mathcal{Y}(\mathrm{i}\sigma+\varepsilon z,\varepsilon)=\frac{1}{\varepsilon^\gamma}\widehat{\mathcal{Y}}_0^+(z)+\frac{\varepsilon^\nu}{\varepsilon^\gamma}\widehat{\mathcal{Y}}_1^+(z,\varepsilon) \tag{2.14}$$

with $\widehat{\mathcal{Y}}_0^+\in\widehat{H}_\delta^{m_0}$ and $\widehat{\mathcal{Y}}_1^+\in\widehat{H}_{\delta,r}^{m_1}$, it is convenient to introduce two systems of coordinates in space and time:

- a first one (Y,ξ), which is the initial one on $\mathbb{R}$, is called the *outer system of coordinates*;
- a second one $(\widehat{Y},z):=(\varepsilon^\gamma Y,(\xi-\mathrm{i}\sigma)/\varepsilon)$ is useful to describe what happens near iσ. It is called *the inner system of coordinates.*

Since this whole step is devoted to the description of $\mathcal{Y}$ near iσ we omit the exponent "+". In what follows $\widehat{\mathcal{Y}}_0^+$, $\widehat{\mathcal{Y}}_1^+$, P_r^+ are respectively denoted by $\widehat{\mathcal{Y}}_0$, $\widehat{\mathcal{Y}}_1$, P_r.

For an orbit $Y(\xi)$ in the outer system of coordinates, let us denote by

$$\widehat{Y}(z)=\varepsilon^\gamma Y(\mathrm{i}\sigma+\varepsilon z)$$

the corresponding orbit in the inner system of coordinates. Conversely, for an orbit $\widehat{Y}(z)$ in the inner system of coordinates, let us denote by

$$Y(\xi)=\frac{1}{\varepsilon^\gamma}\widehat{Y}\left(\frac{\xi-\mathrm{i}\sigma}{\varepsilon}\right)$$

the corresponding orbit in the outer system of coordinates.

We use the same notation for the vector fields: for a vector field F expressed in the outer system of coordinates by $Y\mapsto F(Y,\xi,\varepsilon)$ let us denote by $\widehat{F}$ the corresponding vector fields in the inner system of coordinates given by

$$\widehat{Y}\mapsto\varepsilon^{\gamma+1}F\left(\frac{1}{\varepsilon^\gamma}\widehat{Y}(z),\mathrm{i}\sigma+\varepsilon z,\varepsilon\right)$$

and vice versa. Finally, for a domain Σ of $\mathbb{C}$, in the outer system of coordinates, let us denote by $\hat{\Sigma} = (\Sigma - \mathrm{i}\sigma)/\varepsilon$ the corresponding one in the inner system of coordinates and vice versa.

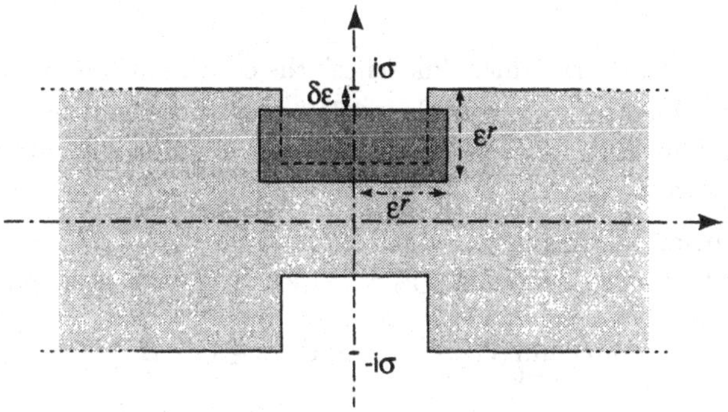

Fig. 2.8. Step 2 : continuation in $\Sigma_{\delta,r,\varepsilon}$

Step 1 gives a holomorphic continuation of $\mathcal{Y}(\xi)$ in $\mathcal{D}_{\sigma,r,\varepsilon}$, i.e. a holomorphic continuation of $\widehat{\mathcal{Y}}(z)$ in $\widehat{\mathcal{D}}_{\sigma,r,\varepsilon}$. Then we must prove the existence of a holomorphic continuation of $\mathcal{Y}(\xi)$ in $\Sigma_{\delta,r,\varepsilon} :=]-\varepsilon^r, \varepsilon^r[\times]\sigma - \varepsilon^r, \sigma - \delta\varepsilon[$ (see Fig. 2.8) which satisfies (2.14), i.e. the existence of a holomorphic continuation of $\widehat{\mathcal{Y}}(z)$ in $\widehat{\Sigma}_{\delta,r,\varepsilon} :=]-1/\varepsilon^{1-r}, 1/\varepsilon^{1-r}[\times]-1/\varepsilon^{1-r}, -\delta[$ (see Fig. 2.9) which satisfies

$$\widehat{\mathcal{Y}}(z,\varepsilon) = \widehat{\mathcal{Y}}_0(z) + \varepsilon^{\nu}\widehat{\mathcal{Y}}_1(z,\varepsilon) \tag{2.15}$$

with $\widehat{\mathcal{Y}}_0 \in \widehat{H}^{m_0}_{\delta}$ and $\widehat{\mathcal{Y}}_1 \in \widehat{H}^{m_1}_{\delta,r}$.

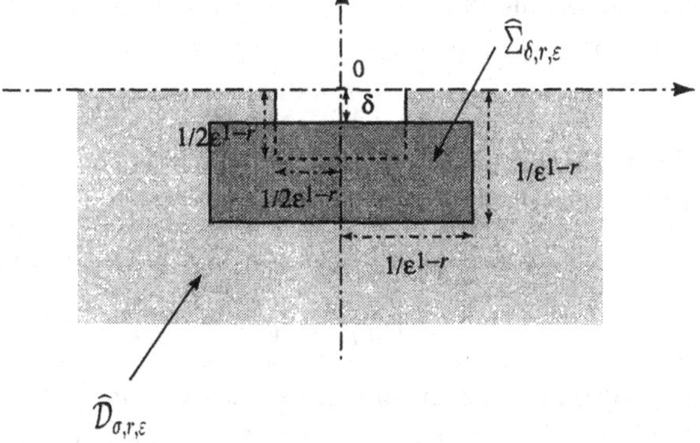

Fig. 2.9. Domains $\widehat{\mathcal{D}}_{\sigma,r,\varepsilon}$ and $\widehat{\Sigma}_{\delta,r,\varepsilon}$

Such a continuation in $\widehat{\Sigma}_{\delta,r,\varepsilon}$ is a holomorphic solution of the full system (2.13) rewritten in the inner system of coordinates

$$\frac{d\widehat{Y}}{dz} = \widehat{F}(\widehat{Y}, z, \varepsilon) := \varepsilon^{\gamma+1} F\left(\frac{1}{\varepsilon^\gamma}\widehat{Y}(z), i\sigma + \varepsilon z, \varepsilon\right), \tag{2.16}$$

which satisfies (2.15) and which "matches" the continuation obtained in $\widehat{\mathcal{D}}_{\sigma,r,\varepsilon}$ at Step 1, i.e. which coincides on $\widehat{\Sigma}_{\delta,r,\varepsilon} \cap \widehat{\mathcal{D}}_{\sigma,r,\varepsilon}$ with the continuation obtained at step 1 (see Fig. 2.9). The existence of such a holomorphic continuation can be proved in two steps.

Step 2.1. Inner system.

The first task is to determine $\widehat{\mathcal{Y}}_0 = \mathrm{P_r}(\mathcal{Y})$. It is characterized by

$$\widehat{\mathcal{Y}}_0 = \lim_{\varepsilon \to 0} \widehat{\mathcal{Y}}(z, \varepsilon) \qquad \text{and} \qquad \widehat{\mathcal{Y}}_0 \in \widehat{H}_\delta^{m_0}.$$

Hence $\widehat{\mathcal{Y}}_0$ can be obtained using the Contraction Mapping Theorem to solve

$$\frac{d\widehat{Y}}{dz} = \widehat{F}_0(\widehat{Y}, z) := \widehat{F}(\widehat{Y}, z, 0) \tag{2.17}$$

in $\widehat{H}_\delta^{m_0}$. This system is called the *"inner system"* and it represents "the relevant part of the system near $i\sigma$": $\widehat{\mathcal{Y}}_0 = \mathrm{P_r}(\mathcal{Y})$ which is the relevant part of $\mathcal{Y}$ for computing oscillatory integral involving it, is obtained as a solution of the inner system.

Step 2.2. Matching of holomorphic continuations

For achieving our purpose we must prove the existence of a solution $\widehat{Y}$ of (2.13) which satisfies

(i) $\widehat{Y}(z,\varepsilon) = \widehat{\mathcal{Y}}_0(z) + \varepsilon^\nu \widehat{\mathcal{Y}}_1(z,\varepsilon)$ where $\widehat{\mathcal{Y}}_0$ has been obtained at Step 2.1,

(ii) $\widehat{\mathcal{Y}}_1(.,\varepsilon)$ is holomorphic in $\widehat{\Sigma}_{\delta,r,\varepsilon}$,

(iii) $\displaystyle\sup_{\substack{\varepsilon \in]0,\varepsilon_0] \\ z \in \widehat{\Sigma}_{\delta,r,\varepsilon}}} \left(|z|^{m_1} |\widehat{\mathcal{Y}}_1(z,\varepsilon)|\right) < +\infty$,

(iv) $\widehat{Y}$ coincides with $\widehat{\mathcal{Y}}$ on $\widehat{\Sigma}_{\delta,r,\varepsilon} \cap \widehat{\mathcal{D}}_{\sigma,r,\varepsilon}$ where $\widehat{\mathcal{Y}}$ is the holomorphic continuation of $\mathcal{Y}$ obtained at step 1.

For that purpose, we can see the full equation as a perturbation of the inner equation

$$\widehat{F}(\widehat{Y}, z, \varepsilon) = \widehat{F}_0(\widehat{Y}, z) + \widehat{F}_1(\widehat{Y}, z, \varepsilon)$$

where the perturbation term is given by $\widehat{F}_1(\widehat{Y}, z, \varepsilon) := \widehat{F}(\widehat{Y}, z, \varepsilon) - \widehat{F}(\widehat{Y}, z, 0)$. In other words, Step 2.1 gives a solution of the inner equation

$$\frac{d\widehat{Y}}{dz} = \widehat{F}_0(\widehat{Y}, z)$$

in $\widehat{\Omega}_\delta$ and we must prove the existence of a solution of the full equation

$$\frac{d\widehat{Y}}{dz} = \widehat{F}_0(\widehat{Y}, z) + \widehat{F}_1(\widehat{Y}, z, \varepsilon)$$

which satisfies (i)-(iv). It amounts to proving the existence of a solution $\widehat{X}_1$ ("$\widehat{X}_1 = \varepsilon^\nu \widehat{\mathcal{Y}}_1$ ") of

$$\frac{d\widehat{X}}{dz} = \widehat{F}_0\left(\widehat{\mathcal{Y}}_0(z) + \widehat{X}(z), z\right) + \widehat{F}_1\left(\widehat{\mathcal{Y}}_0(z) + \widehat{X}(z), z, \varepsilon\right) - \widehat{F}_0\left(\widehat{\mathcal{Y}}_0(z), z\right)$$

in $\widehat{\Sigma}_{\delta,r,\varepsilon}$ which satisfies

(a) $\widehat{X}_1(., \varepsilon)$ is holomorphic in $\widehat{\Sigma}_{\delta,r,\varepsilon}$,

(b) $\displaystyle\sup_{\substack{\varepsilon\in]0,\varepsilon_0] \\ z\in\widehat{\Sigma}_{\delta,r,\varepsilon}}} \left(|z|^{m_1}|\widehat{X}_1(z,\varepsilon)|\right) < M\varepsilon^\nu$,

(c) $\widehat{\mathcal{Y}}_0 + \widehat{X}_1$ coincides with $\widehat{\mathcal{Y}}$ on $\widehat{\Sigma}_{\delta,r,\varepsilon} \cap \widehat{\mathcal{D}}_{\sigma,r,\varepsilon}$ where $\widehat{\mathcal{Y}}$ is the holomorphic continuation of $\mathcal{Y}$ obtained at step 1.

To build $\widehat{X}_1$, we can use the method of continuation along horizontal lines described in Appendix 2.A: we can first see it as a smooth function of $(\tau, \eta) \in \widehat{\Sigma}_{\delta,r,\varepsilon}$ (instead of seeing it as a holomorphic function of $z = \tau + i\eta$), and define it as the unique solution of "the ordinary differential equation along horizontal lines with parameter η"

$$\frac{\partial\widehat{X}}{\partial\tau} = \widehat{F}_0\left(\widehat{\mathcal{Y}}_0(\tau + i\eta) + \widehat{X}(\tau,\eta), \tau + i\eta\right)$$
$$+\widehat{F}_1\left(\widehat{\mathcal{Y}}_0(\tau + i\eta) + \widehat{X}(\tau,\eta), \tau + i\eta, \varepsilon\right) - \widehat{F}_0\left(\widehat{\mathcal{Y}}_0(\tau + i\eta), \tau + i\eta\right).$$

which is equal to $\widehat{\mathcal{Y}}(1/\varepsilon^{1-r} + i\eta)$ for $\tau = 1/\varepsilon^{1-r}$. Using the contraction mapping theorem in a space of functions satisfying

$$\sup_{(\tau,\eta)\in\widehat{\Sigma}_{\delta,r,\varepsilon}} \left(|\tau + i\eta|^{m_1}|\widehat{X}(\tau,\eta,\varepsilon)|\right) < M\varepsilon^\nu.$$

we can prove that $\widehat{X}_1(\tau, \eta, \varepsilon)$ is defined and smooth in $\widehat{\Sigma}_{\delta,r,\varepsilon}$, and satisfies (b).

The lemma of continuation along horizontal lines given in Appendix 2.A ensures that $\widehat{X}_1$ is holomorphic in $\widehat{\Sigma}_{\delta,r,\varepsilon}$ and that $\widehat{\mathcal{Y}}_0 + \widehat{X}_1$ coincides with $\mathcal{Y}$ on $\widehat{\Sigma}_{\delta,r,\varepsilon} \cap \mathcal{D}_{\sigma,r,\varepsilon}$. So (a) and (c) are fulfilled.

Step 3. Continuation near $-i\sigma$: symmetrization

Steps 1 and 2 give a holomorphic continuation of $\mathcal{Y}$ in $\Delta = \mathcal{D}_{\sigma,\frac{1}{2},\varepsilon} \cup \Sigma_{\delta,\frac{1}{2},\varepsilon}$. We still denote by $\mathcal{Y}$ this continuation. Moreover, $\mathcal{Y}$ is real valued on the real axis. Hence the two holomorphic functions $\xi \mapsto \mathcal{Y}(\xi)$ defined on Δ and

$\xi \mapsto \overline{\mathcal{Y}(\overline{\xi})}$ defined on $\overline{\Delta}$ coincide on the real axis, and thus on $\overline{\Delta} \cap \Delta$ because of the isolated zeros theorem. Then, we obtain a holomorphic continuation $\widetilde{\mathcal{Y}}$ of $\mathcal{Y}$ in $\overline{\Delta} \cup \Delta = \mathcal{D}_{\sigma,\frac{1}{2},\varepsilon} \cup \Sigma_{\delta,\frac{1}{2},\varepsilon} \cup \overline{\Sigma_{\delta,\frac{1}{2},\varepsilon}}$ by setting

$$\widetilde{\mathcal{Y}}(\xi) = \begin{cases} \mathcal{Y}(\xi) & \text{for } \xi \in \Delta \\ \overline{\mathcal{Y}(\overline{\xi})} & \text{for } \xi \in \overline{\Delta} \end{cases}$$

The holomorphic continuation being unique, we can omit the "tilde" and we still denote by $\mathcal{Y}$ the continuation obtained in $\overline{\Delta} \cup \Delta$. Finally, since $\xi \mapsto \overline{\xi}$ is an isometry, Steps 1, 2, 3 ensure that $\mathcal{Y}$ belongs to $E^{\gamma,\lambda}_{\sigma,\delta}(m_0, m_1, r, \nu)$ and that $\mathrm{P}_{\mathrm{r}}^{-}(\mathcal{Y})(z) = \overline{\mathrm{P}_{\mathrm{r}}^{+}(\mathcal{Y})(-\overline{z})}$.

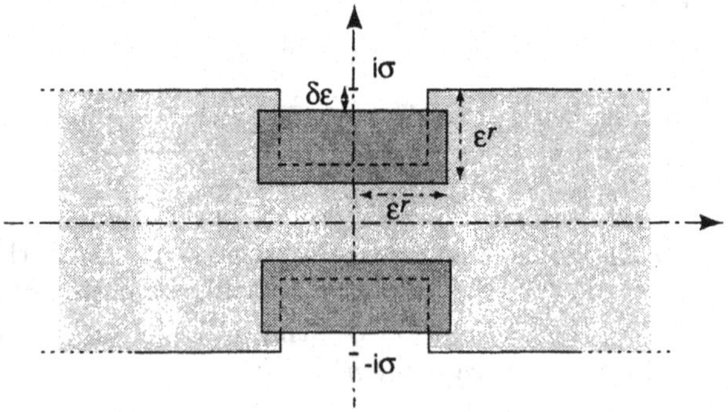

Fig. 2.10. Step 3 : continuation in $\overline{\Sigma_{\delta,r,\varepsilon}}$

Remark 2.1.23. Link with the Matched Asymptotic Expansions Theory. *This strategy of continuation step by step on different subdomains is a rigorous version for this problem of the Matched Asymptotic Expansions theory (M.A.E.)* (The words *outer and inner systems of coordinates*, and *inner system* comes from this theory). Roughly speaking, with the theory of Asymptotic Expansions, one can describe a solution Z of perturbed O.D.E. (or P.D.E.) in the form "$Z = \sum\limits_{n=0}^{N} Z_n + o(Z_N)$" where $Z_{n+1} = o(Z_n)$. The theory of Matched Asymptotic Expansions was introduced to deal with solutions which cannot be described by a unique asymptotic expansion in the whole domain : on one (or several) small subdomain(s) called "inner domains in the language of M.A.E., the asymptotic expansion is no longer well ordered and another asymptotic expansion is required to describe Z in the small subdomain(s). Then, one has to check that the different asymptotic expansions match properly. In our case, we do not describe a solution $\mathcal{Y}$ of a perturbed differential equation with asymptotic expansions, but we decompose it in several holomorphic functions:

in $\mathcal{D}_{\sigma,r,\varepsilon}$ (the *outer domain* in the language of M.A.E.), $\mathcal{Y}$ reads

$$\begin{aligned} \mathcal{Y}(\xi,\varepsilon) &= \text{“leading part on } \mathbb{R} + \text{perturbation term”} \\ &= \mathcal{Y}(\xi,0) + \mathcal{Y}_{\mathrm{p}}(\xi,\varepsilon); \end{aligned} \tag{2.18}$$

and in $\Sigma_{\delta,r,\varepsilon}$ (the *inner domain* in the language of M.A.E.), $\mathcal{Y}$ reads

$$\mathcal{Y}(\xi) = \frac{1}{\varepsilon^\gamma}\widehat{\mathcal{Y}}_0^+\left(\frac{\xi - \mathrm{i}\sigma}{\varepsilon}\right) + \frac{\varepsilon^\nu}{\varepsilon^\gamma}\widehat{\mathcal{Y}}_1^+\left(\frac{\xi - \mathrm{i}\sigma}{\varepsilon}\right). \tag{2.19}$$

Nevertheless, we have to face the same kind of phenomena: the decomposition of $\mathcal{Y}$ in the form (2.18) is not in general relevant in the whole domain. We may have a relative change of size between $\mathcal{Y}(\xi,0)$ and $\mathcal{Y}_{\mathrm{p}}(\xi,\varepsilon)$ when moving from $\mathbb{R}$ to $\mathrm{i}(\sigma - \delta\varepsilon) + \mathbb{R}$: for instance, the function

$$\mathcal{Y}(\xi,\varepsilon) = \frac{1}{\cosh^2(\xi)} + \frac{\varepsilon}{\cosh^4(\xi)}$$

belongs to $E^{3,2}_{\frac{\pi}{2},1}$ and

$$\mathrm{Pr}^+_{E^{3,2}_{\frac{\pi}{2},1}}\left(\frac{1}{\cosh^2(\xi)}\right) = 0, \qquad \mathrm{Pr}^+_{E^{3,2}_{\frac{\pi}{2},1}}\left(\frac{\varepsilon}{\cosh^4(\xi)}\right) = \frac{1}{z^4}.$$

The leading term near $\mathrm{i}\sigma$ comes from the perturbation on $\mathbb{R}$!

For solutions of nonlinear differential equations, this relative change of size between the leading part on $\mathbb{R}$ and the perturbation term when moving from $\mathbb{R}$ to the complex singularity, is in general induced by a similar relative change of size between the leading part of the vector field on $\mathbb{R}$ and the perturbation part. When using the above strategy, the leading part of the solution near $\mathrm{i}\sigma$, i.e. *the principal part* Pr is found as a solution of the *inner system*, which is nothing else than the *leading part of the vector field near* $\mathrm{i}\sigma$. For instance, for all the resonances studied in the third part of this book, we have to face this phenomena of relative change of size, when studying the existence of homoclinic connections to 0. The typical situation is the following:
On $\mathbb{R}$, we use a first decomposition of the vector field

$$\frac{d\mathcal{Y}}{dt} = F(\mathcal{Y}) = N(\mathcal{Y}) + R(\mathcal{Y}), \qquad \mathcal{Y} = \mathcal{Y}_N + \mathcal{Y}_R \ \text{ with } \begin{cases} R(\mathcal{Y}) \ll N(\mathcal{Y}) \\ \mathcal{Y}_R \ll \mathcal{Y}_N \end{cases}$$

where N is the the normal form and R represents the higher order terms, whereas near $\mathrm{i}\sigma$ the above decomposition is no longer relevant because N and R have the same size. So, for catching the principal part $\mathcal{Y}_0^+$ of $\mathcal{Y}$ near $\mathrm{i}\sigma$, we use near $\mathrm{i}\sigma$ a second decomposition of the vector field in the form “inner system+perturbation”,

$$\frac{d\widehat{\mathcal{Y}}}{dz} = \widehat{F}(\widehat{\mathcal{Y}}) = \underbrace{\widehat{N}_0(\widehat{\mathcal{Y}}) + \widehat{R}_0(\widehat{\mathcal{Y}})}_{\widehat{F}_0(\widehat{\mathcal{Y}})} + \underbrace{\widehat{N}_1(\widehat{\mathcal{Y}}) + \widehat{R}_1(\widehat{\mathcal{Y}})}_{\widehat{F}_1(\widehat{\mathcal{Y}})}, \qquad \widehat{\mathcal{Y}} = \widehat{\mathcal{Y}}_0^+ + \varepsilon^\nu \widehat{\mathcal{Y}}_1^+$$

with $\widehat{F}_1(\widehat{\mathcal{Y}}) \ll \widehat{F}_0(\widehat{\mathcal{Y}})$ and $\varepsilon^\nu\widehat{\mathcal{Y}}_1^+ \ll \widehat{\mathcal{Y}}_0^+$ near $\mathrm{i}\sigma$ and $\widehat{\mathcal{Y}} = \varepsilon^\gamma\mathcal{Y}$.

Matched Asymptotic Expansions Theory was precisely designed to deal with perturbed problems, in which such a relative change of size between the unperturbed problem and the perturbation term occurs in some small pathological domains called *inner domains.* The fundamental idea of using an *inner system* to describe properly the solutions in the *inner domain* comes from this theory. A first use of this theory for this kind of problem, was done by Kruskal and Segur in [KS91] which appears as a preprint in 1985.

Remark 2.1.24. The space $E^{\gamma,\lambda}_{\sigma,\delta}$ and the above strategy were designed to deal with perturbed differential equations such that the unperturbed equation has a solution which has poles at $\pm i\sigma$. If the unperturbed problem as a solution with several poles with the same imaginary part, then one has to introduce new spaces similar to $E^{\gamma,\lambda}_{\sigma,\delta}$ which describe the polynomial growth of the functions near each poles. For instance, if the unperturbed problem admits a solution with four poles of the same order at $\pm a \pm \imath\sigma$, with $a, \sigma > 0$ one should introduce the space $E^{\gamma,\lambda}_{\sigma,\delta,a}$ defined as $E^{\gamma,\lambda}_{\sigma,\delta}$ with four "inner domains" instead of two for describing the growth of the functions near each poles (see Fig. 2.11).

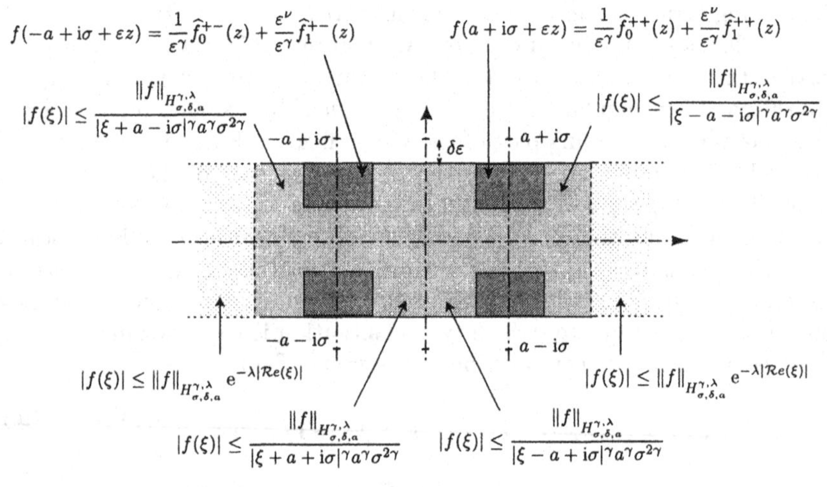

Fig. 2.11. Description of $f \in E^{\gamma,\lambda}_{\sigma,\delta,a}$

2.1.5 Exponential equivalent: an example

Lemma 2.1.21 combined with the strategy explained in the previous subsection enables us to give a final answer to Question 2.1.7

Lemma 2.1.25 (Answer to Question 2.1.7). *Let $\beta^\star(t,\varepsilon,\rho)$ be the solution of*

$$\frac{d\beta}{dt} = 1-\beta^2+\rho\varepsilon^2\left(1-\beta^2\right)\beta^2$$

which vanishes for $t=0$. Observe that $\beta^\star$ is odd and that it is of course real valued on the real axis.

(a) *For $\rho=0$, $\beta^\star=\beta^h(\xi)=\tanh\xi$. Thus it belongs to $E^{1,0}_{\frac{\pi}{2},1}(1,2,\frac{1}{2},\frac{1}{2})$ and $\mathrm{P}^+_{\mathrm{r}}(\beta^h)=1/z$.*

(b) *For ρ_0 arbitrarily fixed, there exist $\delta_1\geq 1$ and $\varepsilon_3>0$ such that for every $\rho\in[-\rho_0,\rho_0]$ and every $\varepsilon\in]0,\varepsilon_3]$, $\beta^\star$ reads*

$$\beta^\star(\xi,\varepsilon,\rho)=\beta^h(\xi)+\gamma^\star(\xi,\varepsilon,\rho) \text{ where } \gamma^\star\in E^{1,\lambda}_{\frac{\pi}{2},\delta_1}(2,2,\tfrac{1}{2},\tfrac{1}{4}) \quad (2.20)$$

with $\lambda\in]0,2[$. Hence, $\beta^\star$ belongs to $E^{1,0}_{\frac{\pi}{2},\delta_1}(1,2,\frac{1}{2},\frac{1}{4})$ and

$$\mathrm{P}^\pm_{\mathrm{r}}(\beta^\star)=\mathrm{P}^\pm_{\mathrm{r}}(\beta^h)+\mathrm{P}^\pm_{\mathrm{r}}(\gamma^\star).$$

Moreover, $\widehat{\beta}^\star_0=\mathrm{P}^+_{\mathrm{r}}(\beta^\star)$ is solution of the inner system

$$\frac{d\widehat{\beta}}{dz}=-\widehat{\beta}^2-\rho\widehat{\beta}^4 \qquad \text{for } z\in\widehat{\Omega}_{\delta_1}, \quad (2.21)$$

and for every $z\in\widehat{\Omega}_{\delta_1}$, $\rho\mapsto\widehat{\beta}^\star_0(z,\rho)$ is holomorphic in $D(0,\rho_0)=\{\rho\in\mathbb{C}/|\rho|<\rho_0\}$

(c) *The oscillatory integral*

$$I(\varepsilon,\rho)=\rho\varepsilon\int_{-\infty}^{+\infty}\mathrm{e}^{\mathrm{i}\omega t/\varepsilon}\left(1-(\beta^\star(t,\varepsilon,\rho))^2\right)\,dt$$

satisfies

$$I(\varepsilon,\rho)=\rho\mathrm{e}^{-\frac{\pi\omega}{2\varepsilon}}\left(\Lambda(\rho)+\underset{\varepsilon\to0}{\mathcal{O}}(\varepsilon^{\frac{1}{4}})\right) \quad (2.22)$$

where $\underset{\varepsilon\to0}{\mathcal{O}}(\varepsilon^{\frac{1}{4}})$ is uniform with respect to $\rho\in[-\rho_0,\rho_0]$ and

$$\Lambda(\rho)=\int_{-\mathrm{i}\eta+\mathbb{R}}-\left(\widehat{\beta}^\star_0(z,\rho)\right)^2\mathrm{e}^{\mathrm{i}\omega z}dz \qquad \text{for any } \eta>\delta_1. \quad (2.23)$$

The function $\rho\mapsto\Lambda(\rho)$ is real analytic in $]-\rho_0,\rho_0[$, and satisfies $\Lambda(0)=2\pi\omega$. Hence, Λ vanishes at most for a finite number of ρ in $[-\rho_0,\rho_0]$.

Because of the particular form of the inner system we can obtain explicit formula for $\widehat{\beta}^\star_0$ and $\Lambda(\rho)$.

Lemma 2.1.26 (Explicit formula for $\widehat{\beta}_0^\star(z,\rho)$ and $\Lambda(\rho)$). *For every $\rho \in]-\rho_0,\rho_0[$,*

$$\widehat{\beta}_0^\star(z,\rho) = \sum_{n\geq 0} a_n \frac{\rho^n}{z^{2n+1}}, \qquad \textit{for } z \in \widehat{\Omega}_{\delta_1},$$

and

$$\Lambda(\rho) = 2\pi\omega \sum_{n\geq 0} \frac{\left(-\rho\omega^2\right)^n}{(2n+1)!} \sum_{p+q=n} a_p a_q \tag{2.24}$$

where the coefficients a_n are positive and given by induction

$$a_0 := 1,$$
$$a_n := \frac{1}{2n-1}\left(\sum_{k=1}^{n-1} a_k a_{n-k} + \sum_{p+q+k+l=n-1} a_p a_q a_k a_l\right) \quad \textit{for } n \geq 1. \tag{2.25}$$

Remark 2.1.27. Lemma 2.1.25 ensures that except for a finite number of $\rho \in [-\rho_0,\rho_0]$, $I(\varepsilon,\rho)$ is exponentially small but does not vanish. Moreover, Lemma 2.1.26 ensures that $\Lambda(\rho) > 0$ for $\rho \in]-\rho_0, 0[$. Thus, the front of our second toy model for $\rho = 0$ does not persists for $\rho \neq 0$ except perhaps for a finite number of ρ in $]0,\rho_0]$. It develops exponentially small oscillations for $t \longrightarrow -\infty$.

For a more general system, the existence of a solution for the inner system in $\widehat{H}_\delta^{mo}$ can be obtained using the Contraction Mapping Theorem as for Lemma 2.1.25. However, in general, an explicit expression of this solution (like the one obtained in Lemma 2.1.26) cannot be obtained.

Remark 2.1.28. For ρ and ε small, (2.22) and (2.24) ensure that

$$I(\varepsilon,\rho) = \rho e^{-\frac{\pi\omega}{2\varepsilon}}\left(2\pi\omega + \underset{\rho\to 0}{O}(\rho) + \underset{\varepsilon\to 0}{O}(\varepsilon^{\frac{1}{4}})\right)$$

So (2.22) generalizes the result obtained with the second exponential Lemma (see (2.5) obtained in Example 2.1.15).

Observe that for ρ and ε small the equivalent of $I(\varepsilon,\rho)$ is given by the Melnikov function $M_e(\varepsilon)$ defined in (2.4). However for ρ of order 1, the equivalent of $I(\varepsilon,\rho)$ has the same exponential degeneracy as the Melnikov function, but the numerical coefficients in front of the exponential are different:

$$I(\varepsilon,\rho) \underset{\varepsilon\to 0}{\sim} \rho\, \Lambda(\rho) e^{-\frac{\pi\omega}{2\varepsilon}}, \qquad M_e(\varepsilon) \underset{\varepsilon\to 0}{\sim} \rho\, 2\pi\omega e^{-\frac{\pi\omega}{2\varepsilon}},$$

and $\Lambda(\rho) = \sum_{n\geq 0} \Lambda_n \rho^n$ with $\Lambda_0 = 2\pi\omega$.

Remark 2.1.29. One could address a more general question : *Compute an equivalent for ε small of the oscillatory integral*

$$I_m(\varepsilon,\rho) = \rho\varepsilon \int_{-\infty}^{+\infty} e^{i\omega t/\varepsilon} \left(1 - (\beta_m^\star)^2\right) \, dt$$

where $\beta_m^\star$ is the solution of

$$\frac{d\beta}{dt} = 1 - \beta^2 + \rho\varepsilon^m \left(1 - \beta^2\right) \beta^2$$

which vanishes for $t = 0$. Question 2.1.7 corresponds to $m = 2$.

$m > 2$ This case can be seen as a particular case of the case $m = 2$, by setting $\rho' = \rho\varepsilon^{m-2}$: $I_m(\varepsilon,\rho) = I_2(\varepsilon,\rho\varepsilon^{m-2})/\varepsilon^{m-2}$. Since for any ρ fixed in $[-\rho_0,\rho_0]$, ρ' still belongs to $[-\rho_0,\rho_0]$, the previous result still holds and

$$\begin{aligned} I_m(\varepsilon,\rho) &= \rho e^{-\frac{\pi\omega}{2\varepsilon}} \left(\Lambda(\rho\varepsilon^{m-2}) + \underset{\varepsilon\to 0}{\mathcal{O}}(\varepsilon^{\frac{1}{4}}) \right) \\ &= \rho e^{-\frac{\pi\omega}{2\varepsilon}} \left(2\pi\omega + \underset{\varepsilon\to 0}{\mathcal{O}}(\varepsilon^{m-2}) + \underset{\varepsilon\to 0}{\mathcal{O}}(\varepsilon^{\frac{1}{4}}) \right) \end{aligned}$$

Hence, for $m > 2$, the equivalent of $I_m(\varepsilon,\rho)$ is given by the Melnikov function.

$m = 2$ It corresponds to the limit case : for ρ small the equivalent of $I(\varepsilon,\rho)$ is given by the Melnikov function $M_e(\varepsilon)$ whereas for ρ of order 1, $I(\varepsilon,\rho)$ has the same exponential degeneracy as $M_e(\varepsilon)$, but the numerical coefficients in front of the exponential are different.

$0 < m < 2$ This case can also be seen as a particular case of the case $m = 2$, by setting $\rho' = \rho/\varepsilon^{2-m}$. However, for a fixed $\rho \in [-\rho_0,\rho_0]$, $\rho' \to +\infty$ for $\varepsilon \to 0$. So in this case, the equivalent of $I(\varepsilon,\rho)$ cannot be deduced from the result obtained for $m = 2$.

For $m \geq 2$, and for any $\beta \in E^{1,0}_{\frac{\pi}{2},\delta}$,

$$1 - \beta^2,\ \rho\varepsilon^m(1-\beta)^2\beta^2 \in E^{2,0}_{\frac{\pi}{2},\delta}$$

whereas for $m < 2$,

$$1 - \beta^2 \in E^{2,0}_{\frac{\pi}{2},\delta} \qquad \text{and} \qquad \rho\varepsilon^m(1-\beta)^2\beta^2 \in E^{4-m,0}_{\frac{\pi}{2},\delta}.$$

Hence, for $m < 2$ the unperturbed system is no longer the relevant part for the computation of $I(\varepsilon,\rho)$ and the Melnikov function is no longer relevant too.

The resonances studied in this book ($0^2 i\omega$, $(i\omega_0)^2 i\omega_1$) lead to perturbed systems (Normal form +higher order terms) analogous to the limit case $m = 2$.

Proof of Lemma 2.1.25.

(a) is left to the reader, and (c) directly follows from (b). Indeed, denote by

$$f := \left(1-(\beta^\star)^2\right) = 1-\left(\beta^h\right)^2 - 2\beta^h\gamma^\star - (\gamma^\star)^2.$$

Then,

$$1-\left(\beta^h(\xi)\right)^2 = \cosh^{-2}(\xi) \in E^{2,2}_{\frac{\pi}{2},1}(2,2,\tfrac{1}{2},\tfrac{1}{2})$$

and

$$\mathrm{P}^+_{\mathrm{r}}\left(1-\left(\beta^h\right)^2\right) = -1/z^2 = -(\widehat{\beta}^h_0)^2.$$

Moreover,

$$-2\beta^h\gamma^\star - (\gamma^\star)^2 \in E^{1+1,\lambda}_{\frac{\pi}{2},\delta_1}(2,2,\tfrac{1}{2},\tfrac{1}{4})$$

and

$$\mathrm{P}^+_{\mathrm{r}}\left(-2\beta^h\gamma^\star - (\gamma^\star)^2\right) = -2\widehat{\beta}^h_0\widehat{\gamma}^\star_0 - (\widehat{\gamma}^\star_0)^2$$

where $\widehat{\beta}^h_0 := \mathrm{P}^+_{\mathrm{r}}(\beta^h)$ and $\widehat{\gamma}^\star_0 := \mathrm{P}^+_{\mathrm{r}}(\gamma^\star)$. Thus, f belongs to $E^{1+1,\lambda}_{\frac{\pi}{2},\delta_1}(2,2,\frac{1}{2},\frac{1}{4})$ and $\mathrm{P}^+_{\mathrm{r}}(f) = -(\widehat{\beta}^h_0 + \widehat{\gamma}^\star_0)^2 = -(\widehat{\beta}^\star_0)^2$. Then, Lemma 2.1.21 gives (2.22) and (2.23). The analyticity of $\rho \mapsto \Lambda(\rho)$ comes from the analyticity of $\widehat{\beta}^\star_0$ with respect to ρ.

There remains (b). We follow the strategy proposed in subsection 2.1.4.

Step 1. Continuation far away from singularities We look for a holomorphic continuation of $\beta^\star$ in $\mathcal{D}_{\frac{\pi}{2},r,\varepsilon}$ (see Fig. 2.7) in the form $\beta^\star(\xi,\varepsilon) = \beta^h(\xi,\varepsilon)+\gamma^\star(\xi,\varepsilon)$. To work with a domain which contains each interval $[0,\xi]$ for every ξ in it, we enlarge a little bit $\mathcal{D}_{\frac{\pi}{2},r,\varepsilon}$ to obtain $\mathcal{D}^\star_{\frac{\pi}{2},r,\varepsilon} \supset \mathcal{D}_{\frac{\pi}{2},r,\varepsilon}$ (see Fig. 2.12 and observe that for $\xi \in \mathcal{D}^\star_{\frac{\pi}{2},r,\varepsilon}$, $|\xi \pm \mathrm{i}\frac{\pi}{2}| \geq \frac{\pi-1}{2\pi}\varepsilon^r$ holds for $0<\varepsilon<1$).

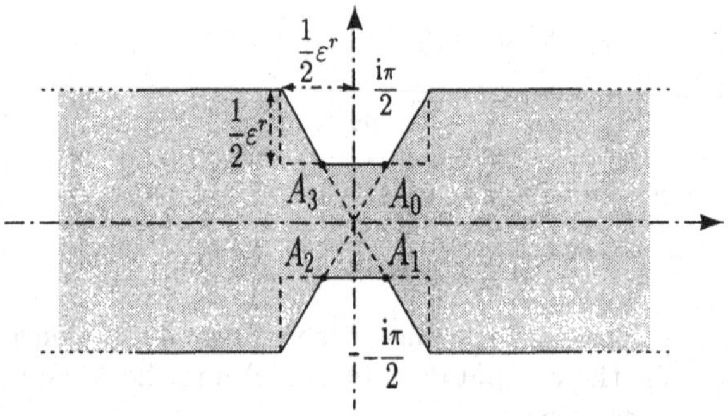

Fig. 2.12. Domain $\mathcal{D}^\star_{\frac{\pi}{2},r,\varepsilon}$. $\left(A_k = \pm\frac{\pi-\varepsilon^r}{2\pi}\varepsilon^r \pm \mathrm{i}\left(\frac{\pi}{2}-\frac{\varepsilon^r}{2}\right)\right)$

We look for $\gamma^\star(\cdot,\varepsilon)$ in the Banach space,

$$\mathcal{H}^{2,\lambda}_\varepsilon(r) := \{f : \mathcal{D}^\star_{\frac{\pi}{2},r,\varepsilon} \to \mathbb{C},\ \text{holomorphic in } \mathcal{D}^\star_{\frac{\pi}{2},r,\varepsilon},\ |f|_{\mathcal{H}^{2,\lambda}_\varepsilon(r)} < +\infty\}$$

with $|f|_{\mathcal{H}^{2,\lambda}_\varepsilon(r)} := \sup_{\xi\in\mathcal{D}^\star_{\frac{\pi}{2},r,\varepsilon}} \left[|f(\xi)| \left(\left|\xi^2 + \frac{\pi^2}{4}\right|^2 \chi_1(\xi) + (1-\chi_1(\xi)) e^{\lambda|\mathcal{R}e(\xi)|} \right) \right]$

and $\lambda \in]0,2[$.

Once more, notice that $\gamma^\star$ is the unique holomorphic function defined in a neighborhood of 0 which satisfies $\gamma^\star = \mathcal{F}(\gamma^\star)$ where

$$\mathcal{F}(\gamma)(\xi) = \cosh^{-2}(\xi) \int_{[0,\xi]} \cosh^2(\zeta)\ G(\gamma(\zeta))\ d\zeta.$$

with $G(\gamma) = -\gamma^2 + \rho\varepsilon^2 \left(1 - (\beta^h+\gamma)^2\right)(\beta^h+\gamma)^2$.

For every $\gamma \in \mathcal{H}^{2,\lambda}_\varepsilon(r)$, $\mathcal{F}(\gamma)$ is clearly holomorphic in $\mathcal{D}^\star{}_{\frac{\pi}{2},r,\varepsilon}$. Rewriting $\mathcal{F}(\gamma)$ as

$$\mathcal{F}(\gamma)(\xi) = \cosh^{-2}(\xi) \left(\int_{[0,\mathcal{R}e(\xi)]} + \int_{[\mathcal{R}e(\xi),\xi]} \right) \cosh^2(\zeta)\ G(\gamma(\zeta))\ d\zeta,$$

one can check that there exists M_{out} such that for every $r \in [0,1]$, $\varepsilon \in]0,1]$, $\rho \in [-\rho_0,\rho_0]$ and every $\gamma,\gamma' \in \mathcal{H}^{2,\lambda}_\varepsilon(r)$, satisfying $|\gamma|_{\mathcal{H}^{2,\lambda}_\varepsilon(r)} \le 1$, $|\gamma'|_{\mathcal{H}^{2,\lambda}_\varepsilon(r)} \le 1$, $\mathcal{F}$ satisfies

$$|\mathcal{F}(\gamma)|_{\mathcal{H}^{2,\lambda}_\varepsilon(r)} \le M_{\text{out}} \left(\frac{|\gamma|^2_{\mathcal{H}^{2,\lambda}_\varepsilon(r)}}{\varepsilon^r} + \varepsilon^{2-r} \left(1 + \frac{|\gamma|_{\mathcal{H}^{2,\lambda}_\varepsilon(r)}}{\varepsilon^r} \right)^4 \right)$$

$$|\mathcal{F}(\gamma) - \mathcal{F}(\gamma')|_{\mathcal{H}^{2,\lambda}_\varepsilon(r)} \le M_{\text{out}}$$
$$\left(\frac{|\gamma|_{\mathcal{H}^{2,\lambda}_\varepsilon(r)} + |\gamma'|_{\mathcal{H}^{2,\lambda}_\varepsilon(r)}}{\varepsilon^r} + \varepsilon^{2-r} \left(1 + \frac{|\gamma|_{\mathcal{H}^{2,\lambda}_\varepsilon(r)}}{\varepsilon^r} \right)^3 \right) |\gamma - \gamma'|_{\mathcal{H}^{2,\lambda}_\varepsilon(r)}$$

These estimates ensure that for any $r \in [0,1[$, there exists $\varepsilon_1(r) > 0$ such that for every $\varepsilon \in]0,\varepsilon_1(r)]$ and every $\rho \in [-\rho_0,\rho_0]$, $\mathcal{F}$ is a contraction mapping from $\mathcal{BH}^{2,\lambda}_\varepsilon(r)[2M_{\text{out}}\varepsilon^{2-r}]$ to itself where $\mathcal{BH}^{2,\lambda}_\varepsilon(r)[d] = \{f \in \mathcal{H}^{2,\lambda}_\varepsilon(r) /\ |f|_{\mathcal{H}^{2,\lambda}_\varepsilon(r)} \le d\}$. We can then conclude

Result of step 1. *For any $r \in [0,1[$, there exists $\varepsilon_1(r) > 0$ such that for every $\varepsilon \in]0,\varepsilon_1(r)]$ and every $\rho \in [-\rho_0,\rho_0]$, $\beta^\star$ has a holomorphic continuation in $\mathcal{D}^\star_{\frac{\pi}{2},r,\varepsilon} \supset \mathcal{D}_{\frac{\pi}{2},r,\varepsilon}$ of the form*

$$\beta^\star = \beta^h + \gamma^\star \qquad \text{where } \gamma^\star \in \mathcal{H}^{2,\lambda}_\varepsilon(r), \text{ and } |\gamma^\star|_{\mathcal{H}^{2,\lambda}_\varepsilon(r)} \le 2M_{\text{out}}\varepsilon^{2-r}.$$

In what follows, we will use this result with $r = \frac{1}{2}$.

Corollary of step 1. *There exists $\varepsilon_2 = \varepsilon_1(\frac{1}{2}) > 0$ such that for every $\varepsilon \in]0, \varepsilon_2]$ and every $\rho \in [-\rho_0, \rho_0]$, $\beta^\star$ has a holomorphic continuation in $\mathcal{D}^*_{\frac{\pi}{2},\frac{1}{2},\varepsilon} \supset \mathcal{D}_{\frac{\pi}{2},\frac{1}{2},\varepsilon}$ of the form*

$$\beta^\star = \beta^h + \gamma^\star \qquad \text{where } \gamma^\star \in \mathcal{H}^{2,\lambda}_\varepsilon(\tfrac{1}{2}), \text{ and } |\gamma^\star|_{\mathcal{H}^{2,\lambda}_\varepsilon(\frac{1}{2})} \le 2M_{\text{out}}\varepsilon^{\frac{3}{2}}.$$

Step 2. Continuation near $\mathrm{i}\frac{\pi}{2}$. This second step is devoted to the description of $\beta^\star$ near $\mathrm{i}\frac{\pi}{2}$. For that purpose, following the strategy described in section 2.1.4, we introduce the *inner system of coordinates* $(\widehat{\beta}, z)$ defined by

$$\beta = \frac{1}{\varepsilon}\widehat{\beta}, \qquad \xi = \mathrm{i}\frac{\pi}{2} + \varepsilon z.$$

In this system of coordinates, (2.1) reads

$$\frac{d\widehat{\beta}}{dz} = \widehat{F}(\widehat{\beta}, \rho, \varepsilon) := \widehat{F}_0(\widehat{\beta}, \rho) + \widehat{F}_1(\widehat{\beta}, \rho, \varepsilon) \tag{2.26}$$

with

$$\widehat{F}_0(\widehat{\beta}, \rho) = \widehat{F}(\widehat{\beta}, \rho, 0) = -\widehat{\beta}^2 - \rho\widehat{\beta}^4, \quad \text{and} \quad \widehat{F}_1(\widehat{\beta}, \rho, \varepsilon) := \varepsilon^2 + \rho\varepsilon^2\widehat{\beta}^2. \tag{2.27}$$

We define the *inner equation* by

$$\frac{d\widehat{\beta}}{dz} = \widehat{F}_0(\widehat{\beta}, \rho) := -\widehat{\beta}^2 - \rho\widehat{\beta}^4. \tag{2.28}$$

Then, we proceed in two sub steps:

- In step 2.1, we build a solution $\widehat{\beta}_0$ of the inner equation which belongs to $\widehat{H}^1_\delta$ for δ sufficiently large. $\widehat{\beta}_0$ is our candidate to be the principal part of $\beta^\star$.
- In step 2.2, we prove that $\widehat{\beta}^\star$ admits in $\widehat{\Sigma}_{\delta,\frac{1}{2},\varepsilon} :=] - \varepsilon^{-1/2}, \varepsilon^{-1/2}[\times] - \varepsilon^{-1/2}, -\delta[$ a holomorphic continuation which reads

$$\widehat{\beta}^\star(z, \rho, \varepsilon) = \widehat{\beta}_0(z, \rho) + \varepsilon^{\frac{1}{4}}\widehat{\beta}^\star_1(z, \rho, \varepsilon) \qquad \text{where } \widehat{\beta}^\star_1 \in \widehat{H}^2_{\delta,\frac{1}{2}}. \tag{2.29}$$

  for sufficiently large δ and small ε. This later formula confirms that $\widehat{\beta}_0 = \mathrm{Pr}(\beta^\star)$ holds.

Step 2.1. Inner equation. The inner equation (2.21) admits for $\rho = 0$, a solution $\widehat{\beta}^h_0 = 1/z$ which is the principal part of $\beta^h(\xi) = \tanh\xi$, i.e. $\mathrm{Pr}(\beta^h) = \widehat{\beta}^h_0 = 1/z$. Thus we look for a solution of the inner equation of the form

$$\widehat{\beta}_0(z, \rho) = \widehat{\beta}^h_0(z) + \widehat{\gamma}_0(z, \rho).$$

Moreover, we want to prove that $\widehat{\gamma}_0$ is holomorphic with respect to ρ. Hence, in this substep we work with $\rho \in D(0, \rho_0) = \{\rho \in \mathbb{C} / \ |\rho| < \rho_0\}$ and we look for $\widehat{\gamma}_0$ in the Banach space $\widehat{\mathcal{H}}^2_\delta$ of functions $\widehat{\gamma}$ satisfying

(a) $\widehat{\gamma} : \widehat{\Omega}_\delta \times D(0,\rho_0) \to \mathbb{C}$,
(b) $\widehat{\gamma}$ is continuous in $\widehat{\Omega}_\delta \times D(0,\rho_0)$,
(c) For every $\rho \in D(0,\rho_0)$, $z \mapsto \widehat{\gamma}(z,\rho)$ is holomorphic in $\widehat{\Omega}_\delta$,
(d) For every $z \in \widehat{\Omega}_\delta$, $\rho \mapsto \widehat{\gamma}(z,\rho)$ is holomorphic in $D(0,\rho_0)$,
(e) $|\widehat{\gamma}|_{\widehat{\mathcal{H}}^2_\delta} := \sup_{\widehat{\Omega}_\delta \times D(0,\rho_0)} \left(|z|^2|\widehat{\gamma}(z,\rho)|\right) + \frac{1}{\sqrt{\delta}} \sup_{\widehat{\Omega}^+_\delta \times D(0,\rho_0)} \left(|z|^{\frac{5}{2}}|\widehat{\gamma}(z,\rho)|\right) < +\infty$,

where $\widehat{\Omega}^+_\delta := \mathbb{R}\times]-\infty,-\delta[$ and $\widehat{\Omega}^+_\delta := [0,+\infty[\times]-\infty,-\delta[$.

Using the variation of constant formula to turn the equation satisfied by $\widehat{\gamma}_0$ into an integral form,

$$\frac{d\widehat{\gamma}}{dz} + 2\widehat{\beta}^h_0\widehat{\gamma} = -\widehat{\gamma}^2 - \rho(\widehat{\beta}^h_0 + \widehat{\gamma})^4$$

we obtain : *if* $\widehat{\gamma}_0 = \widehat{\mathcal{F}}_0(\widehat{\gamma}_0)$ *where*

$$\widehat{\mathcal{F}}_0(\widehat{\gamma}) := \frac{1}{z^2}\int_0^{+\infty} (z+t)^2 \left(\widehat{\gamma}^2 + \rho(\widehat{\beta}^h_0 + \widehat{\gamma})^4\right)(z+t)\, dt, \tag{2.30}$$

then $\widehat{\beta}_0 := \widehat{\beta}^h_0 + \widehat{\gamma}_0$ *is a solution of the inner equation (2.21).*

The theorems of continuity and differentiability under the symbol $\int$ derived from Lebesgue's dominated convergence theorem ensure that for every $\widehat{\gamma} \in \widehat{\mathcal{H}}^2_\delta$ satisfying $|\widehat{\gamma}|_{\widehat{\mathcal{H}}^2_\delta} \le 1$, $\widehat{\mathcal{F}}_0(\widehat{\gamma})$ is continuous in $\widehat{\Omega}_\delta \times D(0,\rho_0)$ and holomorphic with respect to z (resp. ρ) in $\widehat{\Omega}_\delta$ (resp. in $D(0,\rho_0)$). The estimates

$$\left|\widehat{\mathcal{F}}_0(\widehat{\gamma})\right|_{\widehat{\mathcal{H}}^2_\delta} \le M_{\text{in}} \left(\frac{|\widehat{\gamma}|^2_{\widehat{\mathcal{H}}^2_\delta} + 1}{\delta}\right)$$

$$\left|\widehat{\mathcal{F}}_0(\widehat{\gamma}) - \widehat{\mathcal{F}}_0(\widehat{\gamma}')\right|_{\widehat{\mathcal{H}}^2_\delta} \le M_{\text{in}} \left(\frac{|\widehat{\gamma}|_{\widehat{\mathcal{H}}^2_\delta} + |\widehat{\gamma}'|_{\widehat{\mathcal{H}}^2_\delta} + 1}{\delta}\right) |\widehat{\gamma} - \widehat{\gamma}'|_{\widehat{\mathcal{H}}^2_\delta}$$

which hold for every $\delta \ge 1$, $\rho \in D(0,\rho_0)$ and for every $\widehat{\gamma}, \widehat{\gamma}' \in \widehat{\mathcal{H}}^2_\delta$ satisfying $|\widehat{\gamma}|_{\widehat{\mathcal{H}}^2_\delta} \le 1$, $|\widehat{\gamma}'|_{\widehat{\mathcal{H}}^2_\delta} \le 1$, ensure that for δ sufficiently large, $\widehat{\mathcal{F}}_0$ is a contraction mapping from $\mathcal{B}\widehat{\mathcal{H}}^2_\delta(2M_{\text{in}}/\delta)$ to itself where $\mathcal{B}\widehat{\mathcal{H}}^2_\delta(d) = \{\widehat{\gamma} \in \widehat{\mathcal{H}}^2_\delta /\ |\widehat{\gamma}|_{\widehat{\mathcal{H}}^2_\delta} \le d\}$. We can then conclude

Result of Step 2.1. *There exists δ_0, such that for every $\rho \in D(0,\rho_0)$ and every $\delta \ge \delta_0$, the inner equation (2.21) admits a solution*

$$\widehat{\beta}_0(z,\rho) = \widehat{\beta}^h_0(z) + \widehat{\gamma}_0(z,\rho)$$

where $\widehat{\gamma}_0 \in \widehat{\mathcal{H}}^2_\delta$ *and*

$$|\widehat{\gamma}_0|_{\widehat{\mathcal{H}}^2_\delta} \le 2M_{\text{in}}/\delta. \tag{2.31}$$

Step 2.2. Matching of holomorphic continuations. From now on, ρ is no longer complex and lies in $[-\rho_0, \rho_0]$. As already explained, this step is devoted to the proof of (2.29). It is equivalent to prove that $\widehat{\beta}^\star(z)$ admits a holomorphic continuation in $\widehat{\Sigma}_{\delta,\frac{1}{2},\varepsilon}$ which reads

$$\widehat{\beta}^\star(z,\rho,\varepsilon) = \widehat{\beta}_0(z,\rho) + \widehat{\beta}_1(z,\rho,\varepsilon)$$

where $\widehat{\beta}_1$ is holomorphic in $\widehat{\Sigma}_{\delta,\frac{1}{2},\varepsilon}$ and satisfies

$$\sup_{z\in\widehat{\Sigma}_{\delta,\frac{1}{2},\varepsilon}} \left(|z|^2|\widehat{\beta}_1(z,\varepsilon)|\right) < M\varepsilon^{\frac{1}{4}}. \tag{2.32}$$

The construction of such a continuation is done using the method of continuation along horizontal lines described in Appendix 2.A. Hence, we must check that Assumptions 2.A.1, 2.A.2, 2.A.3 are fulfilled: we study the complex differential equation (2.26) which has the form

$$\frac{d\widehat{\beta}}{dz} = \widehat{F}_0(\widehat{\beta},\rho) + \widehat{F}_1(\widehat{\beta},\rho,\varepsilon)$$

with $\widehat{F}_0$, $\widehat{F}_1$ given by (2.27) and where z lies in $\mathbb{R}\times]-\varepsilon^{-\frac{1}{2}},-\delta[$ (see Fig. 2.13).

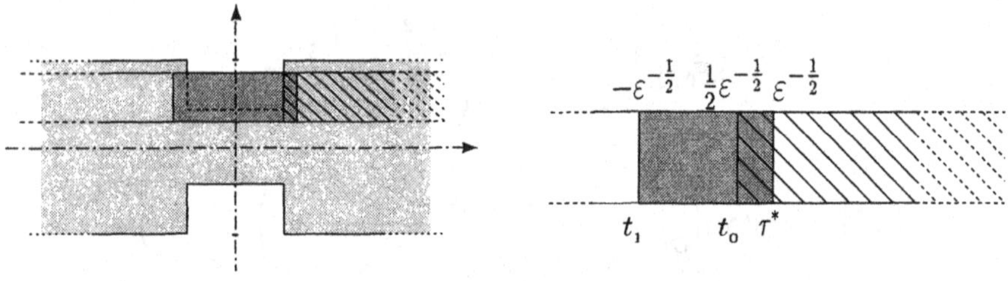

Fig. 2.13. Continuation to $\Sigma_{\delta,\frac{1}{2},\varepsilon}$ along horizontal lines

As. 2.A.1 Step 2.1 ensures the existence of a solution $\widehat{\beta}_0$ of the inner equation

$$\frac{d\widehat{\beta}}{dz} = \widehat{F}_0(\widehat{\beta},\rho)$$

which is holomorphic in $\widehat{\Omega}_\delta \supset \mathbb{R}\times]-\varepsilon^{-\frac{1}{2}},-\delta[$.

As. 2.A.2 Step 1 ensures the existence of a holomorphic continuation of $\beta^\star$ in $\mathcal{D}_{\frac{\pi}{2},\frac{1}{2},\varepsilon}$. This continuation has been obtained in the form $\beta^\star = \beta^h + \gamma^\star$. Rewriting this result in the inner system of coordinates, we can state: the full equation (2.26) admits a holomorphic solution

$$\widehat{\beta}^\star = \widehat{\beta}_0 + \widehat{\beta}_1$$

in $\widehat{\mathcal{D}}_{\frac{\pi}{2},\frac{1}{2},\varepsilon} \supset]\frac{1}{2}\varepsilon^{-\frac{1}{2}}, +\infty[\times] - \varepsilon^{-\frac{1}{2}}, -\delta[$ where

$$\widehat{\beta}_1 = \widehat{\beta}^h + \widehat{\gamma}^\star - \widehat{\beta}_0,$$

with $\widehat{\beta}^h(z) = \varepsilon\beta^h(\mathrm{i}\frac{\pi}{2} + \varepsilon z)$ and $\widehat{\gamma}^\star(z) = \varepsilon\gamma^\star(\mathrm{i}\frac{\pi}{2} + \varepsilon z)$. Thus, Assumption 2.A.2 is fulfilled too.

As. 2.A.3 Following Appendix 2.A, we introduce the equation along horizontal lines

$$\begin{aligned}\frac{\partial\widehat{\beta}}{\partial\tau} = \widehat{F}_0(\widehat{\beta}_0(\tau + \mathrm{i}\eta) + \widehat{\beta}(\tau,\eta)) - \widehat{F}_0(\widehat{\beta}_0(\tau + \mathrm{i}\eta)) \\ + \widehat{F}_1(\widehat{\beta}_0(\tau + \mathrm{i}\eta) + \widehat{\beta}(\tau,\eta)).\end{aligned} \tag{2.33}$$

$\widehat{\beta}_1$ is seen as a function from $\mathbb{R}^2$ to $\mathbb{C}$ and is characterized by the fact that it is the unique solution of (2.33) which is equal to

$$\Delta(\eta) := \widehat{\beta}^h(\varepsilon^{-\frac{1}{2}} + \mathrm{i}\eta) + \widehat{\gamma}^\star(\varepsilon^{-\frac{1}{2}} + \mathrm{i}\eta) - \widehat{\beta}_0(\varepsilon^{-\frac{1}{2}} + \mathrm{i}\eta) \tag{2.34}$$

for $\tau = \varepsilon^{-\frac{1}{2}}$. If we prove that $\widehat{\beta}_1$ has an interval of life which contains $]-\varepsilon^{-\frac{1}{2}}, \varepsilon^{-\frac{1}{2}}]$ for every $\eta \in]-\varepsilon^{-\frac{1}{2}}, -\delta[$, then Assumption 2.A.3 is fulfilled and Lemma 2.A.5 ensures that $\widehat{\beta}_0 + \widehat{\beta}_1$ is a holomorphic continuation of $\widehat{\beta}^\star$ in $\widehat{\Sigma}_{\delta,\frac{1}{2},\varepsilon}$.

To obtain such an estimate on the interval of life of $\widehat{\beta}_1$, we rewrite (2.33) as an integral equation and we solve it using the Contraction Mapping Theorem in an appropriate space of functions defined on $]-\varepsilon^{-\frac{1}{2}}, \varepsilon^{-\frac{1}{2}}]$ satisfying (2.32): we choose the Banach space $C^2_{\delta,\varepsilon}$ of functions $\widehat{\beta}$

$$\begin{aligned}C^2_{\delta,\varepsilon} = \{\widehat{\beta} :] - \varepsilon^{-\frac{1}{2}}, \varepsilon^{-\frac{1}{2}}]\times] - \varepsilon^{-\frac{1}{2}}, -\delta] \to \mathbb{C}, \text{ continuous }, \\ |\widehat{\beta}|_{C^2_{\delta,\varepsilon}} := \sup(|\widehat{\beta}(\tau,\eta)|\ |\tau + \mathrm{i}\eta|^2) < +\infty\}\end{aligned}$$

The equation along horizontal lines (2.33) can be rewritten as

$$\frac{\partial\widehat{\beta}}{\partial\tau} = \widehat{Q}_0(\widehat{\beta}(\tau,\eta),\tau,\eta) + \widehat{Q}_1(\widehat{\beta}(\tau,\eta),\tau,\eta)$$

where

$$\widehat{Q}_0(\widehat{\beta},\tau,\eta) = -2\widehat{\gamma}_0\widehat{\beta} + \widehat{\beta}^2 - \rho\left(\left(\widehat{\beta}^h_0 + \widehat{\gamma}_0 + \widehat{\beta}\right)^4 - \left(\widehat{\beta}^h_0 + \widehat{\gamma}_0\right)^4\right),$$

and

$$\widehat{Q}_1(\widehat{\beta},\tau,\eta)=\varepsilon^2+\rho\varepsilon^2\left(\widehat{\beta}_0^h+\widehat{\gamma}_0+\widehat{\beta}\right)^2.$$

Uniqueness of the Cauchy problem and the variation of constants formula ensure that $\widehat{\beta}_1$ is the unique continuous function of τ with parameter η defined in a neighborhood of $\tau=\varepsilon^{-\frac{1}{2}}$ which satisfies $\widehat{\beta}_1=\widehat{\mathcal{F}}_1(\widehat{\beta}_1)$ where

$$\widehat{\mathcal{F}}_1(\widehat{\beta})(\tau,\eta)=\frac{\Delta(\eta)\left(\varepsilon^{-\frac{1}{2}}+i\eta\right)^2-\int_\tau^{\varepsilon^{-\frac{1}{2}}}(s+i\eta)^2\left(\widehat{Q}_0+\widehat{Q}_1\right)(\widehat{\beta}(s,\eta),s,\eta)\,ds}{(\tau+i\eta)^2}.$$

Rewriting $|\Delta(\eta)|$ as

$$\Delta(\eta)=\left(\widehat{\beta}^h-\widehat{\beta}_0^h+\widehat{\gamma}^\star+\widehat{\gamma}_0\right)(\varepsilon^{-\frac{1}{2}}+i\eta);$$

observing that

$$\sup_{\widehat{\Sigma}_{\delta,\frac{1}{2},\varepsilon}}\left(|z|^2(\widehat{\beta}^h-\widehat{\beta}_0^h)(z)\right)\le M\varepsilon^{\frac{1}{2}}, \tag{2.35}$$

and using the estimates obtained at Step 1 and 2.1

$$\sup_{\widehat{\Omega}_\delta^+\times D(0,\rho_0)}\left(|z|^{\frac{5}{2}}|\widehat{\gamma}_0(z,\rho)|\right)\le\frac{2M_{\text{in}}}{\sqrt{\delta}},\qquad \sup_{\substack{\xi\in\mathcal{D}_{\frac{\pi}{2},\frac{1}{2},\varepsilon}\\|\mathcal{R}e(\xi)|\le 1}}\left(|\widehat{\gamma}^\star(\xi)|\left|\xi^2+\frac{\pi^2}{4}\right|\right)\le 2M_{\text{out}}\varepsilon^{\frac{3}{2}}.$$

we get

$$\left|\frac{\Delta(\eta)\left(\varepsilon^{-\frac{1}{2}}+i\eta\right)^2}{(\tau+i\eta)^2}\right|_{C^2_{\delta,\varepsilon}}\le M\varepsilon^{\frac{1}{4}}.$$

Finally, the estimates

$$\left|\widehat{\mathcal{F}}_1(\widehat{\beta})\right|_{C^2_{\delta,\varepsilon}}\le M_{\text{mat}}\left(|\widehat{\beta}|^2_{C^2_{\delta,\varepsilon}}+\delta^{-1}|\widehat{\beta}|_{C^2_{\delta,\varepsilon}}+\varepsilon^{\frac{1}{4}}\right)$$

$$\left|\widehat{\mathcal{F}}_1(\widehat{\beta})-\widehat{\mathcal{F}}_1(\widehat{\beta}')\right|_{C^2_{\delta,\varepsilon}}\le M_{\text{mat}}\left(|\widehat{\beta}|_{C^2_{\delta,\varepsilon}}+|\widehat{\beta}'|_{C^2_{\delta,\varepsilon}}+\delta^{-1}+\varepsilon^{\frac{3}{2}}\right)\left|\widehat{\beta}-\widehat{\beta}'\right|_{C^2_{\delta,\varepsilon}}$$

which hold for every $\varepsilon\in]0,\varepsilon_2]$, $\delta\in[\delta_0,\varepsilon^{-\frac{1}{2}}$, $\rho\in[-\rho_0,\rho_0]$ and for every $\widehat{\beta},\widehat{\beta}'\in C^2_{\delta,\varepsilon}$ satisfying $|\widehat{\beta}|_{C^2_{\delta,\varepsilon}}\le 1$, $|\widehat{\beta}'|_{C^2_{\delta,\varepsilon}}\le 1$, ensure that for $\delta_1=\max(4M_{\text{mat}},\delta_0)$ and ε sufficiently small, $\widehat{\mathcal{F}}_1$ is a contraction mapping from $\mathcal{B}C^2_{\delta_1,\varepsilon}(2M_{\text{mat}}\varepsilon^{\frac{1}{4}})$ to itself where $\mathcal{B}C^2_{\delta_1,\varepsilon}(d)=\{\widehat{\gamma}\in C^2_{\delta_1,\varepsilon}/\ |\widehat{\gamma}|_{C^2_{\delta_1,\varepsilon}}\le d\}$.

So, Assumption 2.A.3 is fulfilled : $\widehat{\beta}_1$ has an interval of life which contains $]-\varepsilon^{-\frac{1}{2}},\varepsilon^{-\frac{1}{2}}]$ for every $\eta\in]-\varepsilon^{-\frac{1}{2}},-\delta[$. Then, using Lemma 2.A.5 and setting $\widehat{\beta}_1=\varepsilon^{\frac{1}{4}}\widehat{\beta}_1^\star$ we can conclude

Result of Step 2.2. *There exist $\delta_1 \geq \delta_0$ and $\varepsilon_3 \in]0, \varepsilon_2]$, such that for every $\rho \in [-\rho_0, \rho_0]$ and every $\varepsilon \in]0, \varepsilon_3]$, $\widehat{\beta}^\star$ admits a holomorphic continuation in $\widehat{\Sigma}_{\delta_1, \frac{1}{2}, \varepsilon}$ which reads*

$$\widehat{\beta}^\star(z, \rho, \varepsilon) = \widehat{\beta}_0(z, \rho) + \varepsilon^{\frac{1}{4}} \widehat{\beta}_1^\star(z, \rho, \varepsilon)$$

where $\widehat{\beta}_0$ has been obtained at step 2.1 and $\widehat{\beta}_1^\star$ satisfies

$$\sup_{\substack{\varepsilon \in]0, \varepsilon_3], |\rho| \leq \rho_0 \\ z \in \widehat{\Sigma}_{\delta_1, \frac{1}{2}, \varepsilon}}} \left(|z|^2 |\widehat{\beta}_1^\star(z, \rho, \varepsilon)| \right) \leq 2M_{\text{mat}}.$$

Step 3. Continuation near $-\mathrm{i}\frac{\pi}{2}$: symmetrization. Using the argument of symmetry with respect to the real axis given at Step 3 of subsection 2.1.4 ($\beta^\star$ is real valued on the real axis), we can finally conclude

Result of Step 3. *There exist $\delta_1 \geq 1$ and $\varepsilon_3 > 0$, such that for every $\rho \in [-\rho_0, \rho_0]$ and every $\varepsilon \in]0, \varepsilon_3]$, $\widehat{\beta}^\star(\cdot, \varepsilon, \rho)$ admits a holomorphic continuation in*

$$D_\varepsilon := \mathcal{D}_{\frac{\pi}{2}, \frac{1}{2}, \varepsilon} \cup \widehat{\Sigma}_{\delta_1, \frac{1}{2}, \varepsilon} \cup \overline{\widehat{\Sigma}_{\delta_1, \frac{1}{2}, \varepsilon}}.$$

This continuation is still denoted by $\beta^\star$ and it satisfies:

(a) $\beta^\star(\overline{\xi}) = \overline{\beta^\star(\xi)}$ *for every $\xi \in D_\varepsilon$.*

(b) *In $\mathcal{D}_{\frac{\pi}{2}, \frac{1}{2}, \varepsilon}$, $\beta^\star$ reads $\beta^\star = \beta^h + \gamma^\star$ and*

$$\sup_{\substack{\varepsilon \in]0, \varepsilon_3] \\ \xi \in \mathcal{D}_{\frac{\pi}{2}, \frac{1}{2}, \varepsilon}}} \left[|\gamma^\star(\xi, \varepsilon)| \left(\left| \xi^2 + \tfrac{\pi^2}{4} \right| \chi_1(\xi) + (1 - \chi_1(\xi)) \mathrm{e}^{\lambda |\mathcal{R}e(\xi)|} \right) \right] < +\infty$$

(c) *In $\Sigma_{\delta_1, \frac{1}{2}, \varepsilon}$, i.e. for $z \in \widehat{\Sigma}_{\delta_1, \frac{1}{2}, \varepsilon}$, $\beta^\star$ reads*

$$\beta^\star(\mathrm{i}\tfrac{\pi}{2} + \varepsilon z, \varepsilon, \rho) = \frac{1}{\varepsilon} \left(\widehat{\beta}_0^h(z) + \widehat{\gamma}_0(z, \rho) \right) + \frac{\varepsilon^{\frac{1}{4}}}{\varepsilon} \widehat{\beta}_1(z, \rho, \varepsilon)$$

where $\widehat{\beta}_0^h(z) = 1/z \in \widehat{H}^1_{\delta_1}$, $\widehat{\gamma}_0 \in \widehat{H}^2_{\delta_1}$ and $\widehat{\beta}_1^\star \in \widehat{H}^2_{\delta_1, \frac{1}{2}}$.

Moreover, $\widehat{\beta}^h + \widehat{\gamma}_0$ is solution of the inner system (2.21).

Hence, $\beta^\star$ belongs to $\in E^{1,0}_{\frac{\pi}{2}, \delta_1}(1, 1, \frac{1}{2}, \frac{1}{4})$.

Finally, since $\beta^h \in E^{1,0}_{\frac{\pi}{2}, \delta_1}(1, 1, \frac{1}{2}, \frac{1}{2})$, (b),(c) ensure that

$$\gamma^\star = \beta^\star - \beta^h \in E^{1,\lambda}_{\frac{\pi}{2}, \delta_1}(1, 1, \tfrac{1}{2}, \tfrac{1}{2}).$$

Remark. (b) comes from the corollary of step 1 observing that ,

$$\left| \xi^2 + \tfrac{\pi^2}{4} \right| = |\xi - \mathrm{i}\tfrac{\pi}{2}| \, |\xi + \mathrm{i}\tfrac{\pi}{2}| \geq \tfrac{\pi}{2} \, \tfrac{1}{2} \varepsilon^{\frac{1}{2}} \qquad \text{for } \xi \in \mathcal{D}_{\frac{\pi}{2}, \frac{1}{2}, \varepsilon}. \qquad \square$$

Proof of Lemma 2.1.26

Lemma 2.1.25 ensure that $\rho \mapsto \widehat{\beta}(z,\rho)$ is holomorphic in $D(0,2)$. Using the Cauchy integral formula and (2.31) we obtain that

$$\widehat{\beta}_0^\star(z,\rho) = \sum_{n\geq 0} \rho^n \widehat{\beta}_{0,n}^\star(z), \qquad \text{for } z \in \widehat{\Omega}_{\delta_1},\ \rho \in D(0,\rho_0)$$

where $\widehat{\beta}_{0,n}^\star$ is holomorphic in $\widehat{\Omega}_{\delta_1}$,

$$\widehat{\beta}_{0,0} = \widehat{\beta}_0^h(z) = \frac{1}{z} \qquad \text{and } \sup\left(|z|^2 \widehat{\beta}_{0,n}^\star(z)\right) \leq \frac{2M_{\text{in}}}{\delta_1}\frac{1}{(\rho_0)^n}. \tag{2.36}$$

The function $\widehat{\gamma}_0^\star = \widehat{\beta}_0^\star - \beta_0^h = \sum_{n\geq 1} \rho^n \widehat{\beta}_{0,n}^\star$ satisfies

$$\widehat{\gamma}_0 = \widehat{\mathcal{F}}_0(\widehat{\gamma}_0) \tag{2.37}$$

where $\widehat{\mathcal{F}}_0$ is defined by (2.30). Substituting the power expansion of $\widehat{\gamma}_0$ in (2.37), and identifying the power of ρ, we get

$$\widehat{\beta}_{0,n}^\star(z) = -\frac{1}{z^2}\int_0^{+\infty} (z+t)^2 \left(\sum_{p+q=n} \widehat{\beta}_{0,p}^\star \widehat{\beta}_{0,q}^\star + \sum_{p+q+k+l=n-1} \widehat{\beta}_{0,p}^\star \widehat{\beta}_{0,q}^\star \widehat{\beta}_{0,k}^\star \widehat{\beta}_{0,l}^\star \right) (z+t)dt$$

for $n \geq 1$. We can then check by induction that for every $n \geq 0$,

$$\widehat{\beta}_{0,n}^\star(z) = \frac{a_n}{z^{2n+1}} \tag{2.38}$$

holds where the coefficient a_n are given by (2.25). Finally, using (2.23), (2.38), and (2.36)

$$\begin{aligned} \Lambda(\rho) &= -\int_{-i\eta+\mathbb{R}} \left(\widehat{\beta}_0^\star(z,\rho)\right)^2 e^{i\omega z} dz \\ &= -\int_{-i\eta+\mathbb{R}} e^{i\omega z} \sum_{n\geq 0} \rho^n \sum_{p+q=n} \frac{a_p a_q}{z^{2(p+q)+2}}\, dz \\ &= -\sum_{n\geq 0} \rho^n \Big(\sum_{p+q=n} a_p a_q \Big) \int_{-i\eta+\mathbb{R}} \frac{e^{i\omega z}}{z^{2n+2}}\, dz \\ &= -\sum_{n\geq 0} \rho^n \Big(\sum_{p+q=n} a_p a_q \Big) \frac{2\pi i (i\omega)^{2n+1}}{(2n+1)!}. \end{aligned}$$

□

2.2 Bi-frequency oscillatory integrals: partial complexification of time

The previous section gives tools for computing upper bounds and equivalents of mono-frequency oscillatory integrals like the basic example

$$J_1(\varepsilon) = \int_{-\infty}^{+\infty} e^{\frac{i\omega_1 t}{\varepsilon}} g\left(\frac{\varepsilon}{\cosh(t)}\right) dt \underset{\varepsilon\to 0}{\sim} C(g) e^{\frac{-\pi\omega_1}{2\varepsilon}} .$$

which occurs for instance in the $0^2 i\omega$ resonance.

A second basic example of oscillatory integral which occurs in the $(i\omega_0)^2 i\omega_1$ reversible resonance is the following

$$J_2(\varepsilon) = \int_{-\infty}^{+\infty} e^{\frac{i\omega_1 t}{\varepsilon}} g\left(\frac{\varepsilon\cos(\omega_0 t/\varepsilon)}{\cosh(t)}\right) dt$$

which leads to compute for $p \in \mathbb{Z}$ and $n \in \mathbb{N}$

$$\begin{aligned} J_{p,n} &= \int_{-\infty}^{+\infty} e^{\frac{i\omega_1 t}{\varepsilon}} e^{\frac{ip\omega_0 t}{\varepsilon}} \left(\frac{\varepsilon}{\cosh(t)}\right)^n dt \\ &\underset{\varepsilon\to 0}{\sim} \frac{2\pi\varepsilon|\omega_1 + p\omega_0|^{n-1}}{(n-1)!} e^{-\frac{|\omega_1+p\omega_0|\pi}{2\varepsilon}} . \end{aligned}$$

The integral J_2 is the prototype of oscillatory integral which occurs in a system where coexist an oscillatory part, induced by a pair of imaginary eigenvalues of order 1, and a hyperbolic part coming from a set of eigenvalues $\pm\varepsilon \pm i\omega_0$. The resonance between the two frequencies modifies the size of the oscillatory integral. Indeed, for a holomorphic function g, using the residues we get

$$J_2(\varepsilon) = \int_{-\infty}^{+\infty} e^{\frac{i\omega_1 t}{\varepsilon}} g\left(\frac{\varepsilon\cos(\omega_0 t/\varepsilon)}{\cosh(t)}\right) dt \underset{\varepsilon\to 0}{\sim} C'(g) e^{\frac{-\pi\omega_\star}{2\varepsilon}}$$

where $\omega_\star = \min_{p\in\mathbb{Z}} |\omega_1 - p\omega_0|$. The coefficient $\frac{\pi}{2}$ in the exponential comes from the position of the singularity of $(\cosh\xi)^{-1}$ at $i\frac{\pi}{2}$. The methods previously developed for computing the size of mono-frequency oscillatory integrals are based on a complexification of time which determines the holomorphic continuation of solutions and says how this continuation "explodes near singularities". In the first case, the function $H_1 : (\xi,\varepsilon) \mapsto \varepsilon(\cosh\xi)^{-1}$ is bounded near the real axis and explodes near its singularity at $i\frac{\pi}{2}$. In the second case, $H_2 : (\xi,\varepsilon) \mapsto \varepsilon(\cosh\xi)^{-1} \cos(\omega_0\xi/\varepsilon)$ is no longer bounded near the real axis, since for every arbitrary small $\ell > 0$, $\sup_{\varepsilon\in]0,\varepsilon_0]} |H_2(i\ell,\varepsilon)| = +\infty$ because of the oscillation $\left(\cos(i\ell\omega_0/\varepsilon) \underset{\varepsilon\to 0}{\longrightarrow} +\infty\right)$. We are interested in the explosion induced by the singularity and not by the one coming from oscillations. To overcome this difficulty, a *partial complexification* of time is necessary: a function $Y(t,\varepsilon)$ with a rapid oscillation, should be seen as a function of two variables

$$Y(t) = y(t, \omega_0 t/\varepsilon)$$

where $y(\xi, s)$ is holomorphic with respect to $\xi \in \mathbb{C}$ and 2π periodic with respect to s. For our example it amounts to write $H_2(t,\varepsilon) = h_2(t, \omega_0 t/\varepsilon, \varepsilon)$ where $h_2(\xi, s, \varepsilon) = \varepsilon \cos(s)(\cosh\xi)^{-1}$. Now the function h_2 is bounded with respect to (ξ, s, ε) for ξ near the real axis and it explodes at $\xi = i\frac{\pi}{2}$.

In the next subsection, we give three "exponential lemmas" analogous to the one obtained for computing the size of mono-frequency oscillatory integrals, which are adapted to bi-frequency oscillatory integrals.

2.2.1 Exponential upper bounds and equivalent

As for mono-frequency oscillatory integrals we begin with a first lemma which gives rough exponential upper bounds.

Lemma 2.2.1 (First Bi-Frequency Exponential Lemma).
Let ω_0, ω_1, ℓ, λ be four real positive numbers. Define $\omega_\star$ by

$$\omega_\star = \min_{p \in \mathbb{Z}} |\omega_1 - p\omega_0|.$$

Let $\mathcal{B}_\ell$ be the strip in the complex field: $\mathcal{B}_\ell = \{\xi \in \mathbb{C}/|\mathcal{I}m\,(\xi)| < \ell\}$. Let ${}^b\!H_\ell^\lambda$ be the set of functions f satisfying

(a) $f: \ \mathcal{B}_\ell \times \mathbb{R} \times]0,1] \to \mathbb{C}, \quad (\xi, s, \varepsilon) \mapsto f(\xi, s, \varepsilon)$

(b) *$s \mapsto f(\xi, s, \varepsilon)$ is 2π-periodic and of class C^1 in $\mathbb{R}$,*

(c) *$(\xi, s) \mapsto f(\xi, s, \varepsilon)$ is continuous in $\mathcal{B}_\ell \times \mathbb{R}$,*

(d) *$\xi \mapsto f(\xi, s, \varepsilon)$ is holomorphic in $\mathcal{B}_\ell$,*

(e) $\|f\|_{{}^bH_\ell^\lambda} := \sup_{(\xi,s,\varepsilon)\in\mathcal{B}_\ell\times\mathbb{R}\times]0,1]} \left(\left(|f(\xi,s,\varepsilon)| + \left|\frac{\partial f}{\partial s}(\xi,s,\varepsilon)\right| \right) e^{\lambda|\mathcal{R}e(\xi)|} \right) < +\infty.$

Then for every $f \in {}^bH_\ell^\lambda$ and $\varepsilon \in]0,1]$, the bi-oscillatory integral

$$I(f,\varepsilon) = \int_{-\infty}^{+\infty} f(t, \tfrac{\omega_0 t}{\varepsilon}, \varepsilon)\, e^{i\omega_1 t/\varepsilon}\, dt$$

satisfies

$$|I(f,\varepsilon)| \le C(\ell, \omega_0, \omega_1) \frac{\|f\|_{{}^bH_\ell^\lambda}}{\lambda} e^{-\omega_\star \ell/\varepsilon}.$$

Lemma 2.2.1 gives a very efficient way to obtain exponential upper bounds because the membership to ${}^bH_\ell^\lambda$ is stable by addition, multiplication, "composition", which can be summed up

Lemma 2.2.2.

(a) ${}^bH_\ell^\lambda$ *is an algebra.*
(b) *If* $f \in {}^bH_\ell^\lambda$ *and* g *is holomorphic in a domain containing the range of* f *and satisfies* $g(0) = 0$, *then* $g \circ f \in {}^bH_\ell^\lambda$.

Remark 2.2.3. The main difference between mono-and bi-oscillatory integrals, comes from interaction between the two frequencies in the latter case. Indeed, the First Mono-Frequency Exponential Lemma 2.1.1 gives upper bounds of the form

$$|I(f,\varepsilon)| \le \mathrm{e}^{\frac{-\omega_1 \ell}{\varepsilon}} \frac{2\,\|f\|_{H_\ell^\lambda}}{\lambda}$$

whereas the First Bi-Frequency Exponential Lemma gives upper bounds of the form

$$|I(f,\varepsilon)| \le C(\ell)\, \mathrm{e}^{\frac{-\omega_\star \ell}{\varepsilon}} \frac{\|f\|_{{}^bH_\ell^\lambda}}{\lambda}.$$

In the latter case, the interaction between the two frequencies changes the constants in the exponential.

The proof of the First Bi-Frequency Exponential Lemma 2.2.1 and of the following ones are delayed until the end of this subsection. The following lemma gives sharp exponential upper bounds for bi-oscillatory integrals.

Lemma 2.2.4 (Second Bi-Frequency Exponential Lemma). *Let* ω_0, ω_1, γ, λ, σ *be five real positive numbers. Let* δ *be in* $[1, +\infty[$. *Define* $\omega_\star$ *by*

$$\omega_\star = \min_{p\in\mathbb{Z}} |\omega_1 - p\omega_0|.$$

Let ${}^bH_{\sigma,\delta}^{\gamma,\lambda}$ *be the Banach space of functions* f *satisfying*

(a) $$f : \bigcup_{0<\varepsilon<\sigma/\delta} \mathcal{B}_{\sigma-\delta\varepsilon} \times \mathbb{R} \times \{\varepsilon\} \to \mathbb{C}, \quad (\xi, s, \varepsilon) \mapsto f(\xi, s, \varepsilon)$$

(b) $s \mapsto f(\xi, s, \varepsilon)$ *is* 2π*-periodic and of class* C^1 *in* $\mathbb{R}$,
(c) $(\xi, s) \mapsto f(\xi, s, \varepsilon)$ *is continuous in* $\mathcal{B}_{\sigma-\delta\varepsilon} \times \mathbb{R}$,
(d) $\xi \mapsto f(\xi, s, \varepsilon)$ *is holomorphic in* $\mathcal{B}_{\sigma-\delta\varepsilon}$,
(e) $$\|f\|_{{}^bH_{\sigma,\delta}^{\gamma,\lambda}} := \sup_{\substack{0<\varepsilon<\sigma/\delta \\ (\xi,s)\in\mathcal{B}_{\sigma-\delta\varepsilon}\times\mathbb{R}}} \left[\left(|f(\xi,s,\varepsilon)| + \left|\frac{\partial f}{\partial s}(\xi,s,\varepsilon)\right|\right) \left(|\sigma^2 + \xi^2|^\gamma \chi_1(\xi) + (1-\chi_1(\xi))\mathrm{e}^{\lambda|\mathcal{R}e(\xi)|}\right)\right]$$

with $\chi_1(\xi) = \begin{cases} 1 & \text{for } |\mathcal{R}e(\xi)| < 1, \\ 0 & \text{for } |\mathcal{R}e(\xi)| \geq 1. \end{cases}$

Then there exists $C(\gamma, \lambda, \sigma, \delta, \omega_0, \omega_1)$ *such that for every* $f \in {}^b H^{\gamma,\lambda}_{\sigma,\delta}$ *and* $\varepsilon \in]0, \sigma/2\delta]$, *the bi-oscillatory integral*

$$I(f,\varepsilon) = \int_{-\infty}^{+\infty} f(t, \tfrac{\omega_0 t}{\varepsilon}, \varepsilon)\, \mathrm{e}^{\mathrm{i}\omega_1 t/\varepsilon}\, dt$$

satisfies

$$|I(f,\varepsilon)| \leq C\, \frac{1}{\varepsilon^{\gamma-1}} \mathrm{e}^{-\omega_\star \sigma/\varepsilon}\, \|f\|_{{}^b H^{\gamma,\lambda}_{\sigma,\delta}}\,.$$

Unlike ${}^b H^\lambda_\ell$, ${}^b H^{\gamma,\lambda}_{\sigma,\delta}$ *is not an algebra.* However the following lemma enables us to use it in nonlinear differential equations.

Lemma 2.2.5.

(a) *For every* $f \in {}^b H^{\gamma,\lambda}_{\sigma,\delta}$ *and every* $g \in {}^b H^{\gamma',\lambda'}_{\sigma,\delta}$, fg *lies in* ${}^b H^{\gamma+\gamma',\lambda+\lambda'}_{\sigma,\delta}$.

(b) *For every* $f \in {}^b H^{\gamma,\lambda}_{\sigma,\delta}$ *and every* $p \in]0, \gamma[$, $\varepsilon^p f$ *lies in* ${}^b H^{\gamma-p,\lambda}_{\sigma,\delta}$

(c) *For every* $f, g \in {}^b H^{\gamma,\lambda}_{\sigma,\delta}$, $\varepsilon^\gamma fg$ *lies in* ${}^b H^{\gamma,\lambda}_{\sigma,\delta}$.

Finally we define appropriate spaces to obtain equivalent of bi-oscillatory integrals.

Definition 2.2.6. *Let* $\gamma, \lambda \geq 0$, $\sigma > 0$, $\delta \geq 1$, $m_0, m_1 > 1$, $0 \leq r < 1$ *and* $\nu > 0$. *Denote by* ${}^b E^{\gamma,\lambda}_{\sigma,\delta}(m_0, m_1, r, \nu)$ *the set of functions* f *such that*

(a) $f \in {}^b H^{\gamma,\lambda}_{\sigma,\delta}$,

(b) *for every* $\varepsilon \in]0, \varepsilon_0]$ *and every* $z \in \widehat{\Sigma}_{\delta,r,\varepsilon} :=]-\varepsilon^{r-1}, \varepsilon^{r-1}[\times]-\varepsilon^{r-1}, -\delta[$ f *reads*

$$f(\mathrm{i}\sigma + \varepsilon z, s, \varepsilon) = \frac{1}{\varepsilon^\gamma} \widehat{f}_0^+(z,s) + \frac{\varepsilon^\nu}{\varepsilon^\gamma} \widehat{f}_1^+(z,s,\varepsilon)$$

where $\widehat{f}_0^+ \in {}^b\widehat{H}^{m_0}_\delta$ *and* $\widehat{f}_1^+ \in {}^b\widehat{H}^{m_1}_{\delta,r}$ *with*

$$\begin{aligned} {}^b\widehat{H}^{m_0}_\delta = \Big\{ \widehat{f} : \widehat{\Omega}_\delta \times \mathbb{R} \to \mathbb{C}, \quad & \text{where } \widehat{\Omega}_\delta := \mathbb{R} \times]-\infty, -\delta[, \\ & (z,s) \mapsto \widehat{f}(z,s) \text{ continuous in } \widehat{\Omega}_\delta \times \mathbb{R} \\ & z \mapsto \widehat{f}(z,s) \text{ holomorphic in } \widehat{\Omega}_\delta, \\ & s \mapsto \widehat{f}(z,s)\ 2\pi\text{-periodic in } \mathbb{R}, \\ & \big\|\widehat{f}\big\|_{{}^b\widehat{H}^{m_0}_\delta} := \sup_{(z,s)\in\widehat{\Omega}_\delta\times\mathbb{R}} \left(|\widehat{f}(z,s)||z|^{m_0} \right) < +\infty \Big\} \end{aligned}$$

and

$$ {}^b\widehat{H}^{m_1}_{\delta,r} = \Big\{ \widehat{f} : \bigcup_{\varepsilon\in]0,\varepsilon_0]} \left(\widehat{\Sigma}_{\delta,r,\varepsilon} \times \mathbb{R} \times \{\varepsilon\}\right) \to \mathbb{C}, $$

$$ (z,s) \mapsto \widehat{f}(z,s,\varepsilon) \text{ continuous in } \widehat{\Sigma}_{\delta,r,\varepsilon} \times \mathbb{R}, $$

$$ z \mapsto \widehat{f}(z,s,\varepsilon) \text{ holomorphic in } \widehat{\Sigma}_{\delta,r,\varepsilon}, $$

$$ s \mapsto \widehat{f}(z,s,\varepsilon)\ 2\pi\text{-periodic in } \mathbb{R}, $$

$$ \left\|\widehat{f}\right\|_{{}^b\widehat{H}^{m_1}_{\delta,r}} := \sup_{\substack{\varepsilon\in]0,\varepsilon_0] \\ (z,s)\in\widehat{\Sigma}_{\delta,r,\varepsilon}\times\mathbb{R}}} \left(|\widehat{f}(z,s)||z|^{m_1}\right) < +\infty \Big\} $$

(c) *for every* $\varepsilon \in]0,\varepsilon_0]$ *and every* $z \in \widehat{\Sigma}_{\delta,r,\varepsilon}$, f *reads*

$$ f(-\mathrm{i}\sigma - \varepsilon z, s, \varepsilon) = \frac{1}{\varepsilon^\gamma}\widehat{f_0^-}(z,s) + \frac{\varepsilon^\nu}{\varepsilon^\gamma}\widehat{f_1^-}(z,s,\varepsilon) $$

where $\widehat{f}_0 \in {}^b\widehat{H}^{m_0}_\delta$ *and* $\widehat{f}_1 \in {}^b\widehat{H}^{m_1}_{\delta,r}$

Remark 2.2.7. The two functions $\widehat{f}_0$ and $\widehat{f}_1$ are uniquely defined. Indeed, they are given by

$$ \widehat{f_0^\pm}(z,s) := \lim_{\varepsilon\to 0} \varepsilon^\gamma f(\pm(\mathrm{i}\sigma + \varepsilon z), s, \varepsilon) $$

and

$$ \widehat{f_1^\pm}(z) = \lim_{\varepsilon\to 0} \frac{\varepsilon^\gamma f^\pm(\pm(\mathrm{i}\sigma+\varepsilon z), s, \varepsilon) - \widehat{f_0^\pm}(z,s)}{\varepsilon^\nu}. $$

Hence, we can define

Definition 2.2.8.
Denote by $\mathrm{P_r}$ *the operator*

$$ \mathrm{P_r^\pm} : {}^bE^{\gamma,\lambda}_{\sigma,\delta}(m_0,m_1,r,\nu) \to {}^b\widehat{H}^{m_0}_\delta, \qquad f \longmapsto \widehat{f_0^\pm} $$

Lemma 2.2.9 (Third Bi-Frequency Exponential Lemma).
Let $\gamma > 1$, $\delta \geq 1$, $\sigma, \lambda > 0$, $m_0, m_1 > 1$, $0 < r < 1$, $\nu, \omega > 0$.
Let ω_0, ω_1 *be two positive numbers. Define* $\omega_\star$ *by*

$$ \omega_\star = \min_{p\in\mathbb{Z}} |\omega_1 - p\omega_0| $$

and assume that $\omega_\star > 0$, *i.e.* $\frac{\omega_1}{\omega_0} \notin \mathbb{N}$. *For* $\frac{\omega_1}{\omega_0} \notin \mathbb{N} + \frac{1}{2}$, *this minimum is reached for a unique value of* p *denoted by* $p_\star$. *When* $\frac{\omega_1}{\omega_0} \in \mathbb{N} + \frac{1}{2}$, *it is reached for 2 values of* p *and we denote by* $p_\star$ *the smallest one.*

Then, for every function $f \in {}^b E^{\gamma,\lambda}_{\sigma,\delta}(m_0, m_1, r, \nu)$ *the bi-oscillatory integral*

$$I(f,\varepsilon) := \int_{-\infty}^{+\infty} \mathrm{e}^{\mathrm{i}\omega_1 t/\varepsilon} f(t, \tfrac{\omega_0 t}{\varepsilon}, \varepsilon)\, dt$$

satisfies

$$I(f,\varepsilon) = \frac{\mathrm{e}^{-\sigma\omega_\star/\varepsilon}}{\varepsilon^{\gamma-1}} \left(\Lambda + \underset{\varepsilon\to 0}{\mathcal{O}}\left(\varepsilon^{\nu^*}\right)\right)$$

where $\nu^* := \min\big(\nu, (1-r)(m_0-1), (1-r)(\gamma-1)\big)$ *and* Λ *is given by*

(a) *for* $\frac{\omega_1}{\omega_0} \notin \mathbb{N} + \frac{1}{2}$,

$$\Lambda := \int_{-\mathrm{i}\eta+\mathbb{R}} dz\ \mathrm{e}^{\mathrm{i}\omega_\star z} \frac{1}{2\pi} \int_{-\pi}^{\pi} \widehat{f}(z,s) \mathrm{e}^{\mathrm{i}p_\star s} ds \qquad \textit{with } \eta > \delta$$

where $\widehat{f} = \widehat{f_0^+}$ *when* $\omega_1 - p_\star\omega_0 > 0$ *and* $\widehat{f} = \widehat{f_0^-}$ *when* $\omega_1 - p_\star\omega_0 < 0$;

(b) *for* $\omega_1 = (p_\star + \frac{1}{2})\omega_0$,

$$\Lambda := \int_{-\mathrm{i}\eta+\mathbb{R}} dz\ \mathrm{e}^{\mathrm{i}\omega_\star z} \frac{1}{2\pi} \int_{-\pi}^{\pi} \left(\widehat{f_0^+}(z,s) \mathrm{e}^{\mathrm{i}p_\star s} + \widehat{f_0^-}(z,s) \mathrm{e}^{\mathrm{i}(p_\star+1)s}\right) ds$$

with $\eta > \delta$.

This lemma ensures that the relevant parts of $f \in {}^b E^{\gamma,\lambda}_{\sigma,\delta}$ for computing an equivalent of the bi-oscillatory integral $I(f,\varepsilon)$ are $\mathrm{P}^{\pm}_{\Gamma}(f)$. Moreover, it gives a very efficient way to compute equivalents of oscillatory integrals involving solutions of nonlinear differential equations since *the principal part* P_{Γ} *of a sum (resp. of a product) is the sum (resp. the product) of the principal parts.*

Lemma 2.2.10.

(a) ${}^b E^{\gamma,\lambda}_{\sigma,\delta}(m_0, m_1, r, \nu)$ *is a linear space and*

$$\mathrm{P}^{\pm}_{\Gamma} : {}^b E^{\gamma,\lambda}_{\sigma,\delta}(m_0, m_1, r, \nu) \mapsto {}^b \widehat{H}^{m_0}_{\delta}$$

are linear operators.

(b) *The two Banach spaces* ${}^b\widehat{H}^{m_0}_{\delta}$ *and* ${}^b\widehat{H}^{m_1}_{\delta,r}$ *are algebras.*

(c) *For every* $f \in {}^bE_{\sigma,\delta}^{\gamma,\lambda}$ *and every* $g \in {}^bE_{\sigma,\delta}^{\gamma',\lambda'}$, fg *lies in* ${}^bE_{\sigma,\delta}^{\gamma+\gamma',\lambda+\lambda'}$ *and*

$$\mathrm{Pr}^{\pm}_{{}^bE_{\sigma,\delta}^{\gamma+\gamma',\lambda+\lambda'}}(fg) = \mathrm{Pr}^{\pm}_{{}^bE_{\sigma,\delta}^{\gamma,\lambda}}(f)\mathrm{Pr}^{\pm}_{{}^bE_{\sigma,\delta}^{\gamma',\lambda'}}(g).$$

(d) *For every* $f \in {}^bE_{\sigma,\delta}^{\gamma,\lambda}$ *and every* $p \in [0,\gamma[$, $\varepsilon^p f$ *lies in* ${}^bE_{\sigma,\delta}^{\gamma-p,\lambda}$ *and*

$$\mathrm{Pr}^{\pm}_{{}^bE_{\sigma,\delta}^{\gamma-p,\lambda}}(\varepsilon^p f) = \mathrm{Pr}^{\pm}_{{}^bE_{\sigma,\delta}^{\gamma,\lambda}}(f).$$

(e) *For every* $f \in {}^bE_{\sigma,\delta}^{\gamma,\lambda}$ *and every* $g \in {}^bE_{\sigma,\delta}^{\gamma,\lambda}$, $\varepsilon^\gamma fg$ *lies in* ${}^bE_{\sigma,\delta}^{\gamma,\lambda}$ *and*

$$\mathrm{P}^{\pm}_{\mathrm{r}}(\varepsilon^\gamma fg) = \mathrm{P}^{\pm}_{\mathrm{r}}(f)\mathrm{P}^{\pm}_{\mathrm{r}}(g).$$

We now prove the three bi-frequences Exponential Lemmas.

Proof of the First Bi-Frequency Exponential Lemma 2.2.1. Let f be in ${}^bH_\ell^\lambda$. The function $s \mapsto f(\xi, s, \varepsilon)$ is 2π-periodic and of class C^1. Hence, it reads

$$f(\xi,s,\varepsilon) = \sum_{p\in\mathbb{Z}} c_p(\xi,\varepsilon)\mathrm{e}^{-\mathrm{i}ps} \qquad \text{with } c_p(\xi,\varepsilon) := \frac{1}{2\pi}\int_{-\pi}^{\pi} f(\xi,s,\varepsilon)\ \mathrm{e}^{\mathrm{i}ps}\ ds$$

For computing the bi-oscillatory integral

$$I(f,\varepsilon) = \int_{-\infty}^{+\infty} f(t, \tfrac{\omega_0 t}{\varepsilon}, \varepsilon)\ \mathrm{e}^{\mathrm{i}\omega_1 t/\varepsilon}$$

we proceed in several steps.

Step 1. Rewriting of the integral as series. The integral $I(f,\varepsilon)$ reads

$$I(f,\varepsilon) = \int_{-\infty}^{+\infty} F(t)dt$$

with

$$F(t) = \lim_{n\to+\infty} F_n(t) \qquad \text{where} \qquad F_n(t) = \sum_{p=-n}^{n} c_p(t,\varepsilon)\mathrm{e}^{\mathrm{i}(\omega_1 - p\omega_0)t/\varepsilon}.$$

For using the dominated convergence theorem of Lebesgue, we look for g integrable such that $|F_n(t)| \le g(t)$ holds for every n and t. Integrating by parts we get

$$c_p(\xi,\varepsilon) = \frac{\mathrm{i}}{p}\ c'_p(\xi,\varepsilon) \qquad \text{where} \qquad c'_p(\xi,\varepsilon) = \frac{1}{2\pi}\int_{-\pi}^{\pi} \frac{\partial f}{\partial s}(\xi,s,\varepsilon)\ \mathrm{e}^{\mathrm{i}ps}\ ds.$$

Hence, for every $p \neq 0$

$$|c_p(t,\varepsilon)| \leq \frac{e^{-\lambda|t|}}{2p^2} + \frac{e^{\lambda|t|}}{2}|c'_p(t,\varepsilon)|^2.$$

Moreover,

$$|c_0(t)| \leq \|f\|_{{}^bH^\lambda_\ell}\ e^{-\lambda|t|}.$$

Finally, using the Parseval's formula, we get

$$\begin{aligned}
|F_n(t)| &\leq |c_0(t,\varepsilon)| + \sum_{\substack{-n\leq p\leq n\\ p\neq 0}} |c_p(t,\varepsilon)|,\\
&\leq \|f\|_{{}^bH^\lambda_\ell}\ e^{-\lambda|t|} + e^{-\lambda|t|}\sum_{p=1}^{+\infty}\tfrac{1}{p^2} + \tfrac{1}{2}e^{\lambda|t|}\sum_{p\in\mathbb{Z}\setminus\{0\}} |c'_p(t,\varepsilon)|^2,\\
&= \|f\|_{{}^bH^\lambda_\ell}\ e^{-\lambda|t|} + e^{-\lambda|t|}\sum_{p=1}^{+\infty}\tfrac{1}{p^2} + \tfrac{1}{2}e^{\lambda|t|}\ \tfrac{1}{2\pi}\int_{-\pi}^{\pi}\left|\tfrac{\partial f}{\partial s}(t,s,\varepsilon)\right|^2 ds,\\
&= \left(\|f\|_{{}^bH^\lambda_\ell} + \sum_{p=1}^{+\infty}\tfrac{1}{p^2} + \tfrac{1}{2}\|f\|^2_{{}^bH^\lambda_\ell}\right) e^{-\lambda|t|}.
\end{aligned}$$

Then, the dominated convergence theorem of Lebesgue ensures,

$$I(f,\varepsilon) = \sum_{p\in\mathbb{Z}} \int_{-\infty}^{+\infty} c_p(t,\varepsilon)\ e^{i(\omega_1 - p\omega_0)/\varepsilon} dt. \tag{2.39}$$

Step 2. Upper bounds of $I_p(\omega,\varepsilon)$. For $\omega \in \mathbb{R}$, let us define

$$I_p(\omega,\varepsilon) = \int_{-\infty}^{+\infty} c_p(t,\varepsilon)\ e^{i\omega t/\varepsilon} dt.$$

Since $\xi \mapsto c_p(\xi,\varepsilon)$ is holomorphic in $\mathcal{B}_\ell$ and since

$$|c_p(\xi,\varepsilon)| \leq \|f\|_{{}^bH^\lambda_\ell}\ e^{-\lambda|\mathcal{R}e(\xi)|},$$

the First Mono-Frequency Exponential Lemma 2.1.1 ensures

$$|I_p(\omega,\varepsilon)| \leq e^{\frac{-|\omega|\ell}{\varepsilon}}\ \frac{2\,\|f\|_{{}^bH^\lambda_\ell}}{\lambda}, \qquad \text{for } \omega \in \mathbb{R}. \tag{2.40}$$

Step 3. Upper bounds of $I(f,\varepsilon)$. Collecting (2.39) and (2.40) we get

$$
\begin{aligned}
|I(f,\varepsilon)| &= \left|\sum_{p\in\mathbb{Z}} I_p(\omega_1 - p\omega_0, \varepsilon)\right|,\\
&\leq \sum_{p\in\mathbb{Z}} \mathrm{e}^{-|\omega_1 - p\omega_0|\ell/\varepsilon} 2\lambda^{-1} \|f\|_{{}^bH^\lambda_\ell},\\
&\leq 2\lambda^{-1} \|f\|_{{}^bH^\lambda_\ell} \, \mathrm{e}^{-\omega_\star \ell/\varepsilon} \sum_{p\in\mathbb{Z}} \mathrm{e}^{(-|\omega_1 - p\omega_0| + \omega_\star)\ell/\varepsilon},\\
&\leq 2\lambda^{-1} \|f\|_{{}^bH^\lambda_\ell} \, \mathrm{e}^{-\omega_\star \ell/\varepsilon} \sum_{p\in\mathbb{Z}} \mathrm{e}^{(-|\omega_1 - p\omega_0| + \omega_\star)\ell}\\
&\leq 2\lambda^{-1} \|f\|_{{}^bH^\lambda_\ell} \, \mathrm{e}^{-\omega_\star \ell/\varepsilon} \sum_{p\in\mathbb{Z}} \mathrm{e}^{(-|p|\omega_0 + \omega_1 + \omega_\star)\ell}\\
&\leq 2\lambda^{-1} \|f\|_{{}^bH^\lambda_\ell} \, \frac{2\mathrm{e}^{(\omega_1+\omega_\star)\ell}}{1 - \mathrm{e}^{-\omega_0 \ell}} \, \mathrm{e}^{-\omega_\star \ell/\varepsilon}. \;\square
\end{aligned}
$$

Proof of the Second Bi-Frequency Exponential Lemma 2.2.4. Let f be in ${}^bH^{\gamma,\lambda}_{\sigma,\delta}$. As for the previous lemma, we proceed in 3 steps for computing a sharp upper bound of the bi-oscillatory integral $I(f,\varepsilon)$.

Step 1. Rewriting of the integral as series. Using here again the dominated convergence, we check that

$$I(f,\varepsilon) := \int_{-\infty}^{+\infty} f(t, \tfrac{\omega_0 t}{\varepsilon}, \varepsilon)\, \mathrm{e}^{\mathrm{i}\omega_1 t/\varepsilon} = \sum_{p\in\mathbb{Z}} \int_{-\infty}^{+\infty} c_p(t,\varepsilon)\, \mathrm{e}^{\mathrm{i}(\omega_1 - p\omega_0)t/\varepsilon} dt. \tag{2.41}$$

where

$$c_p(\xi,\varepsilon) := \frac{1}{2\pi}\int_{-\pi}^{\pi} f(\xi, s, \varepsilon)\, \mathrm{e}^{\mathrm{i}ps}\, ds.$$

Step 2. Upper bounds of $I_p(\omega,\varepsilon)$. The function $\xi \mapsto c_p(\xi,\varepsilon)$ is holomorphic in $\mathcal{B}_{\sigma-\delta\varepsilon}$ and satisfies

$$\sup_{\substack{0<\varepsilon<\sigma/\delta\\ \xi\in\mathcal{B}_{\sigma-\delta\varepsilon}}} \left[|c_p(\xi,\varepsilon)| \left(|\sigma^2 + \xi^2|^\gamma \chi_1(\xi) + (1-\chi_1(\xi))\mathrm{e}^{\lambda|\mathcal{R}e(\xi)|} \right) \right] \leq \|f\|_{{}^bH^{\gamma,\lambda}_{\sigma,\delta}} < +\infty$$

Hence, the proof of the Second Mono-Frequency Exponential Lemma 2.1.9 ensures that for every $\omega \in \mathbb{R}$ and every $\varepsilon \in]0, \sigma/\delta[$,

$$|I_p(\omega,\varepsilon)| := \left| \int_{-\infty}^{+\infty} c_p(t,\varepsilon)\, \mathrm{e}^{\mathrm{i}\omega t/\varepsilon} dt \right| \leq \frac{M(\gamma,\lambda)}{\varepsilon^{\gamma-1}}\, \mathrm{e}^{\frac{-|\omega|(\sigma-\delta\varepsilon)}{\varepsilon}} \|f\|_{{}^bH^{\gamma,\lambda}_{\sigma,\delta}} \tag{2.42}$$

where

$$M(\gamma,\lambda) = 2\frac{\mathrm{e}^{-\lambda}}{\lambda} + \frac{1}{\sigma^\gamma}\int_{+\infty}^{-\infty} \frac{du}{(1+u^2)^{\gamma/2}}.$$

Step 3. Upper bounds of $I(f,\varepsilon)$. Collecting (2.41) and (2.42) we get that for every $\varepsilon \in]0, \sigma/2\delta[$

$$
\begin{aligned}
|I(f,\varepsilon)| &= \left|\sum_{p\in\mathbb{Z}} I_p(\omega_1 - p\omega_0,\varepsilon)\right|, \\
&\le \sum_{p\in\mathbb{Z}} \frac{1}{\varepsilon^{\gamma-1}} \mathrm{e}^{\frac{-|\omega_1-p\omega_0|(\sigma-\delta\varepsilon)}{\varepsilon}} M(\gamma,\lambda)\,\|f\|_{{}^bH^{\gamma,\lambda}_{\sigma,\delta}}, \\
&\le M(\gamma,\lambda)\,\|f\|_{{}^bH^{\gamma,\lambda}_{\sigma,\delta}} \frac{1}{\varepsilon^{\gamma-1}} \mathrm{e}^{\frac{-\omega_\star(\sigma-\delta\varepsilon)}{\varepsilon}} \sum_{p\in\mathbb{Z}} \mathrm{e}^{\frac{(-|\omega_1-p\omega_0|+\omega_\star)(\sigma-\delta\varepsilon)}{\varepsilon}}, \\
&\le M(\gamma,\lambda)\,\|f\|_{{}^bH^{\gamma,\lambda}_{\sigma,\delta}} \frac{1}{\varepsilon^{\gamma-1}} \mathrm{e}^{\frac{-\omega_\star(\sigma-\delta\varepsilon)}{\varepsilon}} \sum_{p\in\mathbb{Z}} \mathrm{e}^{(-|\omega_1-p\omega_0|+\omega_\star)\sigma/2}, \\
&\le M(\gamma,\lambda)\,\|f\|_{{}^bH^{\gamma,\lambda}_{\sigma,\delta}} \frac{1}{\varepsilon^{\gamma-1}} \mathrm{e}^{\frac{-\omega_\star(\sigma-\delta\varepsilon)}{\varepsilon}} \sum_{p\in\mathbb{Z}} \mathrm{e}^{(-|p|\omega_0+\omega_1+\omega_\star)\sigma/2}, \\
&\le M(\gamma,\lambda)\,\frac{2\mathrm{e}^{(\omega_1+\omega_\star)\sigma/2}}{1-\mathrm{e}^{-\omega_0\sigma/2}}\,\mathrm{e}^{\delta\omega_\star}\,\|f\|_{{}^bH^{\gamma,\lambda}_{\sigma,\delta}} \frac{1}{\varepsilon^{\gamma-1}}\,\mathrm{e}^{-\omega_\star\sigma/\varepsilon}. \ \square
\end{aligned}
$$

Proof of the Third Bi-Frequency Exponential Lemma 2.2.9. Let f be in ${}^bE^{\gamma,\lambda}_{\sigma,\delta}$. To compute an equivalent of the bi-oscillatory integral $I(f,\varepsilon)$ we proceed in four steps.

Step 1 and 2. Rewriting of the integral as series and upper bounds of $I_p(\omega,\varepsilon)$. The two first steps are the same as for the previous lemma. Hence (2.41) and (2.42) still holds for $f\in{}^bE^{\gamma,\lambda}_{\sigma,\delta}$.

Step 3. Splitting of the series in two parts (leading one and remainder). Upper bound of the remainder. We must distinguish two cases. For $\frac{\omega_1}{\omega_0}\notin\mathbb{N}+\frac{1}{2}$, $I(f,\varepsilon)$ reads

$$I(f,\varepsilon) = I_{p_\star}(\omega_1 - p_\star\omega_0) + J(f,\varepsilon) \tag{2.43}$$

whereas for $\omega_1 = (p_\star+\frac{1}{2})\omega_0$,

$$I(f,\varepsilon) = I_{p_\star}(\omega_1 - p_\star\omega_0) + I_{p_\star+1}(\omega_1-(p_\star+1)\omega_0) + J(f,\varepsilon) \tag{2.44}$$

where

$$J(f,\varepsilon) = \sum_{p\in\mathcal{J}} I_p(\omega_1 - p\omega_0)$$

with $\mathcal{J}=\mathbb{Z}\backslash\{p_\star\}$ for $\frac{\omega_1}{\omega_0}\notin\mathbb{N}+\frac{1}{2}$ and $\mathcal{J}=\mathbb{Z}\backslash\{p_\star,p_\star+1\}$ for $\omega_1=(p_\star+\frac{1}{2})\omega_0$.

Observing that there exists $\alpha>0$ such that for every $p\in\mathcal{J}$

$$|\omega_1 - p\omega_0| \ge \omega_\star + \alpha$$

and using the same strategy as in Step 3 of the previous lemma we get

$$|J(f,\varepsilon)| \le M'\ \|f\|_{{}^bH^{\gamma,\lambda}_{\sigma,\delta}} \frac{1}{\varepsilon^{\gamma-1}}\ \mathrm{e}^{-(\omega_\star+\alpha)\sigma/\varepsilon} \tag{2.45}$$

where

$$M' = M(\gamma, \lambda) \, \frac{2e^{(\omega_1+\omega_*+\alpha)\sigma/2}}{1 - e^{-\omega_0\sigma/2}} \, e^{\delta(\omega_*+\alpha)}.$$

Step 4. Computation of the equivalent.

Case 4.1. $\frac{\omega_1}{\omega_0} \notin \mathbb{N} + \frac{1}{2}$. For $\omega_1 - p_*\omega_0 > 0$,

$$I_{p_*}(\omega_1 - p_*\omega_0) = I_{p_*}(+\omega_*) = \int_{-\infty}^{+\infty} c_p(t,\varepsilon) \, e^{+i\omega_* t/\varepsilon} dt$$

whereas for $\omega_1 - p_*\omega_0 < 0$

$$I_{p_*}(\omega_1 - p_*\omega_0) = I_{p_*}(-\omega_*) = \int_{-\infty}^{+\infty} c_p(t,\varepsilon) \, e^{-i\omega_* t/\varepsilon} dt.$$

Moreover, the function $(\xi, \varepsilon) \mapsto c_{p_*}(\xi, \varepsilon)$ which reads

$$c_p(\xi, \varepsilon) := \frac{1}{2\pi} \int_{-\pi}^{\pi} f(\xi, s, \varepsilon) \, e^{ip_* s} \, ds$$

belongs to $E^{\gamma,\lambda}_{\sigma,\delta}(m_0, m_1, r, \nu)$ and

$$\mathrm{P}^{\pm}_{\mathrm{r}}(c_{p_*}) = \frac{1}{2\pi} \int_{-\pi}^{\pi} \widehat{f_0^{\pm}}(\xi, s) \, e^{ip_* s} \, ds.$$

Hence the Third Mono-Frequency Exponential Lemma ensures that

$$I_{p_*}(\omega_1 - p_*\omega_0) = \frac{e^{-\sigma\omega_*/\varepsilon}}{\varepsilon^{\gamma-1}} \left(\Lambda + \underset{\varepsilon\to 0}{\mathcal{O}} \left(\varepsilon^{\nu^*} \right) \right)$$

where $\nu^* := \min\big(\nu, (1-r)(m_0-1), (1-r)(\gamma-1)\big)$ and where Λ is given by

$$\Lambda := \int_{-i\eta+\mathbb{R}} dz \, e^{i\omega_* z} \frac{1}{2\pi} \int_{-\pi}^{\pi} \widehat{f}(z,s) e^{ip_* s} ds \qquad \text{with } \eta > \delta \tag{2.46}$$

where $\widehat{f} = \widehat{f_0^+}$ when $\omega_1 - p_*\omega_0 > 0$ and $\widehat{f} = \widehat{f_0^-}$ when $\omega_1 - p_*\omega_0 < 0$.

Gathering (2.43), (2.45), (2.46), we finally obtain the equivalent of $I(f, \varepsilon)$

$$I(f,\varepsilon) = \frac{e^{-\sigma\omega_*/\varepsilon}}{\varepsilon^{\gamma-1}} \left(\Lambda + \underset{\varepsilon\to 0}{\mathcal{O}} \left(\varepsilon^{\nu^*} \right) \right).$$

Case 4.2. $\omega_1 = (p_* + \frac{1}{2})\omega_0$. Similarly, using (2.43), (2.45) and the First Mono-Frequency Exponential Lemma, we prove

$$I(f,\varepsilon) = \frac{e^{-\sigma\omega_*/\varepsilon}}{\varepsilon^{\gamma-1}} \left(\Lambda + \underset{\varepsilon\to 0}{\mathcal{O}} \left(\varepsilon^{\nu^*} \right) \right)$$

where Λ is given by

$$\Lambda := \int_{-i\eta+\mathbb{R}} dz \, e^{i\omega_* z} \frac{1}{2\pi} \int_{-\pi}^{\pi} \left(\widehat{f_0^+}(z,s) e^{ip_* s} + \widehat{f_0^-}(z,s) e^{i(p_*+1)s} \right) ds$$

with $\eta > \delta$. □

2.A Appendix. Method of continuation along horizontal lines

In this appendix, we describe in a general and abstract way the technique of continuation along horizontal lines, because we need it several times. Knowing a holomorphic continuation of a solution of a perturbed differential equation on a domain, this technique enables us to prove the existence of a holomorphic continuation in a larger domain.

Suppose we study a complex ordinary differential equation of the form

$$\frac{dY}{dz} = F_0(Y,z) + F_1(Y,z) \tag{2.47}$$

where F_0 and F_1 are analytic functions with respect to (Y, z), and z lies in $]T, +\infty[\times]a, b[$ where $a < b$ and $T \in [-\infty, +\infty[$. We make the following assumptions:

Assumption 2.A.1. *There exists a solution $z \mapsto h(z)$ of*

$$\frac{dY}{dz} = F_0(Y,z)$$

defined and holomorphic in $]T, +\infty[\times]a, b[$.

Assumption 2.A.2. *There exists a solution of (2.47) $Y(z) = h(z) + v(z)$, holomorphic in $]t_0, +\infty[\times]a, b[$ with $t_0 > T$*

Let τ^*, t_1 be fixed such that $\tau^* > t_0 > t_1 \geq T$. Our aim is to continue v in $]t_1, +\infty[\times]a, b[$.

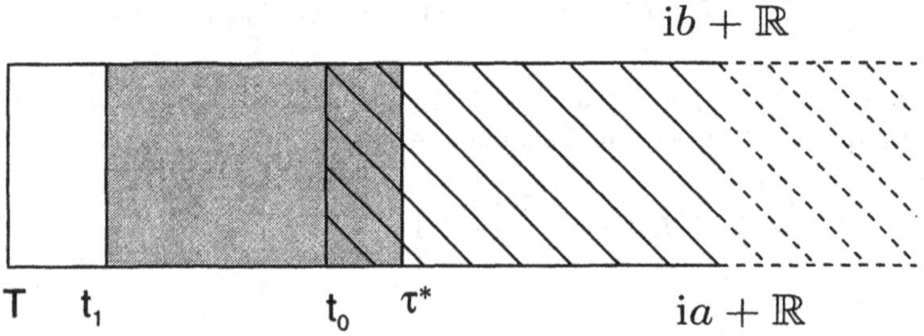

Fig. 2.14. Continuation along horizontal lines

For that purpose, we define the following one parameter family of ordinary differential equation along horizontal lines

$$\begin{aligned}\frac{\partial w}{\partial t} = F_0\big(h(t+\mathrm{i}\eta) + w(t,\eta), t+\mathrm{i}\eta\big) - F_0\big(h(t+\mathrm{i}\eta), t+\mathrm{i}\eta\big) \\ +F_1\big(h(t+\mathrm{i}\eta) + w(t,\eta), t+\mathrm{i}\eta\big) = g(t,\eta,w).\end{aligned} \tag{2.48}$$

Denote by $\Phi(t,\eta,t_\mathrm{i},w_\mathrm{i})$ the flow of this O.D.E: t_i is the initial time and w_i is the initial value ($\Phi(t_\mathrm{i},\eta,t_\mathrm{i},w_\mathrm{i}) = w_\mathrm{i}$). The function g is C^∞ with respect to (t,η,w). Hence, Φ is C^∞ with respect to $(t,\eta,t_\mathrm{i},w_\mathrm{i})$.

Assumption 2.A.3. *Denote by $w^*(t,\eta) = \Phi(t,\eta,\tau^*,v(\tau^*+\mathrm{i}\eta))$. For every $\eta \in]a,b[$, w^* is defined in $]t_1,\tau^*]$: the life interval of w^* contains at least $]t_1,+\infty[$ uniformly with respect to η.*

Remark 2.A.4. This assumption is usually proved using the Contraction Mapping Theorem in appropriate sets of functions defined on $]t_1,\tau^*]$.

Lemma 2.A.5. *Under Assumptions 2.A.1, 2.A.2, 2.A.3, w^* satisfies*

(a) *w^* is defined in $]t_1,+\infty[\times]a,b[$ and $w^*(t,\eta) = v(t+\mathrm{i}\eta)$ for $t > t_0$, and $\eta \in]a,b[$,*
(b) *w^* is holomorphic in $]t_1,\tau^*[\times]a,b[$.*
(c) *$Y = h + w^*$ is solution of (2.47) in $]t_1,+\infty[\times]a,b[$.*

Proof. (a): By definition $w^*(\tau^*,\eta) = \Phi(\tau^*,\eta,\tau^*,v(\tau^*+\mathrm{i}\eta)) = v(\tau^*+\mathrm{i}\eta)$. Moreover, Assumption 2.A.2 ensures that $t \mapsto v(t+\mathrm{i}\eta)$ is solution of (2.48). Thus, $t \mapsto w(t,\eta)$ and $t \mapsto v(t+\mathrm{i}\eta)$ are two solutions of the same O.D.E. which are equal for $t = \tau^*$. Hence, $w^*(t,\eta) = v(t+\mathrm{i}\eta)$ for $t > t_0$, which ensures that w^* is holomorphic in $]t_0,+\infty[\times]a,b[$.

(b): The flow Φ is of class C^∞. Thus w^* has the same regularity and for proving that w^* is holomorphic it is sufficient to prove that w^* satisfies the Cauchy-Riemann equation

$$\frac{\partial w^*}{\partial \eta} = \mathrm{i}\frac{\partial w^*}{\partial t}.$$

We have already proved that $w^*(t,\eta) = v(t+\mathrm{i}\eta)$ for $t > t_0$. By Assumption 2.A.2, v is holomorphic and thus

$$\frac{\partial w^*}{\partial \eta}(\tau^*,\eta) = \mathrm{i}\frac{\partial w^*}{\partial t}(\tau^*,\eta).$$

Moreover we have,

$$\frac{\partial w^*}{\partial t}(t,\eta) = F_0\big(h(t+\mathrm{i}\eta)+w^*(t,\eta),t+\mathrm{i}\eta\big) - F_0\big(h(t+\mathrm{i}\eta),t+\mathrm{i}\eta\big) \\ +F_1\big(h(t+\mathrm{i}\eta)+w^*(t,\eta),t+\mathrm{i}\eta\big),$$

$$\begin{aligned}\frac{\partial^2 w^*}{\partial t\partial\eta}(t,\eta) &= D_Y F_0\big(h(t+\mathrm{i}\eta)+w(t,\eta),t+\mathrm{i}\eta\big).[\mathrm{i}\frac{dh}{dz}(t+\mathrm{i}\eta)+\frac{\partial w^*}{\partial\eta}(t,\eta)] \\ &\quad +\mathrm{i}\frac{\partial F_0}{\partial z}\big(h(t+\mathrm{i}\eta)+w(t,\eta),t+\mathrm{i}\eta\big) \\ &\quad -\mathrm{i}D_Y F_0\big(h(t+\mathrm{i}\eta),t+\mathrm{i}\eta\big).\frac{dh}{dz}(t+\mathrm{i}\eta) \\ &\quad +D_Y F_1\big(h(t+\mathrm{i}\eta)+w(t,\eta)t+\mathrm{i}\eta\big).[\mathrm{i}\frac{dh}{dz}(t+\mathrm{i}\eta)+\frac{\partial w^*}{\partial\eta}(t,\eta)] \\ &\quad +\mathrm{i}\frac{\partial F_1}{\partial z}\big(h(t+\mathrm{i}\eta)+w(t,\eta),t+\mathrm{i}\eta\big) \\ &= G(t,\eta,\frac{\partial w^*}{\partial\eta}),\end{aligned}$$

$$\begin{aligned}\frac{\partial^2 w^*}{\partial t^2}(t,\eta) &= D_Y F_0\big(h(t+\mathrm{i}\eta)+w^*(t,\eta),t+\mathrm{i}\eta\big).[\frac{dh}{dz}(t+\mathrm{i}\eta)+\frac{\partial w^*}{\partial t}(t,\eta)] \\ &\quad +\frac{\partial F_0}{\partial z}\big(h(t+\mathrm{i}\eta)+w(t,\eta),t+\mathrm{i}\eta\big) \\ &\quad -DF_0\big(h(t+\mathrm{i}\eta),t+\mathrm{i}\eta\big).\frac{dh}{dz}(t+\mathrm{i}\eta) \\ &\quad +DF_1\big(h(t+\mathrm{i}\eta)+w^*(t,\eta),t+\mathrm{i}\eta\big).[\frac{dh}{dz}(t+\mathrm{i}\eta)+\frac{\partial w^*}{\partial t}(t,\eta)] \\ &\quad +\frac{\partial F_1}{\partial z}\big(h(t+\mathrm{i}\eta)+w(t,\eta),t+\mathrm{i}\eta\big).\end{aligned}$$

Hence, $\mathrm{i}\frac{\partial^2 w^*}{\partial t^2}(t,\eta) = G(t,\eta,\mathrm{i}\frac{\partial w^*}{\partial t})$. For η fixed in]a,b[, $t\mapsto \mathrm{i}\frac{\partial w^*}{\partial t}$ and $t\mapsto \frac{\partial w^*}{\partial\eta}$ are solutions of the same O.D.E. and are equal for $t=\tau^*$. Therefore,

$$\frac{\partial w^*}{\partial\eta} = \mathrm{i}\frac{\partial w^*}{\partial t} \quad \text{for } \eta\in]a,b[,\ t\in]T,\tau^*].$$

So, w^* is holomorphic.

(c): (b) ensures that w^* is holomorphic and since w^* is solution of (2.48) we have

$$\frac{dw^*}{dz}(t+\mathrm{i}\eta) = \frac{\partial w^*}{\partial t}(t+\mathrm{i}\eta) = (F_0+F_1)(h+v)(t+\mathrm{i}\eta) - \frac{dh}{dz}(t+\mathrm{i}\eta). \ \square$$

Part II

Toolbox for reversible systems studied near resonances

3. Resonances of reversible vector fields

3.1 Definitions and basic properties

We denote by $\mathcal{M}_n(\mathbb{R})$ the set of $n \times n$ real matrices and by $GL_n(\mathbb{R})$ the subset of the invertible ones.

3.1.1 Reversible vector fields

Definition 3.1.1. *Let $V : \mathcal{U} \to \mathbb{R}^n : u \mapsto V(u)$ be a vector field in $\mathbb{R}^n$ defined in an open set $\mathcal{U}$ of $\mathbb{R}^n$. The vector field is said to be reversible when there exists a symmetry $S \in GL_n(\mathbb{R})$, $S^2 = \mathrm{Id}$, which satisfies*

$$SV(u) = -V(Su) \qquad \text{for every } u \in \mathcal{U}.$$

Lemma 3.1.2 (Pair of solutions and reversible solutions). *Let V be a smooth reversible vector field in $\mathbb{R}^n$.*

(a) *If $u(t)$ is a solution of*

$$\frac{du}{dt} = V(u), \tag{3.1}$$

then $Su(-t)$ is also solution of (3.1).

(b) *A solution u of (3.1) is said to be reversible when it satisfies one of the three following properties which are equivalent*

$$u(t) = Su(-t) \text{ holds for every } t \in \mathbb{R}, \tag{3.2}$$

$$\text{there exists } t^* \in \mathbb{R} \text{ such that } u(t^*) = Su(-t^*) \text{ holds}, \tag{3.3}$$

$$u(0) = Su(0). \tag{3.4}$$

Proof. (a) follows directly from Definition 3.1.1 and (b) from uniqueness of the solution of the Cauchy problem. □

Remark 3.1.3. This lemma ensures that solutions of (3.1) are given by pairs except for the particular ones which are reversible (the two solutions coincide in this case).

Remark 3.1.4. Observe that knowing the behavior of a reversible solution at $+\infty$, (3.2) gives its behavior at $-\infty$. It happens to be very useful when studying reversible homoclinic connections. Since $S^2 = \mathrm{Id}$, it is diagonalizable and its eigenvalues are 1 and -1. So, in an appropriate system of coordinates (obtained by a linear transformation), a reversible solution is a solution which has even or odd components depending on the eigenvalues of S.

Lemma 3.1.5 (Reversible fixed points).
Let V be a smooth, reversible vector fields in $\mathbb{R}^n$. Denote by u_0 a reversible fixed point of V : u_0 satisfies $V(u_0) = 0$ and $Su_0 = u_0$.

(a) *Denote by $L := D_uV(u_0)$. Then $LS = -SL$ holds.*
(b) *If σ is an eigenvalue of L, then $-\sigma$ and $\overline{\sigma}$ are also eigenvalues of L. Moreover, the eigenspace and generalized eigenspace belonging to $-\sigma$ (resp. to $\overline{\sigma}$) are the images of the spaces belonging to σ by S (resp. by $u \mapsto \overline{u}$). Hence, three cases may occur*
  (i) *$\sigma \in \mathbb{C} \setminus (\mathbb{R} \cup i\mathbb{R})$ belongs to the spectrum of L. Then $-\sigma$, $\overline{\sigma}$, $-\overline{\sigma}$ are three other eigenvalues of L with the same geometrical and algebraic multiplicity.*
  (ii) *σ is a non zero, real or purely imaginary eigenvalue of L. Then, $-\sigma$ is an eigenvalue of L with the same geometrical and algebraic multiplicity.*
  (iii) *0 is an eigenvalue of L.*
(c) *Denote by $W_s(u_0)$ (resp. $W_u(u_0)$) the stable (resp. unstable) manifold of u_0. Then, $W_s(u_0) = SW_u(u_0)$.*
(d) *There exists a reversible orbit homoclinic to u_0 if and only if $\big(W_s(u_0) \setminus \{u_0\}\big) \cap E^+ \neq \emptyset$ where E^+ is the symmetry space $E^+ := \{u \in \mathbb{R}^n / \; Su = u\}$.*

Remark 3.1.6. For the definition and existence of stable or unstable manifold of a fixed point (non necessarily hyperbolic) see for example [GH83], [Wi90] or [Ke67] and for details see [HPS77].

Remark 3.1.7. Property (d) happens to be very useful when studying reversible homoclinic connections. Indeed, the study of homoclinic connections requires in general the computation of the intersection of the two manifolds W_s and W_u. *For reversible homoclinic connections* it is sufficient to study the intersection of the stable manifold W_s with the fixed, explicitly known linear symmetry space E^+.

Proof. (a): (a) directly follows from Definition 3.1.1 by differentiating.

(b): Property (a) (resp. The fact that L is a real matrix) ensures that if v belongs to $\ker\big((L-\sigma \mathrm{Id})^q\big)$ then $Sv \in \ker\big((L+\sigma \mathrm{Id})^q\big)$ (resp. $Sv \in \ker\big((L-\overline{\sigma}\mathrm{Id})^q\big)$. (b) follows.

(c): (c) comes from the property (a) of Lemma 3.1.2.

(d): If there exists a reversible homoclinic connection h to u_0, then $h(0)$ belongs to $\big(W_s(u_0) \setminus \{u_0\}\big) \cap E^+$ which is consequently not empty. Reciprocally, assume that $\big(W_s(u_0) \setminus \{u_0\}\big) \cap E^+ \neq \emptyset$. Choose $v_0 \in \big(W_s(u_0) \setminus \{u_0\}\big) \cap E^+$ and denote by v the solution of (3.1) such that $v(0) = v_0$. Since $v_0 \in W_s(u_0)$, v satisfies $\lim_{t \to +\infty} v(t) = u_0$. Moreover, since v_0 belongs to E^+, Lemma 3.1.2-(b) ensures that v is reversible, i.e. $Sv(t) = v(-t)$ for every $t \in \mathbb{R}$. Hence, $\lim_{t \to -\infty} v(t) = \lim_{t \to +\infty} v(-t) = \lim_{t \to +\infty} Sv(t) = Su_0 = u_0$, since u_0 is a reversible fixed point. So, v is a reversible homoclinic connection to u_0. □

Lemma 3.1.8 (reversible periodic orbits).

(a) *Let u be a solution of (3.1). Then, there exist $\varphi \in \mathbb{R}$ such that $t \mapsto u(t+\varphi)$ is a reversible periodic function if and only if there exists $t_0, t_1 \in \mathbb{R}$, $t_0 \neq t_1$, such that $u(t_0)$ and $u(t_1)$ belongs to E^+ where E^+ is the symmetry space $E^+ := \{u \in \mathbb{R}^n / \ Su = u\}$.*

(b) *Let u be a reversible periodic solution of (3.1). Denote by $W_s(u)$ (resp. $W_u(u)$) the stable (resp. unstable) manifold of u. Then, $W_s(u) = SW_u(u)$.*

(c) *Let u be a reversible periodic solution of (3.1). There exists a reversible orbit homoclinic to u if and only if $\big(W_s(u) \setminus \{u(t)/t \in \mathbb{R}\}\big) \cap E^+ \neq \emptyset$.*

Remark 3.1.9. Property (a) gives a simple geometric characterization for periodic orbits. As for fixed point, Property (c) gives a simple way to study the existence of reversible homoclinic connections to periodic orbits.

Proof. (a): If $t \mapsto u(t+\varphi)$ is a reversible T-periodic solution, then $u(\varphi)$ and $u(\varphi+T/2)$ belong to E^+. Indeed, define $\widetilde{u}(t) := u(t+\varphi)$. Lemma 3.1.2 ensures that $u(\varphi) = \widetilde{u}(0) = S\widetilde{u}(0) = S(u(\varphi))$ and that $u(\varphi + T/2) = \widetilde{u}(T/2) = \widetilde{u}(T/2 - T) = \widetilde{u}(-T/2) = S(\widetilde{u}(T/2)) = S(u(\varphi + T/2))$.

Reciprocally, assume that u is a solution of (3.1) such that $u(t_0), u(t_1) \in E^+$ for $t_1 > t_0$. Define $T := 2(t_1 - t_0)$, $\varphi := t_0$ and $\widetilde{u}(t) := u(t + t_0)$. Since $u(t_0) \in E^+$, $\widetilde{u}(0) = S\widetilde{u}(0)$ and thus $\widetilde{u}$ is reversible. Hence,

$$\widetilde{u}(-T/2) = S(\widetilde{u}(T/2)) = Su(t_1) = u(t_1) = \widetilde{u}(T/2)$$

since $\widetilde{u}$ is reversible and since $u(t_1)$ belongs to E^+. This last formula and the uniqueness to Cauchy problem ensures that $\widetilde{u}$ is T-periodic.

(b): (b) follows directly from the property (a) of Lemma 3.1.2. Indeed, if h is solution of (3.1) and satisfies $\lim_{t \to +\infty} \big(h(t) - u(t+\varphi)\big) = 0$ then $t \mapsto \widetilde{h}(t) := S(h(-t))$ is also solution of (3.1) and satisfies $\lim_{t \to -\infty} \big(\widetilde{h}(t) - u(t-\varphi)\big) = 0$.

(c): If there exists a reversible homoclinic connection h to u, then $h(0)$ belongs to $\big(W_s(u) \setminus \{u(t)/t \in \mathbb{R}\}\big) \cap E^+$ which is consequently not empty.

Conversely, assume that $\big(W_s(u) \setminus \{u(t)/t \in \mathbb{R}\}\big) \cap E^+ \neq \emptyset$. Choose $v_0 \in \big(W_s(u) \setminus \{u(t)/t \in \mathbb{R}\}\big) \cap E^+$ and denote by v the solution of (3.1) such that

$v(0) = v_0$. Since $v_0 \in W_s(u)$, v satisfies $\lim_{t\to+\infty}(v(t) - u(t+\varphi)) = 0$. Moreover, since v_0 belongs to E^+, Lemma 3.1.2-(b) ensures that v is reversible, i.e. $Sv(t) = v(-t)$ for every $t \in \mathbb{R}$. Hence, $\lim_{t\to-\infty}(v(t) - u(t-\varphi)) = \lim_{t\to+\infty}(v(-t) - u(-t-\varphi)) = \lim_{t\to+\infty} S\big(v(t) - u(t+\varphi)\big) = 0$, since u is a reversible periodic orbit. So, v is a reversible homoclinic connection to u. □

3.1.2 Linear classification and nomenclature of reversible fixed points

One aim of this work is to study the dynamics of

$$\frac{du}{dt} = V(u) \qquad u \in \mathbb{R}^N,$$

for vector fields $V(u)$ reversible with respect to some symmetry S near a reversible fixed point u_0 when the spectrum of the differential $L = D_uV(u_0)$ at the fixed point has purely imaginary eigenvalues (non hyperbolic fixed point). For simplicity we place the fixed point at the origin ($u_0 = 0$). We are specially interested in the existence of periodic orbits and homoclinic connections near the origin. So we are only interested in the dynamics of V up to a linear change of coordinates, and more generally up to a change of coordinates of the form $u = \Phi(u_1)$ with $\Phi(0) = 0$ where Φ is a diffeomorphism near the origin. So a very first step is to perform a linear change of coordinates, $u = P(u_1)$ with $P \in GL(\mathbb{R}^N)$ to obtain an equivalent equation

$$\frac{du_1}{dt} = V_1(u_1) \qquad u \in \mathbb{R}^N,$$

where $S_1 = P^{-1}SP$ and $L_1 = P^{-1}LP = D_{u_1}V_1(0)$ are *simultaneously* as simple as possible. Observe that $V_1(u_1) = P^{-1}V(Pu_1)$ and L_1 are reversible with respect to the symmetry S_1.

Theorem 3.1.10 below gives the classification of the reversible pairs of $N \times N$ matrices, i.e. $\{(L, S) \in \big(\mathcal{M}(\mathbb{R}^N)\big)^2 / S^2 = \mathrm{Id},\ LS = -SL\}$ with respect to the equivalence relation $\sim$ defined by the *simultaneous conjugacy*, i.e.

$$(L, S) \sim (L_1, S_1) \overset{\text{def}}{\Longleftrightarrow} \exists P \in GL(\mathbb{R}^N),\ L_1 = P^{-1}LP \text{ and } S_1 = P^{-1}SP.$$

Moreover it gives for each equivalence class a representative (L, S) where L is a Jordan normal matrix and S is the direct sum of blocks of the form I_p, $-I_q$, and

$$\begin{pmatrix} 0 & \widetilde{S} \\ \widetilde{S} & 0 \end{pmatrix} \qquad \text{where } \widetilde{S} \text{ is a diagonal matrix .}$$

In other words, this theorem says that it is always possible to perform a linear change of coordinates to obtain an equivalent equation governed by a reversible vector field V_1 such that (L_1, S_1) is one of the previous representatives. Then we introduce in Remark 3.1.12, a notation inspired by the notation of Jordan matrices used by V. Arnold in [Ar83] to name each classes.

To state the theorem we introduce the following convenient notation to describe matrices which are blockdiagonal and antiblockdiagonal:

- for a $k \times k$ matrix M and a $p \times p$ matrix N we denote by $M \boxbslash N$ (resp. by $M \boxslash N$) the blockdiagonal matrix (resp. the antiblockdiagonal matrix)

$$M \boxbslash N = \begin{pmatrix} M & 0 \\ 0 & N \end{pmatrix} \qquad M \boxslash N = \begin{pmatrix} 0 & N \\ M & 0 \end{pmatrix}.$$

- for $k_j \times k_j$ matrices M_j we denote by $\underset{j=1}{\overset{p}{\boxbslash}} M_j$ (resp. by $\underset{j=1}{\overset{p}{\boxslash}} M_j$) the blockdiagonal matrix (resp. the antiblockdiagonal matrix)

$$\underset{j=1}{\overset{p}{\boxbslash}} M_j = \begin{pmatrix} M_1 & 0 & 0 \\ 0 & \ddots & 0 \\ 0 & 0 & M_p \end{pmatrix} \qquad \underset{j=1}{\overset{p}{\boxslash}} M_j = \begin{pmatrix} 0 & 0 & M_p \\ 0 & \cdot^{\cdot^{\cdot}} & 0 \\ M_1 & 0 & 0 \end{pmatrix}.$$

For a real matrix L, denote by μ_L its minimal polynomial. It has the form

$$\mu_L(X) := \prod_{j=1}^{P} (X - \lambda_j)^{m(\lambda_j)} \prod_{j=1}^{R} \left((X - \sigma_j)(X - \overline{\sigma_j})\right)^{m(\sigma_j)}.$$

where $\lambda_j \in \mathbb{R}$ and $\sigma_j \in \mathbb{C} \setminus \mathbb{R}$. Denote by $\mathrm{Jor}(L)$ its real Jordan normal form which reads

$$\mathrm{Jor}(L) = \underset{j=1}{\overset{P}{\boxbslash}} \underset{k=1}{\overset{m(\lambda_j)}{\boxbslash}} \underset{\ell=1}{\overset{n_k(\lambda_j)}{\boxbslash}} J_k(\lambda_j) \boxbslash \underset{j=1}{\overset{R}{\boxbslash}} \underset{k=1}{\overset{m(\sigma_j)}{\boxbslash}} \underset{\ell=1}{\overset{n_k(\sigma_j)}{\boxbslash}} \widehat{J}_k(\sigma_j)$$

where $n_k(\xi)$ is the number of Jordan blocks of size k corresponding to the eigenvalue ξ and where $J_k(\lambda)$ is the $k \times k$ matrix

$$J_k(\lambda) = \begin{pmatrix} \lambda & 1 & 0 & 0 \\ 0 & \ddots & \ddots & 0 \\ 0 & 0 & \ddots & 1 \\ 0 & 0 & 0 & \lambda \end{pmatrix} \tag{3.5}$$

and $\widehat{J}_k(x + \mathrm{i}y)$ is the $2k \times 2k$ matrix

$$\widehat{J}_k(x + \mathrm{i}y) = \begin{pmatrix} \begin{pmatrix} x & -y \\ y & x \end{pmatrix} & \begin{pmatrix} 1 & 0 \\ 0 & 1 \end{pmatrix} & 0 & 0 \\ 0 & \ddots & \ddots & 0 \\ 0 & 0 & \ddots & \begin{pmatrix} 1 & 0 \\ 0 & 1 \end{pmatrix} \\ 0 & 0 & 0 & \begin{pmatrix} x & -y \\ y & x \end{pmatrix} \end{pmatrix}.$$

Theorem 3.1.10 (Classification of real reversible matrices).

(a) *Let (L, S) be two $N \times N$ matrices such that S is a symmetry $S^2 = \mathrm{Id}$ and $SL = -LS$ holds. Then the minimal polynomial of L reads*

$$\mu_L(X)=X^m\prod_{j=1}^{P}(X^2-\lambda_j^2)^{m(\lambda_j)}\prod_{j=1}^{Q}(X^2+\omega_j^2)^{m(\mathrm{i}\omega_j)}\prod_{j=1}^{R}((X^2-\sigma_j^2)(X^2-\overline{\sigma}_j^2))^{m(\sigma_j)}$$

where $\lambda_j, \omega_j > 0$ and $\sigma_j \in \mathbb{C}$ with $\mathcal{R}e\,(\sigma_j)\,, \mathcal{I}m\,(\sigma_j) > 0$. The numbers m, P, Q, R may be equal to 0. So the normal Jordan form of L reads

$$\mathrm{Jor}(L) = \mathop{\boxbslash}_{k=1}^{m}\left(\mathop{\boxbslash}_{\ell=1}^{n_k(0)} J_k(0)\right) \boxbslash \mathop{\boxbslash}_{j=1}^{P}\mathop{\boxbslash}_{k=1}^{m(\lambda_j)}\mathop{\boxbslash}_{\ell=1}^{n_k(\lambda_j)}\left(J_k(\lambda_j) \boxbslash J_k(-\lambda_j)\right)$$
$$\boxbslash \mathop{\boxbslash}_{j=1}^{Q}\mathop{\boxbslash}_{k=1}^{m(\mathrm{i}\omega_j)}\mathop{\boxbslash}_{\ell=1}^{n_k(\mathrm{i}\omega_j)} \widehat{J}_k(\mathrm{i}\omega_j) \boxbslash \mathop{\boxbslash}_{j=1}^{R}\mathop{\boxbslash}_{k=1}^{m(\sigma_j)}\mathop{\boxbslash}_{\ell=1}^{n_k(\sigma_j)}\left(\widehat{J}_k(\sigma_j) \boxbslash \widehat{J}_k(-\sigma_j)\right)$$

(b) $(L, S) \sim (\mathrm{Jor}(L), \mathrm{Sor}(L, S))$ *where* $\mathrm{Sor}(L, S)$ *is the matrix*

$$\mathrm{Sor}(L, S) = \mathop{\boxbslash}_{k=1}^{m}\left(\mathop{\boxbslash}_{\ell=1}^{p_k(L,S)} S_k \boxbslash \mathop{\boxbslash}_{\ell=1}^{q_k(L,S)} (-S_k)\right) \boxbslash \mathop{\boxbslash}_{j=1}^{P}\mathop{\boxbslash}_{k=1}^{m(\lambda_j)}\mathop{\boxbslash}_{\ell=1}^{n_k(\lambda_j)}\left(S_k \boxslash S_k\right)$$
$$\boxbslash \mathop{\boxbslash}_{j=1}^{Q}\mathop{\boxbslash}_{k=1}^{m(\mathrm{i}\omega_j)}\mathop{\boxbslash}_{\ell=1}^{n_k(\mathrm{i}\omega_j)} \widehat{S}_k \boxbslash \mathop{\boxbslash}_{j=1}^{R}\mathop{\boxbslash}_{k=1}^{m(\sigma_j)}\mathop{\boxbslash}_{\ell=1}^{n_k(\sigma_j)}\left(\overline{S}_k \boxslash \overline{S}_k\right)$$

where S_k (resp. $\widehat{S}_k$, $\overline{S}_k$) is the $k \times k$ (resp. $2k \times 2k$) diagonal matrix given by

$$S_k := \mathrm{diag}(1, -1, 1, -1, \cdots), \quad \widehat{S}_k = \mathop{\boxbslash}_{j=0}^{k-1}(-1)^j S_2, \quad \overline{S}_k = \mathop{\boxbslash}_{j=0}^{k-1}(-1)^j I_2, \quad (3.6)$$

with $p_k(L, S)$ and $q_k(L, S)$ are given by inverse induction

$$p_N(L,S) = \dim\big(\ker(L) \cap \mathrm{im}(L^{N-1}) \cap \ker(S - \mathrm{Id})\big),$$
$$q_N(L,S) = \dim\big(\ker(L) \cap \mathrm{im}(L^{N-1}) \cap \ker(S + \mathrm{Id})\big),$$

where $\mathrm{im}(L)$ *is the image (or the range) of L and*

$$p_k(L,S) = \dim\big(\ker(L) \cap \mathrm{im}(L^{k-1}) \cap \ker(S - \mathrm{Id})\big) - \sum_{i=k+1}^{N} p_i(L,S),$$
$$q_k(L,S) = \dim\big(\ker(L) \cap \mathrm{im}(L^{k-1}) \cap \ker(S + \mathrm{Id})\big) - \sum_{i=k+1}^{N} q_i(L,S).$$

(c) *So, $(L, S) \sim (L_1, S_1)$ if and only if* $\mathrm{Jor}(L) = \mathrm{Jor}(L_1)$ *and*

$$\big(p_k(S,L), q_k(S,L)\big) = \big(p_k(S_1,L_1), q_k(S_1,L_1)\big), \qquad \text{for } 1 \le k \le N.$$

The proof of this theorem is given in appendix 3.A

Remark 3.1.11. This theorem ensures that for reversible matrices $L, L_1 \in GL(\mathbb{R}^N)$ (i.e. invertible), $(L,S) \sim (L_1,S_1)$ holds if and only if $\mathrm{Jor}(L) = \mathrm{Jor}(L_1)$. In other words, for the reversible matrices which have the same invertible Jordan normal form there is only one possible symmetry up to simultaneous conjugacy. In this case, the theorem gives for each Jordan block the corresponding symmetry.

However, for the reversible matrices which have the same non invertible Jordan normal form, there are several possible symmetries which leads to different equivalence classes for the simultaneous conjugacy. Each class is characterized by a non invertible Jordan normal form and by a sequence of positive integers (p_k, q_k) with $1 \le k \le N$ which satisfy $p_k + q_k = n_k(0)$ where $n_k(0)$ is the number of 0-Jordan block of size k in the Jordan normal form: for a class characterized by (p_k, q_k), the symmetry corresponding to the direct sum of $n_k(0)$ 0-Jordan blocks of size k, i.e. $\boxbslash_{\ell=1}^{n_k(0)} J_k(0)$, is given by $\left(\boxbslash_{\ell=1}^{p_k} S_k\right) \boxbslash \left(\boxbslash_{\ell=1}^{q_k}(-S_k)\right)$.

Remark 3.1.12. Nomenclature of classes of reversible pairs up to simultaneous conjugacy. For denoting each class, we must introduce a name which describes the Jordan normal matrix and the sequence (p_k, q_k), i.e. the structure of the symmetry corresponding to the 0-Jordan blocks. For denoting Jordan normal matrices we can use the nomenclature introduced by V. Arnold in [Ar83]: A Jordan block $J_k(\lambda)$ corresponding to a real eigenvalue is denoted by λ^k and a Jordan block $\widehat{J}_k(\sigma)$ corresponding to a pair of complex conjugate eigenvalues $(\sigma, \overline{\sigma})$ is denoted by $\sigma^k\overline{\sigma}^k$. Then a general Jordan normal matrix is denoted by the formal product of the name of its Jordan blocks. Hence, $\lambda_0\lambda_0\lambda_1^3\sigma^2\overline{\sigma}^2$ represents the Jordan normal matrix $J_1(\lambda_0) \boxbslash J_1(\lambda_0) \boxbslash J_3(\lambda_1) \boxbslash \widehat{J}_2(\sigma)$.

For reversible Jordan normal matrices, we know that if $\lambda \in \mathbb{R} \cup i\mathbb{R} \setminus \{0\}$ is an eigenvalue, then $-\lambda$ is also an eigenvalue with the same Jordan blocks. So we can omit the product of the name of the blocks corresponding to $-\lambda$. Similarly, we simply write σ^k instead of $\sigma^k\overline{\sigma}^k(-\sigma^k)(-\overline{\sigma}^k)$. So in what follows, $\lambda\lambda^2\sigma^3$ represents the reversible Jordan normal matrix

$$J_1(\lambda) \boxbslash J_1(-\lambda) \boxbslash J_2(\lambda) \boxbslash J_2(-\lambda) \boxbslash \widehat{J}_3(\sigma) \boxbslash \widehat{J}_3(-\sigma).$$

To obtain an abbreviated name, instead of writing $\lambda\lambda\lambda\lambda^2\lambda^2$, we write $3.\lambda(2.\lambda^2)$

Finally for denoting a class up to simultaneous conjugacy we must add to the name of the Jordan normal form something which represents the sequence (p_k, q_k) $1 \le k \le N$ (as already explained, this sequence describe the symmetry corresponding to the 0 Jordan blocks). For that purpose we incorporate the sequence in the name of the 0 Jordan block : we denote by $(p.0^{k+})(q.0^{k-})$ the class of the pair

$$\left(\mathop{\boxed{\diagdown}}\limits_{\ell=1}^{p+q} J_k(0), \mathop{\boxed{\diagdown}}\limits_{\ell=1}^{p} S_k \;\boxed{\diagdown}\; \mathop{\boxed{\diagdown}}\limits_{\ell=1}^{q}(-S_k)\right)$$

Moreover, when p (resp. q) is equal to 0 we do not write the term $(p.0^{k+})$ (resp. $(q.0^{k-})$). Hence, for example, we denote by $2.0^{+}0^{2-}(\mathrm{i}\omega)^2$ the class of the reversible pairs which are simultaneously conjugated to the pair (L, S) with

$$L = \begin{pmatrix} 0 & 0 & 0 & 0 & 0 & 0 \\ 0 & 0 & 0 & 0 & 0 & 0 \\ 0 & 0 & 0 & 1 & 0 & 0 \\ 0 & 0 & 0 & 0 & 0 & 0 \\ 0 & 0 & 0 & 0 & 0 & -\omega \\ 0 & 0 & 0 & 0 & \omega & 0 \end{pmatrix} \qquad S = \begin{pmatrix} 1 & 0 & 0 & 0 & 0 & 0 \\ 0 & 1 & 0 & 0 & 0 & 0 \\ 0 & 0 & -1 & 0 & 0 & 0 \\ 0 & 0 & 0 & 1 & 0 & 0 \\ 0 & 0 & 0 & 0 & 1 & 0 \\ 0 & 0 & 0 & 0 & 0 & -1 \end{pmatrix}.$$

3.1.3 Families of reversible vector fields and resonances

Definition 3.1.13 (Reversible families of vector fields). *Let $V : \mathcal{U} \times \Lambda \to \mathbb{R}^n : (u, \underline{\lambda}) \mapsto V(u, \underline{\lambda})$ be a p-parameters family of vector fields in $\mathbb{R}^n$ defined in an open set $\mathcal{U}$ of $\mathbb{R}^n$ with parameter $\underline{\lambda}$ lying in an open set Λ of $\mathbb{R}^p$. The family is said to be reversible when there exists a symmetry $S \in GL_n(\mathbb{R})$, $S^2 = \mathrm{Id}$, which satisfies*

$$SV(u, \underline{\lambda}) = -V(Su, \underline{\lambda}) \qquad \textit{for every } u \in \mathcal{U} \textit{ and } \underline{\lambda} \in \Lambda.$$

In other words the symmetry is the same for all $\underline{\lambda} \in \Lambda$.

Definition 3.1.14. (Fixed points of a reversible family of vector fields) *A fixed point u_0 of a family of vector fields $V(u, \underline{\lambda})$ with $u \in \mathcal{U}$, $\underline{\lambda} \in \Lambda$ is a point u_0 such that $V(u_0, \underline{\lambda}) = 0$ holds for every $\underline{\lambda} \in \Lambda$.*

When studying the dynamics of

$$\frac{du}{dt} = V(u, \underline{\lambda}) \qquad u \in \mathbb{R}^n,$$

for a reversible family $V(u, \underline{\lambda})$ of vector fields near a reversible fixed point u_0 where the spectrum of the differential $L(\underline{\lambda}) = D_u V(u_0, \underline{\lambda})$ at the fixed point has purely imaginary eigenvalues (non hyperbolic fixed point) for some critical value of the parameter $\underline{\lambda}_c$, a first question is the behavior of the eigenvalues of $L(\underline{\lambda})$ for $\underline{\lambda}$ close to $\underline{\lambda}_c$. As already mentioned the eigenvalues of $L(\underline{\lambda})$ are given by pairs $(\sigma, -\sigma)$ for $\sigma \in (\mathbb{R} \cup \mathrm{i}\mathbb{R}) \setminus \{0\}$ and by quadruplets $(\sigma, -\sigma, \overline{\sigma}, -\overline{\sigma})$ for $\sigma \in \mathbb{C}$ such that $\mathcal{R}e\,(\sigma)\,\mathcal{I}m\,(\sigma) \neq 0$. Moreover, $(\sigma, -\sigma)$

(resp. $(\sigma, -\sigma, \overline{\sigma}, -\overline{\sigma})$) have generalized eigenspaces with the same Jordan decomposition. So, if $\sigma(\underline{\lambda})$ is a simple eigenvalue for $\underline{\lambda} = \underline{\lambda}_c$ it cannot leave the imaginary axis for $\underline{\lambda}$ close to $\underline{\lambda}_c$. So bifurcation of the spectrum can only occur for multiple purely imaginary eigenvalues. So, we define

Definition 3.1.15 (Resonant reversible fixed points). *A reversible fixed point of a reversible family of vector fields $V(u, \underline{\lambda})$ is said to be resonant for $\underline{\lambda} = \underline{\lambda}_c$ when the spectrum of $D_u V(u_0, \underline{\lambda}_c)$ contains multiple purely imaginary eigenvalues.*

Remark 3.1.16 (Nomenclature of reversible fixed points). We name such a fixed point using the nomenclature introduced at the previous subsection. So for instance, we say that u_0 is a $0^{2+}\mathrm{i}\omega$ resonant fixed point of $V(u, \underline{\lambda})$ for $\underline{\lambda} = \underline{\lambda}_c$ or equivalently that $V(u, \underline{\lambda})$ admits a $0^{2+}\mathrm{i}\omega$ resonance at u_0 for $\underline{\lambda} = \underline{\lambda}_c$ if

- $V(u_0, \underline{\lambda}) = 0$ for $\underline{\lambda}$ close to $\underline{\lambda}_c$,
- the spectrum of $D_u V(u_0, \underline{\lambda}_c)$ is $\{0, \pm \mathrm{i}\omega\}$ where 0 is double non semi simple eigenvalue and $\pm \mathrm{i}\omega$ are simple eigenvalues.
- $S(\varphi_0) = +\varphi_0$ where φ_0 is any eigenvector corresponding to 0.

When we say for instance that $V(u, \underline{\lambda})$ admits a $0^{2+}\mathrm{i}\omega$ resonance without mentioning the fixed point and the critical value of the parameter we always assume that the fixed point is placed at the origin, and that the critical value of the parameter is equal to 0.

3.2 Global normal forms associated with resonances

As already mentioned, one aim of this work is to study the dynamics of

$$\frac{dX}{dt} = V(X, \underline{\lambda}), X \in \mathbb{R}^n,$$

for a reversible family of vector fields near a resonant fixed point placed at the origin and for a critical value of the parameter $\underline{\lambda}_c$ also equal to 0. Let us denote by T_k the Taylor expansion of the vector field up to terms of order $k \geq 2$, and by R_k the higher order terms,

$$V(X, \underline{\lambda}) = T_k(X, \underline{\lambda}) + R_k(X, \underline{\lambda}), \qquad R_k(X, \underline{\lambda}) = \mathcal{O}(|X|^{k+1}).$$

A natural idea is to begin with the study of the truncated system

$$\frac{dX}{dt} = T_k(X, \underline{\lambda}) \tag{3.7}$$

and then to show that the orbits obtained for the truncated system persist under the perturbation induced by the higher order terms. Unfortunately,

T_k is in general too complicated (even for $k = 2$) to determine whether the truncated system (3.7) admits for instance a homoclinic connection to 0 or periodic orbits. So, a second idea is to perform a change of variables $X = \Phi(Y)$, to obtain an equivalent equation near the fixed point

$$\frac{dY}{dt} = \widetilde{V}(Y, \underline{\lambda}), Y \in \mathbb{R}^n,$$

governed by a new reversible family of vector fields, the principal part of which being as simple as possible. The theory of normal forms is precisely designed for that purpose. We only recall here the theorem giving the global characterization of normal form. See [IA92] for a good introduction to this theory. Normal form Theory for reversible systems can be also found in [Se86].

3.2.1 The fully critical case

We begin with a first theorem giving the normal form for fixed points where the differential has all its eigenvalues on the imaginary axis.

Theorem 3.2.1 (Normal form theorem for critical spectrum). *Let $V : \mathcal{U} \times \Lambda \to \mathbb{R}^n : (u, \underline{\lambda}) \mapsto V(u, \underline{\lambda})$ be a p-parameters family of smooth vector fields in $\mathbb{R}^n$ defined in an open set $\mathcal{U}$ of $\mathbb{R}^n$ with parameter $\underline{\lambda}$ lying in an open set Λ of $\mathbb{R}^p$.*

Assume that 0 is a fixed point of the vector field $V(X, 0)$, i.e. $V(0,0) = 0$ holds and that the spectrum of the differential at the origin $L_0 := D_X V(0,0)$ is fully critical, i.e. that all the eigenvalues of L_0 are on the imaginary axis.

(a) *Then, for any integer $k \geq 2$, there are two open sets $\widetilde{\mathcal{U}}$, $\widetilde{\Lambda}$, $0 \in \widetilde{\mathcal{U}} \subset \mathcal{U}$, $0 \in \widetilde{\Lambda} \subset \Lambda$, and there are polynomials smoothly dependent in $\underline{\lambda}$, $\Phi(\cdot, \underline{\lambda}) : \mathbb{R}^n \to \mathbb{R}^n$ and $N(\cdot, \underline{\lambda}) : \mathbb{R}^n \to \mathbb{R}^n$ of degree $\leq k$, with $\Phi(0,0) = 0$, $D_Y\Phi(0,0) = 0$ and $N(0,0) = 0, D_Y N(0,0) = 0$ such that for*

$$X = Y + \Phi(Y, \underline{\lambda}) \qquad \text{and } (Y, \underline{\lambda}) \in \widetilde{\mathcal{U}} \times \widetilde{\Lambda},$$

the equation

$$\frac{dX}{dt} = V(X, \underline{\lambda}),$$

becomes

$$\frac{dY}{dt} = L_0 Y + N(Y, \underline{\lambda}) + R(Y, \underline{\lambda})$$

with

$$N(e^{tL_0^*}Y, \underline{\lambda}) = e^{tL_0^*}N(Y, \underline{\lambda}) \tag{3.8}$$

for all $Y, \underline{\lambda}$ and t and

$$R(Y, \underline{\lambda}) = o(|Y|^k).$$

(b) *If there is a unitary map T on $\mathbb{R}^n$ with $V(TX, \underline{\lambda}) = TV(X, \underline{\lambda})$ for all X and $\underline{\lambda}$, then for all Y and $\underline{\lambda}$,*

$$\Phi(TY, \underline{\lambda}) = T\Phi(Y, \underline{\lambda}),\ N(TY, \underline{\lambda}) = TN(Y, \underline{\lambda}) \text{ and } R(TY, \underline{\lambda}) = TR(Y, \underline{\lambda}).$$

(c) *If the family of vector fields is reversible, i.e. if there is a symmetry $S \in GL(\mathbb{R}^n)$, $S^2 = I_n$ with $V(SX, \underline{\lambda}) = -SV(X, \underline{\lambda})$ for all X and $\underline{\lambda}$, then for all Y and $\underline{\lambda}$,*

$$\Phi(SY, \underline{\lambda}) = S\Phi(Y\underline{\lambda}),\ N(SY, \underline{\lambda}) = -SN(Y, \underline{\lambda}) \text{ and } R(SY, \underline{\lambda}) = -SR(Y, \underline{\lambda}).$$

Remark 3.2.2. Observe that for this theorem we do not need that 0 is a fixed point of the whole family of vector fields.

Remark 3.2.3. "$N(\cdot, \underline{\lambda})$ commutes with $e^{tL_0^*}$ for all $\underline{\lambda}$ and t" is equivalent to "$D_Y N(Y, \underline{\lambda}) L_0^* Y = L_0^* N(Y, \underline{\lambda})$ for all Y and $\underline{\lambda}$".

Remark 3.2.4. If 0 is not an eigenvalue of L_0 then the proof is contained in Elphick et al. [1987]. Even when 0 is an eigenvalue of L_0, the theorem still holds after a simple adaptation of the proof above, which uses again the Implicit Function Theorem.

Remark 3.2.5. Property (c) ensures that if the family of vector fields is reversible then the vector field obtained after normal form changes of coordinates is also reversible with respect to the same symmetry.

Remark 3.2.6. Equation (3.8) gives a global characterization of the normal form, and it gives also a way to compute it. Moreover observe that this characterization and property (c) ensure that the normal form N for a reversible family of vector fields, depends only on the linear part $L_0 = D_X V(0,0)$ for the critical value of the parameter and on the symmetry S. We give several examples below.

Remark 3.2.7. (Strategy for studying dynamics near resonant fixed points). When studying, the dynamics of a reversible family of vector fields

$$\frac{dX}{dt} = V(X, \underline{\lambda}) \tag{3.9}$$

near a fixed point placed at the origin

$$V(0, \underline{\lambda}) = 0, \qquad \text{for every } \underline{\lambda},$$

we proceed in four steps:

Step 1. We use the Theorem of classification of reversible matrices 3.1.10 to perform a *linear* change of coordinates $X = P\widetilde{X}$ with $P \in GL(\mathbb{R}^n)$ to obtain an equivalent equation

$$\frac{d\widetilde{X}}{dt} = \widetilde{V}(\widetilde{X}, \underline{\lambda})$$

where $\widetilde{V}$ is a family of vector fields reversible with the symmetry $\widetilde{S}$, where $\widetilde{S}$ and $\widetilde{L}_0 = D_{\widetilde{X}}\widetilde{V}(0,0)$ are in the simple form (Jordan normal form and associated symmetry) given by the Theorem of classification 3.1.10

$$\widetilde{L}_0 = \mathrm{Jor}(L_0) \qquad \text{and} \qquad \widetilde{S} = \mathrm{Sor}(L_0, S).$$

Step 2. We use the normal form theorem 3.2.1 above to perform a second change of coordinates which is *polynomial and close to identity*, $\widetilde{X} = Y + \Phi(Y)$ to obtain a third equivalent equation

$$\frac{dY}{dt} = \widetilde{L}_0 Y + N(Y, \underline{\lambda}) + R(Y, \underline{\lambda})$$

with

$$N(e^{t\widetilde{L}_0^*} Y, \underline{\lambda}) = e^{t\widetilde{L}_0^*} N(Y, \underline{\lambda}) \text{ and } R(Y, \underline{\lambda}) = o(|Y|^k).$$

Step 3. We study the dynamics of the Normal form system

$$\frac{dY}{dt} = \widetilde{L}_0 Y + N(Y, \underline{\lambda})$$

In several examples like the $0^{2+}\mathrm{i}\omega$, $(\mathrm{i}\omega_0)^2$, $(\mathrm{i}\omega_0)^2\mathrm{i}\omega_1$ resonances, the normal form system is integrable and so the study of its dynamics is not too difficult.

Step 4. We have to determine which of the previously found orbits persist for the full system, i.e. when the normal form system is perturbed by higher order terms denoted by R. This work is devoted to the question of persistence of homoclinic connections to fixed points or to periodic orbits.

The two first steps of this strategy ensure that there is a uniquely determined normal form system for all the reversible families of vector fields admitting the same resonance at the origin.

Example 3.2.8 (the $(\mathrm{i}\omega)^2$ resonance).

Let $V : \mathbb{R}^4 \times \mathbb{R} \to \mathbb{R}^4 : (X, \mu) \mapsto V(X, \mu)$ be a one-parameter family of smooth vector fields in $\mathbb{R}^4$. Assume that 0 is a $(\mathrm{i}\omega)^2$ resonant fixed point of the family.

Step 1. The Theorem of classification 3.1.10 ensures that up to a linear change of coordinates, we can assume that $L_0 = DV(0,0)$ and the symmetry reads

$$L_0 = \begin{pmatrix} 0 & -\omega & 1 & 0 \\ \omega & 0 & 0 & 1 \\ 0 & 0 & 0 & -\omega \\ 0 & 0 & \omega & 0 \end{pmatrix} \text{ and } S = \begin{pmatrix} 1 & 0 & 0 & 0 \\ 0 & -1 & 0 & 0 \\ 0 & 0 & -1 & 0 \\ 0 & 0 & 0 & 1 \end{pmatrix}$$

Step 2. Using the characterization of the Normal Form given by Theorem 3.2.1 one can check that the normal form system reads

$$\frac{dz_1}{dt} = \mathrm{i}\omega z_1 + z_2 + \mathrm{i}z_1 P(u, v, \mu)$$
$$\frac{dz_2}{dt} = \mathrm{i}\omega z_2 + \mathrm{i}z_2 P(u, v, \mu) + z_1 Q(u, v, \mu)$$

with $Y = (a_1, b_1, a_2, b_2)$, $z_1 = a_1 + ib_1$, $z_2 = a_2 + ib_2$ and where $u = |z_1|^2$, $v = \mathrm{i}(z_1\overline{z_2} - z_2\overline{z_1})/2$, and P and Q are real polynomials in (u, v) such that $P(0,0,0) = Q(0,0,0) = 0$. For the explicit computation of this normal form see [ETBCI87]. Moreover, one can check that the normal form system admits the following two first integrals

$$\rho = \mathrm{i}(z_1\overline{z_2} - z_2\overline{z_1})/2 \text{ and } K = |z_2|^2 - \int_0^u Q(s, \rho, \mu)\, ds.$$

Step 3 an 4. The above system was found by [IMD89] in the context of bifurcating bounded steady flows in a cylindrical (unbounded) domain. This case also occurs in water-wave problems, and allows in particular, to prove the existence of solitary waves [IK90]. The proof of persistence of quasi-periodic solutions for the untruncated system is made in [IL90], and the persistence of homoclinic solutions to periodic solutions or to the origin is made in [IP93].

Example 3.2.9 (the $0^2\mathrm{i}\omega$ resonances).
Let $V : \mathbb{R}^4 \times \mathbb{R} \to \mathbb{R}^4 : (X, \mu) \mapsto V(X, \mu)$ be a one-parameter family of smooth vector fields in $\mathbb{R}^4$. Assume that 0 is a $0^2\mathrm{i}\omega$ resonant fixed point of the family.

Step 1. The Theorem of classification 3.1.10 ensures that there are only two possible symmetries up to a linear change of coordinates. So, we can assume that $L_0 = DV(0,0)$ reads

$$L_0 = \begin{pmatrix} 0 & 1 & 0 & 0 \\ 0 & 0 & 0 & 0 \\ 0 & 0 & 0 & -\omega \\ 0 & 0 & \omega & 0 \end{pmatrix}$$

and that the symmetry has the form

$$S = \mathrm{diag}(\ \ 1, -1, 1, -1) \qquad \text{for the } 0^{2+}\mathrm{i}\omega \text{ resonance,}$$
$$S = \mathrm{diag}(-1, \ \ 1, 1, -1) \qquad \text{for the } 0^{2-}\mathrm{i}\omega \text{ resonance.}$$

Step 2. Using the characterization of the Normal Form given by Theorem 3.2.1 one can check that the normal form system for the $0^{2+}i\omega$ resonance reads

$$\begin{aligned}\frac{d\alpha}{dx} &= \beta,\\ \frac{d\beta}{dx} &= \Phi(\alpha, A^2+B^2, \mu),\\ \frac{dA}{dx} &= -B\Psi(\alpha, A^2+B^2, \mu),\\ \frac{dB}{dx} &= A\Psi(\alpha, A^2+B^2, \mu),\end{aligned} \tag{3.10}$$

where Φ,Ψ are real polynomials satisfying

$$\Phi(0,0,\mu)=0,\ D_Y\Phi(0,0,0)=0, \qquad \Psi(0,0,\mu)=\omega+\mathcal{O}(\mu).$$

The normal form system for the $O^{2-}i\omega$ resonance reads

$$\begin{aligned}\frac{d\alpha}{dx} &= \beta+\alpha^2\ N_0(\alpha^2, A^2+B^2, \mu),\\ \frac{d\beta}{dx} &= \alpha\ \beta N_0(\alpha^2, A^2+B^2, \mu)+\alpha\ N_1(\alpha^2, A^2+B^2, \mu),\\ \frac{dA}{dx} &= -B\Psi(\alpha^2, A^2+B^2, \mu)+\alpha A\Phi(\alpha^2, A^2+B^2, \mu),\\ \frac{dB}{dx} &= A\Psi(\alpha^2, A^2+B^2, \mu)+\alpha B\Phi(\alpha^2, A^2+B^2, \mu),\end{aligned} \tag{3.11}$$

where N_0, N_1, Φ, Ψ are real polynomials.

The normal form system for the $0^{2+}i\omega$ resonance is integrable with the two first integrals

$$\begin{aligned}&A^2+B^2=\rho^2,\\ &\beta^2-\hat{\Phi}(\alpha,\rho^2,\mu)=H, \qquad \partial_\alpha\hat{\Phi}=2\Phi.\end{aligned}$$

So the phase portrait of the normal form system can be obtained with no difficulty in this case. For the $0^{2-}i\omega$ resonance the situation is not so clear (see [IA92] chap. 1).

Step 3. and 4. The $0^{2+}i\omega$ resonance appears in water waves problem. The explicit computation of the normal form and the proof the persistence of quasi periodic solutions are given in [IK92]. The persistence of homoclinic connections to exponentially small periodic orbit and the generic non persistence of homoclinic connection to 0 are proved in Chapter 7. The infinite dimensional case and the water wave problem are studied in Chapter 8.

3.2.2 The general case

We now give a second theorem which gives normal forms for fixed points where the eigenvalues of the differential are not necessarily all on the imaginary axis. Usually, this theorem is not used to study dynamics of vector fields close to fixed points when one is interested in periodic solutions, homoclinic connections, and more generally in bounded solutions close to the fixed point, since all these orbits lie on a center manifold. So usually, one use the Center Manifold Theorem to obtain a reduced equation on a center manifold which has a linear part with only critical eigenvalues. For such reduced equations the previous Normal Form Theorem 3.2.1 is sufficient. See [IA92] for a pedagogical explanation of this approach. Such an approach can only be used when a C^k smoothness of the vector field is sufficient to make the proofs, because the center manifold reduction destroys *analyticity* and C^∞ smoothness. As already explained, one of our aims is to study the existence of homoclinic connections near resonant fixed points. For vector fields, for which the exponential tools given in chapter 2, are required to study the persistence of homoclinic connection, analyticity is fundamental. So, center manifold reduction cannot be used, and so, the second Normal Form Theorem below is necessary. This theorem also works in infinite dimensions. See [IL90] for an example in infinite dimensions where C^∞ smoothness is required to study the persistence of quasi periodic orbits using the Hard Implicit Function Theorem. See also chapter 8 for a second example in infinite dimensions where analyticity is required to use the tools of chapter 2.

Theorem 3.2.10. (Normal form theorem)
Let $V : \mathcal{U}\times\Lambda : (u,\underline{\lambda}) \mapsto V(u,\underline{\lambda})$ be a p-parameters family of smooth vector fields in $\mathbb{R}^n$ defined in an open set $\mathcal{U}$ of $\mathbb{R}^n$ with parameter $\underline{\lambda}$ lying in an open set Λ of $\mathbb{R}^p$.
Assume that 0 is a fixed point of the vector field $V(X,0)$, i.e. $V(0,0)=0$ holds. Denote by E_{c} (resp. E_{h}) the vector space spanned by the eigenvectors and generalized eigenvectors of $L_0 = D_X V(0,0)$ corresponding to the eigenvalues with 0 real parts (resp. with non zero real parts).

$$\mathbb{R}^n = E_{\mathrm{c}} \bigoplus E_{\mathrm{h}}.$$

Moreover let us denote $L_{0\mathrm{c}} = L_0|_{E_{\mathrm{c}}}$ and $L_{0\mathrm{h}} = L_0|_{E_{\mathrm{h}}}$

(a) *Then, for any integer $k \geq 2$ there are two open sets $\widetilde{\mathcal{U}}, \widetilde{\Lambda}$, $0 \in \widetilde{\mathcal{U}} \subset \mathcal{U}$, $0 \in \widetilde{\Lambda} \subset \Lambda$ and there are two polynomials smoothly dependent on $\underline{\lambda}$ $\Phi(\cdot,\cdot,\underline{\lambda}) : E_{\mathrm{c}} \times E_{\mathrm{h}} \to \mathbb{R}^n$ and $N(\cdot,\underline{\lambda}) : E_{\mathrm{c}} \to E_{\mathrm{c}}$ of degree $\leq k$ with $\Phi(0,0,0) = 0$, $D_{Y_{\mathrm{c}},Y_h}\Phi(0,0,0) = 0$ and $N(0,0) = 0, D_{Y_{\mathrm{c}}}N(0,0) = 0$ such that for*

$$X = Y_{\mathrm{c}} + Y_{\mathrm{h}} + \Phi(Y_{\mathrm{c}}, Y_{\mathrm{h}}, \underline{\lambda}) \qquad \text{and } (Y_{\mathrm{c}}, Y_{\mathrm{h}}) \in \widetilde{\mathcal{U}},\ \underline{\lambda} \in \widetilde{\Lambda}$$

the equation

$$\frac{dX}{dt} = V(X, \underline{\lambda}), X \in \mathbb{R}^n,$$

becomes

$$\begin{aligned}\frac{dY_c}{dt} &= L_{0c}Y_c + N(Y_c, \underline{\lambda}) + R_c(Y_c, Y_h, \underline{\lambda}) \\ \frac{dY_h}{dt} &= L_{0h}Y_h + M(Y_c, Y_h, \underline{\lambda}) + R_h(Y_c, Y_h, \underline{\lambda})\end{aligned}$$

with

$$N(e^{tL_0^*}Y_c, \underline{\lambda}) = e^{tL_0^*}N(Y_c, \underline{\lambda}) \tag{3.12}$$

for all $Y, \underline{\lambda}$ *and* t *and*

$$\begin{aligned}M(Y_c, Y_h, \underline{\lambda}) &= O\Big(|Y_h|(|\underline{\lambda}| + |Y_c| + |Y_h|)\Big), \\ R_c(Y_c, Y_h, \underline{\lambda}) &= O\Big(|Y_h|^2 + |Y_c|^{k+1}\Big), \\ R_h(Y_c, Y_h, \underline{\lambda}) &= O(|Y_c|^{k+1}).\end{aligned}$$

(b) *If there is a unitary map* T *on* $\mathbb{R}^n$ *with* $V(TX, \underline{\lambda}) = TV(X, \underline{\lambda})$ *for all* X *and* $\underline{\lambda}$, *then for all* Y *and* $\underline{\lambda}$,

$$\begin{aligned}&\Phi(TY_c, TY_h, \underline{\lambda}) = T\Phi(Y_c, Y_h, \underline{\lambda}),\ N(TY_c, \underline{\lambda}) = TN(Y_c, \underline{\lambda}) \\ &R_{c \text{ or } h}(TY_c, TY_h, \underline{\lambda}) = TR_{c \text{ or } h}(Y_c, Y_h, \underline{\lambda}).\end{aligned}$$

(c) *If the family of vector fields is reversible, i.e. if there is a symmetry* $S \in GL(\mathbb{R}^n)$, $S^2 = I_n$ *with* $V(SX, \underline{\lambda}) = -SV(X, \underline{\lambda})$ *for all* X *and* $\underline{\lambda}$, *then for all* Y *and* $\underline{\lambda}$,

$$\begin{aligned}&\Phi(SY_c, SY_h, \underline{\lambda}) = S\Phi(Y_c, Y_h, \underline{\lambda}),\ N(SY_c, \underline{\lambda}) = -SN(Y_c, \underline{\lambda}) \\ &R_{c \text{ or } h}(SY_c, SY_h, \underline{\lambda}) = -SR_{c \text{ or } h}(Y_c, Y_h, \underline{\lambda}).\end{aligned}$$

The proof of this theorem follows from the indication given in [ETBCI87].

3.A Appendix. Proof of theorem of classification of reversible matrices

To prove Theorem 3.1.10, it is more convenient to reformulate it in terms of linear maps rather than in terms of matrices. So, let u and s be two linear maps in $\mathbb{R}^N$ whose matrices in the canonical basis of $\mathbb{R}^N$ are respectively L and S. Then for proving (b) of Theorem 3.1.10 it is equivalent to prove that there exists an appropriate basis of $\mathbb{R}$ where the matrices of u and s, called respectively M_u and M_s are given by M_u=Jor(L) and M_s=Sor(L, S). Since

$$\mathbb{R}^N = \ker(u^m) \quad \oplus \bigoplus_{j=1}^{P} \ker\left((u^2-\lambda_j^2)^{m(\lambda_j)}\right) \oplus \bigoplus_{j=1}^{Q} \ker\left((u^2+\omega_j^2)^{m(\mathrm{i}\omega_j)}\right)$$
$$\oplus \bigoplus_{j=1}^{R} \Big(\ker\Big((u^2 - 2\mathcal{R}e\,(\sigma_j)\,u + |\sigma_j|^2)^{m(\sigma_j)} \Big)$$
$$\oplus \ker\big(u^2 + 2\mathcal{R}e\,(\sigma_j)\,u + |\sigma_j|^2\big)^{m(\sigma_j)} \Big),$$

we can split the proofs in several parts and find appropriate basis for each kernel. So it is sufficient to prove the following lemmas

Lemma 3.A.1. *Let u, s be two linear maps of $\mathbb{R}^N$ and assume that*

– *u is nilpotent of index m, i.e. $u^m = 0$, $u^{m-1} \neq 0$.*
– *s is a symmetry ($s^2 = \mathrm{Id}$) which anticommutes with u, i.e $us = -su$.*

(a) *Then, there exists a basis of $\mathbb{R}^N$ in which the matrices of u and s are respectively given by $M_u = \boxbslash_{p=1}^{r} J_{k_p}(0)$ and $M_s = \boxbslash_{p=1}^{r} \varepsilon_p S_{k_p}$ with $\varepsilon_p = \pm 1$, $k_r = m$, $k_{p+1} \leq k_p$ and where J_k and S_k are respectively given by (3.5) and (3.6).*

(b) *There exists a basis of $\mathbb{R}^N$ in which the matrices of u and s are respectively given by*

$$M_u = \boxbslash_{k=1}^{m} \boxbslash_{\ell=1}^{n_k(0)} J_k(0) \qquad M_s = \boxbslash_{k=1}^{m} \left(\boxbslash_{\ell=1}^{p_k(0)} S_k \boxbslash \boxbslash_{\ell=1}^{q_k(0)} (-S_k) \right).$$

where (p_k, q_k) are given by the inverse induction

$$\sum_{i=k}^{N} p_i = \dim\Big(\mathrm{im}(u^{k-1}) \cap \ker(u) \cap \ker(s - \mathrm{Id})\Big),$$

$$\sum_{i=k}^{N} q_i = \dim\Big(\mathrm{im}(u^{k-1}) \cap \ker(u) \cap \ker(s + \mathrm{Id})\Big).$$

Lemma 3.A.2. *Let $\lambda > 0$ and let u, s be two linear maps of $\mathbb{R}^N$ and assume that*

– *$u^2 - \lambda^2$ is nilpotent of index m.*
– *s is a symmetry ($s^2 = \mathrm{Id}$) which anticommutes with u, i.e $us = -su$.*

Then, there exists a basis of $\mathbb{R}^N$ in which the matrices of u and s are respectively given by

$$M_u = \mathop{\boxbslash}_{k=1}^{m} \mathop{\boxbslash}_{\ell=1}^{n_k} (J_k(\lambda) \boxbslash J_k(-\lambda)), \qquad M_s = \mathop{\boxbslash}_{k=1}^{m} \mathop{\boxbslash}_{\ell=1}^{n_k} (S_k \boxslash S_k),$$

where n_k is the number of Jordan blocks of size k corresponding to the eigenvalue λ.

Lemma 3.A.3. *Let $\omega > 0$ and u, s be two linear maps of $\mathbb{R}^N$ and assume that*

– $u^2 + \omega^2$ is nilpotent of index m.
– s is a symmetry ($s^2 = \mathrm{Id}$) which anticommutes with u, i.e $us = -su$.

Then, there exists a basis of $\mathbb{R}^N$ in which the matrices of u and s are respectively given by

$$M_u = \mathop{\boxbslash}_{k=1}^{m} \mathop{\boxbslash}_{\ell=1}^{n_k} \widehat{J}_k(\mathrm{i}\omega), \qquad M_s = \mathop{\boxbslash}_{k=1}^{m} \mathop{\boxbslash}_{\ell=1}^{n_k} \widehat{S}_k, \tag{3.13}$$

where n_k is the number of complex Jordan blocks of size k corresponding to the eigenvalues $\pm\mathrm{i}\omega$.

Lemma 3.A.4. *Let $\sigma \in \mathbb{C}$ with $\mathcal{R}e\,(\sigma)\,\mathcal{I}m\,(\sigma) \neq 0$. Let u, s be two linear maps of $\mathbb{R}^N$ and assume that*

– $(u^2 - \mathcal{R}e\,(\sigma)\,u + |\sigma|^2)(u^2 + \mathcal{R}e\,(\sigma)\,u + |\sigma|^2)$ is nilpotent of index m.
– s is a symmetry ($s^2 = \mathrm{Id}$) which anticommutes with u, i.e $us = -su$.

Then, there exists a basis of $\mathbb{R}^N$ in which the matrices of u and s are respectively given by

$$M_u = \mathop{\boxbslash}_{k=1}^{m} \mathop{\boxbslash}_{\ell=1}^{n_k} (\widehat{J}_k(\sigma) \boxbslash \widehat{J}_k(-\sigma)), \qquad M_s = \mathop{\boxbslash}_{k=1}^{m} \mathop{\boxbslash}_{\ell=1}^{n_k} (\overline{S}_k \boxslash \overline{S}_k), \tag{3.14}$$

where n_k is the number of complex Jordan blocks of size k corresponding to the eigenvalue σ.

Proof of Lemma 3.A.2. Observe that if $(e_j)_{0 \leq j \leq k-1}$ is a family of vectors satisfying

$$u(e_0) = \lambda e_0, \qquad u(e_j) = \lambda e_j + e_{j-1} \text{ for } j \geq 1,$$

then $e'_j = (-1)^j s(e_j)$ is a family of vectors satisfying

$$u(e'_0) = -\lambda e'_0, \qquad u(e'_j) = -\lambda e'_j + e'_{j-1} \text{ for } j \geq 1.$$

Lemma 3.A.2 follows readily. □

Proof of Lemma 3.A.4. The proof of this lemma follows from the previous observation and from the following one: if $(e_j)_{0 \leq j \leq k-1}$ is a family of complex vectors satisfying

$$u(e_0) = \sigma e_0, \qquad u(e_j) = \sigma e_j + e_{j-1} \text{ for } j \geq 1,$$

then $e''_j = \overline{e_j}$ is a family of vectors satisfying

$$u(e''_0) = \overline{\sigma} e''_0, \qquad u(e''_j) = \overline{\sigma} e''_j + e''_{j-1} \text{ for } j \geq 1.$$

So, let $e_j^{k,\ell}$, $1 \leq k \leq m$, $1 \leq \ell \leq n_k$, $0 \leq j \leq k-1$ be a complex basis of $ker(u-\sigma)^m$ in which the matrices of u is under Jordan normal form with Jordan blocks ordered by size, i.e. for $1 \leq k \leq m$ and $1 \leq \ell \leq n_k$,

$$u(e_0^{k,\ell}) = \sigma e_0^{k,\ell}, \qquad u(e_j^{k,\ell}) = \sigma e_j^{k,\ell} + e_{j-1}^{k,\ell} \text{ for } 1 \leq j \leq k-1.$$

Then, $\overline{e_j^{k,\ell}}$, $(-1)^j s(e_j^{k,\ell})$, $\overline{(-1)^j s(e_j^{k,\ell})}$ are respectively basis of $ker(u-\overline{\sigma})^m$, $ker(u+\sigma)^m$, $ker(u+\overline{\sigma})^m$. Finally, in the real basis

$$(x_j^{k,\ell}, y_j^{k,\ell}, (-1)^j s(x_j^{k,\ell}), (-1)^j s(y_j^{k,\ell}))$$

with $x_j^{k,\ell} = (e_j^{k,\ell} + \overline{e_j^{k,\ell}})/2$ and $y_j^{k,\ell} = (\overline{e_j^{k,\ell}} - e_j^{k,\ell})/2\mathrm{i}$, the matrices of u and s are given by (3.14). □

Proof of Lemma 3.A.3. For proving this Lemma it is sufficient to prove the existence of a complex basis of $\ker(u - \mathrm{i}\omega)^m$ in which the matrix of u is under Jordan normal form with Jordan blocks ordered by size, i.e. for $1 \leq k \leq m$ and $1 \leq \ell \leq n_k$,

$$u(e_0^{k,\ell}) = \mathrm{i}\omega e_0^{k,\ell}, \qquad u(e_j^{k,\ell}) = \mathrm{i}\omega e_j^{k,\ell} + e_{j-1}^{k,\ell} \text{ for } 1 \leq j \leq k-1.$$

which satisfies also

$$\overline{e_j^{k,\ell}} = (-1)^j s(e_j^{k,\ell}). \tag{3.15}$$

Indeed, if such a complex basis exists, then in the real basis $(x_j^{k,\ell}, y_j^{k,\ell})$ with $x_j^{k,\ell} = (e_j^{k,\ell} + \overline{e_j^{k,\ell}})/2$ and $y_j^{k,\ell} = (\overline{e_j^{k,\ell}} - e_j^{k,\ell})/2\mathrm{i}$, the matrices of u and s are given by (3.13). Then, for building such a complex basis of $\ker(u - \mathrm{i}\omega)^m$ satisfying (3.15), we first consider a basis $(f_j^{k,\ell})$ of $\ker(u - \mathrm{i}\omega)^m$ in which the matrix of u is under Jordan normal form with Jordan blocks ordered by size. We observe that for any family of real numbers $(\theta^{k,\ell})_{\substack{1 \leq k \leq m \\ 1 \leq \ell \leq n_k}}$ the family

$$\widetilde{f}_j^{k,\ell}(\theta^{k,\ell}) := \mathrm{e}^{\mathrm{i}\theta^{k,\ell}} f_j^{k,\ell} + \mathrm{e}^{-\mathrm{i}\theta^{k,\ell}} (-1)^j s(\overline{f_j^{k,\ell}})$$

is also a family of vectors which satisfies

$$u(\widetilde{f}_0^{k,\ell}) = \mathrm{i}\omega \widetilde{f}_0^{k,\ell}, \qquad u(\widetilde{f}_j^{k,\ell}) = \mathrm{i}\omega \widetilde{f}_j^{k,\ell} + \widetilde{f}_{j-1}^{k,\ell} \text{ for } 1 \le j \le k-1.$$

Moreover this family also satisfies (3.15). So the only thing to prove is that it is possible to choose an appropriate family of real numbers $(\theta^{k,\ell})_{\substack{1\le k\le m\\ 1\le \ell\le n_k}}$ so that the family of vectors $(\widetilde{f}_j^{k,\ell})$ is linearly independent. To prove this property, we proceed in several steps.

Step 1. By contradiction we check that it is possible to choose $\theta^{1,1}$ so that the family $(\widetilde{f}_0^{1,1}(\theta^{1,1}), f_0^{1,2}, \cdots, f_0^{m,n_m})$ are linearly independent. Otherwise, for all θ there would exist complex numbers $\mu^{k,\ell}(\theta)$ such that

$$\widetilde{f}_0^{1,1}(\theta^{1,1}) = \sum_{\substack{1\le k\le m, 1\le \ell \le n_k - 1\\ (k,\ell)\neq(1,1)}} \mu(\theta) f_0^{k,\ell}.$$

Then, the two previous formula obtained for $\theta = \pi/2$ and $\theta = \pi$ imply that the family $(f_0^{k,\ell})$, $1 \le k \le m$, $1 \le \ell \le n_k$ is not linearly independent, which is false.

Step 2. Similarly, we prove by induction that it is possible to choose $\theta^{1,2}, \cdots, \theta^{m,n_m}$ such that the vectors $\widetilde{f}_0^{k,\ell}(\theta^{k,\ell})$, $1 \le k \le m$, $1 \le \ell \le n_k$ are linearly independent.

Step 3. Finally, assume that the vectors $\widetilde{f}_j^{k,\ell}(\theta^{k,\ell}), 1 \le k \le m$, $1 \le \ell \le n_k$, $0 \le j \le k-1$ are not linearly independent

$$\sum_{k=1}^{m}\sum_{\ell=1}^{n_k}\sum_{j=0}^{k-1} \mu_j^{k,\ell} \widetilde{f}_j^{k,\ell}(\theta_j^{k,\ell}) = 0.$$

and denote by p the largest integer j such that there exists (k,ℓ) with $\mu_j^{k,\ell} \neq 0$. Then,

$$0 = (u - \mathrm{i}\omega)^p \left(\sum_{k=1}^{m}\sum_{\ell=1}^{n_k}\sum_{j=0}^{k-1} \mu_j^{k,\ell} \widetilde{f}_j^{k,\ell}(\theta_j^{k,\ell}) \right) = \sum_{k,\ell} \mu_p^{k,\ell} \widetilde{f}_0^{k,l}$$

and step 2 ensures that $\mu_p^{k,\ell} = 0$ for all (k,ℓ) which leads to a contraction. Hence, the complex vectors $e_j^{k,\ell} = \widetilde{f}_j^{k,\ell}(\theta_j^{k,\ell})$ is a basis of $ker(u - \mathrm{i}\omega)$ in which the matrix of u is under Jordan normal form with Jordan blocks ordered by size and which also satisfies (3.15). □

Proof of Lemma 3.A.1.

(a): To prove this first part of Lemma 3.A.1 we proceed in two main steps:

1. We begin by building a m-dimensional subspace F of $\mathbb{R}^N$ stable by u and s which reads $F := \text{vect}(x, u(x), \cdots, u^{m-1}(x))$ with $s(x) = (-1)^{m-1}\varepsilon_1 x$ and $\varepsilon_1 = \pm 1$. Then, in the basis $x_j = u^{m-1-j}(x)$ with $0 \le j \le m-1$ the matrices of u and s are respectively given by $J_m(0)$ and $\varepsilon_1 S_m$.
2. We build a supplementary subspace F' of F ($\mathbb{R}^N = F \bigoplus F'$) which is also stable by u and s.

Then, the proof of Lemma 3.A.1-(a) follows by induction and applying the result of Step 1 to $u|_{F'}$ and $s|_{F'}$.

Step 1. By hypothesis $\text{im}(u^{m-1}) \neq \{0\}$ and it is stable under s since u and s anticommute. So, since $s|_{\text{im}(u^{m-1})}$ is also a symmetry, there exists necessarily $y \neq 0$ in $\text{im}(u^{m-1})$ such that $s(y) = \varepsilon_1 y$ and $\varepsilon_1 = \pm 1$. Moreover, there exists z such that $y = u^{m-1}(z)$. Then we check that $x = \left(z + \varepsilon_1(-1)^{m-1}s(z)\right)/2$ satisfies

$$s(x) = (-1)^{m-1}\varepsilon_1 x, \qquad u^{m-1}(x) = y.$$

Denote by μ_x the minimal polynomial of x. Since u is nilpotent of index m then μ_x divides X^m. Moreover, $u^{m-1}(x) = y \neq 0$. Thus, $\mu_x = X^m$ and the vectors $(x, u(x), \cdots, u^{m-1}(x))$ are linearly independent. We denote by F the m dimensional vector space spanned by these vectors. F is stable under u and s since $\mu_x = X^m$ and since $s(u^k(x)) = (-1)^{m-1-k}\varepsilon_1 u^k(x)$.

Step 2. s is diagonalizable and $(x, u(x), \cdots, u^{m-1}(x))$ is a family of linearly independent eigenvectors of s. Hence it is possible to add to this family other 0-eigenvectors of s to obtain a basis $(e_j)_{1\le j\le N}$ of eigenvectors of $\mathbb{R}^N$ with $e_j = u^{j-1}(x)$ for $1 \le j \le m$. Denote by $(e_j^*)_{1\le j\le N}$ the associated dual basis, by

$$H = \operatorname*{vect}_{i\in\mathbb{N}}\left({}^t\!u^i(e_m^*)\right)$$

and by $F' := H^\perp$ the orthogonal space of H for the duality. Our aim is to prove that $\mathbb{R}^N = F \bigoplus F'$ and that F' is stable under u and s. We proceed in several substeps.

Step 2.1. Stability of F'. We first observe that H is stable under ${}^t\!u$ and thus $F' = H^\perp$ is stable under u. Secondly since $(e_j)_{1\le j\le N}$ is a basis of eigenvectors of s, then $(e_j^*)_{1\le j\le N}$ is a basis of eigenvectors of ${}^t\!s$. In particular, ${}^t\!s(e_m^*) = \varepsilon e_m^*$ with $\varepsilon = \pm 1$. Hence, since ${}^t\!u$ and ${}^t\!s$ anticommute

$${}^t\!s\left({}^t\!u^i(e_m^*)\right) = (-1)^i\, {}^t\!u^i(s(e_m^*)) = (-1)^i \varepsilon\, {}^t\!u^i(e_m^*).$$

Thus H is stable under ${}^t\!s$ and $F' = H^\perp$ is stable under s.

Step 2.2. dimensions of H and $H^\perp$. Since u is nilpotent of index m, ${}^t\!u^m = 0$. Hence the minimal polynomial $\mu_{e_m^*}$ of e_m^* divides X^m. Moreover,

$$\langle {}^t\!u^{m-1}(e_m^*), e_1\rangle = \langle e_m^*, u^{m-1}(x)\rangle = \langle e_m^*, e_m\rangle = 1.$$

Thus ${}^t u^{m-1}(e_m^*) \neq 0$ and so $\mu_{e_m^*} = X^m$. Hence $({}^t u^i(e_m^*))_{0 \le i \le m-1}$ is a basis of H. Finally,

$$\dim(H) = m, \qquad \dim(H^\perp) = N - m.$$

Step 2.3. $F \cap H^\perp = \{0\}$. Let z be in $F \cap H^\perp$. Step 2.2 ensures that since $z \in F$ it reads

$$z = \sum_{i=0}^{m-1} \alpha_i u^i(x) = \sum_{i=1}^{m} \alpha_i e_i.$$

Moreover since $z \in H^\perp$, for $0 \le k \le m-1$

$$\begin{aligned} 0 &= \langle {}^t u^{m-1-k}(e_m^*), z \rangle \\ &= \sum_{i=0}^{k} \alpha_i \langle e_m^*, u^{m-k+i-1}(x) \rangle \\ &= \sum_{j=m-k}^{m} \alpha_{j-m+k} \langle e_m^*, u^{j-1}(x) \rangle \\ &= \alpha_k \langle e_m^*, e_m \rangle = \alpha_k. \end{aligned}$$

Hence $z = 0$ and $F \cap H^\perp = \{0\}$.

Steps 2.2, 2.3, ensure that $\mathbb{R}^N = F \bigoplus H^\perp$ and Step 2.1 ensures that F and $H^\perp$ are stable under u and s.

(b):. The proof of (b) follows from (a) by ordering the vectors of the basis in an appropriate way. The determination of (p_k, q_k) is made by inverse induction counting the numbers of eigenvalues of s equal to ± 1 in the space $F_k := \mathrm{im}(u^{k-1}) \cap \ker(u)$. □

4. Analytic description of periodic orbits bifurcating from a pair of simple purely imaginary eigenvalues

4.1 Real periodic orbits: explicit form

When studying the existence of homoclinic connections to periodic orbits, a preliminary task is to study the existence of periodic orbits. We are more precisely interested in the dynamics of one parameter families of vector fields $V(X,\mu)$ near a resonant fixed point placed at the origin, such that even after bifurcation, the differential $DV(0,\mu)$ still admits a pair of simple, purely imaginary eigenvalues. Indeed, in such families of vector fields, there coexist an oscillatory part with frequency of order one induced by the pair of simple eigenvalues and a "slow" hyperbolic part induced by a set of hyperbolic eigenvalues with small real parts and, as explained in the introduction, this configuration of spectrum leads to exponentially small phenomena which are the object of this book.

So this chapter is devoted to the existence of reversible periodic orbits bifurcation from a pair of purely imaginary eigenvalues. The following theorem due to Devaney ensures, for reversible vector fields, the existence of such periodic orbits provided that some "non resonance criteria" hold.

Theorem 4.1.1 (Devaney). *Let V be a smooth reversible vector field in $\mathbb{R}^{2n}$ such that the origin is a fixed point. Assume that the spectrum of the differential $DV(0)$ at the origin reads $\{\pm i\omega,\ \pm\lambda_2,\cdots,\pm\lambda_n\}$ where $\pm i\omega$, $\omega>0$ is a pair of simple purely imaginary eigenvalues. Moreover assume that none of the ratio $\lambda_k/i\omega$ is an integer. Then, the equation*

$$\frac{dX}{d\tau}=V(X)$$

admits near the origin a smooth one-parameter family of T_k-periodic orbits P_k with $k\in[0,k_0]$ such that

$$\lim_{k\to 0}\left(\sup_{\tau\in\mathbb{R}}(|P_k(\tau)|)\right)=0,\qquad \lim_{k\to 0}(T_k)=\frac{2\pi}{\omega}.$$

Moreover, if the vector field is analytic then P_k depends also analytically on k.

The proof of this theorem can be found in [De76] or in [Se86]. An analogous theorem due to Lyapunov holds for Hamiltonian system. It can be found in [SM71]. In both cases, the proof is based on the Implicit Functions Theorem.

When one wants to apply the above theorem to reversible families $V(X,\mu)$ of vector fields admitting a resonance at the origin such that the spectrum of $DV(0,\mu)$ admits a pair of simple purely imaginary eigenvalues, a first difficulty is that the theorem does not always apply for $\mu = 0$. For instance, it does not apply to the $0^2 i\omega$ resonance for $\mu = 0$, and it applies to the $(i\omega_0)^2 i\omega_1$ resonance for $\mu = 0$ if and only if ω_0/ω_1 is not an integer.

Even if it can be applied for $\mu > 0$, a second difficulty arises when complexifying time. Indeed, just after bifurcation, the spectrum of the differential at the origin $DV(0,\mu)$ can be split in three parts

P 1. A pair of purely imaginary eigenvalues $\pm i\omega(\mu)$, with $\omega(\mu) > 0$,

P 2. A set of eigenvalues which are very close to the imaginary axis. For instance $\pm\sqrt{\mu}$ for the $0^2 i\omega$ resonance, and $\pm\sqrt{\mu} \pm i\omega_0$ for the $(i\omega_0)^2 i\omega_1$ resonance.

P 3. A set of eigenvalues $\lambda_2, .., \lambda_p$ bounded away from the imaginary axis, i.e. $\mathcal{R}e\,(\lambda_k(\mu)) > \delta > 0$ holds for any k and any μ close to 0.

So, Devaney's Theorem 4.1.1 always applies for $\mu > 0$ and, for analytic vector fields it ensures the existence of a one parameter family of reversible periodic orbits of the form

$$P_{k,\mu}(\tau) = \sum_{n \geq 1} k^n p_n(\tau,\mu) \tag{4.1}$$

where p_n is a periodic function. However, the behavior of $p_n(\tau,\mu)$ when μ tends to 0 cannot be obtained by this theorem when it does not apply for $\mu = 0$. This leads to a second difficulty when complexifying time : when studying the existence of homoclinic connections to exponentially small periodic orbits, using the exponential tools developed in chapter 2, we need to extend and to control the size of such periodic orbits in a complex strip of the form $\mathcal{B}_{c/\sqrt{\mu}} = \{\xi \in \mathbb{C},\ |\mathcal{I}m\,(\xi)\,| < c/\sqrt{\mu}\}$ with $c > 0$. For that purpose, since $\cos(i\omega\ell/\sqrt{\mu}) \underset{\mu\to 0}{\longrightarrow} +\infty$ for any $\ell > 0$, we need to know very precisely how $p_n(\tau,\mu)$ depends on τ and μ. So we need a more precise theorem than Devaney's one which gives this dependence.

This dependence can be obtained by redoing the proof of existence of periodic orbits with the method of undetermined coefficients with majorization of the power series rather than with a direct application of the analytic implicit function theorem as for Devaney's theorem 4.1.1.

Theorem 4.1.2 below gives such a dependence. It is stated in the form under which we will use it in chapters 7 and 9, i.e. with rescaled time and parameter (for instance $t = \sqrt{\mu}\tau$ and $\nu = \sqrt{\mu}$ for the $0^2 i\omega$ and the $(i\omega_0)^2 i\omega_1$ resonances).

Theorem 4.1.2. *Let us consider the reversible system*

$$\frac{dY}{dt} = L_\nu Y + Q(Y,\nu) \tag{4.2}$$

where $Y = (A, B, Y_{\rm h}) \in \mathbb{R}^{2+d}$, $\nu \in]0, \nu_0]$ *and*

(i) $L_\nu = \begin{bmatrix} 0 & -\frac{\omega_{0,\nu}}{\nu} & 0 \\ \frac{\omega_{0,\nu}}{\nu} & 0 & 0 \\ 0 & 0 & \frac{L_{\rm h}(\nu)}{\nu} \end{bmatrix}$ *where* $\omega_{0,\nu} = \omega + \nu\ a(\nu)$,

with $m_0 = \inf_{\nu\in]0,\nu_0]} \omega_{0,\nu} > 0$, $|a|_0 = \sup_{\nu\in]0,\nu_0]} |a(\nu)| < +\infty$ *and where* $L_{\rm h}(\nu)$ *is a* $d \times d$ *matrix satisfying for every* $\nu \in]0, \nu_0]$

$$\|L_{\rm h}(\nu)\| \le M_{\rm h}, \tag{4.3}$$

$$\left\| \left({\rm Id} - \exp\left(\frac{2\pi L_{\rm h}(\nu)}{\omega_{0,\nu}} \right) \right)^{-1} \right\| \le \frac{C_{\rm h}}{\nu}. \tag{4.4}$$

(ii) Q *satisfies* $Q(0,\nu) = D_Y Q(0,\nu) = 0$; *there exists* $r_0 > 0$ *such that* $Y \mapsto Q(Y,\nu)$ *is analytic in* $D(0,r_0) := \{Y \in \mathbb{C}^{2+d}, |Y| < r_0\}$, *for all* $\nu \in]0, \nu_0]$ *and*

$$\sup_{|Y|<r_0,\ \nu\in]0,\nu_0]} |Q(Y,\nu)| < +\infty.$$

Then, there exists $k_0 > 0$ *such that for all* ν *in* $]0, \nu_0]$ *the system* (4.2) *admits a one parameter family of reversible periodic solutions* $(Y_{k,\nu}(t))_{k\in[0,k_0]}$ *with*

$$Y_{k,\nu}(t) = \widehat{Y}(\underline{\omega}_{k,\nu} t, k, \nu)$$

and

$$\text{(a)} \begin{cases} \widehat{Y}(s,k,\nu) = \sum_{n\ge1} k^n\ \widehat{Y}_n(s,\nu), \\ \widehat{Y}_1 = (\cos(s), \sin(s), 0), \quad \widehat{Y}_n \in \tilde{\mathcal{P}}_{n,R}, \text{ for } n \ge 2 \\ (k_0)^n \left\| \widehat{Y}_n \right\|_{\mathcal{P}1} \le C_{\widehat{Y}} \text{ for every } \nu \in]0,\nu_0], n \ge 1. \end{cases}$$

with $\tilde{\mathcal{P}}_{n,R} = \Big\{ \widehat{Y} : \mathbb{R} \mapsto \mathbb{R}^{2+d},\ \widehat{Y}(s) = a_0 + \sum_{p=1}^{n} a_p \cos(ps) + b_p \sin(ps),$

$$Sa_p = a_p,\ Sb_p = -b_p \Big\} \cap \left\{ \widehat{Y} /\ \int_0^{2\pi} \langle \widehat{Y}(s), \widehat{Y}_1(s) \rangle ds = 0 \right\}$$

and $\left\| \widehat{Y} \right\|_{\mathcal{P}1} := \sup_{s\in[0,2\pi]} |\widehat{Y}(s)| + \sup_{s\in[0,2\pi]} \left| \frac{d\widehat{Y}}{ds}(s) \right|.$

(b) $$\begin{cases} \underline{\omega}_{k,\nu} = \dfrac{\omega_{0,\nu}}{\nu} + \widehat{\omega}(k,\nu), \\ \widehat{\omega}(k,\nu) = \sum_{n\geq 2} \widehat{\omega}_{2n}(\nu)\; k^{2n} \quad \textit{for every } k \in [0,k_0],\ \nu \in]0,\nu_0], \\ |\widehat{\omega}(k,\nu)| \leq C_\omega k^2 \textit{ for every } \nu \in]0,\nu_0],\ k \in [0,k_0]. \end{cases}$$

(c) $\widehat{\omega}(-k,\nu) = \widehat{\omega}(k,\nu), \qquad \widehat{Y}(s+\pi,-k,\nu) = \widehat{Y}(s,k,\nu).$

(d) *If we assume that there exists* $\alpha \geq 0$ *such that*

$$\sup_{|Y|<r_0,\ \nu\in]0,\nu_0]} |\nu^{-\alpha} Q_A(Y,\nu)| < +\infty, \qquad \sup_{|Y|<r_0,\ \nu\in]0,\nu_0]} |\nu^{-\alpha} Q_A(Y,\nu)| < +\infty \quad (4.5)$$

where $Q = (Q_A, Q_B, Q_\mathrm{h})$, *then we get that* $|\widehat{\omega}(k,\nu)| \leq C_\omega k^2 \nu^\alpha$.

The detailed proof of this theorem is given in Appendix 4.A.

Remark 4.1.3. Observe that no hypothesis on the smoothness of L_ν and $Q(.,\nu)$ with respect to ν is made.

Remark 4.1.4. In Devaney's Theorem 4.1.1 the "non resonance criteria" on the eigenvalues comes from the fact that for building the family of periodic orbits by the Implicit Function Theorem, one needs to invert the matrix $\mathrm{Id} - \exp(2\pi L_\mathrm{h}/\omega)$ where L_h is the restriction of $DV(0)$ to the direct sum of the eigenspaces and generalized eigenspaces corresponding to the eigenvalues $\pm\lambda_2, \cdots, \pm\lambda_n$. Here we do not assume that the non resonance criteria holds for $\nu = 0$, but hypothesis (4.4) controls the size of

$$\left(\mathrm{Id} - \mathrm{e}^{\frac{2\pi L_\mathrm{h}(\nu)}{\omega_{0,\nu}}}\right)^{-1}$$

when ν tends to 0.

Remark 4.1.5. Observe that $C_{\widehat{Y}}$, C_ω are independent of k, ν. We have constructed a particular family of reversible periodic solutions $Y_{k,\nu}$ of (4.2). Since (4.2) is autonomous, we can deduce from $Y_{k,\nu}$, other periodic solutions by an arbitrary shift in time. Among these periodic solutions, $Y_{k,\nu}(t)$ and $Y_{k,\nu}(t + \pi/\underline{\omega}_{k,\nu})$ are the only ones which are reversible. .

Remark 4.1.6. The Theorem of classification of reversible matrices ensures that S reads $S : (A, B, Y_\mathrm{h}) \mapsto (A, -B, S_\mathrm{h} Y_\mathrm{h})$ where $S_\mathrm{h} \in GL_d(\mathbb{R})$ is a symmetry which anticommutes with $L_\mathrm{h}(\nu)$ and $Q_\mathrm{h}(\cdot,\nu)$.

4.2 Complexification of the periodic orbits

The description of the periodic orbits given by the above theorem, enables us to complexify time: for each n, $\widehat{Y}_n$ is a trigonometric polynomial of degree n. Thus it can be extended by analyticity to obtain an entire function. It remains now to find the radius of convergence of the power series $\sum k^n|\widehat{Y}_n(\xi)|$ for $\xi \in \mathbb{C}$. The following proposition gives a lower bound of this radius.

Proposition 4.2.1. *There exists $C'_{\widehat{Y}}$ such that*

$$k^n \left| \frac{d^j \widehat{Y}_n}{d\xi^j}(\xi,\nu) \right| \leq C'_{\widehat{Y}} n^j \sqrt{n+1} \left(\frac{k}{k_0}\right)^n e^{n|\mathcal{I}m(\xi)|},$$

holds for every $j \geq 0$, $k \in [0,k_0]$, $\nu \in [0,\nu_0]$, $\xi \in \mathbb{C}$, and $n \geq 1$.

The proof of this proposition is given in appendix 4.B.

Proposition 4.2.1 ensures that taking $(k/k_0)e^{|\mathcal{I}m(\xi)|} < 1$, one can define $\widehat{Y}(\xi,k,\nu)$. Since $Y_{k,\nu}(\xi) = \widehat{Y}(\underline{\omega}_{k,\nu}\xi,k,\nu)$, $Y_{k,\nu}(\xi)$ is well defined if $(k/k_0)e^{\underline{\omega}_{k,\nu}|\mathcal{I}m(\xi)|} < 1$. But $\underline{\omega}_{k,\nu}$ has a "bad" dependence on ν ($\underline{\omega}_{k,\nu} = w/\nu + \cdots$). So one way to keep $(k/k_0)e^{\underline{\omega}_{k,\nu}|\mathcal{I}m(\xi)|} < 1$ is to keep ξ in a complex strip $\mathcal{B}_\ell = \{\xi \in \mathbb{C}/|\mathcal{I}m(\xi)| < \ell\}$ and to choose k in $[0, k_0 e^{-\ell\underline{\omega}_{k,\nu}}]$. Then, $Y_{k,\nu}(\xi)$ is well defined in $\mathcal{B}_\ell$, as well as its derivatives.

Theorem 4.1.2 and Proposition 4.2.1 will be used in the second part of this book to prove the existence of Homoclinic connections to exponentially small periodic orbits for the $0^2 i\omega$ and the $(i\omega_0)^2 i\omega_1$ resonances using the Exponential Lemmas 2.1.1, 2.2.1 which require complexification of time in complex strips $\mathcal{B}_\ell$.

4.3 Analytic conjugacy to circles

The very precise description given by Theorem 4.1.2 of the reversible periodic orbits bifurcating from a pair of simple purely imaginary eigenvalues, enables us to prove that there exists an analytic diffeomorphism Θ_ν close to identity, such that setting $Y = \Theta_\nu(Y')$ the resulting equation admits a one parameter family of reversible periodic orbits which are circles.

Theorem 4.3.1. *Under the hypothesis (i) and (ii) of Theorem 4.1.2, there exists $k_1 \in]0,k_0[$ and an analytic diffeomorphism Θ_ν close to identity such that setting $Y = \Theta_\nu(Y')$, the equation (4.2) is equivalent to*

$$\frac{dY'}{dt} = L_\nu Y' + Q'(Y',\nu) \tag{4.6}$$

with

(a) *For* $k \in]0, k_1[$, $\nu \in]0, \nu_0]$, *the periodic orbits* $Y_{k,\nu}$ *of (4.2) reads* $Y_{k,\nu} = \Theta_\nu(Y'_{k,\nu})$ *where* $Y'_{k,\nu}(t) = (k\cos(\underline{\omega}_{k,\nu}t), k\sin(\underline{\omega}_{k,\nu}t), 0)$ *is a periodic solution of (4.6);*

(b) $S\Theta_\nu(Y) = \Theta_\nu(SY)$;

(c) $\Theta_\nu(A', B', Y'_h) = (A', B', Y'_h) + \Theta^1_\nu(A', B')$,
where $\Theta^1_\nu : (] - k_1, k_1[)^2 \mapsto \mathbb{R}^{2+d}$ *satisfies*

$$|\Theta^1_\nu(A', B')| \leq C(|A'|^2 + |B'|^2), \quad |D\Theta^1_\nu(A', B')| \leq C(|A'|^2 + |B'|^2)^{\frac{1}{2}}$$

for $\nu \in]0, \nu_0]$, *and* $(|A'|^2 + |B'|^2)^{\frac{1}{2}} < k_1$.

(d) Q' *is analytic, reversible and satisfies* $Q'(0, \nu) = DQ'(0, \nu) = 0$.

The proof of this theorem is given in Appendix 4.C

4.A Appendix. Proof of Theorem 4.1.2

4.A.1 Rewriting of the system as an implicit equation

We start by rewriting System (4.2) as an implicit equation $\mathcal{G}_\nu(u, k) = 0$ (see Lemma 4.A.5) where $\mathcal{G}_\nu$ is holomorphic with respect to u, k, but not with respect to ν. Moreover, $\mathcal{G}_\nu$ is not defined for $\nu = 0$. We could apply for each given ν the Analytical Implicit Function Theorem at the point $(u, k) = (0, 0)$ to obtain a reversible holomorphic periodic solution $u(k, \nu)$, but we would not know how u depends on ν. Then, to determine the dependence of u on ν we use the method of undetermined coefficients with majorization of power series. So, we rewrite System (4.2) and we introduce Banach spaces to be able to apply the Analytical Implicit Function Theorem. First, using the analyticity of the vector field we check that the following lemma holds:

Lemma 4.A.1. *System* (4.2) *is equivalent to*

$$\frac{dY}{dt} = L_\nu Y + \sum_{n \geq 2} Q_n(Y^{(n)}, \nu), \tag{4.7}$$

with

(a) $Q_n(Y^{(n)}, \nu) = Q_n(\underbrace{Y, \cdots, Y}_{\text{n times}}, \nu)$
is a n-linear symmetric operator from $(\mathbb{R}^{2+d})^n$ *to* $\mathbb{R}^{2+d}$.

(b) *There exist* C_0, $r > 0$, *such that*

$$|Q_n(Y_1, \cdots, Y_n, \nu)| \leq C_0 \frac{|Y_1|}{r} \cdots \frac{|Y_n|}{r}$$

for every $\nu \in [0, \nu_0]$, $n \geq 2$, $Y_i \in \mathbb{R}^{2+d}$, $1 \leq i \leq n$.

(c) *If* Q *satisfies (4.5), then there exists* C_1 *such that*

$$|Q_n^A(Y_1, \cdots, Y_n, \nu)| \leq C_1 \nu^\alpha \frac{|Y_1|}{r} \cdots \frac{|Y_n|}{r},$$

$$|Q_n^B(Y_1, \cdots, Y_n, \nu)| \leq C_1 \nu^\alpha \frac{|Y_1|}{r} \cdots \frac{|Y_n|}{r},$$

for every $\nu \in [0, \nu_0]$, $n \geq 2$, $Y_i \in \mathbb{R}^{2+d}$, $1 \leq i \leq n$.

(d) *Reversibility properties read*

$$-SL_\nu = L_\nu S \qquad \text{and } Q_n(SY_1, \cdots, SY_n, \nu) = -SQ_n(Y_1, \cdots, Y_n, \nu).$$

for every $\nu \in [0, \nu_0]$, $n \geq 2$, $Y_i \in \mathbb{R}^{2+d}$, $1 \leq i \leq n$.

We look for periodic solutions, but we do not know the period. We expect it to be near $\frac{2\pi\nu}{\omega_{0,\nu}}$, which is the case for the periodic solutions of the linear system. To work in the set of 2π-periodic functions we set

$$s = \underline{\omega}_{k,\nu} t, \quad \underline{\omega}_{k,\nu} = \frac{\omega_{0,\nu}}{\nu} + \widehat{\omega}, \quad Y(t) = \widehat{Y}(s) = \widehat{Y}(\underline{\omega}_{k,\nu} t). \tag{4.8}$$

So, we now look for $(\widehat{Y}(s), \widehat{\omega})$ and we want to know how $\widehat{Y}(s)$ depends on ν.

We define all the Banach spaces needed to find the 2π-periodic solutions:

Definition 4.A.2.

$$\mathcal{P}^j_{2\pi} := \{f : \mathbb{R} \longrightarrow \mathbb{R} /\ 2\pi\text{-periodic of class } C^j\},$$

$$\mathcal{P}^1_R := \{\widehat{Y} /\ \widehat{Y} \in (\mathcal{P}^1_{2\pi})^{2+d}, \widehat{Y} \ \text{reversible}\},$$

$$\mathcal{P}^0_{AR} := \{\widehat{Y} /\ \widehat{Y} \in (\mathcal{P}^0_{2\pi})^{2+d}, \widehat{Y} \ \text{antireversible}\},$$

$$\tilde{\mathcal{P}}^1_R := \mathcal{P}^1_R \cap \{\widehat{Y} /\ \int_0^{2\pi} \langle \widehat{Y}(s), \widehat{Y}_1(s) \rangle ds = 0\},$$

$$\tilde{\mathcal{U}}^1_R := \{u = (\widehat{Y}, \widehat{\omega}) /\ \widehat{Y} \in \tilde{\mathcal{P}}^1_R, \widehat{\omega} \in \mathbb{R}\},$$

where $\widehat{Y}_1(s) = (\cos(s), \sin(s), 0)$. *We norm these spaces with*

$$\left\|\widehat{Y}\right\|_{\mathcal{P}^1} := \sup_{s\in[0,2\pi]} |\widehat{Y}(s)| + \sup_{s\in[0,2\pi]} \left|\frac{d\widehat{Y}}{ds}(s)\right|, \text{ for } \mathcal{P}^1_R,\ \tilde{\mathcal{P}}^1_R,$$

$$\left\|\widehat{Y}\right\|_{\mathcal{P}^0} := \sup_{s\in[0,2\pi]} |\widehat{Y}(s)|, \text{ for } \mathcal{P}^0_{AR},$$

$$\|u\|_{\mathcal{U}^1} := \left\|\widehat{Y}\right\|_{\mathcal{P}^1} + |\widehat{\omega}|, \text{ for } \tilde{\mathcal{U}}^1_R.$$

We define now finite-dimensional subspaces of the previously defined ones. These are the "trigonometric polynomial" subspaces (n indicates trigonometric polynomials of degree less than n):

Definition 4.A.3.

$$\mathcal{P}_n := \left\{\widehat{Y} : \mathbb{R} \mapsto \mathbb{R}^{2+d},\ \widehat{Y}(s) = a_0 + \sum_{p=0}^{n} a_p \cos(ps) + b_p \sin(ps), \right\}$$

$$\mathcal{P}_{n,R} := \left\{\widehat{Y} \in \mathcal{P}_n,\ Sa_p = a_p,\ Sb_p = -b_p\right\}$$

$$\mathcal{P}_{n,AR} := \left\{\widehat{Y} \in \mathcal{P}_n,\ Sa_p = -a_p,\ Sb_p = b_p\right\}$$

$$\tilde{\mathcal{P}}_{n,R} := \mathcal{P}_{n,R} \cap \{\widehat{Y} / \int_0^{2\pi} \langle \widehat{Y}(s), \widehat{Y}_1(s)\rangle ds = 0\},$$

$$\tilde{\mathcal{U}}_{n,R} := \tilde{\mathcal{P}}_{n,R} \times \mathbb{R}.$$

Now considering the form of one parameter families of reversible periodic solutions for the truncated system, we look for 2π-periodic solutions of the form

$$\widehat{Y}(s,k) = k\widehat{Y}_1(s) + k\widehat{Z}(s,k) \text{ with } \widehat{Z} \text{ in } \tilde{\mathcal{P}}^1_R. \tag{4.9}$$

Our aim is to expand $\widehat{Y}(s,k)$ in powers of k: $\widehat{Y}(s,k) = \sum_{n\geq 1} k^n \widehat{Y}_n(s)$. We choose for the first term $\widehat{Y}_1$ the periodic solution of the linear system. $\widehat{Z}$ is considered as a perturbation. We can now prove

Lemma 4.A.4. *For every* $\nu \in]0, \nu_0]$, k *satisfying* $|k| < \frac{1}{2}r$ *where* r *is introduced in Lemma* 4.A.1 *and* $\widehat{Z} \in \tilde{\mathcal{P}}^1_R$ *satisfying* $\left\|\widehat{Z}\right\|_{\mathcal{P}^1} < 1$, *System* (4.2) *is equivalent to*

$$\mathcal{L}_\nu(\widehat{Z},\widehat{\omega}) = -\widehat{\omega}\frac{d\widehat{Z}}{ds} + \sum_{p\geq 1} k^p \sum_{m=0}^{p+1} C_{p+1}^m \mathcal{Q}_{p+1}(\widehat{Y}_1^{(p+1-m)}, \widehat{Z}^{(m)}, \nu)$$

with

$$\mathcal{L}_\nu(\widehat{Z},\widehat{\omega}) = \frac{\omega_{0,\nu}}{\nu}\frac{d\widehat{Z}}{ds} - L_\nu \widehat{Z} + \widehat{\omega}\frac{d\widehat{Y}_1}{ds}, \quad C_p^m = \frac{p!}{m!(p-m)!}.$$

Proof. Using Lemma 4.A.1, we verify that for every k satisfying $|k| < \frac{r}{2}$ and $\widehat{Z} \in \tilde{\mathcal{P}}_R^1$ satisfying $\left\|\widehat{Z}\right\|_{\mathcal{P}^1} < 1$, the power series $\sum_{n\geq 2} k^n \mathcal{Q}_n((\widehat{Y}_1 + \widehat{Z})^{(n)}, \nu)$ is absolutely convergent since

$$\begin{aligned} |k^n \mathcal{Q}_n((\widehat{Y}_1 + \widehat{Z})^{(n)}, \nu)| &= |k^n \sum_{m=o}^{p+1} C_n^m \mathcal{Q}_n(\widehat{Y}_1^{(n-m)}, \widehat{Z}^{(m)}, \nu)\ | \\ &\leq k^n \sum_{m=o}^{n} C_n^m C_0 \frac{1}{r^n} = C_0 (\frac{2k}{r})^n. \end{aligned}$$

The conclusion of the proof follows. □

Finally, we can rewrite System (4.2) as the following implicit equation:

Lemma 4.A.5. *For every given $\nu \in]0, \nu_0]$, k satisfying $|k| < \dfrac{r}{2}$ and $u \in \tilde{\mathcal{U}}_R^1$ satisfying $\|u\|_{\mathcal{U}^1} < 1$, System (4.2) reads*

$$\mathcal{G}_\nu(k,u) = 0 \quad \textit{for } u = (\widehat{Z},\widehat{\omega}), \tag{4.10}$$

where

$$\mathcal{G}_\nu(k,u) = \sum_{p+q\geq 1} k^p \mathcal{G}_{pq}[u^{(q)}] = \mathcal{L}_\nu(u) + k\mathcal{G}_{10} + \sum_{p+q\geq 2} k^p \mathcal{G}_{pq}[u^{(q)}],$$

$$\mathcal{G}_{01}[u] = \mathcal{L}_\nu(u), \quad \mathcal{G}_{02}[u^{(2)}] = \widehat{\omega}\frac{d\widehat{Z}}{ds}, \quad \mathcal{G}_{0q}[u^{(q)}] = 0 \ \ \forall q \geq 3,$$

$$\forall p \geq 1, \ k^p \mathcal{G}_{pq}[u^{(q)}] = \begin{cases} -C_{p+1}^q k^p \mathcal{Q}_{p+1}(\widehat{Y}_1^{(p+1-q)}, \widehat{Z}^{(q)}, \nu) & if\, q \leq p+1, \\ 0 & if\, q \geq p+2. \end{cases}$$

Moreover,

(a) *for every $(p,q) \in \mathbb{N}^2$, $\mathcal{G}_{pq}$ is a q-linear operator from $(\tilde{\mathcal{U}}_R^1)^q$ to $\mathcal{P}_{AR}^0$,*

(b) *there exists* C_2 *such that*

$$\left\| k^p \mathcal{G}_{pq}[u_1, \cdots, u_q] \right\|_{\mathcal{P}^0} \leq C_2 \left\| u_1 \right\|_{\mathcal{U}^1} \cdots \left\| u_q \right\|_{\mathcal{U}^1} \left(\frac{2k}{r} \right)^p$$

holds for every $\nu \in]0, \nu_0]$, k *satisfying* $|k| < \frac{1}{2}r$, $(p,q) \neq (0,1)$, *and* $u_i \in \tilde{\mathcal{U}}^1_R$ *satisfying* $\left\| u_i \right\|_{\mathcal{U}^1} < 1$, $1 \leq i \leq q$.

Recall that we look for $u = (\widehat{Z}, \widehat{\omega})$ and that we want to know how u depends on ν. As explained earlier, our strategy is now to apply Cauchy's method of undetermined coefficients with majorization of the power series to obtain an explicit solution. We proceed in three steps:

- As usual, we start by studying $\mathcal{L}_\nu = \frac{\partial \mathcal{G}}{\partial u}(0,0)$.
- We look for u in the form $u(s,k,\nu) = \sum_{n\geq 1} k^n u_n(s,\nu)$. We put $\sum_{n \geq 1} k^n u_n$ in (4.10) and we identify the coefficients of the powers of k. This step gives the necessary form of $(u_n)_{n\geq 1}$ (u_n is obtained by induction).
- Finally, using a majorizing power series, we find a lower bound of the radius of convergence of the power series $\sum_{n\geq 1} k^n u_n$ where u_n has been obtained at the previous step.

4.A.2 Study of $\mathcal{L}_\nu$

Lemma 4.A.6.

(a) *For all* ν *in* $]0, \nu_0]$, $\mathcal{L}_\nu$ *is a bounded operator from* $\tilde{\mathcal{U}}^1_R$ *to* $\mathcal{P}^0_{AR}$.
(b) *There exists* C_3, *such that* $|||\mathcal{L}_\nu||| \leq C_3/\nu$ *for every* $\nu \in]0, \nu_0]$.
(c) *For all* ν *in* $]0, \nu_0]$, $\mathcal{L}_\nu$ *is an isomorphism from* $\tilde{\mathcal{U}}^1_R$ *to* $\mathcal{P}^0_{AR}$.

Proof. (a),(b). For all $\widehat{Z}$ in $\tilde{\mathcal{P}}^1_R$, the functions $d\widehat{Y}_1/ds$, $d\widehat{Z}/ds$, $L_\nu \widehat{Z}$ are antireversible. Thus $\mathcal{L}_\nu(u)$ is antireversible. Moreover,

$$\left\| \mathcal{L}_\nu(u) \right\|_{\mathcal{P}^0} \leq \frac{2|\omega_{0,\nu}| + \|L_{\mathrm{h}}\|)}{\nu} \left\| u \right\|_{\mathcal{U}^1} \leq \frac{2(\omega + |a|_0 \nu_0 + M_{\mathrm{h}})}{\nu} \left\| u \right\|_{\mathcal{U}^1}.$$

(c): Let $f \in \mathcal{P}^0_{AR}$. We look for u in $\tilde{\mathcal{U}}^1_R$ satisfying $\mathcal{L}_\nu(u) = f$, which is equivalent to

$$\widehat{Z}(s) = \mathrm{e}^{s\nu L_\nu/\omega_{0,\nu}} \widehat{Z}(0) + \int_0^s \mathrm{e}^{(s-\tau)\nu L_\nu/\omega_{0,\nu}} \left[\frac{\nu f(\tau)}{\omega_{0,\nu}} - \frac{\nu \widehat{\omega}}{\omega_{0,\nu}} \frac{d\widehat{Y}_1}{ds} \right] d\tau. \quad (4.11)$$

$\widehat{Z}$ is 2π-periodic if and only if $\widehat{Z}(0) = \widehat{Z}(2\pi)$, which is equivalent to

$$\left[\mathrm{Id} - \mathrm{e}^{2\pi\nu L_\nu/\omega_{0,\nu}}\right] \widehat{Z}(0) = \int_0^{2\pi} \mathrm{e}^{(2\pi-\tau)\frac{\nu L_\nu}{\omega_{0,\nu}}} \left[\frac{\nu f(\tau)}{\omega_{0,\nu}} - \frac{\nu\widehat{\omega}}{\omega_{0,\nu}}\frac{d\widehat{Y}_1}{ds}\right] d\tau. \quad (4.12)$$

Computing explicitly $\mathrm{e}^{sL_\nu/\omega_{0,\nu}}$, we find that (4.12) is equivalent to

$$0 = \frac{\nu}{\omega_{0,\nu}}\int_0^{2\pi} \left[\cos(\tau)f_A(\tau) + \sin(\tau)f_B(\tau)\right] d\tau, \quad (4.13)$$

$$0 = \frac{\nu}{\omega_{0,\nu}}\int_0^{2\pi} \left[-\sin(\tau)f_A(\tau) + \cos(\tau)f_B(\tau)\right] d\tau - \frac{2\pi\nu\widehat{\omega}}{\omega_{0,\nu}} \quad (4.14)$$

$$\widehat{Z}_\mathrm{h}(0) = \left(\mathrm{Id} - \mathrm{e}^{\frac{2\pi L_\mathrm{h}(\nu)}{\omega_{0,\nu}}}\right)^{-1} \int_0^{2\pi} \mathrm{e}^{\frac{(2\pi-s)L_\mathrm{h}(\nu)}{\omega_{0,\nu}}} \frac{\nu f_\mathrm{h}(s)}{\omega_{0,\nu}} ds. \quad (4.15)$$

Equation (4.14) gives $\widehat{\omega}$ and equation (4.15) gives $\widehat{Z}_\mathrm{h}(0)$. Equation (4.13) is automatically satisfied because f lies in $\mathcal{P}^0_{AR}$. Thus f is 2π-periodic; f_A is odd; f_B is even.

Now, $\widehat{Z}$ is reversible if and only if $\widehat{Z}(0)$ reads $\widehat{Z}(0) = (\widehat{A}(0), 0, \widehat{Z}_\mathrm{h}(0))$ where $S_\mathrm{h}\widehat{Z}_\mathrm{h}(0) = \widehat{Z}_\mathrm{h}(0)$. So we must check that the necessary value of $\widehat{Z}_\mathrm{h}$ given by (4.15) satisfies this last condition. For that purpose, we use that S_h anticommutes with $L_\mathrm{h}(\nu)$ and f_h is antireversible with respect to S_h. We get

$$\begin{aligned} S_\mathrm{h}\widehat{Z}_\mathrm{h}(0) &= -\left(\mathrm{Id} - \mathrm{e}^{\frac{-2\pi L_\mathrm{h}(\nu)}{\omega_{0,\nu}}}\right)^{-1} \int_0^{2\pi} \mathrm{e}^{\frac{(s-2\pi)L_\mathrm{h}(\nu)}{\omega_{0,\nu}}} \frac{\nu f_\mathrm{h}(-s)}{\omega_{0,\nu}} ds, \\ &= \left(\mathrm{Id} - \mathrm{e}^{\frac{2\pi L_\mathrm{h}(\nu)}{\omega_{0,\nu}}}\right)^{-1} \int_0^{2\pi} \mathrm{e}^{\frac{sL_\mathrm{h}(\nu)}{\omega_{0,\nu}}} \frac{\nu f_\mathrm{h}(-s)}{\omega_{0,\nu}} ds, \\ &= \left(\mathrm{Id} - \mathrm{e}^{\frac{2\pi L_\mathrm{h}(\nu)}{\omega_{0,\nu}}}\right)^{-1} \int_0^{2\pi} \mathrm{e}^{\frac{(2\pi-s)L_\mathrm{h}(\nu)}{\omega_{0,\nu}}} \frac{\nu f_\mathrm{h}(s)}{\omega_{0,\nu}} ds = \widehat{Z}_\mathrm{h}(0). \end{aligned}$$

Finally, the last condition to be satisfied is $\int_0^{2\pi} \langle \widehat{Z}(s), \widehat{Y}_1(s)\rangle ds = 0$ which is equivalent to

$$2\pi\widehat{A}(0) = -\int_0^{2\pi}\int_0^s \langle \mathrm{e}^{\frac{(s-\tau)L_\nu}{\omega_{0,\nu}}} g(\tau)d\tau,\ \widehat{Y}_1(s)\rangle ds \ \text{ with } g(\tau) = \frac{\nu f(\tau)}{\omega_{0,\nu}} - \frac{\nu\widehat{\omega}}{\omega_{0,\nu}}\frac{d\widehat{Y}_1}{ds}.$$

Computing explicitly the exponential and the scalar product we get

$$\begin{aligned}2\pi\widehat{A}(0) &= -\int_0^{2\pi}\int_0^s (g_A(\tau)\cos\tau + g_B(\tau)\sin\tau)\, d\tau ds,\\ &= -\int_0^{2\pi} (2\pi-\tau)(g_A(\tau)\cos\tau + g_B(\tau)\sin\tau)\, d\tau,\\ &= \int_0^{2\pi} \tau(g_A(\tau)\cos\tau + g_B(\tau)\sin\tau)\, d\tau\end{aligned}$$

since g_A is odd and g_B even and both are 2π periodic. Finally, since $\langle \frac{d\widehat{Y}_1}{ds}, \widehat{Y}_1\rangle = 0$,

$$\widehat{A}(0) = \frac{1}{2\pi}\int_0^{2\pi} \tau\langle g(\tau), \widehat{Y}_1(\tau)\rangle d\tau = \frac{\nu}{2\pi\omega_{0,\nu}}\int_0^{2\pi} \tau\langle f(\tau), \widehat{Y}_1(\tau)\rangle d\tau.$$

This equation gives $\widehat{A}(0)$. So, there exists a unique u in $\tilde{\mathcal{U}}_R^1$ satisfying $\mathcal{L}_\nu(u) = f$. □

The proof of Lemma 4.A.6 gives the explicit form of $\mathcal{L}_\nu^{-1}$ which is rewritten in the next lemma.

Lemma 4.A.7. *Let $f \in \mathcal{P}_{AR}^0$, $f = (f_A, f_B, f_{\rm h})$. Set $u = (\widehat{Z}, \widehat{\omega}) = \mathcal{L}_\nu^{-1}(f)$, with $\widehat{Z} = (\widehat{A}, \widehat{B}, \widehat{Z}_{\rm h}) \in \tilde{\mathcal{U}}_R^1$. Then*

$$\widehat{\omega} = \frac{1}{2\pi}\int_0^{2\pi} \langle f(s), \frac{d\widehat{Y}_1}{ds}(s)\rangle ds,$$

Moreover,

$$\begin{aligned}\widehat{A}(s) &= \widehat{A}(0)\cos(s) + \widetilde{A}(s),\\ \widehat{B}(s) &= \widehat{A}(0)\lambda\sin(s) + \widetilde{B}(s),\\ \widehat{Z}_{\rm h}(s) &= e^{sL_{\rm h}(\nu)/\omega_{0,\nu}}\widehat{Z}_{\rm h}(0) + \tfrac{\nu}{\omega_{0,\nu}}\int_0^s e^{(s-\tau)L_{\rm h}(\nu)/\omega_{0,\nu}} f_{\rm h}(\tau)\, d\tau.\end{aligned}$$

where

$$\widetilde{A}(s) = \frac{\nu}{\omega_{0,\nu}}\int_0^s [\cos(s-\tau)f_A(\tau) - \sin(s-\tau)f_B(\tau)]\, d\tau + \frac{\widehat{\omega}\nu}{\omega_{0,\nu}} s\sin(s),$$

$$\widetilde{B}(s) = \frac{\nu}{\omega_{0,\nu}}\int_0^s [\sin(s-\tau)f_A(\tau) + \cos(s-\tau)f_B(\tau)]\, d\tau - \frac{\widehat{\omega}\nu}{\omega_{0,\nu}} s\cos(s),$$

$$\widehat{A}(0) = \frac{\nu}{2\pi\omega_{0,\nu}}\int_0^{2\pi} \tau\langle f(\tau), \widehat{Y}_1(\tau)\rangle d\tau,$$

and

$$\widehat{Z}_{\rm h}(0) = \frac{\nu}{\omega_{0,\nu}} \left(\mathrm{Id} - \mathrm{e}^{\frac{2\pi L_{\rm h}(\nu)}{\omega_{0,\nu}}} \right)^{-1} \int_0^{2\pi} \mathrm{e}^{\frac{(2\pi - s)L_{\rm h}(\nu)}{\omega_{0,\nu}}} \frac{\nu f_{\rm h}(s)}{\omega_{0,\nu}} f_{\rm h}(s) ds.$$

Now using the explicit form of $\mathcal{L}_\nu^{-1}$ we can prove

Lemma 4.A.8. *There exists C_4 such that $\left\|\mathcal{L}_\nu^{-1}(f)\right\|_{\mathcal{U}^1} \leq C_4 \|f\|_{\mathcal{P}^0}$ for every $\nu \in]0, \nu_0]$, and $f \in \mathcal{P}^0_{AR}$.*

Proof. Denote $u = (\widehat{Z}, \widehat{\omega}) = \mathcal{L}_\nu^{-1}(f)$. Firstly, Lemma 4.A.7 ensures that $|\widehat{\omega}| \leq \|f\|_{\mathcal{P}^0}$. Secondly, using Lemma 4.A.7 and the assumptions on $L_{\rm h}(\nu)$ and $\omega_{0,\nu}$ given in Theorem 4.1.2-(i) we check that there exists C such that $\left\|\widehat{Z}\right\|_{\mathcal{P}^0} \leq C \|f\|_{\mathcal{P}^0}$ holds for $\nu \in]0, \nu_0]$. Now, to estimate $\left\|\frac{d\widehat{Z}}{ds}\right\|_{\mathcal{P}^0}$, recall that

$$\mathcal{L}_\nu(u) = f \iff \frac{d\widehat{Z}}{ds} = \frac{\nu L_\nu}{\omega_{0,\nu}} \widehat{Z} - \frac{\nu \widehat{\omega}}{\omega_{0,\nu}} \frac{d\widehat{Y}_1}{ds} + \frac{\nu f}{\omega_{0,\nu}}.$$

The conclusion of the proof follows. □

To conclude the study of $\mathcal{L}_\nu$ we study the image of the "trigonometric polynomial subspaces" under $\mathcal{L}_\nu$. More precisely,

Lemma 4.A.9. *$\mathcal{L}_\nu$ is an isomorphism from $\tilde{\mathcal{U}}_{n,R}$ to $\mathcal{P}_{n,AR}$.*

Proof. For all $u = (\widehat{Z}, \widehat{\omega})$ in $\tilde{\mathcal{P}}_{n,R}$, $\frac{d\widehat{Z}}{ds}$, $L_\nu \widehat{Z}$, $\frac{d\widehat{Y}_1}{ds}$ lie in $\mathcal{P}_{n,AR}$. Thus $\mathcal{L}_\nu(u)$ lies in $\mathcal{P}_{n,AR}$. Moreover $\mathcal{L}_\nu$ is injective in $\tilde{\mathcal{U}}^1_R \supset \tilde{\mathcal{U}}_{n,R}$. So $\mathcal{L}_\nu$ is injective from $\tilde{\mathcal{U}}_{n,R}$ to $\mathcal{P}_{n,AR}$. To obtain that $\mathcal{L}_\nu$ is surjective from $\tilde{\mathcal{U}}_{n,R}$ to $\mathcal{P}_{n,AR}$ we use a dimensional argument. Observe that 0 is not an eigenvalue of L_ν. Then, the theorem of classification of reversible matrices 3.1.10, ensures that d is even and that

$$\dim(\ker(S - \mathrm{Id})) = \dim(\ker(S + \mathrm{Id})) = \frac{d+2}{2}.$$

Then,

$$\dim(\mathcal{P}_{n,R}) = \dim(\mathcal{P}_{n,AR}) = \frac{2+d}{2}\,(2n+1).$$

The map $\widehat{Z} \longrightarrow \int_0^{2\pi} \langle \widehat{Z}, \widehat{Y}_1 \rangle$ is a linear form in $\mathcal{P}_{n,R}$ which is not identically equal to zero because $\widehat{Y}_1$ lies in $\mathcal{P}_{n,R}$. Thus $\dim(\tilde{\mathcal{P}}_{n,R}) = \dim(\mathcal{P}_{n,R}) - 1$.

Finally $\dim(\tilde{\mathcal{U}}_{n,R}) = \dim(\tilde{\mathcal{P}}_{n,R} \times \mathbb{R}) = \dim \mathcal{P}_{n,R} = \dim(\mathcal{P}_{n,AR})$. So, $\mathcal{L}_\nu$ is an isomorphism from $\tilde{\mathcal{U}}_{n,R}$ to $\mathcal{P}_{n,AR}$. □

4.A.3 Expansion in powers of k

Recall that we look for a solution $u = (\widehat{Z}, \widehat{\omega})$ of (4.2) rewritten as $\mathcal{G}_\nu(k,u) = 0$. (See (4.10)). We look for u in the form $u(s,k,\nu) = \sum_{n\geq 1} k^n u_n(s,\nu)$. More precisely, we look for $k_0 < \frac{r}{2}$, $(\widehat{Z}_n(s,\nu))_{n\geq 1} \in (\tilde{\mathcal{P}}_{n,R})^{\mathbb{N}^*}$ and $(\widehat{\omega}_n(\nu))_{n\geq 1} \in (\mathbb{R})^{\mathbb{N}^*}$ such that

(a) $\sum_{n\geq 1} \left\| \widehat{Z}_n(s,\nu) \right\|_{\mathcal{P}^1} |k|^n$ converges for every $k \in [-k_0, k_0]$, and $\nu \in]0, \nu_0]$.

(b) $\sum_{n\geq 1} |\widehat{\omega}_n(\nu)||k|^n$ converges for every $k \in [-k_0, k_0]$, and $\nu \in]0, \nu_0]$.

(c) $\|u\|_{\mathcal{U}^1} < 1$ where $u = \sum_{n\geq 1} k^n u_n$ and $u_n = (\widehat{Z}_n, \widehat{\omega}_n)$.

(d) $\mathcal{G}_\nu(k, u(s,k,\nu)) = 0$.

We proceed in two steps:

- We put $\sum_{n\geq 1} k^n u_n$ in (4.10) and we identify the coefficients of powers of k. This step gives the necessary form of $(u_n)_{n\geq 1}$. (u_n is obtained by induction in Lemma 4.A.10).
- Then, using a majorizing power series, we find a lower bound for the radius of convergence of the power series $\sum_{n\geq 1} k^n u_n$ where the u_n have been obtained at the previous step (Proposition 4.A.11).

4.A.3.1 Identification of the coefficients. Proceeding just as explained and using the form of $\mathcal{G}_\nu$ given in Lemma 4.A.5 we obtain

Lemma 4.A.10. *$\mathcal{G}_\nu(k,u) = 0$ has at most one solution given by $u = \sum_{n\geq 1} k^n u_n$ where $u_n \in \tilde{\mathcal{U}}^1_R$. The sequence $(u_n)_{n\geq 1}$ is given by induction:*

$$u_1 = -\mathcal{L}_\nu^{-1}(Q_2(\widehat{Y}_1, \widehat{Y}_1, \nu)), \tag{4.16}$$

$$u_n = -\mathcal{L}_\nu^{-1} \sum_{\substack{p+q\geq 2 \\ p\leq n \\ q\leq n}} \sum_{\substack{i_1+\ldots+i_q=n-p \\ i_l\geq 1}} \mathcal{G}_{pq}[u_{i_1}, \ldots, u_{i_q}]). \tag{4.17}$$

The formal power series $u = \sum_{n\geq 1} k^n u_n$ is a true solution if and only if its radius of convergence is not equal to 0 and if $\|u\|_{\mathcal{U}^1} < 1$.

4.A.3.2 Majorizing power series. Using Lemmas 4.A.8 and 4.A.5-(b) we check that

$$u_1 = \mathcal{L}_\nu^{-1}(\mathcal{G}_{10}) \quad \Rightarrow |k| \left\|u_1\right\|_{\mathcal{U}^1} \leq C_4 C_2 \frac{2|k|}{r},$$

$$|k|^n \left\|u_n\right\|_{\mathcal{U}^1} \leq C_4 C_2 \sum_{\substack{p+q\geq 2\\ p\leq n\\ q\leq n}} \sum_{\substack{i_1+\ldots+i_q=n-p\\ i_l\geq 1}} (\frac{2|k|}{r})^p |k|^{i_1} \left\|u_{i_1}\right\|_{\mathcal{U}^1} \cdots |k|^{i_q} \left\|u_{i_q}\right\|_{\mathcal{U}^1}.$$

Now we replace inequalities by equalities and denote $\left\|u_n\right\|_{\mathcal{U}^1}$ by γ_n, $\eta = \sum_{n\geq 1} \gamma_n |k|^n$; we obtain a majorizing power series, the coefficients of which are given by induction. Let M be C_2C_4. M is independent of ν and

$$|k|\gamma_1 = M\frac{2|k|}{r},$$

$$|k|^n \gamma_n = M \sum_{\substack{p+q\geq 2\\ p\leq n\\ q\leq n}} \sum_{\substack{i_1+\ldots+i_q=n-p\\ i_l\geq 1}} (\frac{2|k|}{r})^p |k|^{i_1}\gamma_{i_1} \cdots |k|^{i_q}\gamma_{i_q}.$$

Thus η formally satisfies the equation

$$\eta = M\left[\frac{2|k|}{r} + \sum_{p+q\geq 2} (\frac{2|k|}{r})^q \eta^p\right].$$

But when $|k| < \frac{r}{2}$ and $\eta < 1$ this equation reads

$$\eta = \frac{M}{(1-\frac{2|k|}{r})(1-\eta)} - M - M\eta. \tag{4.18}$$

This equation must have a solution $\eta = \eta(|k|)$ satisfying

(i) $\eta(0) = 0$,
(ii) $\eta(|k|)$ is defined for $|k| \in [0, k_0]$ with $k_0 < \frac{r}{2}$,
(iii) $|\eta(|k|)| < 1$ for $k \in [0, k_0]$.

But (4.18) is equivalent to

$$\eta^2 - \frac{\eta}{1+M} + \frac{M}{1+M}\frac{2|k|/r}{1-2|k|/r} = 0,$$

which has real solutions if and only if $|k| \leq k_0 \leq \frac{r}{2}$ where $k_0 = \dfrac{r}{2(1+2M)^2}$. The function η is then given by

$$\eta = \frac{1}{2(1+M)}\left(1 - \frac{\sqrt{1-|k|/k_0}}{\sqrt{1-2|k|/r}}\right),$$

which satisfies (i), (ii), (iii). Observe that k_0 does not depend on ν since C_2 and C_4 do not depend on ν. We can then conclude:

Proposition 4.A.11. *There exists k_0 such that for all ν in $]0, \nu_0]$ and for all k in $[-k_0, k_0]$ the full system rewritten as $\mathcal{G}(k, \nu) = 0$ admits a solution*

$$u(s, k, \nu) = \sum_{n \geq 1} k^n u_n(s, n)$$

satisfying

(a) *For every $n \geq 1$, $u_n(\cdot, \nu) \in \tilde{\mathcal{U}}^1_R$.*
(b) *The power series is absolutely convergent for k in $[-k_0, k_0]$.*
(c) *There exists C_5 such that $(k_0)^n \|u_n(\cdot, \nu)\|_{\mathcal{U}^1} \leq C_5$ holds for every $\nu \in]0, \nu_0]$.*
(d) $\|u\|_{\mathcal{U}^1} < 1$.

Statement (c) holds because the majorizing power series does not depend on ν.

4.A.3.3 Supplementary information about u_n, $\widehat{\omega}$. Using the explicit definition of $\mathcal{G}_{pq}$ and the definition by induction of u_n we can prove that u_n is a trigonometric polynomial. More precisely,

Lemma 4.A.12. *For every $n \geq 1$, $u_n \in \tilde{\mathcal{U}}_{n+1,R}$.*

Proof. We prove this lemma by induction. $u_1 = -\mathcal{L}_\nu^{-1}(Q_2(\widehat{Y}_1, \widehat{Y}_1, \nu))$. Moreover, $\widehat{Y}_1 = (0, 0, \cos(s), \sin(s))$, Q_2 is 2-linear, thus $Q_2(\widehat{Y}_1, \widehat{Y}_1, \nu)$ lies in $\mathcal{P}_{2,AR}$. Lemma 4.A.9 ensures that $\mathcal{L}_\nu$ is an isomorphism from $\tilde{\mathcal{U}}_{n,R}$ to $\mathcal{P}_{n,AR}$. Thus u_1 lies in $\tilde{\mathcal{U}}_{2,R}$.

Now, assume that for every $p \leq n-1$, $u_p \in \tilde{\mathcal{U}}_{p+1,R}$. The explicit form of $\mathcal{G}_{pq}$ given in Lemma 4.A.5, and Lemma 4.A.10 ensure that

$$\begin{aligned} \mathcal{L}_\nu(u_n) &= -\sum_{p=1}^{n-1} \frac{d\widehat{Z}_p}{ds}\widehat{\omega}_{n-p} + Q_{n+1}(\widehat{Y}_1^{(n+1)}, \nu) \\ &\quad + \sum_{p=1}^{n-1}\sum_{q=1}^{p+1} C^q_{p+1} \sum_{\substack{i_1+\dots+i_q=n-p \\ i_l \geq 1}} Q_{p+1}(\widehat{Y}_1^{(p+1-q)}, \widehat{Z}_{i_1}, \dots, \widehat{Z}_{i_q}, \nu). \end{aligned}$$

By induction hypothesis, $\dfrac{d\widehat{Z}_p}{ds} \in \mathcal{P}_{p+1,AR} \subset \mathcal{P}_{n,AR} \subset \mathcal{P}_{n+1,AR}$ for every $p \leq n-1$. $\widehat{Y}_1 = (0, 0, \cos(s), \sin(s))$, and Q_{n+1} is $(n+1)$-linear; thus $Q_{n+1}(\widehat{Y}_1(n+1), \nu)$ lies in $\mathcal{P}_{n+1,AR}$.
In the same way, $Q_{p+1}(\widehat{Y}_1^{(p+1-q)}, \widehat{Z}_{i_1}, \dots, \widehat{Z}_{i_q}, \nu)$ lies in $\mathcal{P}_{N,AR}$ with N given by

$$N = p+1-q+(i_1+1)+\ldots.+(i_q+1) = p+1-q+(i_1+\ldots.+i_q)+q = n+1.$$

Thus $\mathcal{L}_\nu(u_n)$ lies in $\mathcal{P}_{n+1,AR}$. Finally, u_n lies in $\tilde{\mathcal{U}}_{n+1,R}$. □

To complete the proof of Theorem 4.1.2 we now prove

Lemma 4.A.13. *Let $u(s,k,\nu) = (\widehat{Z}(s,k,\nu), \widehat{\omega}(k,\nu))$ denote the solution of the full system* (4.2) *given by Proposition* 4.A.11.

(a) $\widehat{\omega}(k,\nu) = \widehat{\omega}(-k,\nu)$ *and* $-\widehat{Z}(s+\pi,-k,\nu) = \widehat{Z}(s,k,\nu)$.

(b) *If Q satisfies the extra hypothesis (4.5), then there exists C_6 such that* $|\widehat{\omega}(k,\nu)| \leq C_6 k^2 \nu^\alpha$ *for every* $k \in [-k_0, k_0]$ *and* $\nu \in]0, \nu_0]$.

Proof. (a): Denote by $v(s,k,\nu) = \big(-\widehat{Z}(s+\pi,-k,\nu), \widehat{\omega}(-k,\nu)\big)$. Observe that

$$\mathcal{G}_\nu(v,-k) = -\mathcal{G}_\nu(u,k) = 0$$

where $\mathcal{G}_\nu$ is defined in Lemma 4.A.5. Proposition 4.A.11 ensures that

$$u(s,k,\nu) = \sum_{k\geq 1} k^n u_n(s,\nu) \qquad \text{with } u_n(s,\nu) = (\widehat{Z}_n(s,\nu), \omega_n(\nu)) \in \tilde{\mathcal{U}}^1_R.$$

Thus,

$$v(s,k,\nu) = \sum_{k\geq 1} k^n v_n(s,\nu) \text{ with } v_n(s,\nu) = (-1)^n(-\widehat{Z}_n(s+\pi,\nu), \omega_n(\nu)) \in \tilde{\mathcal{U}}^1_R.$$

Finally, the uniqueness of the solution given by Lemma 4.A.10 ensures that $v(s,k,\nu) = u(s,k,\nu)$, and thus

$$\widehat{\omega}(-k,\nu) = \widehat{\omega}(k,\nu), \qquad -\widehat{Z}(s+\pi,-k,\nu) = \widehat{Z}(s,k,\nu).$$

(b): Lemmas 4.A.4 and 4.A.7 ensure that

$$\begin{aligned} \widehat{\omega} &= \frac{1}{2\pi}\int_0^{2\pi} \langle \sum_{p\geq 1} k^p \sum_{m=0}^{p+1} C^m_{p+1} Q_{p+1}(\widehat{Y}_1^{(p+1-m)}, \widehat{Z}^{(m)}, \nu), \frac{d\widehat{Y}_1}{ds}\rangle ds \\ &\quad -\frac{\widehat{\omega}}{2\pi}\int_0^{2\pi} \langle \frac{d\widehat{Y}_1}{ds}, \frac{d\widehat{Z}}{ds}\rangle ds. \end{aligned}$$

Integrating by parts we find that

$$\int_0^{2\pi} \langle \frac{d\widehat{Y}_1}{ds}, \frac{d\widehat{Z}}{ds}\rangle\, ds = \int_0^{2\pi} \langle \widehat{Y}_1, \widehat{Z}\rangle ds = 0,$$

because $\widehat{Z}$ lies in $\tilde{\mathcal{U}}^1_R$. Thus,

$$\widehat{\omega} = \frac{1}{2\pi}\int_0^{2\pi} \langle f, \frac{d\widehat{Y}_1}{ds}\rangle ds \quad \text{with } f = \sum_{p\geq 1} k^p \sum_{m=0}^{p+1} C^m_{p+1} Q_{p+1}(\widehat{Y}_1^{(p+1-m)}, \widehat{Z}^{(m)}, \nu).$$

The frequency $\widehat{\omega}$ is even with respect to k. Thus

$$\widehat{\omega} = \frac{1}{2\pi}\int_0^{2\pi} \langle g, \frac{d\widehat{Y}_1}{ds}\rangle ds \quad \text{with } g = \sum_{\substack{p\geq 2\\ p \text{ even}}} k^p \sum_{m=0}^{p+1} C_{p+1}^m Q_{p+1}(\widehat{Y}_1^{(p+1-m)}, \widehat{Z}^{(m)}, \nu).$$

so that

$$|\widehat{\omega}| = |\frac{1}{2\pi}\int_0^{2\pi} [-g_A(s)\sin(s) + g_B(s)\cos(s)]ds| \leq \|g_A\|_{\mathcal{P}0} + \|g_B\|_{\mathcal{P}0}.$$

Lemma 4.A.1 ensures that $Q_{p+1}^A(\widehat{Y}_1^{(p+1-m)}, \widehat{Z}^{(m)}, \nu) \leq \frac{C_1\nu^\alpha}{r^{p+1}}|\widehat{Z}|^m$. Moreover, $\left\|\widehat{Z}\right\|_{\mathcal{P}0} \leq \|u\|_{\mathcal{U}^1} < 1$. Thus

$$\sum_{m=0}^{p+1} C_{p+1}^m \left\|Q_{p+1}^A(\widehat{Y}_1^{(p+1-m)}, \widehat{Z}^{(m)}, \nu)\right\|_{\mathcal{P}0} \leq (\frac{2}{r})^{p+1} C_1\nu^\alpha.$$

Finally,

$$\|g_A\|_{\mathcal{P}0} \leq \left[\frac{8C_1}{r^3}\sum_{p=0}^{+\infty}(\frac{2k_0}{r})^p\right] k^2\nu^\alpha.$$

The power series is convergent because $k_0 < \frac{r}{2}$. The same inequality holds for g_B. □

4.B Appendix. Proof of Proposition 4.2.1

We start by recalling some material about Fourier series and norms.

Lemma 4.B.1. *For $f \in \mathcal{P}_n$ ($\mathcal{P}_n$ is defined in Definition 4.A.3) let*

$$\begin{aligned} N_1(f) &= |a_0| + \sum_{p=1}^n |a_p| + |b_p|, \\ N_2(f) &= \sqrt{\frac{1}{2\pi}\int_0^{2\pi} |f|^2} = \sqrt{a_0^2 + \frac{1}{2}\sum_{p=1}^n |a_p|^2 + |b_p|^2} \\ \|f\|_0 &= \sup_{s\in[0,2\pi]} |f(s)|. \end{aligned}$$

Then

$$N_1(f) \leq \sqrt{2(2n+1)}N_2(f) \leq \sqrt{2(2n+1)}\,\|f\|_0\,.$$

Now we compare the size of a trigonometric polynomial at a complex point ξ, $f(\xi)$ and the size of the polynomial on the real axis $\|f\|_0$.

Lemma 4.B.2. *The following estimates hold for every $f \in \mathcal{P}_n$, $\xi \in \mathbb{C}$ and $j \geq 0$*

$$\left|\frac{d^j f}{d\xi^j}(\xi)\right| \leq n^j \sqrt{2(2n+1)}\ \|f\|_0\, \mathrm{e}^{n|\mathcal{I}m(\xi)|}.$$

Proof.

$$\begin{aligned}\left|\frac{d^j f}{d\xi^j}(\xi)\right| &\leq \sum_{p=0}^{n} |p^j a_p| \mathrm{e}^{p|\mathcal{I}m(\xi)|} \leq n^j N_1(f) \mathrm{e}^{n|\mathcal{I}m(\xi)|} \\ &\leq n^j \sqrt{2(2n+1)}\ \|f\|_0\, \mathrm{e}^{n|\mathcal{I}m(\xi)|}. \ \square\end{aligned}$$

Now we have enough material to control the size of $|\widehat{Y}_n(\xi,\nu)|$ in the complex plane.

Proof of Proposition 4.2.1. Applying Lemma 4.B.2 to $\widehat{Y}_n(\xi,\nu)$ we get

$$\begin{aligned} k^n \left|\frac{d^j \widehat{Y}_n(\xi,\nu)}{d\xi^j}\right| &\leq k^n n^j \sqrt{2(2n+1)} \left\|\widehat{Y}_n\right\|_0 \mathrm{e}^{n|\mathcal{I}m(\xi)|} \\ &\leq (k_0)^n \left\|\widehat{Y}_n\right\|_{\mathcal{P}1} \left(\frac{k}{k_0}\right)^n n^j \sqrt{2(2n+1)} \mathrm{e}^{n|\mathcal{I}m(\xi)|}. \end{aligned}$$

Applying Theorem 4.1.2 we finally get

$$k^n \left|\frac{d^j \widehat{Y}_n(\xi,\nu)}{d\xi^j}\right| \leq 2C_{\widehat{Y}}\ n^j \sqrt{n+1} \left(\frac{k}{k_0}\right)^n \mathrm{e}^{n|\mathcal{I}m(\xi)|}. \ \square$$

4.C Proof of Theorem 4.3.1

We perform a first linear change of coordinates $\mathsf{Y} = \Psi(Y)$ where Ψ is the nonsingular linear operator

$$\Psi : \mathbb{R}^{2+d} \to \Delta \times \mathbb{R}^d : (A, B, Y_{\mathrm{h}}) \mapsto (C = A + \mathrm{i}B, \widetilde{C} = A - \mathrm{i}B, Y_{\mathrm{h}}).$$

where $\Delta = \{(C,\widetilde{C}) \in \mathbb{C}^2, \widetilde{C} = \overline{C}\}$. Observe that Ψ can be extended to an isomorphism of $\mathbb{C}^{2+d}$. We equip $\Delta \times \mathbb{R}^d$ with the inner product

$$\langle \mathsf{Y}, \mathsf{Y}' \rangle = \frac{1}{2}\overline{C}C' + \frac{1}{2}\overline{\widetilde{C}}\widetilde{C}' + \langle Y_{\mathrm{h}}, Y_{\mathrm{h}}' \rangle_{\mathbb{R}^d}.$$

This way Ψ is an isometry. We can reformulate the result of Theorem 4.1.2 in this new system of coordinates.

Lemma 4.C.1. *Performing the linear change of coordinates* $\mathsf{Y} = \Psi(Y)$, *(4.2) is equivalent to*

$$\frac{d\mathsf{Y}}{dt} = \mathsf{L}_\nu \mathsf{Y} + \mathsf{Q}(\mathsf{Y}, \nu) \tag{4.19}$$

where $\mathsf{L}_\nu(C, \widetilde{C}, Y_\mathrm{h}) = (\mathrm{i}\omega_{0,\nu} C, -\mathrm{i}\omega_{0,\nu}\widetilde{C}, \frac{1}{\nu} L_\mathrm{h}(\nu) Y_\mathrm{h})$ *and where* Q *also satisfies hypothesis (ii) of Theorem 4.1.2. Moreover, (4.19) admits a one parameter family of periodic orbits* $\mathsf{Y}_{k,\nu}$ *given by*

$$\mathsf{Y}_{k,\nu}(t) = \widehat{\mathsf{Y}}(\underline{\omega}_{k,\nu} t, k, \nu)$$

where

$$\widehat{\mathsf{Y}}(s, k, \nu) = \sum_{n \geq 1} k^n \, \widehat{\mathsf{Y}}_n(s, \nu), \qquad \mathsf{Y}_1 = (\mathrm{e}^{\mathrm{i}s}, \mathrm{e}^{-\mathrm{i}s}, 0),$$

and for $n \geq 2$, $\widehat{\mathsf{Y}}_n$ *reads*

$$\widehat{\mathsf{Y}}_n(s, \nu) = \sum_{p=-n}^{n} (\widehat{c}_{n,p}(\nu) \ \mathrm{e}^{\mathrm{i}ps}, \widehat{c}_{n,p}(\nu) \ \mathrm{e}^{-\mathrm{i}ps}, \widehat{Y}_{\mathrm{h},n,p}(\nu) \ \mathrm{e}^{\mathrm{i}ps}),$$

with

(a) $\widehat{c}_{n,p}(\nu) \in \mathbb{R}$, $\quad \widehat{c}_{n,1}(\nu) = 0$, $\quad \widehat{Y}_{\mathrm{h},n,-p}(\nu) = \overline{\widehat{Y}_{\mathrm{h},n,p}(\nu)} = S_\mathrm{h} \widehat{Y}_{\mathrm{h},n,p}(\nu)$.

(b) $k_0^n \left(\sup_{[0,2\pi]} |\widehat{\mathsf{Y}}_n(s,\nu)| + \sup_{[0,2\pi]} \left| \frac{d\widehat{\mathsf{Y}}_n}{ds}(s,\nu) \right| \right) \leq C_{\widehat{\mathsf{Y}}}$ *for* $\nu \in]0, \nu_0]$, $n \geq 1$.

(c) $\widehat{c}_{n,p}(\nu) = 0$ *and* $\widehat{Y}_{\mathrm{h},n,p}(\nu) = 0$ *if* $n + p$ *is odd.*

Remark 4.C.2. Statement (c) is equivalent to Theorem 4.1.2-(c) and for this new system of coordinates, the symmetry of reversiblity S reads $\mathsf{S}(C, \widetilde{C}, Y_\mathrm{h}) = (\widetilde{C}, C, S_\mathrm{h} Y_\mathrm{h})$ (see Remark 4.1.6).

Lemma 4.C.3. *Denote* $D_2(0, r) = \{(C, \widetilde{C}) \in \mathbb{C}^2, |C|^2 + |\widetilde{C}|^2 < r^2\}$ *and let* $\theta_\nu = \mathrm{Id} + \theta_\nu^1$ *where*

$$\theta_\nu^1 : D_2(0, k_0) \times \mathbb{C}^d \to \mathbb{C}^{2+d} : (C', \widetilde{C}', Y_\mathrm{h}') \mapsto \sum_{n \geq 2} \theta_{\nu,n}^1(C', \widetilde{C}')$$

with

$$\theta_{\nu,n}^1(C', \widetilde{C}') = \sum_{\substack{p=-n \\ p+n \text{ even}}}^{n} \left(\widehat{c}_{n,p} C'^{\frac{n+p}{2}} \widetilde{C}'^{\frac{n-p}{2}}, \widehat{c}_{n,p} C'^{\frac{n-p}{2}} \widetilde{C}'^{\frac{n+p}{2}}, \widehat{Y}_{\mathrm{h},n,p} C'^{\frac{n+p}{2}} \widetilde{C}'^{\frac{n-p}{2}} \right).$$

(a) *Then, for any* $k_0' \in]0, k_0[$, *there exists* $M(k_0') > 0$ *such that for* $\nu \in]0, \nu_0]$ *and* $\mathsf{Y}' \in D_2(0, k_0') \times \mathbb{C}^d$,

$$|\theta_\nu^1(\mathsf{Y}')| \leq M(k_0')(|C'|^2 + |\widetilde{C}'|^2), \quad |D\theta_\nu^1(\mathsf{Y}')| \leq M(k_0')(|C'|^2 + |\widetilde{C}'|^2)^{\frac{1}{2}}.$$

(b) $\theta_\nu(k\mathrm{e}^{\mathrm{i}s}, k\mathrm{e}^{-\mathrm{i}s}, 0) = \widehat{\mathsf{Y}}(s, k, \nu)$ *for* $k \in [0, k_0[$ *and* $\nu \in]0, \nu_0]$.

(c) $\theta_\nu(\Delta \times \mathbb{R}^d) \subset \Delta \times \mathbb{R}^d$ *and* $\mathsf{S}\theta_\nu(\mathsf{Y}') = \theta_\nu(\mathsf{S}\mathsf{Y}')$.

(d) *There exists* $k_1 < k_0$ *and a family of open sets* $\Omega_\nu \subset \mathbb{C}^2$, $0 < \nu \leq \nu_0$, *containing the origin, such that for every* $\nu \in]0, \nu_0]$, θ_ν *is an analytic diffeomorphism from* $D_2(0, k_1) \times \mathbb{C}^d$ *to* $\Omega_\nu \times \mathbb{C}^d$.

Remark 4.C.4. Observe that $\theta_\nu^1(\mathsf{Y}')$ only depends on $(C', \widetilde{C}')$. Moreover, Theorem 4.3.1 readily follows from this lemma by setting $\Theta_\nu = \Psi \circ \theta_\nu \circ \Psi^{-1} = \mathrm{Id} + \Psi \circ \theta_\nu^1 \circ \Psi^{-1}$.

Proof. (a): We first check that θ_ν is well defined for $\mathsf{Y}' \in D_2(0, k_0') \times \mathbb{C}^d$ with $k_0' < k_0$. Using the Plancherel Formula and Lemma 4.C.1 we get, for $n \geq 2$,

$$\begin{aligned}
|\theta_{\nu,n}^1(C', \widetilde{C}')|^2 &\leq (2n+1)(|C'|^2 + |\widetilde{C}'|^2)^n \sum_{p=-n}^{n} (|\widehat{c}_{n,p}|^2 + |\widehat{Y}_{\mathrm{h},n,p}|^2) \\
&\leq (2n+1)(|C'|^2 + |\widetilde{C}'|^2)^n \frac{1}{2\pi} \int_0^{2\pi} |\widehat{Y}_n(s, \nu)|^2 \\
&\leq (C_{\widehat{Y}})^2 (2n+1) \left(\frac{|C'|^2 + |\widetilde{C}'|^2}{k_0^2} \right)^n.
\end{aligned}$$

Thus, for $k_0' < k_0$ and every $(C', \widetilde{C}') \in D_2(0, k_0')$, θ_ν^1 is well defined and

$$|\theta_\nu^1(C', \widetilde{C}')| \leq \frac{C_{\widehat{Y}}}{k_0^2} (|C'|^2 + |\widetilde{C}'|^2) \sum_{n \geq 2} \sqrt{2n+1} \left(\frac{k_0'}{k_0} \right)^{n-2}$$

Similarly, we check that for $k_0' < k_0$ and every $(C', \widetilde{C}') \in D_2(0, k_0')$,

$$|D\theta_\nu^1(C', \widetilde{C}')| \leq M(k_0') \, (|C'|^2 + |\widetilde{C}'|^2)^{\frac{1}{2}}$$

with

$$M(k_0') = \frac{C_{\widehat{Y}}}{k_0^2} \sum_{n \geq 2} n\sqrt{2n+1} \left(\frac{k_0'}{k_0} \right)^{n-2}.$$

(b): Observing that

$$(k\mathrm{e}^{\mathrm{i}s})^{\frac{n+p}{2}} (k\mathrm{e}^{-\mathrm{i}s})^{\frac{n-p}{2}} = k^n \mathrm{e}^{\mathrm{i}ps}, \qquad (k\mathrm{e}^{\mathrm{i}s})^{\frac{n-p}{2}} (k\mathrm{e}^{-\mathrm{i}s})^{\frac{n+p}{2}} = k^n \mathrm{e}^{-\mathrm{i}ps},$$

Lemma 4.C.1 ensures that $\theta_\nu(ke^{is}, ke^{-is}, 0) = \widehat{Y}(s,k,\nu)$ for $k \in [0, k_0[$ and $\nu \in]0, \nu_0]$.

(c): Using Lemma 4.C.1-(a) we obtain

$$
\begin{aligned}
&S\theta^1_{\nu,n}(C', \widetilde{C}') \\
&= \left(\sum_{\substack{p=-n \\ p+n \text{ even}}}^{n} \widehat{c}_{n,p} C'^{\frac{n-p}{2}} \widetilde{C}'^{\frac{n+p}{2}}, \sum_{\substack{p=-n \\ p+n \text{ even}}}^{n} \widehat{c}_{n,p} C'^{\frac{n+p}{2}} \widetilde{C}'^{\frac{n-p}{2}}, \sum_{\substack{p=-n \\ p+n \text{ even}}}^{n} \widehat{Y}_{\mathrm{h},n,-p} C'^{\frac{n+p}{2}} \widetilde{C}'^{\frac{n-p}{2}} \right), \\
&= \left(\sum_{\substack{p=-n \\ p+n \text{ even}}}^{n} \widehat{c}_{n,p} C'^{\frac{n-p}{2}} \widetilde{C}'^{\frac{n+p}{2}}, \sum_{\substack{p=-n \\ p+n \text{ even}}}^{n} \widehat{c}_{n,p} C'^{\frac{n+p}{2}} \widetilde{C}'^{\frac{n-p}{2}}, \sum_{\substack{q=-n \\ q+n \text{ even}}}^{n} \widehat{Y}_{\mathrm{h},n,q} C'^{\frac{n-q}{2}} \widetilde{C}'^{\frac{n+q}{2}} \right) \\
&= \sum_{\substack{p=-n \\ p+n \text{ even}}}^{n} \left(\widehat{c}_{n,p} C'^{\frac{n-p}{2}} \widetilde{C}'^{\frac{n+p}{2}}, \widehat{c}_{n,p} C'^{\frac{n+p}{2}} \widetilde{C}'^{\frac{n-p}{2}}, \widehat{Y}_{\mathrm{h},n,p} C'^{\frac{n-p}{2}} \widetilde{C}'^{\frac{n+p}{2}} \right), \\
&= \theta^1_{\nu,n}(\widetilde{C}', C').
\end{aligned}
$$

Thus, $S\theta_\nu(Y') = SY' + S\theta^1_\nu(C', \widetilde{C}') = SY' + \theta^1_\nu(\widetilde{C}', C') = \theta_\nu(SY')$. Similarly, we check that $\theta_\nu(\Delta \times \mathbb{R}^d) \subset \Delta \times \mathbb{R}^d$.

(d): We proceed in several steps.

Step 1. $\theta_\nu = \mathrm{Id} + \sum_{n \geq 2} \theta^1_{\nu,n}$ holds; $\theta^1_{\nu,n}$ is an analytic function and the computation made for (d) ensure that the series is normally convergent for $Y' \in D_2(0, k_0') \times \mathbb{C}^d$ for $k_0' < k_0$. Hence, Θ_ν is an analytic function on $D_2(0, k_0) \times \mathbb{C}^d$.

Step 2. We denote $\theta^1_\nu = (\theta^1_{\nu,C}, \theta^1_{\nu,\widetilde{C}}, \theta^1_{\nu,\mathrm{h}})$ and we introduce

$$\phi_\nu : D_2(0, k_0) \to \mathbb{C}^2 : (C', \widetilde{C}') \mapsto (C' + \theta^1_{\nu,C}(C', \widetilde{C}'), \widetilde{C}' + \theta^1_{\nu,\widetilde{C}}(C', \widetilde{C}')).$$

where. Since ϕ_ν is an analytic function which is uniformly bounded for $\nu \in]0, \nu_0]$ and for $(C', \widetilde{C}') \in D_2(0, k_0')$ with $0 < k_0' < k_0$ and which satisfies $D\phi_\nu(0) = \mathrm{Id}$, the analytic Inverse Function Theorem ensures that there there exits $k_1 \in]0, k_0[$ which does not depend on ν and a family of open sets $\mathcal{U}_\nu$ containing the origin such that ϕ_ν is an analytical diffeomorphism from $D_2(0, k_1)$ to $\mathcal{U}_\nu$.

Step 3. So, θ_ν is a one to one map from $D_2(0, k_1) \times \mathbb{C}^d$ onto $\mathcal{U}_\nu \times \mathbb{C}^d$ whose inverse is explicitly given by

$$\theta_\nu^{-1}(Y) = (\phi_\nu^{-1}(C, \widetilde{C}), Y_\mathrm{h} - \theta^1_{\nu,\mathrm{h}} \circ \phi_\nu^{-1}(C, \widetilde{C})).$$

where $Y = (C, \widetilde{C}, Y_\mathrm{h})$ and $\theta^1_\nu = (\theta^1_{\nu,C}, \theta^1_{\nu,\widetilde{C}}, \theta^1_{\nu,\mathrm{h}})$. This explicit formula for θ_ν^{-1} ensures that θ_ν is an analytical diffeomorphism from $D_2(0, k_1) \times \mathbb{C}^d$ onto $\mathcal{U}_\nu \times \mathbb{C}^d$. □

5. Constructive Floquet Theory for periodic matrices near a constant one

When looking for a homoclinic connection to a periodic orbit, one has to study the system linearized around the periodic orbit for determining the dimension of the stable and unstable manifolds of the periodic orbit and the Floquet exponents which give the speed of attraction of an orbit which tends to a periodic orbit in infinite time. Usually, the determination of the Floquet exponents is very difficult because it requires the knowledge of the fundamental matrix after one period. However when one studies periodic matrices close to a constant one, one can determine Floquet exponents using perturbation theory. Such families of periodic matrices close to a constant one, typically occur when linearizing around a one parameter family of periodic orbits near a fixed point, like the one obtained by bifurcation from a pair of simple purely imaginary eigenvalues (see chapter 4). So this chapter is devoted to such a "constructive Floquet theory" for periodic matrices near a constant one.

More precisely, for any continuous, $n \times n$ complex matrix function A , we denote by R_A the fundamental matrix, i.e. the unique $n \times n$ matrix function R such that

$$\frac{dR}{dt} = A(t)R, \qquad R(0) = \mathrm{Id}.$$

When A is T-periodic, Floquet theory ensures that R_A reads

$$R_A(t) = Q(t)\mathrm{e}^{t\Lambda}$$

where Q is a T-periodic invertible matrix and Λ is a constant matrix. The eigenvalues of Λ are called Floquet exponents of A determined up to $\frac{2n\pi \mathrm{i}}{T}$, $n \in \mathbb{Z}$. When A is obtained by linearization around a periodic orbit, these exponents characterize the stability properties of the periodic orbit. Their explicit computation is quite difficult in general since Λ is simply characterized by the relation $R(T) = \mathrm{e}^{T\Lambda}$. However, when A is close to a constant matrix, perturbation theory enables us to compute Floquet exponents. We study in this chapter, a linear time dependent differential equation

$$\frac{dY}{dt} = L_0 Y + M(t, \underline{\lambda})Y, \qquad Y \in \mathbb{C}^n, \underline{\lambda} \in \mathbb{R}^m \tag{5.1}$$

where L_0 is $n \times n$ complex matrix and where M is a smooth, T-periodic matrix function which satisfies

$$M(t,0) = 0, \qquad \text{for } t \in \mathbb{R}.$$

Floquet theory ensures that, for every $\underline{\lambda} \in \mathbb{R}^m$, there exists a T-periodic, invertible matrix $Q(\cdot, \underline{\lambda})$, such that performing the change of coordinates $Y(t) = Q(t, \underline{\lambda})$, (5.1) is equivalent to

$$\frac{dZ}{dt} = L(\underline{\lambda})Z$$

where $L(\underline{\lambda})$ is a constant matrix, the eigenvalues of which are Floquet exponents. Since $L_0 + M(t, \underline{\lambda})$ is close to L_0 for $\underline{\lambda}$ close to 0, one can hope that $L(\underline{\lambda})$ and $Q(\cdot, \underline{\lambda})$ are respectively close to L_0 and Id. The following theorem ensures that this is only true if the eigenvalues of L_0 are non resonant. In the resonant case, L and Q are not close to L_0 and Id.

5.1 Constructive Floquet Theory in the non resonant case

Theorem 5.1.1 (Floquet Theory in the non resonant case). *Let L_0 be a $n \times n$ matrix and let $M : (t, \underline{\lambda}) \mapsto M(t, \underline{\lambda})$ be a matrix function of class C^r with $r > 0$, which is T-periodic with respect to t for any small $\underline{\lambda} \in \mathbb{R}^m$ and which satisfies $M(t,0) = 0$ for $t \in \mathbb{R}$. Denote $\sigma_1, \cdots, \sigma_n$ the eigenvalues of L_0 and assume that they are not resonant, i.e.*

$$\sigma_j - \sigma_k \notin \frac{2\pi \mathrm{i}\mathbb{Z}}{T} \setminus \{0\}, \qquad \text{for } 1 \le j \le n,\ 1 \le k \le n. \tag{5.2}$$

(a) *Then, for any sufficiently small $\underline{\lambda}$, there exists an invertible T-periodic, $n \times n$ matrix $Q(\cdot, \underline{\lambda})$ such that setting $Y(t) = Q(t, \underline{\lambda})Z(t)$, (5.1) is equivalent to*

$$\frac{dZ}{dt} = L(\underline{\lambda})Z \tag{5.3}$$

with

$$L(\underline{\lambda}) = L_0 + L_1(\underline{\lambda}), \qquad L_1(0) = 0, \qquad L_0^* L_1(\underline{\lambda}) - L_1(\underline{\lambda}) L_0^* = 0,$$

and

$$Q(t, \underline{\lambda}) = \mathrm{Id} + Q_1(t, \underline{\lambda}), \qquad Q_1(t,0) = 0$$

where $\underline{\lambda} \mapsto L_1(\underline{\lambda})$ and $(t, \underline{\lambda}) \mapsto Q_1(t, \underline{\lambda})$ are functions of class C^r and $t \mapsto Q_1(t, \underline{\lambda})$ is of class C^{r+1}.

(b) *If L_0 and M are reversible with respect to some orthogonal symmetry S, i.e.*

$$L_0 S = -S L_0, \qquad SM(-t,\underline{\lambda}) = -M(t,\underline{\lambda})S, \qquad S = S^* = S^{-1},$$

then

$$L_1(\underline{\lambda})S = -SL_1(\underline{\lambda}), \qquad SQ_1(-t,\underline{\lambda}) = Q_1(t,\underline{\lambda})S.$$

(c) *If L_0 and M are real matrices, then L_1 and Q_1 are also real matrices.*

(d) *If $(t,\underline{\lambda}) \mapsto M(t,\underline{\lambda})$ is an analytic functions then $\underline{\lambda} \mapsto L(\underline{\lambda})$ and $(t,\underline{\lambda}) \mapsto$ are also analytic functions.*

The proof of this theorem is given in Appendix 5.A. It is based on a rewriting of the equation satisfied by L_1, Q_1 as an implicit equation, solved by the Implicit Function Theorem. So It is possible to compute explicitly the Taylor expansion of L_1 and Q_1 in powers of $\underline{\lambda}$.

Moreover, we can prove that when M is "an analytic matrix function computed around circles", then Q_1 is also an "analytic function computed around circles".

Lemma 5.1.2 ("Analytic matrices computed around circles"). *Let L_0 satisfy the hypothesis of Theorem 5.1.1 and let*

$$M(t,\rho) = \check{M}(\rho e^{i\omega t}, \rho e^{-i\omega t}) \text{ with } \rho \in \mathbb{R},\ \omega > 0.$$

where $\check{M} : \mathbb{C}^2 \to \mathcal{M}_n(\mathbb{C})$ be an analytic function on $(D(0,\rho_1))^2$ with $\rho_1 > 0$, where $D(0,r) = \{z \in \mathbb{C}, |z| < r\}$ satisfying $\check{M}(0,0) = D\check{M}(0,0)$, i.e.

$$\check{M}(C,\widetilde{C}) = \sum_{p+q\geq 2} \check{M}_{p,q} C^p \widetilde{C}^q, \qquad \text{with } |\check{M}_{p,q}| \leq \frac{m_1}{(\rho_1)^{p+q}}.$$

Then there exist $\rho_2 \in]0,\rho_1]$ and m_2 such that for $t \in \mathbb{R}$ and $\rho \in]-\rho_2,\rho_2[$, the two matrix functions given by Theorem 5.1.1 $L_1(\rho)$ and $Q_1(t,\rho)$ read

$$L_1(\rho) = \sum_{k\geq 1} \rho^{2k} L_{1,2k}, \qquad L_0^* L_{1,2k} - L_0^* L_{1,2k} = 0, \qquad |L_{1,2k}| \leq \frac{m_2}{(\rho_2)^{2k}},$$

and $Q_1(t,\rho) = \check{Q}_1(\rho e^{-i\omega t}, \rho e^{-i\omega t})$ with

$$\check{Q}_1(C,\widetilde{C}) = \sum_{p+q\geq 2} \check{Q}_{p,q} C^p \widetilde{C}^q, \qquad |\check{Q}_{p,q}|(1+|p-q|\omega) \leq \frac{m_2}{(\rho_2)^{p+q}}.$$

The proof of this lemma is given in appendix 5.B

Remark 5.1.3. This type of matrix M typically occurs when linearizing analytic vector fields around periodic orbits which are circles, like the one obtained in Theorem 4.3.1.

Remark 5.1.4. The above description of M and Q_1 appears to be invaluable when complexifying time. Indeed, if ξ lies in a strip $\mathcal{B}_\ell = \{\xi \in \mathbb{C}, |\mathcal{I}m\,(\xi)\,| < \ell\}$, then the above description ensures that $M(\xi, \rho)$ and $Q_1(\xi, \rho)$ are both well defined provided that $\rho < \rho_2 \mathrm{e}^{-\ell\omega}$.

5.2 Constructive Floquet Theory in the resonant case

The resonant case can be deduced from the non resonant one using sufficiently many linear T-periodic transformation V of the type given by Lemma 5.2.1 below. Observe that consequently, in the resonant case, Q and L are not close to Id and L_0.

Lemma 5.2.1. *Let L_0 be a $n \times n$ matrix and denote σ_j, $1 \le j \le \ell$ its eigenvalues ($\sigma_j \neq \sigma_k$ for $j \neq k$) .*
Let $M : (t, \underline{\lambda}) \mapsto M(t, \underline{\lambda})$ be a matrix function of class C^r with $r > 0$, which is T-periodic with respect to t for any small $\underline{\lambda} \in \mathbb{R}^m$ and which satisfies $M(t, 0) = 0$ for $t \in \mathbb{R}$.

Then, there exists an invertible matrix function V, such that the transformation $Y(t) = V(t)\widetilde{Y}(t)$ transforms (5.1) into an equivalent system

$$\frac{d\widetilde{Y}}{dt} = \widetilde{L}_0\widetilde{Y} + \widetilde{M}(t, \underline{\lambda}), \tag{5.4}$$

where $\widetilde{L}_0$ has for eigenvalues $\sigma_1 - \frac{2i\pi}{T}, \sigma_2, \cdots, \sigma_\ell$ and where $\widetilde{M}$ satisfies the same properties as M.

Proof. We first assume that L_0 is under Jordan normal form

$$L_0 = \begin{bmatrix} L_{01} & 0 \\ 0 & L_{02} \end{bmatrix}$$

where L_{01} is the $n_1 \times n_1$ matrix equal to the direct sum of all the Jordan block corresponding to the eigenvalue σ_1,

$$L_{01} = \begin{bmatrix} \sigma_1 & \delta_2 & 0 & 0 \\ 0 & \sigma_1 & \ddots & 0 \\ & \ddots & \ddots & \delta_{n_1} \\ 0 & & 0 & \sigma_1 \end{bmatrix}$$

δ_i being 0 or 1. Then, denote by U the matrix

$$U(t) = \begin{bmatrix} e^{\frac{2i\pi t}{T}} I_{n_1} & 0 \\ 0 & I_{n-n_1} \end{bmatrix}.$$

Clearly U is invertible with $U^{-1}(t) = U(-t)$ and setting $Y(t) = U(t)\widetilde{Y}(t)$, (5.1) is equivalent to (5.4) with

$$\widetilde{L}_0 = \begin{bmatrix} L_{01} - \frac{2\pi i}{T} I_{n_1} & 0 \\ 0 & L_{02} \end{bmatrix}, \qquad \widetilde{M}(t, \underline{\lambda}) = U(-t)M(t, \underline{\lambda})U(t).$$

Thus, $\widetilde{L}_0$ and $\widetilde{M}$ have the required properties.

In case L_0 is not under Jordan normal form, the transformation U can be replaced by PU, where P is chosen so that $P^{-1}L_0P$ is under Jordan normal form. Setting $V = PU$ the lemma is proved. □

5.A Appendix. Proof of Theorem 5.1.1

5.A.1 Writing of the equation as an implicit equation

We look for (Q_1, L_1) such that setting $Y = (\mathrm{Id} + Q_1(\cdot, \underline{\lambda}))Z$, equation (5.1) is equivalent to

$$\frac{dZ}{dt} = (L_0 + L_1(\underline{\lambda}))Z.$$

To obtain the equation satisfied by L_1 and Q_1, we perform the linear change of coordinate $Y = (\mathrm{Id} + Q_1(\cdot, \underline{\lambda}))Z$ in (5.1)

$$\frac{d(Z + Q_1(t, \underline{\lambda})Z)}{dt} = L_0(Z + Q_1(t, \underline{\lambda})Z) + M(t, \underline{\lambda})(Z + Q_1(t, \underline{\lambda})Z)$$

and we prescribe that $\frac{dZ}{dt} = (L_0 + L_1(\underline{\lambda}))$ holds. Hence, we get

$$L_1 Z + \frac{dQ_1}{dt} Z + Q_1 L_0 Z + Q_1 L_1 Z = L_0 Q_1 Z + MZ + MQ_1 Z.$$

So we can conclude

Lemma 5.A.1. *If $u = (Q_1, L_1)$ satisfy*

$$\mathcal{L}(u) = \mathcal{N}(u, \underline{\lambda}) \tag{5.5}$$

where

$$\mathcal{L}(u) = L_1 + Q_1 L_0 - L_0 Q_1 + \frac{dQ_1}{dt}, \quad \mathcal{N}(u, \underline{\lambda}) = -Q_1 L_1 + M(\cdot, \underline{\lambda}) + M(\cdot, \underline{\lambda})Q_1$$

and if $\mathrm{Id} + Q_1(t, \underline{\lambda})$ is invertible for $t \in \mathbb{R}$, then (5.1) is equivalent to (5.3).

The idea is then to solve this equation using the implicit function Theorem in appropriate Banach Spaces for $u = (Q_1, L_1)$ and $\underline{\lambda}$ close to 0. For that purpose we introduce the following notation and definitions

Definition 5.A.2. *Denote by $\mathcal{M}_n(\mathbb{C})$ the set of the complex $n \times n$ matrices equipped with the inner product $\langle A, B\rangle = \operatorname{tr}(A^* B)$ and let $|\cdot|$ be the corresponding norm which is a norm of algebra, i.e. $|AB| \leq |A|\,|B|$.*

Definition 5.A.3. *Let us define the Banach spaces*

$$\begin{aligned}
\mathcal{P}_T^r(\mathcal{M}_n) &= \{Q : \mathbb{R} \to \mathcal{M}_n(\mathbb{C}), T\text{-periodic, of class } C^r\},\\
\mathcal{P}_T^{r,\perp}(\mathcal{M}_n) &= \mathcal{P}_T^r(\mathcal{M}_n) \cap \left\{Q : \mathbb{R} \to \mathcal{M}_n, \textstyle\int_0^T Q(t)dt \in (\ker \mathcal{A}_0)^\perp\right\}\\
\mathcal{U}_T^{r,\perp}(\mathcal{M}_n) &= \{u = (Q, L),\ Q \in \mathcal{P}_T^{r,\perp}(\mathcal{M}_n),\ L \in \ker \mathcal{A}_0^*\},
\end{aligned}$$

normed with

$$\begin{aligned}
\|Q\|_{\mathcal{P}_T^r} &= \sum_{j=0}^{r} \sup_{[0,T]} \left|\frac{d^j Q(t)}{dt^j}\right| && \text{for } \mathcal{P}_T^r(\mathcal{M}_n), \mathcal{P}_T^{r,\perp}(\mathcal{M}_n),\\
\|u\|_{\mathcal{U}_T^{r,\perp}} &= \|Q\|_{\mathcal{P}_T^r} + |L| && \text{for } \mathcal{P}_T^{r,\perp}(\mathcal{M}_n).
\end{aligned}$$

where $\mathcal{A}_0$ is the linear operator given by

$$\mathcal{A}_0 : \mathcal{M}_n(\mathbb{C}) \to \mathcal{M}_n(\mathbb{C}) : P \mapsto L_0 P - P L_0.$$

5.A.2 Study of $\mathcal{L}(u)$

As usual, the crucial point is the study of the linear part (classical result).

Lemma 5.A.4. *Let $\mathcal{A}_0 : \mathcal{M}_n(\mathbb{C}) \to \mathcal{M}_n(\mathbb{C}) : P \mapsto L_0 P - P L_0$.*

(a) *When $\mathcal{M}_n(\mathbb{C})$ is equipped with the inner product introduced in the previous definition, the adjoint $\mathcal{A}_0^*$ of $\mathcal{A}_0$ is given by $\mathcal{A}_0^*(P) = L_0^* P - P L_0^*$.*

(b) *The eigenvalues of $\mathcal{A}_0$ are the numbers $\sigma_j - \sigma_k$, $1 \leq j \leq \ell$, $1 \leq k \leq \ell$ where $\sigma_1, \cdots, \sigma_\ell$ are the eigenvalues of L_0.*

Proof. (a):

$$\begin{aligned}
\langle Q, \mathcal{A}_0 P\rangle &= \operatorname{tr}(Q^* L_0 P) - \operatorname{tr}(Q^* P L_0)\\
&= \operatorname{tr}(Q^* L_0 P) - \operatorname{tr}(L_0 Q^* P)\\
&= \langle L_0^* Q - Q L_0^*, P\rangle.
\end{aligned}$$

(a): Let us denote $(x_{j,k})_{\substack{1\le j\le \ell\\ 0\le k\le n_j-1}}$ a basis of $\mathbb{C}^n$ for which L_0 is under Jordan normal form, i.e.

$$\begin{aligned}\mathcal{A}_0 x_{j,0} &= \sigma_j x_{j,0} && \text{for } 1\le j\le \ell,\\ \mathcal{A}_0 x_{j,k} &= \sigma_j x_{j,k} + \delta_{j,k}\, x_{j,k-1} && \text{for } 1\le j\le \ell,\ 1\le k\le n_j-1\end{aligned}$$

where $\delta_{j,k} = 0$ or 1. Similarly, since the transpose ${}^tL_0$ of L_0 has the same eigenvalues as L_0 with Jordan blocks of the same size, we can find a basis $(y_{j,k})_{\substack{1\le j\le \ell\\ 0\le k\le n_j-1}}$ for which ${}^tL_0$ is under Jordan normal form, i.e.

$$\begin{aligned}\mathcal{A}_0 y_{j,0} &= \sigma_j y_{j,0} && \text{for } 1\le j\le \ell\\ \mathcal{A}_0 y_{j,k} &= \sigma_j y_{j,k} + \delta'_{j,k} y_{j,k-1} && \text{for } 1\le j\le q,\ 1\le k\le n_j-1\end{aligned}$$

where $\delta'_{j,k} = 0$ or 1. Then, the family of matrices $(x_{j,k}\,{}^ty_{p,q})_{\substack{1\le j,p\le \ell\\ 0\le k,q\le n_j-1}}$ is a basis of $\mathcal{M}_n(\mathbb{C})$. Moreover,

$$\mathcal{A}_0(x_{j,k}\,{}^ty_{p,q}) = (\sigma_j-\sigma_p)x_{j,k}\,{}^ty_{p,q} + \delta_{p,q}x_{j,k}\,{}^ty_{p,q-1} + \delta_{j,k}x_{j,k-1}\,{}^ty_{p,q}\,.$$

So, if we order the basis $x_{j,k}\,{}^ty_{p,q}$ with the lexicographical order for (j,k,p,q), we observe that for this basis of $\mathcal{M}_n(\mathbb{C})$ the matrix of $\mathcal{A}_0$ is an upper triangular matrix with coefficient $\sigma_j-\sigma_p$ on the diagonal. So, the numbers $\sigma_j-\sigma_p$ are the eigenvalues of $\mathcal{A}_0$. □

The study of the operator $\mathcal{L}$ defined in Lemma 5.A.1 will be essentially based on the following technical lemma.

Lemma 5.A.5. *Let $T>0$ and let $\mathcal{A}$ be a linear operator on a $\mathbb{C}$-Hilbert space E of finite dimension. Denote by $\lambda_1,\cdots,\lambda_q$ the eigenvalues of $\mathcal{A}$. Assume*

$$\lambda_j \notin \frac{2\pi i\mathbb{Z}}{T}\setminus\{0\},\qquad \textit{for } 1\le j\le q.$$

Then, the operator $\mathcal{B}=\int_0^T \exp(s\mathcal{A})ds$ is an isomorphism of E. Moreover,

$$\mathcal{B}(\ker\mathcal{A}^*)\bigoplus \mathrm{im}\mathcal{A} = E,\qquad \mathcal{B}(\ker\mathcal{A})\bigoplus(\ker\mathcal{A})^\perp = E.$$

and

$$\ker\mathcal{A} = \ker(\mathrm{Id}-\exp(T\mathcal{A})),\qquad \mathrm{im}\mathcal{A} = \mathrm{im}(\mathrm{Id}-\exp(T\mathcal{A})).$$

where im$\mathcal{A}$ *is the image (or the range) of $\mathcal{A}$.*

Proof. Step 1. First assume that in an appropriate basis, the matrix $M_\mathcal{A}$ of $\mathcal{A}$, reads $M_\mathcal{A} = \Lambda\mathrm{Id} + J_q$ where J_q is the $q\times q$ matrix given by

$$J_q = \begin{bmatrix} \lambda & 1 & 0 & 0 \\ 0 & \ddots & \ddots & 0 \\ 0 & 0 & \ddots & 1 \\ 0 & 0 & 0 & \lambda \end{bmatrix}.$$

If $\lambda = 0$, then the matrix $M_{\mathcal{B}}$ of $\mathcal{B}$ in this basis reads

$$M_{\mathcal{B}} = \int_0^T \left(I_q + sJ_q + \cdots + \frac{s^{q-1}(J_q)^{q-1}}{(q-1)!} \right) ds = TI_q + \frac{T^2 J_q}{2} + \cdots + \frac{T^q (J_q)^{q-1}}{q!}$$

which is a nonsingular, upper triangular matrix. Similarly, if $\lambda \notin \frac{2\pi i \mathbb{Z}}{T}$, then the matrix $M_{\mathcal{B}}$ of $\mathcal{B}$ in this basis reads

$$\begin{aligned} M_{\mathcal{B}} &= \int_0^T e^{\lambda s} \big(I_q + sJ_q + \cdots + \tfrac{s^{q-1}}{(q-1)!} (J_q)^{q-1} \big) ds \\ &= (e^{\lambda T} - 1) I_q + \int_0^T s e^{\lambda s} ds \; J_q + \cdots + \int_0^T \tfrac{s^{q-1}}{(q-1)!} e^{\lambda s} ds \; (J_q)^{q-1} \end{aligned}$$

which is a nonsingular, upper triangular matrix, since $(e^{T\lambda} - 1) \neq 0$. So, $\mathcal{B}$ is invertible. The general case, readily follows from this particular case, considering a basis for which the matrix of $\mathcal{A}$ is under Jordan normal form.

Step 2. Denote $F = \mathcal{B}(\ker \mathcal{A}^*)$. Since $\mathcal{B}$ is invertible,

$$\dim F = \dim(\ker \mathcal{A}^*) = \dim(\mathrm{im}(\mathcal{A})^\perp) = \dim E - \dim(\mathrm{im} \mathcal{A}).$$

Hence, to show that $E = F \bigoplus \mathrm{im} \mathcal{A}$, it is sufficient to show that $F \cap \mathcal{I}m\,(\mathcal{A}) = \{0\}$. So, let x be in $F \cap \mathrm{im} \mathcal{A}$. Thus, $x = \mathcal{B} z$ with $z \in \ker \mathcal{A}^* = (\mathrm{im} \mathcal{A})^\perp$ and

$$0 = \langle x, z \rangle = \int_0^T \langle \exp(s\mathcal{A}) z, z \rangle ds = \int_0^T \langle z, \exp(s\mathcal{A}^*) z \rangle ds = \int_0^T \langle z, z \rangle ds = T |z|^2$$

since for $z \in \ker \mathcal{A}^*$, $\exp(s\mathcal{A}^*) z = z$. Hence $z = 0$ and $x = 0$.

Step 3. $\mathcal{B}|_{\ker \mathcal{A}} = T\mathrm{Id}$. Thus, $\mathcal{B}(\ker \mathcal{A}) = \ker \mathcal{A}$ and thus

$$\mathcal{B}(\ker \mathcal{A}) \bigoplus (\ker \mathcal{A})^\perp = E.$$

Step 4. Observing that

$$\mathcal{A}\mathcal{B} = \int_0^T \mathcal{A} \exp(s\mathcal{A}) ds = \mathrm{Id} - \exp(T\mathcal{A})$$

we deduce that $\mathrm{im}\mathcal{A} = \mathrm{im}(\mathrm{Id} - \exp(T\mathcal{A}))$ since $\mathcal{B}$ is an isomorphism of E.

Step 5. Applying the result of the two previous steps to $\mathcal{A}^*$, whose eigenvalues are $\overline{\lambda_1}, \cdots, \overline{\lambda_q}$ and thus also satisfy the same non resonance criteria as the one of $\mathcal{A}$, we get

$$\ker \mathcal{A} = (\mathrm{im} \mathcal{A}^*)^\perp = (\mathrm{im}(\mathrm{Id} - \exp(T\mathcal{A}^*)))^\perp = \ker(\mathrm{Id} - \exp(T\mathcal{A})). \; \square$$

We conclude this subsubsection devoted to the study of $\mathcal{L}$ by

Lemma 5.A.6. *For any $k \geq 0$, $\mathcal{L}$ is a Banach isomorphism from the space $\mathcal{U}_T^{k+1,\perp}(\mathcal{M}_n)$ to $\mathcal{P}_T^k(\mathcal{M}_n)$.*

Proof. $\mathcal{L}$ is clearly a linear bounded operator from $\mathcal{U}_T^{k+1,\perp}(\mathcal{M}_n)$ to $\mathcal{P}_T^k(\mathcal{M}_n)$. conversely, let N be in $\mathcal{P}_T^k(\mathcal{M}_n)$. We look for $u = (Q_1, L_1)$ in $\mathcal{U}_T^{k+1,\perp}(\mathcal{M}_n)$ such that $\mathcal{L}(u) = N$ which is equivalent to

$$Q_1(t) = \exp(t\mathcal{A}_0)Q_1(0) + \int_0^t \exp((t-s)\mathcal{A}_0).(N(s) - L_1)ds.$$

This formula ensures that Q_1 is of class C^{k+1} since N is of class C^k. Moreover, Q_1 is T-periodic if

$$\begin{aligned}(\mathrm{Id} - \exp(T\mathcal{A}_0)).Q_1(0) &= \int_0^T \exp((T-s)\mathcal{A}_0).(N(s) - L_1)\, ds,\\ &= \int_0^T \exp((T-s)\mathcal{A}_0).N(s)ds - \mathcal{B}_0.L_1\end{aligned} \tag{5.6}$$

where

$$\mathcal{B}_0.L_1 = \int_0^T \exp(s\mathcal{A}_0).L_1\, ds.$$

Equation (5.6) as a solution if and only the right hand side belongs to $\mathrm{im}(\mathrm{Id} - \exp(T\mathcal{A}_0))$. Lemma 5.A.4 and the non resonance criteria (5.2) ensure that Lemma 5.A.5 holds for $\mathcal{A}_0$. Thus,

$$\mathrm{im}(\mathrm{Id} - \exp(T\mathcal{A}_0)) = \mathrm{im}\mathcal{A}_0, \qquad \ker(\mathrm{Id} - \exp(T\mathcal{A}_0)) = \ker\mathcal{A}_0$$

and

$$\mathcal{B}_0(\ker\mathcal{A}_0^*)\bigoplus \mathrm{im}\mathcal{A}_0 = \mathcal{M}_n(\mathbb{C}), \qquad \mathcal{B}_0(\ker\mathcal{A}_0)\bigoplus(\ker\mathcal{A}_0)^\perp = \mathcal{M}_n(\mathbb{C}).$$

Thus there exists a unique $\widetilde{L}_1 \in \ker\mathcal{A}_0^*$ such that (5.6) has at least one solution. Moreover, for $L_1 = \widetilde{L}_1$, there exists a unique $\widetilde{Q}_{10} \in (\ker\mathcal{A}_0)^\perp$ which is solution of (5.6) and all the solutions of (5.6) are given by

$$Q_1(0) = \widetilde{Q}_{10} + Q_{11} \qquad \text{where } Q_{11} \in \ker\mathcal{A}_0.$$

Finally, $\int_0^T Q_1(s)ds \in (\ker\mathcal{A}_0)^\perp$ holds if

$$\mathcal{B}_0.(\widetilde{Q}_{10} + Q_{11}) + \int_0^T\int_0^t \exp((t-s)\mathcal{A}_0).(N(s) - \widetilde{L}_1)ds \in (\ker\mathcal{A}_0)^\perp. \tag{5.7}$$

Since $\mathcal{B}_0(\ker\mathcal{A}_0)\bigoplus(\ker\mathcal{A}_0)^\perp = \mathcal{M}_n(\mathbb{C})$ there exists a unique Q_{11} and thus a unique $Q_1(0) = \widetilde{Q}_{10} + Q_{11}$ such that (5.7) holds. This shows that for any $N \in \mathcal{P}_T^k(\mathcal{M}_n)$ there exists a unique $u = (Q_1, L_1) \in \mathcal{U}_T^{k+1,\perp}(\mathcal{M}_n)$ such that $\mathcal{L}(u) = N$. □

5.A.3 The implicit equation

We have now enough material to prove Theorem 5.1.1.

(a): Observing that $\mathcal{N}$ is a C^r function from $\mathcal{U}_T^{r+1,\perp}(\mathcal{M}_n)\times\mathbb{R}^m$ to $\mathcal{P}_T^r(\mathcal{M}_n)$ such that $\mathcal{N}(0,0)=0$, and since $\mathcal{L}$ is an isomorphism from $\mathcal{U}_T^{r+1,\perp}(\mathcal{M}_n)\times\mathbb{R}^m$ to $\mathcal{P}_T^r(\mathcal{M}_n)$, the implicit function theorem ensure that (5.5) has a unique solution which reads $u=\varphi(\underline{\lambda})$ where $\varphi:\mathbb{R}^m\to\mathcal{U}_T^{r+1,\perp}(\mathcal{M}_n)$ is of class C^r and satisfies $\varphi(0)=0$. This completes the proof of (a).

(b): If M and L_0 are reversible with respect to some orthogonal symmetry S, then if $u=(Q_1,L_1)\in\mathcal{U}_T^{r+1,\perp}(\mathcal{M}_n)$ is a solution of the implicit equation (5.5), then $v=(SQ_1(-t)S,-SL_1S)$ is another solution which also lies in $\mathcal{U}_T^{r+1,\perp}(\mathcal{M}_n)$ since $S:\mathcal{M}_n(\mathbb{C})\to\mathcal{M}_n(\mathbb{C}):P\mapsto SPS$ is a self adjoint isomorphism which anticommutes with $\mathcal{A}_0$ and $\mathcal{A}_0^*$. The uniqueness of the solution given by the implicit function theorem ensures that $v=u$.

(c): If M, L_0 and S are real matrices, then if $u=(Q_1,L_1)\in\mathcal{U}_T^{r+1,\perp}(\mathcal{M}_n)$ is a solution of the implicit equation (5.5), then $v=(\overline{Q_1},\overline{L_1})$ is another solution which also lies in $\mathcal{U}_T^{r+1,\perp}(\mathcal{M}_n)$ since $\mathcal{R}:\mathcal{M}_n(\mathbb{C})\to\mathcal{M}_n(\mathbb{C}):P\mapsto\overline{P}$ is a semilinear isomorphism which commutes with $\mathcal{A}_0$ and $\mathcal{A}_0^*$. The uniqueness of the solution given by the implicit function theorem ensures that $v=u$. □

5.B Proof of Lemma 5.1.2

Step 1. Let us denote $\underline{\lambda}=(\rho,\widetilde{\rho})\in\mathbb{C}^2$ and $\widetilde{M}(t,\underline{\lambda}):(D(0,\rho_1))^2\to\mathcal{M}_n(\mathbb{C})$,

$$\widetilde{M}(t,\underline{\lambda})=\check{M}(\rho e^{-i\omega t},\widetilde{\rho}e^{-i\omega t}).$$

As for Theorem 5.1.1, we introduce the implicit equation

$$\mathcal{F}(u,\underline{\lambda}):=\mathcal{L}(u)-\mathcal{N}(u,\underline{\lambda})=0, \tag{5.8}$$

where $u=(Q_1,L_1)$ and

$$\mathcal{L}(u)=L_1+Q_1L_0-L_0Q_1+\frac{dQ_1}{dt},\quad \mathcal{N}(u,\underline{\lambda})=-Q_1L_1+\widetilde{M}(\cdot,\underline{\lambda})+\widetilde{M}(\cdot,\underline{\lambda})Q_1.$$

$\mathcal{F}$ can be seen as an analytic function from $\mathcal{U}_T^{r+1,\perp}(\mathcal{M}_n)\times(D(0,\rho_1))^2$ to $\mathcal{P}_T^r(\mathcal{M}_n)$ with $T=2\pi/\omega$. Using the Analytic Implicit Function Theorem at the origin, we get that there exists $\rho_2\le\rho_1$ and a unique $\widetilde{u}(\underline{\lambda})$ such that

$$\mathcal{F}(\widetilde{u},\underline{\lambda})=0$$

for $\underline{\lambda}$ close to 0. Moreover, $\underline{\lambda}\mapsto\widetilde{u}(\underline{\lambda})$ is an analytic function on $(D(0,\rho_2))^2$, i.e.

$$\widetilde{u}(\underline{\lambda})=\widetilde{u}(\rho,\widetilde{\rho})=\sum_{p+q\ge1}\widetilde{u}_{p,q}\,\rho^p\widetilde{\rho}^q$$

with

$$\widetilde{u}_{p,q} \in \mathcal{U}_T^{r+1,\perp}(\mathcal{M}_n), \qquad \|\widetilde{u}_{p,q}\|_{\mathcal{U}_T^{r+1,\perp}} \leq \frac{m_2}{(\rho_2)^{p+q}}. \tag{5.9}$$

Moreover, for small $|\rho|$, $\rho \in \mathbb{R}$, the function $u(\rho) = (Q_1(\cdot,\rho), L_1(\rho))$ given by Theorem 5.1.1, are found as the unique solution of $\mathcal{F}(u,\rho,\rho) = 0$ in the space $\mathcal{U}_T^{r+1,\perp}(\mathcal{M}_n)$. Hence,

$$u(\rho) = \widetilde{u}(\rho,\rho).$$

Step 2. Observing that for any $\varphi \in \mathbb{R}$ and every $t \in \mathbb{R}$

$$\widetilde{M}(\rho,\widetilde{\rho},t+\varphi) = \check{M}(\rho e^{i(t+\varphi)}, \widetilde{\rho}e^{-i(t+\varphi)}) = \widetilde{M}(\rho e^{i\varphi}, e^{-i\varphi}\widetilde{\rho}, t)$$

we get that if $\widetilde{v} = (Q_1(t), L_1)$ is a solution of $\mathcal{F}(\widetilde{v},\rho,\widetilde{\rho}) = 0$ in $\mathcal{U}_T^{r+1,\perp}(\mathcal{M}_n)$, then $\widetilde{w} = (Q_1(t+\varphi), L_1)$ is a solution of $\mathcal{F}(\widetilde{w},\rho e^{i\varphi},\widetilde{\rho}e^{-i\varphi}) = 0$ in $\mathcal{U}_T^{r+1,\perp}(\mathcal{M}_n)$. Thus, for every $\varphi \in \mathbb{R}$,

$$\widetilde{L}_1(\rho,\widetilde{\rho}) = \widetilde{L}_1(\rho e^{i\varphi}, e^{-i\varphi}\widetilde{\rho}), \qquad \widetilde{Q}_1(\rho,\widetilde{\rho},t+\varphi) = \widetilde{Q}_1(\rho e^{i\varphi}, e^{-i\varphi}\widetilde{\rho}, t), \tag{5.10}$$

where $\widetilde{u}(\rho,\widetilde{\rho}) = \big(\widetilde{L}_1(\rho,\widetilde{\rho}), \widetilde{Q}_1(\rho,\widetilde{\rho},t)\big)$ is the unique solution of

$$\mathcal{F}(u,\rho,\widetilde{\rho}) = 0$$

given by the Implicit Function Theorem for $|\rho|, |\widetilde{\rho}|$ close to 0. From (5.9) we get that $(\widetilde{Q}_1, \widetilde{L}_1)$ reads

$$\widetilde{Q}_1(\rho,\widetilde{\rho},t) = \sum_{p+q\geq 1} \widetilde{Q}_{1,p,q}(t)\, \rho^p \widetilde{\rho}^q, \qquad \widetilde{L}_1(\rho,\widetilde{\rho}) = \sum_{p+q\geq 1} \widetilde{L}_{1,p,q}\, \rho^p \widetilde{\rho}^q,$$

where $\widetilde{Q}_{1,p,q} \in \mathcal{P}_T^{r+1,\perp}(\mathcal{M}_n)$. Then, (5.10) ensures that

$$\widetilde{Q}_{1,p,q}(t+\varphi) = \widetilde{Q}_{1,p,q}(t)\, e^{i(p-q)\varphi}, \qquad \widetilde{L}_{1,p,q} = \widetilde{L}_{1,p,q}\, e^{i(p-q)\varphi}, \tag{5.11}$$

for every $\varphi \in \mathbb{R}$. Hence, $\widetilde{L}_{1,p,q} = 0$ for $p \neq q$ and $\widetilde{L}_1$ reads

$$\widetilde{L}_1(\rho,\widetilde{\rho}) = \sum_{k\geq 1} \widetilde{L}_{1,k,k}\rho^k \widetilde{\rho}^k. \tag{5.12}$$

Moreover, since $\widetilde{Q}_{1,p,q}$ is of class C^1 and T-periodic we can expand it in Fourier series

$$\widetilde{Q}_{1,p,q}(t) = \sum_{k\in\mathbb{Z}} \widetilde{\mathrm{Q}}_{1,p,q,k}\, e^{ik\omega t}$$

and (5.10) ensures that for every $\varphi \in \mathbb{R}$,

$$\widetilde{\mathrm{Q}}_{1,p,q,k}\, e^{i(p-q)\varphi} = \widetilde{\mathrm{Q}}_{1,p,q,k}\, e^{ik\varphi}.$$

So, $\widetilde{\mathrm{Q}}_{1,p,q,k} = 0$ for $k \neq p-q$ and $\widetilde{Q}_1$ reads

$$\widetilde{Q}_1(\rho, \widetilde{\rho}, t) = \sum_{p+q \geq 1} \widetilde{\mathrm{Q}}_{1,p,q,p-q} \, \mathrm{e}^{\mathrm{i}\omega(p-q)t} \rho^p \widetilde{\rho}^q, \tag{5.13}$$

Finally grouping together (5.9), (5.11) ,(5.12), (5.13), we get that for $\rho \in \mathbb{R}$,

$$u(\rho) = \big(Q_1(\rho, t), L_1(\rho)\big) = \big(\widetilde{Q}_1(\rho, \rho, t), \widetilde{L}_1(\rho, \rho)\big)$$

reads

$$L_1(\rho) = \sum_{k \geq 1} \rho^{2k} L_{1,2k}, \qquad \widetilde{Q}_1(\rho, \rho, t) = \sum_{p+q \geq 1} \widetilde{\mathrm{Q}}_{1,p,q,p-q} \, (\rho \mathrm{e}^{\mathrm{i}\omega t})^p (\rho \mathrm{e}^{\mathrm{i}\omega t})^q,$$

with

$$\widetilde{L}_{1,k,k} \in \ker \mathcal{A}_0^*, \qquad |\widetilde{L}_{1,k,k}| \leq \frac{m_2}{(\rho_2)^{2k}}, \qquad |\widetilde{\mathrm{Q}}_{1,p,q,p-q}|(1+\omega|p-q|) \leq \frac{m_2}{(\rho_2)^{p+q}}.$$

This completes the proof of Lemma 5.1.2. □

6. Inversion of affine equations around reversible homoclinic connections

When studying the persistence of homoclinic connections h to 0 of a perturbed system of the form

$$\frac{dY}{dt} = N(Y, \varepsilon) + R(Y, \varepsilon)$$

with

$$\frac{dh}{dt} = N(h, \varepsilon),$$

a natural idea is to look for a homoclinic solution of the full system under the form $Y = h + v$ where v is a perturbation term and to rewrite the equation for v in the form

$$\mathcal{L}(v) = \mathcal{N}(v, \varepsilon) \tag{6.1}$$

where

$$\mathcal{L}(v) = \frac{dv}{dt} - D_v N(h, \varepsilon).v,$$
$$\mathcal{N}(v, \varepsilon) = N(h + v, \varepsilon) - N(h, \varepsilon) - D_v N(h, \varepsilon).v + R(h + v, \varepsilon).$$

Then, for solving (6.1) in a space of functions v which satisfies $v \underset{t \to \pm\infty}{\longrightarrow} 0$ there are two usual methods which are essentially equivalent:

1. One can try to invert the linear map $\mathcal{L}$ in an appropriate space of functions which go to 0 at $\pm\infty$. If one succeeds in this inversion, then one can rewrite (6.1) as a fixed point equation

   $$v = \mathcal{L}^{-1}\mathcal{N}(v, \varepsilon)$$

   and try to solve it using the Contraction Mapping Theorem.
2. One can try to solve (6.1) using the Lyapunov Schmidt Method (see [Ha78] for a good introduction to this method). This method is based on the Implicit Function Theorem, and once more the crucial step is the inversion of $\mathcal{L}$ in appropriate Banach spaces.

So for both cases, the crucial point of the analysis is the inversion of affine equations with linear part obtained by linearization around homoclinic connections. When $\mathcal{L}$ is not invertible, then $\mathcal{N}(v,\varepsilon)$ must belong to the range of $\mathcal{L}$. This leads to so called compatibility or solvability conditions "the principal part" of which are nothing else than a generalized Melnikov function for systems of dimension greater than 2. The tools given in chapter 2, were developed for the study of these solvability conditions in the hyperdegenerated case when the principal part is given by an oscillatory integral which is exponentially small.

This chapter is devoted to such affine equations for reversible systems with linear part obtained by linearization around homoclinic connections. Sections 6.1, 6.2 and 6.3 deal with the linear homogeneous equations, i.e. with the kernel of $\mathcal{L}$. Sections 6.4 and 6.5 are devoted to affine equations and in particular to the solvability condition required to belong to the range of $\mathcal{L}$.

6.1 Explicit computation of a basis of solutions of the linear homogeneous equation

For solving affine equations of the form

$$\frac{dv}{dt} = A(t)v + f(t), \qquad v \in \mathbb{R}^n$$

where $A(t)$ is a $n \times n$ time dependent matrix, the first step is to build a basis of solutions of the linear homogeneous equation

$$\frac{dv}{dt} = A(t)v. \tag{6.2}$$

In general, it is not possible to compute explicitly all the solutions of such an equation. So in general, to obtain the behavior at infinity of a basis of solutions, one has to use mathematical tools such as the Theory of Dichotomies (see [Co78] for a good introduction). However, when (6.2) comes from the linearization around an explicitly known solution h of a vector field N, i.e.

$$\frac{dh}{dt} = N(h(t)), \qquad \text{and} \qquad A(t) = DN(h(t)), \tag{6.3}$$

one can find one or several solution and sometimes all the solutions of the homogeneous equations

$$\frac{dv}{dt} = DN(h(t)).v \tag{6.4}$$

using several "tricks":

1. Observe that $p(t) = \dfrac{dh}{dt}$ is a first solution of (6.4) (differentiate (6.3)). Moreover, since
$$p(t) = \frac{dh}{dt} = N(h(t)),$$
we deduce that
   - when h is reversible, i.e. $Sh(t) = h(-t)$, then p is antireversible, i.e. $Sp(t) = -p(-t)$.
   - when h is a homoclinic connection to 0, then p also tends to 0 at infinity with at least the same speed as h since
$$|p(t)| = |N(h(t))| \leq \sup_{Y \in K} (\|DN(Y)\|)\ |h(t)|.$$
   where K is any compact set containing the origin and the graph of h.
2. More generally, if h belongs to a smooth m-parameter family of solutions $H(t, \varphi_1, \cdots, \varphi_m)$, i.e. $h(t) = H(t, \varphi_1^0, \cdots, \varphi_m^0)$ with
$$\frac{\partial H}{\partial t} = N\big(H(t, \varphi_1, \cdots, \varphi_m)\big) \text{ for any } \varphi_1, \cdots, \varphi_m,$$
then the m functions
$$p_j(t) = \frac{\partial H}{\partial \varphi_j}(t, \varphi_1^0, \cdots, \varphi_m^0) \qquad \text{for } 1 \leq j \leq m,$$
are solutions of the homogeneous equation (6.4). However, they are not always linearly independent. The case 1 , is a particular case of this one with a one parameter family of solutions given by $H(t, \varphi) = h(t + \varphi)$.
3. It happens frequently when N is a normal form that $DN(h(t))$ is block diagonal. So the problem of the inversion can be solved for each block separately, which leads to problems of smaller dimension which are in general, easier to solve.
4. Another helpful observation is that the Wronskian determinant $W(t)$ of a basis of solutions $(b_1(t), \cdots, b_n(t))$ of the homogeneous equation (6.4) defined by
$$W(t) = \det(b_1(t), \cdots, b_n(t))$$
is always explicitly known. Indeed it is given by
$$W(t) = W(0) \exp\left(\int_0^t \operatorname{tr}(DN(h(s)))\ ds\right)$$
where $\operatorname{tr}(A)$ is the trace of a matrix A. This observation happens to be useful when one knows explicitly $n-1$ independent solutions and wants to compute a last one. For $n-1$ vectors x_i, $1 \leq i \leq n-1$, denote by $x_1 \wedge \cdots \wedge x_{n-1}$ the unique vector y of $\mathbb{R}^n$ such that for every $z \in \mathbb{R}^n$,
$$\det(x_1, \cdots, x_{n-1}, z) = \langle y, z \rangle$$
where $\langle \cdot, \cdot \rangle$ is the usual scalar product in $\mathbb{R}^n$. We then have:

Lemma 6.1.1. *Let $b_1(t), \cdots, b_{n-1}(t)$, $n-1$ be linearly independent solutions of a linear time dependent equation*

$$\frac{dv}{dt} = A(t)v$$

where $A(t)$ is a continuous $n \times n$ matrix function defined on $\mathbb{R}$. Then the last solution b such that the Wronskian determinant satisfies

$$W(0) := \det(b_1(0), \cdots, b_{n-1}(0), b(0)) = 1,$$

is given by

$$b(t) = \sum_{i=1}^{n-1} \alpha_i(t)\ b_i(t) + \alpha(t)\ r(t)$$

where $r(t) = b_1(t) \wedge \cdots \wedge b_{n-1}(t)$,

$$\alpha(t) = \frac{W(t)}{\langle r(t), r(t)\rangle}, \qquad W(t) = \exp\left(\int_0^t \operatorname{tr}(A(s))ds\right),$$

and

$$\alpha_i(t) = \int_0^t \alpha(s)\langle b_i^*(s), \left(Ar - \frac{dr}{dt}\right)(s)\rangle ds$$

with

$$b_i^*(t) = (-1)^{n-i} \frac{b_1(t) \wedge \cdots \wedge b_{i-1}(t) \wedge b_{i+1}(t) \cdots \wedge b_{n-1}(t) \wedge r(t)}{\langle r(t), r(t)\rangle}.$$

6.2 Complex singularities of solutions of the homogeneous equation

In the previous section we recalled the classical methods for computing a real basis of solution of a linear equation obtained by linearization of a system around a homoclinic connection to 0. The solutions of the basis obtained are essentially characterized by their reversibility properties and by their behavior at infinity.

When complexifying time for using the exponential tools given in chapter 2, other important information required on the solutions of the basis is the location and the nature of their singularities in the complex field. The theory of Fuchs enables us to compute another basis of solutions characterized by their singularities. Then, the difficulty is to find the link between the two bases. See for illustration, Example 6.2.10 below for the $0^{2+}i\omega$ resonance.

6.2.1 Theory of Fuchs for linear systems

A first part of the theory of Fuchs deals with systems of the form

$$\frac{dv}{d\xi} = A(\xi)v \tag{6.5}$$

where A is an $n \times n$ matrix with at most an isolated singularity at some point ξ_0, but which is otherwise single valued and holomorphic near ξ_0. A second part deals with equation of nth order. It is given in the next paragraph. *We only recall here the main result of the theory of Fuchs. For complete proofs and example see for instance* [CL55] *Chap. 4 and 5.*

A first result gives a rather qualitative description of the fundamental matrix of (6.5) when A has an arbitrary isolated singularity.

Theorem 6.2.1. *Let $\xi \mapsto A(\xi)$ be a single valued, holomorphic $n \times n$ matrix function in the punctured disk $\dot{D}(\xi_0, r) := \{\xi \in \mathbb{C}/0 < |\xi - \xi_0| < r\}$.*

(a) *Since $\dot{D}(\xi_0, r)$ is not a simply connected domain, the solutions of (6.5) are not necessary single valued. However, the problem can be considered in a simply connected domain if the domain is allowed to be many-sheeted: the Contraction Mapping theorem ensures the existence of an analytic fundamental matrix on Ω given by*

$$\Omega : 0 < \rho < r, \ -\infty < \theta < +\infty, \qquad \text{and } \xi - \xi_0 = \rho e^{i\theta}.$$

(b) *Moreover, every fundamental matrix Ψ of (6.5) has the form*

$$\Psi(\xi) = Q(\xi) e^{[\log(\xi - \xi_0)]B}$$

where Q is a single valued holomorphic $n \times n$ matrix function in $\dot{D}(\xi_0, r)$ and B is a constant matrix.

Remark 6.2.2. Denote by $J := Jor(B)$ the Jordan normal form of B. J and B are linked by $B = P^{-1}JP$ where $P \in GL_n(\mathbb{C})$. Then, $\Psi(\xi) = Q(\xi)P^{-1}e^{\log(\xi)J}P$, and $\widetilde{\Psi} = \Psi P^{-1}$ is another fundamental matrix of (6.5) which reads $\widetilde{\Psi}(\xi) = \widetilde{Q}(\xi)e^{\log(\xi - \xi_0)J}$ where $\widetilde{Q}$ is a single valued holomorphic $n \times n$ matrix function in $\dot{D}(\xi_0, r)$. Thus $\widetilde{\Psi}$ reads

$$\widetilde{\Psi}(\xi) = \widetilde{Q}(\xi) \sum_{1 \le j \le q} \sum_{k=1}^{m(\lambda_j)} \frac{(J - \lambda_j I_n)^k}{k!} (\log(\xi - \xi_0))^k \Pi_j$$

where λ_j, $1 \le j \le q$ are the eigenvalues of B (or J) and Π_j are the projectors on the generalized eigenspace corresponding to λ_j. Hence, for each eigenvalue λ_j of B (or J), equation (6.5) always admits a solution of the form $v(\xi) = (\xi - \xi_0)^{\lambda_j}$.

When A has only a pole of order one, a more precise description of the fundamental matrix can be obtained.

Theorem 6.2.3. *Let A be a $n \times n$ matrix function given by*

$$A(\xi) = \xi^{-1}R + \sum_{m=0}^{+\infty} \xi^m A_m \tag{6.6}$$

where $R \neq 0$ and A_m are constant matrices, and where the power series converges for $|\xi| < r$, $r > 0$.

(a) *If R has eigenvalues which do not differ by positive integers, then (6.5) has a fundamental matrix Ψ of the form*

$$\Psi(\xi) = Q(\xi)\mathrm{e}^{\log(\xi)R} \qquad \text{for } 0 < |\xi| < r_0, \text{ with } r_0 > 0,$$

where $Q(\xi)$ is a convergent power series

$$Q(\xi) = \sum_{m=0}^{+\infty} \xi^m Q_m, \qquad Q_0 = I_n.$$

(b) *In the general case, equation (6.5) has a fundamental matrix of the form*

$$\Psi(\xi) = Q(\xi)\mathrm{e}^{\log(\xi)\widehat{R}} \qquad \text{for } 0 < |\xi| < c, \text{ with } c > 0,$$

where $Q(\xi)$ is a convergent power series

$$Q(\xi) = \sum_{m=0}^{+\infty} \xi^m Q_m$$

and where $\widehat{R}$ is a constant matrix with eigenvalues which do not differ by positive integers.

Remark 6.2.4. The general case (b) can be deduced from the particular case (a) using sufficiently many transformation V of the type given by Lemma 6.2.5 below. We give the proof of this lemma because it gives the explicit form of V and enables to compute explicitly $\widehat{R}$ as a function of R.

Lemma 6.2.5. *Let A be a $n \times n$ matrix function given by (6.6). Let λ_j, $1 \leq j \leq q$, be the eigenvalues of R.*
Then, there exists a matrix function V of ξ invertible for $\xi \neq 0$ and linear in ξ, such that the transformation $v = V\widetilde{v}$ transforms (6.5) into an equivalent system

$$\frac{d\widetilde{v}}{d\xi} = \widetilde{A}(\xi)\widetilde{v}, \qquad \textit{with } \widetilde{A}(\xi) = \xi^{-1}\widetilde{R} + \sum_{m=0}^{+\infty} \xi^m \widetilde{A}_m \tag{6.7}$$

where $\widetilde{R}$ *has for eigenvalues* $\lambda_1 - 1, \lambda_2, \cdots, \lambda_q$.

Proof. We first assume that R is under Jordan normal form

$$R = \begin{bmatrix} R_1 & 0 \\ 0 & R_2 \end{bmatrix}$$

where R_1 is the $p_1 \times p_1$ matrix equal to the direct sum of all the Jordan block corresponding to the eigenvalue λ_1,

$$R_1 = \begin{bmatrix} \lambda_1 & \delta_2 & 0 & 0 \\ 0 & \lambda_1 & \ddots & 0 \\ & \ddots & \ddots & \delta_{p_1} \\ 0 & & 0 & \lambda_1 \end{bmatrix}$$

δ_i being 0 or 1. Then, denote by U the matrix

$$U = \begin{bmatrix} \xi I_{p_1} & 0 \\ 0 & I_{n-p_1} \end{bmatrix}.$$

Clearly U is invertible for $\xi \neq 0$ and

$$U^{-1} = \begin{bmatrix} \xi^{-1} I_{p_1} & 0 \\ 0 & I_{n-p_1} \end{bmatrix}.$$

Then, setting $v = U\widetilde{v}$, (6.5) is equivalent to (6.7) with

$$\begin{aligned} \widetilde{A}(\xi) &= \xi^{-1}U^{-1}RU - U^{-1}\frac{dU}{d\xi} + \sum_{m=0}^{+\infty} \xi^m U^{-1} A_m U \\ &= \xi^{-1}\widetilde{R} + \sum_{m=0}^{+\infty} \xi^m \widetilde{A}_m \end{aligned}$$

where

$$\widetilde{R} = \begin{bmatrix} R_1 - I_{p_1} & A_{12} \\ 0 & R_2 \end{bmatrix} \qquad \text{with } A_0 = \begin{bmatrix} A_{11} & A_{12} \\ A_{21} & A_{22} \end{bmatrix}$$

where A_{11} is the block of A_0 of length and width p_1. Thus, $\widetilde{R}$ has the required properties.

In case R is not under Jordan normal form, the transformation U can be replaced by PU, where P is chosen so that $P^{-1}RP$ is under Jordan normal form. Setting $V = PU$ the lemma is proved. □

Theorem 6.2.3 gives the form of the fundamental matrix for (6.5) when A has at most a pole of order one at the origin. However, it does not give explicit solutions. When one wants to compute *explicitly* the fundamental matrix the following theorem 6.2.6 appears to be very useful. Indeed it ensures that when A has at most a pole of order one at the origin , *any "formal solution"* of (6.5) is in fact *a true solution.* So, one can compute explicitly true solutions by looking for formal solutions with an a priori form given for instance by theorem 6.2.3. The coefficients of such solutions will be given by induction. Then, Theorem 6.2.6 will ensure that all the series involved in any formal solutions are convergent.

Theorem 6.2.6. *Let A be a $n \times n$ matrix function given by*

$$A(\xi) = \xi^{-1}R + \sum_{m=0}^{+\infty} \xi^m A_m$$

where $R \neq 0$ and A_m are constant matrices, and where the power series converges for $|\xi| < r$, $r > 0$. Then, if v is a formal solution of (6.5) of the form

$$v(\xi) = \sum_{j=0}^{J}\sum_{k=0}^{K}\left(\sum_{\ell=-m_{jk}}^{+\infty} a_{jk\ell}\,\xi^{\ell}\right)\xi^{\lambda_j}(\log\xi)^k$$

where $J, K, m_{jk} \in \mathbb{N}$ and λ_j, $a_{jk\ell} \in \mathbb{C}$, then formal series occurring in v are convergent in $0 < |\xi| < r$.

Remark 6.2.7. This theorem is no longer true when A has a pole of order 2. This leads to serious difficulties. See for instance [CL55] Chap.5 for more details.

6.2.2 Theory of Fuchs for equation of the nth order

This paragraph is devoted to the study of an equation of nth order

$$\sum_{m=0}^{n} a_{n-m}(\xi)\frac{d^m\varphi}{d\xi^m} = 0, \qquad \text{with } a_0(\xi) \equiv 1. \tag{6.8}$$

where the a_k are single-valued and holomorphic in a punctured neighborhood of a point ξ_0. The study of such equations can be deduced from the study of the usual associated first order linear system. With this method the detailed description of the fundamental matrix given by Theorem 6.2.3 can only be obtained for equations of the nth order (6.8) if the coefficients a_k have at most a pole of order 1 at ξ_0. However, using another associated system, one can apply the result of Theorem 6.2.3 to a far more large class of equations of nth order:

Lemma 6.2.8. *Let a_k, $0 \le k \le n$ be n single valued, holomorphic functions defined on a punctured neighborhood of ξ_0 such that*

$$a_0(\xi) \equiv 1, \qquad a_k(\xi) = (\xi - \xi_0)^{-k} b_k(\xi) \text{ for } 1 \le k \le n.$$

where b_k are holomorphic functions. To every holomorphic function φ defined on a punctured neighborhood of ξ_0, we associate a vector function $\widehat{\varphi}$ in $\mathbb{C}^n$ with components $\widehat{\varphi}_1, \cdots, \widehat{\varphi}_n$ given by

$$\widehat{\varphi}_k(\xi) = (\xi - \xi_0)^{k-1} \frac{d^{k-1}\varphi}{d\xi^{k-1}} \qquad \text{for } 1 \le k \le n.$$

Then, φ is solution of (6.8) if and only if $\widehat{\varphi}$ is solution of the first order system

$$\frac{dv}{d\xi} = A(\xi)v \tag{6.9}$$

with

$$A(\xi) = (\xi - \xi_0)^{-1} \begin{bmatrix} 0 & 1 & 0 & 0 & \cdots & 0 \\ 0 & 1 & 1 & 0 & \cdots & 0 \\ 0 & 0 & 2 & 1 & \cdots & 0 \\ 0 & 0 & 0 & 3 & \cdots & 0 \\ \cdot & \cdot & \cdot & \cdot & \cdots & \cdot \\ 0 & 0 & 0 & 0 & \cdots & 1 \\ -b_n & -b_{n-1} & \cdot & \cdot & \cdots & (n-1) - b_1 \end{bmatrix} \tag{6.10}$$

Proof. The proof directly follows from

$$(\xi - \xi_0)\widehat{\varphi}_k' = (k-1)\widehat{\varphi}_k + \widehat{\varphi}_{k+1} \qquad \text{for } 1 \le k \le n-1,$$
$$(\xi - \xi_0)\widehat{\varphi}_n' = (n-1)\widehat{\varphi}_n + (\xi - \xi_0)^n \widehat{\varphi}^{(n)}. \square$$

Remark 6.2.9. As already explained, under the hypothesis of Lemma 6.2.8 the explicit computation of a fundamental basis can be performed by considering (6.9), (6.10) and using Theorem 6.2.3. If (6.9) is rewritten in the form

$$\frac{dv}{d\xi} = \left[(\xi - \xi_0)^{-1} R + \sum_{m=0}^{+\infty} (\xi - \xi_0)^m \, A_m \right] v$$

where R, A_m are constant matrices, and if the coefficients b_k reads

$$b_k(\xi) = \sum_{m=0}^{+\infty} b_{k,m} \xi^m$$

then the eigenvalues of R are the roots of the equation

$$\lambda(\lambda-1)\cdots(\lambda-n+1) + b_{1,0}\lambda(\lambda-1)\cdots(\lambda-n+2) + \cdots + b_{n-1,0}\lambda + b_{n,0} = 0.$$

This equation is called the *indicial equation* for (6.8) relative to the singular point ξ_0.

So, if for instance the eigenvalues $\lambda_1, \cdots, \lambda_n$ of R are distinct and do not differ by positive integers then a basis of solution of (6.8) is given by

$$\varphi_k(\xi) = (\xi - \xi_0)^{\lambda_k} p_k(\xi)$$

where p_k are single valued holomorphic functions in a neighborhood of ξ_0 satisfying $p_k(\xi_0) \neq 0$. For more complicated cases, we must proceed as explained for Theorem 6.2.3.

Example 6.2.10. (Linearization around the homoclinic connection to 0 of the normal form of order 2 for the $0^{2+}i\omega$ resonance)

In Example 3.2.9 we gave the general normal form system of order k for the $0^{2+}i\omega$ resonance. The normal form system of order 2 reads

$$\begin{aligned}
\frac{d\widetilde{\alpha}}{d\tau} &= \widetilde{\beta}, \\
\frac{d\widetilde{\beta}}{d\tau} &= n_1\mu\widetilde{\alpha} + n_2\widetilde{\alpha}^2 + n_3(\widetilde{A}^2 + \widetilde{B}^2), \\
\frac{d\widetilde{A}}{d\tau} &= -\widetilde{B}(\omega + \mu\omega_1 + m\widetilde{\alpha}), \\
\frac{d\widetilde{B}}{d\tau} &= \widetilde{A}(\omega + \mu\omega_1 + m\widetilde{\alpha}).
\end{aligned}$$

where μ is the bifurcation parameter close to 0 and ω_1, m, n_1, n_2, n_3 are real fixed numbers. In this system of coordinates the symmetry of reversibility reads

$$S(\widetilde{\alpha}, \widetilde{\beta}, \widetilde{A}, \widetilde{B}) = (\widetilde{\alpha}, -\widetilde{\beta}, \widetilde{A}, -\widetilde{B}).$$

Assuming that $n_1 \neq 0$, $n_2 \neq 0$, the normal form system admits a reversible homoclinic connection $\widetilde{h}$ for $n_1\mu > 0$ given by

$$\widetilde{h} = \left(\frac{-3\nu^2}{2n_2} \frac{1}{\cosh^2(\frac{1}{2}\nu\tau)}, \frac{3\nu^3 \tanh(\frac{1}{2}\nu\tau)}{2n_2 \cosh^2(\frac{1}{2}\nu\tau)}, 0, 0 \right) \qquad \text{with } \nu = \sqrt{n_1\mu}.$$

In order to work with a homoclinic connection which does not depend on the bifurcation parameter we perform a scaling for time, space, and parameter

$$t = \nu\tau,\ \widetilde{\alpha} = -\frac{3\nu^2}{2n_2}\alpha,\ \widetilde{\beta} = -\frac{3\nu^3}{2n_2}\beta,\ \widetilde{A} = \nu^2 A,\ \widetilde{B} = \nu^2 B,\ \nu = \sqrt{n_1\mu}.$$

After scaling, the normal form system of order 2 reads

$$\frac{dY}{dt} = N(Y, \nu)$$

with

$$Y = (\alpha, \beta, A, B), \quad \text{and} \quad N(Y, \nu) = \begin{pmatrix} \beta \\ \alpha - \frac{3}{2}\alpha^2 + d(A^2 + B^2) \\ -B(\frac{\omega}{\nu} + a\nu + b\nu\alpha) \\ A(\frac{\omega}{\nu} + a\nu + b\nu\alpha) \end{pmatrix}$$

where a, b, d are three real constants given by

$$a = \frac{\omega_1}{n_1}, \; b = -\frac{3m}{2n_2}, \; d = -\frac{2n_2 n_3}{3}.$$

This system admits a unique (up to time shift) homoclinic connection h to 0, given by $h(t) = (\alpha^h(t), \beta^h(t), 0, 0)$ with

$$\alpha^h(t) = \cosh^{-2}(\tfrac{1}{2}t), \qquad \beta^h(t) = -\cosh^{-2}(\tfrac{1}{2}t)\tanh(\tfrac{1}{2}t). \tag{6.11}$$

Step 1. Computation of a real basis of solutions characterized by reversibility properties and behavior at infinity

This orbit is reversible. Using the "tricks" given in Section 6.1 we are able to compute explicitly a basis of solutions of the equation linearized around h

$$\frac{dv}{dt} = DN(h(t), \nu).v \tag{6.12}$$

with

$$DN(h(t)) = \begin{bmatrix} 0 & 1 & 0 & 0 \\ 1 - 3\alpha^h & 0 & 0 & 0 \\ 0 & 0 & 0 & -(\frac{\omega}{\nu} + a\nu + b\nu\alpha^h) \\ 0 & 0 & (\frac{\omega}{\nu} + a\nu + b\nu\alpha^h) & 0 \end{bmatrix}.$$

Indeed a first solution is given by $p = \frac{dh}{dt} = (p_\alpha, p_\beta, 0, 0)$ with

$$p_\alpha(t) = \beta^h(t) = -\frac{\tanh(\frac{1}{2}t)}{\cosh^2(\frac{1}{2}t)},$$

$$p_\beta(t) = \alpha^h(t) - \tfrac{3}{2}(\alpha^h(t))^2 = \frac{1}{\cosh^2(\frac{1}{2}t)} - \frac{3}{2}\frac{1}{\cosh^4(\frac{1}{2}t)}.$$

and it satisfies

$$Sp(-t) = -p(-t), \qquad \sup_{t \in \mathbb{R}}(|p(t)|e^{|t|}) < +\infty.$$

A second solution is obtained using the "Wronskian trick" for the first diagonal block. Denote q the solution of (6.12) which satisfies $q(0) = (1, 0, 0, 0)$.

Observe that since $DN(h)$ is block diagonal, $q(t) = (q_\alpha(t), q_\beta(t), 0, 0)$ holds for $t \in \mathbb{R}$. So we work with the reduce linear time dependent system

$$\frac{dq}{dt} = A(t)q \qquad \text{with } A(t) = \begin{bmatrix} 0 & 1 \\ 1 - 3\alpha^h(t) & 0 \end{bmatrix} \tag{6.13}$$

For computing q it suffices to observe that $\operatorname{tr}(A(t)) = 0$. So the Wronskian of the reduced system is constant. The Wronskian equation reads

$$W(0) = \frac{1}{2} = W(t) = p_\alpha(t)q_\beta(t) - p_\beta(t)q_\alpha(t) = p_\alpha(t)\frac{dq_\alpha}{dt}(t) - p_\beta(t)q_\alpha(t).$$

Thus, since $p(t)$ is explicitly known, this equation enables us to compute q explicitly. We obtain

$$\begin{aligned} q_\alpha(t) &= \frac{1}{\cosh^2(\frac{1}{2}t)} - \frac{15}{16}\frac{t\tanh(\frac{1}{2}t)}{\cosh^2(\frac{1}{2}t)} - \frac{1}{8}\cosh(t)\tanh^2(\tfrac{1}{2}t) - \tanh^2(\tfrac{1}{2}t), \\ q_\beta(t) &= \frac{dq_\alpha}{dt}. \end{aligned}$$

Moreover, q is reversible and satisfies

$$\sup_{t\in\mathbb{R}} |q(t)|e^{-|t|} < +\infty.$$

The two last solution are found explicitly because of the special form of the second diagonal block of $DN(h(t))$. Denote by r_+, r_- the solutions of (6.12) satisfying $r_+(0) = (0,0,1,0)$ and $r_- = (0,0,0,1)$. Then r_+, r_- are given by

$$r_+(t) = \big(0, 0, \cos\psi_\nu(t), \sin\psi_\nu(t)\big), \qquad r_-(t) = \big(0, 0, -\sin\psi_\nu(t), \cos\psi_\nu(t)\big)$$

where $\psi_\nu(t) = (\omega/\nu + a\nu)t + 2b\nu\tanh(\frac{1}{2}t)$ and they satisfy

$$|r_+(t)| = |r_-(t)| = 1 \qquad \text{for } t \in \mathbb{R}.$$

Finally, r_+ is reversible whereas r_- is antireversible.

Remark. This first step ensures that when $\mathcal{L} = \dfrac{d}{dt} - DN(h,\nu)$ is restricted to the space of reversible functions which decay to 0 at $\pm\infty$, its kernel is equal to $\{0\}$.

Step 2. Computation of a Fuchs basis near $i\pi$

At the previous step we computed a real basis of solutions of (6.13). This basis is characterized by reversibility properties and behavior at infinity of the solutions.

Now we are interested in a second basis adapted when complexifying time and in particular when t is close to $i\pi$. Since all the computations of the previous step are explicit, we readily check that (p, q, r_+, r_-) are holomorphic in $\mathbb{C} \setminus \{in\pi, n \in \mathbb{Z}\}$ and that

– p_α, q_α (resp. p_β, q_β) admit poles of order 3 at $in\pi$, (resp. poles of order 4).
– r_+, r_- have essential singularities at $in\pi$.

In what follows we show how the Theory of Fuchs gives "a better basis" near $i\pi$. Indeed this theory ensures the existence of a basis (p, u, r_+, r_-) where u which reads $u = (u_\alpha, u_\beta, 0, 0)$ is such that u_α has a zero of order 4 and u_β, a zero of order 3.

The theory of Fuchs is local, and it may be difficult to link the real basis characterized by its behavior at infinity and the local basis given by Theory of Fuchs. Here, since the computations of Step 1 are fully explicit, there is no difficulty, and we know explicitly the link between (p, u) and (p, q) (see (6.16)). These two bases happen to be very useful when studying the persistence of homoclinic connection h when the normal form is perturbed by higher order terms. The first basis is useful when describing behavior at infinity (we look for homoclinic connection to 0). The second is useful for describing the behavior near $i\pi$ (the study of the persistence of h requires the exponential tool developed in chapter 2). Since the link between the two bases is explicit there is no difficulty to link the two behaviors (see chapter 7 for details).

So, as previously explained our aim is now to compute a basis of solutions of (6.12) which have a good behavior near $i\pi$. More precisely we are interested in a basis of the 2 dimensional system (6.13) which corresponds to the first diagonal block of $DN(h)$. Observe from (6.13) that A is holomorphic in $\mathbb{C} \setminus \{in\pi, n \in \mathbb{Z}\}$ $(\alpha^h(\xi) = \cosh^{-2}(\frac{1}{2}\xi))$.

To study singularities at $i\pi$ it is more convenient to set $\xi = i\pi + \zeta$ and $\widetilde{\alpha}(\zeta) = \alpha(i\pi + \zeta)$. With this notation, system (6.13) can be rewritten as an equation of order 2

$$\widetilde{\alpha}''(\zeta) + \zeta^{-2} b_2(\zeta)\widetilde{\alpha}(\zeta) = 0 \qquad \text{with } b_2(\zeta) = -\zeta^2\Big(1 + \frac{3}{\sinh^2(\zeta/2)}\Big). \tag{6.14}$$

Following Lemma 6.2.8, setting $\widehat{\varphi}(\zeta) = (\alpha(\zeta), \zeta\alpha'(\zeta))$, this equation is associated with the first order system

$$\frac{d\widehat{\varphi}}{d\zeta} = \frac{1}{\zeta}\begin{bmatrix} 0 & 1 \\ -b_2(\zeta) & 1 \end{bmatrix}\widehat{\varphi} = \Big[\zeta^{-1}R + \sum_{m=0}^{+\infty} \zeta^m A_m\Big]\widehat{\varphi}. \tag{6.15}$$

Then, Remark 6.2.9 ensures that the eigenvalues of R are the solutions of the indicial equation

$$0 = \lambda(\lambda - 1) + b_{2,0} \qquad \text{where } b_{2,0} = b_2(0) = -12.$$

Thus the eigenvalues of R are -3 and 4. They differ by positive integer. Then, Theorem 6.2.3 and Lemma 6.2.5 ensure that (6.15) admits a fundamental matrix Ψ of the form

$$\Psi(\zeta) = Q(\zeta)e^{\log(\zeta)\widehat{R}}$$

where Q is an holomorphic function in a neighborhood of 0 and

$$\widehat{R} = \begin{bmatrix} -3 & \delta \\ 0 & -3 \end{bmatrix} \qquad \text{with } \delta \in \mathbb{R}.$$

Hence, (6.15) admits a basis of solution of the form

$$\widehat{\varphi}_1(\zeta) = \zeta^{-3}\widehat{\psi}_1(\zeta), \qquad \widehat{\varphi}_2(\zeta) = \delta\zeta^{-3}\log(\zeta)\widehat{\psi}_1(\zeta) + \zeta^{-3}\widehat{\psi}_2(\zeta)$$

where $\widehat{\psi}_i : \mathbb{C} \to \mathbb{C}^2$, $i = 1, 2$ are two holomorphic functions in a neighborhood of 0. Then, Theorem 6.2.6 ensures that all formal solutions of (6.15) or equivalently of (6.14) are in fact true solutions. So, for computing true solutions of (6.14), it suffices to look for them as formal solutions of the form

$$\widetilde{\alpha}_1(\zeta) = \sum_{n=-3}^{+\infty} \alpha_{1n}\,\zeta^n, \qquad \widetilde{\alpha}_2(\zeta) = \delta\log(\zeta)\sum_{n=-3}^{+\infty} \alpha_{1n}\,\zeta^n + \sum_{n=-3}^{+\infty} \alpha_{2n}\,\zeta^n.$$

Looking for a formal solution $\widetilde{\alpha}_1$ with the above form we get, in fact two independent formal solutions. Since, all formal solutions are true solutions, this simply ensures that $\delta = 0$. So, a basis of true solutions reads

$$\widetilde{\alpha}_1(\zeta) = \sum_{n=-2}^{+\infty} a_{1,n}\,\zeta^{2n+1}, \qquad \widetilde{\alpha}_2(\zeta) = \sum_{n=2}^{+\infty} a_{2,n}\,\zeta^{2n}$$

with

$$a_{1,-2} = 1, \qquad a_{1,-1} = a_{1,0} = 0 \qquad a_{1,n+1} = \frac{\sum\limits_{p+q=n} a_{1,p}\,b_q}{12 - (2n+2)(2n+3)}$$

$$a_{2,2} = 1, \qquad a_{2,n+1} = \frac{\sum\limits_{p+q=n} a_{2,p}\,b_q}{12 - (2n+1)(2n+2)}$$

where

$$b_2(\zeta) = -\zeta^2\left(1 + \frac{3}{\sinh^2(\zeta/2)}\right) = -\left(12 + \sum_{q=0}^{+\infty} b_q\,\zeta^{2q+2}\right), \qquad (b_0 = 0).$$

Observe that the first solution $\widetilde{\alpha}_1$ has a pole of order 3 and is odd whereas the second $\widetilde{\alpha}_2$ has a zero of order 4 and is even.

Coming back to our original problem coordinates, this analysis ensure the existence of a pair of solutions $(\mathfrak{p}, \mathfrak{q})$ of (6.12) which read $\mathfrak{p} = (\mathfrak{p}_\alpha, \mathfrak{p}_\beta, 0, 0)$ and $\mathfrak{q} = (\mathfrak{q}_\alpha, \mathfrak{q}_\beta, 0, 0)$ where

$$\mathfrak{p}_\alpha(\xi) = \widetilde{\alpha}_1(\xi - i\pi) \underset{\xi \to i\pi}{\sim} \frac{1}{(\xi - i\pi)^3} \qquad \mathfrak{p}_\beta = \frac{d\mathfrak{p}_\alpha}{d\xi},$$

$$\mathfrak{q}_\alpha = \widetilde{\alpha}_2(\xi - i\pi) \underset{\xi \to i\pi}{\sim} (\xi - i\pi)^4 \qquad \mathfrak{q}_\beta = \frac{d\mathfrak{q}_\alpha}{d\xi}.$$

Finally, using the evenness properties of the two pairs with respect to $\xi - i\pi$, we finally get

$$p(\xi) = 8\mathfrak{p}(\xi), \qquad q(\xi) = \frac{15}{16} i\pi p(\xi) + u(\xi) \tag{6.16}$$

where $u = (u_\alpha, u_\beta, 0, 0) = \dfrac{1}{112}\mathfrak{q}$ reads

$$\begin{aligned} u_\alpha(\xi) &= -\frac{15}{16}(\xi - i\pi)\frac{\tanh\frac{1}{2}\xi}{\cosh^2\frac{1}{2}\xi} - \frac{1}{8}\cosh\xi + \frac{15}{8}\frac{1}{\cosh^2\frac{1}{2}\xi} - \frac{3}{4}, \\ u_\beta(\xi) &= \frac{du_\alpha}{d\xi}(\xi). \end{aligned}$$

6.3 Linearization around homoclinic connections

This section is devoted to affine equations obtained by linearization around a reversible homoclinic connection generated by a pair of opposite, real, eigenvalues $\pm\lambda_0$, $\lambda_0 > 0$ or by four conjugate eigenvalues $\pm i\omega_0 \pm \lambda_0$, $\omega_0, \lambda_0 > 0$. The following lemmas give in each case, a particular basis of solutions characterized by their smoothness, reversibility properties and by their behavior at infinity. Moreover, in each case, a similar set of informations is also given for the dual basis which is required when solving the corresponding affine equations using the variation of constants formula. We first introduce some Banach spaces necessary to the description of the basis.

Definition 6.3.1.

(a) *Let I be an open interval of $\mathbb{R}$. For $k, n \in \mathbb{N}$ and $\lambda \in \mathbb{R}$, let $C^k_\lambda(I, \mathbb{R}^n)$ the Banach set given by*

$$C^k_\lambda(I, \mathbb{R}^n) := \{f : I \to \mathbb{R}^n, f \text{ is of class } C^k, |f|_{C^k_\lambda(I,\mathbb{R}^n)} < +\infty\}$$

where $|f|_{C^k_\lambda(I,\mathbb{R}^n)} = \sum_{j=0}^{k} \sup_{t\in\mathbb{R}} \left|\dfrac{d^j f(t)}{dt^j}\right| e^{\lambda|t|}$.

(b) *For $\ell > 0$ and $\lambda \in \mathbb{R}$, let $\mathcal{H}^\lambda_\ell(\mathbb{C}^n)$ be the Banach set given by*

$$\mathcal{H}^\lambda_\ell(\mathbb{C}^n) := \{ f : \mathcal{B}_\ell \to \mathbb{C}^n, \text{ holomorphic in } \mathcal{B}_\ell, \; |f|_{\mathcal{H}^\lambda_\ell(\mathbb{C}^n)} := \sup_{\xi\in\mathcal{B}_\ell} |f(\xi)| e^{\lambda|\mathcal{R}e(\xi)|} < +\infty\}$$

where $\mathcal{B}_\ell := \{\xi \in \mathbb{C}, \; |\mathcal{I}m(\xi)| < \ell\}$.

(c) *For a symmetry S and a vector space $\mathcal{X}$ of functions from $\mathbb{R}$ or $\mathbb{C}$ to $\mathbb{R}^n$ or $\mathbb{C}^n$, let us introduce the four subspaces of $\mathcal{X}$ spanned by the functions which are reversible or antireversible with respect to S, or reversible or antireversible with respect to transpose ${}^tS$ of S i.e.*

$$\begin{aligned}\mathcal{X}|_{\mathrm{R}} &= \mathcal{X} \cap \{f, Sf(-t) = f(t) \text{ for any } t\},\\ \mathcal{X}|_{\mathrm{AR}} &= \mathcal{X} \cap \{f, Sf(-t) = -f(t) \text{ for any } t\}.\\ \mathcal{X}|_{\mathrm{R}^*} &= \mathcal{X} \cap \{f, {}^tSf(-t) = f(t) \text{ for any } t\},\\ \mathcal{X}|_{\mathrm{AR}^*} &= \mathcal{X} \cap \{f, {}^tSf(-t) = -f(t) \text{ for any } t\}.\end{aligned}$$

Remark 6.3.2. The last definitions of $\mathcal{X}|_{\mathrm{R}^*}$ and $\mathcal{X}|_{\mathrm{AR}^*}$ are useful to describe the symmetry properties of dual bases.

We begin with a technical lemma which is very helpful to bound bases of solutions

Lemma 6.3.3. *Let A be a $n \times n$ diagonalizable matrix. Denote by*

$$\lambda = \max_{\sigma \in \mathrm{Sp}(A)} \mathcal{R}e\,(\sigma), \qquad \omega = \max_{\sigma \in \mathrm{Sp}(A)} |\mathcal{I}m\,(\sigma)|$$

where $\mathrm{Sp}(A)$ is the spectrum of A.

If $\xi \mapsto B(\xi)$ is bounded on $\Omega \subset \{\xi \in \mathbb{C}/\mathcal{R}e\,(\xi) \geq 0\}$, then there exists M_{AB} such that

$$\|\exp(\xi A + B(\xi))\| \leq M_{AB} e^{\lambda \mathcal{R}e(\xi) + \omega |\mathcal{I}m(\xi)|}, \qquad \text{for } \xi \in \Omega.$$

Remark 6.3.4. Observe that ω is non negative whereas λ may be negative.

Proof. On the set of the $n \times n$ matrices, all the norms are equivalent. So we can prove this lemma, with a suitable choice of the norm for matrices. Hence, we use the following norm, which is a norm of algebra

$$\|M\| := |||P^{-1}MP||| := \sup_{x \neq 0} \frac{|P^{-1}MPx|}{|x|} \qquad \text{with } |x|^2 = \sum_{i=1}^{n} |x_i|^2$$

where P is an invertible matrix such that $D = P^{-1}AP$ is diagonal. Let us denote $D = \mathrm{diag}(\sigma_1, \cdots, \sigma_n)$. By definition, for every j, $1 \leq j \leq n$, $\mathcal{R}e\,(\sigma_j) \leq \lambda$ and $|\mathcal{I}m\,(\sigma_j)| \leq \omega$. Moreover, for every pair (X, Y) of $n \times n$ matrices

$$\exp(X + Y) = \lim_{k \to +\infty} \left(\exp(X/k)\exp(Y/k)\right)^k.$$

Thus, for every pair (X, Y) of $n \times n$ matrices, there exists N such that

$$\|\exp(X+Y)\| \leq 2\big\|\exp(X/N)\big\|^N \exp(\|Y\|).$$

Hence, for every $\xi \in \Omega$, there exists N_ξ such that

$$\|\exp(\xi A + B(\xi))\| \leq 2\big\|\exp(\xi A/N_\xi)\big\|^{N_\xi} \exp(M_B) \quad \text{where } M_B = \sup_{\xi\in\Omega} \|B(\xi)\|.$$

Finally, observing that for every $\tau \geq 0$ and $\eta \in \mathbb{R}$,

$$\|\mathrm{e}^{(\tau+\mathrm{i}\eta)A}\| = \||\mathrm{e}^{(\tau+\mathrm{i}\eta)D}\|| = \sup_{(x_1,\cdots,x_n)\in\mathbb{C}^n} \frac{\left(\sum_{j=1}^{n} |\mathrm{e}^{(\tau+\mathrm{i}\eta)\sigma_j} x_j|^2\right)^{\frac{1}{2}}}{\left(\sum_{j=1}^{n} |x_j|^2\right)^{\frac{1}{2}}} \leq \mathrm{e}^{\lambda\tau+\omega|\eta|},$$

we get that for $\xi \in \Omega$,

$$\|\mathrm{e}^{\xi A+B(\xi)}\| \leq 2\big(\mathrm{e}^{(\lambda\mathcal{R}e(\xi)+\omega|\mathcal{I}m(\xi)|)/N_\xi}\big)^{N_\xi} \mathrm{e}^{M_B} \leq 2\mathrm{e}^{M_B}\mathrm{e}^{\lambda\mathcal{R}e(\xi)+\omega|\mathcal{I}m(\xi)|} \;\square.$$

Moreover, for determining the properties of dual bases, we introduce the following notation

Definition 6.3.5 (real and complex dual basis).

(a) *For $x, y \in \mathbb{C}^n$, denote by*

$$\langle x, y\rangle_* = \sum_{j=1}^{n} x_j y_j = \langle \overline{x}, y\rangle$$

where $\langle\cdot,\cdot\rangle$ is the canonical inner product in $\mathbb{C}^n$.

(b) *For any basis $(e_1, \cdots, e_n)$ of $\mathbb{R}^n$ (resp. of $\mathbb{C}^n$) we define its dual basis $(e_1^*, \cdots, e_n^*)$ as the unique set of vectors of $\mathbb{R}^n$ (resp. of $\mathbb{C}^n$) such that*

$$\langle e_i, e_j\rangle_* = 1 \text{ if } i = j \text{ and } \langle e_i, e_j\rangle_* = 0 \text{ otherwise.}$$

Here, it is more convenient to identify $(\mathbb{C}^n)^*$ with $\mathbb{C}^n$ with $y \mapsto \langle y, \cdot\rangle_*$ rather than with $y \mapsto \langle y, \cdot\rangle$, because when working with bases $(e_1(\xi), \cdots, e_n(\xi))$ which are holomorphic functions of ξ, $\xi \mapsto e_i^*(\xi)$ still are holomorphic functions whereas with the classical identification with the standard inner product the dual basis is not holomorphic but its conjugate is.

Lemma 6.3.6. *For any set of $n-1$ vectors $(x_1, \cdots, x_{n-1})$ of $\mathbb{R}^n$ (resp. of $\mathbb{C}^n$), we define $x_1 \wedge \cdots \wedge x_{n-1}$ as the unique vector of $\mathbb{R}^n$ (resp. of $\mathbb{C}^n$) such that*

$$\det(x_1, \cdots, x_{n-1}, z) = \langle x_1 \wedge \cdots \wedge x_{n-1}, z\rangle_*, \quad \textit{for every } z \in \mathbb{R}^n \text{ (resp. } \mathbb{C}^n).$$

(a) $(x_1, \cdots, x_{n-1}) \mapsto x_1 \wedge \cdots \wedge x_{n-1}$ *is* $n-1$ *linear in* $\mathbb{R}^n$ *or* $\mathbb{C}^n$.

(b) *For any* $n \times n$ *real or complex invertible matrix* A,

$$Ax_1 \wedge \cdots \wedge Ax_{n-1} = \det A \; ({}^tA)^{-1}(x_1 \wedge \cdots \wedge x_{n-1})$$

where ${}^tA$ *is the transpose of* A.

(c) $|x_1 \wedge \cdots \wedge x_{n-1}| \leq M_n |x_1| \cdots |x_{n-1}|$.

(d) *For any basis* $(e_1, \cdots, e_n)$ *of* $\mathbb{R}^n$ *or* $\mathbb{C}^n$ *the dual basis* $(e_1^*, \cdots, e_n^*)$ *is explicitly given by*

$$e_j^* = \frac{e_1 \wedge \cdots \wedge e_{i-1} \wedge e_{i+1} \wedge \cdots \wedge e_n}{\det(e_1, \cdots, e_n)(-1)^{n-j}}.$$

We begin with homoclinic connections generated by a pair of opposite eigenvalues.

Lemma 6.3.7. *Let* N *be a smooth vector field in* $\mathbb{R}^2$ *reversible with respect to some symmetry* S. *Assume that*

(i) *the origin is a fixed point and the spectrum of the differential* $DN(0)$ *of* N *at the origin is* $\{\pm\lambda_0\}, \lambda_0 > 0$;

(ii) *the vector field* N *admits a reversible homoclinic connection* h *to* 0.

(a) *Then, the equation linearized around* h

$$\frac{dv}{dt} - DN(h).v = 0 \tag{6.17}$$

admits a basis of solutions (p, q) *which belongs to* $C^k_{\lambda_0}(\mathbb{R}, \mathbb{R}^2)|_{\mathrm{AR}} \times C^k_{-\lambda_0}(\mathbb{R}, \mathbb{R}^2)|_{\mathrm{R}}$ *for every* $k \geq 0$.

(b) *The dual basis* (p^*, q^*) *of* (p, q) *belongs to the space* $C^k_{-\lambda_0}(\mathbb{R}, \mathbb{R}^2)|_{\mathrm{AR}^*} \times C^k_{\lambda_0}(\mathbb{R}, \mathbb{R}^2)|_{\mathrm{R}^*}$ *for every* $k \geq 0$.

(c) *Moreover, if* N *is holomorphic and if* h *belongs to* $\mathcal{H}_\ell^{\lambda_0}(\mathbb{C}^2)|_{\mathrm{R}}$, *then* (p, q) *belong to* $\mathcal{H}_\ell^{\lambda_0}(\mathbb{C}^2)|_{\mathrm{AR}} \times \mathcal{H}_\ell^{-\lambda_0}(\mathbb{C}^2)|_{\mathrm{R}}$ *and* (p^*, q^*) *belongs to* $\mathcal{H}_\ell^{-\lambda_0}(\mathbb{C}^2)|_{\mathrm{AR}^*} \times \mathcal{H}_\ell^{\lambda_0}(\mathbb{C}^2)|_{\mathrm{R}^*}$

Proof. A first solution of (6.17) is given by $p = \dfrac{dh}{dt}$. Thus p is smooth and antireversible since h is reversible. So $Sp(0) = -p(0)$ holds. The Theorem of classification of reversible matrices 3.1.10 ensures that the spectrum of S is $\{+1, -1\}$. Hence, there exists a unique $q_0 \in \mathbb{R}^2$ such that

$$Sq_0 = q_0, \qquad \det(p_0, q_0) = 1.$$

We can define q as the unique solution of (6.17) such that $q(0) = q_0$. Then q is smooth and reversible. Moreover q can be computed explicitly using lemma 6.1.1. The end of the proof readily follows observing that $|h(t)| \underset{t\mapsto+\infty}{\sim} Ce^{-\lambda_0 t}$ with $C > 0$ and that the dual basis is explicitly given by

$$p^* = \frac{1}{w}(q_\beta, -q_\alpha),\ q^* = \frac{1}{w}(-p_\beta, p_\alpha) \text{ with } p = (p_\alpha, p_\beta),\ q = (q_\alpha, q_\beta)$$

where $w(t) = det(p(t), q(t)) = \exp\left(\int_0^t \operatorname{tr}(DN(h(s))ds\right)$. □

Lemma 6.3.8. *Let N be a smooth vector field in $\mathbb{R}^4$ reversible with respect to some symmetry S. Assume that*

(i) *the origin is a fixed point and the spectrum of the differential $DN(0)$ of N at the origin is $\{\pm i\omega_0 \pm \lambda_0\}, \omega_0, \lambda_0 > 0$;*
(ii) *N commutes with a smooth one parameter family of matrices R_ω, $\omega \in \mathbb{R}$, such that $R_0 = \mathrm{Id}$ and $J := \frac{dR}{d\omega}(0)$ is reversible.*
(iii) *the vector field N admits a reversible homoclinic connection to 0 $h \in C^0_{\lambda_0}(\mathbb{R}, \mathbb{R}^4)|_{\mathrm{R}}$ such that $\frac{dh(0)}{dt}$ and $Jh(0)$ are independent.*

(a) *Then, the equation linearized around h*

$$\frac{dv}{dt} - DN(h).v = 0 \tag{6.18}$$

admits a basis of solutions (p_0, p_1, q_0, q_1) which belongs to the space $\left(C^k_{\lambda_0}(\mathbb{R}, \mathbb{R}^4)|_{\mathrm{AR}}\right)^2 \times \left(C^k_{-\lambda_0}(\mathbb{R}, \mathbb{R}^4)|_{\mathrm{R}}\right)^2$ for every $k \geq 0$.

(b) *The dual basis $(p_0^*, p_1^*, q_0^*, q_1^*)$ belongs to the space $\left(C^k_{\lambda_0}(\mathbb{R}, \mathbb{R}^4)|_{\mathrm{AR}^*}\right)^2 \times \left(C^k_{-\lambda_0}(\mathbb{R}, \mathbb{R}^4)|_{\mathrm{R}^*}\right)^2$ for every $k \geq 0$.*
(c) *Moreover, if N is holomorphic and if h belongs to $\mathcal{H}^{\lambda_0}_\ell(\mathbb{C}^4)|_{\mathrm{R}}$, then (p_0, p_1, q_0, q_1) belong to $\left(\mathcal{H}^{\lambda_0}_\ell(\mathbb{C}^4)|_{\mathrm{AR}}\right)^2 \times \left(\mathcal{H}^{-\lambda_0}_\ell(\mathbb{C}^4)|_{\mathrm{R}}\right)^2$ and $(p_0^*, p_1^*, q_0^*, q_1^*)$ belong to $\left(\mathcal{H}^{-\lambda_0}_\ell(\mathbb{C}^4)|_{\mathrm{AR}^*}\right)^2 \times \left(\mathcal{H}^{\lambda_0}_\ell(\mathbb{C}^4)|_{\mathrm{R}^*}\right)^2$.*

Remark 6.3.9. Examples of such vector fields satisfying (i), (ii) and (iii) are given by the normal forms of any order corresponding to the $(i\omega)^2$ resonance after bifurcation (see Example 3.2.8) where the family R_ω is a group of rotations.

Proof. (a): Observing that for any $\omega \in \mathbb{R}$, $R_\omega h$ is also an homoclinic connection for N and differentiating respectively with respect to t and ω we get that $p_0 = \frac{dh}{dt}$ and $p_1 = Jh$ are two smooth antireversible solutions of (6.18).

Moreover,

$$|p_0|_{C^0_{\lambda_0}(\mathbb{R},\mathbb{R}^4)} \leq \sup_{X\in K} \|DN(X)\| \, |h|_{C^0_{\lambda_0}(\mathbb{R},\mathbb{R}^4)}, \quad |p_1|_{C^0_{\lambda_0}(\mathbb{R},\mathbb{R}^4)} \leq \|J\| \, |h|_{C^0_{\lambda_0}(\mathbb{R},\mathbb{R}^4)}$$

where K is any compact set containing the range of h. The successive upper bounds of the norms $|p_j|_{C^k_{\lambda_0}(\mathbb{R},\mathbb{R}^4)}$ for $k \geq 1$, can be obtained by induction differentiating $k-1$ times the equation (6.18).

The Theorem of classification of reversible matrices 3.1.10 ensures that $\dim(\ker S \pm \mathrm{Id}) = 2$. Moreover, $(p_0(0), p_1(0))$ is a basis of $\ker(S+\mathrm{Id})$. Then, choosing (q_{00}, q_{10}) any basis of $\ker(S-\mathrm{Id})$, we define (q_0, q_1) as the unique solutions of (6.18) such that $(q_0(0), q_1(0)) = (q_{00}, q_{10})$. Hence q_0, q_1 are smooth and reversible and (p_0, p_1, q_0, q_1) is a full basis of solutions of (6.18).

Observing that $q_j(t) = \exp\left(\int_0^t DN(h(s))ds\right) q_j(0)$ and using Lemma 6.3.3 with $A = DN(0)$ and $B(t) = \int_0^t (DN(h(s)) - DN(0))ds$ which is bounded on $\mathbb{R}$, we get that

$$|q_j|_{C^0_{-\lambda_0}(\mathbb{R},\mathbb{R}^4)} \leq \max(1, \|S\|) \, |q_j|_{C^0_{-\lambda_0}(\mathbb{R}^+,\mathbb{R}^4)} < +\infty.$$

The successive upper bounds of the norms $|q_j|_{C^k_{\lambda_0}(\mathbb{R},\mathbb{R}^4)}$ for $k \geq 1$ can be obtained by induction differentiating $k-1$ times the equation (6.18).

(b): Lemma 6.3.6 gives explicit formulas for the dual basis. Thus,

$$p_0^* = -\frac{p_1 \wedge q_0 \wedge q_1}{w}$$

where $w(t) = \det(p_0(t), p_1(t), q_0(t), q_1(t)) = w(0) \exp\left(\int_0^t \mathrm{tr}(DN(h(s)))\, ds\right)$. Since $\mathrm{tr}(DN(h)) = \mathrm{tr}(DN(h) - DN(0)) \underset{t\to\pm\infty}{=} \mathcal{O}(\mathrm{e}^{-\lambda_0|t|})$, the integral in the exponential converges and $m = \inf_{\mathbb{R}}(w) > 0$. Hence

$$|p_0|_{C^0_{-\lambda_0}(\mathbb{R},\mathbb{R}^4)} \leq \frac{M_3}{m} |p_0|_{C^0_{\lambda_0}(\mathbb{R},\mathbb{R}^4)} |q_0|_{C^0_{-\lambda_0}(\mathbb{R},\mathbb{R}^4)} |q_1|_{C^0_{-\lambda_0}(\mathbb{R},\mathbb{R}^4)}.$$

The explicit formula giving the derivative $\dot{p}_0^* = \frac{dp_0^*}{dt}$ of p_0

$$\dot{p}_0^* = -\frac{\dot{p}_1 \wedge q_0 \wedge q_1 + p_1 \wedge \dot{q}_0 \wedge q_1 + \dot{p}_1 \wedge q_0 \wedge q_1}{w} + \frac{\mathrm{tr}(DN(h))}{w}(p_1 \wedge q_0 \wedge q_1)$$

ensures that $|p_0^*|_{C^1_{-\lambda_0}(\mathbb{R},\mathbb{R}^4)} < +\infty$. The successive upper bounds of the norms $|p_0^*|_{C^k_{\lambda_0}(\mathbb{R},\mathbb{R}^4)}$ for $k \geq 2$ can be obtained by induction differentiating $k-2$ times the above formula.

Finally, since N and h are reversible, the Wronskian determinant w is an even function. Then, the explicit formula giving p_0^* and Lemma 6.3.6-(b) ensure that

$$p_0^*(-t) = \frac{Sp_1(t) \wedge Sq_0(t) \wedge Sq_1(t)}{w(t)} = -{}^tSp_0^*(t)$$

since p_1 is antireversible; q_0, q_1 are reversible and $\det(S) = 1$.
Similar proofs hold for p_1^*, q_0^*, q_1^*.

(c): the proof is very similar to the previous one. So, the details are left to the reader. □.

The previous lemma gives a description of a basis of solutions of the linearized equation around a reversible homoclinic connection induced by a set of conjugate eigenvalues $\pm i\omega_0 \pm \lambda_0$. However, when the frequency ω_0 depends on a small parameter $\varepsilon \in]0,1]$,

$$\underline{\omega}_0(\varepsilon) = \frac{\omega_0}{\varepsilon} + \underline{\omega}_{01}(\varepsilon), \qquad \text{with } \omega_0 > 0, \ \sup_{\varepsilon \in]0,1]} |\underline{\omega}_{01}(\varepsilon)| < +\infty,$$

one needs a more precise description of the basis. Indeed the rapid oscillation due to the high frequency, leads to very bad estimates when complexifying time, since for every $\ell > 0$, $\cos(i\omega\ell/\varepsilon) \underset{\varepsilon \to 0}{\longrightarrow} +\infty$. The following lemma enables us to control this bad behavior of the basis, since it ensures that the rapid oscillation can factored out of the basis.

This factorization will enable us to complexify partially the time when solving the corresponding affine equations (see section 6.5). This partial complexification of time is the crucial point of the study of the existence of reversible homoclinic connections near a $(i\omega_0)^2 i\omega_1$ resonant fixed point. Indeed, the rescaled normal form systems of any order corresponding to the $(i\omega_0)^2 i\omega_1$ resonance admit a reversible homoclinic connection induced by a set of conjugate eigenvalues $\pm i\frac{\omega_0(\varepsilon)}{\varepsilon} \pm 1$. The differential of the normal form at the homoclinic connection is block diagonal. A first block, corresponding to the set of eigenvalues $\pm i\frac{\omega_0(\varepsilon)}{\varepsilon} \pm 1$ is four dimensional and the corresponding basis of solutions directly follows from the Lemma 6.3.11 below.

Definition 6.3.10. *For $\ell > 0$ and $\lambda \in \mathbb{R}$, let $H_\ell^\lambda(\mathbb{C}^n)$ be the set of functions f satisfying*

(a) $f : \mathcal{B}_\ell \times]0,1] \longrightarrow \mathbb{C}^n$ *where* $\mathcal{B}_\ell = \{\xi \in \mathbb{C} / |\mathcal{I}m(\xi)| < \ell\}$,
(b) $\xi \mapsto f(\xi, \varepsilon)$ *is holomorphic in* $\mathcal{B}_\ell$,
(c) $\|f\|_{H_\ell^\lambda(\mathbb{C}^n)} := \sup_{\varepsilon \in]0,1], \xi \in \mathcal{B}_\ell} (|f(\xi,\varepsilon)| e^{\lambda |\mathcal{R}e(\xi)|}) < +\infty$,

Observe that functions lying in $H_\ell^\lambda(\mathbb{C}^n)$ depend on ε and that the norm is uniform with respect to ε.

Lemma 6.3.11. *Let* $N : \mathbb{R}^4 \times]0,1] \to \mathbb{R}^4 : (Y, \varepsilon) \mapsto N(Y, \varepsilon)$ *be a family of smooth vector fields reversible with respect to some symmetry* S. *Let* $\underline{\omega}_0(\varepsilon) = \omega_0 + \varepsilon \underline{\omega}_{01}(\varepsilon)$ *with* $|\underline{\omega}_{01}|_0 = \sup_{\varepsilon \in]0,1]} |\underline{\omega}_{01}(\varepsilon)| < +\infty$. *Assume that*

(i) $N(Y, \varepsilon) = L(\varepsilon)Y + Q(Y, \varepsilon)$ *where*

$$L(\varepsilon) = \tfrac{\underline{\omega}_0(\varepsilon)}{\varepsilon} J + \Lambda \text{ with } \begin{cases} P^{-1}JP = \mathrm{diag}(\mathrm{i}, \mathrm{i}, -\mathrm{i}, -\mathrm{i}) \\ P^{-1}\Lambda P = \mathrm{diag}(\lambda_0, -\lambda_0, \lambda_0, -\lambda_0), \ \lambda_0 > 0 \end{cases}$$

where P *is an invertible matrix and where* $Q(0, \varepsilon) = DQ(0, \varepsilon) = 0$ *and* Q *is holomorphic on some open set* $\Omega \subset \mathbb{C}^4$ *containing the origin and* $\sup_{\varepsilon \in]0,1],\, Y \in K} \|D^n Q(Y, \varepsilon)\| < \infty$ *for all* $n \in \mathbb{N}$ *and all compact set* $K \subset \Omega$.

(ii) $NR_\omega = R_\omega N$ *for* $\omega \in \mathbb{R}$ *where* $R_\omega = \exp(J\omega)$.

(iii) *the vector field* N *admits a reversible homoclinic connection* h *to 0 such that* $\frac{dh(0)}{dt}$ *and* $Jh(0)$ *are independent and such that*

$$\check{h}(\xi, \varepsilon) = R_{\frac{-\omega_0 \xi}{\varepsilon}} h(\xi, \varepsilon) \in H_\ell^{\lambda_0}(\mathbb{C}^4).$$

(a) *Then, the equation linearized around* h

$$\frac{dv}{dt} - DN(h).v = 0 \tag{6.19}$$

admits a basis of solutions (p_0, p_1, q_0, q_1) *such that*

$$\check{p}_j = R_{\frac{-\omega_0 \xi}{\varepsilon}} p_j \in H_\ell^{\lambda_0}(\mathbb{C}^4)|_{\mathrm{AR}}, \qquad \check{q}_j = R_{\frac{-\omega_0 \xi}{\varepsilon}} q_j \in H_\ell^{-\lambda_0}(\mathbb{C}^4)|_{\mathrm{R}}.$$

(b) *The dual basis* $(p_0^*, p_1^*, q_0^*, q_1^*)$ *of* (p_0, p_1, q_0, q_1) *satisfies*

$$\check{p}_j^* = {}^t R_{\frac{\omega_0 \xi}{\varepsilon}}\, p_j^* \in H_\ell^{-\lambda_0}(\mathbb{C}^4)|_{\mathrm{AR}^*}, \qquad \check{q}_j^* = {}^t R_{\frac{\omega_0 \xi}{\varepsilon}}\, q_j^* \in H_\ell^{\lambda_0}(\mathbb{C}^4)|_{\mathrm{R}^*}.$$

Proof. (a):
Step 1. Preliminary remarks. Since N is holomorphic and commutes with R_ω for $\omega \in \mathbb{R}$ then it also commutes with R_ω for $\omega \in \mathbb{C}$. Moreover, $L(\varepsilon)S = -SL(\varepsilon)$ holds and the necessary form of the symmetry S given by the Theorem of classification of reversible matrices 3.1.10 ensures that $SJ = -JS$ and thus that $SR_\omega = R_{-\omega}S$.

Step 2. The two first solutions p_0 and p_1. Observing that for any $\omega \in \mathbb{R}$, $R_\omega h$ is also an homoclinic connection of N and differentiating respectively with respect to t and ω we get that $p = \dfrac{dh}{dt}$ and $p' = Jh$ are two antireversible solutions of (6.19) holomorphic on $\mathcal{B}_\ell$. However, when differentiating h

$$p = \frac{dh}{d\xi} = R_{\frac{\omega_0\xi}{\varepsilon}}\left(\tfrac{\omega_0\xi}{\varepsilon} J\check{h} + \frac{d\check{h}}{d\xi}\right) = \tfrac{\omega_0\xi}{\varepsilon} p' + R_{\frac{\omega_0\xi}{\varepsilon}} \frac{d\check{h}}{d\xi}$$

we observe that p has a bad behavior with respect to ε because of the term $\frac{\omega_0\xi}{\varepsilon}$ in front of p'. So we choose for the two first solutions

$$p_0 := -\tfrac{\omega_0\xi}{\varepsilon} p' + p = R_{\frac{\omega_0\xi}{\varepsilon}} \frac{d\check{h}}{d\xi}, \qquad p_1 := p' = Jh.$$

which are antireversible and holomorphic on $\mathcal{B}_\ell$ and satisfy

$$\check{p}_0 := R_{\frac{-\omega_0\xi}{\varepsilon}} p_0 = \frac{d\check{h}}{d\xi}, \qquad \check{p}_1 := R_{\frac{-\omega_0\xi}{\varepsilon}} p_1 = J\check{h}.$$

Then, for $\xi \in \mathcal{B}_\ell$

$$\begin{aligned}
|\check{p}_0(\xi,\varepsilon)| &= \left|\frac{d}{d\xi}\left(R_{\frac{-\omega_0\xi}{\varepsilon}} h(\xi,\varepsilon)\right)\right| \\
&= \left|(\underline{\omega}_{01}(\varepsilon)J + \Lambda)\check{h}(\xi,\varepsilon) + Q(\check{h}(\xi,\varepsilon),\varepsilon)\right| \\
&\le \left(|\underline{\omega}_{01}|_0 \|J\| + \|\Lambda\|\right) \|\check{h}\|_{H_\ell^{\lambda_0}(\mathbb{C}^4)} e^{-\lambda_0|\mathcal{R}e(\xi)|} \\
&\quad + \sup_{\varepsilon\in]01],X\in K} \|D^2Q(X)\| \, \|\check{h}\|^2_{H_\ell^{\lambda_0}(\mathbb{C}^4)} e^{-2\lambda_0|\mathcal{R}e(\xi)|}
\end{aligned}$$

where K is any compact set containing the image of $\check{h}$. Hence, $\check{p}_0$ belongs to $H_\ell^{\lambda_0}(\mathbb{C}^4)|_{\text{AR}}$. Similarly, we deduce from $\check{p}_1 = J\check{h}$ that $\check{p}_1$ belongs to $H_\ell^{\lambda_0}(\mathbb{C}^4)|_{\text{AR}}$.

Step 3. The two last solutions q_0 and q_1. The Theorem of classification of reversible matrices 3.1.10 ensures that $\dim(\ker S \pm \text{Id}) = 2$. Moreover, $(p_0(0), p_1(0))$ is a basis of $\ker(S + \text{Id})$. Then, choosing (q_{00}, q_{10}) to be any basis of $\ker(S - \text{Id})$, we define (q_0, q_1) as the unique solutions of (6.19) such that $(q_0(0), q_1(0)) = (q_{00}, q_{10})$. Hence q_0, q_1 are smooth and reversible and (p_0, p_1, q_0, q_1) is a full basis of solutions of (6.19). Observing that

$$\frac{d\check{q}_j}{d\xi} = (\underline{\omega}_{01}(\varepsilon)J + \Lambda)\check{q}_j + DQ(\check{h},\varepsilon).\check{q}_j \tag{6.20}$$

and thus that $q_j(\xi,\varepsilon) = \exp\left((\underline{\omega}_{01}(\varepsilon)J + \Lambda)\xi + \int_{[0,\xi]} DQ(\check{h}(s,\varepsilon))ds\right) q_j(0)$ and using Lemma 6.3.3 with $A = \underline{\omega}_{01}(\varepsilon)J + \Lambda$ and $B(\xi) = \displaystyle\int_{[0,\xi]} DQ(\check{h}(s,\varepsilon))ds$ which is bounded on $\mathcal{B}_\ell$, we get that $\|\check{q}_j\|_{H_\ell^{\lambda_0}(\mathbb{C}^4)} < +\infty$ and thus that $\check{q}_j$ belongs to $H_\ell^{\lambda_0}(\mathbb{C}^4)|_{\text{R}}$.

(b): Lemma 6.3.6 gives explicit formulas for the dual basis. Thus,

$$p_0^* = -\frac{p_1 \wedge q_0 \wedge q_1}{w} = -\frac{R_{\frac{\omega_0 \xi}{\varepsilon}} \check{p}_1 \wedge R_{\frac{\omega_0 \xi}{\varepsilon}} \check{q}_0 \wedge R_{\frac{\omega_0 \xi}{\varepsilon}} \check{q}_1}{w} = -\,{}^t R_{\frac{-\omega_0 \xi}{\varepsilon}} \frac{\check{p}_1 \wedge \check{q}_0 \wedge \check{q}_1}{w}$$

since $\det(R_\omega) = 1$ for every $\omega \in \mathbb{C}$ and where

$$w = \det(p_0, p_1, q_0, q_1) = \det(\check{p}_0, \check{p}_1, \check{q}_0, \check{q}_1) = \exp\left(\int_{[0,\xi]} \operatorname{tr}(DQ(\check{h}(s,\varepsilon),\varepsilon))ds\right).$$

Thus, $\check{p}_0^* = -\frac{\check{p}_1 \wedge \check{q}_0 \wedge \check{q}_1}{w}$. As for Lemma 6.3.11-(b), we deduce from this explicit formula that $\check{p}_0^* \in H_\ell^{-\lambda_0}(\mathbb{C}^4)$ and that $\check{p}_0^*$ is antireversible with respect to ${}^tS$. Similar proofs hold for p_1^*, q_0^*, q_1^*. □.

6.4 Affine equations with real and complexified time

The three previous sections deal with homogeneous equations of the form

$$\mathcal{L}(v) = 0 \qquad \text{with } \mathcal{L}(v) = \frac{dv}{dt} - A(t)v.$$

Sections 6.1 and 6.3 are devoted to the kernel of $\mathcal{L}$ when A converges exponentially quickly to a constant matrix (this situation typically occurs when linearizing around a homoclinic connection to 0). This section is now devoted to the range of $\mathcal{L}$ and in particular to the solvability conditions required on a function f for belonging to the range of $\mathcal{L}$ when $\mathcal{L}$ is restricted to reversible functions which go to 0 at infinity. We give three lemmas which correspond to different linear operators $A(t)$:

- The first one typically occurs after linearization around a reversible homoclinic connection to 0 induced by a pair of opposite eigenvalues.
- The second one occurs for instance, when using the second Normal Form Theorem 3.2.10 (with non fully critical spectrum). It corresponds to the part of the spectrum which is bounded away from the imaginary axis. This second Normal Form Theorem is typically used when the center manifold theorem cannot be used because one needs analyticity of the vector field to use the exponential tools developed in Chapter 2.
- The last one occurs when a pair of simple purely imaginary eigenvalues stay on the imaginary axis even after bifurcation. This is precisely the case for the $0^{2+}\mathrm{i}\omega$ and the $(\mathrm{i}\omega_0)^2\mathrm{i}\omega_1$ resonances.

Of course these three cases may occur simultaneously when A is block diagonal. This is the case for the $0^{2+}\mathrm{i}\omega$ resonance (see Example 6.2.10). In such a case, each affine equation corresponding to each diagonal block can be solved separately.

Lemma 6.4.1. *Let A be a smooth 2×2 matrix function which is reversible with some symmetry S, i.e. $SA(-t) = -A(t)S$ holds for every $t \in \mathbb{R}$. Assume that the homogeneous equation*

$$\mathcal{L}(v) = 0 \qquad \text{with } \mathcal{L}(v) = \frac{dv}{dt} - A(t)v$$

admits a basis of solution $(p,q) \in C^k_{\lambda_0}(\mathbb{R},\mathbb{R}^2)|_{\mathrm{AR}} \times C^k_{-\lambda_0}(\mathbb{R},\mathbb{R}^2)|_{\mathrm{R}}$ for every $k \geq 0$ where $\lambda_0 > 0$ and such that its dual basis (p^*,q^*) belongs to $C^k_{-\lambda_0}(\mathbb{R},\mathbb{R}^2)|_{\mathrm{AR}^*} \times C^k_{\lambda_0}(\mathbb{R},\mathbb{R}^2)|_{\mathrm{R}^*}$.

(a) *Then for any $\lambda \in]0,\lambda_0[$, and every function $f \in C^0_\lambda(\mathbb{R},\mathbb{R}^2)|_{\mathrm{AR}}$, there exists a unique solution v of $\mathcal{L}(v) = f$ which tends to 0 at $\pm\infty$ and which is reversible. This solution is explicitly given by*

$$\mathcal{F}(f)(t) = \int_{[0,t]} \langle p^*(s), f(s)\rangle_* ds\; p(t) - \int_0^{+\infty} \langle q^*(t+s), f(t+s)\rangle_* ds\; q(t) \quad (6.21)$$

(b) *If (p,q) belongs to $\mathcal{H}^{\lambda_0}_\ell(\mathbb{C}^2)|_{\mathrm{AR}} \times \mathcal{H}^{-\lambda_0}_\ell(\mathbb{C}^2)|_{\mathrm{R}}$ and if (p^*,q^*) belongs to $\mathcal{H}^{-\lambda_0}_\ell(\mathbb{C}^2)|_{\mathrm{AR}^*} \times \mathcal{H}^{\lambda_0}_\ell(\mathbb{C}^2)|_{\mathrm{R}^*}$ then $\mathcal{F}$ is a bounded linear operator from $\mathcal{H}^{\lambda}_\ell(\mathbb{C}^2)|_{\mathrm{AR}}$ to $\mathcal{H}^{\lambda}_\ell(\mathbb{C}^2)|_{\mathrm{R}}$.*

(c) *For any $\lambda \in]0,\lambda_0[$ and any $k \geq 0$, $\mathcal{F}$ is a bounded linear operator from $C^k_\lambda(\mathbb{R},\mathbb{R}^2)|_{\mathrm{AR}}$ to $C^{k+1}_\lambda(\mathbb{R},\mathbb{R}^2)|_{\mathrm{R}}$.*

Remark 6.4.2. This lemma ensures that $\mathcal{L}$ is a Banach isomorphism from $C^{k+1}_\lambda(\mathbb{R},\mathbb{R}^2)|_{\mathrm{R}}$ onto $C^k_\lambda(\mathbb{R},\mathbb{R}^2)|_{\mathrm{AR}}$ and from $\mathcal{H}^{\lambda}_\ell(\mathbb{C}^2)|_{\mathrm{R}}$ onto $\mathcal{H}^{\lambda}_\ell(\mathbb{C}^2)|_{\mathrm{AR}}$. In both cases $\mathcal{L}^{-1} = \mathcal{F}$. So, in both cases there is no solvability condition required on f for belonging to the range of $\mathcal{L}$.

Remark 6.4.3. An example of such a linear operator occurs when considering the first diagonal block of the matrix obtained by linearization around the homoclinic connection to 0 of the normal form of order 2 for the $0^{2+}\mathrm{i}\omega$ resonance. More generally, it occurs after linearization around a homoclinic connection to 0 induced by a pair of real opposite eigenvalues (see Lemma 6.3.7).

Lemma 6.4.4. *Let A_{h} be a $n \times n$ matrix. Assume that $m(A_{\mathrm{h}}) = \min_{\sigma \in \mathrm{Sp}(A_{\mathrm{h}})} |\mathcal{R}e\,(\sigma)| > 0$.*

Let $\gamma_\pm : [0,1] \to \mathbb{C}$ be two rectangular paths with side parallel to the axis such that : $\gamma_-(t) = -\gamma_+(t)$; all the eigenvalues of A_{h} of positive real parts are in the interior of γ_+; $\mathcal{R}e\,(\gamma(t)) \geq \delta$, for $t \in [0,1]$ with $0 < \delta < m(A_{\mathrm{h}})$ (see Fig. 6.1).

Denote by E^+ (resp. E^-) the direct sum of all the generalized eigenspaces corresponding to eigenvalues of A with positive real parts (resp. negative real parts). Then $\mathbb{R}^n = E^+ \bigoplus E^-$ and the projectors on $E^\pm$ parallel to $E^\mp$ are given by

$$\pi^\pm = \int_{\gamma_\pm} (\lambda - A_\mathrm{h})^{-1}\, d\lambda.$$

Denote by $A_\mathrm{h}^\pm : \mathbb{R}^n \to E^\pm \subset \mathbb{R}^n$ the two matrices $A_\mathrm{h}^\pm = A_\mathrm{h}\pi^\pm$.

(a) *Let $f \in C_\lambda^0(\mathbb{R}, \mathbb{R}^n)$ with $\lambda > 0$. Then, there exists a unique solution of*

$$\mathcal{L}_\mathrm{h}(v) = f \qquad \text{with } \mathcal{L}_\mathrm{h}(v) = \frac{dv}{dt} - A_\mathrm{h} v \tag{6.22}$$

which is bounded on $\mathbb{R}$. This solution is given by

$$\begin{aligned} \mathcal{F}_\mathrm{h}(f) &= -\int_t^{+\infty} \mathrm{e}^{(t-s)A_\mathrm{h}^+} f(s)ds + \int_{-\infty}^t \mathrm{e}^{(t-s)A_\mathrm{h}^-} f(s)ds, \\ &= \int_{-\infty}^{+\infty} K_\mathrm{h}(s) f(t-s)ds \end{aligned} \tag{6.23}$$

where $K_\mathrm{h}(s) = -\mathrm{e}^{sA_\mathrm{h}^+}$ for $s < 0$ and $K_\mathrm{h}(s) = \mathrm{e}^{sA_\mathrm{h}^-}$ for $s > 0$.

(b) *If A_h satisfies $SA_\mathrm{h} = -A_\mathrm{h}S$ and if $f \in C_\lambda^0(\mathbb{R}, \mathbb{R}^n)$ is antireversible, then $\mathcal{F}_\mathrm{h}(f)$ is reversible.*

(c) *For any λ, $0 < \lambda < m(A_\mathrm{h})$, $\mathcal{F}_\mathrm{h}$ is a bounded linear operator from $\mathcal{H}_\ell^\lambda(\mathbb{C}^n)$ to $\mathcal{H}_\ell^\lambda(\mathbb{C}^n)$.*

(d) *For any λ, $0 < \lambda < m(A_\mathrm{h})$, $\mathcal{F}_\mathrm{h}$ is a bounded linear operator from $C_\lambda^k(\mathbb{R}, \mathbb{R}^n)$ to $C_\lambda^{k+1}(\mathbb{R}, \mathbb{R}^n)$.*

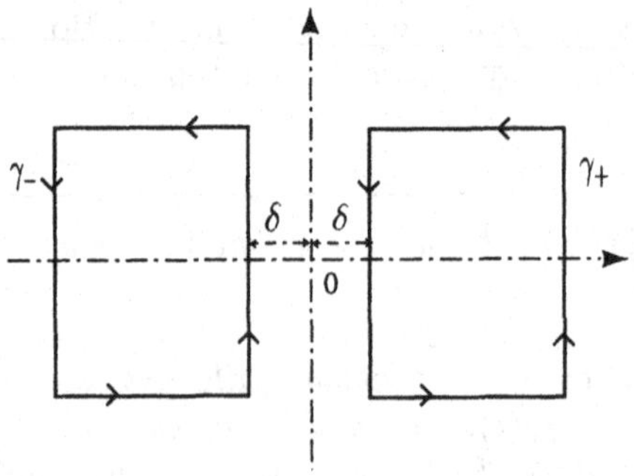

Fig. 6.1. Paths $\gamma_\pm$

Remark 6.4.5. This Lemma ensures that $\mathcal{L}_{\mathrm{h}}$ is a Banach isomorphism from $C_\lambda^{k+1}(\mathbb{R},\mathbb{R}^n)$ onto $C_\lambda^k(\mathbb{R},\mathbb{R}^n)$ and from $\mathcal{H}_\ell^\lambda(\mathbb{C}^n)$ onto $\mathcal{H}_\ell^\lambda(\mathbb{C}^n)$. When A_{h} is reversible then $\mathcal{L}_{\mathrm{h}}$ is also a Banach isomorphism from $C_\lambda^{k+1}(\mathbb{R},\mathbb{R}^n)|_{\mathrm{R}}$ onto $C_\lambda^k(\mathbb{R},\mathbb{R}^n)|_{\mathrm{AR}}$ and from $\mathcal{H}_\ell^\lambda(\mathbb{C}^n)|_{\mathrm{R}}$ to $\mathcal{H}_\ell^\lambda(\mathbb{C}^n)|_{\mathrm{AR}}$. In each cases, $\mathcal{L}_{\mathrm{h}}^{-1} = \mathcal{F}_{\mathrm{h}}$. Here again, there is no solvability conditions for belonging to the range of $\mathcal{L}_{\mathrm{h}}$. The same result holds in infinite dimensions when A is an unbounded linear operator (see the general result given in section 8.2 of chapter 8).

Lemma 6.4.6. *Let $\omega > 0$ and let A_ω be the 2×2 matrix reversible with respect to S given by*

$$A_\omega = \begin{bmatrix} 0 & -\omega \\ \omega & 0 \end{bmatrix}, \qquad S = \begin{bmatrix} 1 & 0 \\ 0 & -1 \end{bmatrix}.$$

(a) *Let $f \in C_\lambda^0(\mathbb{R},\mathbb{R}^2)$ with $\lambda > 0$. Then, there exists a unique solution of*

$$\mathcal{L}_\omega(v) = f \qquad \text{with } \mathcal{L}_\omega = \frac{dv}{dt} - A_\omega v \tag{6.24}$$

which tends to 0 at $+\infty$.
This solution is given by

$$\begin{aligned} \mathcal{F}_\omega(f) &= -\int_t^{+\infty} \langle r_+(s), f(s)\rangle ds \; r_+(t) - \int_t^{+\infty} \langle r_-(s), f(s)\rangle ds \; r_-(t), \\ &= \int_0^{+\infty} K_\omega(s).f(t+s)ds \end{aligned} \tag{6.25}$$

where $r_+(t) = (\cos t, \sin t)$, $r_-(t) = (-\sin t, \cos t)$ and

$$K_\omega(s) = -\begin{bmatrix} \cos(\omega s) & \sin(\omega s) \\ -\sin(\omega s) & \cos(\omega s) \end{bmatrix}.$$

(b) *For any $f \in C_\lambda^0(\mathbb{R},\mathbb{R}^2)$ with $\lambda > 0$, $\mathcal{F}(f) \underset{t\to-\infty}{\longrightarrow} 0$ if and only if f satisfies the two solvability conditions*

$$\int_{-\infty}^{+\infty} \langle r_+(s), f(s)\rangle ds = 0, \qquad \int_{-\infty}^{+\infty} \langle r_-(s), f(s)\rangle ds = 0.$$

(c) *For any $f = (f_0, f_1) \in C_\lambda^0(\mathbb{R},\mathbb{R}^2)|_{\mathrm{AR}}$ with $\lambda > 0$, the following conditions are equivalent*

(i) $\mathcal{F}_\omega(f)$ *is reversible,*

(ii) $\mathcal{F}_\omega(f) \underset{t\to-\infty}{\longrightarrow} 0$,

(iii) $\int_{-\infty}^{+\infty} e^{i\omega t}(f_1(t) + i f_0(t))dt = 0$,

(iv) $\int_{-\infty}^{+\infty} \langle r_-(s), f(s)\rangle ds = 0,$

(v) $\int_0^{+\infty} \langle r_-(s), f(s)\rangle ds = 0.$

(d) *The operator* $\mathcal{F}_\omega$ *is a bounded linear operator from* $\mathcal{H}_\ell^\lambda(\mathbb{C}^2)|_{\mathrm{AR}} \cap \{f,\ \int_{-\infty}^{+\infty} \langle r_-(s), f(s)\rangle ds = 0\}$ *to* $\mathcal{H}_\ell^\lambda(\mathbb{C}^2)|_{\mathrm{R}}$.

(e) *For any* $k \geq 0$, $\mathcal{F}_\omega$ *is a bounded linear operator from* $C_\lambda^k(\mathbb{R}, \mathbb{R}^2)|_{\mathrm{AR}} \cap \{f,\ \int_{-\infty}^{+\infty} \langle r_-(s), f(s)\rangle ds = 0\}$ *to* $C_\lambda^{k+1}(\mathbb{R}, \mathbb{R}^2)|_{\mathrm{R}}$.

Remark 6.4.7. This Lemma ensures that $\mathcal{L}_\omega$ is a Banach isomorphism from $C_\lambda^{k+1}(\mathbb{R}, \mathbb{R}^2)|_{\mathrm{R}}$ onto $C_\lambda^k(\mathbb{R}, \mathbb{R}^2)|_{\mathrm{AR}} \cap \{f,\ \int_{-\infty}^{+\infty} \langle r_-(s), f(s)\rangle ds = 0\}$ and from $\mathcal{H}_\ell^\lambda(\mathbb{C}^2)|_{\mathrm{R}}$ onto $\mathcal{H}_\ell^\lambda(\mathbb{C}^2)|_{\mathrm{AR}} \cap \{f,\ \int_{-\infty}^{+\infty} \langle r_-(s), f(s)\rangle ds = 0\}$. In both cases, there is one solvability condition required on a function f for belonging to the range of $\mathcal{L}_\omega$. Observe on (e)(iii) that when ω reads $\omega = \omega_0/\varepsilon$ where ε is close to 0, then this solvability condition is given by an oscillatory integral. The tools described in chapter 2, have been precisely developed to study such solvability conditions given by oscillatory integrals.

Although the proofs of these three Lemmas are elementary, we give them because they show precisely where the solvability conditions come from. These solvability conditions are the crucial point of the analysis when studying the persistence of homoclinic connections when a pair of purely imaginary eigenvalues stay on the imaginary axis even after bifurcation.

Proof of Lemma 6.4.1

(a): A basis of solutions of $\mathcal{L}(v) = 0$ is given by (p, q). Using the variation of constants formula, we get that any solution v of $\mathcal{L}(v) = f$ reads

$$v(t) = \left(a_0 + \int_0^t \langle p^*(s), f(s)\rangle_* ds\right) p(t) + \left(b_0 + \int_0^t \langle q^*(s), f(s)\rangle_* ds\right) q(t).$$

If $v \underset{t \to +\infty}{\longrightarrow} 0$, then necessarily

$$\langle q^*(t), v(t)\rangle_* = b_0 + \int_0^t \langle q^*(s), f(s)\rangle_* \underset{t \to +\infty}{\longrightarrow} 0.$$

Thus

$$b_0 = -\int_0^{+\infty} \langle q^*(s), f(s)\rangle_* ds.$$

Now if v is reversible, then $0 = Sv(0) - v(0) = 2a_0 p(0)$, since p is antireversible and q is reversible. So, since $(p(0), q(0))$ is a basis of $\mathbb{R}^2$, $p(0) \neq 0$ and $a_0 = 0$.

Thus, there exists at most one solution which tends to 0 at $+\infty$ and which is reversible. This solution is given by $\mathcal{F}(f)$ where $\mathcal{F}(f)$ is defined in (6.21).

Conversely, if f is antireversible, then the two functions $s \mapsto \langle p^*(s), f(s)\rangle_*$ and $s \mapsto \langle q^*(s), f(s)\rangle_*$ are respectively even and odd. Thus, for any $b_0 \in \mathbb{R}$, the function

$$t \mapsto \int_0^t \langle p^*(s), f(s)\rangle_* ds\ p(t) + \left(b_0 + \int_0^t \langle q^*(s), f(s)\rangle_* ds\right) q(t).$$

is reversible because p is antireversible and q reversible. Hence $\mathcal{F}(f)$ is reversible. Moreover, for any $t \geq 0$

$$\begin{aligned}|\mathcal{F}(f)(t)| \leq & \int_0^t e^{(\lambda_0-\lambda)|s|}\, |f|_{C^0_\lambda}\, |p^*|_{C^0_{-\lambda_0}}\, ds\ |p|_{C^0_{\lambda_0}}\, e^{-\lambda_0|t|} \\ & + \int_0^{+\infty} e^{-(\lambda_0+\lambda)|s+t|}\, |f|_{C^0_\lambda}\, |q^*|_{C^0_{\lambda_0}}\, |q|_{C^0_{-\lambda_0}}\, e^{\lambda_0|t|} \\ & \leq \left(\tfrac{1}{\lambda_0-\lambda} + \tfrac{1}{\lambda_0+\lambda}\right) (|p|_{C^0_{\lambda_0}}\, |p^*|_{C^0_{-\lambda_0}} + |q^*|_{C^0_{\lambda_0}}\, |q|_{C^0_{-\lambda_0}})\ |f|_{C^0_\lambda}\, e^{-\lambda t}.\end{aligned}$$

So, $\mathcal{F}(f) \underset{t\to+\infty}{\longrightarrow} 0$ holds. Moreover, since $\mathcal{F}(f)$ is reversible, this last estimate also ensures that $\mathcal{F}(f) \underset{t\to-\infty}{\longrightarrow} 0$ and more precisely that

$$|\mathcal{F}(f)|_{C^0_\lambda} \leq \left(\frac{2\lambda_0}{\lambda_0^2-\lambda^2}\right) (|p|_{C^0_{\lambda_0}}\ |p^*|_{C^0_{-\lambda_0}} + |q^*|_{C^0_{\lambda_0}}\ |q|_{C^0_{-\lambda_0}})\ |f|_{C^0_\lambda}\,.$$

(b): The integral expression of $\mathcal{F}(f)$ and the theorem of derivation under the integral derived from Lebesgue's dominated convergence theorem, ensure that $\mathcal{F}_\omega(f)$ is holomorphic in $\mathcal{B}_\ell$ for any $f \in \mathcal{H}^\lambda_\ell(\mathbb{C}^2)$. As previously we check that

$$|\mathcal{F}(f)|_{\mathcal{H}^\lambda_\ell(\mathbb{C}^2)} \leq M(|p|_{C^0_{\lambda_0}}\ |p^*|_{C^0_{-\lambda_0}} + |q^*|_{C^0_{\lambda_0}}\ |q|_{C^0_{-\lambda_0}})\ |f|_{\mathcal{H}^\lambda_\ell(\mathbb{C}^2)}\,.$$

(c): The proof is very similar to (b) and thus left to the reader. □

Proof of Lemma 6.4.4

(a): Any solution u of (6.22) reads

$$u(t) = e^{tA_\mathrm{h}}u_0 + \int_0^t e^{(t-s)A_\mathrm{h}f(s)ds} \qquad \text{with } u_0 \in \mathbb{R}^n.$$

Observing that $A_\mathrm{h}^\pm = A_\mathrm{h}\pi^\pm = \pi^\pm A_\mathrm{h}$, $e^{tA_\mathrm{h}^\pm} = e^{tA_\mathrm{h}}\pi^\pm = \pi^\pm e^{tA_\mathrm{h}}$ and projecting the above equation on E^+ we get,

$$e^{-tA_\mathrm{h}^+}u(t) = \pi^+(u_0) + \int_0^t e^{-sA_\mathrm{h}^+}f(s).$$

So, since $e^{-tA_h^+} \underset{t\to+\infty}{\longrightarrow} 0$, if u is bounded necessarily

$$\pi^+(u_0) = -\int_0^{+\infty} e^{-sA_h^+} f(s)ds.$$

Similarly, we get that if u is bounded necessarily

$$\pi^-(u_0) = -\int_{-\infty}^{0} e^{-sA_h^+} f(s)ds.$$

Hence, there is at most a solution u which is bounded on $\mathbb{R}$ and it is given by $u = \mathcal{F}_h(f)$ where $\mathcal{F}_h$ is given by (6.23). Conversely, For checking that $\mathcal{F}_h(f)$ is bounded on $\mathbb{R}$ for $f \in C_0^0(\mathbb{R}, \mathbb{R}^n)$, we observe that for any $\delta \in]0, m(A_h)[$, there exists $M > 0$ such that

$$\begin{aligned} |e^{tA_h^+} x| &\leq M e^{-|t|\delta} |x| \qquad \text{for } t < 0,\ x \in \mathbb{R}^n, \\ |e^{tA_h^-} x| &\leq M e^{-|t|\delta} |x| \qquad \text{for } t > 0,\ x \in \mathbb{R}^n. \end{aligned} \tag{6.26}$$

Then it readily follows that $|\mathcal{F}_h(f)|_{C_0^0(\mathbb{R},\mathbb{R}^n)} \leq 2M\delta^{-1} |f|_{C_0^0(\mathbb{R},\mathbb{R}^n)}$

(b): If A_h is reversible with respect to the symmetry S, we check that

$$S\pi^+ = \int_{\gamma^+} S(\lambda - A_h)^{-1} = \int_{\gamma^+} (\lambda + A_h)^{-1} S = \int_{\gamma^-} (\lambda - A_h)^{-1} S = \pi^- S$$

and consequently $Se^{tA_h^+} = e^{-tA_h^-} S$ holds for $t \in \mathbb{R}$. Hence, if the function f is continuous, bounded and antireversible and if A_h is reversible, then $\mathcal{F}_h(f)$ is reversible. Indeed,

$$S\mathcal{F}_h(f)(t) = \int_{-\infty}^{+\infty} SK_h(s) f(t-s)ds = \int_{-\infty}^{+\infty} K_h(-s) f(-t+s)ds$$

$$= \int_{-\infty}^{+\infty} K_h(s) f(-t-s)ds = \mathcal{F}(f)(-t).$$

Thus $\mathcal{F}_h(f)$ is reversible.

(c): The expression of $\mathcal{F}_h(f)$ as a convolution product with kernel K_h and the theorem of derivation under the integral derived from Lebesgue's dominated convergence theorem, ensure that $\mathcal{F}_h(f)$ is holomorphic in $\mathcal{B}_\ell$ for any $f \in \mathcal{H}_\ell^\lambda(\mathbb{C}^n)$. Moreover, for a given $\lambda \in]0, m(A_h)[$, we can choose δ such that $0 < \lambda < \delta < m(A_h)$. Then, (6.26) ensures that

$$\begin{aligned} |\mathcal{F}_h(f)| &\leq \int_{-\infty}^{+\infty} M e^{-\delta|u|} |f|_{\mathcal{H}_\ell^\lambda(\mathbb{C}^n)} e^{-\lambda|\mathcal{R}e(\xi-u)|} du \\ &\leq e^{-\lambda|\mathcal{R}e(\xi)|} \int_{-\infty}^{+\infty} M e^{(-\delta+\lambda)|u|} |f|_{\mathcal{H}_\ell^\lambda(\mathbb{C}^n)} du \\ &\leq \frac{2M}{\delta-\lambda} |f|_{\mathcal{H}_\ell^\lambda(\mathbb{C}^n)} e^{-\lambda|\mathcal{R}e(\xi)|}. \end{aligned}$$

Hence, we get that for every $f \in \mathcal{H}_\ell^\lambda(\mathbb{C}^n)$,

$$|\mathcal{F}_{\mathrm{h}}(f)|_{\mathcal{H}_\ell^\lambda(\mathbb{C}^n)} \leq \frac{2M}{m(A_{\mathrm{h}}) - \lambda} |f|_{\mathcal{H}_\ell^\lambda(\mathbb{C}^n)}.$$

(d): The proof is very similar to (c) and thus left to the reader. □

Proof of Lemma 6.4.6

(a): A basis of solutions of $\mathcal{L}_\omega(v) = 0$ is given by (r_+, r_-). This basis is an orthonormal basis and using the variation of constants formula, we get that any solution u of (6.24) reads

$$u(t) = \left(u_0^+ + \int_0^t \langle r_+(s), f(s)\rangle ds\right) r_+(t) + \left(u_0^- + \int_0^t \langle r_-(s), f(s)\rangle ds\right) r_-(t).$$

If $u \underset{t\to+\infty}{\longrightarrow} 0$, then necessarily

$$\langle u(t), r_\pm(t)\rangle = u_0^\pm + \int_0^t \langle r_\pm(s), f(s)\rangle \underset{t\to+\infty}{\longrightarrow} 0.$$

Thus

$$u_0^\pm = -\int_0^{+\infty} \langle r_\pm(s), f(s)\rangle ds \qquad \text{and } u = \mathcal{F}_\omega(f)$$

where $\mathcal{F}_\omega(f)$ is given by (6.25). Conversely, one checks easily that $\mathcal{F}_\omega(f)$ is well defined for $f \in C_\lambda^0(\mathbb{R}, \mathbb{R}^2)$ and that it satisfies $\mathcal{F}_\omega(f) \underset{t\to+\infty}{\longrightarrow} 0$.

(b): (b) readily follows from (6.25), recalling that (r_+, r_-) is an orthonormal basis.

(c): (c) follows from (b) observing that when f is antireversible, $s \mapsto \langle r_+(s), f(s)\rangle$ is odd and thus the first solvability condition

$$\int_{-\infty}^{+\infty} \langle r_+(s), f(s)\rangle ds = 0$$

is automatically satisfied.

(d): The integral expression of $\mathcal{F}_\omega(f)$ with kernel K_ω and the theorem of derivation under the integral derived from Lebesgue's dominated convergence theorem, ensure that $\mathcal{F}_\omega(f)$ is holomorphic in $\mathcal{B}_\ell$ for any $f \in \mathcal{H}_\ell^\lambda(\mathbb{C}^2)$. Moreover, (c) ensures that since f is antireversible and satisfies (c)(iv), then $\mathcal{F}_\omega(f)$ is reversible. Then, for any $\xi \in \mathcal{B}_\ell$ such that $\mathcal{R}e\,(\xi) \geq 0$,

$$\begin{aligned}
|\mathcal{F}_\omega(f)| &\leq \int_0^{+\infty} |||K_\omega(s)||| \, |f|_{\mathcal{H}_\ell^\lambda(\mathbb{C}^2)} \, \mathrm{e}^{-\lambda|\mathcal{R}e(\xi+s)|} ds \\
&\leq \mathrm{e}^{-\lambda \mathcal{R}e(\xi)} \int_0^{+\infty} \mathrm{e}^{-\lambda t} \, |f|_{\mathcal{H}_\ell^\lambda(\mathbb{C}^2)} \, ds \leq \frac{1}{\lambda} |f|_{\mathcal{H}_\ell^\lambda(\mathbb{C}^2)} \, \mathrm{e}^{-\lambda \mathcal{R}e(\xi)}.
\end{aligned}$$

Hence, since $\mathcal{F}_\omega(f)$ is reversible, this last estimate ensures that for every $f \in \mathcal{H}_\ell^\lambda(\mathbb{C}^2)$,

$$|\mathcal{F}_\omega(f)|_{\mathcal{H}_\ell^\lambda(\mathbb{C}^2)} \le \frac{1}{\lambda} |f|_{\mathcal{H}_\ell^\lambda(\mathbb{C}^2)}.$$

(e): The proof is very similar to (d) and thus left to the reader. □

6.5 Affine equations with partially complexified time

Sections 6.1 and 6.3 deal with homogeneous equations of the form

$$\mathcal{L}(v) = 0 \qquad \text{with } \mathcal{L}(v) = \frac{dv}{dt} - A(t)v,$$

i.e. with the kernel of $\mathcal{L}$ when A converges exponentially quickly to a constant matrix (this situation typically occurs when linearizing around a homoclinic connection to 0).

The previous section is devoted to the range of $\mathcal{L}$ and in particular to the solvability conditions required on a function f for belonging to the range of $\mathcal{L}$ when $\mathcal{L}$ is restricted to different sets of reversible functions which tends to 0 at infinity. More precisely, we have studied the range of $\mathcal{L}$ when $\mathcal{L}$ is restricted to $C_\lambda^k(\mathbb{R}, \mathbb{R}^n)|_{\mathrm{R}}$ and to $\mathcal{H}_\ell^\lambda(\mathbb{C}^n)|_{\mathrm{R}}$ for different matrices $A(t)$ typically obtained by linearization around reversible homoclinic connections.

This section is now devoted to affine equations with partially complexified time, i.e. to the study of the solvability conditions that a function $\mathrm{g} : \mathcal{B}_\ell \times \mathbb{R} \to \mathbb{C}^n$, 2π-periodic with respect to the second variable, should satisfy so that, there exists a reversible solution v of

$$\mathcal{L}(v)(t) = \mathrm{g}(t, \tfrac{\omega_0 t}{\varepsilon}) \qquad \text{with } \mathcal{L}(v) = \frac{dv}{dt} - A(t)v$$

which tends to 0 at $\pm\infty$ and which reads

$$v(t) = \mathrm{v}(t, \tfrac{\omega_0 t}{\varepsilon})$$

where $\mathrm{v} : \mathcal{B}_\ell \times \mathbb{R} \to \mathbb{C}^n$ is 2π-periodic with respect to the second variable. Such solutions of affine equations are required, when one needs to use the First Bi-frequency Exponential Lemma 2.2.1 given in chapter 2. We give three propositions which correspond to different linear operators $A(t)$:

- The first and the second ones respectively typically occur after linearization around a reversible homoclinic connection to 0 induced by a pair of double semi-simple eigenvalues $\pm\lambda_0$ and a set of conjugate eigenvalues $\pm \mathrm{i}\frac{\omega_0(\varepsilon)}{\varepsilon} \pm \lambda_0$.
- The third one occurs when a pair of simple purely imaginary eigenvalues stay on the imaginary axis even after bifurcation. This is precisely the case for the $0^{2+}\mathrm{i}\omega$ and the $(\mathrm{i}\omega_0)^2\mathrm{i}\omega_1$ resonances.

Of course these cases may occur simultaneously when A is block diagonal. For instance, the second and the third one occur for the $(i\omega_0)^2 i\omega_1$ resonance. In such a case, each affine equation corresponding to each diagonal block can be solved separately.

Definition 6.5.1. *Let $\lambda \in \mathbb{R}$, $\ell > 0$ and let ${}^b\mathcal{H}_\ell^\lambda(\mathbb{C}^n)$ be the set of functions f satisfying*

(a) $f : \mathcal{B}_\ell \times \mathbb{R} \to \mathbb{C}^n$, *where* $\mathcal{B}_\ell = \{\xi \in \mathbb{C},\ |\mathcal{I}m\,(\xi)| < \ell\}$

(b) $s \mapsto f(\xi, s)$ *is 2π-periodic and of class C^1 in $\mathbb{R}$,*

(c) $(\xi, s) \mapsto f(\xi, s)$ *and* $(\xi, s) \mapsto \frac{\partial f}{\partial s}(\xi, s)$ *are continuous in $\mathcal{B}_\ell \times \mathbb{R}$,*

(d) $\xi \mapsto f(\xi, s)$ *is holomorphic in $\mathcal{B}_\ell$,*

(e) $$|f|_{{}^b\mathcal{H}_\ell^\lambda} := \sup_{(\xi,s)\in\mathcal{B}_\ell\times\mathbb{R}} \left(\left(|f(\xi,s)| + |\frac{\partial f}{\partial s}(\xi,s)|\right) e^{\lambda|\mathcal{R}e(\xi)|} \right) < +\infty.$$

For a symmetry S and $\omega > 0$, let us introduce the four subspaces of ${}^b\mathcal{H}_\ell^\lambda(\mathbb{C}^n)$ spanned by the functions which are real on the real axis and which are reversible or antireversible with respect to S or reversible or antireversible with respect to the transpose ${}^tS$ of S , i.e.

$$\begin{aligned}
{}^b\mathcal{H}_\ell^\lambda(\mathbb{C}^n)|_{\mathrm{R},\mathbb{R}} &= {}^b\mathcal{H}_\ell^\lambda(\mathbb{C}^n) \cap \{f, Sf(-\xi,-s) = f(\xi,s)\} \cap {}^b\mathcal{H}|_{\mathbb{R}},\\
{}^b\mathcal{H}_\ell^\lambda(\mathbb{C}^n)|_{\mathrm{AR},\mathbb{R}} &= {}^b\mathcal{H}_\ell^\lambda(\mathbb{C}^n) \cap \{f, Sf(-\xi,-s) = -f(\xi,s)\} \cap {}^b\mathcal{H}|_{\mathbb{R}},\\
{}^b\mathcal{H}_\ell^\lambda(\mathbb{C}^n)|_{\mathrm{R}^*,\mathbb{R}} &= {}^b\mathcal{H}_\ell^\lambda(\mathbb{C}^n) \cap \{f, {}^tSf(-\xi,-s) = f(\xi,s)\} \cap {}^b\mathcal{H}|_{\mathbb{R}},\\
{}^b\mathcal{H}_\ell^\lambda(\mathbb{C}^n)|_{\mathrm{AR}^*,\mathbb{R}} &= {}^b\mathcal{H}_\ell^\lambda(\mathbb{C}^n) \cap \{f, {}^tSf(-\xi,-s) = -f(\xi,s)\} \cap {}^b\mathcal{H}|_{\mathbb{R}}
\end{aligned}$$

where ${}^b\mathcal{H}|_{\mathbb{R}} = \{f : \mathcal{B}_\ell \times \mathbb{R} \to \mathbb{C}^n,\ f(t, \frac{\omega_0 t}{\varepsilon}) \in \mathbb{R}^n \text{ for any } t \in \mathbb{R}, \varepsilon \in]0,1]\}$.

Proposition 6.5.2. *Let $\ell, \omega_0 > 0$ and $\check{A}_\mathrm{h} : \mathcal{B}_\ell \times]0,1] \to \mathcal{M}_4(\mathbb{C})$ a 4×4 matrix valued function, reversible with respect to some real symmetry S (i.e. $S\check{A}_\mathrm{h}(\xi,\varepsilon) = -\check{A}_\mathrm{h}(-\xi,\varepsilon)S$) which is real for $(\xi,\varepsilon) \in \mathbb{R}\times]0,1]$ and holomorphic with respect to ξ.*

Assume that the homogeneous equation

$$\check{\mathcal{L}}_{\mathrm{h},\varepsilon}(v) = 0, \qquad \text{with } \check{\mathcal{L}}_{\mathrm{h},\varepsilon}(v) = \frac{dv}{dt} - \check{A}_\mathrm{h}(t,\varepsilon)v$$

admits a basis of solutions $(\check{p}_0, \check{p}_1, \check{q}_0, \check{q}_1)$ lying in $\left(H_\ell^{\lambda_0}(\mathbb{C}^4)|_{\mathrm{AR},\mathbb{R}}\right)^2 \times \left(H_\ell^{-\lambda_0}(\mathbb{C}^4)|_{\mathrm{R},\mathbb{R}}\right)^2$ and such that its dual basis $(\check{p}_0^*, \check{p}_1^*, \check{q}_0^*, \check{q}_1^*)$ belongs to $\left(H_\ell^{-\lambda_0}(\mathbb{C}^4)|_{\mathrm{AR}^*,\mathbb{R}}\right)^2 \times \left(H_\ell^{\lambda_0}(\mathbb{C}^4)|_{\mathrm{R}^*,\mathbb{R}}\right)^2$ where $\lambda_0, \omega_0 > 0$.

(a) *For every* $\mathrm{g} \in {}^b\mathcal{H}_\ell^\lambda(\mathbb{C}^4)|_{\mathrm{AR,R}}$ *with* $\lambda > \lambda_0$*, there exists a unique* v_g *reversible and which tends to* 0 *at* $+\infty$ *such that*

$$\check{\mathcal{L}}_{\mathrm{h},\varepsilon}(v_\mathrm{g})(t) = \mathrm{g}(t, \tfrac{\omega_0 t}{\varepsilon}). \tag{6.27}$$

(b) *Moreover, for every* $\varepsilon \in]0,1]$*, there exists a linear operator* $\check{\mathrm{T}}_{\mathrm{h},\varepsilon}$ *which maps* ${}^b\mathcal{H}_\ell^\lambda(\mathbb{C}^4)|_{\mathrm{AR,R}}$ *to* ${}^b\mathcal{H}_\ell^{\lambda_0}(\mathbb{C}^4)|_{\mathrm{R,R}}$ *for any* $\lambda > \lambda_0$*, such that*

$$v_\mathrm{g}(t) = \check{\mathrm{T}}_{\mathrm{h},\varepsilon}(\mathrm{g})(t, \tfrac{\omega_0 t}{\varepsilon}) \qquad \text{for every } \mathrm{g} \in {}^b\mathcal{H}_\ell^\lambda(\mathbb{C}^4)|_{\mathrm{AR,R}} \text{ with } \lambda > \lambda_0.$$

(c) *The family of operators* $\check{\mathrm{T}}_{\mathrm{h},\varepsilon}$ *is uniformly bounded, i.e. there exists* $M_{\check{\mathrm{T}}_\mathrm{h}}$ *such that*

$$\left|\check{\mathrm{T}}_{\mathrm{h},\varepsilon}(\mathrm{g})\right|_{{}^b\mathcal{H}_\ell^{\lambda_0}} \leq M_{\check{\mathrm{T}}_\mathrm{h}} |\mathrm{g}|_{{}^b\mathcal{H}_\ell^\lambda} \qquad \text{for } \mathrm{g} \in {}^b\mathcal{H}_\ell^\lambda(\mathbb{C}^4)|_{\mathrm{AR,R}},\ \varepsilon \in]0,1].$$

Remark 6.5.3. This Lemma says that when g is "partially complexified", i.e. when $\mathrm{g} \in {}^b\mathcal{H}_\ell^\lambda(\mathbb{C}^4)|_{\mathrm{AR,R}}$, then $v_\mathrm{g}(t) = \check{\mathcal{L}}_{\mathrm{h},\varepsilon}^{-1}(\mathrm{g}(t, \frac{\omega_0 t}{\varepsilon}))$ has the same form and belongs to ${}^b\mathcal{H}_\ell^{\lambda_0}(\mathbb{C}^4)|_{\mathrm{R,R}}$.

We deduce from the above proposition, a second one which will be used for the study of the $(\mathrm{i}\omega_0)^2\mathrm{i}\omega_1$ resonance.

Proposition 6.5.4. *Let* $\ell > 0$ *and* $A_\mathrm{h} : \mathcal{B}_\ell \times]0,1] \to \mathcal{M}_4(\mathbb{C})$ *a* 4×4 *matrix valued function, reversible with respect to some real symmetry* S *(i.e.* $SA_\mathrm{h}(\xi,\varepsilon) = -A_\mathrm{h}(-\xi,\varepsilon)S$*) which is real for* $(\xi,\varepsilon) \in \mathbb{R}\times]0,1]$ *and holomorphic with respect to* ξ*.*

Let J *be a* 4×4 *real reversible matrix such that* $\pm\mathrm{i}$ *are double semi-simple eigenvalues and denote* $\mathsf{A}_\mathrm{h}(\xi,s,\varepsilon) := R_s A_\mathrm{h}(\xi,\varepsilon) R_{-s}$ *where for* $\omega \in \mathbb{C}$*,* $R_\omega = \exp(\omega J)$*.*

Assume that the homogeneous equation

$$\mathcal{L}_{\mathrm{h},\varepsilon}(v) = 0, \qquad \text{with } \mathcal{L}_{\mathrm{h},\varepsilon}(v) = \frac{dv}{dt} - \mathsf{A}_\mathrm{h}(t, \tfrac{\omega_0 t}{\varepsilon}, \varepsilon)v$$

admits a basis of solutions (p_0, p_1, q_0, q_1) *such that*

$$\check{p}_j = R_{\frac{-\omega_0\xi}{\varepsilon}} p_j \in H_\ell^{\lambda_0}(\mathbb{C}^4)|_{\mathrm{AR,R}}, \quad \check{q}_j = R_{\frac{-\omega_0\xi}{\varepsilon}} q_j \in H_\ell^{-\lambda_0}(\mathbb{C}^4)|_{\mathrm{R,R}}$$

and such that its dual basis $(p_0^*, p_1^*, q_0^*, q_1^*)$ *satisfies*

$$\check{p}_j^* = {}^tR_{\frac{\omega_0 \xi}{\varepsilon}}\, p_j^* \in H_\ell^{-\lambda_0}(\mathbb{C}^4)|_{\mathrm{AR}^*,\mathbb{R}}, \quad \check{q}_j^* = {}^tR_{\frac{\omega_0 \xi}{\varepsilon}}\, q_j^* \in H_\ell^{\lambda_0}(\mathbb{C}^4)|_{\mathrm{R}^*,\mathbb{R}}.$$

where $\lambda_0, \omega_0 > 0$.

(a) *For every* $\mathrm{g} \in {}^b\mathcal{H}_\ell^\lambda(\mathbb{C}^4)|_{\mathrm{AR},\mathbb{R}}$ *with* $\lambda > \lambda_0$, *there exists a unique* v_g *reversible and which tends to 0 at* $+\infty$ *such that*

$$\mathcal{L}_{\mathrm{h},\varepsilon}(v_\mathrm{g})(t) = \mathrm{g}(t, \tfrac{\omega_0 t}{\varepsilon}). \tag{6.28}$$

(b) *Moreover, for every* $\varepsilon \in]0,1]$ *there exists a linear operator* $\mathsf{T}_{\mathrm{h},\varepsilon}$ *which maps* ${}^b\mathcal{H}_\ell^\lambda(\mathbb{C}^4)|_{\mathrm{AR},\mathbb{R}}$ *to* ${}^b\mathcal{H}_\ell^{\lambda_0}(\mathbb{C}^4)|_{\mathrm{R},\mathbb{R}}$ *for any* $\lambda > \lambda_0$, *such that*

$$v_\mathrm{g}(t) = \mathsf{T}_{\mathrm{h},\varepsilon}(\mathrm{g})(t, \tfrac{\omega_0 t}{\varepsilon}) \qquad \textit{for every } \mathrm{g} \in {}^b\mathcal{H}_\ell^\lambda(\mathbb{C}^4)|_{\mathrm{AR},\mathbb{R}} \textit{ with } \lambda > \lambda_0.$$

(c) *The family of operators* $\mathsf{T}_{\mathrm{h},\varepsilon}$ *is uniformly bounded, i.e. there exists* $M_{\mathsf{T}_\mathrm{h}}$ *such that*

$$|\mathsf{T}_{\mathrm{h},\varepsilon}(\mathrm{g})|_{{}^b\mathcal{H}_\ell^{\lambda_0}} \le M_{\mathsf{T}_\mathrm{h}}\, |\mathrm{g}|_{{}^b\mathcal{H}_\ell^\lambda} \qquad \textit{for } \mathrm{g} \in {}^b\mathcal{H}_\ell^\lambda(\mathbb{C}^4)|_{\mathrm{AR},\mathbb{R}},\ \varepsilon \in]0,1].$$

Remark 6.5.5. This Lemma says that when g is "partially complexified", i.e. when $\mathrm{g} \in {}^b\mathcal{H}_\ell^\lambda(\mathbb{C}^4)|_{\mathrm{AR},\mathbb{R}}$, then $v_\mathrm{g}(t) = \mathcal{L}_{\mathrm{h},\varepsilon}^{-1}(\mathrm{g}(t, \frac{\omega_0 t}{\varepsilon}))$ has the same form and belongs to ${}^b\mathcal{H}_\ell^{\lambda_0}(\mathbb{C}^4)|_{\mathrm{R},\mathbb{R}}$.

Remark 6.5.6. An example of such a linear operator occurs when considering the first diagonal block of the matrix obtained by linearization around the homoclinic connection to 0 of the normal form of any order for the $(\mathrm{i}\omega_0)^2\mathrm{i}\omega_1$ resonance. More generally, it occurs after linearization around a homoclinic connection to 0 induced by a set of conjugate eigenvalues $\pm\mathrm{i}\frac{\omega_0(\varepsilon)}{\varepsilon} \pm \lambda_0$ (see Lemma 6.3.11).

Proposition 6.5.7. *Let* $\omega_0, \omega_1, \ell > 0$ *and* $\underline{\omega}_1(\varepsilon) = \omega_1 + \varepsilon\underline{\omega}_{11}(\varepsilon)$ *with* $|\underline{\omega}_{11}|_0 = \sup_{\varepsilon \in]0,1]} |\underline{\omega}_{11}(\varepsilon)| < +\infty$. *Let*

$$A_c(\xi,\varepsilon) = \begin{pmatrix} 0 & -\frac{\underline{\omega}_1(\varepsilon)}{\varepsilon} - \psi(\xi,\varepsilon) \\ \frac{\underline{\omega}_1(\varepsilon)}{\varepsilon} + \psi(\xi,\varepsilon) & 0 \end{pmatrix}, \qquad S = \begin{pmatrix} 1 & 0 \\ 0 & -1 \end{pmatrix}$$

where $\psi : \mathcal{B}_\ell \times]0,1] \to \mathbb{C}$ *is holomorphic and even with respect to* $\xi \in \mathcal{B}_\ell$ *and* $\sup_{(\xi,\varepsilon) \in \mathcal{B}_\ell \times]0,1]} \left(|\psi(\xi,\varepsilon)| + \left| \int_{[0,\xi]} \psi(s,\varepsilon) ds \right| \right) < +\infty$.

(a) *Then, the linear homogeneous equation*

$$\mathcal{L}_{c,\varepsilon}(v) = 0, \qquad \text{where } \mathcal{L}_{c,\varepsilon}(v) := \frac{dv}{dt} - A_c(t,\varepsilon)v$$

admits a basis of solutions (r_+, r_-) *which reads* $r_+ = (\cos\phi, \sin\phi)$, $r_- = (-\sin\phi, \cos\phi)$, *where* $\phi(t,\varepsilon) = \frac{\omega_1(\varepsilon)}{\varepsilon}t + \int_{[0,t]} \psi(s,\varepsilon)ds$.

(b) *For every function* g, *if there exists* v *reversible which tends to* 0 *at* $+\infty$ *such that* $\mathcal{L}_{c,\varepsilon}(v) = g$, *then* g *is antireversible and*

$$\int_0^{+\infty} \langle r_-(t,\varepsilon), g(t)\rangle_* \, dt = 0.$$

(c) *Conversely, denote*

$${}^b\mathcal{H}_\ell^\lambda(\mathbb{C}^2)|^\perp_{AR,\mathbb{R}} = {}^b\mathcal{H}_\ell^\lambda(\mathbb{C}^2)|_{AR,\mathbb{R}} \cap \{f : \mathbb{R}^2 \to \mathbb{R}^2, \int_0^{+\infty} \langle r_-(t), f(t, \tfrac{\omega_0 t}{\varepsilon})\rangle_* \, dt = 0\}.$$

For every $\mathrm{g} \in {}^b\mathcal{H}_\ell^\lambda(\mathbb{C}^2)|^\perp_{AR,\mathbb{R}}$ *with* $\lambda > 0$, *there exists a unique* v_{g} *reversible and which tends to 0 at* $+\infty$ *such that*

$$\mathcal{L}_{c,\varepsilon}(v_{\mathrm{g}})(t) = \mathrm{g}(t, \tfrac{\omega_0 t}{\varepsilon}).$$

(d) *Moreover, for every* $\varepsilon \in]0,1]$, *there exists a linear operator* $\mathsf{T}_{c,\varepsilon}$ *which maps* ${}^b\mathcal{H}_\ell^\lambda(\mathbb{C}^2)|^\perp_{AR,\mathbb{R}}$ *to* ${}^b\mathcal{H}_\ell^\lambda(\mathbb{C}^2)|_{R,\mathbb{R}}$ *for any* $\lambda > 0$, *such that*

$$v_{\mathrm{g}}(t) = \mathsf{T}_{c,\varepsilon}(\mathrm{g})(t, \tfrac{\omega_0 t}{\varepsilon}) \text{ for every } \mathrm{g} \in {}^b\mathcal{H}_\ell^\lambda(\mathbb{C}^2)|^\perp_{AR,\mathbb{R}} \text{ with } \lambda > 0.$$

(e) *The family of operators* $\mathsf{T}_{c,\varepsilon}$ *is uniformly bounded, i.e. there exists* M_{T_c} *such that*

$$|\mathsf{T}_{c,\varepsilon}(\mathrm{g})|_{{}^b\mathcal{H}_\ell^\lambda} \le M_{\mathsf{T}_c} |\mathrm{g}|_{{}^b\mathcal{H}_\ell^\lambda} \qquad \text{for } \mathrm{g} \in {}^b\mathcal{H}_\ell^\lambda(\mathbb{C}^2)|^\perp_{AR,\mathbb{R}},\ \varepsilon \in]0,1].$$

Remark 6.5.8. This Lemma says that when g is "partially complexified", i.e. when $\mathrm{g} \in {}^b\mathcal{H}_\ell^\lambda(\mathbb{C}^2)|_{AR,\mathbb{R}}$, then g should satisfy one solvability condition for belonging to the range of $\mathcal{L}_{c,\varepsilon}$. If g satisfies such a condition, then $v_{\mathrm{g}}(t) = \mathcal{L}_{c,\varepsilon}^{-1}(\mathrm{g}(t, \frac{\omega_0 t}{\varepsilon}))$ has the same form and belongs to ${}^b\mathcal{H}_\ell^{\lambda_0}(\mathbb{C}^2)|_{R,\mathbb{R}}$. Observe that the solvability condition required on g is a bi-oscillatory integral. The Bi-frequency Exponential Lemmas given in Chapter 2 have been developed to study such solvability conditions.

Remark 6.5.9. An example of such a linear operator occurs when considering the second diagonal block of the matrix obtained by linearization around the homoclinic connection to 0 of the normal form of any order for the $(i\omega_0)^2 i\omega_1$ resonance. More generally, it occurs when a pair of simple purely imaginary eigenvalues stays on the imaginary axis after bifurcation.

The proofs of the three previous proposition are given in the next three subsections.

Proof of Proposition 6.5.2

Let g be in ${}^b\mathcal{H}_\ell^\lambda(\mathbb{C}^4)|_{\mathrm{AR,R}}$ with $\lambda > \lambda_0$. We look for solutions of the equation (6.27) which are reversible and which go to 0 at $+\infty$.

If $\check{\mathcal{L}}_{\mathrm{h},\varepsilon}(v) = 0$ then v reads $v(t) = a_0\check{p}_0(t) + a_1\check{p}_1(t) + b_0\check{q}_0(t) + b_1\check{q}_1(t)$ where a_0, a_1, b_0, b_1 are four real constants. On one hand, if v is reversible then $a_0 = a_1 = 0$ since $\check{p}_0, \check{p}_1$ are independent and antireversible and $\check{q}_0, \check{q}_1$ are reversible. On the other hand, if v tends to 0 at $+\infty$, then $b_0 = b_1 = 0$ since $b_j = \langle \check{q}_j^*(t), v(t)\rangle_*$ and $v, \check{q}_j^*$ go to 0 at $+\infty$. *Hence there exists at most one solution of (6.27) which is reversible and which tends to 0 at $+\infty$.*

The rest of this subsection is devoted to the proof of the existence of such a solution of the form $v_{\mathrm{g}} = \mathsf{v}(t, \frac{\omega_0 t}{\varepsilon})$ with $\mathsf{v} \in {}^b\mathcal{H}_\ell^{\lambda_0}|_{\mathrm{R,R}}$. For that purpose we introduce the following linear partial differential equation

$$\check{\mathsf{L}}_{\mathrm{h},\varepsilon}(\mathsf{v})(\xi, s) = \mathrm{g}(\xi, s) \tag{6.29}$$

where $\check{\mathsf{L}}_{\mathrm{h},\varepsilon}(\mathsf{v})(\xi, s) = \dfrac{\partial \mathsf{v}}{\partial \xi} + \dfrac{\omega_0}{\varepsilon}\dfrac{\partial \mathsf{v}}{\partial s} - \check{A}_{\mathrm{h}}(\xi, \varepsilon)\mathsf{v}$ and we check

Lemma 6.5.10. *If* v *is a solution of (6.29), then* $v(t) = \mathsf{v}(t, \frac{\omega_0 t}{\varepsilon})$ *is a solution of (6.27).*

Remark 6.5.11. The above associated P.D.E. is not the unique possible one. The above lemma is also true for any P.D.E. of the form $\check{\mathsf{L}}_{\mathrm{h},\varepsilon}(\mathsf{v})(\xi, s) = \mathrm{g}(\xi, s) + \widetilde{\mathrm{g}}(\xi, s)$ where $\widetilde{\mathrm{g}}(t, \frac{\omega_0 t}{\varepsilon}) = 0$ holds for every $t \in \mathbb{R}$. For instance such a condition is fulfilled by any $\widetilde{\mathrm{g}}(\xi, s) = G(\xi - \frac{\varepsilon s}{\omega_0})$ where G is holomorphic and satisfies $G(0) = 0$. In what follows, we explain which function $\widetilde{\mathrm{g}}$ should be chosen to obtain the desired result. We first check that

Lemma 6.5.12. *The functions* $(\xi, s) \mapsto \check{p}_j(\xi, \varepsilon)$, $(\xi, s) \mapsto \check{q}_j(\xi, \varepsilon)$ *are solutions of the linear homogeneous equation* $\check{\mathsf{L}}_{\mathrm{h},\varepsilon}(\mathsf{v}) = 0$.

For obtaining a solution of (6.27) in the form $v_{\mathrm{g}}(t) = \mathsf{v}(t, \frac{\omega_0 t}{\varepsilon}) = \check{\mathsf{T}}_{\mathrm{h},\varepsilon}(\mathrm{g})(t, \frac{\omega_0 t}{\varepsilon})$ we proceed in two steps:

1. We first obtain formally an integral formula for $\check{\mathsf{T}}_{\mathrm{h},\varepsilon}(\mathrm{g})$.

2. We check carefully that the proposed formula is really a solution of (6.27) of the form $v(t) = \mathrm{v}(t, \frac{\omega_0 t}{\varepsilon})$ with $\mathrm{v} \in {}^b\mathcal{H}_\ell^{\lambda_0}|_{\mathbb{R},\mathbb{R}}$.

Step 1. Formal obtaining of an integral formula for $\check{T}_{h,\varepsilon}(g)$. The integral is obtained using a variation of constants formula with the basis of solutions given by Lemma 6.5.12. Moreover, we also explain, how (6.29) should be modified to obtain a new associated P.D.E. (see Remark 6.5.11) which has all the desired symmetry properties. Performing the change of coordinates

$$\tau = s - \tfrac{\omega_0 \xi}{\varepsilon}, \quad \xi = \xi, \qquad \mathrm{v}(\xi, s) = \widetilde{\mathrm{v}}(\xi, \tau) = \widetilde{\mathrm{v}}(\xi, s - \tfrac{\omega_0 \xi}{\varepsilon}),$$

the equation (6.29) is equivalent to

$$\frac{\partial \widetilde{\mathrm{v}}}{\partial \xi} - \check{A}_h(\xi, \varepsilon)\widetilde{\mathrm{v}} = \mathrm{g}(\xi, \tau + \tfrac{\omega_0 \xi}{\varepsilon}).$$

Using the variation of constants formula, we get

$$\begin{aligned} \widetilde{\mathrm{v}}(\xi,\tau) = \quad & \sum_{j=0}^{1} \left(a_j(\tau) + \int_{[0,\xi]} \langle \check{p}_j^*(\zeta), g(\zeta, \tfrac{\omega_0 \zeta}{\varepsilon} + \tau) \rangle_* \, d\zeta \right) \check{p}_j(\xi) \\ & + \sum_{j=0}^{1} \left(b_j(\tau) + \int_{[0,\xi]} \langle \check{q}_j^*(\zeta), g(\zeta, \tfrac{\omega_0 \zeta}{\varepsilon} + \tau) \rangle_* \, d\zeta \right) \check{q}_j(\xi) \end{aligned}$$

For obtaining an integral formula which is well defined, i.e. such that $\mathrm{g}(\zeta, \tau + \frac{\omega_0 \zeta}{\varepsilon})$ is well defined it is necessary that

$$\tau + \tfrac{\omega_0 \zeta}{\varepsilon} \in \mathbb{R} \;\Leftrightarrow\; s + \tfrac{\omega_0}{\varepsilon}(\zeta - \xi) \in \mathbb{R} \;\Leftrightarrow\; \zeta \in \xi + \mathbb{R}. \qquad (6.30)$$

This can be achieved by an appropriate choice of the constants $a_j(\tau), b_j(\tau)$. Indeed, using one of the two paths drawn in Figure 6.2

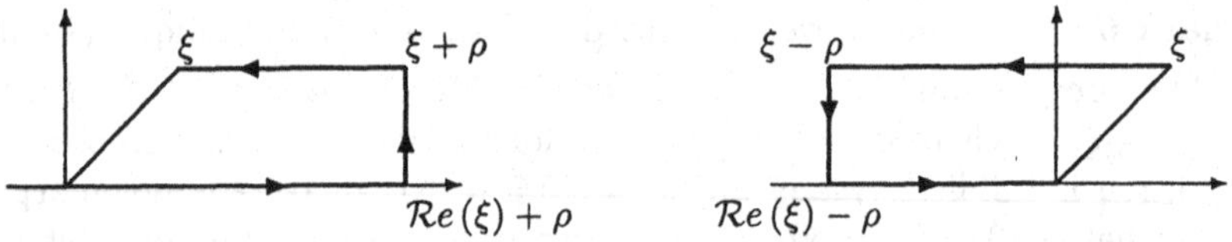

Fig. 6.2. Paths $\Gamma_\rho^\pm$

we check that any of the following natural choice for the constants

$$c = -\int_0^{+\infty} \;\Rightarrow\; c + \int_{[0,\xi]} = -\int_{\xi+\mathbb{R}^+}, \qquad \text{(C1)}$$

$$c = \int_{-\infty}^{0} \;\Rightarrow\; c + \int_{[0,\xi]} = \int_{\xi+\mathbb{R}^-}, \qquad \text{(C2)}$$

$$c = -\frac{1}{2}\int_0^{+\infty} + \frac{1}{2}\int_{-\infty}^{0} \;\Rightarrow\; c + \int_{[0,\xi]} = -\frac{1}{2}\int_{\xi+\mathbb{R}^+} + \frac{1}{2}\int_{\xi+\mathbb{R}^-}. \qquad \text{(C3)}$$

ensures that (6.30) is fulfilled. To determine which of the previous choices, we shall make for a_j, b_j we must keep in mind that the integrals should converge and that we look for a solution v which satisfies

$$\sup_{\mathcal{B}_\ell \times \mathbb{R}} |\mathsf{v}(\xi, s)| \mathrm{e}^{\lambda_0 |\mathcal{R}e(\xi)|} < +\infty, \qquad S\mathsf{v}(-\xi, -s) = \mathsf{v}(\xi, s). \tag{6.31}$$

With the choice (C3) for a_j we obtain that

$$\check{\mathsf{T}}_{p_j,\varepsilon}(\mathbf{g})(\xi, s) := \left(-\frac{1}{2}\int_{\xi+\mathbb{R}^+} + \frac{1}{2}\int_{\xi+\mathbb{R}^-}\right) \langle \check{p}_j^*(\zeta, \varepsilon), g(\zeta, \tfrac{\omega_0 \zeta}{\varepsilon} + \tau)\rangle_* d\zeta \; \check{p}_j(\xi, \varepsilon)$$

reads

$$\check{\mathsf{T}}_{p_j,\varepsilon}(\mathbf{g})(\xi, s) = \check{\mathsf{U}}_{p_j,\varepsilon}(\mathbf{g})(\xi, s) + S\check{\mathsf{U}}_{p_j,\varepsilon}(\mathbf{g})(-\xi, -s)$$

where

$$\check{\mathsf{U}}_{p_j,\varepsilon}(\mathbf{g})(\xi, s) = -\frac{1}{2}\int_0^{+\infty} \langle \check{p}_j^*(\xi + t, \varepsilon), g(\xi + t, s + \tfrac{\omega_0 t}{\varepsilon})\rangle_* dt \; \check{p}_j(\xi, \varepsilon)$$

and that

$$|\check{\mathsf{U}}_{p_j,\varepsilon}(\mathbf{g})(\xi, s)| \leq \|\check{p}_j^*\|_{H_\ell^{-\lambda_0}} \|\check{p}_j\|_{H_\ell^{\lambda_0}} |\mathbf{g}|_{{}^b\mathcal{H}_\ell^\lambda} \frac{\mathrm{e}^{-\lambda|\mathcal{R}e(\xi)|}}{2(\lambda - \lambda_0)} \qquad \text{for } \mathcal{R}e\,(\xi) \geq 0,$$

$$|\check{\mathsf{U}}_{p_j,\varepsilon}(\mathbf{g})(\xi, s)| \leq \|\check{p}_j^*\|_{H_\ell^{-\lambda_0}} \|\check{p}_j\|_{H_\ell^{\lambda_0}} |\mathbf{g}|_{{}^b\mathcal{H}_\ell^\lambda} \frac{\mathrm{e}^{-\lambda_0|\mathcal{R}e(\xi)|}}{\lambda - \lambda_0} \qquad \text{for } \mathcal{R}e\,(\xi) \leq 0.$$

So, (6.31) is fulfilled by $\check{\mathsf{T}}_{p_j,\varepsilon}(\mathbf{g})$ for the choice (C3) for a_j.

Although, the situation seems very similar for b_j, it is in fact more tricky. Indeed, here again we shall choose b_j such that (6.31) is fulfilled. With the choice (C1) for b_j we get

$$\begin{aligned}\check{\mathsf{T}}_{q_j,\varepsilon}(\mathbf{g})(\xi, s) &:= -\int_{\xi+\mathbb{R}^+} \langle \check{q}_j^*(\zeta, \varepsilon), g(\zeta, \tfrac{\omega_0 \zeta}{\varepsilon} + \tau)\rangle_* d\zeta \; \check{q}_j(\xi, \varepsilon) \\ &= -\int_0^{+\infty} \langle \check{q}_j^*(\xi + t, \varepsilon), g(\xi + t, s + \tfrac{\omega_0 t}{\varepsilon})\rangle_* dt \; \check{q}_j(\xi, \varepsilon)\end{aligned}$$

and that

$$|\check{\mathsf{T}}_{q_j,\varepsilon}(\mathbf{g})(\xi, s)| \leq \|\check{q}_j^*\|_{H_\ell^{-\lambda_0}} \|\check{q}_j\|_{H_\ell^{\lambda_0}} |\mathbf{g}|_{{}^b\mathcal{H}_\ell^\lambda} \frac{\mathrm{e}^{-\lambda|\mathcal{R}e(\xi)|}}{\lambda + \lambda_0}, \qquad \text{for } \mathcal{R}e\,(\xi) \geq 0,$$

$$|\check{\mathsf{T}}_{q_j,\varepsilon}(\mathbf{g})(\xi, s)| \leq \|\check{q}_j^*\|_{H_\ell^{-\lambda_0}} \|\check{q}_j\|_{H_\ell^{\lambda_0}} |\mathbf{g}|_{{}^b\mathcal{H}_\ell^\lambda} \frac{2\mathrm{e}^{+\lambda_0|\mathcal{R}e(\xi)|}}{\lambda + \lambda_0}, \qquad \text{for } \mathcal{R}e\,(\xi) \leq 0.$$

So, with this choice we obtain a very bad estimate for $\mathcal{R}e\,(\xi) \leq 0$. With the choice (C2) the estimate is then bad for $\mathcal{R}e\,(\xi) \geq 0$ and with the choice (C3), although the solution is reversible, we obtain bad estimates for both $\mathcal{R}e\,(\xi) \geq 0$ and $\mathcal{R}e\,(\xi) \leq 0$. To overcome this difficulty, we determine under which condition $\check{\mathsf{T}}_{q_j,\varepsilon}(\mathsf{g})$ obtained with the choice (C1) is reversible. We check that

$$S\check{\mathsf{T}}_{q_j,\varepsilon}(\mathsf{g})(-\xi,-s) = \check{\mathsf{T}}_{q_j,\varepsilon}(\mathsf{g})(\xi,s) + \mathsf{d}_{q_j,\varepsilon}(\mathsf{g})(\xi,s)\check{q}_j(\xi,\varepsilon) \tag{6.32}$$

where the defect of symmetry $\mathsf{d}_{q_j,\varepsilon}(\mathsf{g})$ is given by

Lemma 6.5.13. *For* $\mathsf{g} \in {}^b\mathcal{H}_\ell^\lambda(\mathbb{C}^4)|_{\mathrm{AR,R}}$, *let* $\mathsf{d}_{q_j,\varepsilon}(\mathsf{g})$ *be given by*

$$\mathsf{d}_{q_j,\varepsilon}(\mathsf{g})(\xi,s) = \int_{-\infty}^{+\infty} \langle \check{q}_j^*(\xi+t,\varepsilon), g(\xi+t, s+\tfrac{\omega_0 t}{\varepsilon})\rangle_* \, dt.$$

Then, for every $(\xi,s) \in \mathcal{B}_\ell \times \mathbb{R}$ *and* $\mathsf{g} \in {}^b\mathcal{H}_\ell^\lambda(\mathbb{C}^4)|_{\mathrm{AR,R}}$,

$$\mathsf{d}_{q_j,\varepsilon}(\mathsf{g})(\xi,s) = \theta_{q_j,\varepsilon}(\mathsf{g})(\xi - \tfrac{\varepsilon s}{\omega_0})$$

where $\theta_{q_j,\varepsilon}$ *is an odd,* $\frac{2\pi\varepsilon}{\omega_0}$*-periodic function holomorphic on* $\mathcal{B}_\ell$, *real valued on the real axis and explicitly given by*

$$\theta_{q_j,\varepsilon}(\mathsf{g})(\zeta) = \int_{-\infty}^{+\infty} \langle \check{q}_j^*(\zeta+u,\varepsilon), g(\zeta+u, \tfrac{\omega_0 u}{\varepsilon})\rangle_* \, du.$$

Moreover, $\mathsf{d}_{q_j,\varepsilon}$ *is a linear operator from* ${}^b\mathcal{H}_\ell^\lambda(\mathbb{C}^4)|_{\mathrm{AR,R}}$ *to* ${}^b\mathcal{H}_\ell^0(\mathbb{C})$ *which is uniformly bounded with respect to* ε, *i.e. there exists* $M_{\mathsf{d}_{q_j}}$ *such that*

$$|\mathsf{d}_{q_j,\varepsilon}(\mathsf{g})|_{{}^b\mathcal{H}_\ell^0} \le M_{\mathsf{d}_{q_j}} |\mathsf{g}|_{{}^b\mathcal{H}_\ell^\lambda} \qquad \text{for } \mathsf{g} \in {}^b\mathcal{H}_\ell^\lambda(\mathbb{C}^4)|_{\mathrm{AR,R}},\ \varepsilon \in]0,1].$$

Proof. Perform the change of variable $u = t - \frac{\varepsilon s}{\omega_0}$ to obtain the explicit formula for $\theta_{q_j,\varepsilon}$. Its properties readily follows. Using the theorem of derivation under the integral derived from Lebesgue's dominated convergence theorem, we prove that $\mathsf{d}_{q_j,\varepsilon}$ maps ${}^b\mathcal{H}_\ell^\lambda(\mathbb{C}^4)|_{\mathrm{AR,R}}$ to ${}^b\mathcal{H}_\ell^0(\mathbb{C})$ with

$$|\mathsf{d}_{q_j,\varepsilon}(\mathsf{g})|_{{}^b\mathcal{H}_\ell^0} \le \frac{4\|\check{q}_j^*\|_{H_\ell^{\lambda_0}}}{\lambda+\lambda_0} |\mathsf{g}|_{{}^b\mathcal{H}_\ell^\lambda} \qquad \text{for } \mathsf{g} \in {}^b\mathcal{H}_\ell^\lambda(\mathbb{C}^4)|_{\mathrm{AR,R}},\ \varepsilon \in]0,1].$$

To obtain a term $\check{\mathsf{T}}_{q_j,\varepsilon}(\mathsf{g})$ with no symmetry defect, i.e. with $\mathsf{d}_{q_j,\varepsilon}(\mathsf{g}) = 0$, we shall work with a modified g, i.e with a modified associated P.D.E. given by

$$\check{\mathsf{L}}_{\mathsf{h},\varepsilon}(\mathsf{g}) = \mathsf{g} + \sum_{j=0}^{1} \Pi_{q_j,\varepsilon}(\mathsf{g}) \tag{6.33}$$

where

$$\Pi_{q_j,\varepsilon}(\mathsf{g})(\xi,s) = -\frac{1}{\sqrt{\pi}} \mathrm{e}^{-\xi^2} \mathsf{d}_{q_j,\varepsilon}(\mathsf{g})(\xi,s)\, \check{q}_j(\xi,\varepsilon)$$

Observe that since $\Pi_{q_j,\varepsilon}(t, \frac{\omega_0 t}{\varepsilon}) = 0$ for every t, the property of Lemma 6.5.10 is still true: if v is a solution of (6.33), then $v(t) = \mathsf{v}(t, \frac{\omega_0 t}{\varepsilon})$ is a solution of (6.27). Moreover, we have modified the equation in the direction $\check{q}_j$ which ensures that the previous computations made for $\check{p}_j$ are still true, since $\check{\mathsf{T}}_{p_j,\varepsilon}(\mathsf{g} + \sum_{j=0}^{1} \Pi_{q_j,\varepsilon}(\mathsf{g})) = \check{\mathsf{T}}_{p_j,\varepsilon}(\mathsf{g})$. So, we propose the following integral formula for $\check{\mathsf{T}}_{\mathsf{h},\varepsilon}$

Definition 6.5.14. *Let*

$$\check{\mathsf{T}}_{\mathsf{h},\varepsilon}(\mathsf{g}) = \sum_{j=0}^{1} \check{\mathsf{T}}_{p_j,\varepsilon}(\mathsf{g}) + \check{\mathsf{T}}_{q_j,\varepsilon}(\mathsf{g} + \Pi_{q_j,\varepsilon}(\mathsf{g}))$$

where $\Pi_{q_j,\varepsilon}$ is defined in (6.33) and where

$$\check{\mathsf{T}}_{p_j,\varepsilon}(\mathsf{f})(\xi, s) = \check{\mathsf{U}}_{p_j,\varepsilon}(\mathsf{f})(\xi, s) + S\check{\mathsf{U}}_{p_j,\varepsilon}(\mathsf{f})(-\xi, -s)$$

with

$$\check{\mathsf{U}}_{p_j,\varepsilon}(\mathsf{f})(\xi, s) = -\frac{1}{2}\int_0^{+\infty} \langle \check{p}_j^*(\xi + t, \varepsilon), g(\xi + t, s + \tfrac{\omega_0 t}{\varepsilon})\rangle_* dt \; \check{p}_j(\xi, \varepsilon)$$

and

$$\check{\mathsf{T}}_{q_j,\varepsilon}(\mathsf{f})(\xi, s) = -\int_0^{+\infty} \langle \check{q}_j^*(\xi + t, \varepsilon), g(\xi + t, s + \tfrac{\omega_0 t}{\varepsilon})\rangle_* dt \; \check{q}_j(\xi, \varepsilon).$$

Step 2. Study of $\check{\mathsf{T}}_{\mathsf{h},\varepsilon}$. Our aim is now to check that the above integral formula proposed for $\check{\mathsf{T}}_{\mathsf{h},\varepsilon}(\mathsf{g})$ is well defined for $\mathsf{g} \in {}^b\mathcal{H}_\ell^\lambda(\mathbb{C}^4)|_{\mathrm{AR},\mathbb{R}}$ and that it is a solution of (6.33) which lies in ${}^b\mathcal{H}_\ell^{\lambda_0}(\mathbb{C}^4)|_{\mathrm{R},\mathbb{R}}$. We proceed in three substeps. We begin with

Lemma 6.5.15.

(a) *$\check{\mathsf{T}}_{p_j,\varepsilon}$ is a linear operator from ${}^b\mathcal{H}_\ell^\lambda(\mathbb{C}^4)|_{\mathrm{AR},\mathbb{R}}$ to ${}^b\mathcal{H}_\ell^{\lambda_0}(\mathbb{C}^4)|_{\mathrm{R},\mathbb{R}}$ which is uniformly bounded with respect to ε, i.e. there exists $M_{\check{\mathsf{T}}_{p_j}}$ such that*

$$\left|\check{\mathsf{T}}_{p_j,\varepsilon}(\mathsf{g})\right|_{{}^b\mathcal{H}_\ell^{\lambda_0}} \le M_{\check{\mathsf{T}}_{p_j}} |\mathsf{g}|_{{}^b\mathcal{H}_\ell^\lambda} \qquad \text{for } \mathsf{g} \in {}^b\mathcal{H}_\ell^\lambda(\mathbb{C}^4)|_{\mathrm{AR},\mathbb{R}},\ \varepsilon \in]0,1].$$

(b) $\check{\mathsf{L}}_{\mathsf{h},\varepsilon}(\check{\mathsf{T}}_{p_j,\varepsilon}(\mathsf{g}))(\xi, s) = \langle \check{p}_j^*(\xi, \varepsilon), \mathsf{g}(\xi, s)\rangle_* \check{p}_j(\xi, \varepsilon)$.

Proof. **(a):** As already mentioned, we check without any difficulty that for $\mathrm{g} \in {}^{b}\mathcal{H}_{\ell}^{\lambda}(\mathbb{C}^4)|_{\mathrm{AR},\mathbb{R}}$, $\check{\mathsf{U}}_{p_j,\varepsilon}(\mathrm{g})$ belongs to ${}^{b}\mathcal{H}_{\ell}^{\lambda_0}(\mathbb{C}^4)|_{\mathrm{R},\mathbb{R}}$; is real on the real axis and satisfies for every $\varepsilon \in]0,1]$

$$\left|\check{\mathsf{U}}_{p_j,\varepsilon}(\mathrm{g})\right|_{{}^{b}\mathcal{H}_{\ell}^{\lambda_0}} \leq \frac{1}{\lambda-\lambda_0} \|\check{p}_j^*\|_{H_\ell^{-\lambda_0}} \|\check{p}_j\|_{H_\ell^{\lambda_0}} |\mathrm{g}|_{{}^{b}\mathcal{H}_{\ell}^{\lambda}}\,.$$

Thus $\check{\mathsf{T}}_{p_j,\varepsilon}(\mathrm{g})$ is reversible and belongs to ${}^{b}\mathcal{H}_{\ell}^{\lambda_0}(\mathbb{C}^4)|_{\mathrm{R},\mathbb{R}}$.
(b): Denote $\check{\mathsf{U}}_{p_j,\varepsilon}(\mathrm{g})(\xi,s) = \mathsf{u}(\xi,s)\check{p}_j(\xi,\varepsilon)$. Since $\check{\mathsf{L}}_{\mathrm{h},\varepsilon}(\check{p}_j(\xi,\varepsilon)) = 0$, we obtain

$$\check{\mathsf{L}}_{\mathrm{h},\varepsilon}(\check{\mathsf{U}}_{p_j,\varepsilon}(\mathrm{g}))(\xi,s) = \left(\frac{\partial}{\partial\xi} + \frac{\omega_0}{\varepsilon}\frac{\partial}{\partial s}\right)(\mathsf{u}(\xi,s))\,\check{p}_j.$$

Then differentiating under the sign $\int$ we get

$$\begin{aligned}\left(\tfrac{\partial}{\partial\xi} + \tfrac{\omega_0}{\varepsilon}\tfrac{\partial}{\partial s}\right)(\mathsf{u}(\xi,s)) &= -\frac{1}{2}\int_0^{+\infty} \frac{d}{dt}\left(\langle \check{p}_j^*(\xi+t,\varepsilon), g(\xi+t, s+\tfrac{\omega_0 t}{\varepsilon})\rangle_*\right) dt\\ &= \tfrac{1}{2}\langle \check{p}_j^*(\xi,\varepsilon), \mathrm{g}(\xi,s)\rangle_*.\end{aligned}$$

Thus

$$\check{\mathsf{L}}_{\mathrm{h},\varepsilon}(\check{\mathsf{U}}_{p_j,\varepsilon}(\mathrm{g}))(\xi,s) = \frac{1}{2}\langle \check{p}_j^*(\xi,\varepsilon), \mathrm{g}(\xi,s)\rangle_*\,\check{p}_j(\xi,\varepsilon).$$

Similarly, since $S\check{\mathsf{U}}_{p_j,\varepsilon}(\mathrm{g})(-\xi,-s) = -\mathsf{u}(-\xi,-s)\check{p}_j(\xi,\varepsilon)$, we check

$$\begin{aligned}\check{\mathsf{L}}_{\mathrm{h},\varepsilon}(S\check{\mathsf{U}}_{p_j,\varepsilon}(\mathrm{g})(-\xi,-s)) &= \left(\tfrac{\partial}{\partial\xi} + \tfrac{\omega_0}{\varepsilon}\tfrac{\partial}{\partial s}\right)(\mathsf{u})(-\xi,-s)\,\check{p}_j(\xi,\varepsilon)\\ &= \frac{1}{2}\langle \check{p}_j^*(-\xi,\varepsilon), \mathrm{g}(-\xi,-s)\rangle_*\,\check{p}_j(\xi,\varepsilon)\\ &= \frac{1}{2}\langle -{}^{t}S\check{p}_j^*(-\xi,\varepsilon), -S\mathrm{g}(\xi,s)\rangle_*\,\check{p}_j(\xi,\varepsilon)\\ &= \frac{1}{2}\langle \check{p}_j^*(\xi,\varepsilon), \mathrm{g}(\xi,s)\rangle_*\,\check{p}_j(\xi,\varepsilon).\end{aligned}$$

Hence,

$$\check{\mathsf{L}}_{\mathrm{h},\varepsilon}(\check{\mathsf{U}}_{p_j,\varepsilon}(\mathrm{g}))(\xi,s) = \langle \check{p}_j^*(\xi,\varepsilon), \mathrm{g}(\xi,s)\rangle_*\,\check{p}_j(\xi,\varepsilon). \;\square$$

Lemma 6.5.16.

(a) *$\Pi_{q_j,\varepsilon}$ is a linear operator from ${}^{b}\mathcal{H}_{\ell}^{\lambda}(\mathbb{C}^4)|_{\mathrm{AR},\mathbb{R}}$ to ${}^{b}\mathcal{H}_{\ell}^{\lambda}(\mathbb{C}^4)|_{\mathrm{AR},\mathbb{R}}$ which is uniformly bounded with respect to ε, i.e. there exists $M_{\Pi_{q_j}}$ such that*

$$|\Pi_{q_j,\varepsilon}(\mathrm{g})|_{{}^{b}\mathcal{H}_{\ell}^{\lambda}} \leq M_{\Pi_{q_j}} |\mathrm{g}|_{{}^{b}\mathcal{H}_{\ell}^{\lambda}} \qquad \textit{for } \mathrm{g} \in {}^{b}\mathcal{H}_{\ell}^{\lambda}(\mathbb{C}^4)|_{\mathrm{AR},\mathbb{R}},\ \varepsilon \in]0,1].$$

(b) *For every* $\mathsf{f} \in \mathcal{H}_\ell^\lambda(\mathbb{C}^4)|_{\mathrm{AR,R}}$, $s \mapsto \check{\mathsf{T}}_{q_j,\varepsilon}(\mathsf{f})(\xi,s)$ *is* 2π*-periodic and of class* C^1 *on* $\mathbb{R}$*;* $\xi \mapsto \check{\mathsf{T}}_{q_j,\varepsilon}(\mathsf{f})(\xi,s)$ *is holomorphic on* $\mathcal{B}_\ell$*;* $(\xi,s) \mapsto \check{\mathsf{T}}_{q_j,\varepsilon}(\mathsf{f})(\xi,s)$ *and* $(\xi,s) \mapsto \dfrac{\partial \check{\mathsf{T}}_{q_j,\varepsilon}(\mathsf{f})}{\partial s}(\xi,s)$ *are continuous on* $\mathcal{B}_\ell \times \mathbb{R}$*. Moreover, there exists* $M^+_{\check{\mathsf{T}}_{q_j}}$ *such that for every* $\varepsilon \in]0,1]$ *and* $\mathsf{f} \in \mathcal{H}_\ell^\lambda(\mathbb{C}^4)|_{\mathrm{AR,R}}$

$$|\check{\mathsf{T}}_{q_j,\varepsilon}(\mathsf{f})(\xi,s)| + \left|\frac{\partial \check{\mathsf{T}}_{q_j,\varepsilon}(\mathsf{f})}{\partial s}(\xi,s)\right| \le M^+_{\check{\mathsf{T}}_{q_j}} \,\mathrm{e}^{-\lambda|\mathcal{R}e(\xi)|}\, |\mathsf{f}|_{{}^b\mathcal{H}_\ell^\lambda}$$

for $\xi \in \mathcal{B}_\ell$, $\mathcal{R}e\,(\xi) \ge 0$, $s \in \mathbb{R}$.

(c) $\mathsf{g} \mapsto \check{\mathsf{T}}_{q_j,\varepsilon}(\mathsf{g} + \Pi_{q_j,\varepsilon}(\mathsf{g}))$ *is a linear operator from* ${}^b\mathcal{H}_\ell^\lambda(\mathbb{C}^4)|_{\mathrm{AR,R}}$ *to* ${}^b\mathcal{H}_\ell^\lambda(\mathbb{C}^4)|_{\mathrm{R,R}}$ *which is uniformly bounded with respect to* ε*, i.e. there exists* $M_{\check{\mathsf{T}}_{q_j}}$ *such that*

$$\left|\check{\mathsf{T}}_{q_j,\varepsilon}(\mathsf{g} + \Pi_{q_j,\varepsilon}(\mathsf{g}))\right|_{{}^b\mathcal{H}_\ell^\lambda} \le M_{\check{\mathsf{T}}_{q_j}} \,|\mathsf{g}|_{{}^b\mathcal{H}_\ell^\lambda} \quad \text{for } \mathsf{g} \in {}^b\mathcal{H}_\ell^\lambda(\mathbb{C}^4)|_{\mathrm{AR,R}},\ \varepsilon \in]0,1].$$

(d) *For every* $\mathsf{f} \in {}^b\mathcal{H}_\ell^\lambda(\mathbb{C}^4)|_{\mathrm{AR,R}}$,

$$\check{\mathsf{L}}_{\mathrm{h},\varepsilon}(\check{\mathsf{T}}_{q_j,\varepsilon}(\mathsf{f}))(\xi,s) = \langle \check{q}_j^*(\xi,\varepsilon), \mathsf{f}(\xi,s)\rangle_* \,\check{q}_j(\xi,\varepsilon).$$

Proof. (a): Statement (a) directly follows from Lemma 6.5.13 observing that for every $\ell > 0$, $\lambda > 0$ and $\xi \in \mathcal{B}_\ell$,

$$|\mathrm{e}^{-\xi^2}| \le \mathrm{e}^{\ell^2 - |\mathcal{R}e(\xi)|^2} \le \mathrm{e}^{\ell^2+\lambda^2/4}\mathrm{e}^{-\lambda|\mathcal{R}e(\xi)|}.$$

(b): The smoothness properties of $\check{\mathsf{T}}_{q_j,\varepsilon}(\mathsf{f})$ are obtained using the theorem of derivation under the integral derived from Lebesgue's dominated convergence theorem. $\check{\mathsf{T}}_{q_j,\varepsilon}(\mathsf{f})$ is clearly real valued on the real axis. Moreover we get that for every $\mathsf{f} \in \mathcal{H}_\ell^\lambda(\mathbb{C}^4)|_{\mathrm{AR,R}}$, $s \in \mathbb{R}$, $\xi \in \mathcal{B}_\ell$ with $\mathcal{R}e\,(\xi) \ge 0$

$$\begin{aligned}|\check{\mathsf{T}}_{q_j,\varepsilon}(\mathsf{f})(\xi,s)| &+ \left|\frac{\partial \check{\mathsf{T}}_{q_j,\varepsilon}(\mathsf{f})}{\partial s}(\xi,s)\right| \\ &\le \frac{2}{\lambda+\lambda_0}\|\check{q}_j^*\|_{H_\ell^{-\lambda_0}}\|\check{q}_j\|_{H_\ell^{\lambda_0}}\,|\mathsf{f}|_{{}^b\mathcal{H}_\ell^\lambda}\,\mathrm{e}^{-\lambda|\mathcal{R}e(\xi)|}.\end{aligned}$$

(c): (a), (b) ensure that for every $\mathsf{g} \in {}^b\mathcal{H}_\ell^\lambda(\mathbb{C}^4)|_{\mathrm{AR,R}}$, $\mathsf{g} + \Pi_{q_j,\varepsilon}(\mathsf{g})$ also belongs to ${}^b\mathcal{H}_\ell^\lambda(\mathbb{C}^4)|_{\mathrm{AR,R}}$, and that $\check{\mathsf{T}}_{q_j,\varepsilon}(\mathsf{g} + \Pi_{q_j,\varepsilon}(\mathsf{g}))$ has all the desired properties to belong to ${}^b\mathcal{H}_\ell^\lambda(\mathbb{C}^4)|_{\mathrm{R,R}}$ except the reversibility and the estimate for

$\mathcal{R}e\,(\xi) \le 0$. If we prove that $\check{\mathsf{T}}_{q_j,\varepsilon}(\mathsf{g} + \Pi_{q_j,\varepsilon}(\mathsf{g}))$ is reversible then the estimates for $\mathcal{R}e\,(\xi) \le 0$ follows automatically. We first compute $\check{\mathsf{T}}_{q_j,\varepsilon}(\Pi_{q_j,\varepsilon}(\mathsf{g}))$. Since

$$\begin{aligned}\Pi_{q_j,\varepsilon}(\mathsf{g})(\xi+t, s+\tfrac{\omega_0 t}{\varepsilon}) &= -\frac{\mathrm{e}^{-(\xi+t)^2}}{\sqrt{\pi}}\theta_{q_j,\varepsilon}(\mathsf{g})\left(\xi+t-\tfrac{\varepsilon}{\omega_0}\left(s+\tfrac{\omega_0 t}{\varepsilon}\right)\right)\check{q}_j(\xi+t,\varepsilon)\\ &= -\theta_{q_j,\varepsilon}(\mathsf{g})(\xi - \tfrac{\varepsilon s}{\omega_0})\,\frac{\mathrm{e}^{-(\xi+t)^2}}{\sqrt{\pi}}\check{q}_j(\xi+t,\varepsilon),\end{aligned}$$

we get

$$\begin{aligned}\check{\mathsf{T}}_{q_j,\varepsilon}(\Pi_{q_j,\varepsilon}(\mathsf{g})) &= -\int_0^{+\infty}\langle \check{q}_j^*(\xi+t,\varepsilon), \Pi_{q_j,\varepsilon}(\mathsf{g})(\xi+t, s+\tfrac{\omega_0 t}{\varepsilon})\rangle_* \,dt\ \check{q}_j(\xi,\varepsilon)\\ &= \theta_{q_j,\varepsilon}(\mathsf{g})(\xi - \tfrac{\varepsilon s}{\omega_0})\int_0^{+\infty}\frac{\mathrm{e}^{-(\xi+t)^2}}{\sqrt{\pi}}dt\ \check{q}_j(\xi,\varepsilon).\end{aligned}$$

Hence,

$$\check{\mathsf{T}}_{q_j,\varepsilon}(\mathsf{g} + \Pi_{q_j,\varepsilon}(\mathsf{g}))(\xi,s) = \check{\mathsf{T}}_{q_j,\varepsilon}(\mathsf{g})(\xi,s) + \alpha(\xi)\mathsf{d}_{q_j,\varepsilon}(\xi,s)\ \check{q}_j(\xi,\varepsilon)$$

where

$$\alpha(\xi) = \frac{1}{\sqrt{\pi}}\int_0^{+\infty}\mathrm{e}^{-(\xi+t)^2}dt.$$

We have already proved (see (6.32)) that for $\mathsf{g} \in {}^b\mathcal{H}_\ell^\lambda(\mathbb{C}^4)|_{\mathrm{AR,R}}$

$$S\check{\mathsf{T}}_{q_j,\varepsilon}(\mathsf{g})(-\xi,-s) = \check{\mathsf{T}}_{q_j,\varepsilon}(\mathsf{g})(\xi,s) + \mathsf{d}_{q_j,\varepsilon}(\xi,s)\ \check{q}_j(\xi,\varepsilon).$$

Thus, since $\theta_{q_j,\varepsilon}$ is odd, $\mathsf{d}_{q_j,\varepsilon}(-\xi,-s) = \theta_{q_j,\varepsilon}(-\xi + \frac{\varepsilon s}{\omega_0}) = -\mathsf{d}_{q_j,\varepsilon}(\xi,s)$, we get

$$\begin{aligned}S\check{\mathsf{T}}_{q_j,\varepsilon}(\mathsf{g}+\Pi_{q_j,\varepsilon}(\mathsf{g}))(-\xi,-s) &= (\check{\mathsf{T}}_{q_j,\varepsilon}(\mathsf{g})(\xi,s) + \mathsf{d}_{q_j,\varepsilon}(\xi,s)\ \check{q}_j(\xi,\varepsilon))\\ &\qquad -\alpha(-\xi)\mathsf{d}_{q_j,\varepsilon}(\xi,s)\ S\check{q}_j(-\xi,\varepsilon)\\ &= \check{\mathsf{T}}_{q_j,\varepsilon}(\mathsf{g})(\xi,s)+(1-\alpha(-\xi))\mathsf{d}_{q_j,\varepsilon}(\xi,s)\ \check{q}_j(\xi,\varepsilon).\end{aligned}$$

Finally, observing that

$$\alpha(\xi) + \alpha(-\xi) = \frac{1}{\sqrt{\pi}}\int_{-\infty}^{+\infty}\mathrm{e}^{-(z+t)^2}dt = \frac{1}{\sqrt{\pi}}\int_{-\infty}^{+\infty}\mathrm{e}^{-t^2}dt = 1$$

we obtain that $\check{\mathsf{T}}_{q_j,\varepsilon}(\mathsf{g} + \Pi_{q_j,\varepsilon}(\mathsf{g}))$ is reversible

$$\begin{aligned}S\check{\mathsf{T}}_{q_j,\varepsilon}(\mathsf{g} + \Pi_{q_j,\varepsilon}(\mathsf{g}))(-\xi,-s) &= \check{\mathsf{T}}_{q_j,\varepsilon}(\mathsf{g})(\xi,s) + \alpha(\xi)\mathsf{d}_{q_j,\varepsilon}(\xi,s)\ \check{q}_j(\xi,\varepsilon)\\ &= \check{\mathsf{T}}_{q_j,\varepsilon}(\mathsf{g} + \Pi_{q_j,\varepsilon}(\mathsf{g}))(\xi,s).\end{aligned}$$

This symmetry property combined with the estimate obtained previously for $\mathcal{R}e\,(\xi) \ge 0$ enables us to obtain the desired estimates on $\mathcal{B}_\ell$,

$$\left|\check{\mathsf{T}}_{q_j,\varepsilon}(\mathrm{g}+\Pi_{q_j,\varepsilon}(\mathrm{g}))\right|_{{}^b\mathcal{H}_\ell^\lambda} \le \max(1,\|S\|)M^+_{\check{\mathsf{T}}_{q_j}} \left|\mathrm{g}+\Pi_{q_j,\varepsilon}(\mathrm{g})\right|_{{}^b\mathcal{H}_\ell^\lambda}$$
$$\le \max(1,\|S\|)M^+_{\check{\mathsf{T}}_{q_j}} (1+M_{\Pi_{q_j}}) \left|\mathrm{g}\right|_{{}^b\mathcal{H}_\ell^\lambda}.$$

which holds for every $\mathrm{g} \in {}^b\mathcal{H}_\ell^\lambda(\mathbb{C}^4)|_{\mathrm{AR,R}}$ and $\varepsilon \in]0,1]$. Hence, $\mathrm{g} \mapsto \check{\mathsf{T}}_{q_j,\varepsilon}(\mathrm{g}+\Pi_{q_j,\varepsilon}(\mathrm{g}))$ is a linear operator from $\mathcal{H}_\ell^\lambda(\mathbb{C}^4)|_{\mathrm{AR,R}}$ to $\mathcal{H}_\ell^\lambda(\mathbb{C}^4)|_{\mathrm{R,R}}$ which is uniformly bounded with respect to ε.

(d): The proof is the same as the one for $\check{p}_j$ given in Lemma 6.5.15. □

We can finally conclude that

Lemma 6.5.17. *$\check{\mathsf{T}}_{\mathrm{h},\varepsilon}$ is a linear operator from the space $\mathcal{H}_\ell^\lambda(\mathbb{C}^4)|_{\mathrm{AR,R}}$ to $\mathcal{H}_\ell^{\lambda_0}(\mathbb{C}^4)|_{\mathrm{R,R}}$ which satisfies*

$$\check{\mathsf{L}}_{\mathrm{h},\varepsilon}(\check{\mathsf{T}}_{\mathrm{h},\varepsilon}(\mathrm{g})) = \mathrm{g} + \sum_{j=0}^{1} \Pi_{q_j,\varepsilon}(\mathrm{g}) \qquad \text{for } \mathrm{g} \in {}^b\mathcal{H}_\ell^\lambda(\mathbb{C}^4)|_{\mathrm{AR,R}}$$

and which is uniformly bounded with respect to ε, i.e. there exists $M_{\check{\mathsf{T}}_\mathrm{h}}$ such that

$$\left|\check{\mathsf{T}}_{\mathrm{h},\varepsilon}(\mathrm{g})\right|_{{}^b\mathcal{H}_\ell^{\lambda_0}} \le M_{\check{\mathsf{T}}_\mathrm{h}} \left|\mathrm{g}\right|_{{}^b\mathcal{H}_\ell^\lambda} \quad \text{for } \mathrm{g} \in {}^b\mathcal{H}_\ell^\lambda(\mathbb{C}^4)|_{\mathrm{AR,R}},\ \varepsilon \in]0,1]. \tag{6.34}$$

Remark 6.5.18. The Proof of Proposition 6.5.2 follows directly from this Lemma since $\Pi_{q_j,\varepsilon}(\mathrm{g})(t,\frac{\omega_0 t}{\varepsilon}) = 0$ (see Lemmas 6.5.10 and 6.5.13 and Remark 6.5.11).

Proof. Lemmas 6.5.15 and 6.5.16 ensure that (6.34) holds with

$$M_{\check{\mathsf{T}}_\mathrm{h}} = \sum_{j=0}^{1} M_{\check{\mathsf{T}}_{p_j}} + M_{\check{\mathsf{T}}_{q_j}}$$

and that

$$\begin{aligned}
\check{\mathsf{L}}_{\mathrm{h},\varepsilon}(\check{\mathsf{T}}_{\mathrm{h},\varepsilon}(\mathrm{g}))(\xi,s) &= \sum_{j=0}^{1} \langle \check{p}_j^*(\xi,\varepsilon), \mathrm{g}(\xi,s)\rangle_* \check{p}_j(\xi,\varepsilon) \\
&\quad + \sum_{j=0}^{1} \langle \check{q}_j^*(\xi,\varepsilon), (\mathrm{g}+\Pi_{q_j,\varepsilon}(\mathrm{g}))(\xi,s)\rangle_* \check{q}_j(\xi,\varepsilon) \\
&= \mathrm{g} + \sum_{j=0}^{1} \Pi_{q_j,\varepsilon}(\mathrm{g})
\end{aligned}$$

since

$$\langle \check{p}_k^*, \Pi_{q_j,\varepsilon}(\mathrm{g})\rangle_* = 0 \quad \text{for } j,k = 0,1,$$
$$\langle \check{q}_k^*, \Pi_{q_j,\varepsilon}(\mathrm{g})\rangle_* = 0 \quad \text{for } j,k = 0,1,\ k \neq j. \ \square$$

Proof of Proposition 6.5.4

Proposition 6.5.4 directly follows from Proposition 6.5.2. Indeed, for $\mathrm{g} \in {}^b\mathcal{H}_\ell^\lambda(\mathbb{C}^4)|_{\mathrm{AR,R}}$, the equation

$$\frac{dv}{dt} - \mathrm{A_h}(t, \tfrac{\omega_0 t}{\varepsilon}, \varepsilon) = \mathrm{g}(t, \tfrac{\omega_0 t}{\varepsilon}) \tag{6.35}$$

is equivalent to

$$\frac{dw}{dt} - \check{A}_\mathrm{h}(t,\varepsilon)w = R_{-\frac{\omega_0 t}{\varepsilon}}\mathrm{g}(t, \tfrac{\omega_0 t}{\varepsilon})$$

where

$$w(t) = R_{-\frac{\omega_0 t}{\varepsilon}} v(t), \qquad \check{A}_\mathrm{h}(t,\varepsilon) = A_\mathrm{h}(t,\varepsilon) - \tfrac{\omega_0}{\varepsilon} J.$$

Since J is reversible and has $\pm \mathrm{i}$ for double semi-simple eigenvalues, we get that

$$SR_s = R_{-s}S \qquad \text{for } s \in \mathbb{R}, \qquad M_R := \sup_{s\in\mathbb{R}} \|\mathbb{R}_s\| < +\infty.$$

Thus $(\xi, s) \mapsto R_{-s}\mathrm{g}(\xi, s)$ also belongs ${}^b\mathcal{H}_\ell^\lambda(\mathbb{C}^4)|_{\mathrm{AR,R}}$ and

$$|R_s \mathrm{g}|_{{}^b\mathcal{H}_\ell^\lambda} \le M_R(1 + \|J\|)\, |\mathrm{g}|_{{}^b\mathcal{H}_\ell^\lambda}\,.$$

Observe that $(\check{p}_0, \check{p}_1, \check{q}_0, \check{q}_1)$ is a basis of solutions of

$$\frac{dw}{dt} - \check{A}_\mathrm{h}(t,\varepsilon)W = 0$$

which belongs to $\left(H_\ell^{\lambda_0}(\mathbb{C}^4)|_{\mathrm{AR,R}}\right)^2 \times \left(H_\ell^{-\lambda_0}(\mathbb{C}^4)|_{\mathrm{R,R}}\right)^2$ such that its dual basis $(\check{p}_0^*, \check{p}_1^*, \check{q}_0^*, \check{q}_1^*)$ lies in $\left(H_\ell^{-\lambda_0}(\mathbb{C}^4)|_{\mathrm{AR^*,R}}\right)^2 \times \left(H_\ell^{\lambda_0}(\mathbb{C}^4)|_{\mathrm{R^*,R}}\right)^2$. Hence, we deduce from Proposition 6.5.4 that (6.35) has a unique solution v_g which is reversible and which tends to 0 at $+\infty$. Moreover, Proposition 6.5.4 also ensure that

$$v_\mathrm{g}(t) = \mathrm{T}_{\mathrm{h},\varepsilon}(\mathrm{g})(t, \tfrac{\omega_0 t}{\varepsilon})$$

where

$$\mathrm{T}_{\mathrm{h},\varepsilon}(\mathrm{g})(\xi, s) = R_s \check{\mathrm{T}}_{\mathrm{h},\varepsilon}(R_{-s}\mathrm{g}(\xi, s))$$

is a linear operator from ${}^b\mathcal{H}_\ell^\lambda(\mathbb{C}^4)|_{\mathrm{AR,R}}$ to ${}^b\mathcal{H}_\ell^{\lambda_0}(\mathbb{C}^4)|_{\mathrm{R,R}}$ such that

$$|\mathrm{T}_{\mathrm{h},\varepsilon}(\mathrm{g})|_{{}^b\mathcal{H}_\ell^{\lambda_0}} \le M_{\check{\mathrm{T}}_\mathrm{h}}(1 + \|J\|)^2 M_R^2\, |\mathrm{g}|_{{}^b\mathcal{H}_\ell^{\lambda_0}}\,.$$

Proof of Proposition 6.5.7

The basis of solution (r_+, r_-) of $\mathcal{L}_{c,\varepsilon}(v) = 0$, readily follows from the explicit form of A_c. Moreover, observe that $r_\pm^* = r_\pm$.

Then, if g is such that there exists v which is reversible, which tends to 0 at $+\infty$ and satisfies $\mathcal{L}_{c,\varepsilon}(v) = g$, using the variation of constant formula we get that

$$v(t) = \left(a_0 + \int_0^t \langle r_+(s,\varepsilon), g(s)\rangle_* ds\right) r_+(t,\varepsilon) + \left(b_0 + \int_0^t \langle r_-(s,\varepsilon), g(s)\rangle_* ds\right) r_-(t,\varepsilon).$$

On one hand, since v and r_+ are reversible and r_- is antireversible, we get that $b_0 = 0$. On the other hand, v tends to 0 at $+\infty$ and r_- is bounded, thus $\langle r_-(t,\varepsilon), v(t)\rangle_* \underset{t\to+\infty}{\longrightarrow} 0$ and

$$b_0 = -\int_0^{+\infty} \langle r_-(s,\varepsilon), g(s)\rangle_* ds.$$

Hence,

$$\int_0^{+\infty} \langle r_-(s,\varepsilon), g(s)\rangle_* ds = 0.$$

The proofs of statements (c), (d), (e) are very similar to the proof of Proposition 6.5.2. So we give only an outline of the proof. The details are left to the reader. As for Proposition 6.5.2, to look for solution of

$$\frac{dv}{dt} - A_c(t,\varepsilon) = \mathrm{g}(t, \tfrac{\omega_0 t}{\varepsilon}) \tag{6.36}$$

of the form $v(t) = \mathrm{v}(t, \frac{\omega_0 t}{\varepsilon})$ we introduce an associated P.D.E. where the function g has to be modified to obtain the desired properties. So we introduce

$$\frac{\partial \mathrm{v}}{\partial \xi} + \frac{\omega_0}{\varepsilon}\frac{\partial \mathrm{v}}{\partial s} - A_c(\xi,\varepsilon).\mathrm{v} = \mathrm{g} + \Pi_{+,\varepsilon}(\mathrm{g}) + \Pi_{-,\varepsilon}(\mathrm{g}) \tag{6.37}$$

where $\Pi_{+,\varepsilon}$ and $\Pi_{-,\varepsilon}$ are given by

Lemma 6.5.19. *For* $\mathrm{g} \in {}^b\mathcal{H}_\ell^\lambda(\mathbb{C}^2)|_{\mathrm{AR,R}}$, *let* $\Pi_{\pm,\varepsilon}(\mathrm{g})$ *be given by*

$$\Pi_{\pm,\varepsilon}(\mathrm{g})(\xi,s) = -\frac{\mathrm{e}^{-\xi^2}}{\sqrt{\pi}}\,\mathrm{d}_{\pm,\varepsilon}(\xi,s)\, r_\pm(t,\varepsilon)$$

$$\mathrm{d}_{\pm,\varepsilon}(\mathrm{g})(\xi,s) = \int_{-\infty}^{+\infty} \langle r_\pm(\xi+t,\varepsilon), \mathrm{g}(\xi+t, s+\tfrac{\omega_0 t}{\varepsilon})\rangle_*\, dt.$$

(a) *Then, for every* $(\xi, s) \in \mathcal{B}_\ell \times \mathbb{R}$ *and* $\mathrm{g} \in {}^b\mathcal{H}_\ell^\lambda(\mathbb{C}^4)|_{\mathrm{AR,R}}$,

$$\mathrm{d}_{\pm,\varepsilon}(\mathrm{g})(\xi, s) = \theta_{\pm,\varepsilon}(\mathrm{g})(\xi - \tfrac{\varepsilon s}{\omega_0})$$

where $\theta_{\pm,\varepsilon}$ *is a* $\frac{2\pi\varepsilon}{\omega_0}$*-periodic function holomorphic on* $\mathcal{B}_\ell$*, real valued on the real axis and explicitly given by*

$$\theta_{\pm,\varepsilon}(\mathrm{g})(\zeta) = \int_{-\infty}^{+\infty} \langle r_\pm(\zeta + u, \varepsilon), \mathrm{g}(\xi + u, \tfrac{\omega_0 u}{\varepsilon})\rangle_* \, du.$$

Moreover, $\theta_{+,\varepsilon}(\mathrm{g})$ *is odd and* $\theta_{-,\varepsilon}(\mathrm{g})$ *is even.*

(b) $\Pi_{+,\varepsilon} + \Pi_{-,\varepsilon}$ *is a linear operator from* ${}^b\mathcal{H}_\ell^\lambda(\mathbb{C}^2)|_{\mathrm{AR,R}}^\perp$ *to* ${}^b\mathcal{H}_\ell^\lambda(\mathbb{C}^2)|_{\mathrm{AR,R}}$ *which is uniformly bounded with respect to* ε*, i.e. there exists* $M_{\Pi_\pm}$ *such that*

$$|\Pi_{+,\varepsilon}(\mathrm{g}) + \Pi_{-,\varepsilon}(\mathrm{g})|_{{}^b\mathcal{H}_\ell^\lambda} \le M_{\Pi_\pm} \, |\mathrm{g}|_{{}^b\mathcal{H}_\ell^\lambda} \text{ for } \mathrm{g} \in {}^b\mathcal{H}_\ell^\lambda(\mathbb{C}^4)|_{\mathrm{AR,R}},\ \varepsilon \in]0,1].$$

Remark 6.5.20. Observe that $\theta_+(\mathrm{g})$ is odd. So $\theta_+(\mathrm{g})(0) = 0$, whereas $\theta_-(\mathrm{g})$ is even. However, $\theta_-(\mathrm{g})(0)$ is also equal to 0 because g satisfies the solvability condition

$$\int_0^{+\infty} \langle r_-(t,\varepsilon), \mathrm{g}(t, \tfrac{\omega_0 t}{\varepsilon})\rangle_* \, dt = 0$$

since $\mathrm{g} \in {}^b\mathcal{H}_\ell^\lambda(\mathbb{C}^2)|_{\mathrm{AR,R}}^\perp$. Thus, $\Pi_{+,\varepsilon}(\mathrm{g})(t, \frac{\omega_0 t}{\varepsilon}) = \Pi_{-,\varepsilon}(\mathrm{g})(t, \frac{\omega_0 t}{\varepsilon}) = 0$, and each solution v of (6.37) gives a solution v of (6.36) with $v(t) = \mathrm{v}(t, \frac{\omega_0 t}{\varepsilon})$.

The proof of Lemma 6.5.19 directly follows from the definition $\mathrm{d}_{\pm,\varepsilon}$ and from the explicit formula giving $\Pi_+ + \Pi_-$,

$$\Pi_+(\mathrm{g}) + \Pi_-(\mathrm{g}) = \int_0^{+\infty} K(t,\xi,\varepsilon).\mathrm{g}(\xi + t, s + \tfrac{\omega_0 t}{\varepsilon})dt$$

where

$$K(t,\xi,\varepsilon) = \begin{bmatrix} \cos\Psi(t,\xi,\varepsilon) & \sin\Psi(t,\xi,\varepsilon) \\ -\sin\Psi(t,\xi,\varepsilon) & \cos\Psi(t,\xi,\varepsilon) \end{bmatrix}$$

with

$$\Psi(t,\xi,\varepsilon) = \phi(\xi + t, \varepsilon) - \phi(\xi,\varepsilon) = \frac{\omega_{01}(\varepsilon)}{\varepsilon} t + \int_{[0,\xi+t]} \psi(u,\varepsilon)du - \int_{[0,\xi]} \psi(u,\varepsilon)du$$

whose imaginary part is bounded on $\mathcal{B}_\ell \times \mathbb{R} \times]0,1]$ so that K is also bounded on $\mathcal{B}_\ell \times \mathbb{R} \times]0,1]$.

Now, let us introduce the operator $\mathsf{T}_{c,\varepsilon}$ given by

$$\begin{aligned}\mathsf{T}_{c,\varepsilon} &= (\mathsf{T}_{+,\varepsilon}+\mathsf{T}_{+,\varepsilon})\left(\mathsf{g}+\Pi_{+,\varepsilon}(\mathsf{g})+\Pi_{-,\varepsilon}(\mathsf{g})\right)\\ &= \mathsf{T}_{+,\varepsilon}(\mathsf{g}+\Pi_{+,\varepsilon}(\mathsf{g}))+\mathsf{T}_{-,\varepsilon}(\mathsf{g}+\Pi_{-,\varepsilon}(\mathsf{g}))\end{aligned}$$

where

$$\mathsf{T}_{\pm,\varepsilon}(\mathsf{f})(\xi,s)=-\int_0^{+\infty}\left\langle r_\pm(\xi+t,\varepsilon),\mathsf{g}(\xi+t,s+\tfrac{\omega_0 t}{\varepsilon})\right\rangle_{\!*}dt\; r_\pm(\xi,\varepsilon).$$

As for Proposition 6.5.2, our aim is now to prove that $\mathsf{T}_{c,\varepsilon}$ is solution of (6.37). For that purpose we first prove

Lemma 6.5.21.

(a) *For every* $\mathsf{f}\in\mathcal{H}_\ell^\lambda(\mathbb{C}^2)|_{\mathrm{AR},\mathbb{R}}$, $s\mapsto\mathsf{T}_{\pm,\varepsilon}(\mathsf{f})(\xi,s)$ *is* 2π*-periodic and of class* C^1 *on* $\mathbb{R}$; $\xi\mapsto\mathsf{T}_{\pm,\varepsilon}(\mathsf{f})(\xi,s)$ *is holomorphic on* $\mathcal{B}_\ell$; $(\xi,s)\mapsto\mathsf{T}_{\pm,\varepsilon}(\mathsf{f})(\xi,s)$ *and* $(\xi,s)\mapsto\dfrac{\partial\mathsf{T}_{\pm,\varepsilon}(\mathsf{f})}{\partial s}(\xi,s)$ *are continuous on* $\mathcal{B}_\ell\times\mathbb{R}$. *Moreover, there exists* $M^+_{\mathsf{T}_\pm}$ *such that for every* $\varepsilon\in]0,1]$ *and* $\mathsf{f}\in\mathcal{H}_\ell^\lambda(\mathbb{C}^2)|_{\mathrm{AR},\mathbb{R}}$

$$|(\mathsf{T}_{+,\varepsilon}+\mathsf{T}_{-,\varepsilon})(\mathsf{f})(\xi,s)|+\left|\frac{\partial(\mathsf{T}_{+,\varepsilon}+\mathsf{T}_{-,\varepsilon})(\mathsf{f})}{\partial s}(\xi,s)\right|\le M^+_{\mathsf{T}_\pm}\mathrm{e}^{-\lambda|\mathcal{R}e(\xi)|}\,|\mathsf{f}|_{{}^b\mathcal{H}_\ell^\lambda}$$

for $\xi\in\mathcal{B}_\ell$, $\mathcal{R}e\,(\xi)\ge 0$, $s\in\mathbb{R}$.

(b) $S\mathsf{T}_{\pm,\varepsilon}(\mathsf{g}+\Pi_{\pm,\varepsilon}(\mathsf{g}))(-\xi,-s)=\mathsf{T}_{\pm,\varepsilon}(\mathsf{g}+\Pi_{\pm,\varepsilon}(\mathsf{g}))(\xi,s)$ *for* $\mathsf{g}\in{}^b\mathcal{H}_\ell^\lambda(\mathbb{C}^2)|^\perp_{\mathrm{AR},\mathbb{R}}$.

(c) $\mathsf{g}\mapsto\mathsf{T}_{c,\varepsilon}(\mathsf{g}):=\mathsf{T}_{+,\varepsilon}(\mathsf{g}+\Pi_{+,\varepsilon}(\mathsf{g}))+\mathsf{T}_{-,\varepsilon}(\mathsf{g}+\Pi_{-,\varepsilon}(\mathsf{g}))$ *is a linear operator from* ${}^b\mathcal{H}_\ell^\lambda(\mathbb{C}^2)|^\perp_{\mathrm{AR},\mathbb{R}}$ *to* ${}^b\mathcal{H}_\ell^\lambda(\mathbb{C}^2)|_{\mathrm{R},\mathbb{R}}$ *which is uniformly bounded with respect to* ε*, i.e. there exists* M_{T_c} *such that*

$$|\mathsf{T}_{c,\varepsilon}(\mathsf{g})|_{{}^b\mathcal{H}_\ell^\lambda}=|\mathsf{T}_{+,\varepsilon}(\mathsf{g}+\Pi_{+,\varepsilon}(\mathsf{g}))+\mathsf{T}_{-,\varepsilon}(\mathsf{g}+\Pi_{-,\varepsilon}(\mathsf{g}))|_{{}^b\mathcal{H}_\ell^\lambda}\le M_{\mathsf{T}_c}\,|\mathsf{g}|_{{}^b\mathcal{H}_\ell^\lambda}\quad(6.38)$$

for $\mathsf{g}\in{}^b\mathcal{H}_\ell^\lambda(\mathbb{C}^2)|^\perp_{\mathrm{AR},\mathbb{R}}$, $\varepsilon\in]0,1]$.

(d) *For every* $\mathsf{f}\in{}^b\mathcal{H}_\ell^\lambda(\mathbb{C}^2)|_{\mathrm{AR},\mathbb{R}}$,

$$\mathsf{L}_{c,\varepsilon}(\mathsf{T}_{\pm,\varepsilon}(\mathsf{f}))(\xi,s)=\left\langle r_\pm(\xi,\varepsilon),\mathsf{f}(\xi,s)\right\rangle_{\!*}\,r_\pm(\xi,\varepsilon).$$

(e) *For every* $\mathsf{g}\in{}^b\mathcal{H}_\ell^\lambda(\mathbb{C}^2)|^\perp_{\mathrm{AR},\mathbb{R}}$,

$$\mathsf{L}_{c,\varepsilon}(\mathsf{T}_{c,\varepsilon}(\mathsf{g}))(\xi,s)=\mathsf{g}+\Pi_{+,\varepsilon}(\mathsf{g})+\Pi_{-,\varepsilon}(\mathsf{g}).$$

The proof of this Lemma is very similar to the ones of Lemmas 6.5.16 and 6.5.17. For the estimate 6.38, we use the explicit formula giving $(\mathsf{T}_{+,\varepsilon}+\mathsf{T}_{-,\varepsilon})(\mathsf{f})$. The details are left to the reader.

The Proof of Proposition 6.5.7 follows directly from this Lemma since $\Pi_{\pm,\varepsilon}(\mathbf{g})(t, \frac{\omega_0 t}{\varepsilon}) = 0$. □

Part III

Applications to homoclinic orbits near resonances in reversible systems

7. The $0^{2+}i\omega$ resonance

7.1 Introduction

This chapter is devoted to the $0^{2+}i\omega$ resonance in $\mathbb{R}^4$: we study *analytic* one parameter families of vector fields in $\mathbb{R}^4$,

$$\frac{du}{dx} = \mathcal{V}(u, \mu), \qquad u \in \mathbb{R}^4,\ \mu \in [-\mu_0, \mu_0],\ \mu_0 > 0 \tag{7.1}$$

near a fixed point placed at the origin , i.e.

$$\mathcal{V}(0, \mu) = 0 \qquad \text{for } \mu \in [-\mu_0, \mu_0]. \tag{H1}$$

In addition, the family is supposed to be *reversible*, i.e. there exists a reflection $S \in GL_4(\mathbb{R})$ such that for every u and μ,

$$\mathcal{V}(Su, \mu) = -S\mathcal{V}(u, \mu) \tag{H2}$$

holds. We assume that the origin is a $0^2i\omega$ resonant fixed point, i.e. that the spectrum of the differential at the origin $D_u\mathcal{V}(0,0)$ is $\{\pm i\omega, 0\}$ with $\omega > 0$ and where 0 is a double non semi-simple eigenvalue, and we denote by $(\varphi_0, \varphi_1, \varphi_+, \varphi_-)$ a basis of eigenvectors and generalized eigenvectors

$$D_u\mathcal{V}(0,0)\varphi_0 = 0, \qquad D_u\mathcal{V}(0,0)\varphi_1 = \varphi_0, \qquad D_u\mathcal{V}(0,0)\varphi_\pm = \pm i\omega\varphi_\pm, \tag{H3}$$

and by $(\varphi_0^*, \varphi_1^*, \varphi_+^*, \varphi_-^*)$ the corresponding dual basis.

The Theorem of classification of reversible matrices 3.1.10 ensures that there exists only two types of such vector fields up to linear change of coordinates. Indeed hypothesis (H2) and (H3) imply that $S\varphi_0 = \pm\varphi_0$ holds. The vector fields corresponding to $S\varphi_0 = \varphi_0$ are said to admit a $0^{2+}i\omega$ resonance at the origin, the other ones are said to admit a $0^{2-}i\omega$ resonance. *In this chapter, we only study vector fields admitting a $0^{2+}i\omega$ resonance due to their physical interest.* So we assume that

$$S\varphi_0 = \varphi_0. \tag{H4}$$

Two final generic hypotheses on the linear and the quadratic parts of the vector field are made. The first one concerns the linear part of the vector field,

$$c_{10} = \langle \varphi_1^*, D^2_{\mu,u}\mathcal{V}(0,0)\varphi_0\rangle \neq 0. \tag{H5}$$

It ensures that the bifurcation described at Figure 7.1 really occurs: we do not study the hyper-degenerated case when the double eigenvalue 0 stays at 0 for $\mu \neq 0$.
The last hypothesis,

$$c_{20} := \langle \varphi_1^*, D^2_{uu}\mathcal{V}(0,0)[\varphi_0, \varphi_0]\rangle \neq 0 \tag{H6}$$

ensures that, in some sense, the quadratic part of the vector field is not degenerated.

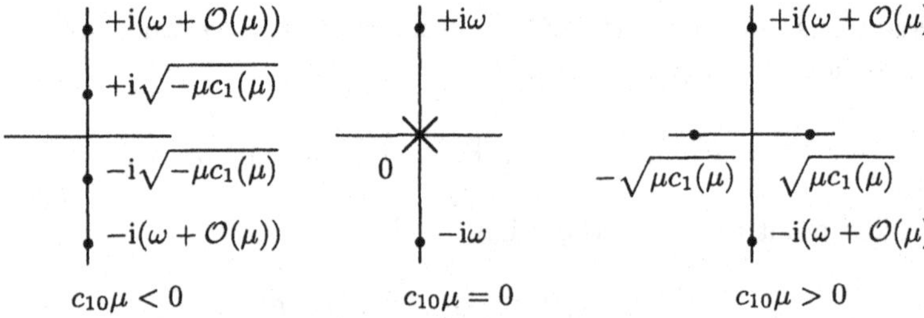

Fig. 7.1. Spectrum of $D_u\mathcal{V}(0,\mu)$ for different values of $c_{10}\mu$ ($c_1(\mu) = c_{10} + \mathcal{O}(\mu)$).

Finally, since we are interested in the existence of homoclinic connections, we only study the "half bifurcation" corresponding to

$$c_{10}\mu > 0 \tag{H7}$$

for which the differential $D_u\mathcal{V}(0,\mu)$ admits eigenvalues with non zero real parts. This chapter is devoted to the proof of the following theorem:

Theorem 7.1.1. *Let $\mathcal{V}(\cdot,\mu)$ be an analytic, reversible one parameter family of vector fields in $\mathbb{R}^4$ admitting a non degenerate $0^{2+}i\omega$ resonance at the origin, i.e. satisfying Hypothesis (H1),$\cdots$,(H6).*

Then, there exist five constants $\sigma, \kappa_3, \kappa_2, \kappa_1 > 0$, $\kappa_0 \geq 0$ such that for $|\mu|$ small enough with $c_{10}\mu > 0$ the vector field $\mathcal{V}(\cdot,\mu)$ admits near the origin

(a) *a one parameter family of periodic orbits $p_{\kappa,\mu}$ of arbitrary small size $\kappa \in [0, \kappa_3|\mu|]$;*

(b) *for every $\kappa \in \left[\kappa_1|\mu|e^{-\frac{\omega(\pi-\sigma(c_{10}\mu)^{\frac{3}{10}})}{\sqrt{c_{10}\mu}}}, \kappa_2|\mu|\right]$, a pair of reversible homoclinic connections to $p_{\kappa,\mu}$ with one loop;*

(c) *for every $\kappa \in \left[0, \kappa_0|\mu|e^{-\frac{\pi\omega}{\sqrt{c_{10}\mu}}}\right[$, no reversible homoclinic connections to $p_{\kappa,\mu}$ with one loop. Generically (with respect to $\mathcal{V}$), κ_0 is positive, i.e. $\kappa_0 > 0$.*

More precise versions of this theorem are given below: see Theorem 7.1.4 for the periodic orbits, Theorems 7.1.7, 7.1.18 and Remark 7.1.19 for the existence of homoclinic connections to periodic orbits and Theorem 7.1.11 and Remark 7.1.19 for the non existence.

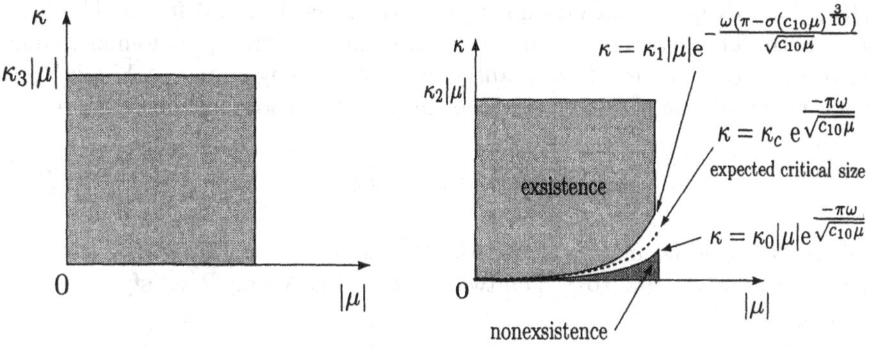

Periodic orbits $p_{\kappa,\mu}$ Reversible homoclinic connections to $p_{\kappa,\mu}$

Fig. 7.2. Domain of existence of the reversible periodic orbits $p_{\kappa,\mu}$ and domains of existence and nonexistence of reversible homoclinic connections to $p_{\kappa,\mu}$

Figure 7.2 shows in the parameter plane $(\kappa, |\mu|)$ the domain of existence and non existence stated in Theorem 7.1.1. As for our first toy model 1.1 given in the introduction of the book, it appears a critical size $K_c(\mu)$ for the existence of reversible homoclinic connections to the periodic orbits. This critical size is exponentially small

$$\kappa_0|\mu|e^{-\frac{\pi\omega}{\sqrt{c_{10}\mu}}} \le K_c(\mu) \le \kappa_1|\mu|e^{-\frac{\omega(\pi-\sigma(c_{10}\mu)^{\frac{3}{10}})}{\sqrt{c_{10}\mu}}}.$$

We expect that the critical $K_c(\mu)$ size is generically given by

$$K_c(\mu) = \kappa_c\, e^{-\frac{\pi\omega}{\sqrt{c_{10}\mu}}}.$$

7.1.1 Full system, normal form and scaling

For studying the dynamics of (7.1) near the origin we follow the strategy explained in Remark 3.2.7 :

Step 1. Linear change of coordinates. The Theorem of classification of reversible matrices 3.1.10 ensures that up to a linear change of coordinates, we can assume that $L_0 = D_u\mathcal{V}(0,0)$ and the symmetry S read

$$L_0 = \begin{pmatrix} 0 & 1 & 0 & 0 \\ 0 & 0 & 0 & 0 \\ 0 & 0 & 0 & -\omega \\ 0 & 0 & \omega & 0 \end{pmatrix}, \qquad S = \begin{pmatrix} 1 & 0 & 0 & 0 \\ 0 & -1 & 0 & 0 \\ 0 & 0 & 1 & 0 \\ 0 & 0 & 0 & -1 \end{pmatrix}.$$

Step 2. Normal form and scaling. The previous linear change of coordinates leads to a simple linear part. We now wish to obtain a quadratic part as simple as possible. The theory of normal form was precisely designed for that purpose.

Step 2.1. Polynomial change of coordinates: normal form. The Normal Form Theorem 3.2.1 ensures that we can perform a polynomial change of coordinates close to identity, analytically depending on μ, $u = \widetilde{Y} + \Phi(\widetilde{Y}, \mu)$ such that equation (7.1) is equivalent in a neighborhood of the origin to

$$\frac{d\widetilde{Y}}{dx} = \widetilde{N}(\widetilde{Y}, \mu) + \widetilde{R}(\widetilde{Y}, \mu) \tag{7.2}$$

where $\widetilde{N}$ is the normal form of order 2 and $\widetilde{R}$ represents the new terms of order greater or equal to 3. The two vector fields $\widetilde{N}$ and $\widetilde{R}$ satisfy

$$\widetilde{N}(\widetilde{Y}, \mu) = \begin{pmatrix} \widetilde{\beta} \\ c_1(\mu)\mu\widetilde{\alpha} + c_2(\mu)\widetilde{\alpha}^2 + c_3(\mu)(\widetilde{A}^2 + \widetilde{B}^2) \\ -\widetilde{B}\ (\omega + \mu\omega_1(\mu) + m(\mu)\widetilde{\alpha}) \\ \widetilde{A}\ (\omega + \mu\omega_1(\mu) + m(\mu)\widetilde{\alpha}) \end{pmatrix},$$

$$\widetilde{R}(\widetilde{Y}, \mu) := \begin{pmatrix} 0 \\ \widetilde{R}_\beta(\widetilde{Y}, \mu) \\ \widetilde{R}_A(\widetilde{Y}, \mu) \\ \widetilde{R}_B(\widetilde{Y}, \mu) \end{pmatrix} = \mathcal{O}(|\widetilde{Y}|^3),$$

where $\widetilde{Y} = (\widetilde{\alpha}, \widetilde{\beta}, \widetilde{A}, \widetilde{B})$, and $c_1, c_2, c_3, \omega_1, m$ are analytic function of μ. Moreover hypothesis (H5) and (H6) ensure that $c_{10} = c_1(0) \neq 0$ and $c_{20} = c_2(0) \neq 0$. The first component of the rest $\widetilde{R}$ can always be chosen equal to 0. This can be done by a change of coordinates of the type $\widetilde{\beta}' = \widetilde{\beta} + \widetilde{R}_\alpha(\widetilde{\alpha}, \widetilde{\beta}, \widetilde{A}, \widetilde{B}, \mu)$.

Reversibility is preserved by the two changes of coordinates. So system (7.2) has the following symmetry properties:

$$S\widetilde{N}(\widetilde{Y}, \mu) = -\widetilde{N}(S\widetilde{Y}, \mu), \qquad S\widetilde{R}(\widetilde{Y}, \mu) = -\widetilde{R}(S\widetilde{Y}, \mu)$$

with $S(\widetilde{\alpha}, \widetilde{\beta}, \widetilde{A}, \widetilde{B}) = (\widetilde{\alpha}, -\widetilde{\beta}, \widetilde{A}, -\widetilde{B})$. Moreover, analyticity is preserved too. Thus, near the origin, $\widetilde{R}(\widetilde{Y}, \mu)$ reads

$$\widetilde{R}(\widetilde{Y}, \mu) = \sum_{(\overline{m}, \ell) \in \widetilde{\mathcal{I}}} \widetilde{a}_{\overline{m}, \ell}\ \widetilde{Y}^{\overline{m}}\ \mu^\ell,$$

where

$$\begin{aligned} \overline{m} &= (m_\alpha, m_\beta, m_A, m_B) \in \mathbb{N}^4, \qquad |\overline{m}| = m_\alpha + m_\beta + m_A + m_B, \\ \widetilde{\mathcal{I}} &= \{(\overline{m}, \ell) \in \mathbb{N}^4 \times \mathbb{N} /\ |\overline{m}| \geq 3\}, \end{aligned}$$

and

$$\begin{aligned}\widetilde{a}_{\overline{m},\ell} &= (0, \widetilde{a}_{\beta,\overline{m},\ell}, \widetilde{a}_{A,\overline{m},\ell}, \widetilde{a}_{B,\overline{m},\ell}) \in \{0\} \times \mathbb{R}^3, \\ \widetilde{Y}^{\overline{m}} &= \widetilde{\alpha}^{m_\alpha} \ \widetilde{\beta}^{m_\beta} \ \widetilde{A}^{m_A} \ \widetilde{B}^{m_B}.\end{aligned}$$

We denote by $(\widetilde{r}_0, \mu_0)$ (where $\widetilde{r}_0 = (\widetilde{r}_\alpha, \widetilde{r}_\beta, \widetilde{r}_A, \widetilde{r}_B) \in (\mathbb{R}_+^*)^4$) the "polyradius" of convergence of this power series, i.e. there exists M_R such that

$$|\widetilde{a}_{\overline{m},\ell}| \le \frac{M_R}{\widetilde{r}_0^{\overline{m}} \mu_0^\ell} = \frac{M_R}{\widetilde{r}_\alpha^{m_\alpha} \widetilde{r}_\beta^{m_\beta} \widetilde{r}_A^{m_A} \widetilde{r}_B^{m_B} \mu_0^\ell} \quad \text{for } (\overline{m}, \ell) \in \widetilde{\mathcal{I}}. \tag{7.3}$$

Step 2.2. Scaling. The last technical transformation of the system before studying dynamics is a scaling in space and time. We are interested in the existence of homoclinic connections. For $c_{10}\mu > 0$ the truncated system

$$\frac{d\widetilde{Y}}{dx} = \widetilde{N}(\widetilde{Y}, \mu)$$

admits a unique —up to a time shift— homoclinic connection $\widetilde{h}$ to 0 which depends on μ, whereas for $c_{10}\mu < 0$ there is none. Observe that $c_{10}\mu > 0$ corresponds to the half bifurcation where the differential $D_u\mathcal{V}(0, \mu)$ generates "oscillatory" dynamics with a frequency of order 1 — due to two simple, opposite eigenvalues lying on the imaginary axis— and a "slow" hyperbolic dynamics — due to the real eigenvalues $\pm\sqrt{c_1(\mu)\mu}$. The homoclinic connection $\widetilde{h}$ to 0 comes from the two real roots, and its two components along the oscillatory part of the truncated system are equal to 0. Since we want to use perturbation theory to study the persistence of $\widetilde{h}$, we wish $\widetilde{h}$ not to depend on the bifurcation parameter μ. This can be achieved by performing the scaling

$$\widetilde{\alpha} = -\frac{3}{2c_{20}}\nu^2\alpha, \quad \widetilde{\beta} = -\frac{3}{2c_{20}}\nu^3\beta, \quad \widetilde{A} = \nu^2 A, \quad \widetilde{B} = \nu^2 B, \quad x = t/\nu \tag{7.4}$$

where $\nu^2 = c_{10}\mu$. Setting $Y = (\alpha, \beta, A, B)$, we obtain a rescaled full system equivalent to our initial equation (7.1) which reads

$$\frac{dY}{dt} = N(Y, \nu) + R(Y, \nu) \tag{7.5}$$

with

$$N(Y,\nu) = \begin{pmatrix} \beta \\ \alpha - \frac{3}{2}\alpha^2 - c(A^2 + B^2) \\ -B(\frac{\omega}{\nu} + a\nu + b\nu\alpha) \\ A(\frac{\omega}{\nu} + a\nu + b\nu\alpha) \end{pmatrix}, \quad R(Y,\nu) = \begin{pmatrix} 0 \\ R_\beta(Y,\nu) \\ R_A(Y,\nu) \\ R_B(Y,\nu) \end{pmatrix}.$$

where a, b, c are real constants given by $a = \omega_1(0)/c_{10}$, $b = m(0)/c_{10}$, $c = -2c_3(0)c_{20}/3$. We can now write

$$\begin{aligned} R_\beta(Y,\nu) &= \sum_{(\overline{m},\ell)\in\mathcal{I}} a_{\beta,\overline{m},\ell}\, Y^{\overline{m}}\, \nu^{2|\overline{m}|+m_\beta+2\ell-4}, \\ R_A(Y,\nu) &= \sum_{(\overline{m},\ell)\in\mathcal{I}} a_{A,\overline{m},\ell}\, Y^{\overline{m}}\, \nu^{2|\overline{m}|+m_\beta+2\ell-3}, \\ R_B(Y,\nu) &= \sum_{(\overline{m},\ell)\in\mathcal{I}} a_{B,\overline{m},\ell}\, Y^{\overline{m}}\, \nu^{2|\overline{m}|+m_\beta+2\ell-3}. \end{aligned} \tag{7.6}$$

where

$$\mathcal{I} = \{(\overline{m},\ell) \in \mathbb{N}^4 \times \mathbb{N} / \ |\overline{m}| + \ell \geq 3, |\overline{m}| > 0\},$$

since we incorporate in R the terms of order $(\mu|\widetilde{Y}|^2 + \mu^2|\widetilde{Y}|)$ coming from $\widetilde{N}$. Estimate (7.3) ensures that

$$|a_{\overline{m},\ell}| \leq \frac{M_R}{r_0^{\overline{m}}\nu_0^{2\ell}} \quad \text{for } (\overline{m},\ell) \in \mathcal{I} \tag{7.7}$$

with $\nu_0 = \sqrt{c_{10}\mu_0}$ and $r_0 = (r_\alpha, r_\beta, r_A, r_B) = (\frac{2}{3}|c_{20}|\widetilde{r}_\alpha, \frac{2}{3}|c_{20}|\widetilde{r}_\beta, \widetilde{r}_A, \widetilde{r}_B)$.

The symmetry properties are now

$$SN(Y,\nu) = -N(SY,\nu), \quad SR(Y,\nu) = -R(SY,\nu), \tag{7.8}$$

with $S(\alpha,\beta,A,B) = (\alpha,-\beta,A,-B)$. Using the power expansion of R and the bound of $a_{\overline{m},\ell}$ given by (7.6) and (7.7) we check that

Lemma 7.1.2. *For every $d > 0$, there exist two positive constants $\nu_R(d)$, $M(d)$ such that for every $Y \in \mathbb{C}^4$ satisfying $|Y| \leq d$, and every $\nu \in]0,\nu_R(d)]$, $R(Y,\nu)$ is well defined and satisfies*

$$|R_\beta(Y,\nu)| \leq M\nu^2|Y|, \quad |R_A(Y,\nu)| \leq M\nu^3|Y|, \quad |R_B(Y,\nu)| \leq M\nu^3|Y|.$$

Moreover, for every $(Y,Y') \in (\mathbb{C}^4)^2$ satisfying $|Y| \leq d$, $|Y'| \leq d$ and every $\nu \in]0,\nu_R(d)]$, the following estimates hold:

$$\begin{aligned} |R_\beta(Y,\nu) - R_\beta(Y',\nu)| &\leq M\nu^2|Y-Y'|, \\ |R_A(Y,\nu) - R_A(Y',\nu)| &\leq M\nu^3|Y-Y'|, \\ |R_B(Y,\nu) - R_B(Y',\nu)| &\leq M\nu^3|Y-Y'|. \end{aligned}$$

Finally we obtain that

$$D_Y(N+R)(0,\nu) = \begin{pmatrix} 0 & 1 & 0 & 0 \\ 1+\mathcal{O}(\nu^2) & 0 & 0 & \\ 0 & 0 & 0 & -\dfrac{\omega_{0,\nu}}{\nu} \\ 0 & 0 & \dfrac{\omega_{0,\nu}}{\nu} & 0 \end{pmatrix}. \tag{7.9}$$

where $\omega_{0,\nu} = \omega + a\nu^2 + \mathcal{O}(\nu^4)$.

The study of the existence of homoclinic connections requires the exponential tools developed in Chapter 2. Thus a complexification of time is necessary. So, we shall study the following complex differential equation:

$$\frac{dY}{d\xi} = N(Y,\nu) + R(Y,\nu) \qquad \text{with } \xi = t + i\eta \in \mathbb{C}. \tag{7.10}$$

This equation makes sense because $N(\cdot,\nu)$ is a holomorphic function in $\mathbb{C}^4$ and because $R(\cdot,\nu)$ is holomorphic in some neighborhood of 0 in $\mathbb{C}^4$. Observe that the symmetry properties (7.8) are still true when Y lies in $\mathbb{C}^4$.

For studying the dynamics of (7.1) near the origin, we have performed a polynomial change of coordinates followed by a scaling to obtain an equivalent system (7.5) for which the "linear+quadratic" part N is as simple as possible. So, following strategy explained in Remark 3.2.7, the next step is the study of the truncated system (7.11). This study is done in the next subsection. The rest of the chapter is then devoted to *the problem of the persistence for the full system of the homoclinic connections obtained for the truncated system.*

7.1.2 Truncated system

The truncated system

$$\frac{dY}{dt} = N(Y,\nu) \tag{7.11}$$

happens to be integrable, which makes its study far more easy. This property of integrability, is not only true for the normal form of order 2, but it also holds for normal forms of any order. For the normal form system of order 2, two first integrals are

$$A^2 + B^2 = k^2, \qquad \beta^2 = \alpha^2 - \alpha^3 - 2ck^2\alpha + H. \tag{7.12}$$

The truncated system admits near the origin a one parameter family of reversible periodic orbits $Y^*_{k,\nu}$ of size $k \geq 0$ explicitly given by

$$Y^*_{k,\nu}(t) = \begin{pmatrix} \alpha^*(k) = \dfrac{1-\sqrt{1-6k^2c}}{3} \\ 0 \\ k\cos(\omega_0^*(k,\nu)t) \\ k\sin(\omega_0^*(k,\nu)t) \end{pmatrix}$$

with $\omega_0^*(k,\nu) = \dfrac{\omega}{\nu} + a\nu + b\nu\alpha^*(k)$. In other words, the truncated system admits, for each ν, periodic solutions of arbitrary small size. Moreover, each periodic orbits admits a two-parameters family of solutions homoclinic to itself

$$h_{k,\varphi}(t) = \begin{pmatrix} \alpha_{h_k}(t) \\ \beta_{h_k}(t) \\ k\cos[\omega_0^*(k)(t+\varphi) + 2b\nu\eta(k)\tanh(\frac{1}{2}\eta(k)t)] \\ k\sin[\omega_0^*(k)(t+\varphi) + 2b\nu\eta(k)\tanh(\frac{1}{2}\eta(k)t)] \end{pmatrix}$$

with

$$\begin{aligned}
\eta(k) &= \sqrt{1-3\beta_0^*(k)} = (1-6k^2c)^{1/4}, \\
\alpha_{h_k}(t) &= \alpha^*(k) + \eta^2(k)\frac{1}{\cosh^2\left(\frac{1}{2}\eta(k)t\right)}, \\
\beta_{h_k}(t) &= -\eta^3(k)\tanh\left(\tfrac{1}{2}\eta(k)t\right)\frac{1}{\cosh^2\left(\frac{1}{2}\eta(k)t\right)}.
\end{aligned}$$

Among these homoclinic solutions there are two one-parameter families of *reversible* solutions

$$h_{k,0}(t),\ h_{k,\frac{\pi}{\omega_0^*(k)}}(t).$$

Observe that

$$h_{k,0} \underset{\mathcal{R}e(t)\to+\infty}{\longrightarrow} Y_{k,\nu}^*\left(t+\frac{2b\nu\eta(k)}{\omega_0^*(k)}\right).$$

Thus, the phase shift between $h_{k,0}$ and $Y_{k,\nu}^*$ at $+\infty$ is

$$\varphi = \frac{2b\nu\eta(k)}{\omega_0^*(k)} + \frac{2n\pi}{\omega_0^*(k)},\quad n\in\mathbb{Z}. \tag{7.13}$$

In the same way, the phase shift between $h_{k,\frac{\pi}{\omega_0^*(k)}}$ and $Y_{k,\nu}^*$ at $+\infty$ is

$$\varphi = \frac{2b\nu\eta(k)}{\omega_0^*(k)} + \frac{\pi}{\omega_0^*(k)} + \frac{2n\pi}{\omega_0^*(k)},\quad n\in\mathbb{Z}. \tag{7.14}$$

And finally, the truncated system admits a unique reversible homoclinic connection h to 0 explicitly given by

$$h := (\alpha_h,\beta_h,0,0) = \left(\frac{1}{\cosh^2\frac{1}{2}t}, \frac{-\tanh\frac{1}{2}t}{\cosh^2\frac{1}{2}t}, 0, 0\right). \tag{7.15}$$

The orbit h appears to be the limit case for $k\longrightarrow 0$ since

$$Y_{0,\nu}^* = 0,\qquad h_{0,0}, = h_{0,\frac{\pi}{\omega_0^*(0)}} = h.$$

Observe that the homoclinic connection h is induced by the pair of real eigenvalues of the linear part whereas the periodic orbits are induced by the pair of simple eigenvalues which stay on the imaginary axis after bifurcation.

Remark 7.1.3. The truncated system also admits two other class of solutions : elliptic periodic orbits and quasiperiodic orbits. However, this book is devoted to the problem of the persistence of homoclinic connections. So, we do not study the persistence of these two class of solutions. For more details about these two class of solution we refer to [IK92].

There remains the problem of the persistence for the full system (7.5) of the previously found periodic and homoclinic orbits. In other words, do the previously found orbits persist when the normal form system is perturbed by the higher order terms?

7.1.3 Statement of the results of persistence for periodic solutions

For the periodic solutions the answer is yes : the full system admits a one parameter family of reversible periodic orbits $Y_{k,\nu}$ of arbitrary small size $k \geq 0$ given by Theorem 7.1.4 below. This result follows directly from the general Theorem 4.1.2 given in chapter 4 which ensures for reversible systems the existence of a one parameter family of periodic orbits bifurcating from a pair of simple purely imaginary eigenvalues. This theorem gives a detailed description of the periodic orbits which is very useful when complexifying time for studying the persistence of homoclinic connections to exponentially small periodic orbits.

Theorem 7.1.4. *There exist* $k_0 > 0$, $\nu_1 > 0$ *such that for all* ν *in* $]0, \nu_1]$ *the real full system* (7.5) *admits a one parameter family of periodic solutions* $\left(Y_{k,\nu}(t)\right)_{k\in[0,k_0]}$ *with*

$$Y_{k,\nu}(t) = \widehat{Y}(\underline{\omega}_{k,\nu}t, k, \nu)$$

and

$$\text{(a)} \begin{cases} \widehat{Y}(s,k,\nu) = \sum\limits_{n\geq 1} k^n\, \widehat{Y}_n(s,\nu), \\ \widehat{Y}_n \in \tilde{\mathcal{P}}_{n,R}, \ \widehat{Y}_1 = (0,0,\cos(s),\sin(s)), \\ (k_0)^n \left\|\widehat{Y}_n\right\|_{\mathcal{P}1} \leq C_{\widehat{Y}} \ \text{for every } \nu \in]0,\nu_1], \ n \geq 1. \end{cases}$$

$$\text{with } \tilde{\mathcal{P}}_{n,R} = \Big\{\widehat{Y} : \mathbb{R} \mapsto \mathbb{R}^4, \ \widehat{Y}(s) = a_0 + \sum_{p=0}^{n} a_p \cos(ps) + b_p \sin(ps),$$

$$Sa_p = a_p, \ Sb_p = -b_p\Big\} \cap \Big\{\widehat{Y} / \int_0^{2\pi} \langle \widehat{Y}(s), \widehat{Y}_1(s)\rangle ds = 0\Big\}$$

$$\text{and } \left\|\widehat{Y}\right\|_{\mathcal{P}1} := \sup_{s\in[0,2\pi]} |\widehat{Y}(s)| + \sup_{s\in[0,2\pi]} \left|\frac{d\widehat{Y}}{ds}(s)\right|.$$

$$\text{(b)} \begin{cases} \underline{\omega}_{k,\nu} = \dfrac{\omega_{0,\nu}}{\nu} + \widehat{\omega}(k,\nu), \ \text{where } \omega_{0,\nu} = \omega + a\nu^2 + \mathcal{O}(\nu^4), \\ \widehat{\omega}(k,\nu) = \sum\limits_{n\geq 1} \widehat{\omega}_n(\nu)\, k^n, \\ \widehat{\omega}(k,\nu) \leq C_\omega k\nu \ \text{for every } \nu \in]0,\nu_1], \ k \in [0,k_0]. \end{cases}$$

Remark 7.1.5. Observe that $C_{\widehat{Y}}$, C_ω are independent of k, ν. We have constructed a particular family of reversible periodic solutions $Y_{k,\nu}$ of (7.5). Since (7.5) is autonomous, we can deduce other periodic solutions from $Y_{k,\nu}$ by an arbitrary shift in time. Among these periodic solutions, $Y_{k,\nu}(t), Y_{k,\nu}(t+$

$\pi/\underline{\omega}_{k,\nu}$) are the only ones which are reversible. In addition, we can check that $\underline{\omega}_{k,\nu}$ is even with respect to k.

Figure 7.3 shows in the (ν, k) plane the domain of existence of hyperbolic periodic orbits of size k respectively for the truncated system and for the full system.

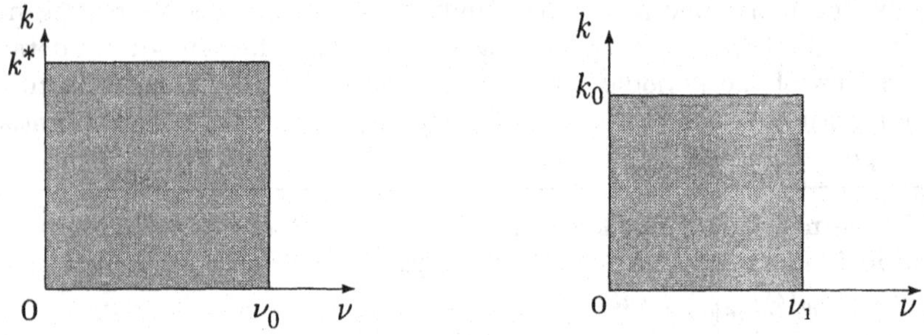

Truncated system ($\rho = 0$) Full system (7.16) ($\rho \neq 0$)

Fig. 7.3. Domain of existence of reversible periodic orbits of size k for the truncated system and domain of persistence obtained in this chapter

7.1.4 Statement of the results of persistence for homoclinic connections

For the homoclinic connections the situation is far more intricate. One way to make up one's mind is to formulate the problem in terms of stable manifolds and to illustrate it geometrically. We denote by $W_s^*(k,\nu)$ the stable manifold of the periodic solution $Y_{k,\nu}^*$ of the truncated system. The manifold $W_s^*(k,\nu)$ looks like a two dimensional cylinder in $\mathbb{R}^4$ (one dimension for time and one dimension for the phase shift).The radius of this cylinder is essentially the size of the periodic orbit, i.e. k. The fundamental remark, is that *for a reversible system, a reversible periodic solution admits a reversible homoclinic connection to itself if and only if the intersection between its stable manifold and the symmetry space* $\mathcal{P}_+ = \{Y/SY = Y\}$ *is not empty* (see Lemma 3.1.8). In our case, $\mathcal{P}_+$ is a 2 dimensional plane in $\mathbb{R}^4$. For $k > 0$, the intersection consists of two points which lead to the two reversible homoclinic connections to the periodic orbit $Y_{k,\nu}^*$. For $k = 0$, there is a unique point of intersection, which leads to the unique reversible homoclinic connection to 0 of the truncated system.

To illustrate the situation , we draw a picture in $\mathbb{R}^3$ (we cannot draw in $\mathbb{R}^4$!). On this picture $W_s^*(k,\nu)$ still looks like a 2 dimensional cylinder, whereas the symmetry plane is now a symmetry line. So, Figure 7.4 represents the intersection between $W_s^*(k,\nu)$ and $\mathcal{P}_+$ in two case : $k > 0$ and $k = 0$ for a fixed value of ν.

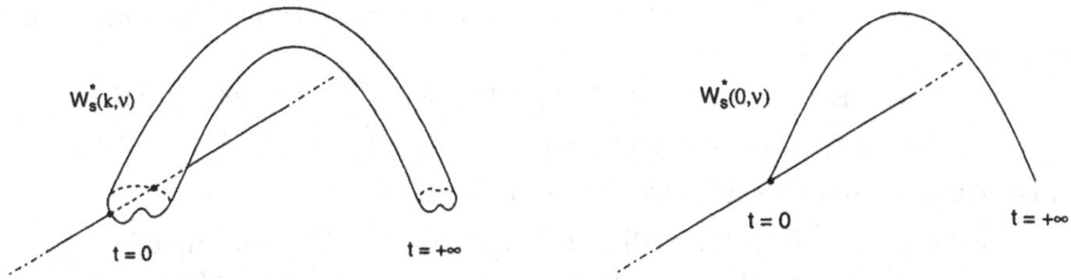

$k > 0$ $\qquad$ $k = 0$

Fig. 7.4. Stable manifold of the periodic solution $Y^*_{k,\nu}$ of the truncated system

Let us denote by $W_s(k,\nu)$ the stable manifold of the periodic solution $Y_{k,\nu}$ of the full system. The stable manifold $W_s(k,\nu)$ is obtained by perturbation of the stable manifold $W^*_s(k,\nu)$ of the periodic solution of the truncated system. On Figure 7.4 we can observe that for $k = 0$ the situation is not robust: there might not exist any reversible homoclinic connection to 0 for the full system, whereas for k large enough, the two points of intersection between $W^*_s(k,\nu)$ and $\mathcal{P}_+$ should persist, i.e. there should exist two reversible homoclinic connections to $Y_{k,\nu}$. The natural question is then the following:

Question 7.1.6.

(a) *Determine, for a fixed value of the bifurcation parameter ν, the smallest size $k_c(\nu)$ of a periodic solution which admits a reversible homoclinic connection to itself : is it 0 or not? And if it is not 0 we would like to compute k_c with respect to ν.*
(b) *If $k_c(\nu) > 0$, what is the behavior in the past of stable manifold $W_s(0,\nu)$ of 0?*

A more analytical approach complements usefully the geometrical approach and points out that we have to face the same kind of difficulty as for our toy model (1.2) given in the introduction of the book: the truncated system can be seen as a system of two "uncoupled" two dimensional systems, which have very different behaviors. Using the two firsts integrals (7.12), the first two dimensional system can be written

$$\begin{pmatrix} \dot\alpha \\ \dot\beta \end{pmatrix} = \begin{pmatrix} \beta \\ \alpha - \dfrac{3}{2}\alpha^2 - ck^2 \end{pmatrix},$$

and the second system reads

$$\begin{pmatrix} \dot A \\ \dot B \end{pmatrix} = \begin{pmatrix} -B(\dfrac{\omega}{\nu} + a\nu + b\nu\alpha) \\ A(\dfrac{\omega}{\nu} + a\nu + b\nu\alpha) \end{pmatrix}.$$

The first system is "hyperbolic and slow". The differential at the origin admits two simple real eigenvalues ± 1. For $k = 0$, it admits a reversible connection to 0 given by (α_h, β_h).

The second is "oscillatory with high frequency". The differential at the origin admits two purely imaginary eigenvalues $\pm i(\frac{\omega}{\nu} + a\nu)$. The solutions of this system oscillate rapidly along the circle $A^2 + B^2 = k^2$.

For the homoclinic orbit h, these two systems are really uncoupled because $k^2 = A^2 + B^2 = 0$ holds for h. Now, when we add the rest "R", i.e. higher order terms, we "couple" these two systems. The question is then to determine whether or not the homoclinic orbit h survives the oscillations induced by the eigenvalues $\pm i(\frac{\omega}{\nu} + a\nu)$? So, as for our first toy model, we expect that the answer is no. Moreover, we expect the appearance of a *critical size* $k_c(\nu)$ *which is exponentially small* such that for $k \geq k_c(\nu)$ there exists reversible homoclinic connections to periodic orbits of size k, and for $k < k_c(\nu)$ there should not exist any reversible homoclinic connections to periodic orbits of size k and in particular there should not exist any homoclinic connections to 0.

For our toy model (1.2) the problem of the persistence was solved by explicit computations. Here, no explicit computation are possible for the full system. So we use the "exponential tools" developed in chapter 2 for studying such systems.

7.1.4.1 Persistence of homoclinic connections to exponentially small periodic orbits. Using the First Mono frequency Exponential Lemma 2.1.1 we prove the persistence of homoclinic connections to exponentially small periodic orbits.

Theorem 7.1.7. *For all ℓ, $0 < \ell < \pi$ and all λ, $0 < \lambda < 1$ there exist M, ν_4 such that for all ν in $]0, \nu_4]$ the full System* (7.5) *admits two reversible solutions of the form*

$$Y(t) = h(t) + v(t) + Y_{k,\nu}(t + \varphi \tanh \tfrac{1}{2}t)$$

where

$$|h(t)| \leq M e^{-|t|}, \qquad |v(t)| \leq M\nu^2 e^{-\lambda|t|} \qquad \text{for } t \in \mathbb{R},$$

and where $Y_{k,\nu}$ is a periodic solution of the full system satisfying

$$\forall t \in \mathbb{R}, \ |Y_{k,\nu}(t)| \leq M\nu^2 e^{-\ell\omega/\nu}.$$

A complete description of the periodic solution and of its period is given in Theorem 7.1.4. The proof of this theorem is given in Section 7.3.

Remark 7.1.8. It is possible to obtain such a reversible homoclinic connection to $Y_{k,\nu}$ with

$$|v(t)| \leq M\nu^2 \mathrm{e}^{-\lambda_\nu |t|} \qquad \text{for } t \in \mathbb{R}, \qquad \text{with } \lambda_\nu = 1 + \mathcal{O}(\nu^2).$$

The proof is more intricate since it requires the computation of the Floquet exponents of the periodic orbit $Y_{k,\nu}$. These exponents can be estimated since $Y_{k,\nu}$ is close to a fixed point (see Chapter 5).

Remark 7.1.9. To come back to our original Equation (7.1) it suffices to perform the scaling (7.4) and the polynomial change of coordinates given by the Normal Form Theorem 3.2.1. The solution has the same form and the oscillations at infinity are less than $M\nu^4 \mathrm{e}^{-\ell\omega/\nu}$.

Remark 7.1.10. This Theorem gives a first partial answer to Question 7.1.6: it ensures that $k_c(\nu)$ is at least exponentially small.

7.1.4.2 Generic non persistence of homoclinic connection to 0.

A second partial answer which completes the previous one is given by the following theorem which ensures the generic non persistence of homoclinic connection to 0. To state the theorem we introduce in equation (7.5) an extra parameter $\rho \in [0, 1]$

$$\frac{dY}{dt} = N(Y, \nu) + \rho R(Y, \nu). \tag{7.16}$$

For $\rho = 0$, we get the truncated system which admits a homoclinic connection h to 0. Now, we would like to understand what happens for $\rho \neq 0$ and in particular for $\rho = 1$ which corresponds to the full system (7.5).

Theorem 7.1.11. *There exists a real analytic function defined on [0,1]*

$$\Lambda(\rho) = \sum_{n \geq 1} \rho^n \Lambda_n$$

and $d_s > 0$ such that for any $\rho \in [0, 1]$ with $\Lambda(\rho) \neq 0$, and for ν small enough, the full system does not admit any reversible homoclinic connection Y to 0, which principal part is h, i.e. such that

$$\sup_{t \in \mathbb{R}} |Y(t) - h(t)| \leq d_s. \tag{7.17}$$

The first coefficient Λ_1 is explicitly known in terms of Bessel functions and in terms of the coefficients of the power expansion of the rest R (see Remark 7.1.15 for the explicit expression of Λ_1).

Corollary 7.1.12. *If $\Lambda_1 \neq 0$ (which is generically satisfied), there is at most a finite number of values of ρ for which the full system admits, for ν small enough, a reversible homoclinic connection to 0 which principal part is h.*

The proof of this Theorem is given in Section 7.4.

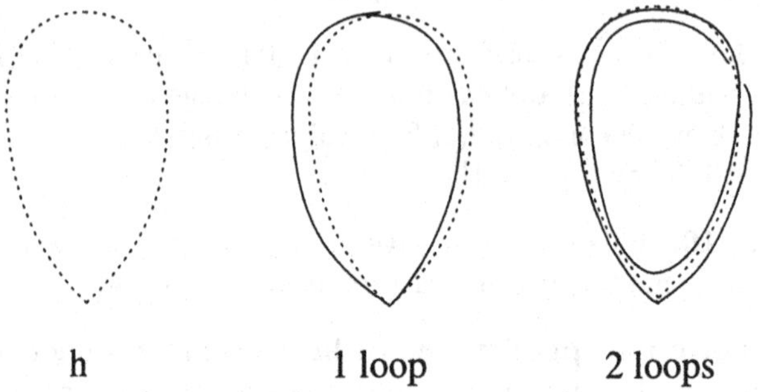

Fig. 7.5. Homoclinic connections with one or more loops

Remark 7.1.13. Equation (7.17) characterizes the homoclinic connections with one loop (see Figure 7.5). So, Theorem 7.1.11 ensures the generic non persistence of homoclinic connections to 0 with one loop whereas Theorem 7.1.7 ensures that there always exist homoclinic orbits with one loop connecting an exponentially small periodic orbit to itself. The question of the existence of homoclinic connections to 0 or to periodic orbits with more than one loop is not studied here.

Remark 7.1.14. For $\rho = 0$, $\Lambda(\rho) = 0$. Indeed, the truncated system does admit a reversible homoclinic connection to 0.

Remark 7.1.15. The coefficient Λ_1 is explicitly given by

$$\Lambda_1 = \sum_{\substack{m_\alpha, m_\beta \in \mathbb{N}^2 \\ m_\alpha + m_\beta \geq 3}} b_{m_\alpha, m_\beta} \, \Lambda_{m_\alpha, m_\beta} \tag{7.18}$$

with

$$\Lambda_{m_\alpha, m_\beta} = 2\pi \, (-1)^{m_\beta + E(\frac{m_\beta+1}{2})} \sum_{n=0}^{+\infty} \frac{(-2b)^n \, (2\omega)^{2m_\alpha + 3m_\beta + n - 1}}{(2m_\alpha + 3m_\beta + n - 1)! \, n!},$$

and

$$b_{m_\alpha, m_\beta} = a_{A,(m_\alpha, m_\beta, 0,0),0} + a_{B,(m_\alpha, m_\beta,0,0),0}$$

where $a_{A,\overline{m},\ell}$, $a_{B,\overline{m},\ell}$ are defined in (7.6) and where for $x \in \mathbb{R}$, $E(x)$ is the largest integer smaller or equal to x (see Proposition 7.4.8 for the proof of this result). Observe that $\Lambda_{m_\alpha, m_\beta}$ can be rewritten using Bessel functions,

$$\Lambda_{m_\alpha,m_\beta}=2\pi\ (-1)^{m_\beta+E(\frac{m_\beta+1}{2})}(2\omega)^{2m_\alpha+3m_\beta-1}\sum_{n=0}^{+\infty}\frac{(-4b\omega)^n}{(2m_\alpha+3m_\beta+n-1)!\ n!},$$

Using the Bessel function$J_\lambda(z)$ and the modified Bessel function $I_\lambda(z)$

$$J_\lambda(z)=\left(\frac{z}{2}\right)^\lambda\sum_{n=0}^{+\infty}\frac{\left(-\frac{1}{4}z^2\right)^n}{n!\ \Gamma(\lambda+n+1)},\quad I_\lambda(z)=\left(\frac{z}{2}\right)^\lambda\sum_{n=0}^{+\infty}\frac{\left(\frac{1}{4}z^2\right)^n}{n!\ \Gamma(\lambda+n+1)},$$

we obtain

– for $b=0$, $\Lambda_{m_\alpha,m_\beta}=2\pi\ \varepsilon_{m_\alpha,m_\beta}\dfrac{(2\omega)^{2m_\alpha+3m_\beta-1}}{(2m_\alpha+3m_\beta-1)!}$

– for $b<0$, $\Lambda_{m_\alpha,m_\beta}=2\pi\varepsilon_{m_\alpha,m_\beta}\left(\dfrac{-\omega}{b}\right)^{\frac{2m_\alpha+3m_\beta-1}{2}}I_{2m_\alpha+3m_\beta-1}(4\sqrt{-b\omega})$,

– for $b>0$, $\Lambda_{m_\alpha,m_\beta}=2\pi\ \varepsilon_{m_\alpha,m_\beta}\left(\dfrac{\omega}{b}\right)^{\frac{2m_\alpha+3m_\beta-1}{2}}J_{2m_\alpha+3m_\beta-1}(4\sqrt{b\omega})$.

with $\varepsilon_{m_\alpha,m_\beta}=(-1)^{m_\beta-1+E\left(\frac{m_\beta+1}{2}\right)}$.

We deduce from this explicit form of $\Lambda_{m_\alpha,m_\beta}$, that for $b\leq 0$, $\Lambda_{m_\alpha,m_\beta}$ never vanishes whereas there exists an infinite number of positive b for which $\Lambda_{m_\alpha,m_\beta}$ vanishes.

Remark 7.1.16. Among the coefficients Λ_n, Λ_1 is the only one which is known explicitly. So we have a generic non existence result, but from a theoretical point of view, for a given system we are not able in general to determine whether $\Lambda(\rho)$ vanishes or not. However, Proposition 7.4.8 gives an integral formula for $\Lambda(\rho)$. This integral does not involves the small parameter ν. It only involves explicitly known terms and the solution of an *"inner system"* which is deduced from the full system and which does not depend on ν. So this gives a numerical way to compute directly $\Lambda(\rho)$ without having to face the troubles induced by the terms which are exponentially small with respect to ν.

Remark 7.1.17. Conjecture of codimension 1 persistence of homoclinic connections to 0. On Figure 7.4, we see that reversible homoclinic connection to 0 should be of codimension 1. In [Ch2000], A.R. Champneys proposes to use the frequency ω as an extra parameter. He performed numerical computations which suggest that *for $b>0$ there are curves in the (ν,ω) parameter plane along which there exist reversible homoclinic connections to 0.* To prove theoretically such a result, there are two main difficulties:

- For a fixed ρ (typically $\rho = 1$) we can see $\Lambda(\rho)$ as a function of $\omega : \Lambda(\rho, \omega)$. Then, the first problem is to determine for which value of ω, $\Lambda(\rho, \omega)$ vanishes.
- The second problem is to prove a result of the type : If $\Lambda(\rho, \omega_0) = 0$ and $\frac{\partial \Lambda(\rho,\omega_0)}{\partial \omega}(\omega_0) \neq 0$ then there exits a curve $\omega(\nu)$ defined at least for small ν with $\omega(0) = \omega_0$ along which there exists homoclinic connections to 0.

7.1.4.3 Behavior in the past of the stable manifold of 0. Question 7.1.6-(b) remains open: when there is no homoclinic connection to 0, what is the behavior in the past of the stable manifold of 0? One can expect as for Toy Model (1.1), that the stable manifold develops exponentially small oscillations at $-\infty$. This is what happens with the toy model

$$\begin{aligned} \frac{d\widetilde{\alpha}}{dx} &= \widetilde{\beta} \\ \frac{d\beta}{dx} &= c_{10}\mu\widetilde{\alpha} + c_{20}\widetilde{\alpha}^2 \\ \frac{d\widetilde{A}}{dx} &= -\widetilde{B}\ \omega \\ \frac{d\widetilde{B}}{dx} &= \widetilde{A}\ \omega - \rho(2c_{20}/3)\widetilde{\alpha}^3 \end{aligned}$$

which is a particular example of reversible family vector fields admitting a $0^{2+}i\omega$ resonance at the origin. Applying the scaling (7.4), we get

$$\begin{aligned} \frac{d\alpha}{dt} &= \beta \\ \frac{d\beta}{dt} &= \alpha - \tfrac{3}{2}\alpha^2 \\ \frac{dA}{dt} &= -\frac{\omega B}{\nu} \\ \frac{dB}{dt} &= \frac{\omega A}{\nu} + \rho\nu^3\alpha^3 \end{aligned}$$

This systems corresponds to the rescaled normal form of order two with $c = a = b = 0$ perturbed by a unique monomial $(0, 0, 0, \rho\nu^3\alpha^3)$. The solutions of this system can be computed explicitly since the system is only partially coupled. We check that the stable manifold of 0 is explicitly given by $Y_s(t) = (\alpha_s(t), \beta_s(t), A_s(t), B_s(t))$ with

$$\alpha_s(t) = \cosh^{-2}(\tfrac{1}{2}t), \qquad \beta_s(t) = -\cosh^{-2}(\tfrac{1}{2}t)\tanh(\tfrac{1}{2}t),$$

$$Z_s(t) := A_s(t) + iB_s(t) = -i\rho\nu^3 \int_t^{+\infty} e^{i\omega(t-s)/\nu} \frac{ds}{\cosh^6(\frac{1}{2}s)}.$$

and that Y_s develops exponentially small oscillations at $-\infty$

$$Y_s(t) \underset{t\to-\infty}{\sim} (0,0,K(\nu)\sin(\omega t/\nu), -K(\nu)\cos(\omega t/\nu))$$

where

$$K(\nu) := \rho\nu^3 \int_{-\infty}^{+\infty} \mathrm{e}^{\mathrm{i}\omega s/\nu} \frac{ds}{\cosh^6(\frac{1}{2}s)} \underset{\nu\to 0}{\sim} \rho \frac{\pi}{60} \frac{\omega^5}{\nu^2} \mathrm{e}^{-\omega\pi/\nu}.$$

Another example of family of vector fields admitting a $0^{2+}\mathrm{i}\omega$ resonance at the origin is the following

$$\frac{d^4u}{dx^4} + \frac{d^2u}{dx^2} - \varepsilon^2 u + \varepsilon^4 u^2 = 0,$$

Setting $Y = \varepsilon^2 u,\ t = \varepsilon x$ we obtain the equivalent equation

$$\varepsilon^2 \frac{d^4Y}{dt^4} + \frac{d^2Y}{dt^2} - Y + Y^2 = 0,$$

This last equation was studied by Amick and McLeod in [AK89] who proved that the stable manifold of 0, explodes in finite time in the past.

So, there are at least two possible behavior in the past for the stable manifold of 0, when the origin is a $0^{2+}\mathrm{i}\omega$ resonant fixed point of a reversible family of vector fields. At the present time we do not know any general criteria to determine which of these two possible behaviors occurs for a given reversible family of vector fields admitting a $0^{2+}\mathrm{i}\omega$ resonance.

7.1.4.4 Domains of persistence of homoclinic connections. The two previous theorems are devoted to the estimation of the smallest size for a fixed ν of a periodic orbit which admits a reversible homoclinic connection to itself. However, for the truncated system, there also exist reversible homoclinic connections to periodic orbits the size of which is of order one. More precisely, in subsection 7.1.2, we checked that the truncated system (7.11) admits *two one parameter families* $h_{k,0}$, $h_{k,\frac{\pi}{\omega_0^*(k)}}$ of reversible homoclinic connections connecting a periodic orbits of size $k \in [0, k^*]$ with $k^* = \sqrt{1/6c}$ for $c > 0$ and $k^* = +\infty$ for $c \le 0$ where c is defined in (7.5). The following theorem ensures the persistence of the two families of reversible homoclinic connections to $Y_{k,\nu}$ provided that $k \in [K(\ell)\mathrm{e}^{-\ell\omega/\nu}, K_\star[$ for $\nu \in]0, \nu_4(\ell)]$ and $0 < \ell < \pi$.

Theorem 7.1.18. *There exist $K_\star$, $M_\star$ such that for every ℓ, $0 < \ell < \pi$ and every λ, $0 < \lambda < 1$ there exist $K(\ell)$, $\nu_5(\ell)$ such that for every $\nu \in]0, \nu_5(\ell)]$ and every $k \in [K(\ell)\nu^2\mathrm{e}^{-\ell\omega/\nu}, K_\star[$ the full System* (7.5) *admits two reversible solutions of the form*

$$Y(t) = h(t) + v(t) + Y_{k,\nu}(t + \varphi \tanh \tfrac{1}{2}t)$$

where

$$|h(t)| + |v(t)| \le M_\star \mathrm{e}^{-\lambda|t|} \qquad \textit{for } t \in \mathbb{R},$$

. . . .

and where $Y_{k,\nu}$ is a periodic solution of the full system satisfying

$$|Y_{k,\nu}(t)| \le M_\star k \qquad \textit{for } t \in \mathbb{R}.$$

This theorem is a direct consequence of Theorem 7.1.7. Its Proof is given at the end of Section 7.3. Figure 7.6 shows the domains of (non)-existence of reversible homoclinic connections to a periodic orbit of size k proved in this chapter. Compare with the domain of persistence of periodic solutions drawn in figure 7.3.

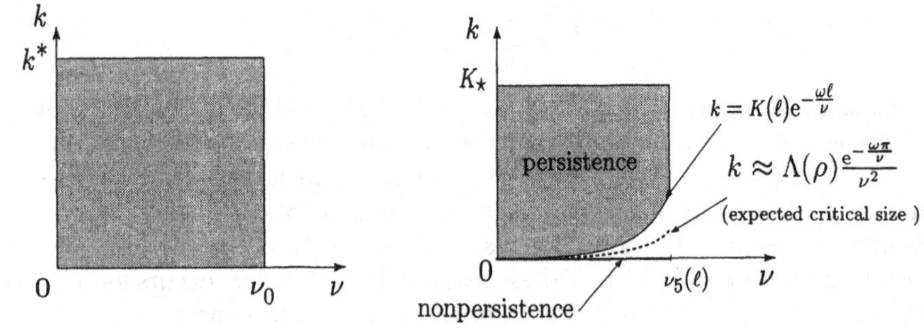

Truncated system ($\rho = 0$) Full system (7.16) ($\rho \neq 0,\ 0 < \ell < \pi$)

Fig. 7.6. Domain of existence of reversible homoclinic connections to a periodic orbit of size k for the truncated system and their domains of persistence and non-persistence obtained in this chapter

Remark 7.1.19. Improved estimates of the critical size for the full system, ($\rho = 1$). In fact, one can obtain sharper lower and upper bounds for the critical size : for any λ, $0 < \lambda < 1$ there exist $\nu_0' > 0$, $\sigma, d_s' > 0$, M', K_2', K_1', such that

(a) For every $\nu \in]0, \nu_0']$ and every $k \in [K_1'\ \mathrm{e}^{-\frac{\omega(\pi - \sigma\nu^{\frac{3}{5}})}{\nu}}, K_2']$ the full System (7.5) admits two reversible solutions of the form

$$Y(t) = h(t) + v(t) + Y_{k,\nu}(t + \varphi \tanh \tfrac{1}{2} t) \tag{7.19}$$

where

$$|h(t)| + |v(t)| \le M' \mathrm{e}^{-\lambda |t|} \qquad \text{for } t \in \mathbb{R},$$

and where $Y_{k,\nu}$ is a periodic solution of the full system satisfying $|Y_{k,\nu}(t)| \le M'k \qquad \text{for } t \in \mathbb{R}$.

(b) For every $\nu \in]0, \nu_0']$ and every $k \in [0, \Lambda(1)\ \mathrm{e}^{\frac{-\pi\omega}{\nu}}[$ the full System (7.5) does not admit any reversible solutions of the form

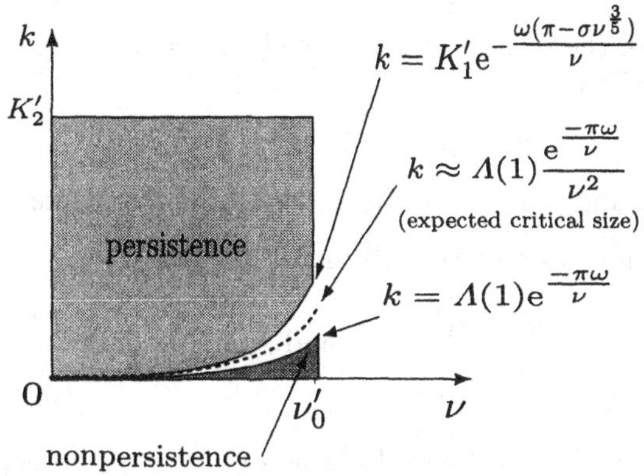

Fig. 7.7. Domains of persistence and non-persistence of reversible homoclinic connections to a periodic orbit of size k for the full system (7.5) ($\rho = 1$)

$$Y(t) = h(t) + v(t) + Y_{k,\nu}(t + \varphi \tanh \tfrac{1}{2}t) \tag{7.20}$$

with

$$\sup_{t\in\mathbb{R}} |v(t)| < d'_s, \qquad \sup_{t\in\mathbb{R}} |v(t)|e^{\lambda|t|} < +\infty. \tag{7.21}$$

These two results, enlarge the domain of existence and non existence previously obtained in Theorems 7.1.11, 7.1.18 (see figure 7.7).

The crucial point of the study of homoclinic connections to the periodic orbit $Y_{k,\nu}$ is that a solution of the full System (7.5) is a reversible homoclinic connection to $Y_{k,\nu}$ with $Y(t) \underset{t\to\pm\infty}{\longrightarrow} Y_{k,\nu}(t \pm \varphi)$ if and only if it satisfies a solvability condition, which reads very roughly speaking,

$$k \sin(\varphi) + I(Y, \nu) = 0$$

where $I(Y, \nu)$ is an *oscillatory integral.* Such an equation admits solutions φ, if and only if $k > |I(Y, \nu)|$. The different result of existence or non existence of homoclinic connections to $Y_{k,\nu}$ corresponds to different estimates of the *oscillatory integral* $I(Y, \nu)$ with the different Mono-frequency Exponential Lemmas introduced in Chapter 2. All these lemmas require a complexification of time. Working in a strip $\mathcal{B}_\ell = \{\xi \in \mathbb{C}, |\mathcal{I}m(\xi)| < \ell\}$ of width $\ell = \pi - \delta\nu$ we obtain that

$$m\Lambda(1)\frac{e^{-\frac{\omega(\pi-\delta\nu)}{\nu}}}{\delta^5\nu^2} \le |I(Y,\nu)| \le M\frac{e^{-\frac{\omega(\pi-\delta\nu)}{\nu}}}{\delta^5\nu^2}.$$

However, to have the periodic orbit $Y_{k,\nu}$ well defined when working in a strip $\mathcal{B}_\ell$ of width $\ell = \pi - \delta\nu$, we need that (see Theorem 7.1.4)

$$k \le k_0 e^{-\frac{\omega(\pi-\delta\nu)}{\nu}}.$$

So, on one hand, the result (a) is obtained for $\delta = \sigma\nu^{-\frac{2}{5}}$, $k = M\mathrm{e}^{-\frac{\omega(\pi-\sigma\nu^{\frac{3}{5}})}{\nu}}$ and σ chosen such that

$$\frac{M}{\sigma^5} \le k_0.$$

On the other hand, the result (b) is obtained for δ fixed, $0 \le \Lambda(1)k \le \mathrm{e}^{-\frac{\omega(\pi-\delta\nu)}{\nu}}$ and ν chosen sufficiently small so that

$$1 < \frac{m\Lambda(1)}{\delta^5\nu^2}, \qquad \text{and} \qquad \frac{M}{\delta^5\nu^2} \ge k_0.$$

Remark 7.1.20. For $c < 0$ (see Equation (7.5)) and the beginning of paragraph 7.1.4.4) the estimates of the critical size can be improved, since the constraint coming from the complexification of time for the periodic orbits can removed. Indeed, in this case, it is possible to build directly the one parameter family of periodic orbits $Y_{k,\nu}$ of size k with the Contraction Mapping Theorem in a space of function f which are periodic, holomorphic in the strip $\mathcal{B}_\pi$, normed with $\|f\| = \sup_{\xi\in\mathcal{B}_\pi,} (|f(\xi)|\mathrm{e}^{-|\mathcal{I}m(\xi)|\omega/\nu})$. Hence for $c < 0$, we obtain that

- there does not exist any reversible homoclinic connection of the form (7.20) satisfying (7.21) for $k \in [0, m\frac{\Lambda(1)}{\nu^2}\,\mathrm{e}^{\frac{-\pi\omega}{\nu}}[$ with $m < 1$;
- there exist reversible homoclinic connections of the form (7.19) for $k \in [K_1''\frac{1}{\nu^2}\,\mathrm{e}^{-\frac{\omega\pi}{\nu}}, K_2']$.

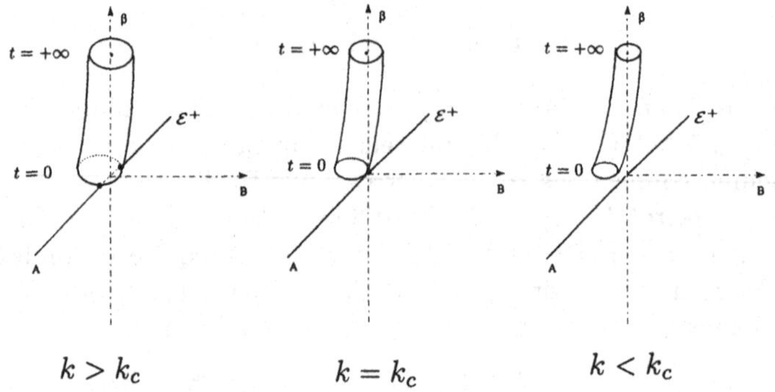

Fig. 7.8. Stable manifold of the periodic orbit $P^+_{k,\varphi}$ for Toy Model (1.1) for different values of k

Remark 7.1.21. In the small tong between the two domains of existence and non existence (see Figure 7.7) nothing is known. A priori, we can not exclude that it may exist in this domain small tongs of existence intertwined with small tong of non existence. Indeed, for our Toy Model (1.1) the stable manifold of a periodic orbit of size k is a circular tube whose cross section are circles of radius equal to k which miss the symmetry line when k is too small (see Figure 7.8). For the $0^{2+}i\omega$ resonance, the stable manifold of the periodic orbits is still a tube, but the cross sections are no longer circles. This may lead to a more subtle mechanism of transition between the domain of existence and non existence of reversible homoclinic connections.

7.2 Proof of the persistence of periodic orbits: explicit form and complexification

This section is devoted to the persistence of periodic solutions. We first prove Theorem 7.1.4 which follows directly from the general Theorem 4.1.2 given in chapter 4 which ensures for reversible systems the existence of a one parameter family of periodic orbits bifurcating from a pair of simple purely imaginary eigenvalues.

Proof of Theorem 7.1.4. The explicit form of N, R and $D_Y(N+R)(0,\nu)$ given respectively by (7.5), (7.6), (7.9) ensures that Equation (7.5) can be rewritten as

$$\frac{dY}{dt} = L_\nu Y + Q(Y,\nu)$$

where

$$\text{(a)}\ \ L_\nu = D_Y(N+R)(0,\nu) = \begin{bmatrix} \dfrac{L_{\mathrm{h}}(\nu)}{\nu} & 0 & 0 \\ 0 & 0 & -\dfrac{\omega_{0,\nu}}{\nu} \\ 0 & \dfrac{\omega_{0,\nu}}{\nu} & 0 \end{bmatrix}$$

$$\text{with } L_{\mathrm{h}}(\nu) = \nu \begin{bmatrix} 0 & 1 \\ \theta(\nu) & 0 \end{bmatrix},\ \theta(\nu) = 1 + \mathcal{O}(\nu^2),\ \ \omega_{0,\nu} = \omega + a\nu^2 + \mathcal{O}(\nu^4).$$

Thus we can choose $\nu_1 \in]0,\nu_0]$ such that

$$\inf_{\nu\in]0,\nu_1]} \omega_{0,\nu} > 0, \quad \sup_{\nu\in]0,\nu_1]} |\nu a + \mathcal{O}(\nu^3)| < +\infty, \quad \sup_{\nu\in]0,\nu_1]} \|L_{\mathrm{h}}(\nu)\| < +\infty$$

Finally, since

$$\left(I_2 - \exp\left(\frac{2\pi L_h(\nu)}{\omega_{0,\nu}}\right)\right)^{-1} = \begin{bmatrix} 1 & (\theta(\nu))^{-\frac{1}{2}}\coth(\eta(\nu)) \\ (\theta(\nu))^{\frac{1}{2}}\coth(\eta(\nu)) & 1 \end{bmatrix}$$

with $\eta(\nu) = \pi\nu(\theta(\nu))^{\frac{1}{2}}/\omega_{0,\nu}$, there exists M such that for every $\nu \in]0,\nu_1]$

$$\left\| \left(I_2 - \exp\left(\frac{2\pi L_h(\nu)}{\omega_{0,\nu}}\right)\right)^{-1} \right\| \leq \frac{M}{\nu}.$$

(b) the nonlinear part Q satisfies $Q(0,\nu) = D_Y Q(0,\nu) = 0$. Moreover, Q is analytic in $D(0,r) = \{Y \in \mathbb{R}^4, |Y| < r\}$ with $0 < r < \min(r_\gamma, r_\beta, r_A, r_B)$ and satisfies

$$\sup_{|Y|<r,\ \nu\in]0,\nu_1]} |Q(Y,\nu)| < +\infty,$$

$$\sup_{|Y|<r,\ \nu\in]0,\nu_1]} |\nu^{-1}Q_A(Y,\nu)| < +\infty, \qquad \sup_{|Y|<r,\ \nu\in]0,\nu_1]} |\nu^{-1}Q_B(Y,\nu)| < +\infty.$$

Thus hypothesis (i) and (ii) of Theorem 4.1.2 are fulfilled. The result follows directly from this last theorem. □

The real full System (7.5) admits a one-parameter family of periodic solutions $Y_{k,\nu}(t)$ given by Theorem 7.1.4 which are real analytic with respect to (t,k). If we succeed in extending $Y_{k,\nu}(t)$ by analyticity into a neighborhood of the real axis in the complex field to obtain $Y_{k,\nu}(\xi)$, then $Y_{k,\nu}(\xi)$ has to be a solution of the complex Equation (7.10). Moreover,

$$Y_{k,\nu}(t) = \widehat{Y}(\underline{\omega}_{k,\nu}t, k, \nu), \text{ and } \widehat{Y}(s,k,\nu) = \sum_{n\geq 1} k^n \widehat{Y}_n(s,\nu), \text{ with } \widehat{Y}_n \in \tilde{\mathcal{P}}_{n,R}.$$

For each n, $\widehat{Y}_n$ is a trigonometric polynomial of degree n. Thus it can be extended by analyticity to obtain an entire function. It remains now to find the radius of convergence of the power series $\sum k^n |\widehat{Y}_n(\zeta)|$.

The following proposition gives a lower bound of this radius.

Proposition 7.2.1. *there exists $C'_{\widehat{Y}}$ such that*

$$k^n \left| \frac{d^j \widehat{Y}_n}{d\zeta^j}(\zeta,\nu) \right| \leq C'_{\widehat{Y}} n^j \sqrt{n+1} \left(\frac{k}{k_0}\right)^n e^{n|\mathcal{I}m(\zeta)|},$$

holds for every $j \geq 0$, $k \in [0,k_0]$, $\nu \in [0,\nu_1]$, $\zeta \in \mathbb{C}$, and $n \geq 1$.

This proposition directly follows from Proposition 4.2.1. □

Remark 7.2.2. Proposition 7.2.1 ensures that taking $(k/k_0)e^{|\mathcal{I}m(\zeta)|} < 1$, one can define $\widehat{Y}(\zeta, k, \nu)$. Since $Y_{k,\nu}(\xi) = \widehat{Y}(\underline{\omega}_{k,\nu}\xi, k, \nu)$, $Y_{k,\nu}(\xi)$ is well defined if $(k/k_0)e^{\underline{\omega}_{k,\nu}|\mathcal{I}m(\xi)|} < 1$. But $\underline{\omega}_{k,\nu}$ has a "bad" dependence on ν ($\underline{\omega}_{k,\nu} = \omega/\nu + \cdots$). So one way to keep $(k/k_0)e^{\underline{\omega}_{k,\nu}|\mathcal{I}m(\xi)|} < 1$ is to keep ξ in a complex strip $\mathcal{B}_\ell = \{\xi \in \mathbb{C}/|\mathcal{I}m(\xi)| < \ell\}$ and to choose k in $[0, k_0 e^{-\ell\underline{\omega}_{k,\nu}}]$. In $\mathcal{B}_\ell$, $Y_{k,\nu}(\xi)$ is then well defined as well as $\frac{dY_{k,\nu}}{d\xi}(\xi)$, $\frac{d^2Y_{k,\nu}}{d\xi^2}(\xi)$. The precise choice of the complex strip and of the different parameters is given in the next section which is devoted to homoclinic connections to exponentially small periodic orbits.

7.3 Proof of the persistence of reversible homoclinic connections to exponentially small periodic solutions

A quite natural idea to find reversible solutions of the real full System (7.5) homoclinic to the periodic orbits $Y_{k,\nu}$ obtained in Theorem 7.1.4 is to compute the stable manifold of $Y_{k,\nu}$ in the form

$$Y(t) = h(t) + Y_{k,\nu}(t + \varphi) + v(t), \quad t \in [0, +\infty], \ \varphi \in \mathbb{R},$$

where $h(t)$ is the unique reversible solution of the truncated system homoclinic to 0. The perturbation v is obtained using the Contraction Mapping Theorem. System (7.5) is reversible and autonomous. Thus, for any solution $Z(t)$ of (7.5), if there exists a time t^* such that $SZ(t^*) = Z(t^*)$, then $Z(t+t^*)$ is a reversible solution of (7.5). So, to find a reversible solution homoclinic to $Y_{k,\nu}$, we must find the intersection between the stable manifold and the symmetry plane $P_+ = \{Y/\ SY = Y\}$. This intersection is given by a scalar equation which can be solved provided that k is of greater size that an integral of the form

$$\int_0^{+\infty} f(t,\nu) \sin\frac{\omega t}{\nu} dt \quad (\nu \longrightarrow 0).$$

Coarsely dominating the integral by $\int_0^{+\infty} |f(t,\nu)|dt$, i.e., not taking into account the oscillations we obtain that k must be greater than ν^n.

We recall here that we want to take k, the size of the periodic orbit, as small as possible. So to get k exponentially small we want to apply the First Exponential Lemma 2.1.1. The trouble is that this lemma only works with integrals of the form

$$\int_{-\infty}^{+\infty} f(t,\nu) e^{i\omega t/\nu} dt.$$

For symmetrizing the integral by evenness we must use the fact that the solution $Y(t)$ is reversible. But this is precisely what we want to know by studying this integral!

To have a correct proof, we do not try to compute the whole stable manifold. We try to find directly a reversible solution homoclinic to the periodic orbit using the Contraction Mapping Theorem in an appropriate set of reversible functions. But now the parameterization of the stable manifold is no longer suitable because the set of functions v

$$\{v/t \mapsto h(t) + Y_{k,\nu}(t+\varphi) + v(t) \ \text{ is reversible}\}$$

is not a vector space. So, we look for $Y(t)$ in the form

$$Y(t) = h(t) + Y_{k,\nu}(t + \varphi \tanh \tfrac{1}{2}t) + v(t), \quad t \in \mathbb{R}, \ \varphi \in \mathbb{R}. \tag{7.22}$$

The choice of $Y_{k,\nu}(t + \varphi \tanh \frac{1}{2}t)$ is motivated by the form of solutions of the truncated system homoclinic to periodic orbits given in Subsection 7.1.2. Observe that the function $t \mapsto h(t) + Y_{k,\nu}(t + \varphi \tanh \frac{1}{2}t)$ is reversible. Now the set of functions $\{v/\ t \mapsto h(t) + Y_{k,\nu}(t + \varphi \tanh \frac{1}{2}t) + v(t)$ is reversible$\}$ is a vector space, since it amounts to having v reversible. So from now on, we look for the reversible perturbation function v and for the phase shift φ. Our strategy to build a solution Y of (7.5) under the form (7.22) is the following:

In subsection 7.3.1, we first rewrite the equation satisfied by v in the form

$$\frac{dv}{dt} - DN(h,\nu).v = g(v, \varphi, k, \nu, t). \tag{7.23}$$

Our aim is to transform this equation into an integral equation and to solve it using the Contraction Mapping Theorem. For that purpose we must invert the linear part of (7.23). The crucial point of our analysis is the following : for an *antireversible* function f which *tends to 0 exponentially at infinity*, there exists a *reversible* function u which *tends to 0 exponentially at infinity* and which satisfies

$$\frac{du}{dt} - DN(h,\nu).u = f$$

if and only if the function f satisfies the solvability condition

$$\int_0^{+\infty} \langle r_-^*(t), f(t) \rangle = 0 \tag{7.24}$$

where the dual vector r_-^* reads

$$r_-^*(t) := (-\sin \psi_\nu(t), \cos \psi_\nu(t), 0, 0) \text{ with } \psi_\nu(t) = \left(\tfrac{\omega}{\nu} + a\nu\right) t + 2b\nu \tanh(\tfrac{1}{2}t).$$

This solvability condition is the typical one which occurs when the linear homogeneous system admits a pair of simple purely imaginary eigenvalues (see Section 6.4). Moreover, observe that the above integral, is an *oscillatory integral* since the dual vector rotates with a frequency of order ω/ν where ν is a small parameter.

In subsection 7.3.5, we prove that for each reversible v tending to 0 exponentially at infinity it is possible to choose $\varphi(v, k, \nu)$ such that $t \mapsto$

$g(v(t), \varphi, k, \nu, t)$ satisfies the solvability condition (7.24) provided that k is greater than an oscillatory integral of the form

$$J(\nu) = \int_{-\infty}^{+\infty} \mathrm{e}^{\frac{i\omega t}{\nu}} G(t)dt.$$

Using the First Exponential Lemma 2.1.1, we are able to obtain an exponentially small upper bound of $J(\nu)$ and thus we can take the size k of the periodic orbit $Y_{k,\nu}$ exponentially small.

So, finally in Subsection 7.3.6, with the above choice of $\varphi(v, k, \nu)$ and with an exponentially small k of the form $k = K\mathrm{e}^{-\frac{\ell\omega}{\nu}}$, $0 < \ell < \pi$, we transform (7.23) into an integral equation which can be solved using the Contraction Mapping Theorem in a space of reversible function tending to 0 at infinity.

Remark 7.3.1. For obtaining a better estimate of $J(\nu)$ which allows us to take $k = K\frac{1}{\nu^2}\mathrm{e}^{-\omega\pi/\nu}$, one should use the Third Exponential Lemma 2.1.21 coupled with the strategy proposed in Subsection 2.1.4. The resulting proof is then far more intricate than the one proposed here which is based on the First Exponential Lemma 2.1.1. The Third Exponential Lemma is used in the next section to prove the generic non persistence of homoclinic connection to 0.

For using Lemma 2.1.1 we have to complexify the problem, i.e., we have to study the complex Equation (7.10). Thus we look for Y in the form

$$Y(\xi) = h(\xi) + Y_{k,\nu}(\xi + \varphi \tanh \tfrac{1}{2}\xi) + v(\xi), \quad \xi \in \mathbb{C}, \ \varphi \in \mathbb{R}.$$

Thus we must restrict ξ, k, φ to have $h(\xi)$, $Y_{k,\nu}(\xi + \varphi \tanh \frac{1}{2}\xi)$ well defined. This is the purpose of the next subsection.

7.3.1 Choice of the parameters

Observe that $h(\xi) = (\alpha_h(\xi), \beta_h(\xi), 0, 0)$, $\tanh \frac{1}{2}\xi$ are meromorphic with their poles in $\mathrm{i}(2n+1)\pi$, $n \in \mathbb{Z}$. To apply Lemma 2.1.1, which requires holomorphic functions (singularities are not allowed), we work in a strip of the complex field $\mathcal{B}_\ell$ with $\ell < \pi$:

Let ℓ be fixed, $0 < \ell < \pi$. Denote $\mathcal{B}_\ell := \{\xi \in \mathbb{C} / \ \xi = t + \mathrm{i}\eta, -\ell < \eta < \ell, \ t \in \mathbb{R}\}$. *From now on, ξ is assumed to lie in $\mathcal{B}_\ell$.*

Now we must choose k, φ to have $Y_{k,\nu}(\xi + \varphi \tanh \frac{1}{2}\xi)$ well defined and holomorphic in $\mathcal{B}_\ell$. $Y_{k,\nu}$ is deduced from $\widehat{Y}$ by

$$Y_{k,\nu}(\xi + \varphi \tanh \tfrac{1}{2}\xi) = \widehat{Y}(\underline{\omega}_{k,\nu}(\xi + \varphi \tanh \tfrac{1}{2}\xi), k, \nu).$$

Proposition 7.2.1 ensures that $Y_{k,\nu}(\xi + \varphi \tanh \frac{1}{2}\xi)$ is well defined if

$$(k/k_0)\mathrm{e}^{|\mathcal{I}m(\underline{\omega}_{k,\nu}(\xi+\varphi\tanh(\xi/2)))|} < 1.$$

So, from now on, we look for φ in $[0, 6\pi\nu/\omega]$ and we set $k = K\mathrm{e}^{-\ell\omega/\nu}$, with $K \in [0, k_0]$.

The following lemma give an upper bound for $|\mathcal{I}m\left(\underline{\omega}_{k,\nu}(\xi + \varphi \tanh \frac{1}{2}\xi)\right)|$. Observe the "bad" dependence on ν coming from $\underline{\omega}_{k,\nu}$.

Lemma 7.3.2. *There exists M_0 such that*

$$|\mathcal{I}m\left(\underline{\omega}_{k,\nu}(K\mathrm{e}^{-\ell\omega/\nu})(\xi + \varphi \tanh \tfrac{1}{2}\xi)\right)| \leq M_0 + \frac{\ell\omega}{\nu}$$

holds for every $\nu \in]0, \nu_1]$, $\varphi \in [0, 6\pi\nu/\omega]$, $K \in [0, k_0]$, and $\xi \in \mathcal{B}_\ell$.

Proof. Observe that for $K \in [0, k_0]$, $k = K\mathrm{e}^{-\ell\omega/\nu} \in [0, k_0]$ which ensures that $\underline{\omega}_{k,\nu}(K\mathrm{e}^{-\ell\omega/\nu})$ is well defined.

By Theorem 7.1.4 we have

$$\underline{\omega}_{k,\nu} = \omega/\nu + a\nu + \mathcal{O}(\nu^3) + \widehat{\omega}(K\mathrm{e}^{-\ell\omega/\nu}, \nu), \quad |\widehat{\omega}| \leq C_\omega \nu K\mathrm{e}^{-\ell\omega/\nu} \leq C_\omega k_0 \nu.$$

Moreover, observing that for all $\xi \in \mathcal{B}_\ell$, $|\mathcal{I}m\left(\tanh \frac{1}{2}\xi\right)| \leq \tan \frac{1}{2}\ell$, we get

$$|\mathcal{I}m\left(\xi + \varphi \tanh \tfrac{1}{2}\xi\right)| \leq \ell + \frac{6\pi\nu}{\omega} \tan \tfrac{1}{2}\ell.$$

Thus,

$$|\mathcal{I}m\left(\underline{\omega}_{k,\nu}(\xi + \varphi \tanh(\tfrac{1}{2}\xi))\right)|$$
$$\leq \frac{\omega\ell}{\nu} + 6\pi \tan \tfrac{1}{2}\ell + [\ell + \frac{6\pi\nu_1}{\omega} \tan \tfrac{1}{2}\ell][|a| + M\nu_1^2 + C_\omega k_0]\nu_1. \square$$

Finally, we can now prove:

Proposition 7.3.3. *From now on let K_1 be fixed such that $K_1 < k_0\mathrm{e}^{-M_0}$. Then, for every $\nu \in]0, \nu_1]$, $\varphi \in [0, 6\pi\nu/\omega]$ and $K \in [0, K_1]$, the functions $Y_{k,\nu}(\xi + \varphi \tanh \frac{1}{2}\xi)$, $\frac{dY_{k,\nu}}{d\xi}(\xi + \varphi \tanh \frac{1}{2}\xi)$, $\frac{d^2Y_{k,\nu}}{d\xi^2}(\xi + \varphi \tanh \frac{1}{2}\xi)$ are well defined and holomorphic in $\mathcal{B}_\ell$.*

Proof. Proposition 7.2.1, and Lemma 7.3.2 ensure that

$$k^n|\widehat{Y}_n(\underline{\omega}_{k,\nu}(\xi + \varphi \tanh \tfrac{1}{2}\xi)| \leq C'_{\widehat{Y}}\sqrt{n+1}\left(\frac{K_1\mathrm{e}^{M_0}}{k_0}\right)^n.$$

But $\frac{K_1\mathrm{e}^{M_0}}{k_0} < 1$ which ensures that the power series converge. We proceed in the same way for $\frac{dY_{k,\nu}}{d\xi}, \frac{d^2Y_{k,\nu}}{d\xi^2}$. $\square$

Definition 7.3.4. *Let λ be fixed such that $0 < \lambda < 1$. We define $\mathcal{H}_\ell^\lambda|_{R,\mathbb{R}}$ the set of functions $v : \mathcal{B}_\ell \mapsto \mathbb{C}^4$, holomorphic, reversible, mapping $\mathbb{R}$ in $\mathbb{R}^4$ and satisfying*

$$|v|_{\mathcal{H}_\ell^\lambda} \stackrel{def}{=} \sup_{\xi \in \mathcal{B}_\ell} [|v(\xi)| e^{\lambda |\mathcal{R}e(\xi)|}] < +\infty.$$

Denote by $\mathcal{BH}_\ell^\lambda|_{R,\mathbb{R}}(\delta\sqrt{\nu})$ the ball of radius $\delta\sqrt{\nu}$ of $\mathcal{H}_\ell^\lambda|_{R,\mathbb{R}}$:

$$\mathcal{BH}_\ell^\lambda|_{R,\mathbb{R}}(\delta\sqrt{\nu}) := \{v \in \mathcal{H}_\ell^\lambda|_{R,\mathbb{R}} / \, |v|_{\mathcal{H}_\ell^\lambda} \leq \delta\sqrt{\nu}\}.$$

We look for v in $\mathcal{BH}_\ell^\lambda|_{R,\mathbb{R}}(\delta\sqrt{\nu})$ with $\delta \leq 1$.

We have chosen parameters to have $h(\xi)$, $Y_{k,\nu}(\xi + \varphi \tanh \frac{1}{2}\xi)$ well defined and holomorphic in $\mathcal{B}_\ell$. Now we compute the equation satisfied by v:

Proposition 7.3.5. *For every $\nu \in]0, \nu_1]$, $\varphi \in [0, 6\pi\nu/\omega]$, $K \in [0, K_1]$ and $\xi \in \mathcal{B}_\ell$, the function $Y(\xi) = h(\xi) + Y_{k,\nu}(\xi + \varphi \tanh \frac{1}{2}\xi) + v(\xi)$ is a solution of the complex differential Equation (7.10) if and only if*

$$\frac{dv}{d\xi} - DN(h,\nu).v = g(v(\xi), \varphi, K, \nu, \xi) \qquad (7.25)$$

with

$$g(v(\xi), \varphi, K, \nu, \xi) = N'(v(\xi), \varphi, K, \nu, \xi) + R'(v(\xi), \varphi, K, \nu, \xi) + \varphi\Phi(\xi, \varphi, K, \nu)$$

$$N'(v, \varphi, K, \nu, \xi) = N(h + Y_{k,\nu} + v, \nu) - N(h,\nu) - N(Y_{k,\nu}, \nu) - DN(h,\nu)v,$$

$$R'(v, \varphi, K, \nu, \xi) = R(h + Y_{k,\nu} + v) - R(Y_{k,\nu}),$$

$$\Phi(\xi, \varphi, K, \nu) = \frac{-1}{2\cosh^2 \frac{1}{2}\xi} \frac{dY_{k,\nu}}{d\xi}(\xi + \varphi \tanh \tfrac{1}{2}\xi),$$

where h, v are evaluated at ξ, and $Y_{k,\nu}$ at $\xi + \varphi \tanh \frac{1}{2}\xi$.

Our aim is now to transform (7.25) into an integral equation. For that purpose we need estimates on $N', R', \varphi\Phi$, as well as a basis for the homogeneous equation

$$\frac{dv}{d\xi} - DN(h,\nu)v = 0. \qquad (7.26)$$

The estimates are given in Subsection 7.3.2, the homogeneous equation is studied in Subsection 7.3.3.

7.3.2 Estimates of $Y_{k,\nu}$, h, N', R', $\varphi\Phi$

The following lemmas give bounds of $Y_{k,\nu}$, h, N', R', $\varphi\Phi$. Their proof are elementary but must be done carefully because we want to know precisely how these different functions in the complex field depend on the small parameter ν. Observe that in each following lemma the constants M_i do not depend on ν, K, φ. The constants M_i depend on ℓ, but ℓ is fixed in this study.

7.3.2.1 Bounds of $Y_{k,\nu}(\xi + \varphi \tanh \frac{1}{2}\xi)$.

Lemma 7.3.6. *There exists M_1 such that for every $\nu \in]0, \nu_1]$, $\varphi \in [0, 6\pi\nu/\omega]$, $K \in [0, K_1]$, and $\xi \in \mathcal{B}_\ell$,*

$$\left|Y_{k,\nu}(\xi + \varphi \tanh \tfrac{1}{2}\xi)\right| \leq M_1 K,$$

$$\left|\frac{dY_{k,\nu}}{d\xi}(\xi + \varphi \tanh \tfrac{1}{2}\xi)\right| \leq M_1 \frac{K}{\nu},$$

$$\left|\frac{d^2Y_{k,\nu}}{d\xi^2}(\xi + \varphi \tanh \tfrac{1}{2}\xi)\right| \leq M_1 \frac{K}{\nu^2},$$

and

$$Y_{k,\nu}(\xi+\varphi \tanh \tfrac{1}{2}\xi) = K e^{-\frac{\ell\omega}{\nu}} \widehat{Y}_1(\underline{\omega}_{k,\nu}(\xi + \varphi \tanh \tfrac{1}{2}\xi)) + O_1(\xi, k, \nu, \varphi),$$

$$\frac{dY_{k,\nu}}{d\xi}(\xi+\varphi \tanh \tfrac{1}{2}\xi) = \frac{\omega}{\nu} K e^{-\frac{\ell\omega}{\nu}} \frac{d\widehat{Y}_1}{d\xi}(\underline{\omega}_{k,\nu}(\xi + \varphi \tanh \tfrac{1}{2}\xi)) + O_2(\xi, k, \nu, \varphi),$$

$$\frac{d^2Y_{k,\nu}}{d\xi^2}(\xi+\varphi \tanh \tfrac{1}{2}\xi) = \frac{\omega^2 K e^{-\frac{\ell\omega}{\nu}}}{\nu^2} \frac{d^2\widehat{Y}_1}{d\xi^2}(\underline{\omega}_{k,\nu}(\xi+\varphi \tanh \tfrac{1}{2}\xi)) + O_3(\xi, k, \nu, \varphi),$$

with

$$\begin{aligned} |O_1(\xi, k, \nu, \varphi)| &\leq M_1 K^2, \\ |O_2(\xi, k, \nu, \varphi)| &\leq M_1 \left(\nu K + \frac{K^2}{\nu}\right), \\ |O_3(\xi, k, \nu, \varphi)| &\leq M_1 \left(K + \frac{K^2}{\nu^2}\right). \end{aligned}$$

Proof. Using Theorem 7.1.4, Proposition 7.2.1 and Lemma 7.3.2 we get

$$\begin{aligned} |Y_{k,\nu}(\xi + \varphi \tanh \tfrac{1}{2}\xi)| &= |\widehat{Y}(\underline{\omega}_{k,\nu}(\xi + \varphi \tanh \tfrac{1}{2}\xi), k, \nu)| \\ &\leq \sum_{n\geq 1} C'_{\widehat{Y}} \sqrt{n+1}\, K^n \frac{e^{-\frac{n\ell\omega}{\nu}}}{(k_0)^n} e^{n\mathcal{I}m\left(\underline{\omega}_{k,\nu}(\xi+\varphi \tanh \frac{1}{2}\xi)\right)} \\ &\leq \frac{C'_{\widehat{Y}} e^{M_0}}{k_0} K \sum_{n\geq 0} \sqrt{n+2}\, \left(\frac{K_1 e^{M_0}}{k_0}\right)^n \\ &\leq M_1 K. \end{aligned}$$

The power series converge because K_1 has been chosen such that $K_1 < e^{-M_0} k_0$ holds. Then we have

$$Y_{k,\nu}(\xi + \varphi \tanh \tfrac{1}{2}\xi) = \widehat{Y}_1(\underline{\omega}_{k,\nu}(\xi + \varphi \tanh \tfrac{1}{2}\xi)) + O_1,$$

with $|O_1| \le C'_{\widehat{Y}}(\frac{e^{M_0}}{k_0})^2 K^2 \sum_{n\ge 0} \sqrt{n+3}\left(\frac{K_1 e^{M_0}}{k_0}\right)^n$.

The proofs for $\frac{dY}{d\xi}$, $\frac{d^2Y}{d\xi^2}$ are done in the same way. □

When $\xi = t$ lies in $\mathbb{R}$ the estimates of the bounds of the periodic solution are far better: in $\mathcal{B}_\ell$ the periodic solution is simply bounded, whereas in $\mathbb{R}$ it is "exponentially bounded".

Lemma 7.3.7. *There exists M_2 such that for every $\nu \in]0,\nu_1]$, $\varphi \in [0, 6\pi\nu/\omega]$, $K \in [0, K_1]$, and $t \in \mathbb{R}$,*

$$\begin{aligned} \left|Y_{k,\nu}(t+\varphi\tanh\tfrac12 t)\right| &\le M_2 K e^{-\ell\omega/\nu}, \\ \left|\frac{dY_{k,\nu}}{dt}(t+\varphi\tanh\tfrac12 t)\right| &\le M_2 \frac{K}{\nu} e^{-\ell\omega/\nu}, \\ \left|\frac{d^2Y_{k,\nu}}{dt^2}(t+\varphi\tanh\tfrac12 t)\right| &\le M_2 \frac{K}{\nu^2} e^{-\ell\omega/\nu}, \end{aligned}$$

$$Y_{k,\nu}(t+\varphi\tanh\tfrac12 t) = K e^{-\frac{\ell\omega}{\nu}} \widehat{Y}_1(\underline{\omega}_{k,\nu}(t+\varphi\tanh\tfrac12 t)) + \tilde{O}_1(t,k,\nu,\varphi),$$

$$\frac{dY_{k,\nu}}{dt}(t+\varphi\tanh\tfrac12 t) = \frac{\omega}{\nu} K e^{-\ell\omega/\nu} \frac{d\widehat{Y}_1}{dt}(\underline{\omega}_{k,\nu}(t+\varphi\tanh\tfrac12 t)) + \tilde{O}_2(t,k,\nu,\varphi),$$

$$\frac{d^2Y_{k,\nu}}{dt^2}(t+\varphi\tanh\tfrac12 t) = \frac{\omega^2}{\nu^2} K e^{-\frac{\ell\omega}{\nu}} \frac{d^2\widehat{Y}_1}{dt^2}(\underline{\omega}_{k,\nu}(t+\varphi\tanh\tfrac12 t)) + \tilde{O}_3(t,k,\nu,\varphi),$$

with

$$\begin{aligned} |\tilde{O}_1(t,k,\nu,\varphi)| &\le M_2 K^2 e^{-2\ell\omega/\nu}, \\ |\tilde{O}_2(t,k,\nu,\varphi)| &\le M_2\left(\nu K e^{-\ell\omega/\nu} + \frac{K^2}{\nu} e^{-2\ell\omega/\nu}\right), \\ |\tilde{O}_3(t,k,\nu,\varphi)| &\le M_2\left(K e^{-\ell\omega/\nu} + \frac{K^2}{\nu^2} e^{-2\ell\omega/\nu}\right). \end{aligned}$$

Proof. The improvement of the estimates comes from the fact that for every $t \in \mathbb{R}$, $\mathcal{I}m\left(\underline{\omega}_{k,\nu}(t+\varphi\tanh\tfrac12 t)\right) = 0$, whereas for every $\xi \in \mathcal{B}_\ell$,

$$\left|\mathcal{I}m\left(\underline{\omega}_{k,\nu}(\xi+\varphi\tanh\tfrac12\xi)\right)\right| \le M_0 + \ell\omega/\nu. \quad \square$$

7.3.2.2 Bound for $h(\xi)$.

Lemma 7.3.8.

(a) *There exists M_3 such that* $\left|\tanh\tfrac12\xi\right| \le M_3$ *holds for every $\xi \in \mathcal{B}_\ell$.*

(b) *There exists M_4 such that* $\dfrac{1}{\left|\cosh^2\tfrac12\xi\right|} \le M_4 e^{-|\mathcal{R}e(\xi)|}$ *holds for every $\xi \in \mathcal{B}_\ell$.*

(c) *There exists* M_5 *such that* $|h(\xi)| \leq M_5 \mathrm{e}^{-|\mathcal{R}e(\xi)|}$ *holds for every* $\xi \in \mathcal{B}_\ell$.

7.3.2.3 Estimates of N', R', $\varphi\Phi$. Let us denote

$$\begin{aligned} h(\xi) &= (\alpha_h(\xi), \beta_h(\xi), 0, 0), \\ Y_{k,\nu}(\xi) &= (\alpha_k(\xi), \beta_k(\xi), A_k(\xi), B_k(\xi)), \\ v(\xi) &= (\alpha(\xi), \beta(\xi), A(\xi), B(\xi)). \end{aligned}$$

Then, the following lemmas hold:

Lemma 7.3.9.

$$N'(v,\varphi,\xi) = \begin{pmatrix} 0 \\ -\frac{3}{2}\alpha^2 - c(A^2+B^2) - 3\alpha_k\alpha - 2c(A_kA + B_kB) - 3\alpha_h\alpha_k \\ -b\nu[B\alpha + \alpha_k B + B_k\alpha + B_k\alpha_h] \\ b\nu[A\alpha + \alpha_k A + A_k\alpha + A_k\alpha_h] \end{pmatrix}$$

where α_h, β_h, α, β, A, B *are evaluated at* ξ, $\alpha_k, \beta_k, A_k, B_k$ *are evaluated at* $\xi + \varphi \tanh \frac{1}{2}\xi$. *Moreover, there exists* M_6 *such that for every* $\nu \in]0,\nu_1]$, $\varphi, \varphi' \in [0, 6\pi\nu/\omega]$, $K \in [0, K_1]$, *and* $\xi \in \mathcal{B}_\ell$, *the following estimates hold:*

(a) $|N'(v,\varphi,\xi)| \leq M_6[|v(\xi)|^2 + K|v(\xi)| + K|h(\xi)|]$,

(b) $|N'(v,\varphi,\xi) - N'(v',\varphi',\xi)| \leq M_6\big[(|v(\xi)| + |v'(\xi)| + K)\ |v'(\xi) - v(\xi)|$
$\qquad + \frac{K}{\nu}(|v(\xi)| + |h(\xi)|)|\varphi' - \varphi|\big]$.

Observe that we have estimated the growth rate of N', $|N'(v,\varphi,\xi) - N'(v',\varphi',\xi)|$, between two different values of v, and also between two different values of φ. Indeed, we look for v and φ and this type of estimate is required by the Contraction Mapping Theorem.

Lemma 7.3.10. *There exists* M_7, *such that for every* $\nu \in]0,\nu_1]$, $\varphi, \varphi' \in [0, 6\pi\nu/\omega]$, $K \in [0, K_1]$, *and* $\xi \in \mathcal{B}_\ell$, *the following estimates hold:*

(a) $|R'(v,\varphi,\xi)| \leq M_7\ \nu^2\big[|v(\xi)| + |h(\xi)|\big]$,

(b) $|R'(v,\varphi,\xi) - R'(v',\varphi',\xi)| \leq M_7\big[\ \nu^2|v'(\xi) - v(\xi)|$
$\qquad + \nu K(|v(\xi)| + |h(\xi)|)\ |\varphi' - \varphi|\ \big]$.

Lemma 7.3.11. *There exists* M_8 *such that for every* $\nu \in]0, \nu_1]$, $\varphi, \varphi' \in [0, 6\pi\nu/\omega]$, $K \in [0, K_1]$, *and* $\xi \in \mathcal{B}_\ell$ *the following estimates hold:*

(a) $|\varphi\Phi(\xi,\varphi)| \leq M_8 K \mathrm{e}^{-|\mathcal{R}e(\xi)|}$,

(b) $|\varphi\Phi(\xi,\varphi) - \varphi'\Phi(\xi,\varphi')| \leq M_8 \dfrac{K}{\nu} \mathrm{e}^{-|\mathcal{R}e(\xi)|}\, |\varphi' - \varphi|$.

The proofs of these three lemmas are given in Appendix 7.A.

7.3.3 Homogeneous equation

7.3.3.1 Basis of solutions. We start the study of the linear part of (7.25) by giving an explicit basis of solutions of (7.26).

Lemma 7.3.12. *Since* h *satisfies* (7.11), *its derivative* $p(\xi) = \dfrac{dh}{d\xi}(\xi)$ *satisfies* (7.26). *Moreover,* p *is antireversible and holomorphic in* $\mathcal{B}_\ell$, $p(\xi) \in \Pi$ *where* $\Pi = \{Y/\ A = B = 0\}$, p *maps* $\mathbb{R}$ *into* $\mathbb{R}^4$, $p(0) = (0, -\frac{1}{2}, 0, 0)$ *and there exists* M_9 *such that*

$$|p(\xi)| \leq M_9 \mathrm{e}^{-|\mathcal{R}e(\xi)|}$$

for every $\xi \in \mathcal{B}_\ell$. *Denote* $p(\xi) = (p_\alpha(\xi), p_\beta(\xi), 0, 0) = \left(\dfrac{d\alpha_h}{d\xi}, \dfrac{d^2\alpha_h}{d\xi^2}, 0, 0\right)$.

Lemma 7.3.13. *Let* q *be the solution of the homogeneous equation* (7.26) *such that* $q(0) = (1,0,0,0)$. *Then* $q(\xi) \in \Pi$. *Thus we denote* $q(\xi) = (q_\alpha(\xi), q_\beta(\xi), 0, 0)$. *The two components* q_α, q_β *are explicitly given by*

$$\begin{aligned} q_\alpha(\xi) &= \frac{1}{\cosh^2 \frac{1}{2}\xi} - \frac{15}{16}\frac{\xi \tanh \frac{1}{2}\xi}{\cosh^2 \frac{1}{2}\xi} - \frac{1}{8}\cosh\xi\ \tanh^2 \tfrac{1}{2}\xi - \tanh^2 \tfrac{1}{2}\xi, \\ q_\beta(\xi) &= \frac{dq_\alpha}{d\xi}(\xi). \end{aligned}$$

Moreover, q *is reversible and holomorphic in* $\mathcal{B}_\ell$, *maps* $\mathbb{R}$ *into* $\mathbb{R}^4$, *and there exists* M_{10} *such that*

$$|q(\xi)| \leq M_{10} \mathrm{e}^{|\mathcal{R}e(\xi)|}$$

for every $\xi \in \mathcal{B}_\ell$.

Proof. Observe that taking $A(0) = B(0) = 0$ for $q(0)$ then for all ξ in $\mathcal{B}_\ell$, $A(\xi) = B(\xi) = 0$ for $q(\xi)$. So we work with the reduced ξ-dependent linear system

$$\frac{dq}{d\xi} = L(\xi)q, \quad \text{with } L(\xi) = \begin{bmatrix} 0 & 1 \\ 1 - 3\alpha_h(\xi) & 0 \end{bmatrix}.$$

To compute q it suffices to observe that $\mathrm{Tr}(L(\xi)) = 0$. So the Wronskian $W(\xi)$ of the reduced system is constant. Since we already know a solution p of this system, we can compute the other one. The Wronskian equation reads

$$\begin{aligned}\forall \xi \in \mathcal{B}_\ell,\quad W(0) = -\tfrac{1}{2} = W(\xi) &= q_\alpha(\xi)p_\beta(\xi) - q_\beta(\xi)p_\alpha(\xi)\\ &= q_\alpha(\xi)p_\beta(\xi) - \frac{dq_\alpha}{d\xi}(\xi)p_\alpha(\xi).\end{aligned}$$

This equation can be explicitly solved to obtain q with the required properties. □

Finally, the System (7.26) admits two other solutions:

Lemma 7.3.14. *Let r_+, r_- be the two solutions of (7.26) satisfying $r_+(0) = (0,0,1,0)$, $r_-(0) = (0,0,0,1)$. Then r_+, r_- are given by*

$$\begin{aligned}r_+(\xi) &= (0,0,\ \ \cos\psi_\nu(\xi), \sin\psi_\nu(\xi)),\\ r_-(\xi) &= (0,0,-\sin\psi_\nu(\xi), \cos\psi_\nu(\xi))\end{aligned}$$

where $\psi_\nu(\xi) = (\omega/\nu + a\nu)\xi + 2b\nu\tanh\frac{1}{2}\xi$. Moreover, r_+, r_- are holomorphic in $\mathcal{B}_\ell$; $r_+(\xi)$, $r_-(\xi) \in \Pi^\perp$; r_+, r_- map $\mathbb{R}$ into $\mathbb{R}^4$; and there exists M_{11} such that

$$|r_+(t)| = |r_-(t)| = 1 \quad \textit{for every } t \in \mathbb{R}$$

$$|r_+(\xi)| = |r_-(\xi)| \leq M_{11}e^{\ell\omega/\nu} \textit{ for every } \nu \in]0,\nu_1] \textit{ and } \xi \in \mathcal{B}_\ell. \tag{7.27}$$

Finally, r_+ is reversible whereas r_- is antireversible.

Proof. We only verify (7.27). It suffices to observe that

$$\forall z \in \mathbb{C},\quad |\cos z| \leq e^{|\mathcal{I}m(z)|},\quad |\sin z| \leq e^{|\mathcal{I}m(z)|},$$

and that for every $\xi \in \mathcal{B}_\ell$, and $\nu \in]0,\nu_1]$,

$$\left|\mathcal{I}m\left((\frac{\omega}{\nu} + a\nu)\xi + 2b\nu\tanh\tfrac{1}{2}\xi\right)\right| \leq \omega\ell/\nu + |a|\nu_1\ell + 2b\nu_1\tan\tfrac{1}{2}\ell. \square$$

7.3.3.2 Dual basis for Equation (7.26). Let us denote by $\langle\cdot,\cdot\rangle$ the inner product in $\mathbb{C}^4$: $\langle Y, Y'\rangle := \overline{\alpha}\alpha' + \overline{\beta}\beta' + \overline{A}A' + \overline{B}B'$. We identify $\mathbb{C}^4$ and $(\mathbb{C}^4)^*$ by $Y \longrightarrow \langle Y,\cdot\rangle_* := \langle\overline{Y},\cdot\rangle$. With this identification, we denote by(p^*, q^*, r_+^*, r_-^*) the dual basis of (p,q,r_+,r_-) (see Definition 6.3.5). The dual basis is given by the following lemma:

Lemma 7.3.15.

(a) $p^* = (2q_\beta, -2q_\alpha, 0, 0)$, $|p^*(\xi)| \leq 2M_{10}e^{|\mathcal{R}e(\xi)|}$, $Sp^*(\xi) = -p^*(-\xi)$.
(b) $q^* = (-2p_\beta, 2p_\alpha, 0, 0)$, $|q^*(\xi)| \leq 2M_9e^{-|\mathcal{R}e(\xi)|}$, $Sq^*(\xi) = q^*(-\xi)$.
(c) $r_+^* = r_+$, $r_+^* \in \Pi^\perp$, $|r_+^*| = |r_+|$, $Sr_+^*(\xi) = r_+^*(-\xi)$.
(d) $r_-^* = r_-$, $r_-^* \in \Pi^\perp$, $|r_-^*| = |r_-|$, $Sr_-^*(\xi) = -r_-^*(-\xi)$.

Since the Wronskian is constant the proof follows. □

7.3.4 Integral equation

For all ξ in $\mathcal{B}_\ell$, (p, q, r_+, r_-) is a basis of $\mathbb{C}^4$. We write $v(\xi)$ in this basis:

$$v(\xi) = a(\xi)p(\xi) + b(\xi)q(\xi) + c(\xi)r_+(\xi) + d(\xi)r_-(\xi).$$

Then v is a solution of (7.25) if and only if

$$\begin{aligned} v(\xi) = & \left[a_0 + \int_{[0,\xi]} \langle p^*(s), g(v(s), \varphi, K, \nu, s)\rangle_* ds \right] p(\xi) \\ & + \left[b_0 + \int_{[0,\xi]} \langle q^*(s), g(v(s), \varphi, K, \nu, s)\rangle_* ds \right] q(\xi) \\ & + \left[c_0 + \int_{[0,\xi]} \langle r_+^*(s), g(v(s), \varphi, K, \nu, s)\rangle_* ds \right] r_+(\xi) \\ & + \left[d_0 + \int_{[0,\xi]} \langle r_-^*(s), g(v(s), \varphi, K, \nu, s)\rangle_* ds \right] r_-(\xi). \end{aligned}$$

Since we require v to be reversible, since p, r_- are antireversible, and since q, r_+ are reversible it follows that $a_0 = d_0 = 0$.

Now, since we require $v(\xi) \underset{|\mathcal{R}e(\xi)| \to +\infty}{\longrightarrow} 0$ and since

$$q^*(\xi) \underset{|\mathcal{R}e(\xi)| \to +\infty}{\longrightarrow} 0, \quad |r_+^*(\xi)| = |r_-^*(\xi)| \le M_{11} e^{\ell\omega/\nu},$$

we necessarily have

$$\begin{aligned} \langle q^*(\xi), v(\xi)\rangle_* & \underset{\mathcal{R}e(\xi) \to +\infty}{\longrightarrow} 0, \\ \langle r_+^*(\xi), v(\xi)\rangle_* & \underset{\mathcal{R}e(\xi) \to +\infty}{\longrightarrow} 0, \\ \langle r_-^*(\xi), v(\xi)\rangle_* & \underset{\mathcal{R}e(\xi) \to +\infty}{\longrightarrow} 0. \end{aligned}$$

This implies

$$\begin{aligned} b_0 & = -\int_0^{+\infty} \langle q^*(t), g(v(t), \varphi, K, \nu, t)\rangle_* dt, \\ c_0 & = -\int_0^{+\infty} \langle r_+^*(t), g(v(t), \varphi, K, \nu, t)\rangle_* dt, \\ d_0 & = -\int_0^{+\infty} \langle r_-^*(t), g(v(t), \varphi, K, \nu, t)\rangle_* dt. \end{aligned}$$

Moreover we choose a_0 which corresponds to a choice of the origin of the time t.

The constant d_0 has to satisfy two conditions. We shall use the phase shift φ to solve both. The estimates of N', R', Φ, ensure that the integrals converge for v in $\mathcal{BH}_\ell^\lambda|_{R,\mathbb{R}}(\delta\sqrt{\nu})$.

Finally to obtain a tractable form of the integral equation we prove

Lemma 7.3.16. *Let $[\xi, +\infty)$ be the half line in the complex plane:*

$$[\xi, +\infty) = \{t + i\eta / \ t \geq \mathcal{R}e\,(\xi)\,,\ \eta = \mathcal{I}m\,(\xi)\}.$$

For the choice of b_0, c_0, d_0 just made and for every $\nu \in]0, \nu_1]$, $\varphi \in [0, 6\pi\nu/\omega]$, $K \in [0, K_1]$, $v \in \mathcal{BH}_\ell^\lambda|_{R,\mathbb{R}}(\delta\sqrt{\nu})$, and $\xi \in \mathcal{B}_\ell$,

$$b_0 + \int_{[0,\xi]} \langle q^*(s), g(v(s), \varphi, K, \nu, s)\rangle_* ds = -\int_{[\xi,+\infty)} \langle q^*(s), g(v(s), \varphi, K, \nu, s)\rangle_* ds,$$

$$c_0 + \int_{[0,\xi]} \langle r_+^*(s), g(v(s), \varphi, K, \nu, s)\rangle_* ds = -\int_{[\xi,+\infty)} \langle r_+^*(s), g(v(s), \varphi, K, \nu, s)\rangle_* ds,$$

$$d_0 + \int_{[0,\xi]} \langle r_-^*(s), g(v(s), \varphi, K, \nu, s)\rangle_* ds = -\int_{[\xi,+\infty)} \langle r_-^*(s), g(v(s), \varphi, K, \nu, s)\rangle_* ds.$$

Proof. Since we work with holomorphic functions, the integrals along the path Γ_3 shown in Fig. 7.9 are equal zero.

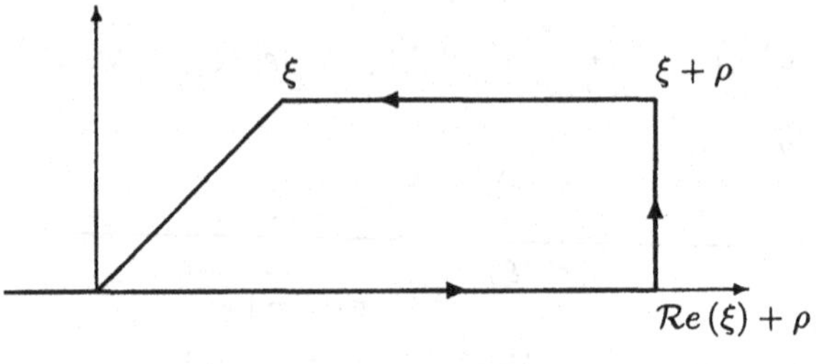

Fig. 7.9. Path Γ_3.

The estimates on N', R', Φ guarantee that for $y^* = q^*,\ r_+^*,\ r_-^*$,

$$I(\rho, y^*) = \int_{[\mathcal{R}e(\xi)+\rho, \xi+\rho]} \langle y^*(s), g(v(s), \varphi, K, \nu, s)\rangle_* ds \underset{\rho\to+\infty}{\longrightarrow} 0. \ \square$$

We can then conclude:

Proposition 7.3.17. *For every* $\nu \in]0, \nu_1]$, $\varphi \in [0, 6\pi\nu/\omega]$, $K \in [0, K_1]$, $v \in \mathcal{BH}_\ell^\lambda|_{R,\mathbb{R}}(\delta\sqrt{\nu})$ *and* $\xi \in \mathcal{B}_\ell$, *if* $v = \mathcal{F}_{K,\varphi,\nu}(v)$ *with*

$$\begin{aligned}(\mathcal{F}_{K,\varphi,\nu}(v))(\xi) &= \int_{[0,\xi]} \langle p^*(s),, g(v(s),\varphi,K,\nu,s)\rangle_* ds\ p(\xi) \\ &- \int_{[\xi,+\infty)} \langle q^*(s), g(v(s),\varphi,K,\nu,s)\rangle_* ds\ q(\xi) \\ &- \int_{[\xi,+\infty)} \langle r_+^*(s), g(v(s),\varphi,K,\nu,s)\rangle_* ds\ r_+(\xi) \\ &- \int_{[\xi,+\infty)} \langle r_-^*(s), g(v(s),\varphi,K,\nu,s)\rangle_* ds\ r_-(\xi)\end{aligned}$$

then v *is a solution of* (7.25).

7.3.5 Choice and control of the phase shift

Of course, we want to prove that $\mathcal{F}_{K,\varphi,\nu}$ is a contraction mapping in $\mathcal{BH}_\ell^\lambda|_{R,\mathbb{R}}(\delta\sqrt{\nu})$ for K, δ, ν small enough. But in order to lie in $\mathcal{BH}_\ell^\lambda|_{R,\mathbb{R}}(\delta\sqrt{\nu})$, $\mathcal{F}_{K,\varphi,\nu}(v)$ must be, in particular, reversible. For a given v in $\mathcal{BH}_\ell^\lambda|_{R,\mathbb{R}}(\delta\sqrt{\nu})$, and for a given phase shift φ, this is generally false.

In this subsection, we prove that, for all v in $\mathcal{BH}_\ell^\lambda|_{R,\mathbb{R}}(\delta\sqrt{\nu})$, there exists a phase shift $\varphi(v,K,\nu)$ such that $\mathcal{F}_{K,\varphi(v,K,\nu),\nu}(v)$ is reversible. Moreover, we have to control $\varphi(v,K,\nu)$ precisely, and we need an estimate of $|\varphi(v,K,\nu) - \varphi(v',K,\nu)|$ to be able to prove that $v \mapsto \mathcal{F}_{K,\varphi(v,K,\nu),\nu}(v)$ is a contraction in $\mathcal{BH}_\ell^\lambda|_{R,\mathbb{R}}(\delta\sqrt{\nu})$.

7.3.5.1 Choice of the phase shift. We start by proving that $\mathcal{F}_{K,\varphi,\nu}(v)$ is reversible if and only if "$d_0 = 0$".

Lemma 7.3.18. *For every* $\nu \in]0, \nu_1]$, $K \in [0, K_1]$, $\varphi \in [0, 6\pi\nu/\omega]$ *and* $v \in \mathcal{BH}_\ell^\lambda|_{R,\mathbb{R}}(\delta\sqrt{\nu})$, $\mathcal{F}_{K,\varphi,\nu}(v)$ *is reversible if and only if* $J(v,\varphi,K,\nu) = 0$, *where*

$$J(v,\varphi,K,\nu) = \int_0^{+\infty} \langle r_-^*(t), g(v(t),\varphi,K,\nu,t)\rangle_* dt. \qquad (7.28)$$

Proof. The symmetry properties of N, R given by (7.8) and the reversibility of h, $Y_{k,\nu}$ ensure that, for all v in $\mathcal{B}_\ell$,

$$\begin{aligned} N'(v(-\xi),\varphi,K,\nu,-\xi) &= -SN'(v(\xi),\varphi,K,\nu,\xi),\\ R'(v(-\xi),\varphi,K,\nu,-\xi) &= -SR'(v(\xi),\varphi,K,\nu,\xi),\\ \varphi\Phi(-\xi,\varphi,K,\nu) &= -S(\varphi\Phi(\xi,\varphi,K,\nu)). \end{aligned}$$

Now using the reversibility of q, r_+, q^*, r_+^* and the antireversibily of p, r_-, p^*, r_-^* we find that

$$\begin{aligned} G(\xi) = & \int_{[0,\xi]} \langle p^*(s), g(v(s),\varphi,K,\nu,s)\rangle_* ds \ \ p(\xi)\\ & + \int_{[0,\xi]} \langle r_-^*(s), g(v(s),\varphi,K,\nu,s)\rangle_* ds \ \ r_-(\xi)\\ & + \Big[b_0 + \int_{[0,\xi]} \langle q^*(s), g(v(s),\varphi,K,\nu,s)\rangle_* \, ds \Big] q(\xi)\\ & + \Big[c_0 + \int_{[0,\xi]} \langle r_+^*(s), g(v(s),\varphi,K,\nu,s)\rangle_* \, ds \Big] r_+(\xi) \end{aligned}$$

is reversible. Then $\mathcal{F}_{K,\varphi,\nu}(v) = G(\xi) + d_0\, r_-(\xi)$ holds. But r_- is antireversible. Thus $\mathcal{F}_{K,\varphi,\nu}(v)$ is reversible if and only if $d_0 = 0$. □

Now we rewrite (7.28) in a more tractable form. We split J into two parts, an explicit dominant part and an exponentially bounded remainder. To achieve this, we split g in two parts, $\varphi\Phi$ being explicitly known (it does not depend on v) gives the dominant part, and $N' + R'$, which depends on v can only be estimated. We conclude from the explicit form of r_-^* that the integral in (7.28) is an oscillatory integral. So, using the First Exponential Lemma 2.1.1 we obtain an exponentially small upper bound. This is why we have complexified the problem: we want to bound an oscillatory integral which depends on the solution of the differential equation v, and Lemma 2.1.1 only works with complex analytic functions. In this way we obtain

Proposition 7.3.19. *There exists M_{15} such that for every $\nu \in]0,\nu_1]$, $K \in [0,K_1]$, $\varphi \in [0, 6\pi\nu/\omega]$, and $v \in \mathcal{BH}_\ell^\lambda|_{\mathbb{R},\mathbb{R}}(\delta\sqrt{\nu})$,*

$$J(v,\varphi,K,\nu) = -\frac{\omega}{\nu} K \mathrm{e}^{-\ell\omega/\nu} \varphi \frac{\sin(\underline{\omega}_{k,\nu}\varphi - 2b\nu)}{\underline{\omega}_{k,\nu}\varphi - 2b\nu} + J_2(v,\varphi,K,\nu)$$

with $|J_2(v,\varphi,K,\nu)| \le M_{15}(\nu^2 + \nu K)\mathrm{e}^{-\ell\omega/\nu}$.

The detailed proof of this proposition is given in Appendix 7.B.

We are now able to prove

Proposition 7.3.20. *There exists $\nu_2 \le \nu_1$ such that for every $\nu \in]0, \nu_2]$, $K \in [8M_{15}\nu^2, K_1]$ and $v \in \mathcal{BH}_\ell^\lambda|_{R,R}(\delta\sqrt{\nu})$, there exists $\varphi(v, K, \nu)$ such that*

$$J(v, \varphi(v, K, \nu), K, \nu) = 0.$$

Moreover, $0 \le \frac{\pi\nu}{8\omega} \le \varphi(v, K, \nu) \le \frac{4\pi\nu}{\omega} \le \frac{6\pi\nu}{\omega}$.

Proof. Let $\varphi_0(K, \nu), \varphi_1(K, \nu)$ be such that

$$\underline{\omega}_{k,\nu}(K\mathrm{e}^{-\ell\omega/\nu})\varphi_0(K, \nu) - 2b\nu = \frac{\pi}{2}, \underline{\omega}_{k,\nu}(K\mathrm{e}^{-\ell\omega/\nu})\varphi_1(K, \nu) - 2b\nu = \frac{3}{2}\pi.$$

Recall that $\underline{\omega}_{k,\nu} = \frac{\omega}{\nu} + a\nu + \mathcal{O}(\nu^3) + \widehat{\omega}(K\mathrm{e}^{-\ell\omega/\nu}, \nu)$ and that $|\widehat{\omega}| \le C_\omega K_1 \nu \mathrm{e}^{-\ell\omega/\nu}$ by Theorem 7.1.4.
Then we can choose $\nu_2 \le \min(\nu_1, \sqrt{K_1/8M_{15}})$ such that

$$|2b\nu| \le \frac{\pi}{4} \le \frac{\pi}{2}, \quad \frac{\omega}{2\nu} \le \underline{\omega}_{k,\nu}(K\mathrm{e}^{-\ell\omega/\nu}) \le \frac{2\omega}{\nu},$$

for every $\nu \in]0, \nu_2]$, and $K \in [0, K_1]$. This choice of ν_2 ensures that

$$0 \le \frac{\pi\nu}{8\omega} \le \varphi_0(K, \nu) \le \frac{2\pi\nu}{\omega} \le \frac{6\pi\nu}{\omega},$$
$$0 \le \frac{\pi\nu}{2\omega} \le \varphi_1(K, \nu) \le \frac{4\pi\nu}{\omega} \le \frac{6\pi\nu}{\omega},$$

for every $\nu \in]0, \nu_2]$, and $K \in [0, K_1]$. Then using Proposition 7.3.19 and choosing ν_2 such that $M_{15}\nu_2 \le \frac{1}{8}$ we get

$$J(v, \varphi_0(K, \nu), K, \nu) \le (-\frac{1}{8}K + M_{15}\nu^2)\mathrm{e}^{-\ell\omega/\nu},$$
$$J(v, \varphi_1(K, \nu), K, \nu) \ge (\frac{5}{24}K - M_{15}\nu^2)\mathrm{e}^{-\ell\omega/\nu},$$

for every $\nu \in]0, \nu_2]$, $K \in [0, K_1]$, and $v \in \mathcal{BH}_\ell^\lambda|_{R,R}(\delta\sqrt{\nu})$.

Finally considering $\varphi \mapsto J(v, \varphi, K, \nu)$, which is continuous, and choosing K in $[8M_{15}\nu^2, K_1]$ we find that there exists $\varphi(v, K, \nu)$ satisfying $\varphi_0(K, \nu) \le \varphi(v, K, \nu) \le \varphi_1(K, \nu)$ such that $J(v, \varphi(v, K, \nu), K, \nu) = 0$. □

Remark 7.3.21. We note that $J(v, \varphi, K, \nu)$ vanishes for φ such that $\underline{\omega}_{k,\nu}\varphi - 2b\nu$ is near π. This is in perfect agreement with the phase shift found for the truncated System (7.14).

Remark 7.3.22. In the same way that we found $\varphi(v, K, \nu)$, we can find another $\varphi'(v, K, \nu)$ such that $J(v, \varphi', K, \nu)$ vanishes. It is such that $\frac{3}{2}\pi < \underline{\omega}_{k,\nu}\varphi' - 2b\nu < \frac{5}{2}\pi$. We note that $\underline{\omega}_{k,\nu}\varphi' - 2b\nu$ is near 2π. This leads to another reversible homoclinic solution as for the truncated System (7.13).

To control $\dfrac{\partial J}{\partial\varphi}(v,\varphi,K,\nu)$ we need an improvement of the previous proposition.

Proposition 7.3.23. *Let γ be in $]0\ \frac{\pi}{2}]$. Then there exists $\nu_2(\gamma)\le\nu_1$ such that for every $\nu\in]0,\nu_2(\gamma)]$, $K\in[\dfrac{8M_{15}\nu^2}{\sin\gamma},K_1]$, and $v\in\mathcal{BH}_\ell^\lambda|_{R,R}(\delta\sqrt{\nu})$, there exists $\varphi(v,K,\nu)$ such that*

$$J(v,\varphi(v,K,\nu),K,\nu)=0.$$

Moreover,

$$0\le\frac{\pi\nu}{8\omega}\le\varphi(v,K,\nu)\le\frac{4\pi\nu}{\omega}\le\frac{6\pi\nu}{\omega},$$
$$\pi-\gamma\le\underline{\omega}_{k,\nu}\varphi(v,K,\nu)-2b\nu\le\pi+\gamma.$$

Proposition 7.3.20 is the particular case of Proposition 7.3.23 with $\gamma=\frac{\pi}{2}$. The proof of Proposition 7.3.23 is the same as that of Proposition 7.3.20 except that here we work with $\varphi_0^\gamma(K,\nu),\varphi_1^\gamma(K,\nu)$ given by

$$\begin{aligned}\underline{\omega}_{k,\nu}(K\mathrm{e}^{-\ell\omega/\nu})\varphi_0^\gamma(K,\nu)-2b\nu &= \pi-\gamma,\\ \underline{\omega}_{k,\nu}(K\mathrm{e}^{-\ell\omega/\nu})\varphi_1^\gamma(K,\nu)-2b\nu &= \pi+\gamma,\end{aligned}$$

and we choose $\nu_2(\gamma)\le\min(\nu_1,\sqrt{\dfrac{K_1\sin\gamma}{8M_{15}}})$ such that

$$|2b\nu|\le\min(\frac{\pi}{4},\gamma),\ \frac{\omega}{2\nu}\le\underline{\omega}_{k,\nu}(K\mathrm{e}^{-\ell\omega/\nu})\le\frac{2\omega}{\nu},$$

for every $\nu\in]0,\nu_2(\gamma)]$, $K\in[0,K_1]$.

7.3.5.2 Estimates of $\varphi(v,K,\nu)$. As explained in the introduction of this subsection we want to control φ and its first derivative. For that purpose, we want to use the Implicit Function Theorem. Thus we need to compute $\dfrac{\partial J}{\partial\varphi}(v,\varphi,K,\nu)$. The following proposition gives a tractable form of $\dfrac{\partial J}{\partial\varphi}(v,\varphi,K,\nu)$.

Proposition 7.3.24. *There exists M_{17} such that for every $\nu\in]0,\nu_1]$, $K\in[0,K_1]$, $\varphi\in[0,6\pi\nu/\omega]$, and $v\in\mathcal{BH}_\ell^\lambda|_{R,R}(\delta\sqrt{\nu})$,*

$$\begin{aligned}\frac{\partial J}{\partial\varphi}(v,\varphi,K,\nu)=&\ \frac{\omega^2}{\nu^2}K\mathrm{e}^{-\frac{\ell\omega}{\nu}}\varphi\left[-\frac{\cos(\underline{\omega}_{k,\nu}\varphi-2b\nu)}{\underline{\omega}_{k,\nu}\varphi-2b\nu}+\frac{\sin(\underline{\omega}_{k,\nu}\varphi-2b\nu)}{(\underline{\omega}_{k,\nu}\varphi-2b\nu)^2}\right]\\ &-\frac{\omega}{\nu}K\mathrm{e}^{-\ell\omega/\nu}\frac{\sin(\underline{\omega}_{k,\nu}\varphi-2b\nu)}{\underline{\omega}_{k,\nu}\varphi-2b\nu}+J_4(v,\varphi,K,\nu),\end{aligned}$$

with $|J_4(v,\varphi,K,\nu)|\le M_{17}K\mathrm{e}^{-\ell\omega/\nu}$.

The proof of this proposition is given in Appendix 7.C. We can now prove

Proposition 7.3.25. *There exist M_{18}, $\nu_3 \le \nu_2$ such that for every $\nu \in]0, \nu_3]$, $K \in [M_{18}\nu^2, K_1]$, and $v \in \mathcal{BH}_\ell^\lambda|_{R,\mathbb{R}}(\delta\sqrt{\nu})$,*

$$J(v, \varphi(v, K, \nu), K, \nu) = 0, \; \frac{\partial J}{\partial \varphi}(v, \varphi(v, K, \nu), K, \nu) \ge \frac{\omega}{32\nu} K e^{-\ell\omega/\nu} > 0.$$

Proof. As in Propositions 7.3.19, 7.3.24 we denote

$$H(0, y) = -\frac{\cos y}{y} + \frac{\sin y}{y^2}, \; G(0, y) = -\frac{\sin y}{y}.$$

Observing that $H(0, \pi) = \frac{1}{\pi}$ and that $G(0, \pi) = 0$ we choose $\gamma \in]0, \frac{\pi}{2}]$ such that for every $y \in [\pi - \gamma, \pi + \gamma]$,

$$H(0, y) > 0, \; \tfrac{\pi}{8} H(0, y) + G(0, y) > \tfrac{1}{16}.$$

Now set $M_{18} = 8M_{15}/\sin\gamma$ and $\nu_3' = \nu_2(\gamma)$. Proposition 7.3.23 ensures that for every $\nu \in]0, \nu_3']$, $K \in [M_{18}\nu^2, K_1]$ and $v \in \mathcal{BH}_\ell^\lambda|_{R,\mathbb{R}}(\delta\sqrt{\nu})$, there exists $\varphi(v, K, \nu)$ such that

$$J(v, \varphi(v, K, \nu), K, \nu) = 0,$$

with $\varphi(v, K, \nu) \ge \pi\nu/8\omega$ and $\pi - \gamma \le \underline{\omega}_{k,\nu}\varphi(v, K, \nu) - 2b\nu \le \pi + \gamma$.

In the following inequalities we denote $y(v, K, \nu) = \underline{\omega}_{k,\nu}\varphi(v, K, \nu) - 2b\nu$. Using Proposition 7.3.24 and observing that $y(v, K, \nu) \in [\pi - \gamma, \pi + \gamma]$ we get

$$\begin{aligned}
\frac{\partial J}{\partial \varphi}(v, \varphi(v, K, \nu), K, \nu) \ge\; & \frac{\omega^2}{\nu^2} K e^{-\ell\omega/\nu} \frac{\pi\nu}{8\omega} \left[-\frac{\cos y(v, K, \nu)}{y(v, K, \nu)} + \frac{\sin y(v, K, \nu)}{y^2(v, K, \nu)} \right] \\
& -\frac{\omega}{\nu} K e^{-\ell\omega/\nu} \frac{\sin y(v, K, \nu)}{y(v, K, \nu)} - M_{17} K e^{-\ell\omega/\nu} \\
\ge\; & K e^{-\ell\omega/\nu} \left(\frac{\omega}{\nu} \left[\tfrac{\pi}{8} H(0, y(v, K, \nu)) \right.\right. \\
& \left.\left. + G(0, y(v, K, \nu)) \right] - M_{17} \right)
\end{aligned}$$

And thus,

$$\frac{\partial J}{\partial \varphi}(v, \varphi(v, K, \nu), K, \nu) \ge K e^{-\ell\omega/\nu} \left(\frac{\omega}{16\nu} - M_{17} \right).$$

To conclude, we choose $\nu_3 \le \nu_3'$ such that $(\frac{\omega}{16\nu} - M_{17}) \ge \frac{\omega}{32\nu}$, for $\nu \in]0, \nu_3]$.
□

Now applying the Implicit Function Theorem we get

Proposition 7.3.26. *For every* $\nu \in]0, \nu_3]$, $K \in [M_{18}\nu^2, K_1]$ *and* $v, v' \in \mathcal{BH}_\ell^\lambda|_{R,\mathbb{R}}(\delta\sqrt{\nu})$, *the function* $\varphi(v, K, \nu)$ *is in* C^1 *with respect to* (v, K) *and*

$$D_v\varphi(v, K, \nu).v' = -\frac{D_vJ(v, \varphi(v, K, \nu), K, \nu).v'}{\dfrac{\partial J}{\partial \varphi}(v, \varphi(v, K, \nu), K, \nu)}.$$

Moreover, there exists M_{20} *such that for every* $\nu \in]0, \nu_3]$, $K \in [M_{18}\nu^2, K_1]$ *and* $v, v' \in \mathcal{BH}_\ell^\lambda|_{R,\mathbb{R}}(\delta\sqrt{\nu})$ *the following inequalities hold:*

(a) $|D_v\varphi(v, K, \nu).v'| \le M_{20}\, |v'|_{\mathcal{H}_\ell^\lambda} \dfrac{\nu^2}{K}(|v|_{\mathcal{H}_\ell^\lambda} + K + \nu)$,

(b) $|\varphi(v, K, \nu) - \varphi(v', K, \nu)| \le M_{20}\dfrac{\nu^2}{K}(\delta\sqrt{\nu} + K + \nu)\, |v - v'|_{\mathcal{H}_\ell^\lambda}$.

The proof of this proposition is given in Appendix 7.D.

We can sum up this subsection devoted to the choice of the phase shift by the following proposition:

Proposition 7.3.27. *There exist* M_{18} *and* $\nu_3 \le \nu_1$ *such that for every* $\nu \in]0, \nu_3]$, $K \in [M_{18}\nu^2, K_1]$ *and* $v \in \mathcal{BH}_\ell^\lambda|_{R,\mathbb{R}}(\delta\sqrt{\nu})$, *there exists* $\varphi(v, K, \nu)$ *such that* $\mathcal{F}_{\varphi(v,K,\nu),K,\nu}(v)$ *is reversible. Moreover, there exists* M_{20} *such that for every* $\nu \in]0, \nu_3]$, $K \in [M_{18}\nu^2, K_1]$ *and* $v, v' \in \mathcal{BH}_\ell^\lambda|_{R,\mathbb{R}}(\delta\sqrt{\nu})$, *the following inequalities hold:*

$$0 \le \frac{\pi\nu}{8\omega} \le \varphi(v, K, \nu) \le \frac{4\pi\nu}{\omega} \le \frac{6\pi\nu}{\omega}, \tag{7.29}$$

$$|\varphi(v, K, \nu) - \varphi(v', K, \nu)| \le M_{20}\frac{\nu^2}{K}(\delta\sqrt{\nu} + K + \nu)\, |v - v'|_{\mathcal{H}_\ell^\lambda}. \tag{7.30}$$

In section 7.3.4 we have rewritten (7.25) as an integral equation $v = \mathcal{F}_{K,\varphi,\nu}(v)$. Since this transformation is an integration, we had four free constants (a_0, b_0, c_0, d_0). But we want v to be reversible and to satisfy $|v|_{\mathcal{H}_\ell^\lambda} \le \delta\sqrt{\nu}$, which implies that $v \underset{\mathcal{R}e(\xi)\to+\infty}{\longrightarrow} 0$. These requirements determined a_0, b_0, c_0 uniquely, but two a priori different values for d_0. (One ensures that $\mathcal{F}(v)$ is bounded and the other ensures that $\mathcal{F}(v)$ is reversible). We have chosen d_0 so that $\mathcal{F}(v)$ is bounded. Then we have an extra scalar equation $J(v, \varphi, K, \nu) = 0$, which ensures that $\mathcal{F}(v)$ is reversible. We have solved this equation for each v in $\mathcal{BH}_\ell^\lambda|_{R,\mathbb{R}}(\delta\sqrt{\nu})$ by an appropriate choice of the phase shift φ. Observe that we could not solve this equation for all (K, ν). We have solved the equation for $K \ge M_{18}\nu^2$, i.e., for $k \ge M_{18}\nu^2 e^{-\ell\omega/\nu}$. Whereas for each ν there exist periodic solutions of arbitrary small size $(Y_{k,\nu}(t))_{k\ge 0}$,

$|Y_{k,\nu}| \le M_2 k$, here the size of the periodic solution has to be chosen larger than $M_{18}\nu^2 e^{-\ell\omega/\nu}$ to meet the extra condition.

7.3.6 Fixed point

Denote

$$\widehat{\mathcal{F}}_{K,\nu}(v) = \mathcal{F}_{K,\varphi(v,K,\nu),\nu}(v) \text{ and } \hat{g}(v(\xi), K, \nu, \xi) = g(v(\xi), \varphi(v, K, \nu), K, \nu, \xi).$$

Our aim is now to prove that $\widehat{\mathcal{F}}_{K,\nu}$ is a contraction mapping in $\mathcal{BH}_\ell^\lambda|_{R,\mathbb{R}}(\delta\sqrt{\nu})$ for K, ν, δ small enough and satisfying $K \ge M_{18}\nu^2$.

Using the estimates of N', R', $\varphi\Phi$, $\varphi(v, K, \nu)$ given in Lemmas 7.3.9, 7.3.10, 7.3.11 and in Proposition 7.3.27 we estimate $\hat{g}$:

Lemma 7.3.28. *There exists M_{21} such that for every $\nu \in]0, \nu_3]$, $K \in [M_{18}\nu^2, K_1]$, $v, v' \in \mathcal{BH}_\ell^\lambda|_{R,\mathbb{R}}(\delta\sqrt{\nu})$, and $\xi \in \mathcal{B}_\ell$, the following estimates hold:*

$$|\hat{g}(v(\xi), K, \nu, \xi)| \le M_{21}[|v(\xi)|^2 + (K + \nu^2)\ |(h + v)(\xi)| + K e^{-|\mathcal{R}e(\xi)|}],$$

$$\begin{aligned} &|\hat{g}(v(\xi), K, \nu, \xi) - \hat{g}(v'(\xi), K, \nu, \xi)| \\ &\quad \le\ M_{21}\Big[\ \big(|v(\xi)| + |v'(\xi)| + K + \nu^2\big)\ |v(\xi) - v'(\xi)| \\ &\qquad + \nu(\delta\sqrt{\nu} + K + \nu)\ \big(|h(\xi)| + |v(\xi)| + e^{-|\mathcal{R}e(\xi)|}\big)\ |v' - v|_{\mathcal{H}_\ell^\lambda}\Big]. \end{aligned}$$

We must point out here that in Lemma 7.3.14 we checked that for every $\xi \in \mathcal{B}_\ell$, $|r_\pm(\xi)| \le M_{11}e^{\ell\omega/\nu}$ holds. So, r_+, r_- are exponentially large in $\mathcal{B}_\ell$. This may lead to a very bad estimate of $\left|\widehat{\mathcal{F}}_{K,\nu}(v)\right|_{\mathcal{H}_\ell^\lambda}$ in Proposition 7.3.17. The following lemma shows that by grouping together the two integrals involving r_+, r_-, i.e., by using the convolution kernel, the exponentially large terms and the exponentially small terms cancel out.

Lemma 7.3.29. *There exists M_{22} such that for all functions*

$$\begin{aligned} f:\ \ \mathcal{B}_\ell \times]0, \nu_3] \times [M_{18}\nu^2, K_1] &\longrightarrow \mathbb{C}^4 \\ (\xi, \nu, K) &\longrightarrow (\cdot, \cdot, f_A(\xi, \nu, K), f_B(\xi, \nu, K)) \end{aligned}$$

satisfying $|f_A|_{\mathcal{H}_\ell^\lambda} < +\infty$, $|f_B|_{\mathcal{H}_\ell^\lambda} < +\infty$, the inequality

. . . .

$$\left| \int_{[\xi,+\infty)} \left[\langle r_+^*(s), f(s,\nu,K) \rangle_* \, r_+(\xi) + \langle r_-^*(s), f(s,\nu,K) \rangle_* r_-(\xi) \right] ds \right| \leq M_{22} \int_{[\xi,+\infty)} \left(|f_A(s,\nu,K)| + |f_B(s,\nu,K)| \right) ds$$

holds for every $\nu \in]0,\nu_3]$, $K \in [M_{18}\nu^2, K_1]$, *and* $\xi \in \mathcal{B}_\ell$.

See the proof of this lemma, given in Appendix 7.E, to enjoy the marvelous cancellation of exploding terms. We have now enough material to prove

Proposition 7.3.30. *For every* $\nu \in]0,\nu_3]$, $K \in [M_{18}\nu^2, K_1]$ *and* $v \in \mathcal{BH}_\ell^\lambda|_{R,\mathbb{R}}(\delta\sqrt{\nu})$, $\widehat{\mathcal{F}}_{K,\nu}(v)$ *is holomorphic in* $\mathcal{B}_\ell$, *is reversible, and maps* $\mathbb{R}$ *in* $\mathbb{R}^4$. *Moreover, there exists* M_{23} *such that for every* $\nu \in]0,\nu_3]$, $K \in [M_{18}\nu^2, K_1]$ *and* $v, v' \in \mathcal{BH}_\ell^\lambda|_{R,\mathbb{R}}(\delta\sqrt{\nu})$, $\widehat{\mathcal{F}}_{K,\nu}$ *satisfies*

$$\left| \widehat{\mathcal{F}}_{K,\nu}(v) \right|_{\mathcal{H}_\ell^\lambda} \leq M_{23}(|v|^2_{\mathcal{H}_\ell^\lambda} + (K+\nu^2)(|v|_{\mathcal{H}_\ell^\lambda} + 1) + K), \tag{7.31}$$

$$\begin{aligned} &\left| \widehat{\mathcal{F}}_{K,\nu}(v) - \widehat{\mathcal{F}}_{K,\nu}(v') \right|_{\mathcal{H}_\ell^\lambda} \\ &\leq M_{23} \big[\, |v|_{\mathcal{H}_\ell^\lambda} + |v'|_{\mathcal{H}_\ell^\lambda} + K + \nu^2 \\ &\qquad + \nu(\delta\sqrt{\nu} + \nu + K)\,(|v|_{\mathcal{H}_\ell^\lambda} + 1)\,\big]\; |v - v'|_{\mathcal{H}_\ell^\lambda}. \end{aligned} \tag{7.32}$$

The proof of Equations (7.31), (7.32) is a simple application of Lemma 7.3.28, which gives the estimates of $\hat{g}$, Lemmas 7.3.12, 7.3.13, 7.3.15 which respectively give the estimates of p, q, p^*, q^*, and of Lemma 7.3.29.

Proof of Theorem 7.1.7 We conclude this section by the proof of Theorem 7.1.7 which ensures the existence of reversible homoclinic connections to exponentially small periodic orbits of the form

$$Y(t) = h(t) + v(t) + Y_{k,\nu}(t + \varphi \tanh \tfrac{1}{2}t)$$

where v is found as fixed point of $\widehat{\mathcal{F}}_{K,\nu}$ and where $\varphi = \varphi(v,K,\nu)$ has been determined in the previous subsection.

Proposition 7.3.30 ensures that $\widehat{\mathcal{F}}_{K,\nu}$ is a contraction mapping from $\mathcal{BH}_\ell^\lambda|_{R,\mathbb{R}}(\delta\sqrt{\nu})$ to $\mathcal{BH}_\ell^\lambda|_{R,\mathbb{R}}(\delta\sqrt{\nu})$ if and only if

$$\nu \in]0, \nu_3], \; K \in [M_{18}\nu^2, K_1],$$
$$M_{23}[\delta^2\nu + (K+\nu^2)\,(1+\delta\sqrt{\nu}) + K] \leq \delta\sqrt{\nu},$$
$$M_{23}[2\delta\sqrt{\nu} + K + \nu^2 + \nu(\delta\sqrt{\nu} + K + \nu)\,(1+\delta\sqrt{\nu})] < 1.$$

We then can choose $K = M_{18}\nu^2$,

$$\delta = \frac{2M_{23}[2(K+\nu^2)+K]}{\sqrt{\nu}} = 2M_{23}[3M_{18}+1]\nu^{3/2}$$

and $\nu_4 \leq \nu_3$ such that for all ν in $]0, \nu_4]$,

$$M_{23}\delta\sqrt{\nu} \leq \tfrac{1}{2}, \; M_{23}[2\delta\sqrt{\nu} + K + \nu^2 + \nu(\delta\sqrt{\nu} + K + \nu)\,(1+\delta\sqrt{\nu})] < 1.$$

This ensures that for all ℓ, $0 < \ell < \pi$ and all λ, $0 < \lambda < 1$ there exist M, ν_4 such that for all ν in $]0, \nu_4]$ the full System (7.5) admits two reversible solutions of the form

$$Y(t) = h(t) + v(t) + Y_{k,\nu}(t + \varphi \tanh \tfrac{1}{2}t)$$

where

$$|h(t)| \leq M\mathrm{e}^{-|t|}, \qquad |v(t)| \leq M\nu^2\mathrm{e}^{-\lambda|t|} \qquad \text{for } t \in \mathbb{R},$$

and where $Y_{k,\nu}$ is a periodic solution of the full system satisfying

$$\forall t \in \mathbb{R}, \; |Y_{k,\nu}(t + \varphi \tanh \tfrac{1}{2}t)| \leq M\nu^2\mathrm{e}^{-\ell\omega/\nu}.$$

This completes the proof of Theorem 2.1.1 □

Remark 7.3.31. For obtaining the reversible homoclinic connections to exponentially small periodic orbits we have first chosen an exponentially small size for the periodic orbit ($k = K\mathrm{e}^{-\omega\ell/\nu}$), and then we have computed the phase shift φ at infinity. It is also possible to first fix the phase shift at infinity, and then to obtain an exponentially small estimate of the size of the periodic orbits. Both approaches give the same results. The last strategy was used by Beale in [Be91] and by Sun in [Su98], for the water wave problem where the resonance ($0^{2+}\mathrm{i}\omega$) occurs for Froude number close to 1 and Bond number less than $\frac{1}{3}$ (see Chapter 8 for more details about the water wave problem).

7.3.7 Proof of Theorem 7.1.18

The last subsection is devoted to the proof of Theorem 7.1.18. For proving Theorem 7.1.7, we have first fixed $\ell \in]0, \pi[$ and thus all the constants M_i, ν_i and K_i occurring in the proof, depend on ℓ. This last Theorem was deduced from proposition 7.3.30 by choosing $K = M_{18}\nu^2$ and ν sufficiently small so that $\mathcal{F}_{K,\nu}$ is a contraction mapping in $\mathcal{BH}_\ell^\lambda|_{R,\mathbb{R}}(d)$ with $d = 2M_{23}[3M_{18}+1]\nu^{\frac{3}{2}}$.

Exactly in the same way, we can obtain the existence of reversible homoclinic connections to $Y_{k,\nu}$ for every $k = K\mathrm{e}^{-\ell\omega/\nu}$ with $K \in [M_{18}\nu^2, M_{18}\nu^{\frac{3}{4}}]$

provided that $\nu \in]0, \nu_4'(\ell)]$. In every cases the perturbation term v lies in $\mathcal{BH}_\ell^\lambda|_{R,\mathbb{R}}(d)$ with $d \leq 2M_{23}[3M_{18}+1]\nu^{\frac{1}{4}} \leq 1$.

Looking carefully at the proof of Proposition 7.3.30 we can obtain a similar proposition which is valid uniformly for any $\ell' \in]0,\ell]$, i.e. the constant ν_3, M_{18}, M_{23}, K_1 are the same for all $\ell' \in]0,\ell]$. So, grouping together the existence result obtained for each $\ell' \in]0,\ell]$ and observing that

$$\bigcup_{0<\ell'<\ell} [M_{18}\nu^2 e^{-\ell'\omega/\nu}, M_{18}\nu^{\frac{3}{4}} e^{-\ell'\omega/\nu}] = [M_{18}\nu^2 e^{-\ell\omega/\nu}, M_{18}\nu^{\frac{3}{4}}[$$

we obtain following corollary

Corollary 7.3.32. *There exists $\mathcal{M}_1$, $\mathcal{M}_2$ such that for every ℓ, $0<\ell<\pi$ and every λ, $0<\lambda<1$ there exist $M_{18}'(\ell)$, $\nu_4'(\ell)$ such that for all $\nu \in]0,\nu_4'(\ell)]$ and all k in the interval $[M_{18}(\ell)\nu^2 e^{-\ell\omega/\nu}, M_{18}(\ell)\nu^{\frac{3}{4}}[$, the full system (7.5) admits two reversible solutions of the form*

$$Y(t) = h(t) + v(t) + Y_{k,\nu}(t + \varphi \tanh \tfrac{1}{2}t)$$

with

$$|h(t)| + |v(t)| \leq \mathcal{M}_1 e^{-\lambda|t|}, \qquad |Y_{k,\nu}(t)| \leq \mathcal{M}_2 k \qquad \text{for } t \in \mathbb{R},$$

where $Y_{k,\nu}$ is the periodic solution of the full system obtained in Theorem 7.1.4.

Remark 7.3.33. $\mathcal{M}_1 = \sup_{t\in\mathbb{R}} |h(t)|e^{-|t|} + 1 < +\infty$ and $\mathcal{M}_2$ can be deduced from Theorem 7.1.4.

Finally, for "$\ell' = 0$", i.e. on the real axis, we can make a similar proof with continuous function on the real line instead of holomorphic function in $\mathcal{B}_\ell$, i.e. looking for v in

$$\mathcal{C}_\lambda^0(\mathbb{R})|_{\mathbb{R}} = \left\{ v : \mathbb{R} \to \mathbb{R}^4, \text{ of class } C^0, \text{ reversible}, \sup_{t\in\mathbb{R}} \left(|v(t)|e^{\lambda|t|} \right) < +\infty \right\}.$$

This way, we obtain the existence of reversible homoclinic connections of the form

$$Y(t) = h(t) + v(t) + Y_{k,\nu}(t + \varphi \tanh \tfrac{1}{2}t)$$

provided that $\nu \in]0, \nu_4'']$ and that $k \in [M\nu, K]$. Finally choosing $\nu_5(\ell) \leq \min(\nu_4'(\ell), \nu_4'')$ such that $M\nu_5 < M_{18}(\ell)\nu_5^{\frac{3}{4}}$ we finally obtain the existence of reversible connections homoclinic to $Y_{k,\nu}$ for every $\nu \in]0, \nu_5(\ell)]$ and every $k \in [M_{18}(\ell)\nu^2 e^{-\ell\omega/\nu}, K]$. This completes the proof of Theorem 7.1.18. □

7.4 Proof of the generic non persistence of homoclinic connections to 0

7.4.1 Introduction

7.4.1.1 Strategy of proof. This section is devoted to the question of the persistence of homoclinic connections to 0 for the full system

$$\frac{dY}{dt} = N(Y,\nu) + R(Y,\nu)$$

where N, R are defined in (7.5), (7.6). For studying this question we introduce in the previous equation an additional parameter $\rho \in [0,1]$

$$\frac{dY}{dt} = N(Y,\nu) + \rho R(Y,\nu) \tag{7.33}$$

For $\rho = 0$, we get the truncated system (7.11) which admits a unique (up to time shift) homoclinic connection h to 0. This homoclinic connection h is reversible and explicitly given by

$$h(t) := (\alpha_h(t), \beta_h(t), 0, 0) = \left(\frac{1}{\cosh^2 \frac{1}{2}t}, \frac{-\tanh \frac{1}{2}t}{\cosh^2 \frac{1}{2}t}, 0, 0 \right).$$

Now, we would like to understand what happens for $\rho \neq 0$ and in particular for $\rho = 1$ which corresponds to the full system (7.5). For determining the existence of reversible homoclinic connection to 0, a classical idea is to use an approach of Melnikov type, which can be summarized as follows:

If the system (7.33) admits a homoclinic connection Y to 0, then the perturbation term $v := Y - h$ satisfies the equation

$$\frac{dv}{dt} - DN(h).v = g(v,t,\rho,\nu) \tag{7.34}$$

with

$$g(v,t,\nu) = Q(v) + \rho R(h+v,\nu)$$

where

$$Q(v,\nu) = N(h+v,\nu) - N(h,\nu) - DN(h,\nu).v.$$

Then, the crucial point of the analysis, is the following : for an *antireversible* function f which *tends to 0 exponentially at infinity*, there exists a *reversible* function u which *tends to 0 exponentially at infinity* and which satisfies

$$\frac{du}{dt} - DN(h,\nu).u = f$$

if and only if the function f satisfies the solvability condition

$$\int_0^{+\infty} \langle r_-^*(t), f(t)\rangle dt = 0$$

where the dual vector r_-^* is given by Lemmas 7.3.15, 7.3.14

$$r_-^*(t) := (0, 0, -\sin\psi_\nu(t), \cos\psi_\nu(t)) \text{ with } \psi_\nu(t) = \left(\tfrac{\omega}{\nu} + a\nu\right)t + 2b\nu\tanh(\tfrac{1}{2}t).$$

This solvability condition is the typical one which occurs when the linear homogeneous system admits a pair of purely imaginary eigenvalues (see Section 6.4). Moreover, observe that the above integral, is an *oscillatory integral* since the dual vector rotates with a frequency of order ω/ν where ν is a small parameter.

Thus, if the full system (7.33) admits a reversible homoclinic connection Y to 0, then Y necessarily satisfies

$$I(Y, \rho, \nu) := \int_0^{+\infty} \langle r_-^*(t), Q(v, \nu) + \rho R(h(t) + v(t), \nu)\rangle dt = 0.$$

where $v = Y - h$. The usual way to study such a solvability condition is to split the integral in two parts $I(Y, \rho, \nu) = M_e(\rho, \nu) + J(\rho, \nu)$ where $M_e(\rho, \nu)$ which only depends on h is explicitly known

$$\begin{aligned} M_e(\rho, \nu) &= \rho \int_0^{+\infty} \langle r^*(t), R(h(t), \nu)\rangle dt \\ &= \frac{\rho}{2}\int_{-\infty}^{+\infty} e^{i((\frac{\omega}{\nu}+a\nu)t+2b\nu\tanh(\frac{1}{2}t))}\,(R_B(h(t), \nu) + iR_A(h(t), \nu))dt \end{aligned}$$

whereas $J(\rho, \nu)$ involves v which is only characterized by the fact that it is a *reversible* solution of (7.33) which *tends to 0 at infinity*. The first part $M_e(\rho, \nu)$ is called the Melnikov function.

When performing such a splitting of the solvability condition, one hopes that the Melnikov function $M_e(\rho, \nu)$ which comes from the leading part of Y on $\mathbb{R}$, is the leading part of the integral $I(Y, \rho, \nu)$. So, we first compute explicitly $M_e(\rho, \nu)$: expanding in power series the essential singularity in the oscillatory term and computing the residues, we get

$$M_e(\rho, \nu) = \frac{\rho}{\nu^2} e^{-\frac{\omega\pi}{\nu}} (\Lambda_1 + \underset{\nu\to 0}{\mathcal{O}}(\nu)).$$

Then, classical perturbation theory enables us to bound J which involves the not explicitly known perturbation term v. This leads to

$$I(Y, \rho, \nu) = \frac{\rho}{\nu^2} e^{-\frac{\omega\pi}{\nu}} (\Lambda_1 + \underset{\nu\to 0}{\mathcal{O}}(\nu)) + \underset{\nu\to 0}{\mathcal{O}}(\rho^2\nu^2).$$

However, such an estimates of $I(Y, \rho, \nu)$ is not accurate enough to determine whether $I(Y, \rho, \nu)$ vanishes or not, since the upper bound of the not explicitly known part J is far bigger than the expected leading part $M_e(\rho, \nu)$ which is exponentially small. This exponential smallness comes from the fact that the solvability condition is given by an oscillatory integral.

The Exponential Tools given in Chapter 2 were precisely developed to study such solvability conditions given by oscillatory integrals. The Third Mono-frequency Exponential Lemma 2.1.21, coupled with the strategy proposed in Subsection 2.1.4 enables us to obtain equivalents of oscillatory integrals involving solutions of nonlinear differential equations.

Using these Exponential tools, we show in this section that if the full system (7.33) admits a reversible homoclinic connection Y to 0 then

$$I(Y,\rho,\nu) = \frac{1}{\nu^2} e^{-\frac{\omega\pi}{\nu}} (\Lambda(\rho) + \underset{\nu\to 0}{O}(\nu^{\frac{1}{4}})),$$

where $\rho \mapsto \Lambda(\rho)$ is a real analytic function on $[0,1]$ such that $\Lambda(\rho) = \Lambda_1\rho + \underset{\rho\to 0}{O}(\rho^2)$. Such an estimate of $I(Y,\rho,\nu)$ ensures that if $\Lambda(\rho) \neq 0$ (which is generically the case), then for ν small enough, $I(Y,\rho,\nu)$ is exponentially small but does not vanish and thus there is no reversible homoclinic connection to 0.

Observe that for a fixed $\rho > 0$, the Melnikov function $M_e(\rho,\nu)$ is not the leading part of the integral $I(Y,\rho,\nu)$ for ν small. The homoclinic connection of the truncated system h and the perturbation term v contributes to the size of $I(Y,\rho,\nu)$ at the same order in ν.

7.4.1.2 Schedule of the proof. In what follows, we will be interested in the holomorphy of the solutions with respect to ρ and we do not want $\rho = 1$ to be on the edge of the disk. So, we work with $\rho \in [0, \rho_s[$ (instead of $[0,1]$) where $\rho_s > 1$ is fixed.

The exact implementation of the above strategy has to be split in several steps:

Step 1. Real stable manifold of 0 and solvability condition for reversible homoclinic connections. In subsection 7.4.2, we prove the following proposition which gives a parameterization of the stable manifold of 0 such that if the full system (7.33) admits a reversible homoclinic connection to 0 then $Y(t) = Y_s(t)$ for $t \geq -1$. Moreover, this proposition ensures that any reversible homoclinic connection to 0 must satisfy the solvability condition (7.35) which is the crucial point of our analysis.

Proposition 7.4.1. *Let $\mathcal{W}_s(\rho,\nu)$ be the stable manifold of 0 for the full system (7.33).*

(a) *Let $\rho_s > 1$ and $\lambda \in]0,1[$. Then, there exist $\nu_{sr} > 0$ and $M_{sr} > 0$ such that for every $\nu \in]0,\nu_{sr}]$, $\rho \in [0,\rho_s[$, the stable manifold $\mathcal{W}_s(\rho,\nu)$ of 0 admits a parameterization of the form*

$$\mathcal{W}_s(\rho,\nu) = \{Y_s(t,\rho,\nu), t \in]-1,+\infty[\}, \quad \textit{with} \quad Y_s = h(t) + v_s(t,\rho,\nu)$$

where

$$\sup_{t\in]-1,+\infty[} \left(|v_s(t,\rho,\nu)|e^{\lambda|t|}\right) < M_{sr}\nu^2$$

holds for every $\nu \in]0,\nu_{sr}]$ and every $\rho \in [0,\rho_s[$.

(b) *Let λ be in $]0,1[$. There exists $d_s > 0$ such that for every $\nu \in]0,\nu_{sr}]$ and every $\rho \in [0,\rho_s[$, if Y is a homoclinic connection to 0 for the full system (7.33) such that*

$$Y \textit{ is reversible,} \tag{P1}$$

$$\sup_{t\in\mathbb{R}} \left(|Y(t)|e^{\lambda|t|}\right) < +\infty, \tag{P2}$$

$$\sup_{t\in\mathbb{R}} |(Y-h)(t)| \le d_s, \tag{P3}$$

then $Y(t) = Y_s(t)$ for every $t \in]-1,+\infty[$.

(c) *If Y is a reversible homoclinic connection to 0 decaying exponentially at infinity, i.e. if Y satisfies* (P1) *and* (P2) *then it satisfies the solvability condition*

$$I(Y,\rho,\nu) = 0 \tag{7.35}$$

where

$$\begin{aligned} I(Y,\rho,\nu) &:= \int_0^{+\infty} \langle r_-^*(t), g(v(t),t,\rho,\nu)\rangle dt \\ &= \frac{1}{2}\int_{-\infty}^{+\infty} e^{i((\frac{\omega}{\nu}+a\nu)t+2b\nu\tanh(\frac{1}{2}t))}\,(g_B + ig_A)(v(t),t,\rho,\nu)dt \end{aligned}$$

where $g = (g_\alpha, g_\beta, g_A, g_B)$ is defined in (7.34) and where $v = Y - h$.

Remark 7.4.2. Denote by v the "perturbation term" $v = Y - h$. (P1) is equivalent to the reversibility of v since h is reversible; (P2) is equivalent to

$$\sup_{t\in\mathbb{R}} \left(|v(t)|e^{\lambda|t|}\right) < +\infty,$$

since h satisfies $\sup_{t\in\mathbb{R}} \left(|h(t)|e^{|t|}\right) < +\infty$; (P3) means that h is the "principal part" of Y on the real axis. These three Properties characterize the homoclinic connections to 0 with only one loop (see Figure 7.5 p. 200).

Step 2. Holomorphic continuation of the stable manifold of 0

As already explained, our aim is to prove that $I(Y, \rho, \nu)$ is exponentially small but that it generically does not vanish. This leads to a contradiction which ensures the generic non persistence of reversible homoclinic connections. To obtain an exponential equivalent of $I(Y, \rho, \nu)$ with the Third Exponential Lemma 2.1.21, we need to prove that a reversible homoclinic connection admits a holomorphic continuation in a space of type $E^{\gamma,\lambda}_{\sigma,\delta}$ described in Definition 2.1.17. For that purpose, we first build a holomorphic continuation of the stable manifold Y_s. However, we are only able to do this for $\mathcal{R}e\,(\xi) \geq -\nu$: we obtain a holomorphic continuation in a space ${}^{s}E^{\gamma,\lambda}_{\sigma,\delta}$ which is defined as $E^{\gamma,\lambda}_{\sigma,\delta}$ except that it contains functions which are only defined on $\mathcal{B}^{s}_{\sigma-\delta\nu,\nu} := \mathcal{B}_{\sigma-\delta\nu} \cap \{\xi \in \mathbb{C},\ \mathcal{R}e\,(\xi) > -\nu\}$ instead of $\mathcal{B}_{\sigma-\delta\nu}$ for $E^{\gamma,\lambda}_{\sigma,\delta}$ (see Fig. 7.11). More precisely, we define a first space ${}^{s}H^{\gamma,\lambda}_{\sigma,\delta}$ analogous to $H^{\gamma,\lambda}_{\sigma,\delta}$ with functions restricted to $\mathcal{R}e\,(\xi) > -\nu$.

Definition 7.4.3. *Let γ, σ, λ be three positive numbers. Let δ be in $[1, +\infty[$. Let ${}^{s}H^{\gamma,\lambda}_{\sigma,\delta}$ be the Banach set of functions satisfying*

$$
{}^{s}H^{\gamma,\lambda}_{\sigma,\delta} := \Big\{ f : \bigcup_{0<\nu<\nu_0} \mathcal{B}^{s}_{\sigma-\delta\nu,\nu} \times \{\nu\} \to \mathbb{C},\ f(\cdot,\nu) \text{ is holomorphic in } \mathcal{B}^{s}_{\sigma-\delta\nu,\nu},\ \|f\|_{{}^{s}H^{\gamma,\lambda}_{\sigma,\delta}} < +\infty \Big\}
$$

where

$$
\mathcal{B}^{s}_{\ell,\nu} := \{\xi \in \mathbb{C}, |\mathcal{I}m\,(\xi)| < \ell,\ \mathcal{R}e\,(\xi) > -\nu\}
$$

and

$$
\|f\|_{{}^{s}H^{\gamma,\lambda}_{\sigma,\delta}} := \sup_{\substack{0<\nu<\nu_0 \\ \xi\in\mathcal{B}^{s}_{\sigma-\delta\nu,\nu}}} \left[(|f(\xi,\nu)| \left(|\sigma^2+\xi^2|^{\gamma}\chi_1(\xi) + (1-\chi_1(\xi))e^{\lambda|\mathcal{R}e(\xi)|} \right) \right]
$$

with $\chi_1(\xi) = 1$ for $|\mathcal{R}e\,(\xi)| < 1$ and $\chi_1(\xi) = 0$ otherwise (see Figure 7.10).

Observe that $|\xi^2+\sigma^2| = |(\xi - i\sigma)(\xi + i\sigma)|$.

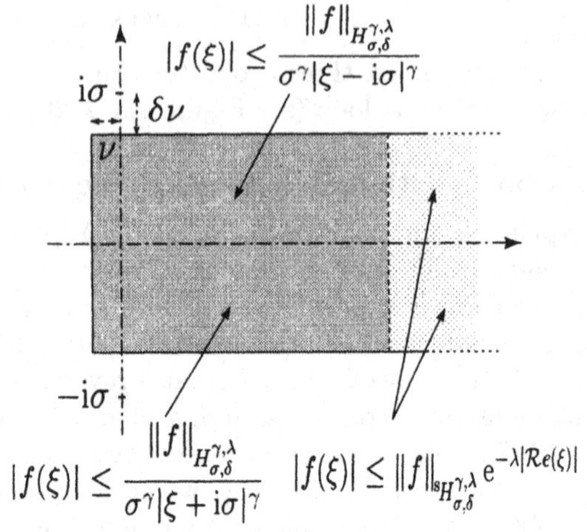

Fig. 7.10. Description of $f \in {}^{\mathrm{s}}H^{\gamma,\lambda}_{\sigma,\delta}$

Then, we define a second space ${}^{\mathrm{s}}E^{\gamma,\lambda}_{\sigma,\delta}$ analogous to $E^{\gamma,\lambda}_{\sigma,\delta}$

Definition 7.4.4. *Let $\gamma, \lambda > 0$, $\sigma > 0$, $\delta \ge 1$, $m_0, m_1 > 1$, $0 \le r < 1$ and $\theta > 0$. Denote by ${}^{\mathrm{s}}E^{\gamma,\lambda}_{\sigma,\delta}$ (m_0, m_1, r, θ) the set of functions f such that*

(a) $f \in {}^{\mathrm{s}}H^{\gamma,\lambda}_{\sigma,\delta}$,

(b) *for every $\nu \in]0, \nu_0]$ and every $z \in \widehat{\Sigma}^{\mathrm{s}}_{\delta,r,\nu} :=]-1, \nu^{r-1}[\times]-\nu^{r-1}, -\delta[$ f reads*

$$f(\mathrm{i}\sigma + \nu z, \nu) = \frac{1}{\nu^\gamma}\widehat{f}_0^+(z) + \frac{\nu^\theta}{\nu^\gamma}\widehat{f}_1^+(z,\nu)$$

where $\widehat{f}_0^+ \in {}^{\mathrm{s}}\widehat{H}^{m_0}_{\delta}$ and $\widehat{f}_1^+ \in {}^{\mathrm{s}}\widehat{H}^{m_1}_{\delta,r}$ with

$${}^{\mathrm{s}}\widehat{H}^{m_0}_{\delta} = \Big\{ \widehat{f} : \widehat{\Omega}^{\mathrm{s}}_\delta \to \mathbb{C},\ \widehat{f} \text{ holomorphic in } \widehat{\Omega}^{\mathrm{s}}_\delta :=]-1, +\infty[\times]-\infty, -\delta[,$$
$$\left\|\widehat{f}\right\|_{{}^{\mathrm{s}}\widehat{H}^{m_0}_{\delta}} := \sup_{z\in\widehat{\Omega}^{\mathrm{s}}_\delta} \left(|\widehat{f}(z)||z|^{m_0}\right) < +\infty \Big\}$$

and

$${}^{\mathrm{s}}\widehat{H}^{m_1}_{\delta,r} = \Big\{ \widehat{f} : \bigcup_{\nu\in]0,\nu_0]} \left(\widehat{\Sigma}^{\mathrm{s}}_{\delta,r,\nu} \times \{\nu\}\right) \to \mathbb{C},\ \widehat{f}(\cdot,\nu) \text{ holomorphic in } \widehat{\Sigma}^{\mathrm{s}}_{\delta,r,\nu},$$
$$\left\|\widehat{f}\right\|_{{}^{\mathrm{s}}\widehat{H}^{m_1}_{\delta,r}} := \sup_{\substack{\nu\in]0,\nu_0]\\ z\in\widehat{\Sigma}^{\mathrm{s}}_{\delta,r,\nu}}} \left(|\widehat{f}(z)||z|^{m_1}\right) < +\infty \Big\},$$

(c) *for every* $\nu \in]0, \nu_0]$ *and every* $z \in -\overline{\widehat{\Sigma}^{\mathrm{s}}_{\delta,r,\nu}}$ f *reads*

$$f(-\mathrm{i}\sigma - \nu z, \nu) = \frac{1}{\nu^\gamma}\widehat{f_0^-}(z) + \frac{\nu^\theta}{\nu^\gamma}\widehat{f_1^-}(z,\nu)$$

with $\widehat{f_0^-} \in {}^{\mathrm{s}}\widehat{H}_\delta^{m_0,-}$ *and* $\widehat{f_1^-} \in {}^{\mathrm{s}}\widehat{H}_{\delta,r}^{m_1,-}$ *where* ${}^{\mathrm{s}}\widehat{H}_\delta^{m_0,-}$ *and* ${}^{\mathrm{s}}\widehat{H}_{\delta,r}^{m_1,-}$ *are defined as* ${}^{\mathrm{s}}\widehat{H}_\delta^{m_0}$ *and* ${}^{\mathrm{s}}\widehat{H}_{\delta,r}^{m_1}$ *with* $-\overline{\widehat{\Omega}^{\mathrm{s}}_\delta}$ *and* $-\overline{\widehat{\Sigma}^{\mathrm{s}}_{\delta,r,\nu}}$ *instead of* $\widehat{\Omega}_\delta$ *and* $\widehat{\Sigma}_{\delta,r,\varepsilon}$.

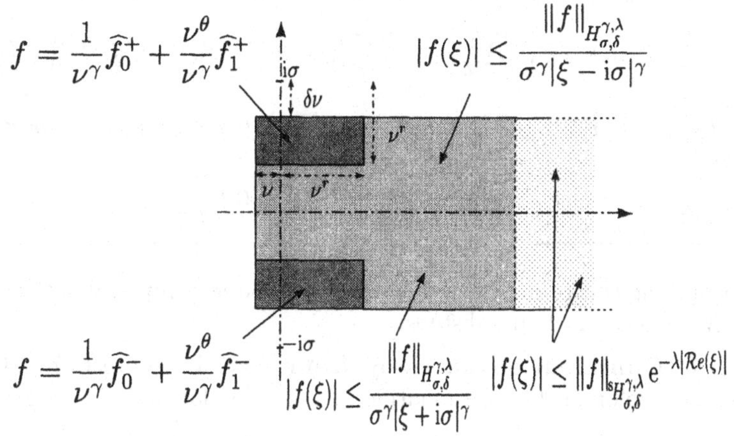

Fig. 7.11. Description of $f \in {}^{\mathrm{s}}E^{\gamma,\lambda}_{\sigma,\delta}$

Remark 7.4.5. Statement (c) of the previous definition is more complicated than Statement (c) of Definition 2.1.17 characterizing $E^{\gamma,\lambda}_{\sigma,\delta}$ since in this later case $\widehat{\Omega}_\delta$ and $\widehat{\Sigma}_{\delta,r,\varepsilon}$ are symmetric about the imaginary axis whereas $\widehat{\Omega}^{\mathrm{s}}_\delta$ and $\widehat{\Sigma}^{\mathrm{s}}_{\delta,r,\nu}$ are not. As in chapter 2, we denote by $\mathrm{P}_{\mathrm{r}}^\pm$ the linear operators $\mathrm{P}_{\mathrm{r}}^\pm : f \mapsto \widehat{f_0^\pm}$. They satisfy the same properties as those given in Remark 2.1.18 and in Lemma 2.1.22. Moreover as in Statement (c) of Lemma 2.1.19, if f is real on the real axis, then $\mathrm{P}_{\mathrm{r}}^-(f)(z) = \overline{\mathrm{P}_{\mathrm{r}}^+(f)(-\overline{z})}$ holds for $z \in -\overline{\widehat{\Omega}^{\mathrm{s}}_\delta}$.

For $\rho = 0$, $Y_{\mathrm{s}} = h$ where h is given by (7.15). Hence the stable manifold Y_{s} admits a holomorphic continuation which lies in

$${}^{\mathrm{s}}E^{2,\lambda}_{\pi,\delta_{\mathrm{s}}}(2,3,\tfrac{1}{2},\tfrac{1}{4}) \times {}^{\mathrm{s}}E^{3,\lambda}_{\pi,\delta_{\mathrm{s}}}(3,4,\tfrac{1}{2},\tfrac{1}{4}) \times {}^{\mathrm{s}}E^{2,\lambda}_{\pi,\delta_{\mathrm{s}}}(2,3,\tfrac{1}{2},\tfrac{1}{4}) \times {}^{\mathrm{s}}E^{2,\lambda}_{\pi,\delta_{\mathrm{s}}}(2,3,\tfrac{1}{2},\tfrac{1}{4})$$

for any $\lambda \in]0,1]$ and any $\delta_{\mathrm{s}} \ge 1$. The following proposition, ensures that it is still true for $\rho \ne 0$ provided that δ_{s} is large enough and that $\lambda \in]0,1[$.

Proposition 7.4.6. *Let Y_s be the parameterization of the stable manifold of 0 for the full system (7.33) obtained in Proposition 7.4.1.*

(a) *There exists $\nu_{sc} \in]0, \nu_{sr}]$, $\delta_s > 1$ such that for every $\nu \in]0, \nu_{sc}]$ and every $\rho \in [0, \rho_s[$, $t \mapsto Y_s(t, \rho, \nu)$ admits a holomorphic continuation in $\mathcal{B}^s_{\pi-\delta_s\nu,\nu}$ still denoted by Y_s which satisfies for every $\rho \in [0, \rho_s[$,*

$$\alpha_s(\cdot, \rho, \cdot),\ A_s(\cdot, \rho, \cdot),\ B_s(\cdot, \rho, \cdot) \in {}^sE^{2,\lambda}_{\pi,\delta_s}\ (2, 3, \tfrac{1}{2}, \tfrac{1}{4}),$$
$$\beta_s(\cdot, \rho, \cdot) \in {}^sE^{3,\lambda}_{\pi,\delta_s}\ (3, 4, \tfrac{1}{2}, \tfrac{1}{4})$$

where $Y_s := (\alpha_s, \beta_s, A_s, B_s)$.

(b) *Denote by $\widehat{Y}^+_{s,0} = (\widehat{\alpha}^+_{s,0}, \widehat{\beta}^+_{s,0}, \widehat{A}^+_{s,0}, \widehat{B}^+_{s,0})$ the principal part of Y_s near $i\pi$*

$$\widehat{\alpha}^+_{s,0} = \mathrm{Pr}^+_{{}^sE^{2,\lambda}_{\pi,\delta_s}}(\alpha_s),\ \widehat{\beta}^+_{s,0} = \mathrm{Pr}^+_{{}^sE^{3,\lambda}_{\pi,\delta_s}}(\beta_s),\ \widehat{A}^+_{s,0} = \mathrm{Pr}^+_{{}^sE^{2,\lambda}_{\pi,\delta_s}}(A_s),\ \widehat{B}^+_{s,0} = \mathrm{Pr}^+_{{}^sE^{2,\lambda}_{\pi,\delta_s}}(B_s).$$

Then, $\rho \mapsto \widehat{Y}^+_{s,0}(z, \rho)$ is holomorphic in $D(0, \rho_s) := \{\rho \in \mathbb{C}, |\rho| < \rho_s\}$ and

$$\widehat{Y}^+_{s,0}(z, 0) = \left(-\frac{4}{z^2}, \frac{8}{z^3}, 0, 0\right).$$

For proving this proposition we follow the strategy proposed in Subsection 2.1.4. We proceed in four substeps:

Step 2.1. Continuation far away from $\pm i\pi$. In Subsection 7.4.3, we determine a holomorphic continuation of Y_s in the domain $\mathcal{D}^s_{\frac{1}{2},\nu}$ where

$$\mathcal{D}^s_{r,\nu} = \{\xi \in \mathbb{C},\ \mathcal{R}e\,(\xi) > -\nu, |\mathcal{I}m\,(\xi)\,| < \pi\} \backslash$$
$$\{\xi \in \mathbb{C}, \mathcal{R}e\,(\xi) \le \tfrac{1}{2}\nu^r, \pi - \tfrac{1}{2}\nu^r \le |\mathcal{I}m\,(\xi)\,|\}$$

(see Figure 7.12). This holomorphic continuation is found in the form $Y_s = h + v_s$ where v_s is a perturbation term small with respect to h in $\mathcal{D}^s_{\frac{1}{2},\nu}$.

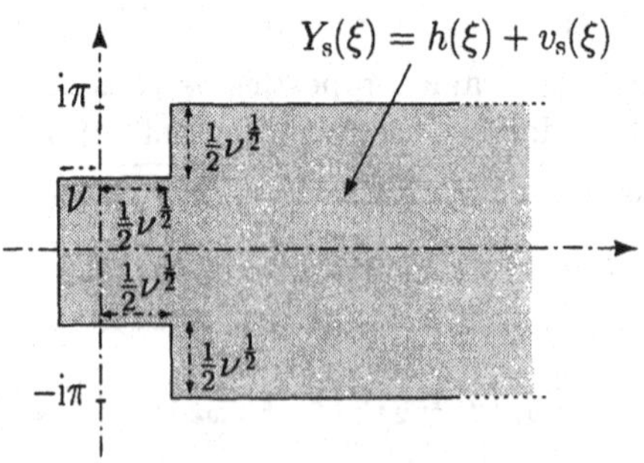

Fig. 7.12. Step 2.1. Holomorphic continuation of Y_s in $\mathcal{D}^s_{\frac{1}{2},\nu}$

Step 2.2. Inner system. For determining a relevant holomorphic continuation of Y_s near $i\pi$, we must first determine the principal part of Y_s near $i\pi$. Following the strategy proposed in Subsection 2.1.4, we introduce in Subsection 7.4.4, the two following system of coordinates in space and time:

- the *outer system of coordinates* (Y, ξ) with $Y = (\alpha, \beta, A, B)$ which is the initial one,
- the *inner system of coordinates* $(\widehat{Y}, z)$ with $\widehat{Y} = (\widehat{\alpha}, \widehat{\beta}, \widehat{A}, \widehat{B})$ where

$$\xi = i\pi + \nu z, \qquad \widehat{\alpha} = \nu^2 \alpha,\ \widehat{\beta} = \nu^3 \beta,\ \widehat{A} = \nu^2 A, \widehat{B} = \nu^2 B,$$

  which is useful for describing the solution near $i\pi$.

We also introduce the *inner system* (7.55) (p. 256) which is the leading part of the system near $i\pi$. Then the stable manifold of 0 for the inner system is a good candidate to be the principal part of Y_s near $i\pi$.

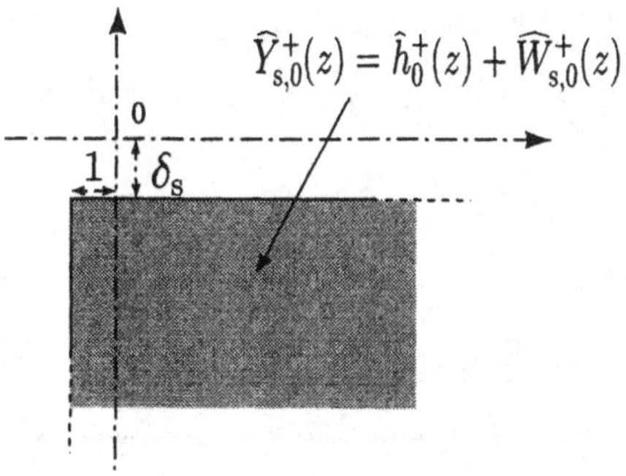

Fig. 7.13. Step 2.2. Stable manifold of the inner system obtained in $\widehat{\Omega}^s_{\delta_s}$ in the inner system of coordinates.

We build a parameterization of this stable manifold, in the inner system of coordinates in the form $z \mapsto \widehat{h}^+_0(z) + \widehat{W}^+_{s,0}(z)$ where $\widehat{h}^+_0$ is the principal part of h near $i\pi$. This parameterization is obtained in the domain $\widehat{\Omega}^s_{\delta_s} :=]-1, +\infty[\times]-\infty, -\delta_s[$ in the inner system of coordinates, i.e. in $\Omega^s_\delta :=]-\nu, +\infty[\times]-\infty, \pi - \delta_s \nu[$ in the outer system of coordinates (see Figure 7.13).

Step 2.3 Continuation near $+i\pi$: matching. Seeing the full system as a perturbation of the inner system, we prove in Subsection 7.4.5 that $\widehat{Y}^+_{s,0}$ is really the principal part of Y_s near $i\pi$. More precisely we prove that in the inner system of coordinates, $\widehat{Y}_s$ admits in $\widehat{\Sigma}^s_{\delta_s, \frac{1}{2}, \nu} =]-1, \nu^{-\frac{1}{2}}[\times]-\nu^{-\frac{1}{2}}, -\delta_s[$ a holomorphic continuation of the form

$$\begin{aligned}\widehat{Y}_s(z, \rho, \nu) &= \widehat{Y}^+_{s,0}(z, \rho) + \widehat{w}_s(z, \rho, \nu) \\ &= \left(\widehat{h}^+_0(z) + \widehat{W}^+_{s,0}(z, \rho)\right) + \widehat{w}_s(z, \rho, \nu) \qquad \text{where } |\widehat{w}_s| = \mathcal{O}(\nu^{\frac{1}{4}}).\end{aligned}$$

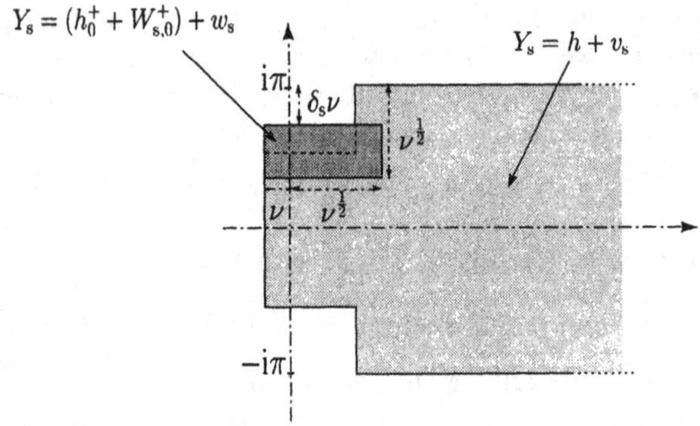

Fig. 7.14. Step 2.3. Decomposition of the holomorphic continuation of Y_s in $\Sigma^s_{\delta_s,\frac{1}{2},\nu}$.

This result expressed in the outer system of coordinates, ensures the existence of a holomorphic continuation of Y_s in $\Sigma^s_{\delta_s,\frac{1}{2},\overline{\nu}} =]-\nu, \nu^{\frac{1}{2}}[\times]\pi - \nu^{\frac{1}{2}}, \pi - \delta_s\nu[$, of the form $Y_s(\xi) = \left(h_0^+(\xi) + W_{s,0}^+(\xi)\right) + w_s(\xi)$, (see Figure 7.14).

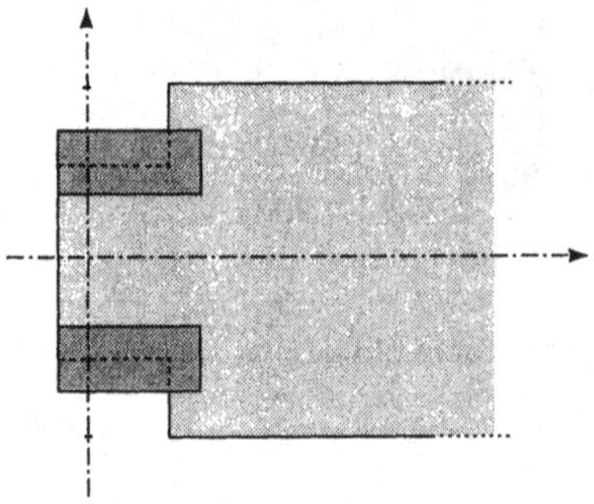

Fig. 7.15. Step 2.4. Holomorphic continuation of Y_s in $\overline{\Sigma^s_{\delta_s,\frac{1}{2},\nu}}$.

Step 2.4. Continuation near $-i\pi$: symmetrization. Finally, in Subsection 7.4.6, using an argument of symmetry about the real axis we obtain a holomorphic continuation of Y_s near $-i\pi$ in the domain $\overline{\Sigma^s_{\delta_s,\frac{1}{2},\nu}}$ (see Figure 7.15).

Remark 7.4.7. The two decompositions of Y_s near and far away from the singularities express that *we have to face a relative change of size of the normal form with respect to the higher order terms.* Indeed, on the real axis $N(h,\nu) = \mathcal{O}(1)$ and $R(h,\nu) = \mathcal{O}(\nu^2)$ whereas near the singularities, for instance for $\xi = i(\pi - \nu)$, $N(h,\nu) = \mathcal{O}(\nu^{-4})$ and $R(h,\nu) = \mathcal{O}(\nu^{-4})$. Hence, on the real axis and far away from the singularities the normal form is the leading part of the system and the holomorphic continuation of the stable manifold Y_s can be found in the form $Y_s = h + v_s$ where h is the homoclinic connection to 0 of the normal form system and where v_s is a perturbation term small with respect to h. However, near the singularities N and R have the same size. So near the singularities , N is no longer the leading part of the system and the decomposition of Y_s in the form $Y_s = h + v_s$ is no longer relevant: h and v have the same size. This is why a second decomposition of the form $Y_s(\xi) = \left(h_0^+(\xi) + W_{s,0}^+(\xi)\right) + w_s(\xi)$, is required to describe Y_s near $i\pi$. Moreover, the principal part of Y_s near $i\pi$ is found as the stable manifold $\widehat{Y}_{s,0}^+ = \widehat{h}_0^+ + \widehat{W}_{s,0}^+$ of 0 for the inner system which is nothing else than the leading part of the system near $i\pi$ (see subsection 7.4.4 for more details).

Step 3. Exponential asymptotics of the solvability condition and generic non persistence. In Subsection 7.4.7, using the description of the holomorphic continuation of Y_s given by Proposition 7.4.6 we can show

Proposition 7.4.8.

(a) *There exists M_{sc} such that for any $\rho \in [0, \rho_s[$ and any $\nu \in]0, \nu_{sc}]$, if the full system (7.33) admits a homoclinic connection satisfying the properties* (P1), (P2), (P3) *defined in Proposition 7.4.1, then the oscillatory integral $I(Y, \rho, \nu)$ defined in Proposition 7.4.1-*(c) *satisfies*

$$I(Y, \rho, \nu) = \frac{1}{\nu^2} e^{-\frac{\omega\pi}{\nu}} \left(\Lambda(\rho) + J(\rho, \nu)\right) \qquad \text{with } |J(\rho, \nu)| \le M_{sc} \nu^{\frac{1}{4}}$$

where

$$\Lambda(\rho) = \mathcal{R}e\left(\int_{-i(\delta_s+1)+\mathbb{R}^+} e^{i\omega z + 4ib/z} (\widehat{g}_{0,B}(z,\rho) + i\widehat{g}_{0,A}(z,\rho))dz\right) \tag{7.36}$$

with

$$\widehat{g}_{0,A}(z,\rho) = -\widehat{B}_{s,0}^+(z,\rho)\ \left(\widehat{\alpha}_{s,0}^+(z,\rho) + 4/z^2\right) + \rho \sum_{|\overline{m}|\ge 3} a_{A,\overline{m},0}\left(\widehat{Y}_{s,0}^+(z,\rho)\right)^{\overline{m}}$$

$$\widehat{g}_{0,B}(z,\rho) = \widehat{A}_{s,0}^+(z,\rho)\left(\widehat{\alpha}_{s,0}^+(z,\rho) + 4/z^2\right) + \rho \sum_{|\overline{m}|\ge 3} a_{B,\overline{m},0}\left(\widehat{Y}_{s,0}^+(z,\rho)\right)^{\overline{m}}$$

where $\widehat{Y}_{s,0}^+ = (\widehat{\alpha}_{s,0}^+, \widehat{\beta}_{s,0}^+, \widehat{A}_{s,0}^+, \widehat{B}_{s,0}^+)$ is defined in Proposition 7.4.6.

(b) *Λ is a real analytic function on $[0, \rho_s[$ which reads $\Lambda(\rho) = \sum_{n\ge 1} \rho^n \Lambda_n$. The first coefficient Λ_1 reads*

$$\Lambda_1 = \sum_{\substack{m_\alpha, m_\beta \in \mathbb{N}^2 \\ m_\alpha + m_\beta \ge 3}} b_{m_\alpha, m_\beta}\ \Lambda_{m_\alpha, m_\beta} \tag{7.37}$$

with

$$\Lambda_{m_\alpha,m_\beta} = 2\pi\ (-1)^{m_\beta+E(\frac{m_\beta+1}{2})} \sum_{n=0}^{+\infty} \frac{(-2b)^n\ (2\omega)^{2m_\alpha+3m_\beta+n-1}}{(2m_\alpha+3m_\beta+n-1)!\ n!},$$

and

$$b_{m_\alpha,m_\beta} = a_{A,(m_\alpha,m_\beta,0,0),0} + a_{B,(m_\alpha,m_\beta,0,0),0}$$

where $a_{A,\overline{m},\ell}$, $a_{B,\overline{m},\ell}$ *are defined in (7.6) and where for* $x \in \mathbb{R}$, $E(x)$ *is the largest integer smaller or equal to* x.

Remark 7.4.9. The reversibility properties of R given by (7.8) ensure that

$$a_{A,(m_\alpha,m_\beta,0,0),0} = 0 \text{ for } m_\beta \text{ even}, \qquad a_{B,(m_\alpha,m_\beta,0,0),0} = 0 \text{ for } m_\beta \text{ odd}.$$

This Proposition finally gives the desired result of generic non persistence for reversible homoclinic connections:

Theorem 7.4.10. *For every* $\rho \in [0, \rho_s[$ *such that* $\Lambda(\rho) \neq 0$ *and every* ν *small enough, the full system (7.33) does not admit any homoclinic connection satisfying* (P1), (P2), (P3) *(see Proposition 7.4.1 p. 234).*

Remark 7.4.11. Theorem 7.1.11 follows from this theorem and from Proposition 7.4.8.

Proof. Let $\rho \in [0, \rho_s[$ be such that $\Lambda(\rho) \neq 0$. Let $\nu^* \in]0, \nu_{sc}]$ be such that $|\Lambda(\rho)| - M_{sc}(\nu^*)^{\frac{1}{4}} > 0$ where M_{sc} is introduced in Proposition 7.4.8. Then, for every $\rho \in [0, \rho_s[$ and every $\nu \in]0, \nu^*]$, Proposition 7.4.1 ensures that if there exists a homoclinic connection Y to 0 satisfying (P1), (P2), (P3), then

$$I(Y, \rho, \nu) = 0.$$

However, Proposition 7.4.8 ensures that in such a case

$$I(Y, \rho, \nu) = \frac{1}{\nu^2} e^{-\frac{\pi\omega}{\nu}} |\Lambda(\rho) + J(\rho, \nu)| > \frac{1}{\nu^2} e^{-\frac{\pi\omega}{\nu}} (|\Lambda(\rho)| - M_{sc}(\nu^*)^{\frac{1}{4}}) > 0.$$

This leads to a contradiction. □

7.4.1.3 Solvability condition and symmetry defects. Looking for reversible homoclinic connections amounts to wondering whether the stable manifold Y_s of 0 can be continued on the whole real axis by reversibility. This way of thinking in terms of continuation by symmetry complements fruitfully our first approach in terms of "solvability conditions". For that purpose we extend the notion of reversibility on $\mathbb{R}$ by the notion of symmetry about the imaginary axis in the complex field.

Definition 7.4.12. *Let $\mathcal{U}^{\mathrm{s}}$ be an open set in $\mathbb{C}$ and let us define $\mathcal{U}^{\mathrm{u}} := -\overline{\mathcal{U}^{\mathrm{s}}}$. Let f_{s} be a holomorphic function defined on $\mathcal{U}^{\mathrm{s}}$.*

f_{s} can be continued by symmetry about the imaginary axis when there exists an holomorphic function f defined on $\mathcal{U}^{\mathrm{s}} \cup \mathcal{U}^{\mathrm{u}}$ such that

$$f|_{\mathcal{U}^{\mathrm{s}}} = f_{\mathrm{s}}, \qquad Sf(-\overline{\zeta}) = \overline{f(\zeta)} \qquad \text{for } \zeta \in \mathcal{U}^{\mathrm{s}} \cup \mathcal{U}^{\mathrm{u}},$$

Remark 7.4.13. The holomorphic continuation of a reversible function which is real valued on the imaginary axis is automatically symmetric about the imaginary axis.

The following Lemma ensures that the knowledge of a holomorphic function on the imaginary axis is sufficient to determine whether this function can be continued by symmetry about the imaginary axis.

Lemma 7.4.14. *Let $\mathcal{U}^{\mathrm{s}}$ be an open set in $\mathbb{C}$ satisfying*

(a) $\mathcal{U}^{\mathrm{s}} \cap \mathrm{i}\mathbb{R} =]\mathrm{i}a, \mathrm{i}b[\quad$ *where* $-\infty \leq a < b \leq +\infty$,
(b) $\mathcal{U}^{\mathrm{s}} \cap \mathcal{U}^{\mathrm{u}}$ *is a simply connected set.*

Let f_{s} be a holomorphic function defined on $\mathcal{U}^{\mathrm{s}}$. Then, f_{s} can be continued by symmetry about the imaginary axis if and only if

$$Sf_{\mathrm{s}}(\mathrm{i}\eta) = \overline{f_{\mathrm{s}}(\mathrm{i}\eta)} \qquad \text{for } \eta \in]a, b[.$$

Proof. $f_{\mathrm{s}}(\zeta)$ and $\overline{f_{\mathrm{s}}(-\overline{\zeta})}$ are two holomorphic functions respectively defined on $\mathcal{U}^{\mathrm{s}}, \mathcal{U}^{\mathrm{u}}$, which are equal on $]\mathrm{i}a, \mathrm{i}b[$ and thus also on $\mathcal{U}^{\mathrm{s}} \cap \mathcal{U}^{\mathrm{u}}$ because of the isolated zeros theorem. □

With this approach, the fundamental remark is the following :*if $\widehat{Y}^{+}_{\mathrm{s},0}$ cannot be continued by symmetry about the imaginary axis then for ν sufficiently small, the full system (7.33) does not admit any homoclinic connection to 0 satisfying properties* (P1), (P2), (P3)*)*. In other words,

Lemma 7.4.15. *Let ρ be fixed in $[0, \rho_{\mathrm{s}}[$. If there exists a sequence $(\nu^*_n)_{n \geq 0}$ with $\nu^*_n \underset{n \to +\infty}{\longrightarrow} 0$, for which the full system (7.33) admits a homoclinic connection $Y_{\nu^*_n}$ satisfying properties* (P1), (P2), (P3), *then necessarily the stable manifold $\widehat{Y}^{+}_{\mathrm{s},0}(\cdot, \rho) = \widehat{h}^{+}_{0} + \widehat{W}^{+}_{\mathrm{s},0}(\cdot, \rho)$ of the inner system can be continued by symmetry about the imaginary axis i.e. it satisfies*

$$S\widehat{Y}^{+}_{\mathrm{s},0}(-\mathrm{i}\eta, \rho) = \overline{\widehat{Y}^{+}_{\mathrm{s},0}(-\mathrm{i}\eta, \rho)}, \qquad \text{for } \eta > \delta_{\mathrm{s}}.$$

This lemma is proved in Appendix 7.J

Then, $I(Y_{\mathrm{s}}, \rho, \nu)$ and $\Lambda(\rho)$ can be seen as two quantities which measures respectively a symmetry defect of the stable manifold Y_{s} of 0 for the full system and a symmetry defect of the stable manifold $\widehat{Y}^+_{s,0}$ of 0 for the inner system. Indeed,

- Proposition 7.4.1 ensures that if Y_s can be continued by reversibility on $\mathbb{R}$ then $I(Y_s, \rho, \nu) = 0$;
- Lemma 7.4.16 below gives necessary and sufficient conditions for $\widehat{Y}^+_{s,0}$ to be continuable by symmetry about the imaginary axis. Moreover this lemma ensures that if $\widehat{Y}^+_{s,0}$ can be continued by symmetry, then necessarily $\Lambda(\rho) = 0$ holds.

So, Theorem 7.4.17 says that generically Y_s has a symmetry defect on $\mathbb{R}$ measured by $I(Y_s, \rho, \nu)$ which is exponentially small with respect to ν. Hence, it is very difficult to detect such a defect. However, Proposition 7.4.6 ensures that near $i\pi$ $\widehat{Y}^+_{s,0}$ is the leading part of Y_s. So the leading part of the symmetry defect of Y_s near $i\pi$ is given by the symmetry defect of $\widehat{Y}^+_{s,0}$, i.e. by $\Lambda(\rho)$. Thus, near $i\pi$ the symmetry defect of Y_s is no longer exponentially small but of order 1 and so far more easy to detect.

Lemma 7.4.16. *Let us define for $\eta > \delta_s$ and $\rho \in [0, \rho_s[$*

$$\Lambda_p(\eta,\rho) = \mathcal{R}e\left(\langle \widehat{\mathbf{p}}^*(-i\eta), \widehat{W}^+_{s,0}(-i\eta,\rho)\rangle_*\right)$$

$$\Lambda_q(\eta,\rho) = \mathcal{I}m\left(\langle \widehat{\mathbf{q}}^*(-i\eta), \widehat{W}^+_{s,0}(-i\eta,\rho)\rangle_*\right)$$

$$\Lambda_+(\eta,\rho)=\mathcal{I}m\left(\langle \widehat{\mathbf{r}}^*_+(-i\eta), \widehat{W}^+_{s,0}(-i\eta,\rho)\rangle_*\right)+\mathcal{R}e\left(\langle \widehat{\mathbf{r}}^*_-(-i\eta), \widehat{W}^+_{s,0}(-i\eta,\rho)\rangle_*\right)$$

$$\Lambda_-(\eta,\rho)=\mathcal{I}m\left(\langle \widehat{\mathbf{r}}^*_+(-i\eta), \widehat{W}^+_{s,0}(-i\eta,\rho)\rangle_*\right)-\mathcal{R}e\left(\langle \widehat{\mathbf{r}}^*_-(-i\eta), \widehat{W}^+_{s,0}(-i\eta,\rho)\rangle_*\right)$$

where $\widehat{\mathbf{p}}^*, \widehat{\mathbf{q}}^*, \widehat{\mathbf{r}}^*, \widehat{\mathbf{r}}^*_+$ are given in Lemma 7.4.40.

(a) *Let ρ be fixed in $[0, \rho_s[$. Then the following statements are equivalent*

(i) *$\widehat{Y}^+_{s,0}(\cdot, \rho)$ can be continued by symmetry about the imaginary axis,*

(ii) *$\widehat{W}^+_{s,0}(\cdot, \rho)$ can be continued by symmetry about the imaginary axis,*

(iii) *For every $\eta \geq \delta_s$,* $\quad \Lambda_p(\eta,\rho) = \Lambda_q(\eta,\rho) = \Lambda_+(\eta,\rho) = \Lambda_-(\eta,\rho) = 0$,

(iv) *There exists $\eta_0 \geq \delta_s$ such that $\Lambda_p(\eta_0,\rho) = \Lambda_q(\eta_0,\rho) = \Lambda_+(\eta_0,\rho) = \Lambda_-(\eta_0,\rho) = 0$.*

(b) *For every $\eta > \delta_s$, the functions $\rho \mapsto \Lambda_p(\eta,\rho)$, $\rho \mapsto \Lambda_q(\eta,\rho)$, $\rho \mapsto \Lambda_+(\eta,\rho)$, $\rho \mapsto \Lambda_-(\eta,\rho)$ are real analytic on $[0, \rho_s[$ and satisfy*

$$\frac{d\Lambda_p}{d\rho}(\eta,0) = \frac{d\Lambda_q}{d\rho}(\eta,0) = \frac{d\Lambda_+}{d\rho}(\eta,0) = 0, \qquad \frac{d\Lambda_-}{d\rho}(\eta,0) = \Lambda_1.$$

where Λ_1 is given in Proposition 7.4.8.

(c) *For every* $\rho \in [0, \rho_s[$, *the function* $\Lambda(\rho)$ *introduced in Proposition 7.4.8 satisfies*

$$\Lambda(\rho) = \Lambda_-(\delta_s + 1, \rho).$$

This lemma is proved in Appendix 7.J

Gathering the results of Lemmas 7.4.15 and 7.4.16, we can prove the following Theorem of non persistence which generalizes Theorem 7.4.10.

Theorem 7.4.17. *Let* $\rho \in [0, \rho_s[$ *be such that there exists* $\eta_0 > \delta_s$ *for which*

$$\Lambda_p(\eta_0, \rho) \neq 0, \text{ or } \Lambda_q(\eta_0, \rho) \neq 0, \text{ or } \Lambda_+(\eta_0, \rho) \neq 0, \text{ or } \Lambda_-(\eta_0, \rho) \neq 0.$$

Then, for all ν *sufficiently small, the full system (7.33) does not admit any homoclinic connection which satisfies properties* (P1), (P2), (P3).

Remark 7.4.18. Because of lemma 7.4.16, this theorem simply says that if the stable manifold of the inner system cannot be continued by symmetry about the imaginary axis then the full system does not admit any reversible homoclinic connection to 0 for ν small.

Proof of Theorem 7.4.17 We make a proof by contradiction: assume that for every $\nu^\#$ there exists $\nu^* < \nu^\#$ such that the full system admits a homoclinic connection which satisfies (P1), (P2), (P3). Then we can build a sequence $(\nu_n^*)_{n\geq 0}$, $\nu_n \underset{n\to+\infty}{\longrightarrow} 0$, such that the full system admits for ν_n^* a homoclinic connection $Y_{\nu_n^*}$ satisfying properties (P1), (P2), (P3). Then, lemmas 7.4.15 and 7.4.16 ensure that for every $\eta > \delta_s$,

$$\Lambda_p(\eta, \rho) = \Lambda_q(\eta, \rho) = \Lambda_+(\eta, \rho) = \Lambda_-(\eta, \rho) = 0.$$

Here is the contradiction. □

7.4.2 Local stable manifold of 0

This Subsection is devoted to the proof of Proposition 7.4.1.

7.4.2.1 Proof of Proposition 7.4.1-(a). We begin by computing a parameterization of the form $Y_s(t) = h(t) + v_s(t)$ where $t \in]-1, +\infty[$. We choose $]-1, +\infty[$ because for later use we need an interval which contains strictly $[0, +\infty[$ (any interval $]-T, +\infty[$ with $T > 0$ could be chosen). The form of the parameterization expresses that on the real axis, the normal form is the relevant part of the full system (7.33). It will be seen later that near the singularities of h the normal form is no longer the relevant part of the system and that another decomposition of Y_s must be used.

So, *we fix* $\lambda \in]0,1[$, and we look for the perturbation term v_s as a solution of (7.34) which lies in the Banach space C^0_λ defined by

Definition 7.4.19.

$$C^0_\lambda = \{v :]-1,+\infty[\longrightarrow \mathbb{R}^4, \text{ continuous, } |v|_{C^0_\lambda} := \sup_{t\in]-1,+\infty[} |v|e^{\lambda|t|} < +\infty\}.$$

Let us denote by $BC^0_\lambda(d)$ *the ball of radius* d *of* C^0_λ

$$BC^0_\lambda(d) = \{v \in C^0_\lambda,\ |v|_{C^0_\lambda} \le d\}.$$

Our aim is now to transform (7.34) into an integral equation which can be solved with the Contraction Mapping Theorem. We first check

Lemma 7.4.20. *There exists* $\nu_1 > 0$ *such that for every* $\nu \in]0,\nu_1]$, $\rho \in [0,\rho_s[$, $v \in BC^0_0(1)$ *and every* $t \in]-1,+\infty[$, $g(v(t),t,\rho,\nu)$ *is well defined. Moreover, there exists* $\mathcal{M}_1$ *such that for every* $\nu \in]0,\nu_1]$, $\rho \in [0,\rho_s[$, $(v,v') \in (BC^0_0(1))^2$, *and every* $t \in]-1,+\infty[$,

$$|g(v(t),t,\rho,\nu)| \le \mathcal{M}_1\big(|v(t)|^2 + \rho\nu^2(|h(t)| + |v(t)|)\big),$$
$$|g(v(t),t,\rho,\nu) - g(v'(t),t,\rho,\nu)| \le \mathcal{M}_1\big(|v(t)| + |v'(t)| + \rho\nu^2\big)|v(t) - v'(t)|.$$

The proof readily follows from the explicit formula giving $Q(v,\nu)$

$$Q(v,\nu) = \begin{pmatrix} 0 \\ -\frac{3}{2}\alpha^2 - c(A^2 + B^2) \\ -\nu b\alpha B \\ \nu b\alpha A \end{pmatrix}$$

and from Lemma 7.1.2 which gives the size of R. □

Then to transform (7.34) into an integral equation, we use the variation of constants formula with the basis (p,q,r_+,r_-) of the homogeneous equation

$$\frac{dv}{dt} - DN(h,\nu).v = 0$$

given by Lemmas 7.3.12, 7.3.13, 7.3.14. Let us write $v(t)$ in this basis

$$v(t) = a(t)p(t) + b(t)q(t) + c(t)r_+(t) + d(t)r_-(t).$$

Then v is a solution of (7.34) if and only if

$$\begin{aligned} v(t) = \quad & \left[a_0 + \textstyle\int_0^t \langle p^*(s), g(v(s),s,\rho,\nu)\rangle_* \, ds\,\right] p(t) \\ & + \left[b_0 + \textstyle\int_0^t \langle q^*(s), g(v(s),s,\rho,\nu)\rangle_* \, ds\,\right] q(t) \\ & + \left[c_0 + \textstyle\int_0^t \langle r_+^*(s), g(v(s),s,\rho,\nu)\rangle_* \, ds\,\right] r_+(t) \\ & + \left[d_0 + \textstyle\int_0^t \langle r_-^*(s), g(v(s),s,\rho,\nu)\rangle_* \, ds\,\right] r_-(t). \end{aligned} \quad (7.38)$$

If we require that $v(t) \underset{t\to+\infty}{\longrightarrow} 0$ and recalling that

$$q^*(t) \underset{t\to+\infty}{\longrightarrow} 0, \qquad |r_+^*(t)| = |r_-^*(t)| = 1,$$

we necessarily have

$$\langle q^*(t), v(t)\rangle_* \underset{t\to+\infty}{\longrightarrow} 0, \langle r_+^*(t), v(t)\rangle_* \underset{t\to+\infty}{\longrightarrow} 0, \langle r_-^*(t), v(t)\rangle_* \underset{t\to+\infty}{\longrightarrow} 0.$$

This implies

$$b_0 = -\int_0^{+\infty} \langle q^*(t), g(v(t), t, \rho, \nu)\rangle_* \, dt, \tag{7.39}$$

$$c_0 = -\int_0^{+\infty} \langle r_+^*(t), g(v(t), t, \rho, \nu)\rangle_* \, dt, \tag{7.40}$$

$$d_0 = -\int_0^{+\infty} \langle r_-^*(t), g(v(t), t, \rho, \nu)\rangle_* \, dt. \tag{7.41}$$

Moreover we choose $a_0 = 0$, which corresponds to a choice of the origin of the time t.

and we can conclude that

Lemma 7.4.21. *For every $\nu \in]0, \nu_1]$, $\rho \in [0, \rho_s[$, $v \in BC_\lambda^0(1)$ if $v = \mathcal{F}(v)$ with*

$$\begin{aligned}(\mathcal{F}(v))(t) = & \int_0^t \langle p^*(s), g(v(s), s, \rho, \nu)\rangle_* \, ds \; p(t) \\ & - \int_t^{+\infty} \langle q^*(s), g(v(s), s, \rho, \nu)\rangle_* \, ds \; q(t) \\ & - \int_t^{+\infty} \langle r_+^*(s), g(v(s), s, \rho, \nu)\rangle_* \, ds \; r_+(t) \\ & - \int_t^{+\infty} \langle r_-^*(s), g(v(s), s, \rho, \nu)\rangle_* \, ds \; r_-(t)\end{aligned}$$

then v is a solution of (7.34).

Then, using Lemmas 7.3.12, 7.3.13, 7.3.14, 7.3.15, which give the size of (p, q, r_+, r_-) and (p^*, q^*, r_+^*, r_-^*), and Lemma 7.4.20 which gives the size of g we obtain

Lemma 7.4.22. *There exists $\mathcal{M}_2$ such that for every $\nu \in]0, \nu_1]$, $\rho \in [0, \rho_s[$ and every $v, v' \in BC_\lambda^0(1)$,*

(a) $|\mathcal{F}(v)|_{C_\lambda^0} \le \mathcal{M}_2 \left(|v|_{C_\lambda^0}^2 + \rho\nu^2\right)$

(b) $|\mathcal{F}(v) - \mathcal{F}(v')|_{C_\lambda^0} \le \mathcal{M}_2 \left(|v|_{C_\lambda^0} + |v'|_{C_\lambda^0} + \rho\nu^2\right) |v - v'|_{C_\lambda^0}$

We can then conclude

Lemma 7.4.23. *There exists ν_2 such that for every $\nu \in]0, \nu_2]$ and every $\rho \in [0, \rho_s[$, the full system (7.33) admits a solution of the form $Y_s(t, \rho, \nu) = h(t) + v_s(t, \rho, \nu)$ where v_s is the unique fixed point of $\mathcal{F}$ in $BC^0_\lambda(d)$ with $d = 2\mathcal{M}_2 \rho\nu^2$.*

Remark 7.4.24. Statement (a) of Proposition 7.4.1 readily follows from this Lemma.

Proof. Lemma 7.4.22 ensures that $\mathcal{F}$ is a contraction mapping in $BC^0_\lambda(d)$ if

$$d \leq 1, \qquad \mathcal{M}_2\,(d^2 + \rho\nu^2) \leq d, \qquad \mathcal{M}_2\,(2d + \rho\nu^2) < 1.$$

Choosing $d = 2\mathcal{M}_2\rho\nu^2$, and

$$\nu_2 < \min\left(\nu_1, (2\mathcal{M}_2\rho_s)^{-1/2}, ((4\mathcal{M}_2^2 + \mathcal{M}_2)\rho_s)^{-1/2}\right),$$

ensures that these three conditions are fulfilled for every $\nu \in]0, \nu_2]$ and every $\rho \in [0, \rho_s[$. □

For finishing this first paragraph devoted to the building of the stable manifold, we give a uniqueness result for the fixed point of $\mathcal{F}$ which will be used later:

Lemma 7.4.25. *There exists d_s, $\nu_3 \in]0, \nu_2]$ such that for every $\nu \in]0, \nu_3]$, and every $\rho \in [0, \rho_s[$, $\mathcal{F}$ admits at most one fixed point v satisfying*

$$|v|_o := \sup_{t\in[0,+\infty[} |v| \leq d_s, \tag{7.42}$$

$$|v|_\lambda := \sup_{t\in[0,+\infty[} |v|e^{\lambda|t|} < +\infty, \tag{7.43}$$

Proof. Like for Lemma 7.4.22, using Lemmas 7.3.12, 7.3.13, 7.3.14, 7.3.15, which give the size of (p, q, r_+, r_-) and (p^*, q^*, r_+^*, r_-^*), and Lemma 7.4.20 which gives the size of g we obtain that

$$|\mathcal{F}(v) - \mathcal{F}(v')|_\lambda \leq M\,(|v|_o + |v'|_o + \rho\nu^2)\,|v - v'|_\lambda$$

holds for v, v' satisfying , $|v|_o \leq 1$, $|v'|_o \leq 1$, $|v|_\lambda < +\infty$, $|v'|_\lambda < +\infty$.

So, we choose d_s, $\nu_3 \in]0, \nu_2]$ small enough to have

$$d_s \leq 1, \qquad d_s < \frac{1}{4M}, \qquad \rho_s\nu_3^2 M < \frac{1}{2}.$$

Hence, $\mathcal{F}$ is a contraction which ensures the uniqueness of the fixed point. □

7.4.2.2 Proof of Proposition 7.4.1-(b),(c). Let Y be a homoclinic connection satisfying the Properties (P1), (P2). Since, $v := Y - h$ is a solution of (7.34) which tends to 0 at $+\infty$, it satisfies (7.38) where b_0, c_0, d_0 are respectively given by (7.39),(7.40), (7.41). Moreover v is reversible. Thus,

$$0 = Sv(0) - v(0) = a_0(Sp(0) - p(0)) + b_0(Sq(0) - q(0)) \\ +c_0(Sr_+(0) - r_+(0)) + d_0(Sr_-(0) - r_-(0)).$$

Hence

$$a_0 = d_0 = 0 \tag{7.44}$$

since q, r_+ are reversible, and since p, r_- are antireversible and independent.

Thus v is a fixed point of $\mathcal{F}$ and it satisfies the solvability condition (7.35) which comes from the fact that d_0 has two necessary values given on one hand by (7.41) and on the other hand by (7.44) . This proves (c).

Now to prove (b) we assume that Y also satisfies (P3) where d_s is defined in Lemma 7.4.25. Moreover, Lemma 7.4.23 ensures that $|v_s|_0 \leq |v_s|_\lambda \leq |v_s|_{C^0_\lambda} \leq 2\mathcal{M}_2\rho_s\nu^2$. Then, we choose $\nu_{sr} \in]0, \nu_3[$ small enough to have $2\mathcal{M}_2\rho_s\nu_{sr}^2 < d_s$. For $\nu \in]0, \nu_{sr}]$ and $\rho \in [0, \rho_s[$, v and v_s are two fixed point of $\mathcal{F}$ defined on $[0, +\infty[$ which satisfy (7.42) and (7.43). Then Lemma 7.4.25 ensures that v and v_s are equal on $[0, +\infty[$. Finally, uniqueness of the solution of the Cauchy problem ensures that they coincide on $]-1, +\infty[$. □

7.4.3 Holomorphic continuation of the stable manifold of 0 far away from the singularities

Subsections 7.4.3, 7.4.4, 7.4.5 are devoted to the construction of the holomorphic continuation of the stable manifold Y_s of 0. We proceed in several steps following the strategy proposed in Subsection 2.1.4. We begin in this subsection, by a continuation far away from the singularities.

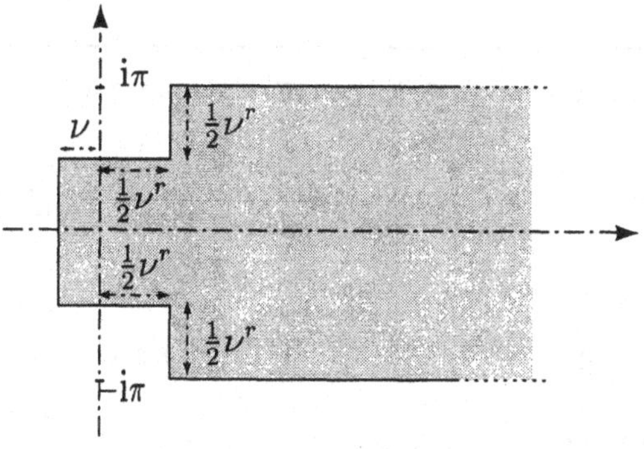

Fig. 7.16. Domain $\mathcal{D}^s_{r,\nu}$

More precisely, we want to show that Y_s admits a holomorphic continuation in $\mathcal{D}^s_{r,\nu}$ for $r \in]0,1[$, where $\mathcal{D}^s_{r,\nu}$ is given by

$$\mathcal{D}^s_{r,\nu} = \{\xi \in \mathbb{C},\ \mathcal{R}e\,(\xi) > -\nu, |\mathcal{I}m\,(\xi)\,| < \pi\}\backslash \{\xi \in \mathbb{C}, \mathcal{R}e\,(\xi) \le \tfrac{1}{2}\nu^r, \pi - \tfrac{1}{2}\nu^r \le |\mathcal{I}m\,(\xi)\,|\}$$

and is drawn in Figure 7.16.

For working in a domain which contains the intervals $[0,\xi]$ and $[\xi,+\infty) := \xi + \mathbb{R}^+$ for all the points ξ inside it, we work with the domain $\mathcal{D}^*_{r,\nu} \supset \mathcal{D}^s_{r,\nu}$ defined as the interior of the polygon whose vertices are the points of affix

$$-\nu + \mathrm{i}(\pi - \tfrac{1}{2}\nu^r),\ \left(\frac{\pi - \frac{1}{2}\nu^r}{\pi}\right)\tfrac{1}{2}\nu^r + \mathrm{i}(\pi - \tfrac{1}{2}\nu^r),\ \mathrm{i}\pi + \tfrac{1}{2}\nu^r,$$
$$-\mathrm{i}\pi + \tfrac{1}{2}\nu^r,\ \left(\frac{\pi - \frac{1}{2}\nu^r}{\pi}\right)\tfrac{1}{2}\nu^r - \mathrm{i}(\pi - \tfrac{1}{2}\nu^r), -\nu - \mathrm{i}(\pi - \tfrac{1}{2}\nu^r).$$

$\mathcal{D}^*_{r,\nu}$ is drawn in figure 7.17.

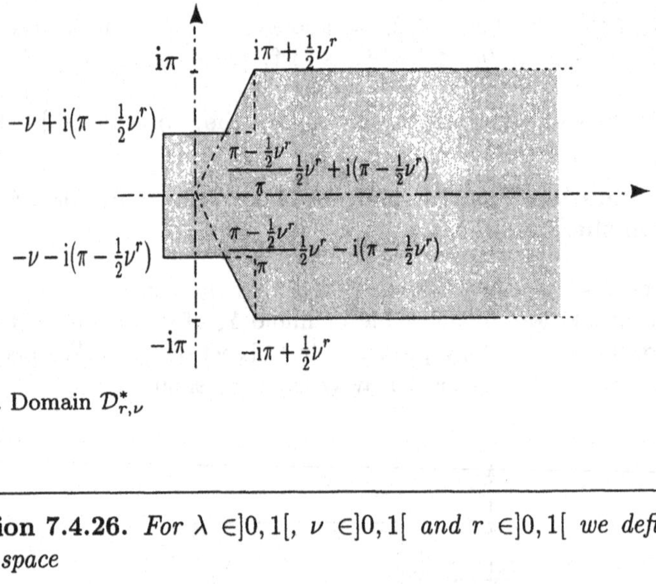

Fig. 7.17. Domain $\mathcal{D}^*_{r,\nu}$

Definition 7.4.26. *For $\lambda \in]0,1[$, $\nu \in]0,1[$ and $r \in]0,1[$ we define the Banach space*

$${}^s\mathcal{H}^{\gamma,\lambda}_{\nu}\,(r) = \{f : \mathcal{D}^*_{r,\nu} \to \mathbb{C} \text{ holomorphic},\ f|_{]-\nu,+\infty[} \in \mathbb{R},\ |f|_{{}^s\mathcal{H}^{\gamma,\lambda}_{\nu}(r)} < +\infty\}$$

with

$$|f|_{{}^s\mathcal{H}^{\gamma,\lambda}_{\nu}(r)} := \sup_{\xi \in \mathcal{D}^*_{r,\nu}} \left[(|f(\xi,\nu)| \left(|\pi^2 + \xi^2|^{\gamma}\chi_1(\xi) + (1-\chi_1(\xi))\mathrm{e}^{\lambda|\mathcal{R}e(\xi)|}\right)\right]$$

with $\chi_1(\xi) = 1$ if $\mathcal{R}e\,(\xi) < 1$ and $\chi_1(\xi) = 0$ otherwise (see Figure 7.18).

We can then define the Banach space

$$^{s}\mathsf{H}_{\nu}^{\lambda}(r) = {}^{s}\mathcal{H}_{\nu}^{3,\lambda}(r) \times {}^{s}\mathcal{H}_{\nu}^{4,\lambda}(r) \times {}^{s}\mathcal{H}_{\nu}^{3,\lambda}(r) \times {}^{s}\mathcal{H}_{\nu}^{3,\lambda}(r)$$

normed with

$$|v|_{^{s}\mathsf{H}_{\nu}^{\lambda}(r)} := |\alpha|_{^{s}\mathcal{H}_{\nu}^{3,\lambda}(r)} + |\beta|_{^{s}\mathcal{H}_{\nu}^{4,\lambda}(r)} + |A|_{^{s}\mathcal{H}_{\nu}^{3,\lambda}(r)} + |B|_{^{s}\mathcal{H}_{\nu}^{3,\lambda}(r)} \tag{7.45}$$

where $v = (\alpha, \beta, A, B)$.

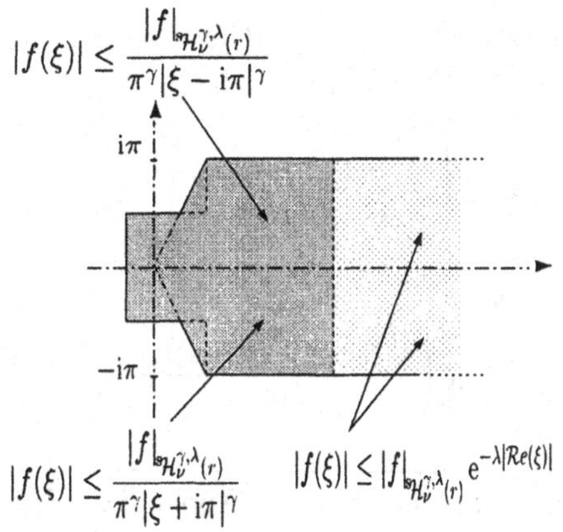

Fig. 7.18. Description of $f \in {}^{s}\mathcal{H}_{\nu}^{\gamma,\lambda}(r)$

We look for an holomorphic continuation of v_s as a fixed point of $\mathcal{F}$ in the ball $\mathsf{B}\,{}^{s}\mathsf{H}_{\nu}^{\lambda}(r)[d\nu]$ with $d \leq 1$ where

$$\mathsf{B}\,{}^{s}\mathsf{H}_{\nu}^{\lambda}(r)[\delta] = \{v \in {}^{s}\mathsf{H}_{\nu}^{\lambda}(r)\ /\ |v|_{^{s}\mathsf{H}_{\nu}^{\lambda}(r)} \leq \delta\}.$$

For $v \in \mathsf{B}\,{}^{s}\mathsf{H}_{\nu}^{\lambda}(r)[d\nu], \xi \in \mathcal{D}_{r,\nu}^{*}$, $(\mathcal{F}(v))(\xi)$ is given by

$$\begin{aligned}
(\mathcal{F}(v))(\xi) \ = \ & \int_{[0,\xi]} \langle p^{*}(s), g(v(s), s, \rho, \nu)\rangle_{*}\, ds\ p(\xi) \\
& - \int_{[\xi,+\infty)} \langle q^{*}(s), g(v(s), s, \rho, \nu)\rangle_{*}\, ds\ q(\xi) \\
& - \int_{[\xi,+\infty)} \langle r_{+}^{*}(s), g(v(s), s, \rho, \nu)\rangle_{*}\, ds\ r_{+}(\xi) \\
& - \int_{[\xi,+\infty)} \langle r_{-}^{*}(s), g(v(s), s, \rho, \nu)\rangle_{*}\, ds\ r_{-}(\xi)
\end{aligned}$$

where $\langle Y, Y' \rangle_* := \alpha\alpha' + \beta\beta' + AA' + BB'$. Here again, we have identified $\mathbb{C}^4$ and $(\mathbb{C}^4)^*$ with $Y \mapsto \langle Y, \cdot \rangle_*$ (see Definition 6.3.5). The dual basis (p^*, q^*, r_+^*, r_-^*) is given by

$$p^* = (2q_\beta, -2q_\alpha, 0, 0), \; q^* = (-2p_\beta, 2p_\alpha, 0, 0), \; r_+^* = r_+, r_-^* = r_-. \quad (7.46)$$

Lemma 7.4.27. *For every $r \in]0,1[$, there exists $\nu_4(r) > 0$ such that for every $\nu \in]0, \nu_4(r)]$, $\rho \in [0, \rho_s[$, $d \in [0,1]$, and $v \in \mathsf{B}^s\mathsf{H}_\nu^\lambda(r)[d\nu]$, $\mathcal{F}(v)$ is holomorphic in $\mathcal{D}_{r,\nu}^*$ and maps $\mathbb{R}$ in $\mathbb{R}^4$ which is equivalent to*

$$(\mathcal{F}(v))(\overline{\xi}) = \overline{(\mathcal{F}(v))(\xi)}, \quad \text{for } \xi \in \mathcal{D}_{r,\nu}^*.$$

Moreover, there exists M_{out} such that for every $r \in]0,1[$, $\nu \in]0, \nu_4(r)]$, $\rho \in [0, \rho_s[$, $d \in [0,1]$, $v, v' \in \mathsf{B}^s\mathsf{H}_\nu^\lambda(r)[d\nu]$,

$$\text{(a)} \quad |\mathcal{F}(v)|_{{}^s\mathsf{H}_\nu^\lambda(r)} \leq M_{\text{out}} \left(\frac{|v|^2_{{}^s\mathsf{H}_\nu^\lambda(r)}}{\nu^r} + \rho\nu^{2-r} \right),$$

$$\text{(b)} \quad |\mathcal{F}(v) - \mathcal{F}(v')|_{{}^s\mathsf{H}_\nu^\lambda(r)} \leq$$

$$M_{\text{out}} \left(\frac{|v|_{{}^s\mathsf{H}_\nu^\lambda(r)} + |v'|_{{}^s\mathsf{H}_\nu^\lambda(r)}}{\nu^r} + \rho\nu^{2-2r} \right) |v - v'|_{{}^s\mathsf{H}_\nu^\lambda(r)}.$$

The proof of this Lemma is given in Appendix 7.F.

We have now enough material to prove the existence of a holomorphic continuation of v_s in $\mathcal{D}_{r,\nu}^*$.

Proposition 7.4.28. *For every r in $]0,1[$ there exists $\nu_5(r) > 0$ such that for every $\nu \in]0, \nu_5(r)]$, $\rho \in [0, \rho_s[$, Y_s admits a holomorphic continuation in $\mathcal{D}_{r,\nu}^* \supset \mathcal{D}_{r,\nu}^s$ of the form*

$$Y_s = h + v_s.$$

We denote this holomorphic continuation by Y_s too. It satisfies

$$|v_s|_{{}^s\mathsf{H}_\nu^\lambda(r)} = |\alpha_{v_s}|_{{}^s\mathcal{H}_\nu^{3,\lambda}(r)} + |\beta_{v_s}|_{{}^s\mathcal{H}_\nu^{4,\lambda}(r)} + |A_{v_s}|_{{}^s\mathcal{H}_\nu^{3,\lambda}(r)} + |B_{v_s}|_{{}^s\mathcal{H}_\nu^{3,\lambda}(r)} \leq 2M_{\text{out}}\rho\nu^{2-r}$$

for every $r \in]0,1[$, $\nu \in]0, \nu_5(r)]$, and $\rho \in [0, \rho_s[$ where

$$|f|_{{}^s\mathcal{H}_\nu^{\gamma,\lambda}(r)} := \sup_{\xi \in \mathcal{D}_{r,\nu}^*} \left[(|f(\xi,\nu)| \left(|\pi^2 + \xi^2|^\gamma \chi_1(\xi) + (1 - \chi_1(\xi)) e^{\lambda |\mathcal{R}e(\xi)|} \right) \right]$$

with $\chi_1(\xi) = 1$ if $\mathcal{R}e(\xi) < 1$ and $\chi_1(\xi) = 0$ otherwise.

In what follows we will use this proposition with $r = \frac{1}{2}$.

Corollary 7.4.29. *There exists* $\nu_6 := \nu_5(\frac{1}{2}) > 0$ *such that for every* $\nu \in]0, \nu_6]$, $\rho \in [0, \rho_s[$, Y_s *admits a holomorphic continuation in* $\mathcal{D}^*_{\frac{1}{2}r,\nu} \supset \mathcal{D}^s_{\frac{1}{2},\nu}$ *of the form*

$$Y_s = h + v_s.$$

We denote this holomorphic continuation by Y_s *too. It satisfies*

$$|v_s|_{{}^s\mathsf{H}^\lambda_\nu(\frac{1}{2})} = |\alpha_{v_s}|_{{}^s\mathcal{H}^{3,\lambda}_\nu(\frac{1}{2})} + |\beta_{v_s}|_{{}^s\mathcal{H}^{4,\lambda}_\nu(\frac{1}{2})} + |A_{v_s}|_{{}^s\mathcal{H}^{3,\lambda}_\nu(\frac{1}{2})} + |B_{v_s}|_{{}^s\mathcal{H}^{3,\lambda}_\nu(\frac{1}{2})} \le 2M_{\text{out}}\rho\nu^{\frac{3}{2}}$$

for every $\nu \in]0\nu_6]$, *and* $\rho \in [0, \rho_s[$ *where* $v_s := (\alpha_{v_s}, \beta_{v_s}, A_{v_s}, B_{v_s})$.

Proof of Proposition 7.4.28 We proceed in two steps. We first show that $\mathcal{F}$ admits a fixed point v^* in $\mathsf{B}\,{}^s\mathsf{H}^\lambda_\nu(r)[d\nu]$ for d small enough and then, we show that $v^*|_{[-\nu,+\infty[} = v_s$.

Step 1. Lemma 7.4.27 ensures that for $r \in]0, 1]$, $\nu \in]0, \nu_4(r)]$ and $\rho \in [0, \rho_s[$, $\mathcal{F}$ is a contraction mapping in $\mathsf{B}\,{}^s\mathsf{H}^\lambda_\nu(r)[d\nu]$ if

(i) $d \in [0, 1]$,

(ii) $M_{\text{out}}(\frac{d^2\nu^2}{\nu^r} + \rho\nu^{2-r}) \le d\nu$,

(iii) $M_{\text{out}}(\frac{2d\nu}{\nu^r} + \rho\nu^{2-2r}) < 1$.

Setting $d = 2M_{\text{out}}\rho\nu^{1-r}$, (i),(ii),(iii) are equivalent to

$$\rho\nu^{1-r} \le \frac{1}{2M_{\text{out}}}, \qquad \rho\left(4M_{\text{out}}^2 + M_{\text{out}}\right)\nu^{2-2r} < 1.$$

These conditions are satisfied for every $\rho \in [0, \rho_s[$ and $\nu \in]0, \nu_5(r)]$ where $\nu_5(r) > 0$ is chosen such that

$$\nu_5(r) < \min\left[\nu_4(r), \left(\frac{1}{2\rho_s M_{\text{out}}}\right)^{1/(1-r)}, \left(\frac{1}{\rho_s(4M_{\text{out}}^2 + M_{\text{out}})}\right)^{1/(2-2r)}\right].$$

To summarize, this first step ensures that for every r in $]0, 1[$, $\nu \in]0, \nu_5(r)]$ and every $\rho \in [0, \rho_s[$, $\mathcal{F}$ admits a unique fixed point $v^* \in \mathsf{B}\,{}^s\mathsf{H}^\lambda_\nu(r)[2M_{\text{out}}\nu^{2-r}]$.

Step 2. Let us define $\check{v}^* = v^*|_{[0,+\infty[}$. Observe that $\check{v}^*$ maps $\mathbb{R}$ into $\mathbb{R}^4$ and that it is a fixed point of $\mathcal{F}$ too. Moreover, since $e^{\lambda|t|} \ge 1$ and $(\pi^2 + t^2) \ge \pi^2 \ge 1$,

$$\begin{aligned}
\sup_{t\in[0,+\infty[} |\check{v}^*(t)| &= |\check{v}^*|_0 \\
&\le |\alpha^*|_{{}^s\mathcal{H}^{3,\lambda}_\nu(r)} + |\beta^*|_{{}^s\mathcal{H}^{4,\lambda}_\nu(r)} + |A^*|_{{}^s\mathcal{H}^{3,\lambda}_\nu(r)} + |B^*|_{{}^s\mathcal{H}^{3,\lambda}_\nu(r)} \\
&\le |v^*|_{{}^s\mathsf{H}^\lambda_\nu(r)} \\
&\le 2M_{\text{out}}\nu^{2-r}.
\end{aligned}$$

Furthermore v_s is a fixed point of $\mathcal{F}$ too and proposition 7.4.1 ensures that

$$\sup_{t\in[0,+\infty[} |v_s| \le \sup_{t\in]-1,+\infty[} \left(|v_s| e^{\lambda|t|}\right) \le M_{sr}\nu^2.$$

Then, choosing $\nu_5(r)$ small enough to have

$$M_{sr}(\nu_5(r))^2 \le d_s, \qquad 2M_{out}(\nu_5(r))^{2-r} \le d_s,$$

Lemma 7.4.25 ensures that $v^*|_{[0,+\infty[} = \check{v}^* = v_s$. Finally, uniqueness of the solution of the Cauchy problem for equation (7.34) ensures that $v^*|_{]-\nu,+\infty[} = \check{v}^* = v_s$. □

7.4.4 Stable manifold of 0 for the inner system

7.4.4.1 Inner system. As explained in the introduction 7.4.1, our aim is to prove that Y_s admits a holomorphic continuation which belongs to

$${}^{s}E^{2,\lambda}_{\pi,\delta_s}(2,3,\tfrac{1}{2},\tfrac{1}{4})\times {}^{s}E^{3,\lambda}_{\pi,\delta_s}(3,4,\tfrac{1}{2},\tfrac{1}{4})\times {}^{s}E^{2,\lambda}_{\pi,\delta_s}(2,3,\tfrac{1}{2},\tfrac{1}{4})\times {}^{s}E^{2,\lambda}_{\pi,\delta_s}(2,3,\tfrac{1}{2},\tfrac{1}{4}).$$

In the previous subsection, we proved that Y_s admits a holomorphic continuation of the form $Y_s = h + v_s$ in the domain $\mathcal{D}^s_{\frac{1}{2},\nu}$. On this domain v_s is small with respect to h which expresses that the normal form is the relevant part of the full system far away from the singularities.

Now, we want to prove that near the singularities, $Y_s = (\alpha_s, \beta_s, A_s, B_s)$ reads

$$\begin{aligned}
\alpha_s(i\pi+\nu z,\rho,\nu) &= \frac{1}{\nu^2}\widehat{\alpha}^+_{s,0}(z,\rho) + \frac{\nu^{\frac{1}{4}}}{\nu^2}\widehat{\alpha}^+_{s,1}(z,\rho,\nu),\\
A_s(i\pi+\nu z,\rho,\nu) &= \frac{1}{\nu^2}\widehat{A}^+_{s,0}(z,\rho) + \frac{\nu^{\frac{1}{4}}}{\nu^2}\widehat{A}^+_{s,1}(z,\rho,\nu),\\
B_s(i\pi+\nu z,\rho,\nu) &= \frac{1}{\nu^2}\widehat{B}^+_{s,0}(z,\rho) + \frac{\nu^{\frac{1}{4}}}{\nu^2}\widehat{B}^+_{s,1}(z,\rho,\nu),\\
\beta_s(i\pi+\nu z,\rho,\nu) &= \frac{1}{\nu^3}\widehat{\beta}^+_{s,0}(z,\rho) + \frac{\nu^{\frac{1}{4}}}{\nu^3}\widehat{\beta}_{s,1}(z,\rho,\nu)
\end{aligned} \tag{7.47}$$

where

$$\widehat{\alpha}^+_{s,0}(\cdot,\rho,\cdot),\ \widehat{A}^+_{s,0}(\cdot,\rho,\cdot),\ \widehat{B}^+_{s,0}(\cdot,\rho,\cdot) \in {}^{s}\widehat{H}^2_{\delta_s}, \qquad \widehat{\beta}^+_{s,0}(\cdot,\rho,\cdot) \in {}^{s}\widehat{H}^3_{\delta_s},$$
$$\widehat{\alpha}_{s,1}(\cdot,\rho,\cdot),\ \widehat{A}^+_{s,1}(\cdot,\rho,\cdot),\ \widehat{B}^+_{s,1}(\cdot,\rho,\cdot) \in {}^{s}\widehat{H}^3_{\delta_s,\frac{1}{2}}, \qquad \widehat{\beta}_{s,1}(\cdot,\rho,\cdot) \in {}^{s}\widehat{H}^4_{\delta_s,\frac{1}{2}},$$

Following the strategy proposed in 2.1.4, we introduce the two systems of coordinates

- the *outer system of coordinates* (Y,ξ) with $Y = (\alpha,\beta,A,B)$ which is the initial one,

– the *inner system of coordinates* $(\widehat{Y}, z)$ with $\widehat{Y} = (\widehat{\alpha}, \widehat{\beta}, \widehat{A}, \widehat{B})$ where

$$\xi = \mathrm{i}\pi + \nu z, \qquad \widehat{\alpha} = \nu^2\alpha, \ \ \widehat{\beta} = \nu^3\beta, \ \ \widehat{A} = \nu^2 A, \ \ \widehat{B} = \nu^2 B, \tag{7.48}$$

which is useful for describing the solution near $\mathrm{i}\pi$.

Moreover, we introduce the following notation: for an orbit $Z(\xi) = (\alpha(\xi), \beta(\xi), A(\xi), B(\xi))$ in the outer system of coordinates, let us denote by

$$\widehat{Z}(z) = \left(\nu^2\alpha(\mathrm{i}\pi + \nu z),\ \nu^3\beta(\mathrm{i}\pi + \nu z),\ \nu^2 A(\mathrm{i}\pi + \nu z),\ \nu^2 B(\mathrm{i}\pi + \nu z)\right)$$

the corresponding orbit in the inner system of coordinates. Reciprocally, for an orbit $\widehat{Y}(z) = (\widehat{\alpha}(z), \widehat{\beta}(z), \widehat{A}(z), \widehat{B}(z))$ in the inner system of coordinates, let us denote by

$$Y(\xi) = \left(\frac{1}{\nu^2}\widehat{\alpha}\left(\tfrac{\xi - \mathrm{i}\pi}{\nu}\right), \frac{1}{\nu^3}\widehat{\beta}\left(\tfrac{\xi - \mathrm{i}\pi}{\nu}\right), \frac{1}{\nu^2}\widehat{A}\left(\tfrac{\xi - \mathrm{i}\pi}{\nu}\right), \frac{1}{\nu^2}\widehat{B}\left(\tfrac{\xi - \mathrm{i}\pi}{\nu}\right)\right) \tag{7.49}$$

the corresponding orbit in the outer system of coordinates.

We use the same notation for the vector fields: for a vector field f expressed in the outer system of coordinates by

$$\begin{array}{rccl} f: & \mathbb{C}^4 & \to & \mathbb{C}^4 \\ & Y = (\alpha, \beta, A, B) & \mapsto & (f_\alpha(Y), f_\beta(Y), f_A(Y), f_B(Y)), \end{array}$$

let us denote by $\widehat{f}$ the corresponding vector fields in the inner system of coordinates given by

$$\begin{array}{rccl} \widehat{f}: & \mathbb{C}^4 & \to & \mathbb{C}^4 \\ & \widehat{Y} = (\widehat{\alpha}, \widehat{\beta}, \widehat{A}, \widehat{B}) & \mapsto & \begin{pmatrix} \nu^3 f_\alpha(\widehat{\alpha}/\nu^2, \widehat{\beta}/\nu^3, \widehat{A}/\nu^2, \widehat{B}/\nu^2) \\ \nu^4 f_\beta(\widehat{\alpha}/\nu^2, \widehat{\beta}/\nu^3, \widehat{A}/\nu^2, \widehat{B}/\nu^2) \\ \nu^3 f_A(\widehat{\alpha}/\nu^2, \widehat{\beta}/\nu^3, \widehat{A}/\nu^2, \widehat{B}/\nu^2) \\ \nu^3 f_B(\widehat{\alpha}/\nu^2, \widehat{\beta}/\nu^3, \widehat{A}/\nu^2, \widehat{B}/\nu^2) \end{pmatrix} \end{array}$$

and vice versa. Finally, for a domain Σ of $\mathbb{C}$, in the outer system of coordinates, let us denote by $\widehat{\Sigma} = (\Sigma - \mathrm{i}\pi)/\nu$ the corresponding one in the inner system of coordinates and vice versa.

With this notation, the full system (7.33) is equivalent to

$$\frac{d\widehat{Y}}{dz} = \widehat{N}(\widehat{Y}, \nu) + \rho\widehat{R}(\widehat{Y}, \nu) = \widehat{f}(\widehat{Y}, \nu, \rho) \tag{7.50}$$

with

$$\widehat{N}(\widehat{Y}, \nu) = \begin{pmatrix} \widehat{\beta} \\ \nu^2\widehat{\alpha} - \frac{3}{2}\widehat{\alpha}^2 - c(\widehat{A}^2 + \widehat{B}^2) \\ -\widehat{B}\ (\omega + a\nu^2 + b\widehat{\alpha}) \\ \widehat{A}\ (\omega + a\nu^2 + b\widehat{\alpha}) \end{pmatrix},$$

$$\widehat{R}(\widehat{Y},\nu) = \sum_{(\overline{m},\ell)\in\mathcal{I}} a_{\overline{m},\ell}\ \widehat{Y}^{\overline{m}}\ \nu^{2\ell}.$$

Then, for proving that Y_s admits a holomorphic continuation satisfying (7.47) near $\xi = i\pi$, it is equivalent to prove that $\widehat{Y}_s$ admits a holomorphic continuation near $z = 0$ which satisfies

$$\widehat{\alpha}_s = \widehat{\alpha}^+_{s,0} + \nu^{\frac{1}{4}}\widehat{\alpha}^+_{s,1},\quad \widehat{A}_s = \widehat{A}^+_{s,0} + \nu^{\frac{1}{4}}\widehat{A}^+_{s,1},\quad \widehat{B}_s = \widehat{B}^+_{s,0} + \nu^{\frac{1}{4}}\widehat{B}^+_{s,1},$$
$$\widehat{\beta}_s = \widehat{\beta}^+_{s,0} + \nu^{\frac{1}{4}}\widehat{\beta}_{s,1},$$

or equivalently

$$\widehat{Y}_s(z,\rho,\nu) = \widehat{Y}^+_{s,0}(z,\rho) + \nu^{\frac{1}{4}}\widehat{Y}^+_{s,1}(z,\rho,\nu) \tag{7.51}$$

where

$$\widehat{Y}^+_{s,0} = (\widehat{\alpha}^+_{s,0}, \widehat{\beta}^+_{s,0}, \widehat{A}^+_{s,0}, \widehat{B}^+_{s,0}) \qquad \widehat{Y}^+_{s,1} = (\widehat{\alpha}^+_{s,1}, \widehat{\beta}_{s,1}, \widehat{A}^+_{s,1}, \widehat{B}^+_{s,1}).$$

So the first task is to determine the principal part $\widehat{Y}^+_{s,0}$ of Y_s near $i\pi$. Observe that

$$\widehat{Y}^+_{s,0}(z,\rho) = \lim_{\nu\to 0} \widehat{Y}_s(z,\rho,\nu).$$

So following strategy proposed in subsection 2.1.4, our candidate to be the principal part of Y_s near $i\pi$ is the stable manifold of 0 for the *inner system* obtained by setting $\nu = 0$ in (7.50). For that purpose we define

$$\begin{aligned}
\widehat{f}_0(\widehat{Y},\rho) &= \widehat{f}(\widehat{Y},0,\rho), & \widehat{f}_1(\widehat{Y},\nu,\rho) &= \widehat{f}(\widehat{Y},\nu,\rho) - \widehat{f}_0(\widehat{Y},\rho),\\
\widehat{N}_0(\widehat{Y}) &= \widehat{N}(\widehat{Y},0), & \widehat{N}_1(\widehat{Y},\nu) &= \widehat{N}(\widehat{Y},\nu) - \widehat{N}_0(\widehat{Y}),\\
\widehat{R}_0(\widehat{Y}) &= \widehat{R}(\widehat{Y},0), & \widehat{R}_1(\widehat{Y},\nu) &= \widehat{R}(\widehat{Y},\nu) - \widehat{R}_0(\widehat{Y}).
\end{aligned} \tag{7.52}$$

The functions $\widehat{N}_0$, $\widehat{N}_1$, $\widehat{R}_0$, $\widehat{R}_1$ are explicitly given by

$$\widehat{N}_0(\widehat{Y}) = \begin{pmatrix} \widehat{\beta} \\ -\frac{3}{2}\widehat{\alpha}^2 - c(\widehat{A}^2 + \widehat{B}^2) \\ -\widehat{B}\ (\omega + b\widehat{\alpha}) \\ \widehat{A}\ (\omega + b\widehat{\alpha}) \end{pmatrix},\quad \widehat{N}_1(\widehat{Y},\nu) = \begin{pmatrix} 0 \\ \nu^2\widehat{\alpha} \\ -a\widehat{B}\nu^2 \\ a\widehat{A}\nu^2 \end{pmatrix}, \tag{7.53}$$

$$\widehat{R}_0(\widehat{Y}) = \sum_{\overline{m}\in\mathbb{N},\ |\overline{m}|\geq 3} a_{\overline{m},0}\ \widehat{Y}^{\overline{m}},\qquad \widehat{R}_1(\widehat{Y},\nu) = \sum_{\substack{(\overline{m},\ell)\in\mathcal{I}\\ \ell\neq 0}} a_{\overline{m},\ell}\ \widehat{Y}^{\overline{m}}\ \nu^{2\ell}. \tag{7.54}$$

Then, the inner system reads

$$\frac{d\widehat{Y}}{dz} = \widehat{N}_0(\widehat{Y}) + \rho\widehat{R}_0(\widehat{Y}) = \widehat{f}_0(\widehat{Y},\rho) \tag{7.55}$$

Remark 7.4.30. By construction, the inner system does not depend on the bifurcation parameter ν but does not depend on ρ whereas the normal form depends on ν but does not depend on ρ.

Remark 7.4.31. Observe that the scaling (7.48) combined with setting $\nu = 0$ acts like a selector of monomials : some of the monomials of the normal form (those in $\widehat{N}_0$) still are in the principal part near $i\pi$ whereas some others (those in $\widehat{N}_1$) are now treated as perturbation terms. The same thing is true for higher order terms.

Remark 7.4.32. Symmetries. We are only interested in solutions Y of (7.33) which are real on the real axis and which are reversible. The uniqueness of the holomorphic continuation ensures that the holomorphic continuation of such solution satisfies

$$SY(-\xi) = Y(\xi), \qquad Y(\overline{\xi}) = \overline{Y(\xi)}$$

for $\xi \in \mathbb{C}$. These two symmetries imply a third one with respect to the imaginary axis

$$SY(\overline{-\xi}) = \overline{Y(\xi)}.$$

The change of time ($\xi = i\pi + \nu z$) is compatible with this third symmetry. Indeed, for $\xi = i\pi + \nu z$, $-\overline{\xi} = i\pi + \nu(-\overline{z})$ holds. In the inner system of coordinates, this solution satisfies

$$S\widehat{Y}(-\overline{z}) = \overline{\widehat{Y}(z)}.$$

We can sum up this result with

Lemma 7.4.33. *Every reversible solution Y of (7.33) which is real on the real axis satisfies*

$$SY(\overline{-\xi}) = \overline{Y(\xi)},$$

or equivalently

$$S\widehat{Y}(\overline{-z}) = \overline{\widehat{Y}(z)}$$

in the inner set of coordinates.

7.4.4.2 Construction of the stable manifold. This subsection is devoted to the construction of a parameterization of "the stable manifold of 0" for the inner system : we look for a solution $\widehat{Y}^{+}_{\mathrm{s},0}(z,\rho)$ of the inner system which tends to 0 as $\mathcal{R}e\,(z) \to +\infty$. Moreover we are interested in the holomorphy of $\widehat{Y}^{+}_{\mathrm{s},0}$ with respect to ρ. Thus in this subsection we work with $\rho \in D(0,\rho_{\mathrm{s}})$ where $D(0,\rho_{\mathrm{s}}) = \{\rho \in \mathbb{C}/\ |\rho| < \rho_{\mathrm{s}}\}$.

Proposition 7.4.34. *There exists $\delta_1 \geq 1$ such that for every $\delta \geq \delta_1$, the inner system (7.55) admits a solution $\widehat{Y}^{+}_{\mathrm{s},0}(z,\rho)$ continuous with respect to $(z,\rho) \in \widehat{\Omega}^{\mathrm{s}}_{\delta} \times D(0,\rho_{\mathrm{s}})$, holomorphic with respect to z for every fixed $\rho \in D(0,\rho_{\mathrm{s}})$ and holomorphic with respect to ρ for every fixed $z \in \widehat{\Omega}^{\mathrm{s}}_{\delta}$ where $\widehat{\Omega}^{\mathrm{s}}_{\delta}$ is given by $\widehat{\Omega}^{\mathrm{s}}_{\delta} :=]-1,+\infty[\times]-\infty,-\delta[$.*

The function $\widehat{Y}^{+}_{s,0}$ *reads* $\widehat{Y}^{+}_{s,0}(z,\rho) := \widehat{h}^{+}_{0}(z) + \widehat{W}^{+}_{s,0}(z,\rho)$ *where the function* $\widehat{h}^{+}_{0}$ *reads* $\widehat{h}^{+}_{0}(z) = (-4/z^2, 8/z^3, 0, 0)$. *Moreover there exists* M^{+}_{0} *such that for every* $\delta \geq \delta_1$ *and every* $\rho \in D(0,\rho_s)$,

$$\sup_{z\in\widehat{\Omega}^{s}_{\delta}} \left(|z|^{3+\frac{1}{2}} |\widehat{\alpha}_{\widehat{W}^{+}_{s,0}}(z,\rho)|\right) + \sup_{z\in\widehat{\Omega}^{s}_{\delta}} \left(|z|^{4+\frac{1}{2}} |\widehat{\beta}_{\widehat{W}^{+}_{s,0}}(z,\rho)|\right)$$

$$+ \sup_{z\in\widehat{\Omega}^{s}_{\delta}} \left(|z|^{4} |\widehat{A}_{\widehat{W}^{+}_{s,0}}(z,\rho)|\right) + \sup_{z\in\widehat{\Omega}^{s}_{\delta}} \left(|z|^{4} |\widehat{B}_{\widehat{W}^{+}_{s,0}}(z,\rho)|\right) \leq \frac{4M^{+}_{0}|\rho|}{\sqrt{\delta}}$$

where $\widehat{W}^{+}_{s,0}(z,\rho) = (\widehat{\alpha}_{\widehat{W}^{+}_{s,0}}, \widehat{\beta}_{\widehat{W}^{+}_{s,0}}, \widehat{A}_{\widehat{W}^{+}_{s,0}}, \widehat{B}_{\widehat{W}^{+}_{s,0}})$.

Remark 7.4.35. As already mentioned h belongs to

$${}^{s}E^{2,\lambda}_{\pi,\delta_s}(2,3,\tfrac{1}{2},\tfrac{1}{4}) \times {}^{s}E^{3,\lambda}_{\pi,\delta_s}(3,4,\tfrac{1}{2},\tfrac{1}{4}) \times {}^{s}E^{2,\lambda}_{\pi,\delta_s}(2,3,\tfrac{1}{2},\tfrac{1}{4}) \times {}^{s}E^{2,\lambda}_{\pi,\delta_s}(2,3,\tfrac{1}{2},\tfrac{1}{4}).$$

and has for principal part near $i\pi$ the function $\widehat{h}^{+}_{0}$. Observe that $\widehat{h}^{+}_{0}$ is the unique solution (up to time shift) of the truncated inner system

$$\frac{d\widehat{Y}}{dz} = \widehat{N}_0(\widehat{Y})$$

which tends to 0 as $\mathcal{R}e\,(z)$ tends to $+\infty$.

Remark 7.4.36. Since $\widehat{W}^{+}_{s,0}$ is holomorphic with respect to ρ for every $z \in \widehat{\Omega}^{s}_{\delta}$, it can be expanded in power series

$$\widehat{W}^{+}_{s,0}(z,\rho) = \sum_{n\geq 0} \rho^n\, \widehat{W}^{+}_{s,0,n}(z).$$

We check that $\widehat{W}^{+}_{s,0,0} = 0$ and using Cauchy integral for $\widehat{W}^{+}_{s,0}$ with respect to ρ, we prove that $\widehat{W}^{+}_{s,0,n}$ is holomorphic on $\widehat{\Omega}^{s}_{\delta}$. So $\widehat{Y}^{+}_{s,0}$ can be expanded in power series too,

$$\widehat{Y}^{+}_{s,0}(z,\rho) = \sum_{n\geq 0} \rho^n\, \widehat{Y}^{+}_{s,0,n}(z), \tag{7.56}$$

with $\widehat{Y}^{+}_{s,0,0} = \widehat{h}^{+}_{0}$ and $\widehat{Y}^{+}_{s,0,n} = \widehat{W}^{+}_{s,0,n}$ for $n \geq 1$. Using the notation introduced in the previous subsection, we denote by $\widehat{h}$ the transformed of h in the inner system of coordinates

$$\widehat{h} := \left(\widehat{\alpha}_{\widehat{h}}(z), \widehat{\beta}_{\widehat{h}}(z), 0, 0\right) = \left(\frac{-\nu^2}{\sinh^2 \frac{1}{2}\nu z}, \frac{\nu^3 \cosh \frac{1}{2}\nu z}{\sinh^3 \frac{1}{2}\nu z}, 0, 0\right). \tag{7.57}$$

The first component $\widehat{\alpha}_{\widehat{h}}$ has a pole of order 2 and the second $\widehat{\beta}_{\widehat{h}}$ a pole of order 3. Moreover, $\widehat{h}^{+}_{0} = (-4/z^2, 8/z^4, 0, 0)$ is the singular part of $\widehat{h}$ near 0 whereas the singular part of $\widehat{Y}^+_{\mathrm{s},0,n}$ near 0 is unknown in general. However, its general form can be computed in one particular case when the coefficient b in the normal form (7.2) is equal to 0 and when the rest $\widetilde{R}$ consists in a unique monomial, $\widetilde{R} = (0,0,0,\widetilde{\alpha}^3)$ which gives after the scaling (7.4) the rescaled rest $R = (0,0,0,\nu^3\alpha^3)$. This toy model gives an idea of the complexity of the singularities of $\widehat{Y}^+_{\mathrm{s},0,n}$ as n grows: the functions $\widehat{Y}^+_{\mathrm{s},0,n}$ are given by inductions, and for the toy model previously described, it can be showed that every $\widehat{Y}^+_{\mathrm{s},0,n}$ lies in $\mathcal{A}^4$ where $\mathcal{A}$ is the algebra of analytic functions given by

$$\mathcal{A} = \bigcup_{m,q\geq 0} \mathcal{A}_{q,m}, \qquad \mathcal{A}_{q,m} = \left\{ \widehat{f}/\widehat{f}(z) = \sum_{p=0}^{q} \left(\psi_p(z) + \sum_{k=0}^{m} \frac{\lambda_{k,p}}{z^k} \right) (\ln(z))^p \right\}$$

where ψ_p is holomorphic on $\mathbb{C}$. The first functions $\widehat{Y}^+_{\mathrm{s},0,0} = \widehat{h}_0^+$ has a third order pole. The singularities of the following ones $\widehat{Y}^+_{\mathrm{s},0,n}$ are more and more complicated as n grows : $\widehat{Y}^+_{\mathrm{s},0,n}$ lies in $\mathcal{A}_{q_n,m_n}$ and $q_n \underset{n\to+\infty}{\longrightarrow} +\infty$, $m_n \underset{n\to+\infty}{\longrightarrow} +\infty$.

Remark 7.4.37. The spectrum of the linear part of the inner system, i.e. the spectrum of $D_{\widehat{Y}} f_0(0,\rho)$ is $\{0, \pm i\omega\}$ where 0 is a double non semi-simple eigenvalue. Hence, all the eigenvalues have zero real parts. So the stable manifold of 0 can only have a polynomial decay at infinity.

The end of this subsection is devoted to the proof of Proposition 7.4.34: we look for a parameterization of the stable manifold $\widehat{Y}^+_{\mathrm{s},0}$ of 0 for the inner system in the form

$$\widehat{Y}^+_{\mathrm{s},0}(z,\rho) := \widehat{h}_0^+(z) + \widehat{W}^+_{\mathrm{s},0}(z,\rho)$$

where $\widehat{W}^+_{\mathrm{s},0}$ lies in the Banach space ${}^{\mathrm{s}}\widehat{\mathsf{H}}_\delta$ defined by

Definition 7.4.38.

$$\begin{aligned} {}^{\mathrm{s}}\widehat{\mathsf{H}}_\delta := \{\ & \widehat{W} : \widehat{\Omega}^{\mathrm{s}}_\delta \times D(0,2\rho_{\mathrm{s}}) \to \mathbb{C}^4,\ \textit{continuous}\ , \\ & \textit{holomorphic on } \widehat{\Omega}^{\mathrm{s}}_\delta \textit{ for every fixed } \rho \in D(0,2\rho_{\mathrm{s}}), \\ & \textit{holomorphic on } D(0,2\rho_{\mathrm{s}}) \textit{ for every fixed } z \in \widehat{\Omega}^{\mathrm{s}}_\delta, \\ & |\widehat{W}|_{{}^{\mathrm{s}}\widehat{\mathsf{H}}_\delta} < +\infty \} \end{aligned}$$

with

$$\begin{aligned} |\widehat{W}|_{{}^{\mathrm{s}}\widehat{\mathsf{H}}_\delta} = & \sup_{\widehat{\Omega}^{\mathrm{s}}_\delta \times D(0,2\rho_{\mathrm{s}})} \left(|z|^{3+\frac{1}{2}}|\widehat{\alpha}(z,\rho)|\right) + \sup_{\widehat{\Omega}^{\mathrm{s}}_\delta \times D(0,2\rho_{\mathrm{s}})} \left(|z|^{4+\frac{1}{2}}|\widehat{\beta}(z,\rho)|\right) \\ & + \sup_{\widehat{\Omega}^{\mathrm{s}}_\delta \times D(0,2\rho_{\mathrm{s}})} \left(|z|^{4}|\widehat{A}(z,\rho)|\right) + \sup_{\widehat{\Omega}^{\mathrm{s}}_\delta \times D(0,2\rho_{\mathrm{s}})} \left(|z|^{4}|\widehat{B}(z,\rho)|\right). \end{aligned}$$

Let us denote by $\mathsf{B}^{\mathsf{s}}\widehat{\mathsf{H}}_\delta[d]$ *the ball of radius* d *of* ${}^{\mathsf{s}}\widehat{\mathsf{H}}_\delta$

$$\mathsf{B}^{\mathsf{s}}\widehat{\mathsf{H}}_\delta[d] := \{\widehat{W} \in {}^{\mathsf{s}}\widehat{\mathsf{H}}_\delta,\ |\widehat{W}|_{{}^{\mathsf{s}}\widehat{\mathsf{H}}_\delta} \le d\}.$$

Observe that we work here with $\rho \in D(0, 2\rho_s)$. This will be useful for obtaining the estimate of $\widehat{W}^+_{s,0}$ given in Proposition 7.4.34.

The equation satisfied by $\widehat{W}^+_{s,0}$ reads

$$\frac{d\widehat{W}}{dz} - D\widehat{N}_0(\widehat{h}_0^+).\widehat{W} = \widehat{g}_0(\widehat{W}, z, \rho), \tag{7.58}$$

where $\widehat{g}_0(\widehat{W}, z, \rho) := \widehat{Q}_0(\widehat{W}) + \rho\widehat{R}_0(\widehat{h}_0^+ + \widehat{W})$ with

$$\widehat{Q}_0(\widehat{W}) = \begin{pmatrix} 0 \\ -\frac{3}{2}\widehat{\alpha} - c(\widehat{A}^2 + \widehat{B}^2) \\ -b\widehat{\alpha}\widehat{B} \\ b\widehat{\alpha}\widehat{A} \end{pmatrix}. \tag{7.59}$$

Our aim is to transform (7.58) into an integral equation and to solve it using the Contraction Mapping Theorem. For that purpose we need a basis of solution of the linear homogeneous equation

$$\frac{d\widehat{W}}{dz} - D\widehat{N}_0(\widehat{h}_0^+).\widehat{W} = 0. \tag{7.60}$$

Lemma 7.4.39. *A basis of solutions of the homogeneous equation (7.60) is* $(\widehat{\mathbf{p}}, \widehat{\mathbf{q}}, \widehat{\mathbf{r}}_+, \widehat{\mathbf{r}}_-)$ *where*

(a) $\widehat{\mathbf{p}}(z) = (\frac{8}{z^3}, -\frac{24}{z^4}, 0, 0)$ *satisfies* $S\widehat{\mathbf{p}}(-\overline{z}) = -\overline{\widehat{\mathbf{p}}(z)}$,

(b) $\widehat{\mathbf{q}}(z) = (z^4, 4z^3, 0, 0)$ *satisfies* $S\widehat{\mathbf{q}}(-\overline{z}) = \overline{\widehat{\mathbf{q}}(z)}$,

(c) $\widehat{\mathbf{r}}_+ = (0, 0, \cos\widehat{\psi}, \sin\widehat{\psi})$, $\widehat{\mathbf{r}}_- = (0, 0, -\sin\widehat{\psi}, \cos\widehat{\psi})$, $\widehat{\psi}(z) = \omega z + 4b/z$.
The functions $\widehat{\mathbf{r}}_+$, $\widehat{\mathbf{r}}_-$ *satisfy* $S\widehat{\mathbf{r}}_+(-\overline{z}) = \overline{\widehat{\mathbf{r}}_+(z)}$, $S\widehat{\mathbf{r}}_-(-\overline{z}) = -\overline{\widehat{\mathbf{r}}_-(z)}$.

Let us denote by $\langle .,. \rangle$ the inner product in $\mathbb{C}^4$, $\langle Y, Y' \rangle := \overline{\alpha}\alpha' + \overline{\beta}\beta' + \overline{A}A' + \overline{B}B'$. We identify $\mathbb{C}^4$ and $(\mathbb{C}^4)^*$ by $Y \longrightarrow \langle Y, \cdot \rangle_* := \langle \overline{Y}, \cdot \rangle$. With this identification, we denote by $(\widehat{\mathbf{p}}^*, \widehat{\mathbf{q}}^*, \widehat{\mathbf{r}}_+^*, \widehat{\mathbf{r}}_-^*)$ the dual basis of $(\widehat{\mathbf{p}}, \widehat{\mathbf{q}}, \widehat{\mathbf{r}}_+, \widehat{\mathbf{r}}_-)$ (see Definition 6.3.5).

Lemma 7.4.40. *The dual basis* $(\widehat{\mathbf{p}}^*, \widehat{\mathbf{q}}^*, \widehat{\mathbf{r}}_+^*, \widehat{\mathbf{r}}_-^*)$ *is explicitly given by*

$$\widehat{\mathbf{p}}^* = \frac{1}{56}(\widehat{\mathbf{q}}_\beta, -\widehat{\mathbf{q}}_\alpha, 0, 0), \quad \widehat{\mathbf{q}}^* = \frac{1}{56}(-\widehat{\mathbf{p}}_\beta, \widehat{\mathbf{p}}_\alpha, 0, 0), \quad \widehat{\mathbf{r}}_+^* = \widehat{\mathbf{r}}_+, \quad \widehat{\mathbf{r}}_-^* = \widehat{\mathbf{r}}_-.$$

Then using the variation of constants formula we get

Lemma 7.4.41. *There exists* δ_0 *such that for every* $\delta \geq \delta_0$, *every* $\rho \in D(0, 2\rho_s)$ *and every* $\widehat{W} \in \mathsf{B}\,{}^s\widehat{\mathsf{H}}_\delta[1]$, *if* $\widehat{W} = \widehat{\mathcal{F}}_0(\widehat{W})(z,\rho)$ *where*

$$\begin{aligned}(\widehat{\mathcal{F}}_0(\widehat{W}))(z,\rho) \; = \; & - \int_{[z,+\infty)} \langle \widehat{\mathbf{p}}^*(s), \widehat{g}_0(\widehat{W}, s, \rho)\rangle_* ds \; \widehat{\mathbf{p}}(z) \\ & - \int_{[z,+\infty)} \langle \widehat{\mathbf{q}}^*(s), \widehat{g}_0(\widehat{W}, s, \rho)\rangle_* ds \; \widehat{\mathbf{q}}(z) \\ & - \int_{[z,+\infty)} \langle \widehat{\mathbf{r}}_+^*(s), \widehat{g}_0(\widehat{W}, s, \rho)\rangle_* ds \; \widehat{\mathbf{r}}_+(z) \\ & - \int_{[z,+\infty)} \langle \widehat{\mathbf{r}}_-^*(s), \widehat{g}_0(\widehat{W}, s, \rho)\rangle_* ds \; \widehat{\mathbf{r}}_-(z)\end{aligned}$$

then $\widehat{W}$ *is a solution of the inner system (7.55).*

The conditions "$\delta \geq \delta_0$"and "$|\widehat{W}|_{{}^s\widehat{\mathsf{H}}_\delta} \leq 1$" ensure that $\widehat{g}_0(\widehat{W}, s, \rho)$ is well defined. The proof of this lemma is given in Appendix 7.G.

Lemma 7.4.42. *For every* $\delta \geq \delta_0$ *and every* $\widehat{W} \in \mathsf{B}\,{}^s\widehat{\mathsf{H}}_\delta[1]$, $\widehat{\mathcal{F}}_0(\widehat{W})$ *lies in* ${}^s\widehat{\mathsf{H}}_\delta$. *Moreover, there exists* M_0^+ *such that for every* $\delta \geq \delta_0$ *and every* $\widehat{W}$, $\widehat{W}' \in \mathsf{B}\,{}^s\widehat{\mathsf{H}}_\delta[1]$,

(a) $$|\widehat{\mathcal{F}}_0(\widehat{W})|_{{}^s\widehat{\mathsf{H}}_\delta} \leq M_0^+ \left(\frac{|\widehat{W}|^2_{{}^s\widehat{\mathsf{H}}_\delta} + \rho_s}{\sqrt{\delta}} \right),$$

(b) $$|\widehat{\mathcal{F}}_0(\widehat{W}) - \widehat{\mathcal{F}}_0(\widehat{W}')|_{{}^s\widehat{\mathsf{H}}_\delta} \leq M_0^+ \left(\frac{|\widehat{W}|_{{}^s\widehat{\mathsf{H}}_\delta} + |\widehat{W}'|_{{}^s\widehat{\mathsf{H}}_\delta} + \rho_s}{\sqrt{\delta}} \right) |\widehat{W} - \widehat{W}'|_{{}^s\widehat{\mathsf{H}}_\delta}.$$

The proof of this lemma is given in Appendix 7.H.

We have now enough material to prove Proposition 7.4.34: Lemma 7.4.42 ensures that $\widehat{\mathcal{F}}_0$ is a contraction mapping from $\mathsf{B}\,{}^s\widehat{\mathsf{H}}_\delta[d]$ to itself where $\mathsf{B}\,{}^s\widehat{\mathsf{H}}_\delta(d) := \{\widehat{W} \in {}^s\widehat{\mathsf{H}}_\delta \;/\; |\widehat{W}|_{{}^s\widehat{\mathsf{H}}_\delta} \leq d\}$, if the conditions

$$d \le 1, \qquad M_0^+ \left(\frac{d^2+\rho_s}{\sqrt{\delta}}\right) \le d, \qquad M_0^+ \left(\frac{2d+\rho_s}{\sqrt{\delta}}\right) < 1,$$

are fulfilled. Setting $d = 2M_0^+\rho_s/\sqrt{\delta}$ these three conditions are equivalent to

$$\sqrt{\delta} \ge 2M_0^+\sqrt{\rho_s}, \qquad \frac{4(M_0^+)^2}{\delta} + \frac{M_0^+\rho_s}{\sqrt{\delta}} < 1$$

which are fulfilled for every sufficiently large δ.

Thus, $\widehat{\mathcal{F}}_0$ admits a unique fixed point in $\mathsf{B}^s\widehat{\mathsf{H}}_\delta(2M_0^+\rho_s/\sqrt{\delta})$. We denote by $\widehat{W}^+_{s,0}$ this fixed point and we check that for $\rho = 0$, $\widehat{W}^+_{s,0}(z,\rho) \equiv 0$. Moreover, for every $z \in \widehat{\Omega}^s_\delta$, $\rho \mapsto \widehat{W}^+_{s,0}(z,\rho)$ is holomorphic in $D(0, 2\rho_s)$. So, for every $\rho \in D(0,\rho_s)$ and every ρ_0, $\rho_s < \rho_0 < 2\rho_s$ we have

$$\widehat{W}^+_{s,0}(z,\rho) = \rho\int_0^1 \frac{\partial \widehat{W}^+_{s,0}}{\partial\rho}(z,u\rho)\,du = \rho\int_0^1 \frac{1}{2\pi}\int_0^{2\pi} \frac{\widehat{W}^+_{s,0}(z,\rho_0 e^{i\theta})}{(\rho_0 e^{i\theta}-u\rho)^2}\rho_0 e^{i\theta}\,d\theta du.$$

Thus, for every $\rho \in D(0,\rho_s)$ and every ρ_0, $\rho_s < \rho_0 < 2\rho_s$

$$\begin{aligned} N(\widehat{W}^+_{s,0}) = &\sup_{z\in\widehat{\Omega}^s_\delta}\left(|z|^{3+\frac12}|\widehat{\alpha}_{\widehat{W}^+_{s,0}}(z,\rho)|\right) + \sup_{z\in\widehat{\Omega}^s_\delta}\left(|z|^{4+\frac12}|\widehat{\beta}_{\widehat{W}^+_{s,0}}(z,\rho)|\right) \\ &+ \sup_{z\in\widehat{\Omega}^s_\delta}\left(|z|^4|\widehat{A}_{\widehat{W}^+_{s,0}}(z,\rho)|\right) + \sup_{z\in\widehat{\Omega}^s_\delta}\left(|z|^4|\widehat{B}_{\widehat{W}^+_{s,0}}(z,\rho)|\right) \\ \le &\frac{|\rho|\rho_0}{|\rho_0-\rho_s|^2}\frac{M_0^+\rho_s}{\sqrt{\delta}}. \end{aligned}$$

Pushing $\rho \to 2\rho_s$, we finally get the desired estimate

$$N(\widehat{W}^+_{s,0}) \le \frac{4M_0^+|\rho|}{\sqrt{\delta}}.$$

which holds for every $\rho \in D(0,\rho_s)$. □

7.4.5 Holomorphic continuation of the stable manifold of 0 near $i\pi$

In this section ρ is no longer complex and lies in $[0,\rho_s[$. Corollary 7.4.29 ensures that the stable manifold Y_s of the full system (7.33) admits a holomorphic continuation on $\mathcal{D}^s_{\frac12,\nu}$. This holomorphic continuation is computed in the form $Y_s(\xi) = h(\xi) + v_s(\xi)$. This construction is done in the outer system of coordinates (see Figure 7.12 p.238). On the other hand, Proposition 7.4.34 gives a parameterization of the stable manifold $\widehat{Y}^+_{s,0}$ of 0 for the inner system (7.55) which reads

$$\widehat{Y}^+_{s,0}(z,\rho) = \widehat{h}_0^+(z) + \widehat{W}^+_{s,0}(z,\rho), \qquad \text{for } z \in \widehat{\Omega}^s_\delta$$

in the inner system of coordinates (see Figure 7.13 p. 239). To make a matching we must use a unique system of coordinates. Thus, we must rewrite the results obtained in the outer system of coordinates in the inner system of coordinates and vice versa. For that purpose we use the notation introduced in the previous subsection to rewrite in the inner system of coordinates objects defined in the outer system of coordinates and vice versa (see p. 255). Hence we can reformulate the results of Corollary 7.4.29 in the inner system of coordinates : let us denote by $\widehat{Y}_s = \widehat{h} + \widehat{v}_s$ with $\widehat{h}$ given by (7.57) and $\widehat{v}_s$ by

$$\widehat{v}_s(z) = \begin{pmatrix} \alpha_{\widehat{v}_s}(z) \\ \beta_{\widehat{v}_s}(z) \\ A_{\widehat{v}_s}(z) \\ B_{\widehat{v}_s}(z) \end{pmatrix} := \begin{pmatrix} \nu^2 \alpha_{v_s}(i\pi + \nu z) \\ \nu^3 \beta_{v_s}(i\pi + \nu z) \\ \nu^2 A_{v_s}(i\pi + \nu z) \\ \nu^2 B_{v_s}(i\pi + \nu z) \end{pmatrix}$$

where $v_s = (\alpha_{v_s}, \beta_{v_s}, A_{v_s}, B_{v_s})$. Then, $\widehat{Y}_s$ is a solution of the full system rewritten in the inner system of coordinates

$$\frac{d\widehat{Y}}{dz} = \widehat{f}(\widehat{Y}, \nu, \rho) = \widehat{f}_0(\widehat{Y}, \rho) + \widehat{f}_1(\widehat{Y}, \nu, \rho)$$

where $\widehat{f}_0$, $\widehat{f}_1$ are given by (7.52) and where z lies in $\widehat{\mathcal{D}}^s_{\frac{1}{2},\nu}$.

On the other hand, $\widehat{Y}^+_{s,0} = \widehat{h}^+_0 + \widehat{W}^+_{s,0}$ is a solution of the inner system of coordinates

$$\frac{d\widehat{Y}}{dz} = \widehat{f}_0(\widehat{Y}, \rho) \qquad \text{for } z \in \widehat{\Omega}^s_\delta.$$

To obtain a holomorphic continuation of $\widehat{Y}_s$ up to a distance δ of 0 (i.e. a continuation up to a distance $\delta\nu$ of $\pm i\pi$ in the outer system of coordinates), we look for a holomorphic solution of the full system in the form $\widehat{Y}^+_{s,0} + \widehat{w}_s$ which matches $\widehat{Y}_s$ on $\mathcal{D}^s_{\frac{1}{2},\nu}$. More precisely, this section is devoted to the proof of the following proposition:

Proposition 7.4.43. *There exists δ_s and ν_{sc} such that for every $\nu \in]0, \nu_{sc}]$ and every $\rho \in [0, \rho_s[$, $\widehat{Y}_s$ admits a holomorphic continuation in the form*

$$\widehat{Y}_s(z, \rho, \nu) = \widehat{Y}^+_{s,0}(z, \rho) + \widehat{w}_s(z, \rho, \nu),$$

on the domain $\widehat{\Sigma}^s_{\delta_s,\frac{1}{2},\nu}$ where $\widehat{\Sigma}^s_{\delta,r,\nu} =]-1, \frac{1}{\nu^{1-r}}[\times]-\frac{1}{\nu^{1-r}}, -\delta[$ (see Figure 7.19). Moreover there exists M_{in} such that

$$|\widehat{w}_s|_{{}^s\widehat{\mathsf{H}}_{\delta_s,\frac{1}{2},\nu}} = |\alpha_{\widehat{w}_s}|_{{}^s\widehat{\mathcal{H}}^3_{\delta_s,\frac{1}{2},\nu}} + |\beta_{\widehat{w}_s}|_{{}^s\widehat{\mathcal{H}}^4_{\delta_s,\frac{1}{2},\nu}} + |A_{\widehat{w}_s}|_{{}^s\widehat{\mathcal{H}}^3_{\delta_s,\frac{1}{2},\nu}} + |B_{\widehat{w}_s}|_{{}^s\widehat{\mathcal{H}}^3_{\delta_s,\frac{1}{2},\nu}} \leq 4M_{in}\nu^{\frac{1}{4}}$$

holds for every $\nu \in]0, \nu_{sc}]$ and every $\rho \in [0, \rho_s[$, where

$$|\widehat{f}|_{{}^s\widehat{\mathcal{H}}^{m_1}_{\delta,r,\nu}} = \sup_{\widehat{\Sigma}^s_{\delta,r,\nu}} \left(|\widehat{f}(z)||z|^{m_1} \right).$$

Proof. The proof of this theorem is based on the technique of continuation along horizontal lines described in Appendix 2.A. We must check that all the assumptions 2.A.1, 2.A.2, 2.A.3 are fulfilled in this case. We study a complex differential equation

$$\frac{d\widehat{Y}}{dz} = \widehat{f}(\widehat{Y}, \nu, \rho) = \widehat{f}_0(\widehat{Y}, \rho) + \widehat{f}_1(\widehat{Y}, \nu, \rho)$$

where $\widehat{f}_0$, $\widehat{f}_1$ are holomorphic and where z lies in $]-1, +\infty[\times]-1/\nu^{\frac{1}{2}}, -\delta[$ (see figure 7.19).

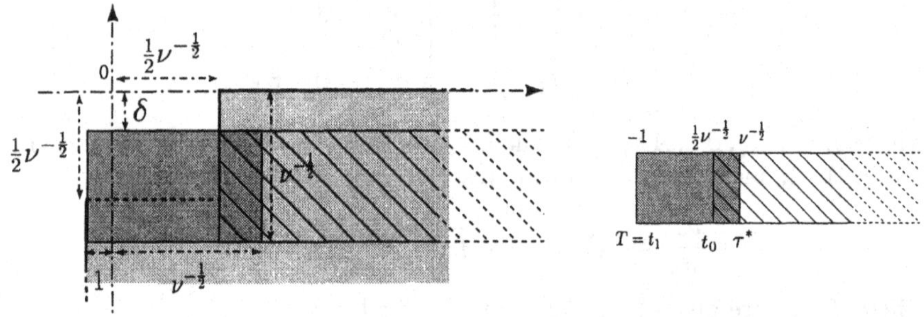

Fig. 7.19. Holomorphic continuation of Y_s in $\widehat{\Sigma}^s_{\delta,\frac{1}{2},\nu}$

Theorem 7.4.34 ensures the existence of a solution $\widehat{Y}^+_{s,0}$ of the inner system

$$\frac{d\widehat{Y}}{dz} = \widehat{f}_0(\widehat{Y}, \rho)$$

which is holomorphic on $\widehat{\Omega}^s_\delta \supset]-1, +\infty[\times]-1/\nu^{\frac{1}{2}}, -\delta[$. So Assumption 2.A.1 is fulfilled.

Corollary 7.4.29 gives a solution Y_s of the full system (7.33) on $\mathcal{D}^s_{\frac{1}{2},\nu}$. Rewriting this result in the inner system of coordinates we can state : the full system (7.50) admits a holomorphic solution $\widehat{Y}_s(z) = \widehat{Y}^+_{s,0}(z) + \widehat{w}_s(z)$ on $]\frac{1}{2}\nu^{-\frac{1}{2}}, +\infty[\times]-\nu^{-\frac{1}{2}}, -\delta[$ with

$$\widehat{w}_s(z) = \widehat{h}(z) + \widehat{v}_s(z) - \widehat{Y}^+_{s,0}(z).$$

So Assumption 2.A.2 is fulfilled too.

Our aim is to prove that $\widehat{Y}_s$ admits a holomorphic continuation on $]-1, +\infty[\times]-1/\nu^{\frac{1}{2}}, -\delta[\supset \widehat{\Sigma}^s_{\delta,\frac{1}{2},\nu}$. As explained in the Appendix 2.A we introduce the equation along horizontal lines: $\widehat{w}_s$ is seen as a function from $\mathbb{R}^2$ to $\mathbb{C}^4$ and satisfies

$$\begin{aligned}\frac{\partial \widehat{w}}{\partial \tau} = \widehat{f}_0\left(\widehat{Y}^+_{s,0}(\tau + i\eta) + \widehat{w}(\tau,\eta)\right) - \widehat{f}_0\left(\widehat{Y}^+_{s,0}(\tau + i\eta)\right) \\ +\widehat{f}_1\left(\widehat{Y}^+_{s,0}(\tau + i\eta) + \widehat{w}(\tau,\eta)\right).\end{aligned} \tag{7.61}$$

It remains to prove assumption 2.A.3 : we must show that the solution $\widehat{w}^*$ of (7.61) which is equal to

$$\widehat{h}(\nu^{-\frac{1}{2}} + i\eta) + \widehat{v}_s(\nu^{-\frac{1}{2}} + i\eta) - \widehat{Y}^+_{s,0}(\nu^{-\frac{1}{2}} + i\eta)$$

for $\tau = \nu^{-\frac{1}{2}}$ and $\eta \in]-1/\nu^{\frac{1}{2}}, -\delta[$, has an interval of life which contains $]-1, +\infty[$ for every $\eta \in]-1/\nu^{\frac{1}{2}}, -\delta[$. If it is the case, Lemma 2.A.5 will ensure that $\widehat{w}^*$ is holomorphic on $]-1, +\infty[\times]-1/\nu^{\frac{1}{2}}, -\delta[$. Denoting by $\widehat{w}_s(\tau + i\eta) = \widehat{w}^*(\tau, \eta)$, it will also ensure that $\widehat{Y}^+_{s,0} + \widehat{w}_s$ is the solution of (7.50) on $]-1, +\infty[\times]-1/\nu^{\frac{1}{2}}, -\delta[$ which is equal to $\widehat{h} + \widehat{v}_s$ on $]\frac{1}{2}\nu^{-\frac{1}{2}}, +\infty[\times]-1/\nu^{\frac{1}{2}}, -\delta[$.

So the only thing to prove is that $\widehat{w}^*$ is defined on $\widehat{\Sigma}^s_{\delta,\frac{1}{2},\nu}$. For that purpose, we rewrite (7.61) as an integral equation and we solve it using the Contraction Mapping Theorem in an appropriate set of functions defined on $\widehat{\Sigma}^s_{\delta,\frac{1}{2},\nu}$.

We rewrite the equation along horizontal lines (7.61) in the form

$$\frac{\partial \widehat{w}}{\partial \tau} - D\widehat{N}_0(\widehat{h}^+_0).\widehat{w} = \widehat{g}_1\left(\widehat{w}, \tau, \eta, \nu, \rho\right)$$

where

$$\begin{aligned}
\widehat{g}_1(\widehat{w}, \tau, \eta, \nu, \rho) &= \widehat{Q}_1(\widehat{w}, \tau, \eta, \nu, \rho) + \widehat{f}_1(\widehat{h}^+_0 + \widehat{W}^+_{s,0} + \widehat{w}, \nu, \rho),\\
\widehat{Q}_1(\widehat{w}, \tau, \eta, \nu, \rho) &= \widehat{f}_0(\widehat{h}^+_0 + \widehat{W}^+_{s,0} + \widehat{w}, \rho) - \widehat{f}_0(\widehat{h}^+_0 + \widehat{W}^+_{s,0}, \rho) - D\widehat{N}_0(\widehat{h}^+_0).\widehat{w}.
\end{aligned}$$

Using lemmas 7.4.39, 7.4.40 and the variation of constant formula we check that $\widehat{w}^*$ is a fixed point of $\widehat{\mathcal{F}}_1$ with

$$\widehat{\mathcal{F}}_1 = \widehat{\mathcal{F}}^{\Delta}_1 + \widehat{\mathcal{F}}_{11}$$

where

$$\begin{aligned}
\widehat{\mathcal{F}}^{\Delta}_1(\tau, \eta) = &\left\langle \widehat{\mathbf{p}}^*(\tfrac{1}{\sqrt{\nu}} + i\eta), \widehat{\Delta}_s(\eta) \right\rangle_* \widehat{\mathbf{p}}(\tau + i\eta) + \left\langle \widehat{\mathbf{q}}^*(\tfrac{1}{\sqrt{\nu}} + i\eta), \widehat{\Delta}_s(\eta) \right\rangle_* \widehat{\mathbf{q}}(\tau + i\eta)\\
&\left\langle \widehat{\mathbf{r}}^*_+(\tfrac{1}{\sqrt{\nu}} + i\eta), \widehat{\Delta}_s(\eta) \right\rangle_* \widehat{\mathbf{r}}_+(\tau + i\eta) + \left\langle \widehat{\mathbf{r}}^*_-(\tfrac{1}{\sqrt{\nu}} + i\eta), \widehat{\Delta}_s(\eta) \right\rangle_* \widehat{\mathbf{r}}_-(\tau + i\eta)
\end{aligned}$$

with

$$\widehat{\Delta}_s(\eta) = \left(\widehat{h} + \widehat{v}_s - \widehat{h}^+_0 - \widehat{W}^+_{s,0}\right)\left(\tfrac{1}{\sqrt{\nu}} + i\eta\right) \tag{7.62}$$

and

$$\begin{aligned}
(\widehat{\mathcal{F}}_{11}(\widehat{w}))(\tau, \eta) = &-\int_\tau^{1/\nu^{\frac{1}{2}}} \left\langle \widehat{\mathbf{p}}^*(s), \widehat{g}_1(\widehat{w}(s, \eta), s, \eta, \nu, \rho) \right\rangle_* ds\ \widehat{\mathbf{p}}(\tau + i\eta)\\
&-\int_\tau^{1/\nu^{\frac{1}{2}}} \left\langle \widehat{\mathbf{q}}^*(s), \widehat{g}_1(\widehat{w}(s, \eta), s, \eta, \nu, \rho) \right\rangle_* ds\ \widehat{\mathbf{q}}(\tau + i\eta)\\
&-\int_\tau^{1/\nu^{\frac{1}{2}}} \left\langle \widehat{\mathbf{r}}^*_+(s), \widehat{g}_1(\widehat{w}(s, \eta), s, \eta, \nu, \rho) \right\rangle_* ds\ \widehat{\mathbf{r}}_+(\tau + i\eta)\\
&-\int_\tau^{1/\nu^{\frac{1}{2}}} \left\langle \widehat{\mathbf{r}}^*_-(s), \widehat{g}_1(\widehat{w}(s, \eta), s, \eta, \nu, \rho) \right\rangle_* ds\ \widehat{\mathbf{r}}_-(\tau + i\eta).
\end{aligned}$$

To show that $\widehat{w}^*$ is defined on $\widehat{\Sigma}^{\mathrm{s}}_{\delta,\frac{1}{2},\nu}$ it is sufficient to prove that $\widehat{\mathcal{F}}_1$ is a contraction mapping in the Banach space ${}^{\mathrm{s}}\widehat{\mathsf{C}}^0_{\delta,\frac{1}{2},\nu}$ given by

Definition 7.4.44. *For $\delta \geq 1$, $\nu \in]0,1[$, and $r \in]0,1[$, we define the Banach space*

$${}^{\mathrm{s}}\widehat{\mathsf{C}}^0_{\delta,r,\nu} := \{\widehat{w} : \widehat{\Sigma}^{\mathrm{s}}_{\delta,r,\nu} \to \mathbb{C}^4, \ \textit{continuous}\ , \ |\widehat{w}|_{{}^{\mathrm{s}}\widehat{\mathsf{C}}^0_{\delta,r,\nu}} < +\infty\}$$

where

$$\begin{aligned} |\widehat{w}|_{{}^{\mathrm{s}}\widehat{\mathsf{C}}^0_{\delta,r,\nu}} = & \sup_{\widehat{\Sigma}^{\mathrm{s}}_{\delta,r,\nu}} \left(|\widehat{\alpha}(\tau,\eta)||\tau + \mathrm{i}\eta|^3\right) + \sup_{\widehat{\Sigma}^{\mathrm{s}}_{\delta,r,\nu}} \left(|\widehat{\beta}(\tau,\eta)||\tau + \mathrm{i}\eta|^4\right) \\ & + \sup_{\widehat{\Sigma}^{\mathrm{s}}_{\delta,r,\nu}} \left(|\widehat{A}(\tau,\eta)||\tau + \mathrm{i}\eta|^3\right) + \sup_{\widehat{\Sigma}^{\mathrm{s}}_{\delta,r,\nu}} \left(|\widehat{B}(\tau,\eta)||\tau + \mathrm{i}\eta|^3\right). \end{aligned}$$

Let us denote by $\mathsf{B}\,{}^{\mathrm{s}}\widehat{\mathsf{C}}^0_{\delta,r,\nu}[d]$ the ball of radius d of ${}^{\mathrm{s}}\widehat{\mathsf{C}}^0_{\delta,r,\nu}$

$$\mathsf{B}\,{}^{\mathrm{s}}\widehat{\mathsf{C}}^0_{\delta,r,\nu}[d] = \{\widehat{w} \in {}^{\mathrm{s}}\widehat{\mathsf{C}}^0_{\delta,r,\nu}, \ |\widehat{w}|_{{}^{\mathrm{s}}\widehat{\mathsf{C}}^0_{\delta,r,\nu}} \leq d\}.$$

The estimates of $\widehat{\mathcal{F}}_1$ are given by

Lemma 7.4.45. *There exists M_{in} and $\delta_2 \geq 1$ such that for every $\nu \in]0,\nu_6]$, $\rho \in [0,\rho_{\mathrm{s}}[$, $\delta \in [\delta_2, 1/\nu^{\frac{1}{2}}]$, and every $\widehat{w}$, $\widehat{w}' \in \mathsf{B}\,{}^{\mathrm{s}}\widehat{\mathsf{C}}^0_{\delta,r,\nu}[1]$,*

$$\text{(a)}\ |\widehat{\mathcal{F}}_1(\widehat{w})|_{{}^{\mathrm{s}}\widehat{\mathsf{C}}^0_{\delta,r,\nu}} \leq M_{\mathrm{in}} \left(\frac{|\widehat{w}|_{{}^{\mathrm{s}}\widehat{\mathsf{C}}^0_{\delta,r,\nu}}}{\delta} + |\widehat{w}|^2_{{}^{\mathrm{s}}\widehat{\mathsf{C}}^0_{\delta,r,\nu}} + \nu^{\frac{1}{4}} \right),$$

$$\text{(b)}\ |\widehat{\mathcal{F}}_1(\widehat{w}) - \widehat{\mathcal{F}}_1(\widehat{w}')|_{{}^{\mathrm{s}}\widehat{\mathsf{C}}^0_{\delta,r,\nu}} \leq M_{\mathrm{in}} \left(\tfrac{1}{\delta} + |\widehat{w}|_{{}^{\mathrm{s}}\widehat{\mathsf{C}}^0_{\delta,r,\nu}} + |\widehat{w}'|_{{}^{\mathrm{s}}\widehat{\mathsf{C}}^0_{\delta,r,\nu}} + \nu \right) |\widehat{w} - \widehat{w}'|_{{}^{\mathrm{s}}\widehat{\mathsf{C}}^0_{\delta,r,\nu}}.$$

The proof of this Lemma is given in Appendix 7.I.

This lemma ensures that $\widehat{\mathcal{F}}_1$ is a contraction mapping from $\mathsf{B}\,{}^{\mathrm{s}}\widehat{\mathsf{C}}^0_{\delta,\frac{1}{2},\nu}[d]$ to itself if

$$M_{\mathrm{in}} \left(\frac{d}{\delta} + d^2 + \nu^{\frac{1}{4}} \right) \leq d, \qquad M_{\mathrm{in}} \left(\frac{1}{\delta} + 2d + \nu \right) < 1, \qquad d \leq 1. \tag{7.63}$$

We choose $d = 4M_{\mathrm{in}}\nu^{\frac{1}{4}}$ and we denote by $\delta_{\mathrm{s}} = \max(\delta_2, 2M_{\mathrm{in}})$. Then, we choose ν_{sc} small enough to have $1/\nu_{\mathrm{sc}}^{\frac{1}{2}} > \delta_2$. Inequalities (7.63) are equivalent to

$$16M_{\mathrm{in}}^2\nu^{\frac{1}{4}} \leq 1, \qquad 8M_{\mathrm{in}}\nu^{\frac{1}{4}} + M_{\mathrm{in}}\nu < \frac{1}{2}, \qquad 4M_{\mathrm{in}}\nu^{\frac{1}{4}} \leq 1$$

which are fulfilled by a choice of a sufficiently small ν_{sc}. This completes the proof of Proposition 7.4.28 □

7.4.6 Symmetrization : holomorphic continuation of the stable manifold of 0 near $-i\pi$

Lemma 7.4.46. *There exists ν_{sc} and $\delta_s \geq 1$ such that for every $\nu \in]0, \nu_{sc}]$ and every $\rho \in [0, \rho_s[$, the stable manifold Y_s of 0 for the full system (7.33) admits a holomorphic continuation in $\mathcal{D}^s_{\frac{1}{2},\nu} \cup \Sigma^s_{\delta_s,\frac{1}{2},\nu} \cup \overline{\Sigma^s_{\delta_s,\frac{1}{2},\nu}}$ (see Figure 7.15 p. 240). Moreover this holomorphic continuation satisfies*

(a) *Y_s is symmetric about the real axis : $Y_s(\overline{\xi}) = \overline{Y_s(\xi)}$ holds for every $\xi \in \mathcal{D}^s_{\frac{1}{2},\nu} \cup \Sigma^s_{\delta_s,\frac{1}{2},\nu} \cup \overline{\Sigma^s_{\delta_s,\frac{1}{2},\nu}}$.*

(b) *On $\mathcal{D}^s_{\frac{1}{2},\nu}$, Y_s reads $Y_s = h + v_s$ and there exists M_{out} such that*

$$|\alpha_{v_s}|_{s\mathcal{H}^{3,\lambda}_\nu(\frac{1}{2})} + |\beta_{v_s}|_{s\mathcal{H}^{4,\lambda}_\nu(\frac{1}{2})} + |A_{v_s}|_{s\mathcal{H}^{3,\lambda}_\nu(\frac{1}{2})} + |B_{v_s}|_{s\mathcal{H}^{3,\lambda}_\nu(\frac{1}{2})} \leq 2M_{out}\rho\nu^{\frac{3}{2}}$$

for every $\nu \in]0, \nu_{sc}]$, and $\rho \in [0, \rho_s[$ where $v_s := (\alpha_{v_s}, \beta_{v_s}, A_{v_s}, B_{v_s})$ and

$$|f|_{s\mathcal{H}^{\gamma,\lambda}_\nu(\frac{1}{2})} = \sup_{\xi \in \mathcal{D}^s_{\frac{1}{2},\nu}} \left[|f(\xi,\nu)| \left(|\pi^2 + \xi^2|^\gamma \chi_1(\xi) + (1 - \chi_1(\xi)) e^{\lambda |\mathcal{R}e(\xi)|} \right) \right]$$

with $\chi_1(\xi) = 1$ if $\mathcal{R}e\,(\xi) < 1$ and $\chi_1(\xi) = 0$ otherwise.

(c) *On $\Sigma^s_{\delta_s,\frac{1}{2},\nu}$, $Y_s = (\alpha_s, \beta_s, A_s, B_s)$ reads $Y_s(\xi) = h_0^+(\xi) + W^+_{s,0}(\xi) + w_s(\xi)$ which is equivalent to*

$$\alpha_s(i\pi + \nu z, \rho, \nu) = \frac{1}{\nu^2}\left(-(4/z^2) + \widehat{\alpha}_{\widehat{W}^+_{s,0}}(z,\rho) \right) + \frac{1}{\nu^2} \widehat{\alpha}_{\widehat{w}_s}(z,\rho,\nu)$$

$$\beta_s(i\pi + \nu z, \rho, \nu) = \frac{1}{\nu^3}\left((8/z^3) + \widehat{\beta}_{\widehat{W}^+_{s,0}}(z,\rho) \right) + \frac{1}{\nu^3} \widehat{\beta}_{\widehat{w}_s}(z,\rho,\nu)$$

$$A_s(i\pi + \nu z, \rho, \nu) = \frac{1}{\nu^2}\left(\widehat{A}_{\widehat{W}^+_{s,0}}(z,\rho) \right) + \frac{1}{\nu^2} \widehat{A}_{\widehat{w}_s}(z,\rho,\nu)$$

$$B_s(i\pi + \nu z, \rho, \nu) = \frac{1}{\nu^2}\left(\widehat{B}_{\widehat{W}^+_{s,0}}(z,\rho) \right) + \frac{1}{\nu^2} \widehat{B}_{\widehat{w}_s}(z,\rho,\nu)$$

for every $z \in \Sigma^s_{\delta_s,\frac{1}{2},\nu}$. Moreover there exist M_0^+ and M_{in} such that

$$|\widehat{\alpha}_{\widehat{W}^+_{s,0}}|_{s\widehat{H}^{3+\frac{1}{2}}_{\delta_s}} + |\widehat{\beta}_{\widehat{W}^+_{s,0}}|_{s\widehat{H}^{4+\frac{1}{2}}_{\delta_s}} + |\widehat{A}_{\widehat{W}^+_{s,0}}|_{s\widehat{H}^4_{\delta_s}} + |\widehat{B}_{\widehat{W}^+_{s,0}}|_{s\widehat{H}^4_{\delta_s}} \leq \frac{4M_0^+\rho}{\sqrt{\delta_s}}$$

$$|\widehat{\alpha}_{\widehat{w}_s}|_{s\widehat{\mathcal{H}}^3_{\delta_s,\frac{1}{2},\nu}} + |\widehat{\beta}_{\widehat{w}_s}|_{s\widehat{\mathcal{H}}^4_{\delta_s,\frac{1}{2},\nu}} + |\widehat{A}_{\widehat{w}_s}|_{s\widehat{\mathcal{H}}^3_{\delta_s,\frac{1}{2},\nu}} + |\widehat{B}_{\widehat{w}_s}|_{s\widehat{\mathcal{H}}^3_{\delta_s,\frac{1}{2},\nu}} \leq 4M_{in}\nu^{\frac{1}{4}}$$

holds for every $\nu \in]0, \nu_{sc}]$ and every $\rho \in [0, \rho_s[$, where

$$|\widehat{f}|_{s\widehat{H}^{m_0}_\delta} = \sup_{\widehat{\Omega}^s_\delta} \left(|\widehat{f}(z)||z|^{m_0} \right), \qquad |\widehat{f}|_{s\widehat{\mathcal{H}}^{m_1}_{\delta,r,\nu}} = \sup_{\widehat{\Sigma}^s_{\delta,r,\nu}} \left(|\widehat{f}(z)||z|^{m_1} \right).$$

Proof. Modulo a rewriting in the outer system of coordinates of the results obtained in the inner system of coordinates, Propositions 7.4.28, 7.4.34, 7.4.43 ensure that the stable manifold of 0 for the full system (7.33) admits a holomorphic continuation in $\mathcal{D}^{\mathrm{s}}_{\frac{1}{2},\nu} \cup \Sigma^{\mathrm{s}}_{\delta_{\mathrm{s}},\frac{1}{2},\nu}$ which satisfies (b),(c). So it remains the problem of the symmetrization.

The two functions $\xi \mapsto Y_{\mathrm{s}}(\xi)$ and $\xi \mapsto \overline{Y_{\mathrm{s}}(\overline{\xi})}$ are respectively holomorphic and defined on $\Delta := \mathcal{D}^{\mathrm{s}}_{\frac{1}{2},\nu} \cup \Sigma^{\mathrm{s}}_{\delta_{\mathrm{s}},\frac{1}{2},\nu}$ and $\overline{\Delta} := \mathcal{D}^{\mathrm{s}}_{\frac{1}{2},\nu} \cup \overline{\Sigma^{\mathrm{s}}_{\delta_{\mathrm{s}},\frac{1}{2},\nu}}$, are equal on $]-\nu, +\infty[$ since Y_{s} is real on the real axis. Then, the isolated zero theorem ensures that these two functions coincide on $\Delta \cap \overline{\Delta}$. Thus we can define a holomorphic continuation of Y_{s} on $\Delta \cup \overline{\Delta}$ given by $Y_{\mathrm{s}}(\xi)$ for $\xi \in \Delta$ and by $\overline{Y_{\mathrm{s}}(\overline{\xi})}$ for $\xi \in \overline{\Delta}$. □

We can now conclude these last four subsections devoted to the study of the holomorphic continuation of Y_{s} by the proof of Proposition 7.4.6.

Proof of Proposition 7.4.6. (a): Lemma 7.4.46 ensures that the stable manifold Y_{s} of the full system (7.33) admits a holomorphic continuation in $\mathcal{D}^{\mathrm{s}}_{\frac{1}{2},\nu} \cup \Sigma^{\mathrm{s}}_{\delta_{\mathrm{s}},\frac{1}{2},\nu} \cup \overline{\Sigma^{\mathrm{s}}_{\delta_{\mathrm{s}},\frac{1}{2},\nu}} \supset \mathcal{B}^{\mathrm{s}}_{\pi-\delta_{\mathrm{s}}\nu,\nu}$.

It remains to check that $\alpha_{\mathrm{s}}(\cdot,\rho,\cdot) \in {}^{\mathrm{s}}E^{2,\lambda}_{\pi,\delta_{\mathrm{s}}}(2,3,\frac{1}{2},\frac{1}{4})$ (see Definition 7.4.4). For $\beta_{\mathrm{s}}, A_{\mathrm{s}}, B_{\mathrm{s}}$ the proof is very similar and thus left to the reader. From Lemma 7.4.46 we get that near $i\pi$, α_{s} reads

$$\alpha_{\mathrm{s}}(i\pi + \nu z, \rho, \nu) = \frac{1}{\nu^2}\widehat{\alpha}^{+}_{\mathrm{s},0}(z,\rho) + \frac{\nu^{\frac{1}{4}}}{\nu^2}\widehat{\alpha}^{+}_{\mathrm{s},1}(z,\rho,\nu)$$

with

$$\widehat{\alpha}^{+}_{\mathrm{s},0}(z,\rho)\Big(-(4/z^2) + \widehat{\alpha}_{\widehat{W}^{+}_{\mathrm{s},0}}(z,\rho)\Big), \qquad \widehat{\alpha}^{+}_{\mathrm{s},1}(z,\rho,\nu) = \nu^{-\frac{1}{4}}\widehat{\alpha}_{\widehat{w}_{\mathrm{s}}}(z,\rho,\nu).$$

where for every $\rho \in [0,\rho_{\mathrm{s}}[$ and every $\nu \in]0,\nu_{\mathrm{sc}}]$, $z \mapsto \widehat{\alpha}_{\widehat{W}^{+}_{\mathrm{s},0}}(z,\rho)$ and $z \mapsto \widehat{\alpha}_{\widehat{w}_{\mathrm{s}}}(z,\rho,\nu)$ are two holomorphic functions respectively defined on $\widehat{\Omega}^{\mathrm{s}}_{\delta_{\mathrm{s}}}$ and $\widehat{\Sigma}^{\mathrm{s}}_{\delta_{\mathrm{s}},\frac{1}{2},\nu}$ which satisfies

$$\sup_{z \in \widehat{\Omega}^{\mathrm{s}}_{\delta_{\mathrm{s}}}} \left(|z|^{3+\frac{1}{2}}|\widehat{\alpha}_{\widehat{W}^{+}_{\mathrm{s},0}}(z,\rho)|\right) \le \frac{4M_0^{+}\rho}{\sqrt{\delta_{\mathrm{s}}}} < +\infty,$$

$$\sup_{\substack{\nu \in]0,\nu_{\mathrm{sc}}] \\ z \in \widehat{\Sigma}^{\mathrm{s}}_{\delta_{\mathrm{s}},\frac{1}{2},\nu}}} \left(\nu^{-\frac{1}{4}}|z|^3|\widehat{\alpha}_{\widehat{w}_{\mathrm{s}}}(z,\rho,\nu)|\right) \le 4M_{\mathrm{in}} < +\infty.$$

Hence, for every $\rho \in [0,\rho_{\mathrm{s}}[$, $\widehat{\alpha}^{+}_{\mathrm{s},0}(\cdot,\rho)$ and $\widehat{\alpha}^{+}_{\mathrm{s},1}(\cdot,\rho,\cdot)$ respectively belongs to $\widehat{H}^2_{\delta_{\mathrm{s}}}$ and $\widehat{H}^3_{\delta_{\mathrm{s}},\frac{1}{2}}$. Thus α_{s} satisfies property (b) of Definition 7.4.4.

The function α_s is real valued on the real axis and thus symmetric about the real axis. So, since it satisfies property (b) of Definition 7.4.4, it automatically satisfies property (c) with $P_r^-(\alpha_s)(z) = \overline{P_r^+(\alpha_s)(-\overline{z})}$.

Lemma 7.4.46 also ensures that far away from the singularities, i.e. in $\mathcal{D}^s_{\frac{1}{2},\nu}$, α_s reads $\alpha_s = \alpha_h + \alpha_{v_s}$ with

$$\alpha_h(\xi) = \frac{1}{\cosh^2(\frac{1}{2}\xi)}, \qquad |\alpha_{v_s}|_{{}^s\mathcal{H}^{3,\lambda}_\nu(\frac{1}{2})} \le 2M_{\text{out}}\rho\nu^{\frac{3}{2}}.$$

where $0 < \lambda < 1$. Thus

$$\|\alpha_h\|_{{}^sH^{2,\lambda}_{\pi,\delta_s}} = \sup_{\substack{\nu\in]0,\nu_{sc}] \\ \xi\in\mathcal{D}^s_{\frac{1}{2},\nu}}} \left[(|\alpha_h(\xi)| \left(|\pi^2+\xi^2|^2\chi_1(\xi) + (1-\chi_1(\xi))e^{\lambda|\mathcal{R}e(\xi)|}\right)\right] < +\infty$$

and

$$\sup_{\substack{\nu\in]0,\nu_{sc}] \\ \xi\in\mathcal{D}^s_{\frac{1}{2},\nu}}} \left[(|\alpha_{v_s}(\xi,\rho,\nu)|\left(|\pi^2+\xi^2|^2\chi_1(\xi) + (1-\chi_1(\xi))e^{\lambda|\mathcal{R}e(\xi)|}\right)\right]$$

$$\le \sup_{\substack{\nu\in]0,\nu_{sc}] \\ \xi\in\mathcal{D}^s_{\frac{1}{2},\nu}}} \left[2M_{\text{out}}\rho\nu^{\frac{3}{2}}\left(|\pi^2+\xi^2|^{-1}\chi_1(\xi) + (1-\chi_1(\xi))e^{\lambda|\mathcal{R}e(\xi)|}\right)\right]$$

$$\le 2M_{\text{out}}\rho_s\left(\frac{\nu_{sc}}{m_0\pi} + \nu_{sc}^{\frac{3}{2}}\right).$$

Hence,

$$\begin{aligned}&\sup_{\substack{\nu\in]0,\nu_{sc}] \\ \xi\in\mathcal{D}^s_{\frac{1}{2},\nu}}} \left[(|\alpha_s(\xi,\rho,\nu)|\left(|\pi^2+\xi^2|^2\chi_1(\xi) + (1-\chi_1(\xi))e^{\lambda|\mathcal{R}e(\xi)|}\right)\right] \\ &\qquad \le \|\alpha_h\|_{{}^sH^{2,\lambda}_{\pi,\delta_s}} + M_{\text{out}}\rho_s\left(\frac{\nu_{sc}}{m_0\pi} + \nu_{sc}^{\frac{3}{2}}\right) < +\infty.\end{aligned} \tag{7.64}$$

Finally, recalling that α_s is real valued on the real axis and thus symmetric about the real axis, we check

$$\|\alpha_s\|_{{}^sH^{2,\lambda}_{\pi,\delta_s}} \le \sup_{\substack{\nu\in]0,\nu_{sc}] \\ \xi\in\mathcal{D}^s_{\frac{1}{2},\nu}\cup\Sigma^s_{\delta_s,\frac{1}{2},\nu}}} \left[(|\alpha_s(\xi,\rho,\nu)|\left(|\pi^2+\xi^2|^2\chi_1(\xi) + (1-\chi_1(\xi))e^{\lambda|\mathcal{R}e(\xi)|}\right)\right]$$

$$\le \|\alpha_h\|_{{}^sH^{2,\lambda}_{\pi,\delta_s}} + M_{\text{out}}\rho_s\left(\frac{\nu_{sc}}{m_0\pi} + \nu_{sc}^{\frac{3}{2}}\right) + \sup_{\substack{\nu\in]0,\nu_{sc}] \\ z\in\widehat{\Sigma}^s_{\delta_s,\frac{1}{2},\nu}}} \left|(4\pi^2+1)(\nu^2z^2)\frac{\widehat{\alpha}_s(z,\rho,\nu)}{\nu^2}\right|$$

$$\le \|\alpha_h\|_{{}^sH^{2,\lambda}_{\pi,\delta_s}} + M_{\text{out}}\rho_s\left(\frac{\nu_{sc}}{m_0\pi} + \nu_{sc}^{\frac{3}{2}}\right) + (4\pi^2+1)\sup_{\substack{\nu\in]0,\nu_{sc}] \\ z\in\widehat{\Sigma}^s_{\delta_s,\frac{1}{2},\nu}}} |-4+z^2(\widehat{\alpha}_{\widehat{W}^+_{s,0}}+\widehat{\alpha}_{\widehat{w}_s})|$$

So,

$$\|\alpha_s\|_{{}^sH^{2,\lambda}_{\pi,\delta_s}} \leq \|\alpha_h\|_{{}^sH^{2,\lambda}_{\pi,\delta_s}} + M_{\text{out}}\rho_s\left(\frac{\nu_{sc}}{m_0\pi}+\nu_{sc}^{\frac{3}{2}}\right)(4\pi^2+1)(4+\frac{4M_0^+\rho}{\sqrt{\delta_s}}+4M_{\text{in}}) <+\infty$$

since for $z \in \widehat{\Sigma}^s_{\delta_s,\frac{1}{2},\nu}$, $|z| \geq |\mathcal{I}m\,(z)\,| \geq \delta_s \geq 1$. So, α_s satisfies property (a) of Definition 7.4.4 and thus, it lies in ${}^sE^{2,\lambda}_{\pi,\delta_s}\,(2,3,\frac{1}{2},\frac{1}{4})$. with $\mathrm{P}^+_{\mathrm{r}}(\alpha_s) = \widehat{\alpha}^+_{s,0} = -(4/z^2) + \widehat{\alpha}_{\widehat{W}^+_{s,0}}$.

(b): Statement (b) of Proposition 7.4.6 directly follows from (a) and from Proposition 7.4.34. □

7.4.7 Exponential asymptotics of the solvability condition

This subsection is devoted to the proof of Proposition 7.4.8 which ensures that if Y is a reversible homoclinic connection to 0 satisfying the properties (P1), (P2), (P3) defined in Proposition 7.4.1, then the oscillatory integral $I(Y,\rho,\nu)$ introduced in Proposition 7.4.1 is exponentially small but does not vanish generically.

7.4.7.1 Proof of Proposition 7.4.8-(a). For a homoclinic connection Y to 0 the full system (7.33), satisfying the properties (P1),(P2),(P3), we want to compute an equivalent of the oscillatory integral $I(Y,\rho,\nu)$ given by

$$I(Y,\rho,\nu) := \frac{1}{2}\int_{-\infty}^{+\infty} \mathrm{e}^{\mathrm{i}((\frac{\omega}{\nu}+a\nu)t+2b\nu\tanh(\frac{1}{2}t))}\,(g_B+\mathrm{i}g_A)(v(t),t,\rho,\nu)dt$$

where $g = (g_\alpha, g_\beta, g_A, g_B)$ is defined in (7.34) and where $v = Y - h$. This oscillatory integral can be rewritten

$$I(Y,\rho,\nu) := \frac{1}{2}\int_{-\infty}^{+\infty} \mathrm{e}^{\frac{\mathrm{i}\omega t}{\nu}}\,f(Y)(t,\rho,\nu)dt$$

where

$$f(Y)(t,\rho,\nu) = \mathrm{e}^{\mathrm{i}(a\nu t+2b\nu\tanh(\frac{1}{2}t))} \times\Big(b\nu\big(\alpha(t)-\alpha_h(t)\big)\big(A(t)-\mathrm{i}B(t)\big)+\rho\big(R_B+\mathrm{i}R_A\big)(Y(t),\nu)\Big). \tag{7.65}$$

with $Y = (\alpha,\beta,A,B)$. We first check

Lemma 7.4.47. *For every $\nu \in]0,\nu_{sc}]$ and every $\rho \in [0,\rho_s[$, the function $t \mapsto f(Y_s)(t,\rho,\nu)$ defined in (7.65) admits a holomorphic continuation in the strip $\mathcal{B}^s_{\pi-\delta_s\nu,\nu}$ which satisfies for every $\rho \in [0,\rho_s[$*

$$f(Y_s)(\cdot,\rho,\cdot) \in {}^sE^{3,\lambda}_{\pi,\delta_s}\,(2,3,\tfrac{1}{2},\tfrac{1}{4})$$

and

$$\mathrm{Pr}^+_{{}^{\mathrm{s}}E^{3,\lambda}_{\pi,\delta_{\mathrm{s}}}}(f(Y_{\mathrm{s}}))(z,\rho) = \mathrm{e}^{\frac{4\mathrm{i}b}{z}}\Big(\Big(-\mathrm{i}\widehat{B}^+_{\mathrm{s},0}(z,\rho)+\widehat{A}^+_{\mathrm{s},0}(z,\rho)\Big)\big(\widehat{\alpha}^+_{\mathrm{s},0}(z,\rho)+\frac{4}{z^2}\big)$$

$$+\rho\sum_{|\overline{m}|\geq 3}(a_{B,\overline{m},0}+\mathrm{i}a_{A,\overline{m},0})\big(\widehat{Y}^+_{\mathrm{s},0}(z,\rho)\big)^{\overline{m}}\Big) \tag{7.66}$$

where $\widehat{Y}^+_{\mathrm{s},0} = (\widehat{\alpha}^+_{\mathrm{s},0}, \widehat{\beta}^+_{\mathrm{s},0}, \widehat{A}^+_{\mathrm{s},0}, \widehat{B}^+_{\mathrm{s},0})$ *is introduced in Proposition (7.4.6) and where* $a_{A,\overline{m},\ell}$ *and* $a_{B,\overline{m},\ell}$ *are given in (7.7).*

Proof. From Proposition 7.4.6 we get that $f(Y_{\mathrm{s}})(\cdot,\rho,\nu)$ admits an holomorphic continuation in $\mathcal{B}^{\mathrm{s}}_{\pi-\delta_{\mathrm{s}}\nu,\nu}$ which reads

$$f(Y_{\mathrm{s}})(\xi,\rho,\nu) = \mathrm{e}^{\mathrm{i}(a\nu\xi+2b\nu\tanh(\frac{1}{2}\xi))}$$
$$\Big(b\nu\big(\alpha_{\mathrm{s}}(\xi)-\alpha_h(\xi)\big)\big(A_{\mathrm{s}}(\xi)-\mathrm{i}B_{\mathrm{s}}(\xi)\big) + \rho(R_B+\mathrm{i}R_A)(Y_{\mathrm{s}}(\xi),\nu)\Big).$$

Step 1. Study of $b\nu(\alpha_{\mathrm{s}}(\xi)-\alpha_h(\xi))(A_{\mathrm{s}}(\xi)-\mathrm{i}B_{\mathrm{s}}(\xi))$. Proposition 7.4.6 also ensures that

$$b\nu\big(\alpha_{\mathrm{s}}(\xi)-\alpha_h(\xi)\big)\big(A_{\mathrm{s}}(\xi)-\mathrm{i}B_{\mathrm{s}}(\xi)\big) \in {}^{\mathrm{s}}E^{2+2-1,\lambda}_{\pi,\delta_{\mathrm{s}}}(2,3,\tfrac{1}{2},\tfrac{1}{4}) \tag{7.67}$$

and

$$\mathrm{Pr}^+_{{}^{\mathrm{s}}E^{3,\lambda}_{\pi,\delta_{\mathrm{s}}}}\Big(b\nu\big(\alpha_{\mathrm{s}}(\xi)-\alpha_h(\xi)\big)\big(A_{\mathrm{s}}(\xi)-\mathrm{i}B_{\mathrm{s}}(\xi)\big)\Big)(z)$$
$$= b(\widehat{\alpha}^+_{\mathrm{s},0}(z)+4/z^2)(\widehat{A}^+_{\mathrm{s},0}(z)-\mathrm{i}\widehat{B}^+_{\mathrm{s},0}(z)).$$

Step 2. Study of R_A. Similarly, we check

$$R_A(Y_{\mathrm{s}}(\xi),\nu) = \sum_{(\overline{m},\ell)\in\mathcal{I}} \Delta_{\overline{m},\ell}$$

where

$$\Delta_{\overline{m},\ell} := a_{A,\overline{m},\ell}\,\alpha_{\mathrm{s}}^{m_\alpha}\beta_{\mathrm{s}}^{m_\beta}A_{\mathrm{s}}^{m_A}B_{\mathrm{s}}^{m_B}\,\nu^{2m_\alpha+3m_\beta+2m_A+2m_B-3}\,\nu^{2\ell}$$

with

$$\alpha_{\mathrm{s}}^{m_\alpha}\beta_{\mathrm{s}}^{m_\beta}A_{\mathrm{s}}^{m_A}B_{\mathrm{s}}^{m_B} \in {}^{\mathrm{s}}E^{2m_\alpha+3m_\beta+2m_A+2m_B,\lambda}_{\pi,\delta_{\mathrm{s}}}(2,3,\tfrac{1}{2},\tfrac{1}{4}).$$

Hence, for every $(\overline{m},\ell)\in\mathcal{I}$, $\Delta_{\overline{m},\ell}\in{}^{\mathrm{s}}E^{3,\lambda}_{\pi,\delta_{\mathrm{s}}}$ and

$$\mathrm{Pr}^+_{{}^{\mathrm{s}}E^{3,\lambda}_{\pi,\delta_{\mathrm{s}}}}(\Delta_{\overline{m},\ell}) = \begin{cases} a_{A,\overline{m},\ell}(\widehat{\alpha}^+_{\mathrm{s},0})^{m_\alpha}(\widehat{\beta}_{\mathrm{s},0})^{m_\beta}(\widehat{A}_{\mathrm{s},0})^{m_A}(\widehat{B}^+_{\mathrm{s},0})^{m_B} & \text{for } \ell=0,\\ 0 & \text{for } \ell\neq 0.\end{cases}$$

Observe that all the monomials of any order in the expansion of R corresponding to $\ell=0$ have a non zero principal part. Modulo the problem of the convergence of the sum, this suggests that

$$R_A(Y_s,\nu) \in {}^s E^{3,\lambda}_{\pi,\delta_s} \qquad \text{with } \mathrm{Pr}^+_{{}^s E^{3,\lambda}_{\pi,\delta_s}}(R_A) = \widehat{R}_{0,A}(\widehat{Y}^+_{s,0}) \tag{7.68}$$

where $\widehat{R}_0 = (0, \widehat{R}_{0,\beta}, \widehat{R}_{0,A}, \widehat{R}_{0,B})$ is defined in (7.52), (7.54). For proving that (7.68) holds, we proceed in 3 Substeps.

Step 2.1. With the notation introduced at the beginning of Subsection 7.4.4, we get that near $i\pi$, R_A reads

$$\begin{aligned} R_A(Y_s,\nu) &= \frac{1}{\nu^3} R_A(\widehat{Y}_s,\nu) \\ &= \frac{1}{\nu^3} R_A(\widehat{h}_0^+ + \widehat{W}^+_{s,0} + \widehat{w}_s,\nu) \\ &= \frac{1}{\nu^3}\widehat{R}_{0,A}(\widehat{h}_0^+ + \widehat{W}^+_{s,0}) + \frac{1}{\nu^3}\Big[\widehat{R}_{0,A}(\widehat{h}_0^+ + \widehat{W}^+_{s,0} + \widehat{w}_s) - \widehat{R}_{0,A}(\widehat{h}_0^+ + \widehat{W}^+_{s,0}) \\ &\qquad + \widehat{R}_{1,A}(\widehat{h}_0^+ + \widehat{W}^+_{s,0} + \widehat{w}_s)\Big]. \end{aligned}$$

The proof of Lemma 7.G.1 (see (7.92)) ensures that for $z \in \widehat{\Omega}^s_{\delta_s}$,

$$|\widehat{R}_{0,A}(\widehat{h}_0^+(z) + \widehat{W}^+_{s,0}(z))| \le \frac{\mathcal{M}_{10}}{|z|^6}$$

and thus $\widehat{R}_{0,A}(\widehat{h}_0^+ + \widehat{W}^+_{s,0}(\cdot,\rho)) \in {}^s\widehat{H}^6_{\delta_s} \subset {}^s\widehat{H}^3_{\delta_s}$.

Similarly the proof of Lemma 7.I.2 (see (7.108) and (7.110)) ensures that for $z \in \widehat{\Sigma}^s_{\delta_s,\frac{1}{2},\nu}$,

$$|\widehat{R}_{0,A}(\widehat{h}_0^+ + \widehat{W}^+_{s,0} + \widehat{w}_s) - \widehat{R}_{0,A}(\widehat{h}_0^+ + \widehat{W}^+_{s,0})| \le \mathcal{M}_{13}\frac{|\widehat{w}_s|_{{}^s\widehat{\mathcal{C}}^0_{\delta_s,\frac{1}{2},\nu}}}{|z|^6} \le \mathcal{M}_{13} 4 M_{\mathrm{in}} \frac{\nu^{\frac{1}{4}}}{|z|^6}$$

and

$$|\widehat{R}_{1,A}(\widehat{h}_0^+ + \widehat{W}^+_{s,0} + \widehat{w}_s)| \le \mathcal{M}_{13}\frac{\nu^2}{|z|^2} \le \sqrt{2}\mathcal{M}_{13}\frac{\nu^{\frac{3}{2}}}{|z|^3}$$

Thus,

$$\widehat{R}_{0,A}(\widehat{h}_0^+ + \widehat{W}^+_{s,0} + \widehat{w}_s) - \widehat{R}_{0,A}(\widehat{h}_0^+ + \widehat{W}^+_{s,0}) + \widehat{R}_{1,A}(\widehat{h}_0^+ + \widehat{W}^+_{s,0} + \widehat{w}_s) \in {}^s\widehat{H}^3_{\delta_s,\frac{1}{2}}.$$

Thus R_A satisfies property (b) of Definition 7.4.4.

Step 2.2. The function $R_A(Y_s,\nu)$ is real valued on the real axis and thus symmetric about the real axis. So, since it satisfies property (b) of Definition 7.4.4, it automatically satisfies property (c) with $\mathrm{P}^-_{\mathrm{r}}(R_A(Y_s,\nu))(z) = \overline{\mathrm{P}^+_{\mathrm{r}}(R_A(Y_s,\nu))(-\overline{z})}$.

Step 2.3. The proof of Lemma 7.F.5 (see (7.77)) ensures that far away from the singularities, i.e. in $\mathcal{D}^s_{\frac{1}{2},\nu}$, $R_A(Y_s,\nu)$ satisfies

$$\begin{aligned}|R_A(Y_s,\nu)| &= |R_A(h+v_s,\nu)| \\ &\leq \mathcal{M}_8\rho\nu^3\left(\frac{\chi_1(\xi)}{|\pi^2+\xi^2|^6}+(1-\chi_1(\xi))e^{\lambda|\mathcal{R}e(\xi)|}\right) \\ &\leq \mathcal{M}_8\rho\left(\frac{\nu_{sc}^{\frac{3}{2}}}{(\pi m_0)^3}\frac{\chi_1(\xi)}{|\pi^2+\xi^2|^3}+\nu_{sc}^3(1-\chi_1(\xi))e^{\lambda|\mathcal{R}e(\xi)|}\right)<+\infty\end{aligned}$$

and thus $\|R_A(Y_s,\nu)\|_{{}^sH^{3,\lambda}_{\pi,\delta_s}} < +\infty$. So, $R_A(Y_s,\nu)$ satisfies property (a) of Definition 7.4.4. Hence,

$$R_A(Y_s,\nu)\in {}^sE^{3,\lambda}_{\pi,\delta_s}(2,3,\tfrac{1}{2},\tfrac{1}{4}),\qquad \Pr_{{}^sE^{3,\lambda}_{\pi,\delta_s}}(R_A(Y_s,\nu))=\widehat{R}_{0,A}(\widehat{Y}^+_{s,0}).$$

For R_B the same analysis holds. So, we finally get

$$\begin{aligned}&\left(R_B(Y_s,\nu)+iR_A(Y_s,\nu)\right)\in {}^sE^{3,\lambda}_{\pi,\delta_s}(2,3,\tfrac{1}{2},\tfrac{1}{4}) \\ &\Pr^+_{{}^sE^{3,\lambda}_{\pi,\delta_s}}(R_B(Y_s,\nu)+iR_A(Y_s,\nu))=(\widehat{R}_{0,B}(\widehat{Y}^+_{s,0})+i\widehat{R}_{0,A}(\widehat{Y}^+_{s,0})).\end{aligned}\tag{7.69}$$

Step 3. Finally, observing that

$$\begin{aligned}&e^{ia\nu+2ib\nu\tanh(\frac{1}{2}\xi)}\in {}^sE^{0,0}_{\pi,\delta_s}(0,0,\tfrac{1}{2},1) \\ &\Pr^{\pm}_{{}^sE^{0,0}_{\pi,\delta_s}}\left(e^{ia\nu+2ib\nu\tanh(\frac{1}{2}\xi)}\right)=e^{\pm\frac{4ib}{z}}\end{aligned}\tag{7.70}$$

and grouping together the previous results (7.70), (7.67), (7.69) we get that $f(Y_s)(\cdot,\rho,\cdot)$ belongs to ${}^sE^{3,\lambda}_{\pi,\delta_s}(2,3,\frac{1}{2},\frac{1}{4})$ with a principal part given by (7.66). □

Lemma 7.4.48. *For every $\nu\in]0,\nu_{sc}]$ and every $\rho\in[0,\rho_s[$, if the full system (7.33) admits a homoclinic connection Y to 0 satisfying the properties (P1),(P2),(P3), then*

(a) *Y admits a holomorphic continuation in the strip $\mathcal{B}_{\pi-\delta_s\nu}$ which is equal to Y_s on the half strip $\mathcal{B}^s_{\pi-\delta_s\nu,\nu}$ (see Figure 7.20) and which is symmetric about the imaginary axis, i.e.*

$$Y(-\overline{\xi})=\overline{SY(\xi)}\tag{7.71}$$

(b) *$f(Y)$ admits a holomorphic continuation in the strip $\mathcal{B}_{\pi-\delta_s\nu}$ which is equal to $f(Y_s)$ on the half strip $\mathcal{B}^s_{\pi-\delta_s\nu,\nu}$ and which satisfies*

$$f(Y)(-\overline{\xi},\rho,\nu)=\overline{f(Y)(\xi,\rho,\nu)}\tag{7.72}$$

Proof. (a): On one hand, Proposition 7.4.1 ensures that $Y(t)=Y_s(t)$ holds for $t\in]-1,+\infty[$. On the other hand, Proposition 7.4.6 ensures that Y_s

admits a holomorphic continuation in $\mathcal{B}^{s}_{\pi-\delta_s\nu,\nu}$. So, Y admits a holomorphic continuation in $\mathcal{B}^{s}_{\pi-\delta_s\nu,\nu}$.

Considering the two holomorphic functions $\xi \mapsto Y(\xi)$ and $\xi \mapsto SY(-\xi)$ respectively defined on the half strips $\mathcal{B}^{s}_{\pi-\delta_s\nu,\nu}$ and $-\mathcal{B}^{s}_{\pi-\delta_s\nu,\nu}$ which coincide on $]-\nu,\nu[$ since by hypothesis Y is reversible on $\mathbb{R}$, we obtain that Y admits a holomorphic continuation on the full strip $\mathcal{B}_{\pi-\delta_s\nu}$ which satisfies $SY(-\xi) = Y(\xi)$ for $\xi \in \mathcal{B}_{\pi-\delta_s\nu}$. Moreover, Y is real valued on the real axis, thus $Y(\overline{\xi}) = \overline{Y(\xi)}$ holds for $\xi \in \mathcal{B}_{\pi-\delta_s\nu}$. Finally combining these two last symmetry properties, we get a third one which reads

$$Y(-\overline{\xi}) = S\overline{Y(\xi)} \qquad \text{for } \xi \in \mathcal{B}_{\pi-\delta_s\nu}.$$

(b): (b) directly follows from (a). □

We can finally prove

Lemma 7.4.49. *There exists M_{sc} such that for every $\nu \in]0,\nu_{sc}]$ and every $\rho \in [0,\rho_s[$, if the full system (7.33) admits a homoclinic connection Y to 0 satisfying the properties (P1),(P2),(P3), then*

$$I(Y,\rho,\nu) = \frac{1}{\nu^2} e^{-\frac{\omega\pi}{\nu}} (\Lambda(\rho) + J(\rho,\nu)) \qquad \text{with } |J(\rho,\nu)| \le M_{sc}\nu^{\frac{1}{4}}$$

where $\Lambda(\rho)$ is given by (7.36).

Remark 7.4.50. Proposition 7.4.8-(a) directly follows from this Lemma.

Proof. We cannot use directly the Third Exponential Lemma 2.1.21 since $f(Y)$ admits a holomorphic continuation in ${}^{s}E^{3,\lambda}_{\pi,\delta_s}$ and not in $E^{3,\lambda}_{\pi,\delta_s}$. So, we redo partially the proof the Third Exponential Lemma here.

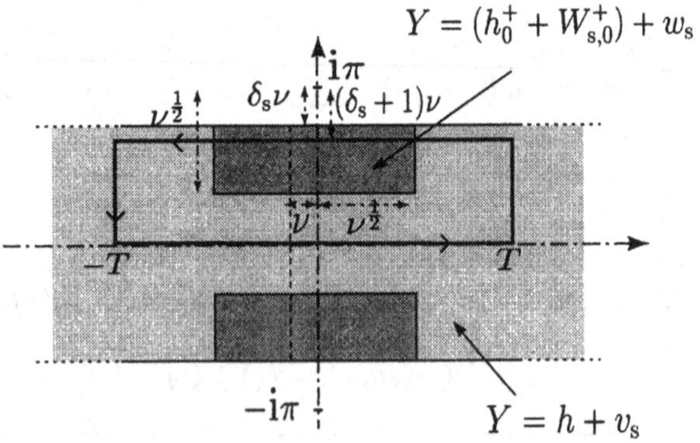

Fig. 7.20. Path Γ_T

Lemma 7.4.48 ensures that $f(Y)$ admits a holomorphic continuation in $\mathcal{B}_{\pi-\delta_s\nu}$. Using the path Γ_T (see Figure 7.20) and pushing $T \to +\infty$ we get that

$$I(Y,\rho,\nu) = \frac{1}{2}\int_{\mathrm{i}(\pi-(\delta_s+1)\nu)+\mathbb{R}} \mathrm{e}^{\frac{\mathrm{i}\omega t}{\nu}} f(Y)(t,\rho,\nu).$$

Then, (7.72) and the reversibility properties of g_A, g_B and h give

$$\begin{aligned} I(Y,\rho,\nu) &= \mathcal{R}e\left(\int_{\mathrm{i}(\pi-(\delta_s+1)\nu)+\mathbb{R}^+} \mathrm{e}^{\frac{\mathrm{i}\omega\xi}{\nu}} f(Y)(\xi,\rho,\nu)d\xi\right) \\ &= \mathcal{R}e\left(\int_{\mathrm{i}(\pi-(\delta_s+1)\nu)+\mathbb{R}^+} \mathrm{e}^{\frac{\mathrm{i}\omega\xi}{\nu}} f(Y_s)(\xi,\rho,\nu)d\xi\right) \end{aligned}$$

since Lemma 7.4.48 ensures that on $\mathcal{B}^s_{\pi-\delta_s\nu,\nu} \supset \mathrm{i}(\pi-(\delta_s+1)\nu)+\mathbb{R}^+$, $f(Y) = f(Y_s)$ holds. Then, using that $f(Y_s) \in {}^s E^{3,\lambda}_{\pi,\delta_s}$ and proceeding as for the proof of the Third Exponential Lemma to estimate the integral along the half line $\mathrm{i}(\pi-(\delta_s+1)\nu)+\mathbb{R}$, we get (see (2.12))

$$I(Y,\rho,\nu) = \frac{1}{\nu^2}\mathrm{e}^{-\frac{\omega\pi}{\nu}}(\Lambda(\rho)+J(\rho,\nu)) \qquad \text{with } |J(\rho,\nu)| \le M_{sc}\nu^{\frac{1}{4}}$$

where

$$\Lambda(\rho) = \mathcal{R}e\left(\int_{-\mathrm{i}(\delta_s+1)+\mathbb{R}^+} \mathrm{e}^{\mathrm{i}\omega z+4\mathrm{i}b/z}(\widehat{g}_{0,B}(z,\rho)+\mathrm{i}\widehat{g}_{0,A}(z,\rho))dz\right)$$

This completes the proof of Lemma 7.4.49. □

7.4.7.2 Proof of Proposition 7.4.8-(b). We first prove

> **Lemma 7.4.51.** *The function Λ introduced in Proposition 7.4.8 is a real analytic function on $[0,\rho_s[$ which reads $\Lambda(\rho) = \sum_{n\ge 1}\rho^n\Lambda_n$.*

Proof. In subsection 7.4.4 the stable manifold of the inner system $\widehat{Y}^+_{s,0}$ is built in the form $\widehat{Y}^+_{s,0} = \widehat{h}^+_0 + \widehat{W}^+_{s,0}$ where $\widehat{W}^+_{s,0}$ was found as a fixed point of $\widehat{\mathcal{F}}_0$ which is defined in Lemma 7.4.41. Hence, for $(z,\rho) \in \widehat{\Omega}^s_{\delta_s} \times [0,\rho_s[$

$$\begin{aligned} \langle \widehat{\mathbf{r}}^*_+(z), \widehat{W}^+_{s,0}(z,\rho)\rangle_* &= -\int_{[z,+\infty)} \langle \widehat{\mathbf{r}}^*_+(\zeta), \widehat{g}_0(\widehat{W}^+_{s,0},\zeta,\rho)\rangle_* d\zeta, \\ &= -\int_{[z,+\infty)} [\cos\left(\omega\left(\zeta+\tfrac{4b}{\zeta}\right)\right)\widehat{g}_{0,A}(\zeta)+\sin\left(\omega\left(\zeta+\tfrac{4b}{\zeta}\right)\right)\widehat{g}_{0,B}(\zeta)]\, d\zeta, \end{aligned}$$

$$\begin{aligned} \langle \widehat{\mathbf{r}}^*_-(z), \widehat{W}^+_{s,0}(z,\rho)\rangle_* &= -\int_{[z,+\infty)} \langle \widehat{\mathbf{r}}^*_-(\zeta), \widehat{g}_0(\widehat{W}^+_{s,0},\zeta,\rho)\rangle_* d\zeta \\ &= \int_{[z,+\infty)} [\sin\left(\omega\left(\zeta+\tfrac{4b}{\zeta}\right)\right)\widehat{g}_{0,A}(\zeta)-\cos\left(\omega\left(\zeta+\tfrac{4b}{\zeta}\right)\right)\widehat{g}_{0,B}(\zeta)]\, d\zeta. \end{aligned}$$

Thus,

$$\Lambda(\rho) = -\mathcal{R}e\left(\mathrm{i}\langle\widehat{\mathbf{r}}^*_+(-\mathrm{i}\eta_s), \widehat{W}^+_{s,0}(-\mathrm{i}\eta_s,\rho)\rangle_* + \langle\widehat{\mathbf{r}}^*_-(-\mathrm{i}\eta_s), \widehat{W}^+_{s,0}(-\mathrm{i}\eta_s,\rho)\rangle_*\right)$$

where $\eta_s = \delta_s + 1$. The Remark 7.4.36 ensures that for $(z,\rho) \in \widehat{\Omega}_\delta \times [0,\rho_s[$, $\widehat{W}^+_{s,0}(z,\rho)$ can be expended in power series of ρ

$$\widehat{W}^+_{s,0}(z,\rho) = \sum_{n\geq 0} \rho^n\, \widehat{W}^+_{s,0,n}(z).$$

Hence, for $\rho \in [0,\rho_s[$

$$\Lambda(\rho) = -\sum_{n\geq 1} \rho^n \mathcal{R}e\left(\mathrm{i}\langle\widehat{\mathbf{r}}^*_+(-\mathrm{i}\eta_s), \widehat{W}^+_{s,0,n}(-\mathrm{i}\eta_s)\rangle_* + \langle\widehat{\mathbf{r}}^*_-(-\mathrm{i}\eta_s), \widehat{W}^+_{s,0,n}(-\mathrm{i}\eta_s)\rangle_*\right).$$

where $\eta_s = \delta_s + 1$. □

Lemma 7.4.52.

$$\Lambda_1 = \sum_{\substack{m_\alpha,m_\beta\in\mathbb{N}^2\\ m_\alpha+m_\beta\geq 3}} b_{m_\alpha,m_\beta}\, \Lambda_{m_\alpha,m_\beta} \tag{7.73}$$

with

$$\Lambda_{m_\alpha,m_\beta} = 2\pi\, (-1)^{m_\beta+E(\frac{m_\beta+1}{2})} \sum_{n=0}^{+\infty} \frac{(-2b)^n\, (2\omega)^{2m_\alpha+3m_\beta+n-1}}{(2m_\alpha+3m_\beta+n-1)!\; n!},$$

and

$$b_{m_\alpha,m_\beta} = a_{A,(m_\alpha,m_\beta,0,0),0} + a_{B,(m_\alpha,m_\beta,0,0),0}$$

where $a_{A,\overline{m},\ell}$, $a_{B,\overline{m},\ell}$ are defined in (7.6) and where for $x \in \mathbb{R}$, $E(x)$ is the largest integer smaller or equal to x.

Remark 7.4.53. The reversibility properties of R ensures that

$$a_{A,(m_\alpha,m_\beta,0,0),0} = 0 \text{ for } m_\beta \text{ even}, \qquad a_{B,(m_\alpha,m_\beta,0,0),0} \qquad \text{for } m_\beta \text{ odd}.$$

Proof. In this Proof, we denote by $\eta_s = \delta_s + 1$.

Step 1. Permutation of $\lim\limits_{\rho\to 0}$ and $\int$. First observe that

$$\Lambda_1 = \frac{d\Lambda}{d\rho}(0) = \lim_{\rho\to 0} \frac{\Lambda(\rho)}{\rho} = \lim_{\rho\to 0} \mathcal{R}e\left(\int_0^{+\infty} G(t,\rho)dt\right)$$

where

$$G(t,\rho) = \mathrm{e}^{\mathrm{i}\omega(t-\mathrm{i}\eta_s)+4\mathrm{i}b/(t-\mathrm{i}\eta_s)}\, \frac{1}{\rho}\left(\widehat{g}_{0,B}(t-\mathrm{i}\eta_s,\rho) + \mathrm{i}\widehat{g}_{0,A}(t-\mathrm{i}\eta_s,\rho)\right).$$

where $\widehat{g}_{0,A}, \widehat{g}_{0,B}$ are given by (7.36). Remark 7.4.36 ensures that

$$\widehat{Y}^{+}_{s,0}(z,\rho) = \sum_{n\geq 0} \rho^n \, \widehat{Y}^{+}_{s,0,n}(z),$$

with $\widehat{Y}^{+}_{s,0,0}(z) = \widehat{h}^{+}_{0}(z) = (-4/z^2, 8/z^3, 0, 0)$. Thus, for every $t \in [0, +\infty]$,

$$\lim_{\rho\to 0} G(t,\rho) = \mathrm{e}^{\mathrm{i}\omega(t-\mathrm{i}\eta_s)+4\mathrm{i}b/(t-\mathrm{i}\eta_s)} \, R_{0,B}\big(\widehat{h}^{+}_{0}(t-\mathrm{i}\eta)\big) + \mathrm{i}R_{0,A}\big(\widehat{h}^{+}_{0}(t-\mathrm{i}\eta)\big).$$

Moreover, Lemma 7.G.1 ensures that

$$\begin{aligned} |G(t,\rho)| &\leq \mathrm{e}^{\eta_s+4|b|} \frac{2\mathcal{M}_{10}}{\rho} \frac{|\widehat{W}^{+}_{s,0}|^2_{s\widehat{H}_\delta} + \rho}{(\eta_s^2+t^2)^3} \\ &\leq \mathrm{e}^{\eta_s+4|b|} \frac{2\mathcal{M}_{10}}{\rho} \frac{(M_0^+)^2\rho^2/\delta_s + \rho}{(\eta_s^2+t^2)^3} \\ &\leq \mathrm{e}^{\eta_s+4|b|} 2\mathcal{M}_{10} \frac{(M_0^+)^2\rho_s/\delta_s + 1}{(\eta_s^2+t^2)^3}. \end{aligned}$$

Using the dominated convergence theorem, we get

$$\begin{aligned} \Lambda_1 &= \lim_{\rho\to 0} \frac{\Lambda(\rho)}{\rho} \\ &= \mathcal{R}e\left(\int_0^{+\infty} \mathrm{e}^{\mathrm{i}\omega(t-\mathrm{i}\eta_s)+4\mathrm{i}b/(t-\mathrm{i}\eta_s)} \left(\mathrm{i}\widehat{R}_{0,A}(\widehat{h}^{+}_{0}(t-\mathrm{i}\eta)) + \widehat{R}_{0,B}(\widehat{h}^{+}_{0}(t-\mathrm{i}\eta))\right) dt \right). \end{aligned}$$

Finally, the symmetry properties of $\widehat{R}_0$ and $\widehat{h}^{+}_{0}$ gives

$$\Lambda_1 = \frac{-1}{2\mathrm{i}} \int_{-\infty}^{+\infty} \mathrm{e}^{\mathrm{i}\omega(t-\mathrm{i}\eta_s)+4\mathrm{i}b/(t-\mathrm{i}\eta_s)} \left(\widehat{R}_{0,A}(\widehat{h}^{+}_{0}(t-\mathrm{i}\eta)) - \mathrm{i}\widehat{R}_{0,B}(\widehat{h}^{+}_{0}(t-\mathrm{i}\eta))\right) dt.$$

Step 2. Permutation of $\int$ and $\sum$. Equations (7.7),(7.6) ensure that

$$\left(\widehat{R}_{0,A}(\widehat{h}^{+}_{0}) - \mathrm{i}\widehat{R}_{0,B}(\widehat{h}^{+}_{0})\right) = \sum_{m_\alpha+m_\beta\geq 3} b_{m_\alpha,m_\beta} (\widehat{h}^{+}_{0})^{\overline{m}} (-\mathrm{i})^{(m_\beta+1)\mathrm{mod}2}.$$

Thus,

$$\mathrm{e}^{\mathrm{i}\omega(t-\mathrm{i}\eta_s)+\frac{4b\mathrm{i}}{t-\mathrm{i}\eta_s}} \left(\widehat{R}_{0,A}(\widehat{h}^{+}_{0}(t-\mathrm{i}\eta_s)) - \mathrm{i}\widehat{R}_{0,B}(\widehat{h}^{+}_{0}(t-\mathrm{i}\eta_s))\right) = \sum_{m_\alpha+m_\beta\geq 3} f_{m_\alpha,m_\beta}(t)$$

where

$$f_{m_\alpha,m_\beta}(t) = \mathrm{e}^{\mathrm{i}\omega(t-\mathrm{i}\eta_s)} \mathrm{e}^{\frac{4b\mathrm{i}}{t-\mathrm{i}\eta_s}} \\ b_{m_\alpha,m_\beta} (-\mathrm{i})^{(m_\beta+1)\mathrm{mod}2} \left(\frac{-4}{(t-\mathrm{i}\eta_s)^2}\right)^{m_\alpha} \left(\frac{8}{(t-\mathrm{i}\eta_s)^3}\right)^{m_\beta}.$$

Using (7.7) we obtain

$$|f_{m_\alpha,m_\beta}(t)| \leq e^{q\eta_s} e^{4|b|} \frac{M_R}{r_\alpha^{m_\alpha} r_\beta^{m_\beta}} \left(\frac{4}{|t-i\eta_s|^2}\right)^{m_\alpha} \left(\frac{8}{|t-i\eta_s|^3}\right)^{m_\beta}.$$

Recalling that $|t - i\eta_s| \geq \delta_s \geq \delta_0 \geq 1$ and that δ_0 has been chosen in Lemma 7.G.1 such that

$$\Theta = \max\left(\frac{4}{\delta_0^2 r_\alpha}, \frac{8}{\delta_0^3 r_\beta}\right) \leq \frac{1}{\delta_0^2} \max\left(\frac{5}{r_\alpha}\frac{8}{r_\beta}\right) \leq \frac{1}{2},$$

we obtain

$$|f_{m_\alpha,m_\beta}(t)| \leq e^{\omega\eta_s} e^{4|b|} M_R \Theta^{m_\alpha+m_\beta-3} \frac{1}{(1+t^2)^3}.$$

Then, the dominated convergence theorem ensures that

$$\Lambda_1 = -\sum_{m_\alpha+m_\beta\geq 3} \frac{b_{m_\alpha,m_\beta}(-i)^{(m_\beta+1)\mathrm{mod}2}}{2i} \int_{-i\eta_s+\mathbb{R}} e^{i\omega z+\frac{4ib}{z}} \left(\frac{-4}{z^2}\right)^{m_\alpha} \left(\frac{8}{z^3}\right)^{m_\beta} dz.$$

Step 3. Expansion of the essential singularity. We cannot compute directly the integral involved in the previous expression of Λ_1 by using the residues, because the term $e^{i\omega z+4bi/z}$ has an essential singularity in 0. To avoid this difficulty we must expand $e^{4bi/z}$ as a power series. Using once again the dominated convergence theorem we obtain

$$\Lambda_1 = -\sum_{m_\alpha+m_\beta\geq 3} \frac{b_{m_\alpha,m_\beta}(-i)^{(m_\beta+1)\mathrm{mod}2}}{2i} \sum_{n\geq 0} \frac{1}{n!} I_{m_\alpha,m_\beta,n}$$

where

$$I_{m_\alpha,m_\beta,n} = \int_{-i\eta_s+\mathbb{R}} e^{i\omega z} \left(\frac{-4}{z^2}\right)^{m_\alpha} \left(\frac{8}{z^3}\right)^{m_\beta} \left(\frac{4ib}{z}\right)^{n} dz$$

Step 4. Computation of the residues. Using the residues with the path Γ_r' drawn in figure 7.21

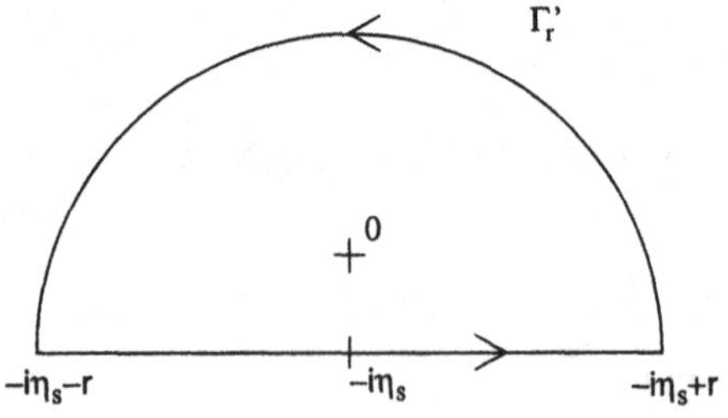

Fig. 7.21. Path Γ_r'

and pushing r to infinity we get

$$I_{m_\alpha,m_\beta,n} = 2\pi\ \mathrm{i}(-1)^{m_\alpha}\ 2^{2m_\alpha+3m_\beta}\frac{(4b\mathrm{i})^n(\mathrm{i}\omega)^{2m_\alpha+3m_\beta+n-1}}{(2m_\alpha+3m_\beta+n-1)!}$$

Hence,

$$\Lambda_1 = \sum_{m_\alpha+m_\beta\geq 3} b_{m_\alpha,m_\beta}\Lambda_{m_\alpha,m_\beta}$$

where

$$\Lambda_{m_\alpha,m_\beta} = \sum_{n\geq 0}\frac{(-\mathrm{i})^{(m_\beta+1)\mathrm{mod}2}}{-2\mathrm{i}}\ \frac{2\pi\mathrm{i}(-1)^{m_\alpha}2^{2m_\alpha+3m_\beta}}{n!}\ \frac{(4b\mathrm{i})^n(\mathrm{i}\omega)^{2m_\alpha+3m_\beta+n-1}}{(2m_\alpha+3m_\beta+n-1)!}$$

$$= 2\pi\ (-1)^{m_\beta+E(\frac{m_\beta+1}{2})}\sum_{n=0}^{+\infty}\frac{(-2b)^n\ (2\omega)^{2m_\alpha+3m_\beta+n-1}}{(2m_\alpha+3m_\beta+n-1)!\ n!}.\ \square$$

7.A Appendix. Proofs of Lemmas 7.3.9, 7.3.10, 7.3.11

Proof of Lemma 7.3.9. The explicit form of N' can be easily computed by using the explicit form of N given by (7.5). Estimate (a) is deduced from the explicit form using Lemma 7.3.6. Now we prove estimate (b):

$$N'(v,\varphi,\xi) - N'(v',\varphi',\xi) = \Delta_1 + \Delta_2$$

where

$$\Delta_1 = N'(v,\varphi,\xi) - N'(v,\varphi',\xi),\quad \Delta_2 = N'(v,\varphi',\xi) - N'(v',\varphi',\xi).$$

Here the explicit form of N' and Lemma 7.3.6 ensure that there exists M such that

$$|\Delta_2| \leq M(|v(\xi)| + |v'(\xi)| + K)|v(\xi) - v'(\xi)|$$

for $\nu\in]0,\nu_1]$, $\varphi'\in[0,6\pi\nu/\omega]$, $K\in[0,K_1]$, and $\xi\in\mathcal{B}_\ell$.

Using Lemmas 7.3.8, 7.3.6 and studying $\varphi\longrightarrow Y_{k,\nu}(\xi+\varphi\tanh\frac{1}{2}\xi)$ we find that

$$|Y_{k,\nu}(\xi+\varphi\tanh\tfrac{1}{2}\xi) - Y_{k,\nu}(\xi+\varphi'\tanh\tfrac{1}{2}\xi)| \leq M_3M_1\frac{K}{\nu}|\varphi'-\varphi|$$

holds for $\nu\in]0,\nu_1]$, $\varphi\in[0,6\pi\nu/\omega]$, $K\in[0,K_1]$, and $\xi\in\mathcal{B}_\ell$. Then using once again the explicit form of N', we conclude that there exists M such that

$$|\Delta_1| \leq M(|v(\xi)| + |h(\xi)| + K)\frac{K}{\nu}|\varphi-\varphi'|$$

for $\nu\in]0,\nu_1]$, $\varphi,\varphi'\in[0,6\pi\nu/\omega]$, $K\in[0,K_1]$, and $\xi\in\mathcal{B}_\ell$.

Proof of Lemma 7.3.10. We start by observing that $h(\xi), Y_{k,\nu}(\xi+\varphi\tanh\frac{1}{2}\xi)$, $v(\xi)$ are uniformly bounded with respect to ξ, φ, K, ν, v. Indeed, we have

$$|h(\xi)| \le M_5 e^{-|\mathcal{R}e(\xi)|} \le M_5,$$
$$|Y_{k,\nu}(\xi + \varphi \tanh \tfrac{1}{2}\xi)| \le M_1 K \le M_1 K_1,$$
$$|v(\xi)| \le |v|_{\mathcal{H}_\ell^\lambda} \le \delta\sqrt{\nu} \le \delta\sqrt{\nu_1}.$$

(a): Now

$$\begin{aligned} R'(v(\xi),\varphi,\xi) &= R\big(h(\xi)+Y_{k,\nu}(\xi+\varphi \tanh \tfrac{1}{2}\xi)+v(\xi)\big) - R(Y_{k,\nu}(\xi+\varphi \tanh \tfrac{1}{2}\xi)) \\ &= \int_0^1 DR(Y_{k,\nu}(\xi+\varphi \tanh \tfrac{1}{2}\xi) + u(h(\xi) + v(\xi))).(h+v)(\xi)\, du. \end{aligned}$$

But R and all its derivate are $\mathcal{O}(\nu^2)$. Thus there exists M_7 such that

$$|R'(v(\xi),\varphi,\xi)| \le M_7 \nu^2 (|v(\xi)| + |h(\xi)|).$$

for every $\nu \in]0,\nu_1]$, $\varphi, \varphi' \in [0, 6\pi\nu\omega]$, $K \in [0, K_1]$, $v \in \mathcal{B}\mathcal{H}_\ell^\lambda|_{R,\mathbb{R}}(\delta\sqrt{\nu})$ and $\xi \in \mathcal{B}_\ell$.

(b): We split R' as we split N':

$$R'(v,\varphi,\xi) - R'(v',\varphi',\xi) = \Delta_1 + \Delta_2$$

where

$$\Delta_1 = R'(v,\varphi,\xi) - R'(v,\varphi',\xi), \quad \Delta_2 = R'(v,\varphi',\xi) - R'(v',\varphi',\xi).$$

The estimate for Δ_2 is done as in (a). We find that there exists M such that

$$|\Delta_2| \le M\nu^2 |v(\xi) - v'(\xi)|$$

holds for every $\nu \in]0,\nu_1]$, $\varphi, \varphi' \in [0, 6\pi\nu/\omega]$, $K \in [0, K_1]$, $v \in \mathcal{B}\mathcal{H}_\ell^\lambda|_{R,\mathbb{R}}(\delta\sqrt{\nu})$, and $\xi \in \mathcal{B}_\ell$.

Denote $\Delta_{\varphi,\varphi'} = Y_{k,\nu}(\xi + \varphi' \tanh \tfrac{1}{2}\xi) - Y_{k,\nu}(\xi + \varphi \tanh \tfrac{1}{2}\xi)$. Then

$$\begin{aligned} \Delta_1 &= \int_0^1 DR(h(\xi) + v(\xi) + Y_{k,\nu}(\xi + \varphi \tanh \tfrac{1}{2}\xi) + u\Delta_{\varphi,\varphi'}).\Delta_{\varphi,\varphi'}\, du \\ &\quad - \int_0^1 DR(Y_{k,\nu}(\xi + \varphi \tanh \tfrac{1}{2}\xi) + u\Delta_{\varphi,\varphi'}).\Delta_{\varphi,\varphi'}\, du \end{aligned}$$

Recalling that $|Y_{k,\nu}(\xi+\varphi \tanh \tfrac{1}{2}\xi) - Y_{k,\nu}(\xi+\varphi' \tanh \tfrac{1}{2}\xi)| \le M_3 M_1 \dfrac{K}{\nu}|\varphi - \varphi'|$, we finally obtain that there exists M' such that

$$|\Delta_1| \le M' K \nu\ (|v(\xi)| + |h(\xi)|)\ |\varphi - \varphi'|$$

holds for $\nu \in]0,\nu_1]$, $\varphi, \varphi' \in [0, 6\pi\nu/\omega]$, $K \in [0, K_1]$, $v \in \mathcal{B}\mathcal{H}_\ell^\lambda|_{R,\mathbb{R}}(\delta\sqrt{\nu})$, and $\xi \in \mathcal{B}_\ell$.

Proof of Lemma 7.3.11

(a): Lemmas 7.3.8, 7.3.6 ensure that

$$|\varphi\Phi(\xi,\varphi,K,\nu)| \le \frac{6\pi}{\omega} M_1 M_4 K \mathrm{e}^{-|\mathcal{R}e(\xi)|}$$

holds for every $\nu \in]0,\nu_1]$, $\varphi \in [0, 6\pi\nu/\omega]$, $K \in [0, K_1]$, $v \in \mathcal{BH}_\ell^\lambda|_{R,\mathbb{R}}(\delta\sqrt{\nu})$, and $\xi \in \mathcal{B}_\ell$.

(b): Here again we split $\varphi\Phi$ in two parts:

$$\varphi\Phi(\xi,\varphi,K,\nu) - \varphi'\Phi(\xi,\varphi',K,\nu) = \Delta_1 + \Delta_2,$$

where

$$\Delta_1 = (\varphi - \varphi')\Phi(\xi,\varphi,K,\nu), \ \Delta_2 = \varphi' \ [\Phi(\xi,\varphi,K,\nu) - \Phi(\xi,\varphi',K,\nu)].$$

The estimate of Δ_1 can be easily deduced from the previous estimate. Then studying $\varphi \longrightarrow \dfrac{dY_{k,\nu}}{d\xi}(\xi + \varphi \tanh \frac{1}{2}\xi)$ we finally obtain that

$$|\varphi\Phi(\xi,\varphi,K,\nu) - \varphi'\Phi(\xi,\varphi',K,\nu)| \le M_1 M_4 \frac{K}{\nu} \mathrm{e}^{-|\mathcal{R}e(\xi)|} \ [1 + \frac{6\pi}{\omega} M_3] \ |\varphi - \varphi'|$$

for $\nu \in]0,\nu_1]$, $\varphi, \varphi' \in [0, 6\pi\nu/\omega]$, $K \in [0, K_1]$, $v \in \mathcal{BH}_\ell^\lambda|_{R,\mathbb{R}}(\delta\sqrt{\nu})$, and $\xi \in \mathcal{B}_\ell$.

7.B Appendix. Proof of Proposition 7.3.19

The proof of Proposition 7.3.19 is essentially based on the following lemma which is a particular form of the First Exponential Lemma 2.1.1.

Lemma 7.B.1. *There exists M_{12} such that for all functions f satisfying*

(a) $$\begin{aligned} f: \ \mathcal{B}_\ell \times]0,\nu_1] &\longrightarrow \mathbb{C}^4 \\ (\xi,\nu) &\longrightarrow (f_\alpha(\xi,\nu), f_\beta(\xi,\nu), f_A(\xi,\nu), f_B(\xi,\nu)), \end{aligned}$$

(b) *f is holomorphic in $\mathcal{B}_\ell$,*

(c) $|f_A|_{\mathcal{H}_\ell^\lambda} < +\infty$, $|f_B|_{\mathcal{H}_\ell^\lambda} < +\infty$,

(d) $Sf(\xi,\nu) = -f(-\xi,\nu)$,

the following inequality holds

$$\left| \int_0^{+\infty} \langle r_-^*(t), f(t,\nu)\rangle_* dt \right| \le M_{12}[|f_A|_{\mathcal{H}_\ell^\lambda} + |f_B|_{\mathcal{H}_\ell^\lambda}] \mathrm{e}^{-\ell\omega/\nu} \quad \text{for } \ \nu \in]0,\nu_1].$$

Proof. The symmetry hypothesis on f is fundamental: f_A is an odd function and f_B is an even function, thus

$$\begin{aligned} \int_0^{+\infty} \langle r_-^*(t), f(t,\nu)\rangle_* dt &= \int_0^{+\infty} [-f_A(t,\nu)\sin\psi_\nu(t) + f_B(t,\nu)\cos\psi_\nu(t)]dt \\ &= \frac{1}{2}\int_{-\infty}^{+\infty} \mathrm{e}^{\mathrm{i}(\frac{\omega}{\nu}+a\nu)t + 2\mathrm{i}b\nu\tanh\frac{1}{2}t}[\mathrm{i}f_A(t,\nu) + f_B(t,\nu)]dt. \end{aligned}$$

The integral now has the appropriate form to apply Lemma 2.1.1. Observing that

$$|e^{ia\nu\xi+2ib\nu\tanh\frac{1}{2}\xi}| \le e^{|a|\nu_1\ell+2|b|\nu\tan\frac{1}{2}\ell} \quad \text{for } \nu \in]0,\nu_1],\ \xi \in \mathcal{B}_\ell,$$

we obtain that

$$\left|(\mathrm{i}f_A(t,\nu)+f_B(t,\nu))e^{ia\nu\xi+2ib\nu\tanh\frac{1}{2}\xi}\right|_{\mathcal{H}_\ell^\lambda} \le e^{|a|\nu_1\ell+2|b|\nu\tan\frac{1}{2}\ell}(|f_A|_{\mathcal{H}_\ell^\lambda}+|f_B|_{\mathcal{H}_\ell^\lambda}).$$

Lemma 2.1.1 gives the result. □

Now we apply this lemma with $f = N' + R'$ to obtain

Lemma 7.B.2. *There exists* M_{13} *such that*

$$\left|\int_0^{+\infty}\langle r_-^*(t), N'(v(t),\varphi,K,\nu,t)+R'(v(t),\varphi,K,\nu,t)\rangle_* dt\right| \le M_{13}(\nu^2+\nu K)e^{-\frac{\ell\omega}{\nu}}$$

for every $\nu \in]0,\nu_1]$, $K \in [0,K_1]$, $\varphi \in [0, 6\pi\nu/\omega]$, *and* $v \in \mathcal{BH}_\ell^\lambda|_{R,\mathbb{R}}(\delta\sqrt{\nu})$.

Proof. $N' + R'$ is holomorphic in $\mathcal{B}_\ell$ and satisfies

$$S(N'+R')(v(\xi),\varphi,K,\nu,\xi) = -(N'+R')(v(-\xi),\varphi,K,\nu,-\xi).$$

Here we use the reversibility of v. To be allowed to symmetrize the reversibility condition of $\mathcal{F}_{\varphi,K,\nu}$, i.e., to obtain an integral from $-\infty$ to $+\infty$, we need that v be reversible. This is why we look for a fixed point in a space of reversible functions, instead of computing the stable manifold of the periodic orbits.

Now using Lemmas 7.3.10, 7.3.8 we obtain

$$|R'(v,\varphi,K,\nu,\xi)|_{\mathcal{H}_\ell^\lambda} \le M_7\nu^2(M_5+|v|_{\mathcal{H}_\ell^\lambda}).$$

Using the explicit form of $[N'(v,\varphi,K,\nu,\xi)]_A$ and Lemmas 7.3.8, 7.3.6 we find that

$$|[N'(v,\varphi,K,\nu,\xi)]_A|_{\mathcal{H}_\ell^\lambda} \le |b|\nu(\delta^2\nu+2M_1K+M_1M_5K).$$

The same inequality holds for $[N'(v,\varphi,K,\nu,\xi)]_B$. Lemma 7.B.1 gives the result. □

Lemma 7.B.3. *There exists* M_{14} *such that for every* $\nu \in]0,\nu_1]$, $K \in [0,K_1]$, $\varphi \in [0,6\pi\nu/\omega]$, *and* $v \in \mathcal{BH}_\ell^\lambda|_{R,\mathbb{R}}(\delta\sqrt{\nu})$,

$$J(v,\varphi,K,\nu) = \frac{\omega}{\nu}Ke^{-\ell\omega/\nu}\varphi G(\widehat{\omega}(Ke^{-\ell\omega/\nu},\nu),\underline{\omega}_{k,\nu}\varphi-2b\nu)+J_1(v,\varphi,K,\nu)$$

where $G(x,y) = -\int_0^{+\infty} \frac{\cos(xt + y\tanh\frac{1}{2}t)}{2\cosh^2\frac{1}{2}t}\,dt$, *and* J_1 *satisfies*

$$|J_1(v,\varphi,K,\nu)| \le M_{14}(\nu^2 + \nu K)\mathrm{e}^{-\ell\omega/\nu}.$$

Proof. For obtaining this result we split J into two parts . One involves $N' + R'$, which have been already studied, and the other one is

$$\int_0^{+\infty} \langle r_-^*(t), \varphi\Phi(t,\varphi,K,\nu)\rangle_* \,dt.$$

For $t \in [0,+\infty[\subset \mathbb{R}$, using Lemma 7.3.7 we have

$$\begin{aligned}
&\langle r_-^*(t), \varphi\Phi(t,\varphi,K,\nu)\rangle_* \\
&\quad = \langle r_-^*(t), \varphi\frac{-1}{2\cosh^2\frac{1}{2}t}\frac{dY_{k,\nu}}{dt}(t+\varphi\tanh\tfrac{1}{2}t)\rangle_* \\
&\quad = \langle r_-^*(t), \frac{-\varphi K\mathrm{e}^{-\ell\omega/\nu}}{2\cosh^2\frac{1}{2}t}\frac{\omega}{\nu}\frac{d\widehat{Y}_1}{dt}(\underline{\omega}_{k,\nu}(t+\varphi\tanh\tfrac{1}{2}t))\rangle_* + \Delta_2 \\
&\quad = -\frac{\omega K\varphi}{\nu\mathrm{e}^{\ell\omega/\nu}}\frac{\cos(\widehat{\omega}(K\mathrm{e}^{-\ell\omega/\nu},\nu)t + (\underline{\omega}_{k,\nu}\varphi - 2b\nu)\tanh\frac{1}{2}t)}{2\cosh^2\frac{1}{2}t} + \Delta_2
\end{aligned}$$

with

$$|\Delta_2| \le \frac{|\varphi||\tilde{O}_2|}{2\cosh^2\frac{1}{2}t} \le \frac{6\pi\nu}{\omega}\frac{1}{2\cosh^2\frac{1}{2}t}M_2\mathrm{e}^{-\ell\omega/\nu}(\nu K + \frac{K^2}{\nu}\mathrm{e}^{-\ell\omega/\nu}).$$

since $|r_-^*(t)| = 1$. Denote $M = \sup_{\nu\in]0,\nu_1]}(\frac{\mathrm{e}^{-\ell\omega/\nu}}{\nu^2})$; then

$$\int_0^{+\infty} |\Delta_2| \le \frac{6\pi M_2(K_1 + K_1^2 M)}{\omega}\nu^2\mathrm{e}^{-\ell\omega/\nu}.$$

Finally, using Lemma 7.B.2 we get $M_{14} = M_{13} + \frac{6\pi M_2(K_1 + K_1^2 M)}{\omega}$. □

Proof of Proposition 7.3.19. First of all, observe that $G(0,y) = -\sin y/y$ and that there exists M_G such that $|G(x,y) - G(0,y)| \le M_G|x|$ for $(x,y) \in \mathbb{R}^2$. Thus $J_2 = J_1 + \Delta$ with

$$\begin{aligned}
|\Delta| &= |G(\widehat{\omega}(K\mathrm{e}^{-\ell\omega/\nu},\nu), \underline{\omega}_{k,\nu}\varphi - 2b\nu) - G(0, \underline{\omega}_{k,\nu}\varphi - 2b\nu)|\frac{\omega}{\nu}K\mathrm{e}^{-\ell\omega/\nu}\varphi \\
&\le \frac{\omega}{\nu}K\mathrm{e}^{-\ell\omega/\nu}\frac{6\pi\nu}{\omega}M_G|\widehat{\omega}(K\mathrm{e}^{-\ell\omega/\nu},\nu)|.
\end{aligned}$$

Finally using the estimate of $\widehat{\omega}$ given by Theorem 7.1.4 we get

$$|\Delta| \le 6\pi M_G C_\omega K_1 K\nu\mathrm{e}^{-\ell\omega/\nu}. \ \square$$

7.C Appendix. Proof of Proposition 7.3.24

Lemma 7.C.1. *There exists M_{16} such that for every $\nu \in]0,\nu_1]$, $K \in [0,K_1]$, $\varphi \in [0,6\pi\nu/\omega]$, and $v \in \mathcal{BH}_\ell^\lambda|_{\mathrm{R,R}}(\delta\sqrt{\nu})$,*

$$\frac{\partial J}{\partial\varphi}(v,\varphi,K,\nu) = \frac{\omega^2}{\nu^2}K\mathrm{e}^{-\ell\omega/\nu}\varphi H(\widehat{\omega}(K\mathrm{e}^{-\ell\omega/\nu},\nu),\underline{\omega}_{k,\nu}\varphi - 2b\nu)$$
$$+\frac{\omega}{\nu}K\mathrm{e}^{-\ell\omega/\nu}G(\widehat{\omega}(K\mathrm{e}^{-\ell\omega/\nu},\nu),\underline{\omega}_{k,\nu}\varphi - 2b\nu) + J_3(v,\varphi,K,\nu)$$

where $H(a,b) = \displaystyle\int_0^{+\infty} \frac{\sin(at + b\tanh\frac{1}{2}t)}{2\cosh^2\frac{1}{2}t}\tanh\tfrac{1}{2}t\, dt$, *and J_3 satisfies* $|J_3(v,\varphi,K,\nu)| \le M_{16}K\mathrm{e}^{-\ell\omega/\nu}$. *($G$ is defined by Lemma* 7.B.3.*)*

Proof.

$$\frac{\partial J}{\partial\varphi} = \int_0^{+\infty}\frac{\partial}{\partial\varphi}[N'(v(t),\varphi,K,\nu,t) + R'(v(t),\varphi,K,\nu,t) + \varphi\Phi(t,\varphi,K,\nu)]\,dt.$$

Here again we split $\partial J/\partial\varphi$ in two parts. We bound the part involving $N' + R'$ and we compute the part involving $\varphi\Phi$. But here we do not use Lemma 7.B.1. Using the explicit form of N' and Lemmas 7.3.7, 7.3.8 we find

$$\left|\left[\frac{\partial N'}{\partial\varphi}(v(t),\varphi,K,\nu,t)\right]_A\right| \le |b|M_2(2\,|v|_{\mathcal{H}_\ell^\lambda}\mathrm{e}^{-\lambda t} + M_5\mathrm{e}^{-t})\;K\mathrm{e}^{-\ell\omega/\nu} \text{ for } t \ge 0.$$

The same inequality holds for $[\dfrac{\partial N'}{\partial\varphi}]_B$. Now we bound $\dfrac{\partial R'}{\partial\varphi}$:

$$\frac{\partial R'}{\partial\varphi}(v(t),\varphi,K,\nu,t) = \tanh\tfrac{1}{2}t$$
$$\int_0^1 D^2R(Y_{k,\nu}(t+\varphi\tanh\tfrac{1}{2}t)+u(h+v)(t)).\left[\frac{dY_{k,\nu}}{dt}(t+\varphi\tanh\tfrac{1}{2}t),(h+v)(t)\right]du.$$

Using Lemmas 7.3.7, 7.3.8 we get for every $t \ge 0$

$$|Y_{k,\nu}(t+\varphi\tanh\tfrac{1}{2}t) + u(h+v)(t)| \le M_2K\mathrm{e}^{-\ell\omega/\nu} + M_5\mathrm{e}^{-t} + |v|_{\mathcal{H}_\ell^\lambda}\mathrm{e}^{-\lambda t}$$
$$\le M_2K_1 + M_5 + \delta\sqrt{\nu},$$
$$\left|[\frac{dY_{k,\nu}}{dt}](t+\varphi\tanh\tfrac{1}{2}t)\right| \le M_2\frac{K}{\nu}\mathrm{e}^{-\ell\omega/\nu},$$
$$|(h+v)(t)| \le M_5\mathrm{e}^{-t} + |v|_{\mathcal{H}_\ell^\lambda}\mathrm{e}^{-\lambda t}.$$

Now recalling that $D^2R = \mathcal{O}(\nu^2)$, we obtain that there exists M such that

$$\frac{\partial R'}{\partial\varphi}(v(t),\varphi,K,\nu,t) \le MM_2\nu(M_5\mathrm{e}^{-t} + |v|_{\mathcal{H}_\ell^\lambda}\mathrm{e}^{-\lambda t})K\mathrm{e}^{-\ell\omega/\nu}$$

for $\nu \in]0, \nu_1]$, $K \in [0, K_1]$, $\varphi \in [0, 6\pi\nu/\omega]$, and $v \in \mathcal{BH}_\ell^\lambda|_{R,\mathbb{R}}(\delta\sqrt{\nu})$. Thus, observing that $|r_-^*(t)| = 1$ for $t > 0$, we conclude that there exists M' such that

$$\int_0^{+\infty} \frac{\partial}{\partial\varphi}[N'(v(t), \varphi, K, \nu, t) + R'(v(t), \varphi, K, \nu, t)]\, dt \leq M' K \mathrm{e}^{-\ell\omega/\nu}.$$

holds for $\nu \in]0, \nu_1]$, $K \in [0, K_1]$, $\varphi \in [0, 6\pi\nu/\omega]$, and $v \in \mathcal{BH}_\ell^\lambda|_{R,\mathbb{R}}(\delta\sqrt{\nu})$.

Now using Lemma 7.3.7 we compute the part coming from $\frac{\partial}{\partial\varphi}[\varphi\Phi]$:

$$\begin{aligned}\frac{\partial}{\partial\varphi}[\varphi\Phi](t, \varphi, K, \nu) = &-\frac{\omega K}{\nu}\mathrm{e}^{-\ell\omega/\nu}\frac{1}{2\cosh^2\frac{1}{2}t}\frac{d\widehat{Y}_1}{dt}(\underline{\omega}_{k,\nu}(t + \varphi\tanh\tfrac{1}{2}t)) \\ &-\varphi\frac{\omega^2}{\nu^2}K\mathrm{e}^{-\ell\omega/\nu}\frac{\tanh\frac{1}{2}t}{2\cosh^2\frac{1}{2}t}\frac{d^2\widehat{Y}_1}{dt^2}(\underline{\omega}_{k,\nu}(t + \varphi\tanh\tfrac{1}{2}t)) + \Delta\end{aligned}$$

with

$$\begin{aligned}\Delta &= \frac{\tilde{O}_2}{2\cosh\frac{1}{2}t} + \varphi\frac{\tanh\frac{1}{2}t}{2\cosh\frac{1}{2}t}\tilde{O}_3, \\ |\Delta| &\leq \frac{M_2}{2\cosh^2\frac{1}{2}t}\,(1 + \frac{6\pi}{\omega})\left[\nu_1 + \frac{K_1}{\nu}\mathrm{e}^{-\ell\omega/\nu}\right] K\mathrm{e}^{-\ell\omega/\nu}.\end{aligned}$$

Thus there exists M'' such that $|\Delta| \leq M''\frac{K\mathrm{e}^{-\ell\omega/\nu}}{2\cosh^2\frac{1}{2}t}$ holds for $\nu \in]0, \nu_1]$, $K \in [0, K_1]$, $\varphi \in [0, 6\pi\nu/\omega]$, and $t \geq 0$. Computing the scalar product we obtain

$$\begin{aligned}\langle r_-^*(t), \frac{d\widehat{Y}_1}{dt}(\underline{\omega}_{k,\nu}(t + \varphi\tanh\tfrac{1}{2}t))\rangle_* &= \cos(\widehat{\omega}t + (\underline{\omega}_{k,\nu}\varphi - 2b\nu)\tanh\tfrac{1}{2}t), \\ \langle r_-^*(t), \frac{d^2\widehat{Y}_1}{dt^2}(\underline{\omega}_{k,\nu}(t + \varphi\tanh\tfrac{1}{2}t))\rangle_* &= -\sin(\widehat{\omega}t + (\underline{\omega}_{k,\nu}\varphi - 2b\nu)\tanh\tfrac{1}{2}t).\end{aligned}$$

We conclude the proof observing that

$$\int_0^{+\infty}\langle r_-^*(t), \Delta\rangle_* \leq \int_0^{+\infty}|\Delta| \leq M''K\mathrm{e}^{-\ell\omega/\nu}.$$

$M_{16} = M' + M''$. □

Proof of Proposition 7.3.24. First of all, observe that

$$\begin{aligned}H(0, y) &= -\frac{\cos y}{y} + \frac{\sin y}{y^2}, \\ |H(x, y) - H(0, y)| &\leq M_H|x| \quad \text{for} \quad (x, y) \in \mathbb{R}^2,\end{aligned}$$

and recall that $G(0, y) = -\sin y/y$ and $|G(x, y) - G(0, y)| \leq M_G|x|$ for $(x, y) \in \mathbb{R}^2$. Thus $J_4 = J_3 + \Delta$ with

$$\begin{aligned}|\Delta| &= [\frac{\omega\varphi}{\nu} H(\widehat{\omega}(K\mathrm{e}^{-\ell\omega/\nu},\nu),\underline{\omega}_{k,\nu}\varphi - 2b\nu) - H(0,\underline{\omega}_{k,\nu}\varphi - 2b\nu)| \\ &\quad +|G(\widehat{\omega}(K\mathrm{e}^{-\ell\omega/\nu},\nu),\underline{\omega}_{k,\nu}\varphi - 2b\nu) - G(0,\underline{\omega}_{k,\nu}\varphi - 2b\nu)|]\frac{\omega}{\nu}K\mathrm{e}^{-\ell\omega/\nu} \\ &\leq \frac{\omega}{\nu}K\mathrm{e}^{-\ell\omega/\nu}\ [6\pi M_H + M_G]|\ \widehat{\omega}(K\mathrm{e}^{-\ell\omega/\nu},\nu)|.\end{aligned}$$

Finally using the estimate of $\widehat{\omega}$ given by Theorem 7.1.4 we get

$$|\Delta| \leq \omega(6\pi M_H + M_G)C_\omega K_1\ K\mathrm{e}^{-\ell\omega/\nu}.\ \square$$

7.D Appendix. Proof of Proposition 7.3.26

We start by proving

Lemma 7.D.1. *There exists M_{19} such that for every $\nu\in]0,\nu_1]$, $K\in[0,K_1]$, $\varphi\in[0,6\pi\nu/\omega]$, and $v,v'\in\mathcal{BH}_\ell^\lambda|_{R,\mathbb{R}}(\delta\sqrt{\nu})$,*

$$|D_vJ(v,K,\varphi,\nu).v'| \leq M_{19}\nu\ (|v|_{\mathcal{H}_\ell^\lambda} + K + \nu)\,|v'|_{\mathcal{H}_\ell^\lambda}\ \mathrm{e}^{-\ell\omega/\nu}.$$

Proof.

$$D_vJ(v,\varphi,K,\nu).v' = \int_0^{+\infty} \langle r_-^*(t), D_v[N'+R'](v(t),\varphi,K,\nu,t).v'(t)\rangle_* dt.$$

Using the explicit form of N' we get

$$|[D_vN'(v,\varphi,K,\nu,\xi).v']_A|_{\mathcal{H}_\ell^\lambda} \leq |b|\nu\,|v'|_{\mathcal{H}_\ell^\lambda}\ (|v|_{\mathcal{H}_\ell^\lambda} + M_1K).$$

The same inequality holds for $|[D_vN'(v,\varphi,K,\nu,\xi).v']_B|_{\mathcal{H}_\ell^\lambda}$.
Now we bound D_vR':

$$D_vR'(v(\xi),\varphi,K,\nu,\xi).v'(\xi) = DR(h(\xi) + Y_{k,\nu}(\xi+\varphi\tanh\tfrac{1}{2}\xi) + v(\xi)).v'(\xi).$$

Using Lemmas 7.3.6, 7.3.8 we get

$$\begin{aligned}|h(\xi) + Y_{k,\nu}(\xi+\varphi\tanh\tfrac{1}{2}\xi) + v(\xi)| &\leq M_5\mathrm{e}^{-|\mathcal{R}e(\xi)|} + M_1K + |v|_{\mathcal{H}_\ell^\lambda}\ \mathrm{e}^{-\lambda|\mathcal{R}e(\xi)|} \\ &\leq M_5 + |v|_{\mathcal{H}_\ell^\lambda} + M_1K.\end{aligned}$$

Recalling that $DR = \mathcal{O}(\nu^2)$ we conclude that there exists M such that

$$|D_vR'(v,\varphi,K,\nu,\xi).v'| \leq M\nu^2\,|v'|_{\mathcal{H}_\ell^\lambda}$$

for $\nu\in]0,\nu_1]$, $K\in[0,K_1]$, $\varphi\in[0,6\pi\nu/\omega]$, and $v,v'\in\mathcal{BH}_\ell^\lambda|_{R,\mathbb{R}}(\delta\sqrt{\nu})$. Finally since

$$f:\ \xi\longrightarrow D_vN'(v(\xi),\varphi,K,\nu,\xi).v'(\xi) + D_vR'(v(\xi),\varphi,K,\nu,\xi).v'(\xi)$$

is holomorphic in $\mathcal{B}_\ell$, and since

$$Sf(-\xi) = -f(-\xi) \text{ for } \xi \in \mathcal{B}_\ell, \quad |f_A|_{\mathcal{H}_\ell^\lambda} < +\infty, \quad |f_B|_{\mathcal{H}_\ell^\lambda} < +\infty,$$

we can apply Lemma 7.B.1, which gives

$$|D_v J(v, K, \nu).v'| \leq M_{12}\ [2|b|\nu(|v|_{\mathcal{H}_\ell^\lambda} + M_1 K) + M\nu^2]\ |v'|_{\mathcal{H}_\ell^\lambda}\ \mathrm{e}^{-\ell\omega/\nu}. \square$$

Proof of Proposition 7.3.26. The expression of $D_v\varphi$ is simply obtained by the Implicit Function Theorem.
(a): Using Proposition 7.3.25 we get

$$\begin{aligned} |D_v\varphi(v, K, \nu).v'| &= \frac{|DvJ(v, \varphi(v, K, \nu), K, \nu).v'|}{|\frac{\partial J}{\partial \varphi}(v, \varphi(v, K, \nu), K, \nu)|} \\ &\leq \frac{M_{19}\nu\ (|v|_{\mathcal{H}_\ell^\lambda} + K + \nu)\,|v'|_{\mathcal{H}_\ell^\lambda}\ \mathrm{e}^{-\ell\omega/\nu}}{\frac{\omega}{32\nu} K \mathrm{e}^{-\ell\omega/\nu}} \\ &\leq \frac{32 M_{19}}{\omega}\frac{\nu^2}{K}(|v|_{\mathcal{H}_\ell^\lambda} + K + \nu)\,|v'|_{\mathcal{H}_\ell^\lambda}. \end{aligned}$$

The proof of (b) follows. $\square$

7.E Appendix. Proof of Lemma 7.3.29

Denote

$$I(\xi) = \int_{[\xi, +\infty)} [\langle r_+^*(s), f(s, K, \nu)\rangle_* \, r_+(\xi) + \langle r_-^*(s), f(s, K, \nu)\rangle_* r_-(\xi)]\ ds.$$

Then, $I(\xi) = (0, 0, I_A(\xi), I_B(\xi))$. because r_+, r_- lie in $\Pi^\perp$. Using the explicit forms of $r_\pm$ given in Lemma 7.3.14 we get

$$\begin{aligned} I_A(\xi) = &-\int_{[\xi, +\infty)} [f_A(s, K, \nu)\cos\psi_\nu(s) + f_B(s, K, \nu)\sin\psi_\nu(s)]\ ds \quad \cos\psi_\nu(\xi) \\ &+ \int_{[\xi, +\infty)} [-f_A(s, K, \nu)\sin\psi_\nu(s) + f_B(s, K, \nu)\cos\psi_\nu(s)]\ ds \quad \sin\psi_\nu(\xi) \\ = &-\int_{[\xi, +\infty)} [f_A(s, K, \nu)\cos(\psi_\nu(s) - \psi_\nu(\xi)) + f_B(s, K, \nu)\sin(\psi_\nu(s) - \psi_\nu(\xi))\ ]ds \\ = &-\int_{\mathcal{R}e(\xi)}^{+\infty} [f_A(\mathrm{i}\mathcal{I}m\,(\xi) + t, K, \nu)\cos\theta + f_B(\mathrm{i}\mathcal{I}m\,(\xi) + t, K, \nu)\sin\theta]dt. \end{aligned}$$

where

$$\theta(t,\nu,\xi) = (\frac{\omega}{\nu} + a\nu)\ (t - \mathcal{R}e\,(\xi)) + 2b\nu[\tanh\tfrac{1}{2}(\mathrm{i}\mathcal{I}m\,(\xi) + t) - \tanh\tfrac{1}{2}\xi].$$

Now observe that $|\cos(z)| \le \mathrm{e}^{|\mathcal{I}m(z)|}$, $|\sin(z)| \le \mathrm{e}^{|\mathcal{I}m(z)|}$ and

$$\begin{aligned} |\mathcal{I}m\,(\theta(t,\nu,\xi))| &\le \left|\mathcal{I}m\,(2b\nu[\tanh\tfrac{1}{2}(\mathrm{i}\mathcal{I}m\,(\xi)+t) - \tanh\tfrac{1}{2}\xi])\right| \\ &\le 4|b|\nu\tan\tfrac{1}{2}\ell \\ &\le 4|b|\nu_3\tan\tfrac{1}{2}\ell. \end{aligned}$$

Thus, there exists M such that

$$|I_A(\xi)| \le M \int_{[\xi,+\infty)} (|f_A(s,K,\nu)| + |f_B(s,K,\nu)|)\ ds.$$

for $\nu \in]0,\nu_3]$, $K \in [M_{18}\nu^2, K_1]$ and $\xi \in \mathcal{B}_\ell$. The same inequality holds for I_B. □

7.F Appendix. Proof of Lemma 7.4.27

To prove Lemma 7.4.27, we need estimates of $h(\xi)$, $p(\xi)$, $q(\xi)$, $r_+(\xi)$, $r_-(\xi)$, g when ξ lies in $\mathcal{D}^*_{r,\nu}$. Since, we use it everywhere in this appendix, we recall that χ_1 is the function given by $\chi_1(\xi) = 1$ if $|\mathcal{R}e\,(\xi)| \le 1$ and $\chi_1(\xi) = 0$ otherwise.

Lemma 7.F.1. *There exists m_0, $\mathcal{M}_3$, $\mathcal{M}_4$, such that for every $r \in [0,1]$, $\nu \in]0,1]$, $\xi \in \mathcal{D}^*_{r,\nu}$,*

(a) $|\xi - \mathrm{i}\pi| \ge m_0\nu^r$, $|\xi + \mathrm{i}\pi| \ge m_0\nu^r$ *and* $|\xi^2 + \pi^2| \ge m_0\pi\nu^r$,

(b) $\left|\tanh\left(\tfrac{1}{2}\xi\right)\right| \le \mathcal{M}_3\left[\frac{1}{|\xi^2+\pi^2|}\chi_1(\xi) + (1-\chi_1(\xi))\right]$,

(c) $\left|\frac{1}{\cosh^2(\frac{1}{2}\xi)}\right| \le \mathcal{M}_4\left[\frac{1}{|\xi^2+\pi^2|^2}\chi_1(\xi) + (1-\chi_1(\xi))\mathrm{e}^{-|\mathcal{R}e(\xi)|}\right]$.

Lemma 7.F.2. *The first solution of the homogeneous equation (7.26) is given by $p = (p_\alpha, p_\beta, 0, 0)$ defined in Lemma 7.3.12. There exists $\mathcal{M}_5$ such that for every $r \in]0,1]$, $\nu \in]0,1]$, $\xi \in \mathcal{D}^*_{r,\nu}$,*

$$|p_\alpha(\xi)| \le \mathcal{M}_5\left[\frac{1}{|\xi^2+\pi^2|^3}\chi_1(\xi) + (1-\chi_1(\xi))\mathrm{e}^{-|\mathcal{R}e(\xi)|}\right]$$

$$|p_\beta(\xi)| \le \mathcal{M}_5\left[\frac{1}{|\xi^2+\pi^2|^4}\chi_1(\xi) + (1-\chi_1(\xi))\mathrm{e}^{-|\mathcal{R}e(\xi)|}\right].$$

Proof. The proof directly follows from the explicit formula giving p defined in Lemma 7.3.12 and from Lemma 7.F.1 . □

The second solution q of the homogeneous equation (7.26), was built to be reversible and to map $\mathbb{R}$ in $\mathbb{R}^4$. Another interesting solution is the one given by Fuchs's theory when we are interested in the singularities at $i\pi$ (see Section 6.2). This theory ensures that there exist two solutions of (7.26) whose first component are respectively equivalent to $(\xi - i\pi)^{-3}$ and $(\xi - i\pi)^4$ when ξ lies near $i\pi$. The first one is already known. Indeed p satisfies

$$p_\alpha(\xi) = -\frac{\tanh(\frac{1}{2}\xi)}{\cosh^2(\frac{1}{2}\xi)} \underset{\xi\to i\pi}{\sim} 8(\xi - i\pi)^{-3}.$$

The following lemma gives the second solution

Lemma 7.F.3. *Lemma 7.3.13 ensures that* $q = (q_\alpha, q_\beta, 0, 0)$ *where* $q_\beta(\xi) = \dfrac{dq_\alpha}{d\xi}$ *and*

$$q_\alpha(\xi) = -\frac{15}{16}\frac{\xi\tanh(\frac{1}{2}\xi}{\cosh^2(\frac{1}{2}\xi)} - \frac{1}{8}\cosh\xi + \frac{15}{8}\frac{1}{\cosh^2(\frac{1}{2}\xi)} - \frac{3}{4},$$

is a solution of (7.26).

Denote by $u = (u_\alpha, u_\beta, 0, 0)$ *where* $u_\alpha(\xi) = q_\alpha(\xi) + \frac{15}{16} i\pi p_\alpha(\xi)$ *and* $u_\beta(\xi) = \dfrac{du_\alpha}{d\xi}$.

Then, u is a solution of (7.26) satisfying:

(a) $u_\alpha(\xi) = -\dfrac{15}{16}(\xi - i\pi)\dfrac{\tanh(\frac{1}{2}\xi)}{\cosh^2(\frac{1}{2}\xi)} - \dfrac{1}{8}\cosh\xi + \dfrac{15}{8}\dfrac{1}{\cosh^2(\frac{1}{2}\xi)} - \dfrac{3}{4}$,

(b) $u_\alpha(\xi) = \dfrac{1}{112}(\xi - i\pi)^4 + \underset{\xi\to i\pi}{\mathcal{O}}\left((\xi - i\pi)^8\right)$,

(c) *there exists* $\mathcal{M}_6$ *such that for every* $r \in [0,1]$, $\nu \in]0,1]$, $\xi \in \mathcal{D}^*_{r,\nu}$ *satisfying* $\mathcal{I}m\,(\xi) \geq 0$,

$$|u_\alpha(\xi)| \leq \mathcal{M}_6\left[|\xi - i\pi|^4\chi_1(\xi) + (1 - \chi_1(\xi))e^{|\mathcal{R}e(\xi)|}\right],$$
$$|u_\beta(\xi)| = \mathcal{M}_6\left[|\xi - i\pi|^3\chi_1(\xi) + (1 - \chi_1(\xi))e^{|\mathcal{R}e(\xi)|}\right].$$

Proof. . The linearity of (7.26) ensures that u is a solution given by (a). Expending u_α in power series we get (b).

(c): For $|\mathcal{R}e\,(\xi)\,| \geq 1$, the estimates comes readily from (a), and for $|\mathcal{R}e\,(\xi)\,| \leq 1$, (b) ensures that the functions $(\xi - i\pi)^{-4}u_\alpha(\xi)$, $(\xi - i\pi)^{-3}u_\alpha(\xi)$ is continuous and thus bounded. □

One can check, using the explicit form of r_+ and r_- given in Lemma 7.3.14, that there exists M such that for every $\alpha \in]0,1]$, $\nu \in]0,1]$, $\xi \in \mathcal{D}^*_{r,\nu}$,

$|r_\pm(\xi)| \le Me^{+\omega\pi/\nu}$. This may lead to a very bad estimate of $\mathcal{F}(v)$. The following lemma shows that grouping together the two integrals involving r_+, and r_-, i.e. using the convolution kernel, the exponentially large terms and the exponentially small ones cancel out.

Lemma 7.F.4. *There exists $\mathcal{M}_7$ such that for every $r \in]0,1[$ and every function*

$$f:\ \mathcal{D}^*_{r,\nu}\times]0,1] \longrightarrow \mathbb{C}^4$$
$$(\xi,\nu) \longrightarrow (.,.,f_A(\xi,\nu),f_B(\xi,\nu))$$

satisfying $|f_A|_{{}^s\mathcal{H}^{3,\lambda}_\nu(r)} < +\infty$, $|f_B|_{{}^s\mathcal{H}^{3,\lambda}_\nu(r)} < +\infty$ the inequality

$$\left| \int_{[\xi,+\infty)} \left[\langle r^*_+(s), f(s,\nu)\rangle_* \, r_+(\xi) + \langle r^*_-(s), f(s,\nu)\rangle_* r_-(\xi)\right] ds \right|$$
$$\le \mathcal{M}_7 \int_{[\xi,+\infty)} (|f_A(s,\nu)| + |f_B(s,\nu)|)\, ds$$

holds for every $\nu \in]0,1]$ and $\xi \in \mathcal{D}^*_{r,\nu}$.

Proof. Denote

$$I(\xi) = -\int_{[\xi,+\infty)} \left[\langle r^*_+(s), f(s,\nu)\rangle_* \, r_+(\xi) + \langle r^*_-(s), f(s,\nu)\rangle_* r_-(\xi)\right] ds.$$

Then, $I(\xi) = (0,0,I_A(\xi),I_B(\xi))$, because r_+, r_- lie in $\Pi^\perp$. Using the explicit forms of $r_\pm$ given in Lemma 7.3.14 we get

$$\begin{aligned}
I_A(\xi) &= -\int_{[\xi,+\infty)} [f_A(s,\nu)\cos\psi_\nu(s) + f_B(s,\nu)\sin\psi_\nu(s)]\, ds \quad \cos\psi_\nu(\xi) \\
&+ \int_{[\xi,+\infty)} [-f_A(s,\nu)\sin\psi_\nu(s) + f_B(s,\nu)\cos\psi_\nu(s)]\, ds \quad \sin\psi_\nu(\xi) \\
&= -\int_{[\xi,+\infty)} [f_A(s,\nu)\cos(\psi_\nu(s)-\psi_\nu(\xi)) + f_B(s,\nu)\sin(\psi_\nu(s)-\psi_\nu(\xi))] ds \\
&= -\int_{\mathcal{R}e(\xi)}^{+\infty} [f_A(\mathrm{i}\mathcal{I}m\,(\xi)+t,\nu)\cos\theta + f_B(\mathrm{i}\mathcal{I}m\,(\xi)+t,\nu)\sin\theta] dt,
\end{aligned}$$

where $\theta(t,\nu,\xi) = (\frac{\omega}{\nu}+a\nu)\,(t-\mathcal{R}e\,(\xi)) + 2b\nu[\tanh(\frac{1}{2}(\mathrm{i}\mathcal{I}m\,(\xi)+t)) - \tanh(\frac{1}{2}\xi)]$.

Now observe that $|\cos(\xi)| \le e^{|\mathcal{I}m(\xi)|}$, $|\sin(\xi)| \le e^{|\mathcal{I}m(\xi)|}$ and that

$$
\begin{aligned}
|\mathcal{I}m\,(\theta)| &\leq 2|b|\nu\left(\left|\tanh(\tfrac{1}{2}(\mathrm{i}\mathcal{I}m\,(\xi)+t))\right|+\left|\tanh(\tfrac{1}{2}\xi)\right|\right)\\
&\leq 4|b|\nu\mathcal{M}_1\left[\frac{\nu^r}{\pi m_0}+1\right]\\
&\leq 4|b|\nu\mathcal{M}_1\left[\frac{1}{\pi m_0}+1\right].
\end{aligned}
$$

since $r\in]0,1[$ and $\nu\in]0,1[$. Thus, there exists M such that

$$
|I_A(\xi)|\leq M\int_{[\xi,+\infty)}(|f_A(s,\nu)|+|f_B(s,\nu)|)\,ds
$$

holds for $\nu\in]0,1]$, $r\in]0,1[$ and $\xi\in\mathcal{D}^*_{r,\nu}$. The same inequality holds for I_B. □

Lemma 7.F.5.

(a) *For every $r\in]0,1[$, there exists $\nu_4(r)$ such that for every $\nu\in]0,\nu_4(r)]$, $\rho\in[0,\rho_s[$, $d\in[0,1]$, $v\in \mathsf{B}{}^s\mathsf{H}^\lambda_\nu(r)[d\nu]$, and $\xi\in\mathcal{D}^*_{r,\nu}$, $g(v,\xi,\rho,\nu)=Q(v,\nu)+\rho R(h(\xi)+v(\xi),\rho,\nu)$ is well defined.*

(b) *Moreover, there exists $\mathcal{M}_8$ such that for every $r\in]0,1[$, $\nu\in]0,\nu_4(r)]$, $\rho\in[0,\rho_s[$, $d\in[0,1]$ and $v\in\mathsf{B}{}^s\mathsf{H}^\lambda_\nu(r)[d\nu]$,*
   (i) $g_\alpha=0$,
   (ii) $|g_\beta(v,\xi,\rho,\nu)|\leq\mathcal{M}_8\left(|v|^2_{{}^s\mathsf{H}^\lambda_\nu(r)}+\rho\nu^2\right)\Upsilon_{6,\lambda}(\xi)$,
   (iii) $|g_A(v,\xi,\rho,\nu)|\leq\mathcal{M}_8\nu\left(|v|^2_{{}^s\mathsf{H}^\lambda_\nu(r)}+\rho\nu^2\right)\Upsilon_{6,\lambda}(\xi)$
   (iv) $|g_B(v,\xi,\rho,\nu)|\leq\mathcal{M}_8\nu\left(|v|^2_{{}^s\mathsf{H}^\lambda_\nu(r)}+\rho\nu^2\right)\Upsilon_{6,\lambda}(\xi)$
   where

$$
\Upsilon_{\gamma,\lambda}(\xi)=\left[\frac{1}{|\xi^2+\pi^2|^\gamma}\chi_1(\xi)+(1-\chi_1(\xi))\mathrm{e}^{-\lambda|\mathcal{R}e(\xi)|}\right].
$$

(c) *There exists $\mathcal{M}_9$ such that for every $r\in]0,1[$, $\nu\in]0,\nu_4(r)]$, $\rho\in[0,\rho_s[$, $d\in[0,1]$, and $v,v'\in\mathsf{B}{}^s\mathsf{H}^\lambda_\nu(r)[d\nu]$,*
   (i) $|g_\beta(v',\xi,\rho,\nu)-g_\beta(v,\xi,\rho,\nu)|\leq$
   $$\mathcal{M}_9\left(|v|_{{}^s\mathsf{H}^\lambda_\nu(r)}+|v'|_{{}^s\mathsf{H}^\lambda_\nu(r)}+\rho\nu^{2-r}\right)|v-v'|_{{}^s\mathsf{H}^\lambda_\nu(r)}\Upsilon_{6,\lambda}(\xi),$$
   (ii) $|g_A(v',\xi,\rho,\nu)-g_A(v,\xi,\rho,\nu)|\leq$
   $$\mathcal{M}_9\nu\left(|v|_{{}^s\mathsf{H}^\lambda_\nu(r)}+|v'|_{{}^s\mathsf{H}^\lambda_\nu(r)}+\rho\nu^{2-r}\right)|v-v'|_{{}^s\mathsf{H}^\lambda_\nu(r)}\Upsilon_{6,\lambda}(\xi),$$
   (iii) $|g_B(v',\xi,\rho,\nu)-g_B(v,\xi,\rho,\nu)|\leq$
   $$\mathcal{M}_9\nu\left(|v|_{{}^s\mathsf{H}^\lambda_\nu(r)}+|v'|_{{}^s\mathsf{H}^\lambda_\nu(r)}+\rho\nu^{2-r}\right)|v-v'|_{{}^s\mathsf{H}^\lambda_\nu(r)}\Upsilon_{6,\lambda}(\xi).$$

Proof. We proceed in three steps.

Step 1. Using the explicit expression of Q given in the proof of Lemma 7.4.20, we check that $Q(v,\nu)$ is well defined and that it satisfies

$$|Q_\beta(v(\xi))| \le \mathcal{M}_8 \, |v|^2_{{}^s\mathsf{H}^\lambda_\nu(r)} \; \Upsilon_{6,\lambda}(\xi),$$
$$|Q_A(v(\xi))| \le \mathcal{M}_8 \nu \, |v|^2_{{}^s\mathsf{H}^\lambda_\nu(r)} \; \Upsilon_{6,\lambda}(\xi),$$
$$|Q_B(v(\xi))| \le \mathcal{M}_8 \nu \, |v|^2_{{}^s\mathsf{H}^\lambda_\nu(r)} \; \Upsilon_{6,\lambda}(\xi),$$

and

$$|Q_\beta(v'(\xi)) - Q_\beta(v(\xi))| \le \mathcal{M}_9 \left(|v|_{{}^s\mathsf{H}^\lambda_\nu(r)} + |v'|_{{}^s\mathsf{H}^\lambda_\nu(r)} \right) |v - v'|_{{}^s\mathsf{H}^\lambda_\nu(r)} \, \Upsilon_{6,\lambda}(\xi),$$
$$|Q_A(v'(\xi)) - Q_A(v(\xi))| \le \mathcal{M}_9 \, \nu \left(|v|_{{}^s\mathsf{H}^\lambda_\nu(r)} + |v'|_{{}^s\mathsf{H}^\lambda_\nu(r)} \right) |v - v'|_{{}^s\mathsf{H}^\lambda_\nu(r)} \, \Upsilon_{6,\lambda}(\xi),$$
$$|Q_B(v'(\xi)) - Q_B(v(\xi))| \le \mathcal{M}_9 \, \nu \left(|v|_{{}^s\mathsf{H}^\lambda_\nu(r)} + |v'|_{{}^s\mathsf{H}^\lambda_\nu(r)} \right) |v - v'|_{{}^s\mathsf{H}^\lambda_\nu(r)} \, \Upsilon_{6,\lambda}(\xi).$$

Step 2. The difficulty comes from the existence and the computation of an upper bound of the rest R. We work only with R_β. The same study can be done with R_A and R_B. The rest $R_\beta(h+w,\nu)$ is given by

$$R_\beta(h+w,\nu) = \sum_{(\overline{m},\ell)\in\mathcal{I}} a_{\beta,\overline{m},\ell} \, (h+w)^{\overline{m}} \, \nu^{2|\overline{m}|+m_\beta+2\ell-4}.$$

Denote by $\Delta_{\overline{m},\ell} := a_{\beta,\overline{m},\ell} \, (h+w)^{\overline{m}} \, \nu^{2|\overline{m}|+m_\beta+2\ell-4}$. Then, estimate (7.7) ensures that

$$|\Delta_{\overline{m},\ell}| \le M_R \frac{|\alpha_h+\alpha|^{m_\alpha}}{r_\alpha^{m_\alpha}} \, \frac{|\beta_h+\beta|^{m_\beta}}{r_\beta^{m_\beta}} \, \frac{|A|^{m_A}}{r_A^{m_A}} \, \frac{|B|^{m_B}}{r_B^{m_B}} \, \frac{\nu^{2|\overline{m}|+m_\beta+2\ell-4}}{\nu_0^{2\ell}}.$$

Using Lemma 7.F.1 we get

$$|\alpha_h(\xi)| \le \mathcal{M}_4 \left[\frac{1}{|\xi^2+\pi^2|^2} \chi_1(\xi) + (1-\chi_1(\xi)) \mathrm{e}^{-|\mathcal{R}e(\xi)|} \right],$$
$$|\beta_h(\xi)| \le \mathcal{M}_4 \mathcal{M}_3 \left[\frac{1}{|\xi^2+\pi^2|^3} \chi_1(\xi) + (1-\chi_1(\xi)) \mathrm{e}^{-|\mathcal{R}e(\xi)|} \right]$$

Moreover, using the definition of $|\cdot|_{{}^s\mathsf{H}^\lambda_\nu(r)}$ given by (7.45) and Lemma 7.F.1 we obtain that for every $d, r, \lambda \in [0,1]$ and every $v \in \mathsf{B}\,{}^s\mathsf{H}^\lambda_\nu\,(r)[d\nu]$,

$$\begin{aligned} |\alpha_h+\alpha| &\le \left[\frac{\chi_1(\xi)}{|\xi^2+\pi^2|^2} + (1-\chi_1(\xi)) \mathrm{e}^{-\lambda|\mathcal{R}e(\xi)|} \right] \left(\mathcal{M}_4 + \frac{|v|_{{}^s\mathsf{H}^\lambda_\nu(r)}}{m_0 \pi \nu^r} \right) \\ &\le \left[\frac{\chi_1(\xi)}{|\xi^2+\pi^2|^2} + (1-\chi_1(\xi)) \mathrm{e}^{-\lambda|\mathcal{R}e(\xi)|} \right] \left(\mathcal{M}_4 + \frac{1}{m_0 \pi} \right). \end{aligned} \tag{7.74}$$

Similarly we check

$$
\begin{aligned}
|\beta_h+\beta| &\leq \left[\frac{\chi_1(\xi)}{|\xi^2+\pi^2|^2}+(1-\chi_1(\xi))\mathrm{e}^{-\lambda|\mathcal{R}e(\xi)|}\right]\left(\frac{\mathcal{M}_4\mathcal{M}_3}{m_0\pi\nu^r}+\frac{1}{m_0^2\pi^2\nu^r}\right),\\
|A| &\leq \left[\frac{1}{|\xi^2+\pi^2|^2}\chi_1(\xi)+(1-\chi_1(\xi))\mathrm{e}^{-\lambda|\mathcal{R}e(\xi)|}\right]\frac{1}{m_0\pi},\\
|B| &\leq \left[\frac{1}{|\xi^2+\pi^2|^2}\chi_1(\xi)+(1-\chi_1(\xi))\mathrm{e}^{-\lambda|\mathcal{R}e(\xi)|}\right]\frac{1}{m_0\pi}.
\end{aligned}
\tag{7.75}
$$

Denote

$$
\begin{aligned}
\Theta := \max\Bigg(&\frac{1}{r_\alpha}\left(\mathcal{M}_4+\frac{1}{m_0\pi}\right),\\
&\frac{1}{r_\beta}\left(\frac{\mathcal{M}_4\mathcal{M}_3}{m_0\pi}+\frac{1}{m_0^2\pi^2}\right),\frac{1}{r_A m_0\pi},\frac{1}{r_B m_0\pi},\frac{1}{\nu_0^2}\Bigg).
\end{aligned}
\tag{7.76}
$$

Then,

$$
|\Delta_{\overline{m},\ell}| \leq M_R\Theta^{|\overline{m}|+\ell}\,\nu^{2(|\overline{m}|+\ell)-4}\left[\frac{\chi_1(\xi)}{|\xi^2+\pi^2|^2}+(1-\chi_1(\xi))\mathrm{e}^{-\lambda|\mathcal{R}e(\xi)|}\right]^{|\overline{m}|}
$$

For $\mathcal{R}e\,(\xi)\geq 1$, and $(\overline{m},\ell)$ such that $|\overline{m}|\geq 1$, $|\overline{m}|+\ell\geq 3$, we get

$$
|\Delta_{\overline{m},\ell}| \leq M_R\Theta^3\nu^2(\nu^2\Theta)^{|\overline{m}|+\ell-3}\mathrm{e}^{-\lambda|\mathcal{R}e(\xi)|}
$$

For $\mathcal{R}e\,(\xi)<1$, denoting by $M := \sup\limits_{\substack{|\mathcal{I}m(\xi)|<\pi\\ \mathcal{R}e(\xi)<1}} |\xi^2+\pi^2|^2$, we get

$$
\begin{aligned}
|\Delta_{\overline{m},\ell}| &\leq M_R\frac{\Theta^{|\overline{m}|+\ell}\,\nu^{2(|\overline{m}|+\ell)-4}}{|\xi^2+\pi^2|^{2(|\overline{m}|+\ell)}}M^\ell\\
&\leq \frac{M_R}{\nu^4}\left(\frac{M\Theta\nu^2}{|\xi^2+\pi^2|^2}\right)^{|\overline{m}|+\ell}\\
&\leq \frac{M_R(M\Theta)^3\nu^2}{|\xi^2+\pi^2|^6}\left(\frac{M\Theta\nu^2}{m_0^2\pi^2\nu^{2r}}\right)^{|\overline{m}|+\ell-3}
\end{aligned}
$$

since $M\geq 1$ and $|\xi^2+\pi^2|\geq m_0\pi\nu^r$.

Then, choosing $\nu_4(r)$ such that

$$
\Theta(\nu_4(r))^2\leq\frac{1}{2},\qquad \frac{M\Theta(\nu_4(r))^{2(1-r)}}{m_0^2\pi^2}\leq\frac{1}{2}
$$

which is possible since $r\in]0,1[$, and denoting by $M'=\max\big(1,(M\Theta)^3\big)$ we finally get

$$|R_\beta(h+v,\nu)| \le \sum_{(\overline{m},\ell)\in\mathcal{I}} \left(\tfrac{1}{2}\right)^{|\overline{m}|+\ell-3} M_R M'\nu^2\left[\frac{\chi_1(\xi)}{|\xi^2+\pi^2|^6} + (1-\chi_1(\xi))e^{-\lambda|\mathcal{R}e(\xi)|}\right]$$
$$\le 2^8 M_R M'\nu^2 \left[\frac{\chi_1(\xi)}{|\xi^2+\pi^2|^6} + (1-\chi_1(\xi))e^{-\lambda|\mathcal{R}e(\xi)|}\right].$$

Proceeding in the same way for R_A and R_B we can finally conclude that for every $r\in]0,1[$, $\nu\in]0,\nu_4(r)]$, $d\in[0,1]$, $v\in \mathsf{B}^s\mathsf{H}^\lambda_\nu(r)[d\nu]$ and $\xi\in\mathcal{D}^*_{r,\nu}$, $R(h(\xi)+v(\xi),\nu)$ is well defined and satisfies

$$\begin{aligned} |R_\beta(h(\xi)+v(\xi),\nu)| &\le 2^8 M_R M'\ \nu^2\ \Upsilon_{6,\lambda}(\xi),\\ |R_A(h(\xi)+v(\xi),\nu)| &\le 2^8 M_R M'\ \nu^3\ \Upsilon_{6,\lambda}(\xi),\\ |R_B(h(\xi)+v(\xi),\nu)| &\le 2^8 M_R M'\ \nu^3\ \Upsilon_{6,\lambda}(\xi), \end{aligned} \tag{7.77}$$

where

$$\Upsilon_{\gamma,\lambda}(\xi) = \left[\frac{\chi_1(\xi)}{|\xi^2+\pi^2|^\gamma} + (1-\chi_1(\xi))e^{-\lambda|\mathcal{R}e(\xi)|}\right].$$

Step 3. We must now bound $R_\beta(h+v,\nu)-R_\beta(h+v',\nu)$. Denote by

$$\widetilde{\Delta}_{\overline{m},\ell} := a_{\beta,\overline{m},\ell}\ \left[(h+v)^{\overline{m}} - (h+v')^{\overline{m}}\right]\ \nu^{2|\overline{m}|+m_\beta+2\ell-4}.$$

The difference $\left[(h+v)^{\overline{m}} - (h+v')^{\overline{m}}\right]$ reads

$$\begin{aligned} &((h+v)^{\overline{m}} - (h+v')^{\overline{m}}) = \\ &\quad \left[(\alpha_h+\alpha)^{m_\alpha} - (\alpha_h+\alpha')^{m_\alpha}\right]\ (\beta_h+\beta)^{m_\beta}\ A^{m_A}\ B^{m_B}\\ &\quad +(\alpha_h+\alpha')^{m_\alpha}\ \left[(\beta_h+\beta)^{m_\beta} - (\beta_h+\beta')^{m_\beta}\right]\ A^{m_A}\ B^{m_B}\\ &\quad +(\alpha_h+\alpha')^{m_\alpha}\ (\beta_h+\beta')^{m_\beta}\ \left[A^{m_A} - A'^{m_A}\right]\ B^{m_B}\\ &\quad +(\alpha_h+\alpha')^{m_\alpha}\ (\beta_h+\beta')^{m_\beta}\ A'^{m_A}\ \left[B^{m_B} - B'^{m_B}\right]. \end{aligned} \tag{7.78}$$

Using (7.74) and (7.75) we get

$$\left|(\alpha_h+\alpha)^{m_\alpha} (\alpha_h+\alpha')^{m_\alpha}\right| = \left|\sum_{k=0}^{m_\alpha-1} (\alpha_h+\alpha)^k\,(\alpha_h+\alpha')^{m_\alpha-1-k}\right| |\alpha-\alpha'|$$
$$\le m_\alpha\left(\mathcal{M}_4+\frac{1}{m_0\pi}\right)^{m_\alpha-1} |\alpha-\alpha'|\ (\Upsilon_{2,\lambda})^{m_\alpha-1},$$

$$\left|(\beta_h+\beta)^{m_\beta} (\beta_h+\beta')^{m_\beta}\right| \le m_\beta\left(\frac{\mathcal{M}_4\mathcal{M}_3}{m_0\pi}+\frac{1}{m_0^2\pi^2}\right)^{m_\beta-1} |\beta-\beta'|\left(\frac{\Upsilon_{2,\lambda}}{\nu^r}\right)^{m_\beta-1},$$

$$|A^{m_A} - A'^{m_A}| \le m_A\left(\frac{1}{m_0\pi}\right)^{m_A-1} |A-A'|\ (\Upsilon_{2,\lambda})^{m_A-1},$$

$$|B^{m_B} - B'^{m_B}| \le m_B\left(\frac{1}{m_0\pi}\right)^{m_B-1} |B-B'|\ (\Upsilon_{2,\lambda})^{m_B-1}.$$

Moreover, using that $\Upsilon_{\gamma+1,\lambda}(\xi) \leq \Upsilon_{\gamma,\lambda}(\xi)/(m_0\pi\nu^r)$ we get,

$$|\alpha-\alpha'| \leq |\alpha-\alpha'|_{{}^s\mathcal{H}_\nu^{3,\lambda}(r)} \Upsilon_{3,\lambda} \leq \frac{|\alpha-\alpha'|_{{}^s\mathcal{H}_\nu^{3,\lambda}(r)}}{m_0\pi\nu^r}\Upsilon_{2,\lambda} \leq \frac{|v-v'|_{{}^s\mathsf{H}_\nu^{\lambda}(r)}}{m_0\pi\nu^r}\Upsilon_{2,\lambda},$$

$$|A-A'| \leq |A-A'|_{{}^s\mathcal{H}_\nu^{3,\lambda}(r)} \Upsilon_{3,\lambda}, \leq \frac{|A-A'|_{{}^s\mathcal{H}_\nu^{3,\lambda}(r)}}{m_0\pi\nu^r}\Upsilon_{2,\lambda} \leq \frac{|v-v'|_{{}^s\mathsf{H}_\nu^{\lambda}(r)}}{m_0\pi\nu^r}\Upsilon_{2,\lambda},$$

$$|B-B'| \leq |B-B'|_{{}^s\mathcal{H}_\nu^{3,\lambda}(r)} \Upsilon_{3,\lambda}, \leq \frac{|B-B'|_{{}^s\mathcal{H}_\nu^{3,\lambda}(r)}}{m_0\pi\nu^r}\Upsilon_{2,\lambda} \leq \frac{|v-v'|_{{}^s\mathsf{H}_\nu^{\lambda}(r)}}{m_0\pi\nu^r}\Upsilon_{2,\lambda},$$

and

$$|\beta-\beta'| \leq |\alpha-\alpha'|_{{}^s\mathcal{H}_\nu^{4,\lambda}(r)} \Upsilon_{4,\lambda} \leq \frac{|\beta-\beta'|_{{}^s\mathcal{H}_\nu^{4,\lambda}(r)}}{(m_0\pi\nu^r)^2}\Upsilon_{2,\lambda} \leq \frac{|v-v'|_{{}^s\mathsf{H}_\nu^{\lambda}(r)}}{(m_0\pi\nu^r)^2}\Upsilon_{2,\lambda},$$

Denote by

$$M'' = \max\left(\frac{1}{m_0\pi}\left(\mathcal{M}_4+\frac{1}{m_0\pi}\right)^{-1}, \frac{1}{(m_0\pi)^2}\left(\frac{\mathcal{M}_4\mathcal{M}_3}{m_0\pi}+\frac{1}{m_0^2\pi^2}\right)^{-1}, 1\right).$$

Then using (7.74), (7.75), (7.76), (7.78) and the above estimates, we obtain

$$\begin{aligned}|\widetilde{\Delta}_{\overline{m},\ell}| &\leq \frac{M_R M''}{\nu^r}|v-v'|_{{}^s\mathsf{H}_\nu^{\lambda}(r)} \; |\overline{m}| \; \Theta^{|\overline{m}|+\ell} \; \frac{\nu^{2|\overline{m}|+m_\beta+2\ell-4}}{\nu^{m_\beta r}} \; (\Upsilon_{2,\lambda})^{|\overline{m}|} \\ &\leq \frac{M_R M'' M'}{\nu^r}|\overline{m}| \; |v-v'|_{{}^s\mathsf{H}_\nu^{\lambda}(r)} \; \nu^{2(|\overline{m}|+\ell)-4} \; \Theta^{|\overline{m}|+\ell} \; (\Upsilon_{2,\lambda})^{|\overline{m}|}\end{aligned}$$

Using the bound of $\nu^{2(|\overline{m}|+\ell)-4} \; \Theta^{|\overline{m}|+\ell} \; (\Upsilon_{2,\lambda})^{|\overline{m}|}$ established at Step 2 we finally get that for every $\nu \in]0,\nu_4(r)]$

$$|R_\beta(h+v,\nu) - R_\beta(h+v',\nu)| \leq \frac{M_R M'' M'}{\nu^r} \; |v-v'|_{{}^s\mathsf{H}_\nu^{\lambda}(r)} \; \nu^2 \Upsilon_{6,\lambda}\Delta'$$

where

$$\Delta' = \sum_{(\overline{m},\ell)\in\mathcal{I}} |\overline{m}| \left(\tfrac{1}{2}\right)^{|\overline{m}|+\ell-3} \leq 8\sum_{(\overline{m},\ell)\in\mathbb{N}^5} |\overline{m}| \left(\tfrac{1}{2}\right)^{|\overline{m}|+\ell} = \frac{8}{1-\frac{1}{2}} \sum_{\overline{m}\in\mathbb{N}^4} |\overline{m}| \left(\tfrac{1}{2}\right)^{|\overline{m}|} = 2^{10}.$$

We can conclude that for every $r \in]0,1[$, $\nu \in]0,\nu_4(r)]$, $d \in [0,1]$ and $v \in \mathsf{B}{}^s\mathsf{H}_\nu^{\lambda}(r)[d\nu]$

$$|R_\beta(h+v,\nu) - R_\beta(h+v',\nu)| \leq 2^{10} M_R M'' M' \, |v-v'|_{{}^s\mathsf{H}_\nu^{\lambda}(r)} \, \nu^{2-r} \; \Upsilon_{6,\lambda}$$

holds. A similar proof ensures that for every $r \in]0,1[$, $\nu \in]0,\nu_4(r)]$, $d \in [0,1]$ and $v \in \mathsf{B}{}^s\mathsf{H}_\nu^{\lambda}(r)[d\nu]$

$$|R_A(h+v,\nu) - R_A(h+v',\nu)| \le 2^{10} M_R M'' M' \, |v-v'|_{{}^s\mathsf{H}_\nu^\lambda(r)} \, \nu^{3-r} \, \Upsilon_{6,\lambda},$$
$$|R_B(h+v,\nu) - R_B(h+v',\nu)| \le 2^{10} M_R M'' M' \, |v-v'|_{{}^s\mathsf{H}_\nu^\lambda(r)} \, \nu^{3-r} \, \Upsilon_{6,\lambda},$$

hold. □

Proof of Lemma 7.4.27 The explicit expression of $\mathcal{F}(w)$ ensures that it is holomorphic in $\mathcal{D}^*_{r,\nu}$ and it maps $\mathbb{R}$ in $\mathbb{R}^4$ which is equivalent to

$$(\mathcal{F}(v))(\overline{\xi}) = \overline{(\mathcal{F}(v))(\xi)}, \quad \text{for } \xi \in \mathcal{D}^*_{r,\nu}.$$

Hence, we can assume that $\mathcal{I}m\,(\xi) \ge 0$ holds, to prove the estimates (a) and (b).

(a): In what follows, we denote by $\mathcal{F} := (\mathcal{F}_\alpha, \mathcal{F}_\beta, \mathcal{F}_A, \mathcal{F}_B)$ the four components of $\mathcal{F}$ and $g = (0, g_\beta, g_A, g_B)$ the four components of g.

Step 1. Bounds of $\mathcal{F}_\alpha$ and $\mathcal{F}_\beta$. For obtaining accurate upper bounds of the two first component of $\mathcal{F}(v)$, the basis of solutions (p,u) of the homogeneous equation (7.26) is more adapted than the basis (p,q). Using the the relationship between u and q given in Lemma 7.F.3 we get

$$(\mathcal{F}_\alpha(v))\,(\xi) = \frac{15}{8} i\pi \; \Delta_1 \; p_\alpha(\xi) - 2\Delta_2(\xi)\; p_\alpha(\xi) - 2\Delta_3(\xi) u_\alpha(\xi),$$
$$(\mathcal{F}_\beta(v))\,(\xi) = \frac{15}{8} i\pi \; \Delta_1 \; p_\beta(\xi) - 2\Delta_2(\xi)\; p_\beta(\xi) - 2\Delta_3(\xi) u_\beta(\xi),$$

where

$$\Delta_1 = \int_0^{+\infty} p_\alpha(t) g_\beta(t) dt, \;\; \Delta_2(\xi) = \int_{[0,\xi]} u_\alpha(t) g_\beta(t) dt, \;\; \Delta_3(\xi) = \int_{[\xi,+\infty)} p_\alpha(t) g_\beta(t) dt.$$

Step 1.1. Estimates of Δ_1. On $[0,+\infty[$, we can compute better estimates of p_α and g_β. Indeed, Lemmas 7.F.2 and 7.F.5 ensure that

$$|p_\alpha(t)| \le \mathcal{M}'_5 \mathrm{e}^{-t}, \qquad |g_\beta(t)| \le \mathcal{M}'_8 \left(|v|^2_{{}^s\mathsf{H}_\nu^\lambda(r)} + \rho\nu^2 \right) \mathrm{e}^{-\lambda t}$$

holds for every $t \in [0,+\infty[$, $r \in]0,1[$, $\nu \in]0,\nu_4(r)]$, $\rho \in [0,\rho_s[$, $d \in [0,1]$, $v \in \mathsf{B}\,{}^s\mathsf{H}_\nu^\lambda(r)[d\nu]$. Hence,

$$|\Delta_1| \le C_1 \left(|v|^2_{{}^s\mathsf{H}_\nu^\lambda(r)} + \rho\nu^2 \right) \tag{7.79}$$

where $C_1 = \dfrac{\mathcal{M}'_5 \mathcal{M}'_8}{1+\lambda}$.

Step 1.2. Estimates of Δ_2. Since we work with functions which are holomorphic in $\mathcal{D}^*_{r,\nu}$, we get

$$\Delta_2 = \Delta_{21} + \Delta_{22}, \text{ where } \Delta_{21}(\xi) = \int_{[0,\mathcal{R}e(\xi)]} u_\alpha(s) g_\beta(s) ds, \;\; \Delta_{22}(\xi) = \int_{[\mathcal{R}e(\xi),\xi]} u_\alpha(s) g_\beta(s) ds.$$

Step 1.2.1 Estimates of Δ_2 for $\xi \in \mathcal{D}^*_{r,\nu}$ such that $-\nu < \mathcal{R}e\,(\xi) \leq 1$ and $\mathcal{I}m\,(\xi) \geq 0$.

• **Bound of Δ_{21}.** Lemmas 7.F.3, 7.F.5 ensure that

$$\begin{aligned} |\Delta_{21}(\xi)| &\leq \frac{\mathcal{M}_6\mathcal{M}_8}{\pi^6}\left(|v|^2_{{}^s\mathsf{H}^\lambda_\nu(r)} + \rho\nu^2\right)\int_0^{|\mathcal{R}e(\xi)|}\frac{dt}{|t-\mathrm{i}\pi|^2} \\ &\leq \frac{\mathcal{M}_6\mathcal{M}_8}{\pi^6}\left(|v|^2_{{}^s\mathsf{H}^\lambda_\nu(r)} + \rho\nu^2\right)\frac{1}{\pi^2}. \end{aligned}$$

since $|\xi^2+\pi^2| \geq \pi|\xi - \mathrm{i}\pi|$ for $\mathcal{I}m\,(\xi) \geq 0$ and $- \leq \nu < \mathcal{R}e\,(\xi) \leq 1$.

• **Bound of Δ_{22}.** Similarly,

$$|\Delta_{22}(\xi)| \leq \frac{\mathcal{M}_6\mathcal{M}_8}{\pi^6}\left(|v|^2_{{}^s\mathsf{H}^\lambda_\nu(r)} + \rho\nu^2\right)I(\xi)$$

holds with

$$I(\xi) = \int_0^{\mathcal{I}m(\xi)}\frac{du}{(\mathcal{R}e\,(\xi))^2 + (u-\pi)^2}.$$

Then, with the help of Figure 7.17 p. 250, we check

– if $\mathcal{R}e\,(\xi) = 0$, then $\pi - \mathcal{I}m\,(\xi) \geq \frac{1}{2}\nu^r$ and

$$I(\xi) = \frac{1}{\pi - \mathcal{I}m\,(\xi)} - \frac{1}{\pi} \leq \frac{2}{\nu^r};$$

– if $\mathcal{R}e\,(\xi) \neq 0$ and $-\nu < \mathcal{R}e\,(\xi) < \dfrac{\pi - \frac{1}{2}\nu^r}{\pi}\frac{1}{2}\nu^r$, then $\pi - \mathcal{I}m\,(\xi) \geq \frac{1}{2}\nu^r$ and

$$\begin{aligned} I(\xi) &= \frac{1}{|\mathcal{R}e\,(\xi)\,|}\left[\arctan\frac{\pi}{|\mathcal{R}e\,(\xi)\,|} - \arctan\frac{\pi - \mathcal{I}m\,(\xi)}{|\mathcal{R}e\,(\xi)\,|}\right] \\ &= \frac{1}{|\mathcal{R}e\,(\xi)\,|}\left[\arctan\frac{|\mathcal{R}e\,(\xi)\,|}{\pi - \mathcal{I}m\,(\xi)} - \arctan\frac{|\mathcal{R}e\,(\xi)\,|}{\pi}\right] \\ &\leq \frac{1}{\pi - \mathcal{I}m\,(\xi)} - \frac{1}{\pi} \leq \frac{2}{\nu^r}; \end{aligned}$$

since $|\arctan x - \arctan y| \leq |x-y|$ for $(x,y) \in \mathbb{R}^2$.

– if $\mathcal{R}e\,(\xi) \geq \frac{\pi - \frac{1}{2}\nu^r}{\pi}\frac{1}{2}\nu^r$, then $\mathcal{R}e\,(\xi) \geq m\nu^r$ with $m = \frac{\pi - \frac{1}{2}}{\pi}\frac{1}{2}$ and

$$I(\xi) = \frac{1}{|\mathcal{R}e\,(\xi)\,|}\left[\arctan\frac{\pi}{|\mathcal{R}e\,(\xi)\,|} - \arctan\frac{\pi - \mathcal{I}m\,(\xi)}{|\mathcal{R}e\,(\xi)\,|}\right] \leq \frac{\pi}{2m\nu^r}.$$

Thus

$$I(\xi) \leq \max\left(2, \tfrac{\pi}{2m}\right)\frac{1}{\nu^r}$$

and

$$|\Delta_{22}(\xi)| \leq \frac{\mathcal{M}_6\mathcal{M}_8}{\pi^6}\left(|v|^2_{{}^s\mathsf{H}^\lambda_\nu(r)} + \rho\nu^2\right)\max\left(2, \tfrac{\pi}{2m}\right)\frac{1}{\nu^r}.$$

We can conclude that for every $r \in]0,1[$, $\nu \in]0,\nu_4(r)]$, $\rho \in [0,\rho_s[$, $d \in [0,1]$, $v \in \mathsf{B}^s\mathsf{H}^\lambda_\nu(r)[d\nu]$ and every $\xi \in \mathcal{D}^*_{r,\nu}$ such that $-\nu < \mathcal{R}e(\xi) \le 1$ and $\mathcal{I}m(\xi) \ge 0$,

$$|\Delta_2(\xi)| \le \frac{\mathcal{M}_6\mathcal{M}_8}{\pi^6}\left(\max\left(2,\tfrac{\pi}{2m}\right)+\frac{1}{\pi^2}\right)\left(|v|^2_{{}^s\mathsf{H}^\lambda_\nu(r)}+\rho\nu^2\right)\frac{1}{\nu^r} \tag{7.80}$$

Step 1.2.2 Estimates of Δ_2 for $\xi \in \mathcal{D}^*_{r,\nu}$ such that $\mathcal{R}e(\xi) \ge 1$ and $\mathcal{I}m(\xi) \ge 0$.

• **Bound of Δ_{21}.** From Lemmas 7.F.3, 7.F.5 we get that

$$|u_\alpha(t)| \le \mathcal{M}'_6 e^t, \qquad |g_\beta(t)| \le \mathcal{M}'_8\left(|v|^2_{{}^s\mathsf{H}^\lambda_\nu(r)}+\rho\nu^2\right)e^{-\lambda t}.$$

holds for every $t \in [0,+\infty[$, $r \in]0,1[$, $\nu \in]0,\nu_4(r)]$, $\rho \in [0,\rho_s[$, $d \in [0,1]$, $v \in \mathsf{B}^s\mathsf{H}^\lambda_\nu(r)[d\nu]$. Hence,

$$\begin{aligned}|\Delta_{21}(\xi)| &\le \mathcal{M}'_6\mathcal{M}'_8\left(|v|^2_{{}^s\mathsf{H}^\lambda_\nu(r)}+\rho\nu^2\right)\int_0^{\mathcal{R}e(\xi)} e^{(1-\lambda)s}ds\\ &\le \frac{\mathcal{M}'_6\mathcal{M}'_8}{1-\lambda}\left(|v|^2_{{}^s\mathsf{H}^\lambda_\nu(r)}+\rho\nu^2\right)e^{(1-\lambda)\mathcal{R}e(\xi)}.\end{aligned}$$

• **Bound of Δ_{22}.** Lemmas 7.F.3 and 7.F.5 ensure that

$$\begin{aligned}|\Delta_{22}(\xi)| &\le \mathcal{M}_6\mathcal{M}_8\left(|v|^2_{{}^s\mathsf{H}^\lambda_\nu(r)}+\rho\nu^2\right)\int_0^{\mathcal{I}m(\xi)} e^{(1-\lambda)\mathcal{R}e(\xi)}du\\ &\le \mathcal{M}_6\mathcal{M}_8\pi\left(|v|^2_{{}^s\mathsf{H}^\lambda_\nu(r)}+\rho\nu^2\right)e^{(1-\lambda)\mathcal{R}e(\xi)}.\end{aligned}$$

We can conclude that for every $r \in]0,1[$, $\nu \in]0,\nu_4(r)]$, $\rho \in [0,\rho_s[$, $d \in [0,1]$, $v \in \mathsf{B}^s\mathsf{H}^\lambda_\nu(r)[d\nu]$ and every $\xi \in \mathcal{D}^*_{r,\nu}$ such that $\mathcal{R}e(\xi) \ge 1$ and $\mathcal{I}m(\xi) \ge 0$,

$$|\Delta_2(\xi)| \le \left(\frac{\mathcal{M}'_6\mathcal{M}'_8}{1-\lambda}+\mathcal{M}_6\mathcal{M}_8\pi\right)\left(|v|^2_{{}^s\mathsf{H}^\lambda_\nu(r)}+\rho\nu^2\right)e^{(1-\lambda)\mathcal{R}e(\xi)}. \tag{7.81}$$

Finally, the estimates of Δ_2 given by (7.80) and (7.81) ensures that

$$|\Delta_2(\xi)| \le C_2\left(|v|^2_{{}^s\mathsf{H}^\lambda_\nu(r)}+\rho\nu^2\right)\left(\frac{\chi_1(\xi)}{\nu^r}+(1-\chi_1(\xi))e^{(1-\lambda)\mathcal{R}e(\xi)}.\right) \tag{7.82}$$

holds for every $r \in]0,1[$, $\nu \in]0,\nu_4(r)]$, $\rho \in [0,\rho_s[$, $d \in [0,1]$, $v \in \mathsf{B}^s\mathsf{H}^\lambda_\nu(r)[d\nu]$ and every $\xi \in \mathcal{D}^*_{r,\nu}$ such that $\mathcal{I}m(\xi) \ge 0$, where

$$C_2 = \max\left(\frac{\mathcal{M}'_6\mathcal{M}'_8}{1-\lambda}+\mathcal{M}_6\mathcal{M}_8\pi, \frac{\mathcal{M}_6\mathcal{M}_8}{\pi^6}\left(\max\left(2,\tfrac{\pi}{2m}\right)+\frac{1}{\pi^2}\right)\right).$$

Step 1.3. Estimates of Δ_3. Now, we want to bound Δ_3. As previously, we proceed in two substeps.

Step 1.3.1 Estimates of Δ_3 for $\xi \in \mathcal{D}^*_{r,\nu}$ such that $\mathcal{R}e\,(\xi) \geq 1$ and $\mathcal{I}m\,(\xi) \geq 0$.

For $\mathcal{R}e\,(\xi) \geq 1$ and $t \in [0 + \infty]$, $\mathcal{R}e\,(\xi + t)$ holds. So, Lemmas 7.F.2 and 7.F.5 ensure that

$$\begin{aligned} |\Delta_3(\xi)| &\leq \mathcal{M}_5 \mathcal{M}_8 \left(|v|^2_{{}^s\mathsf{H}^\lambda_\nu(r)} + \rho\nu^2 \right) \int_0^{+\infty} \mathrm{e}^{-(\lambda+1)\mathcal{R}e(\xi+t)} dt \\ &\leq \frac{\mathcal{M}_5 \mathcal{M}_8}{1+\lambda} \left(|v|^2_{{}^s\mathsf{H}^\lambda_\nu(r)} + \rho\nu^2 \right) \mathrm{e}^{-(\lambda+1)\mathcal{R}e(\xi)} \end{aligned} \tag{7.83}$$

holds for every $t \in [0, +\infty[$, $r \in]0,1[$, $\nu \in]0, \nu_4(r)]$, $\rho \in [0, \rho_s[$, $d \in [0,1]$, $v \in \mathsf{B}^s\mathsf{H}^\lambda_\nu(r)[d\nu]$.

Step 1.3.2 Estimates of Δ_3 for $\xi \in \mathcal{D}^*_{r,\nu}$ such that $-\nu < \mathcal{R}e\,(\xi) \leq 1$ and $\mathcal{I}m\,(\xi) \geq 0$. For $-\nu < \mathcal{R}e\,(\xi) \leq 1$, we split Δ_3 in two parts $\Delta_3 = \int_{\mathcal{R}e(\xi)}^1 + \int_1^{+\infty}$. Using the above estimate for $\int_1^{+\infty}$ and Lemmas 7.F.2 and 7.F.5, we get

$$\begin{aligned} |\Delta_3(\xi)| &= \left| \left(\int_{\mathcal{R}e(\xi)}^1 + \int_1^{+\infty} \right) p_\alpha\,(t + \mathrm{i}\mathcal{I}m\,(\xi))\; g_\beta\,((t + \mathrm{i}\mathcal{I}m\,(\xi))\; dt \right| \\ &\leq \mathcal{M}_5 \mathcal{M}_8 \left(|v|^2_{{}^s\mathsf{H}^\lambda_\nu(r)} + \rho\nu^2 \right) \left(\frac{1}{\pi^9} J(\xi) + \frac{\mathrm{e}^{-(\lambda+1)}}{1+\lambda} \right) \end{aligned}$$

where

$$J(\xi) = \int_{\mathcal{R}e(\xi)}^1 \frac{dt}{(t^2 + (\pi - \mathcal{I}m\,(\xi))^2)^{\frac{9}{2}}}.$$

- **Bound of $J(\xi)$ for $0 \leq \mathcal{R}e\,(\xi) \leq 1$.** In this case we have

$$\begin{aligned} J(\xi) &\leq \frac{1}{((\mathcal{R}e\,(\xi))^2 + (\pi - \mathcal{I}m\,(\xi))^2)^{\frac{7}{2}}} \int_{\mathcal{R}e(\xi)}^1 \frac{dt}{t^2 + (\pi - \mathcal{I}m\,(\xi))^2} \\ &\leq \frac{1}{|\xi - \mathrm{i}\pi|^7} \frac{1}{\pi - \mathcal{I}m\,(\xi)} \left[\arctan \frac{1}{\pi - \mathcal{I}m\,(\xi)} - \arctan \frac{\mathcal{R}e\,(\xi)}{\pi - \mathcal{I}m\,(\xi)} \right] \end{aligned}$$

Then, with the help of Figure 7.17 p. 250, we check

– if $0 \leq \mathcal{R}e\,(\xi) \leq \dfrac{\pi - \frac{1}{2}\nu^r}{\pi} \frac{1}{2}\nu^r$, then $\pi - \mathcal{I}m\,(\xi) \geq \frac{1}{2}\nu^r$ and

$$J(\xi) \leq \frac{\pi}{\nu^r} \frac{1}{|\xi - \mathrm{i}\pi|^7};$$

– if $\mathcal{R}e\,(\xi) \geq \frac{\pi - \frac{1}{2}\nu^r}{\pi} \frac{1}{2}\nu^r$, then $\mathcal{R}e\,(\xi) \geq m\nu^r$ with $m = \frac{\pi - \frac{1}{2}}{\pi} \frac{1}{2}$ and

$$\begin{aligned}J(\xi) &\leq \frac{1}{|\xi - i\pi|^7}\,\frac{1}{(\pi - \mathcal{I}m\,(\xi))}\left[\arctan\frac{1}{\pi - \mathcal{I}m\,(\xi)} - \arctan\frac{\mathcal{R}e\,(\xi)}{\pi - \mathcal{I}m\,(\xi)}\right]\\ &\leq \frac{1}{|\xi - i\pi|^7}\,\frac{1}{(\pi - \mathcal{I}m\,(\xi))}\left[\arctan\frac{\pi - \mathcal{I}m\,(\xi)}{\mathcal{R}e\,(\xi)} - \arctan(\pi - \mathcal{I}m\,(\xi))\right]\\ &\leq \frac{1}{|\xi - i\pi|^7}\left[\frac{1}{\mathcal{R}e\,(\xi)} - 1\right] \leq \frac{1}{|\xi - i\pi|^7}\frac{1}{m\nu^r}.\end{aligned}$$

Thus, for $\xi \in \mathcal{D}^*_{r,\nu}$ such that $0 \leq \mathcal{R}e\,(\xi) \leq 1$ and $\mathcal{I}m\,(\xi) \geq 0$

$$J(\xi) \leq \max\left(\pi, \tfrac{1}{m}\right)\,\frac{1}{\nu^r}\,\frac{1}{|\xi - i\pi|^7} \tag{7.84}$$

- **Bound of $J(\xi)$ for $-\nu < \mathcal{R}e\,(\xi) \leq 0$.** Observing that, in this case

$$\pi - \mathcal{I}m\,(\xi) > \tfrac{1}{2}\nu^r, \qquad 0 \leq \mathcal{R}e\left(-\overline{\xi}\right) \leq \nu \leq 1$$

holds, we get

$$\begin{aligned}J(\xi) &= \int_{\mathcal{R}e(\xi)}^{-\mathcal{R}e(\xi)} \frac{dt}{(t^2 + (\pi - \mathcal{I}m\,(\xi))^2)^{\frac{9}{2}}} + \int_{-\mathcal{R}e(\xi)}^{1} \frac{dt}{(t^2 + (\pi - \mathcal{I}m\,(\xi))^2)^{\frac{9}{2}}}\\ &= 2\int_0^{-\mathcal{R}e(\xi)} \frac{dt}{(t^2 + (\pi - \mathcal{I}m\,(\xi))^2)^{\frac{9}{2}}} + J(-\overline{\xi})\\ &\leq \frac{2}{(\pi - \mathcal{I}m\,(\xi))^7}\int_0^{-\mathcal{R}e(\xi)} \frac{dt}{t^2 + (\pi - \mathcal{I}m\,(\xi))^2} + \max\left(\pi, \tfrac{1}{m}\right)\,\frac{1}{\nu^r}\,\frac{1}{|-\overline{\xi} - i\pi|^7}\\ &\leq \frac{2}{(\pi - \mathcal{I}m\,(\xi))^8}\,\arctan\left(\frac{-\mathcal{R}e\,(\xi)}{\pi - \mathcal{I}m\,(\xi)}\right) + \max\left(\pi, \tfrac{1}{m}\right)\,\frac{1}{\nu^r}\,\frac{1}{|\xi - i\pi|^7}\\ &\leq \pi\,\frac{2}{\nu^r}\,\frac{1}{(\pi - \mathcal{I}m\,(\xi))^7} + \max\left(\pi, \tfrac{1}{m}\right)\,\frac{1}{\nu^r}\,\frac{1}{|\xi - i\pi|^7}.\end{aligned}$$

Finally, for $\xi \in \mathcal{D}^*_{r,\nu}$ satisfying $\mathcal{I}m\,(\xi) \geq 0$ and $-\nu < \mathcal{R}e\,(\xi) \leq 0$, we have

$$\pi - \mathcal{I}m\,(\xi) > \tfrac{1}{2}\nu^r, \qquad \frac{(\mathcal{R}e\,(\xi))^2}{(\pi - \mathcal{I}m\,(\xi))^2} \leq \frac{1}{4}, \qquad |\xi - i\pi| \leq \tfrac{\sqrt{5}}{2}(\pi - \mathcal{I}m\,(\xi)),$$

and

$$J(\xi) \leq \left(2\pi\,\left(\tfrac{\sqrt{5}}{2}\right)^7 + \max\left(\pi, \tfrac{1}{m}\right)\right)\frac{1}{\nu^r}\,\frac{1}{|\xi - i\pi|^7}.$$

We can conclude that for every $r\, \in]0,1[$, $\nu\, \in]0, \nu_4(r)]\, \subset]0,1]$, $\rho \in [0, \rho_s[$, $d \in [0,1]$, $v \in \mathsf{B}^s\mathsf{H}^\lambda_\nu\,(r)[d\nu]$ and every $\xi \in \mathcal{D}^*_{r,\nu}$ such that $-\nu < \mathcal{R}e\,(\xi) \leq 1$ and $\mathcal{I}m\,(\xi) \geq 0$,

$$\begin{aligned}|\Delta_3(\xi)| \leq \mathcal{M}_5\mathcal{M}_8\left(|v|^2_{^s\mathsf{H}^\lambda_\nu(r)} + \rho\nu^2\right)\frac{1}{\nu^r|\xi - i\pi|^7}&\\ \times\left(\tfrac{1}{\pi^9}\left(2\pi\left(\tfrac{\sqrt{5}}{2}\right)^7 + \max\left(\pi, \tfrac{1}{m}\right)\right) + M^7\frac{e^{-(\lambda+1)}}{1 + \lambda}\right)&\end{aligned} \tag{7.85}$$

where $M = \sup_{\substack{0 \le \mathcal{I}m(\xi) < \pi \\ |\mathcal{R}e(\xi)| \le 1}} |\xi - i\pi|$.

Grouping together (7.83) and (7.85) we obtain the following upper bound for $|\Delta_3|$

$$|\Delta_3(\xi)| \le C_3 \left(|v|^2_{{}^s\mathsf{H}^\lambda_\nu(r)} + \rho\nu^2 \right) \left(\frac{\chi_1(\xi)}{\nu^r |\xi - i\pi|^7} + (1-\chi_1(\xi)) e^{-(1+\lambda)\mathcal{R}e(\xi)} \right) \quad (7.86)$$

which holds for every $r \in]0,1[$, $\nu \in]0, \nu_4(r)]$, $\rho \in [0, \rho_s[$, $d \in [0,1]$, $v \in \mathsf{B}\,{}^s\mathsf{H}^\lambda_\nu(r)[d\nu]$ and every $\xi \in \mathcal{D}^*_{r,\nu}$ such that $\mathcal{I}m(\xi) \ge 0$, where

$$C_3 = \max\left(\frac{\mathcal{M}_5 \mathcal{M}_8}{1+\lambda}, \mathcal{M}_5 \mathcal{M}_8 \left[\frac{1}{\pi^9} \left(2\pi \left(\frac{\sqrt{5}}{2} \right)^7 + \max\left(\pi, \frac{1}{m}\right) \right) + M^7 \frac{e^{-(\lambda+1)}}{1+\lambda} \right] \right).$$

Step 1.4 Bound of $\mathcal{F}_\alpha$. Using the estimates of p_α, u_β given by Lemmas 7.F.2, 7.F.3, and the bounds of Δ_i, $i = 1,2,3$ given by (7.79), (7.82), (7.86) we get that

$$|\mathcal{F}_\alpha(v)(\xi)| \le C_\alpha \frac{1}{\nu^r} \left(|v|^2_{{}^s\mathsf{H}^\lambda_\nu(r)} + \rho\nu^2 \right) \left(\frac{\chi_1(\xi)}{|\xi - i\pi|^3} + (1-\chi_1(\xi)) e^{-\lambda\mathcal{R}e(\xi)} \right)$$

holds for every $r \in]0,1[$, $\nu \in]0, \nu_4(r)]$, $\rho \in [0, \rho_s[$, $d \in [0,1]$, $v \in \mathsf{B}\,{}^s\mathsf{H}^\lambda_\nu(r)[d\nu]$ and every $\xi \in \mathcal{D}^*_{r,\nu}$ such that $\mathcal{I}m(\xi) \ge 0$, where

$$C_\alpha = \frac{15\pi}{8} C_1 \mathcal{M}_5 + 2C_2 \mathcal{M}_5 + 2C_3 \mathcal{M}_6.$$

Finally since $|\xi + i\pi|^3 \le (4\pi^2 + 1)^{\frac{3}{2}}$ for $\xi \in \mathcal{D}^*_{r,\nu}$ such $\mathcal{I}m(\xi) \ge 0$ and $|\mathcal{R}e(\xi)| \le 1$ we get that

$$|\mathcal{F}_\alpha(v)(\xi)| \le C'_\alpha \frac{1}{\nu^r} \left(|v|^2_{{}^s\mathsf{H}^\lambda_\nu(r)} + \rho\nu^2 \right) \left(\frac{\chi_1(\xi)}{|\xi^2 + \pi^2|^3} + (1-\chi_1(\xi)) e^{-\lambda\mathcal{R}e(\xi)} \right)$$

with $C'_\alpha = C_\alpha (4\pi^2+1)^{\frac{3}{2}}$ Since $\mathcal{F}(v)$ is symmetric about the real axis, this later estimates ensures

$$|\mathcal{F}_\alpha(v)|_{{}^s\mathcal{H}^{3,\lambda}_\nu(r)} \le C'_\alpha \frac{1}{\nu^r} \left(|v|^2_{{}^s\mathsf{H}^\lambda_\nu(r)} + \rho\nu^2 \right).$$

Step 1.5 Bound of $\mathcal{F}_\beta$. Similarly we obtain

$$|\mathcal{F}_\beta(v)|_{{}^s\mathcal{H}^{4,\lambda}_\nu(r)} \le C'_\beta \frac{1}{\nu^r} \left(|v|^2_{{}^s\mathsf{H}^\lambda_\nu(r)} + \rho\nu^2 \right).$$

Step 2. Bound of $\mathcal{F}_{AB}(v)$. Denote by

$$\mathcal{F}_{AB}(v)(\xi) = \int_{[\xi, +\infty)} \left[\langle r^*_+(s), g(s) \rangle_* \, r_+(\xi) + \langle r^*_-(s), g(s) \rangle_* r_-(\xi) \right] ds.$$

Lemma 7.F.4 ensures that

$$|\mathcal{F}_{AB}(v)(\xi)| \le \mathcal{M}_7 \int_{\mathcal{R}e(\xi)}^{+\infty} (|g_A(\mathrm{i}\mathcal{I}m\,(\xi)+t)| + |g_B(\mathrm{i}\mathcal{I}m\,(\xi)+t)|)\, dt.$$

Step 2.1. Estimates of $\mathcal{F}_{AB}(v)(\xi)$ for $\mathcal{R}e\,(\xi) \ge 1$ and $\mathcal{I}m\,(\xi) \ge 0$ Using Lemma 7.F.5, we obtain that

$$|\mathcal{F}_{AB}(v)(\xi)| \le C^1_{AB} \left(|v|^2_{{}^s\mathsf{H}^\lambda_\nu(r)} + \rho\nu^2 \right) \nu\, \mathrm{e}^{-\lambda \mathcal{R}e(\xi)} \tag{7.87}$$

holds for every $r \in]0,1[$, $\nu \in]0, \nu_4(r)]$, $\rho \in [0, \rho_s[$, $d \in [0,1]$, $v \in \mathsf{B}^s\mathsf{H}^\lambda_\nu\,(r)[d\nu]$ and every $\xi \in \mathcal{D}^*_{r,\nu}$ such that $\mathcal{R}e\,(\xi) \ge 1$ and $\mathcal{I}m\,(\xi) \ge 0$ where

$$C^1_{AB} = \frac{2\mathcal{M}_7\mathcal{M}_8}{\lambda}.$$

Step 2.2. Estimates of $\mathcal{F}_{AB}(v)(\xi)$ for $-\nu < \mathcal{R}e\,(\xi) \le 1$ and $\mathcal{I}m\,(\xi) \ge 0$ Similarly, Lemma 7.F.5, ensures

$$\begin{aligned} |\mathcal{F}_{AB}(v)(\xi)| &= \mathcal{M}_7 \left(\int_{\mathcal{R}e(\xi)}^{1} + \int_{\mathcal{R}e(\xi)}^{+\infty} \right) (|g_A| + |g_B|)(\mathrm{i}\mathcal{I}m\,(\xi)+t)\, dt \\ &\le 2\mathcal{M}_7\mathcal{M}_8 \left(|v|^2_{{}^s\mathsf{H}^\lambda_\nu(r)} + \rho\nu^2 \right) \nu \left(\frac{1}{\pi^6} K(\xi) + \frac{\mathrm{e}^{-\lambda}}{\lambda} \right) \end{aligned}$$

where

$$K(\xi) = \int_{\mathcal{R}e(\xi)}^{1} \frac{dt}{(t^2 + (\pi - \mathcal{I}m\,(\xi)))^{\frac{6}{2}}}.$$

Bounding $K(\xi)$ in the same way as we bounded $J(\xi)$ we obtain that

$$K(\xi) \le \left(2\pi \left(\tfrac{\sqrt{5}}{2} \right)^4 + \max\left(\pi, \tfrac{1}{m}\right) \right) \frac{1}{\nu^r} \frac{1}{|\xi - \mathrm{i}\pi|^4}.$$

Observing that $\dfrac{\nu}{|\xi - \mathrm{i}\pi|} \ge \dfrac{\nu}{m_0 \nu^r} \ge \dfrac{1}{m_0}$.

$$|\mathcal{F}_{AB}(v)(\xi)| \le C^2_{AB} \left(|v|^2_{{}^s\mathsf{H}^\lambda_\nu(r)} + \rho\nu^2 \right) \frac{1}{\nu^r |\xi - \mathrm{i}\pi|^3} \mathrm{e}^{-\lambda \mathcal{R}e(\xi)} \tag{7.88}$$

holds for every $r \in]0,1[$, $\nu \in]0, \nu_4(r)]$, $\rho \in [0, \rho_s[$, $d \in [0,1]$, $v \in \mathsf{B}^s\mathsf{H}^\lambda_\nu\,(r)[d\nu]$ and every $\xi \in \mathcal{D}^*_{r,\nu}$ such that $-\nu < \mathcal{R}e\,(\xi) \le 1$ and $\mathcal{I}m\,(\xi) \ge 0$ where

$$C^2_{AB} = \frac{2\mathcal{M}_7\mathcal{M}_8}{\lambda} \left(\frac{1}{\pi^6} \left(2\pi \left(\tfrac{\sqrt{5}}{2} \right)^4 + \max\left(\pi, \tfrac{1}{m}\right) \right) \frac{1}{m_0} + \frac{\mathrm{e}^{-\lambda}}{\lambda} \right).$$

Gathering estimates (7.88) and (7.87) we finally get

$$|\mathcal{F}_\alpha(v)(\xi)| \leq C_{AB}\left(|v|^2_{{}^s\mathsf{H}^\lambda_\nu(r)} + \rho\nu^2\right)\left(\frac{\chi_1(\xi)}{\nu^r|\xi - \mathrm{i}\pi|^3} + (1-\chi_1(\xi))\mathrm{e}^{-\lambda\mathcal{R}e(\xi)}\right)$$

holds for every $r \in]0,1[$, $\nu \in]0,\nu_4(r)]$, $\rho \in [0,\rho_s[$, $d \in [0,1]$, $v \in \mathsf{B}^s\mathsf{H}^\lambda_\nu(r)[d\nu]$ and every $\xi \in \mathcal{D}^*_{r,\nu}$ such that $\mathcal{I}m(\xi) \geq 0$, where $C_{AB} = \max(C^1_{AB}, C^2_{AB})$. Since $\mathcal{F}$ is symmetric about the real axis and since $0 < \nu < 1$, this later estimate ensures

$$|\mathcal{F}_{AB}(v)|_{{}^s\mathcal{H}^{3,\lambda}_\nu(r)} \leq C_{AB}\frac{1}{\nu^r}\left(|v|^2_{{}^s\mathsf{H}^\lambda_\nu(r)} + \rho\nu^2\right).$$

where $C'_{AB} = C_{AB}(4\pi^2+1)^{\frac{3}{2}})$.

The result of Steps 1 and 2, finally ensures that

$$|\mathcal{F}(v)|_{{}^s\mathsf{H}^\lambda_\nu(r)} \leq M_{\mathrm{out}}\frac{1}{\nu^r}\left(|v|^2_{{}^s\mathsf{H}^\lambda_\nu(r)} + \rho\nu^2\right)$$

holds for every $r \in]0,1[$, $\nu \in]0,\nu_4(r)]$, $\rho \in [0,\rho_s[$, $d \in [0,1]$, $v \in \mathsf{B}^s\mathsf{H}^\lambda_\nu(r)[d\nu]$, where $M_{\mathrm{out}} = \max(C'_\alpha, C'_\beta, C'_{AB})$.

We proceed in the same way for bounding $|\mathcal{F}(v) - \mathcal{F}(v')|_{{}^s\mathsf{H}^\lambda_\nu(r)}$. □

7.G Appendix. Proof of lemma 7.4.41

We must first check that $\widehat{g}_0(\widehat{W}, z, \rho)$ is well defined and we need estimates of $|\widehat{g}_0(\widehat{W}, z, \rho)|$ to prove on one hand that the indefinite integral involved in the definition of $\widehat{\mathcal{F}}_0$ are convergent and on the other hand that $\widehat{\mathcal{F}}_0$ is a contraction mapping.

Lemma 7.G.1. *There exists $\delta_0 \geq 1$ such that for every $\delta \geq \delta_0$, $\rho \in D(0,2\rho_s)$, $z \in \widehat{\Omega}^s_\delta$ and every $\widehat{W} \in \mathsf{B}^s\widehat{\mathsf{H}}_\delta[1]$, $\widehat{g}_0(\widehat{W}, z, \rho)$ is well defined. Moreover there exists $\mathcal{M}_{10}$ such that for every $\delta \geq \delta_0$, $\rho \in D(0,2\rho_s)$, $z \in \widehat{\Omega}^s_\delta$ and every $\widehat{W}, \widehat{W}' \in \mathsf{B}^s\widehat{\mathsf{H}}_\delta[1]$,*

(a) $$|\widehat{g}_0(\widehat{W}, z, \rho)| \leq \mathcal{M}_{10}\left(\frac{|\widehat{W}|^2_{{}^s\widehat{\mathsf{H}}_\delta} + |\rho|}{|z|^6}\right),$$

(b) $$|\widehat{g}_0(\widehat{W}, z, \rho) - \widehat{g}_0(\widehat{W}', z, \rho)| \leq \mathcal{M}_{10}\left(\frac{|\widehat{W}|_{{}^s\widehat{\mathsf{H}}_\delta} + |\widehat{W}'|_{{}^s\widehat{\mathsf{H}}_\delta} + |\rho|}{|z|^6}\right)|\widehat{W} - \widehat{W}'|_{{}^s\widehat{\mathsf{H}}_\delta}.$$

Proof. We proceed in three steps.

Step 1. The explicit expression of $\widehat{Q}_0$ given by (7.59) ensures that there exists M such that for every $\delta \geq 1$, $\widehat{W} \in {}^s\widehat{\mathsf{H}}_\delta$ and every $z \in \widehat{\Omega}^s_\delta$

$$|\widehat{Q}_0(\widehat{W})| \leq M \left(\frac{|\widehat{W}|^2_{{}^s\widehat{H}_\delta}}{|z|^6} \right), \tag{7.89}$$

$$|\widehat{Q}_0(\widehat{W}) - \widehat{Q}_0(\widehat{W}')| \leq M \left(\frac{|\widehat{W}|_{{}^s\widehat{H}_\delta} + |\widehat{W}'|_{{}^s\widehat{H}_\delta}}{|z|^6} \right) |\widehat{W} - \widehat{W}'|_{{}^s\widehat{H}_\delta}. \tag{7.90}$$

since $|z| \geq \delta \geq 1$.

Step 2. The difficulty lies in the definition and in the upper bounds of the rest

$$\widehat{R}_0(\widehat{h}_0^+ + \widehat{W}) = \sum_{|\overline{m}|\geq 3} a_{\overline{m},0}(\widehat{h}_0^+ + \widehat{W})^{\overline{m}}.$$

Estimates (7.7) implies that

$$|a_{\overline{m},0}| \leq \frac{M_R}{r_\alpha^{m_\alpha} r_\beta^{m_\beta} r_A^{m_A} r_B^{m_B}}.$$

Let us denote $\Theta := \max\left(\frac{5}{r_\alpha}, \frac{9}{r_\beta}, \frac{1}{r_A}, \frac{1}{r_B}\right)$. For every $\delta \geq 1$, $\widehat{W} \in \mathrm{B}\,{}^s\widehat{\mathrm{H}}_\delta[1]$ and every $z \in \widehat{\Omega}^s_\delta$, observing that $|z| \geq \delta \geq 1$, we get

$$\left|\frac{\widehat{\alpha}_{\widehat{h}_0^+} + \widehat{\alpha}}{r_\alpha}\right|^{m_\alpha} \leq \left|\frac{4/|z|^2 + |\widehat{W}|_{{}^s\widehat{H}_\delta}/|z|^{3+\frac{1}{2}}}{r_\alpha}\right|^{m_\alpha} \leq \left(\frac{5}{r_\alpha}\right)^{m_\alpha} \frac{1}{|z|^{2m_\alpha}} \leq \frac{\Theta^{m_\alpha}}{|z|^{2m_\alpha}},$$

$$\left|\frac{\widehat{\beta}_{\widehat{h}_0^+} + \widehat{\beta}}{r_\beta}\right|^{m_\beta} \leq \frac{\Theta^{m_\beta}}{|z|^{3m_\beta}},$$

$$\left|\frac{\widehat{A}}{r_A}\right|^{m_A} \leq \left|\frac{|\widehat{W}|_{{}^s\widehat{H}_\delta}}{|z|^4 r_A}\right|^{m_A} \leq \left(\frac{1}{r_A}\right)^{m_A} \frac{1}{|z|^{4m_A}} \leq \frac{\Theta^{m_A}}{|z|^{4m_A}},$$

$$\left|\frac{\widehat{B}}{r_B}\right|^{m_B} \leq \frac{\Theta^{m_B}}{|z|^{4m_B}}. \tag{7.91}$$

Hence, using once again that $|z| \geq \delta \geq 1$ we obtain

$$\begin{aligned} |\widehat{R}_0(\widehat{h}_0^+ + \widehat{W})| &\leq \sum_{|\overline{m}|\geq 3} \frac{M_R\, \Theta^{|\overline{m}|}}{|z|^{2m_\alpha+3m_\beta+4(m_A+m_B)}} \\ &\leq \sum_{|\overline{m}|\geq 3} \frac{M_R\, \Theta^{|\overline{m}|}}{|z|^{2|\overline{m}|}} \\ &\leq \frac{M_R \Theta^3}{|z|^6} \sum_{|\overline{m}|\geq 3} \frac{\Theta^{|\overline{m}|-3}}{|z|^{2(\overline{m}-3)}}. \end{aligned}$$

Let us denote $\delta_0 := \max\left(\sqrt{2\Theta}, 1\right)$. Then for every $\delta \geq \delta_0$, $\widehat{W} \in \mathsf{B}^{\mathsf{s}}\widehat{\mathsf{H}}_\delta[1]$, $z \in \widehat{\Omega}^{\mathsf{s}}_\delta$,

$$\begin{aligned} |\widehat{R}_0(\widehat{h}_0^+ + \widehat{W})| &\leq \frac{M_R \Theta^3}{|z|^6} \sum_{|\overline{m}| \geq 3} \left(\tfrac{1}{2}\right)^{|\overline{m}|-3} \\ &\leq 8 \frac{M_R \Theta^3}{|z|^6} \sum_{|\overline{m}| \geq 0} \left(\tfrac{1}{2}\right)^{|\overline{m}|} \\ &\leq \frac{2^7 M_R \Theta^3}{|z|^6}. \end{aligned} \tag{7.92}$$

Estimates (a) readily follows from (7.89), (7.92).

Step 3. We must now bound $\widehat{R}_0(\widehat{h}_0^+ + \widehat{W}, \nu) - \widehat{R}_0(\widehat{h}_0^+ + \widehat{W}', \nu)$. Let us define

$$\Delta_{\overline{m},0} := a_{\overline{m},0} \left[(\widehat{h}_0^+ + \widehat{W})^{\overline{m}} - (\widehat{h}_0^+ + \widehat{W}')^{\overline{m}}\right].$$

The difference $\left[(\widehat{h}_0^+ + \widehat{W})^{\overline{m}} - (\widehat{h}_0^+ + \widehat{W}')^{\overline{m}}\right]$ reads

$$\begin{aligned} &\left((\widehat{h}_0^+ + \widehat{W})^{\overline{m}} - (\widehat{h}_0^+ + \widehat{W}')^{\overline{m}}\right) = \\ &\quad \left[\left(\widehat{\alpha}_{\widehat{h}_0^+} + \widehat{\alpha}\right)^{m_\alpha} - \left(\widehat{\alpha}_{\widehat{h}_0^+} + \widehat{\alpha}'\right)^{m_\alpha}\right] \left(\widehat{\beta}_{\widehat{h}_0^+} + \widehat{\beta}\right)^{m_\beta} \widehat{A}^{m_A} \, \widehat{B}^{m_B} \\ &\quad + \left(\widehat{\alpha}_{\widehat{h}_0^+} + \widehat{\alpha}'\right)^{m_\alpha} \left[\left(\widehat{\beta}_{\widehat{h}_0^+} + \widehat{\beta}\right)^{m_\beta} - \left(\widehat{\beta}_{\widehat{h}_0^+} + \widehat{\beta}'\right)^{m_\beta}\right] \widehat{A}^{m_A} \, \widehat{B}^{m_B} \\ &\quad + \left(\widehat{\alpha}_{\widehat{h}_0^+} + \widehat{\alpha}'\right)^{m_\alpha} \left(\widehat{\beta}_{\widehat{h}_0^+} + \widehat{\beta}'\right)^{m_\beta} \left[\widehat{A}^{m_A} - \widehat{A}'^{m_A}\right] \widehat{B}^{m_B} \\ &\quad + \left(\widehat{\alpha}_{\widehat{h}_0^+} + \widehat{\alpha}'\right)^{m_\alpha} \left(\widehat{\beta}_{\widehat{h}_0^+} + \widehat{\beta}'\right)^{m_\beta} \widehat{A}'^{m_A} \left[\widehat{B}^{m_B} - \widehat{B}'^{m_B}\right]. \end{aligned} \tag{7.93}$$

Using (7.91) we get

$$\begin{aligned} \frac{\left|\left(\widehat{\alpha}_{\widehat{h}_0^+} + \widehat{\alpha}\right)^{m_\alpha} - \left(\widehat{\alpha}_{\widehat{h}_0^+} + \widehat{\alpha}'\right)^{m_\alpha}\right|}{r_\alpha^{m_\alpha}} &= \frac{|\widehat{\alpha} - \widehat{\alpha}'|}{r_\alpha^{m_\alpha}} \left|\sum_{k=0}^{m_\alpha - 1} \left(\widehat{\alpha}_{\widehat{h}_0^+} + \widehat{\alpha}\right)^k \left(\widehat{\alpha}_{\widehat{h}_0^+} + \widehat{\alpha}'\right)^{m_\alpha - 1 - k}\right| \\ &\leq \frac{|\widehat{\alpha} - \widehat{\alpha}'|}{r_\alpha} m_\alpha \left(\frac{\Theta}{|z|^2}\right)^{m_\alpha - 1}, \\ &\leq \frac{|\widehat{W} - \widehat{W}'|_{{}^{\mathsf{s}}\widehat{\mathsf{H}}_\delta}}{r_\alpha |z|^{3+\frac{1}{2}}} m_\alpha \left(\frac{\Theta}{|z|^2}\right)^{m_\alpha - 1}, \end{aligned}$$

$$\begin{aligned} \frac{\left|\left(\widehat{\beta}_{\widehat{h}_0^+} + \widehat{\beta}\right)^{m_\beta} - \left(\widehat{\beta}_{\widehat{h}_0^+} + \widehat{\beta}'\right)^{m_\beta}\right|}{r_\beta^{m_\beta}} &\leq \frac{|\widehat{\beta} - \widehat{\beta}'|}{r_\beta} m_\beta \left(\frac{\Theta}{|z|^3}\right)^{m_\beta - 1} \\ &\leq \frac{|\widehat{W} - \widehat{W}'|_{{}^{\mathsf{s}}\widehat{\mathsf{H}}_\delta}}{r_\beta |z|^{4+\frac{1}{2}}} m_\beta \left(\frac{\Theta}{|z|^3}\right)^{m_\beta - 1} \end{aligned}$$

$$\frac{\left|\widehat{A}^{m_A} - \widehat{A}'^{m_A}\right|}{r_A^{m_A}} \leq \frac{|\widehat{W} - \widehat{W}'|_{^s\widehat{\mathsf{H}}_\delta}}{r_A|z|^4} m_A \left(\frac{\Theta}{|z|^4}\right)^{m_A-1},$$

$$\frac{\left|\widehat{B}^{m_B} - \widehat{B}'^{m_B}\right|}{r_B^{m_B}} \leq \frac{|\widehat{W} - \widehat{W}'|_{^s\widehat{\mathsf{H}}_\delta}}{r_B|z|^4} m_B \left(\frac{\Theta}{|z|^4}\right)^{m_B-1}.$$

Denoting

$$M = \max\left(r_\alpha^{-1}, r_\beta^{-1}, r_A^{-1}, r_A^{-1}\right)$$

and recalling that $|z| \geq \delta \geq \delta_0 \geq 1$ we obtain

$$\begin{aligned}|\Delta_{\overline{m},0}| &\leq \frac{M_R M \Theta^{|\overline{m}|-1}|\widehat{W} - \widehat{W}'|_{^s\widehat{\mathsf{H}}_\delta}}{|z|^{2m_\alpha+3m_\beta+4(m_A+m_B)}}\left[\frac{m_\alpha + m_\beta}{|z|} + m_A + m_B\right] \\ &\leq M_R M\ |\overline{m}|\ \frac{\Theta^{|\overline{m}|-1}}{|z|^{2|\overline{m}|}}\ |\widehat{W} - \widehat{W}'|_{^s\widehat{\mathsf{H}}_\delta}\end{aligned}$$

We have chosen δ_0 such that $\delta_0 := \max\left(\sqrt{2\Theta}, 1\right)$. Hence,

$$|\Delta_{\overline{m},0}| \leq \frac{M_R M \Theta^2}{|z|^6}\ |\overline{m}|\ \frac{1}{2}^{|\overline{m}|-3}\ |\widehat{W} - \widehat{W}'|_{^s\widehat{\mathsf{H}}_\delta}.$$

Summing these estimates, we obtain

$$\begin{aligned}|\widehat{R}_0(\widehat{h}_0^+ + \widehat{W}) - \widehat{R}_0(\widehat{h}_0^+ + \widehat{W}')| &\leq \frac{M_R M \Theta^2 |\widehat{W} - \widehat{W}'|_{^s\widehat{\mathsf{H}}_\delta}}{|z|^6}\sum_{|\overline{m}|\geq 3}|\overline{m}|\left(\tfrac{1}{2}\right)^{|\overline{m}|-3} \\ &\leq 4\frac{M_R M \Theta^2 |\widehat{W} - \widehat{W}'|_{^s\widehat{\mathsf{H}}_\delta}}{|z|^6}\sum_{|\overline{m}|\geq 0}|\overline{m}|\left(\tfrac{1}{2}\right)^{|\overline{m}|-1} \qquad (7.94) \\ &\leq 2^9 M_R M \Theta^2 \frac{|\widehat{W} - \widehat{W}'|_{^s\widehat{\mathsf{H}}_\delta}}{|z|^6}.\end{aligned}$$

Estimates (b) readily follows from (7.90), (7.94). □

Proof of Lemma 7.4.41 For every $z \in \widehat{\Omega}_\delta^s$, $(\widehat{\mathbf{p}}(z), \widehat{\mathbf{q}}(z), \widehat{\mathbf{r}}_+(z), \widehat{\mathbf{r}}_-(z))$ is a basis of $\mathbb{C}^4$. Let us write $\widehat{W}(z)$ in this basis

$$\widehat{W}(z) = a(z)\widehat{\mathbf{p}}(z) + b(z)\widehat{\mathbf{q}}(z) + c(z)\widehat{\mathbf{r}}_+(z) + d(z)\widehat{\mathbf{r}}_-(z).$$

Let δ, δ^* be such that $\delta^* > \delta \geq \delta_0 \geq 1$. Then $\widehat{W} \in {}^s\widehat{\mathsf{H}}_\delta$ is a solution of (7.58) if and only if

$$\begin{aligned}
\widehat{W}(z) &= \Big(a_0 + \int\limits_{[-i\delta^*,z]} \langle \widehat{\mathbf{p}}^*(s), \widehat{g}_0(\widehat{W}(s),s,\rho)\rangle_* \, ds\Big)\, \widehat{\mathbf{p}}(z) \\
&+ \Big(b_0 + \int\limits_{[-i\delta^*,z]} \langle \widehat{\mathbf{q}}^*(s), \widehat{g}_0(\widehat{W}(s),s,\rho)\rangle_* \, ds\Big)\, \widehat{\mathbf{q}}(z) \\
&+ \Big(c_0 + \int\limits_{[-i\delta^*,z]} \langle \widehat{\mathbf{r}}_+^*(s), \widehat{g}_0(\widehat{W}(s),s,\rho)\rangle_* \, ds\Big)\, \widehat{\mathbf{r}}_+(z) \\
&+ \Big(d_0 + \int\limits_{[-i\delta^*,z]} \langle \widehat{\mathbf{r}}_-^*(s), \widehat{g}_0(\widehat{W}(s),s,\rho)\rangle_* \, ds\Big)\, \widehat{\mathbf{r}}_-(z).
\end{aligned}$$

For $\widehat{W} \in {}^s\widehat{\mathsf{H}}_\delta$, $\widehat{W}(z) \underset{\mathcal{R}e(z)\to+\infty}{\longrightarrow} 0$. Moreover, $\widehat{\mathbf{q}}^*(z) \underset{\mathcal{R}e(z)\to+\infty}{\longrightarrow} 0$ and $\widehat{\mathbf{r}}_+^*(z)$, $\widehat{\mathbf{r}}_-^*(z)$ are bounded for $\mathcal{R}e\,(z) \to +\infty$. Thus setting $z = i\delta^* + \tau$ and pushing $\tau \to +\infty$, we check that necessarily

$$\langle \widehat{\mathbf{q}}^*(z), \widehat{W}(z)\rangle_* \underset{\mathcal{R}e(z)\to+\infty}{\longrightarrow} 0, \langle \widehat{\mathbf{r}}_+^*(z), \widehat{W}(z)\rangle_* \underset{\mathcal{R}e(z)\to+\infty}{\longrightarrow} 0, \langle \widehat{\mathbf{r}}_-^*(z), \widehat{W}(z)\rangle_* \underset{\mathcal{R}e(z)\to+\infty}{\longrightarrow} 0.$$

So we choose:

$$\begin{aligned}
b_0 &= -\int\limits_{[-I\delta^*,z]} \langle \widehat{\mathbf{q}}^*(s), \widehat{g}_0(\widehat{W}(s),s,\rho)\rangle_* \, ds, \\
c_0 &= -\int\limits_{[-i\delta^*,z]} \langle \widehat{\mathbf{r}}_+^*(s), \widehat{g}_0(\widehat{W}(s),s,\rho)\rangle_* \, ds, \\
d_0 &= -\int\limits_{[-i\delta^*,z]} \langle \widehat{\mathbf{r}}_-^*(s), \widehat{g}_0(\widehat{W}(s),s,\rho)\rangle_* \, ds.
\end{aligned}$$

Lemma 7.G.1 ensures that these integrals are convergent. Hence, $\widehat{W} \in {}^s\widehat{\mathsf{H}}_\delta$ is a solution if and only if

$$\begin{aligned}
\widehat{W}(z) &= \Big(a_0 + \int\limits_{[-i\delta^*,z]} \langle \widehat{\mathbf{p}}^*(s), \widehat{g}_0(\widehat{W}(s),s,\rho)\rangle_* \, ds\Big)\, \widehat{\mathbf{p}}(z) \\
&- \Big(\int\limits_{[z,+\infty)} \langle \widehat{\mathbf{q}}^*(s), \widehat{g}_0(\widehat{W}(s),s,\rho)\rangle_* \, ds\Big)\, \widehat{\mathbf{q}}(z) \\
&- \Big(\int\limits_{[z,+\infty)} \langle \widehat{\mathbf{r}}_+^*(s), \widehat{g}_0(\widehat{W}(s),s,\rho)\rangle_* \, ds\Big)\, \widehat{\mathbf{r}}_+(z) \\
&- \Big(\int\limits_{[z,+\infty)} \langle \widehat{\mathbf{r}}_-^*(s), \widehat{g}_0(\widehat{W}(s),s,\rho)\rangle_* \, ds\Big)\, \widehat{\mathbf{r}}_-(z).
\end{aligned}$$

To obtain an integral operator which does not depend on δ^*, we choose

$$a_0 = -\int_{[-i\delta^*,+\infty)} \left\langle \widehat{\mathbf{p}}^*(s), \widehat{g}_0(\widehat{W}(s), s, \rho)\right\rangle_* ds.$$

Here again, lemma 7.G.1 ensures that this integral is convergent. Lemma 7.4.41 follows immediately.

7.H Appendix. Proof of lemma 7.4.42

We begin with two technical lemmas.

Lemma 7.H.1. *There exists $\mathcal{M}_{11}$ such that for every $\delta \geq 1$ and every function*

$$\begin{array}{rccc} \widehat{f}: & \widehat{\Omega}^{\rm s}_{\delta} \times D(0, 2\rho_{\rm s}) & \longrightarrow & \mathbb{C}^4 \\ & (z, \rho) & \longrightarrow & (., ., \widehat{f}_A(z, \rho), \widehat{f}_B(z, \rho)) \end{array}$$

satisfying $\sup_{\widehat{\Omega}^{\rm s}_{\delta} \times D(0,2\rho_{\rm s})} |z^6 \widehat{f}_A| < +\infty$, $\sup_{\widehat{\Omega}^{\rm s}_{\delta} \times D(0,2\rho_{\rm s})} |z^6 \widehat{f}_B| < +\infty$ *the inequality*

$$\left| \int_{[z,+\infty)} \left[\left\langle \widehat{\mathbf{r}}^*_+(s), \widehat{f}(s, \rho)\right\rangle_* \widehat{\mathbf{r}}_+(z) + \left\langle \widehat{\mathbf{r}}^*_-(s), \widehat{f}(s, \rho)\right\rangle_* \widehat{\mathbf{r}}_-(z) \right] ds \right|$$
$$\leq \mathcal{M}_{11} \int_{[z,+\infty)} \left(|\widehat{f}_A(s, \rho)| + |\widehat{f}_B(s, \rho)|\right) ds$$

holds for every $(z, \rho) \in \widehat{\Omega}^{\rm s}_{\delta} \times D(0, 2\rho_{\rm s})$.

Proof. Denote

$$I(z, \rho) = \int_{[z,+\infty)} \left[\left\langle \widehat{\mathbf{r}}^*_+(s), \widehat{f}(s, \rho)\right\rangle_* \widehat{\mathbf{r}}_+(z) + \left\langle \widehat{\mathbf{r}}^*_-(s), \widehat{f}(s, \rho)\right\rangle_* \widehat{\mathbf{r}}_-(z) \right] ds.$$

Using the explicit forms of $\widehat{\mathbf{r}}_\pm$ given in lemmas 7.4.39, 7.4.40 we get $I(z, \rho) = (0, 0, I_A(z, \rho), I_B(z, \rho))$ and

$$\begin{aligned} I_A(z, \rho) &= \int_0^{+\infty} [\widehat{f}_A(z+t, \rho) \cos \widehat{\psi}(z+t) + \widehat{f}_B(z+t, \rho) \sin \widehat{\psi}(z+t)] dt \ \cos \widehat{\psi}(z) \\ &\quad - \int_0^{+\infty} [-\widehat{f}_A(z+t, \rho) \sin \widehat{\psi}(z+t) + \widehat{f}_B(z+t, \rho) \cos \widehat{\psi}(z+t)] dt \ \sin \widehat{\psi}(z) \\ &= \int_0^{+\infty} \Big[\widehat{f}_A(z+t, \rho) \cos \left(\widehat{\psi}(z+t) - \widehat{\psi}(z) \right) \\ &\qquad\qquad + \widehat{f}_B(z+t, \rho) \sin \left(\widehat{\psi}(z+t) - \widehat{\psi}(z) \right) \Big] dt. \end{aligned}$$

Now observe that $|\cos(z)| \leq e^{|\mathcal{I}m(z)|}$, $|\sin(z)| \leq e^{|\mathcal{I}m(z)|}$ and that

$$\left|\mathcal{I}m\left(\widehat{\psi}(z+t)-\widehat{\psi}(z)\right)\right|=\left|\mathcal{I}m\left(\omega t+\frac{4b}{z+t}-\frac{4b}{z}\right)\right|\leq 4|b|\left(\frac{1}{|z+t|}+\frac{1}{|z|}\right)\leq 8|b|$$

since $|z+t| \geq |\mathcal{I}m\,(z)\,| \geq \delta \geq 1$ and $|z| \geq |\mathcal{I}m\,(z)\,| \geq \delta \geq 1$. Thus,

$$|I_A(z)| \leq e^{8|b|}\int_0^{+\infty}(|\widehat{f}_A(z+t,\rho)|+|\widehat{f}_B(z+t,\rho)|)\,dt$$

holds for $(z,\rho) \in \widehat{\Omega}^{s}_{\delta} \times D(0,2\rho_s)$. The same inequality holds for I_B. □

Lemma 7.H.2. *For every $\delta \geq 1$, $z \in \widehat{\Omega}^{s}_{\delta}$ and every $m \geq 2$*

$$J_m(z) := \int_0^{+\infty}\frac{dt}{|z+t|^m} \leq \frac{\pi}{2}\left(\frac{1+\left(\sqrt{2}\right)^{m-1}}{|z|^{m-1}}\right).$$

Proof. We distinguish two cases:

Case 1: $\mathcal{R}e\,(z) \geq 0$. Let us denote $z = \tau - i\eta$ with $\tau \geq 0$ and $\eta \geq \delta$. For every $t \geq 0$, $|z+t| \geq |z|$ holds. Hence,

$$J_m(z) \leq \frac{1}{|z|^{m-2}}J_2(z). \tag{7.95}$$

The integral J_2 can be explicitly computed

$$J_2(z) = \begin{cases} \dfrac{\pi}{2\eta} = \dfrac{\pi}{2|z|} & \text{for } \tau = 0 \\ \dfrac{1}{\eta}\arctan\dfrac{\eta}{\tau} & \text{for } \tau > 0 \end{cases} \tag{7.96}$$

Studying the function

$$\begin{array}{rccl} f: &]0,+\infty[& \to & \mathbb{R}, \\ & x & \mapsto & \sqrt{1+x^2}\,\arctan\frac{1}{x}. \end{array}$$

we obtain that $f(x) \leq \frac{\pi}{2}$, for $x > 0$. Thus, setting $x = \eta/\tau$, we deduce that

$$\frac{1}{\eta}\arctan\frac{\eta}{\tau} \leq \frac{\pi}{2}\frac{1}{\sqrt{\tau^2+\eta^2}} \qquad \text{for } \tau > 0,\ \eta \geq \delta. \tag{7.97}$$

Hence, (7.96),(7.97) ensures that

$$J_2(z) \leq \frac{\pi}{2|z|} \qquad \text{for } z \in \widehat{\Omega}^{s}_{\delta},\ \mathcal{R}e\,(z) \geq 0. \tag{7.98}$$

Finally, (7.95) and (7.98) gives

$$J_m(z) \leq \frac{\pi}{2|z|^{m-1}} \qquad \text{for } z \in \widehat{\Omega}_\delta^{\rm s},\ \mathcal{R}e\,(z) \geq 0. \tag{7.99}$$

Case 2: $\mathcal{R}e\,(z) \in [-1, 0]$. Let us denote $z = -\tau - i\eta$ with $\tau \in [0,1]$, $\eta \geq \delta$.

$$J_m(z) = \int_0^{2\tau} \frac{dt}{((t-\tau)^2+\eta^2)^{\frac{m}{2}}} + \int_{2\tau}^{+\infty} \frac{dt}{((t-\tau)^2+\eta^2)^{\frac{m}{2}}}.$$

Setting $u = t - \tau$ in the first integral and $t' = t - 2\tau$ in the second one, we obtain

$$J_m(z) = J_m^-(z) + J_m(\tau - i\eta).$$

where

$$J_m^-(z) = \int_{-\tau}^{\tau} \frac{du}{(u^2+\eta^2)^{\frac{m}{2}}}.$$

Estimate (7.99) ensures that

$$J_m(\tau - i\eta) \leq \frac{\pi}{2}\frac{1}{|\tau - i\eta|^{m-1}} = \frac{\pi}{2}\frac{1}{|-\tau - i\eta|^{m-1}} = \frac{\pi}{2}\frac{1}{|z|^{m-1}} \tag{7.100}$$

Setting $u = \eta x$ we get

$$J_m^-(z) = \frac{2}{\eta^{m-1}} \int_0^{\tau/\eta} \frac{dx}{(x^2+1)^{\frac{m}{2}}} \leq \frac{2}{\eta^{m-1}} \int_0^{1} \frac{dx}{(x^2+1)^{\frac{m}{2}}} = \frac{\pi}{2\eta^{m-1}}.$$

For $0 \leq \tau \leq 1$, $\eta \geq \delta \geq 1$, $0 \leq \tau/\eta \leq 1$ holds. Thus, $\sqrt{1+\tau^2/\eta^2} \leq \sqrt{2}$ which gives $\eta^{-1} \leq \sqrt{2}/|z|$.Thus,

$$\left|J_m^-(z)\right| \leq \frac{\pi\left(\sqrt{2}\right)^{m-1}}{2|z|^{m-1}}. \tag{7.101}$$

Finally (7.100), (7.101) and (7.99) ensure that

$$J_m(z) \leq \frac{\pi}{2}\left(\frac{1+\left(\sqrt{2}\right)^{m-1}}{|z|^{m-1}}\right) \qquad \text{for } z \in \widehat{\Omega}_\delta^{\rm s}. \ \square$$

Proof of Lemma 7.4.42. Using the theorem of derivation under the integral coming from the theorem of dominated convergence of Lebesgue, we check that for every $\widehat{W} \in \mathsf{B}^{\rm s}\widehat{\mathsf{H}}_\delta[1]$, $\widehat{\mathcal{F}}_0(\widehat{W})$ is continuous with respect to $(z,\rho) \in \widehat{\Omega}_\delta^{\rm s} \times D(0, 2\rho_{\rm s})$, holomorphic with respect to z for every fixed $\rho \in D(0, 2\rho_{\rm s})$ and holomorphic with respect to ρ for every fixed $z \in \widehat{\Omega}_\delta^{\rm s}$. We have now to prove estimates (a) and (b).

Estimate (a). We proceed in two steps. A first one for the two first components of $\widehat{\mathcal{F}}_0$ and the second one for the two last ones.

Step 1. Let us denote

$$\begin{aligned}
(\widehat{\mathcal{F}}_{0,\alpha\beta}(\widehat{W}))(z,\rho) &:= -\int_{[z,+\infty)} \langle \widehat{\mathbf{p}}^*(s), \widehat{g}_0(\widehat{W}(s),s,\rho)\rangle_* ds\ \widehat{\mathbf{p}}(z) \\
&\quad -\int_{[z,+\infty)} \langle \widehat{\mathbf{q}}^*(s), \widehat{g}_0(\widehat{W},s,\rho)\rangle_* ds\ \widehat{\mathbf{q}}(z) \\
&= \Delta_1(z,\rho)\widehat{\mathbf{p}}(z) + \Delta_2(z,\rho)\widehat{\mathbf{q}}(z)
\end{aligned}$$

with

$$\Delta_1(z,\rho) = \int_0^{+\infty} \frac{(z+t)^4}{56}\, \widehat{g}_{0,\beta}(\widehat{W}(z+t,\rho), z+t, \rho)\ dt\ \ \widehat{\mathbf{p}}(z),$$

$$\Delta_2(z,\rho) = \int_0^{+\infty} \frac{-1}{7(z+t)^3}\, \widehat{g}_{0,\beta}(\widehat{W}(z+t,\rho), z+t, \rho)\ dt\ \ \widehat{\mathbf{q}}(z).$$

Using lemmas 7.4.39, 7.4.40, 7.G.1 and 7.H.2, we obtain

$$\begin{aligned}
|\Delta_1(z,\rho)| &\le \mathcal{M}_{10}\int_0^{+\infty} \frac{|z+t|^4}{56}\, \frac{|\widehat{W}|^2_{{}^s\widehat{\mathsf{H}}_\delta} + 2\rho_s}{|z+t|^6}\, dt \\
&\le \frac{\mathcal{M}_{10}}{56}\left(|\widehat{W}|^2_{{}^s\widehat{\mathsf{H}}_\delta} + 2\rho_s\right) J_2(z) \\
&\le \frac{\pi\mathcal{M}_{10}}{112}\ (1+\sqrt{2})\ \frac{|\widehat{W}|^2_{{}^s\widehat{\mathsf{H}}_\delta} + 2\rho_s}{|z|}.
\end{aligned}$$

Similarly

$$|\Delta_2(z,\rho)| \le \frac{\pi\mathcal{M}_{10}}{14}\ \left(1+\left(\sqrt{2}\right)^8\right)\ \frac{|\widehat{W}|^2_{{}^s\widehat{\mathsf{H}}_\delta} + 2\rho_s}{|z|^8}.$$

Finally, there exists M such that

$$\begin{aligned}
\sup_{\widehat{\Omega}^s_\delta \times D(0,2\rho_s)} \left(|z|^{3+\frac{1}{2}} |\widehat{\mathcal{F}}_{0,\alpha}(\widehat{W})(z,\rho)|\right) &\le M \sup_{\widehat{\Omega}^s_\delta \times D(0,2\rho_s)} |z|^{3+\frac{1}{2}} \left(\frac{8}{|z|^{3+1}} + \frac{|z|^4}{|z|^8}\right)\left(|\widehat{W}|^2_{{}^s\widehat{\mathsf{H}}_\delta} + \rho_s\right) \\
&\le 9M \frac{|\widehat{W}|^2_{{}^s\widehat{\mathsf{H}}_\delta} + 2\rho_s}{\sqrt{\delta}} \le 18M \frac{|\widehat{W}|^2_{{}^s\widehat{\mathsf{H}}_\delta} + \rho_s}{\sqrt{\delta}}.
\end{aligned}$$

and

$$\begin{aligned}
\sup_{\widehat{\Omega}^s_\delta \times D(0,2\rho_s)} \left(|z|^{4+\frac{1}{2}} |\widehat{\mathcal{F}}_{0,\beta}(\widehat{W})(z,\rho)|\right) &\le M \sup_{\widehat{\Omega}^s_\delta \times D(0,2\rho_s)} |z|^{4+\frac{1}{2}} \left(\frac{24}{|z|^{4+1}} + \frac{4|z|^3}{|z|^8}\right)\left(|\widehat{W}|^2_{{}^s\widehat{\mathsf{H}}_\delta} + 2\rho_s\right) \\
&\le 28M \frac{|\widehat{W}|^2_{{}^s\widehat{\mathsf{H}}_\delta} + 2\rho_s}{\sqrt{\delta}} \le 56M \frac{|\widehat{W}|^2_{{}^s\widehat{\mathsf{H}}_\delta} + \rho_s}{\sqrt{\delta}}.
\end{aligned}$$

Step 2. Using lemma 7.H.1 we get

$$\left| \int_{[z,+\infty)} \left[\langle \widehat{\mathbf{r}}_+^*(s), \widehat{g}_0(\widehat{W}(s,\rho),s,\rho) \rangle_* \widehat{\mathbf{r}}_+(z) + \langle \widehat{\mathbf{r}}_-^*(s), \widehat{g}_0(\widehat{W}(s,\rho),s,\rho) \rangle_* \widehat{\mathbf{r}}_-(z) \right] ds \right|$$

$$\leq \mathcal{M}_{11} \int_{[z,+\infty)} \left(|\widehat{g}_{0,A}(\widehat{W}(s,\rho),s,\rho)| + |\widehat{g}_{0,B}(\widehat{W}(s,\rho),s,\rho)| \right) ds$$

$$\leq \mathcal{M}_{11} \int_0^{+\infty} \frac{|\widehat{W}|^2_{{}^s\widehat{\mathsf{H}}_\delta} + 2\rho_s}{|z+t|^6} dt$$

$$\leq \frac{\pi \mathcal{M}_{11}}{2} \left(1 + (\sqrt{2})^5 \right) \frac{|\widehat{W}|^2_{{}^s\widehat{\mathsf{H}}_\delta} + 2\rho_s}{|z|^5}.$$

Finally,

$$\sup_{\widehat{\Omega}^s_\delta \times D(0,2\rho_s)} \left| z^4 \widehat{\mathcal{F}}_{0,A}(\widehat{W})(z,\rho) \right| \leq \frac{\pi \mathcal{M}_{11}}{2} \left(1 + (\sqrt{2})^5 \right) \frac{|\widehat{W}|^2_{{}^s\widehat{\mathsf{H}}_\delta} + 2\rho_s}{\delta}$$

$$\leq \pi \mathcal{M}_{11} \left(1 + (\sqrt{2})^5 \right) \frac{|\widehat{W}|^2_{{}^s\widehat{\mathsf{H}}_\delta} + \rho_s}{\sqrt{\delta}}.$$

The same inequality holds $\widehat{\mathcal{F}}_{0,B}$.

Estimates (b) The proof is very similar to the one of (a), and is left to the reader. □

7.I Appendix. Proof of lemma 7.4.45

Lemma 7.I.1. *There exists $\mathcal{M}_{12}$ such that for every $\nu \in]0, \nu_6]$, $\rho \in [0, \rho_s[$ and every $\delta \in [\delta_0, 1/\nu^{\frac{1}{2}}[$,*

$$|\widehat{\mathcal{F}}_1^\Delta|_{{}^s\widehat{\mathsf{C}}^0_{\delta,\frac{1}{2},\nu}} \leq \mathcal{M}_{12} \nu^{\frac{1}{4}}.$$

Proof. We procceed in 3 steps.

Step 1. Estimates of $\widehat{\Delta}_s = (\widehat{\alpha}_{\widehat{\Delta}_s}, \widehat{\beta}_{\widehat{\Delta}_s}, \widehat{A}_{\widehat{\Delta}_s}, \widehat{B}_{\widehat{\Delta}_s})$.

Step 1.1. Estimates of $\widehat{\alpha}_{\widehat{\Delta}_s}$. The first component $\widehat{\alpha}_{\widehat{\Delta}_s}$ reads

$$\widehat{\alpha}_{\widehat{\Delta}_s}(\eta) = \left(\widehat{\alpha}_{\widehat{h}} + \widehat{\alpha}_{\widehat{v}_s} - \widehat{\alpha}_{\widehat{h}_0^+} - \widehat{\alpha}_{\widehat{W}^+_{s,0}} \right) \left(1/\nu^{\frac{1}{2}} + i\eta \right).$$

On one hand scaling (7.48), corollary 7.4.29 ensure that

$$\left|\widehat{\alpha}_{\widehat{v_s}}\left(\nu^{-\frac{1}{2}}+\mathrm{i}\eta\right)\right| = \nu^2\left|\alpha_{v_s}(\mathrm{i}\pi+\mathrm{i}\eta\nu+\nu^{\frac{1}{2}})\right| \le \nu^2\frac{2M_{\text{out}}\ \rho\ \nu^{\frac{3}{2}}}{\pi^3|\mathrm{i}\eta\nu+\nu^{\frac{1}{2}}|^3} \le \frac{2M_{\text{out}}\rho_s}{\pi^3}\ \nu^2.$$

On the other hand, recalling that $\delta \ge 1$, Proposition 7.4.34 gives

$$\left|\widehat{\alpha}_{\widehat{W}^+_{s,0}}\left(1/\nu^{\frac{1}{2}}+\mathrm{i}\eta\right)\right| \le \frac{4M_0^+\rho}{\sqrt{\delta}}\frac{1}{\left|1/\nu^{\frac{1}{2}}+\mathrm{i}\eta\right|^{3+\frac{1}{2}}} \le 4M_0^+\rho_s\ \nu^{\frac{1}{2}(3+\frac{1}{2})}.$$

Moreover,

$$\begin{aligned}\left(\widehat{\alpha}_{\widehat{h}}-\widehat{\alpha}_{\widehat{h}_0^+}\right)\left(1/\nu^{\frac{1}{2}}+\mathrm{i}\eta\right) &= \nu^2\alpha_h\left(\mathrm{i}\pi+\nu^{\frac{1}{2}}+\mathrm{i}\eta\nu\right)-\widehat{\alpha}_{\widehat{h}_0^+}\left(1/\nu^{\frac{1}{2}}+\mathrm{i}\eta\right)\\ &= \nu^2 f(\nu^{\frac{1}{2}}+\mathrm{i}\eta\nu)\end{aligned}$$

where $f(u) = -\sinh^{-2}(\frac{1}{2}u)+4u^{-2}$. Observing that

$$f(u) = \frac{1}{3}+\underset{u\to 0}{\mathcal{O}}(u^2), \qquad |\eta| \le 1/\nu^{\frac{1}{2}}, \qquad |\nu^{\frac{1}{2}}+\mathrm{i}\eta\nu| \le \sqrt{2}\nu^{\frac{1}{2}},$$

we obtain that there exists M such that for every $\nu \in]0,\nu_6]$,

$$\left|\widehat{\alpha}_{\widehat{h}}\left(1/\nu^{\frac{1}{2}}+\mathrm{i}\eta\right)-\widehat{\alpha}_{\widehat{h}_0^+}\left(1/\nu^{\frac{1}{2}}+\mathrm{i}\eta\right)\right| \le M\nu^2.$$

We conclude that there exists M_α such that for every $\nu \in]0,\nu_6]$ and every $\eta \in]-1/\nu^{\frac{1}{2}},-\delta[$,

$$|\widehat{\alpha}_{\widehat{\Delta}_s}(\eta)| \le M_\alpha\ \nu^{\frac{1}{2}(3+\frac{1}{2})}. \tag{7.102}$$

Step 1.2. Estimates of $\widehat{\beta}_{\widehat{\Delta}_s}$. Similarly, we prove that there exists M_β such that for every $\nu \in]0,\nu_6]$ and every $\eta \in]-1/\nu^{\frac{1}{2}},-\delta[$,

$$|\widehat{\beta}_{\widehat{\Delta}_s}(\eta)| \le M_\beta\ \nu^{\frac{1}{2}(4+\frac{1}{2})}. \tag{7.103}$$

Step 1.3. Estimates of $\widehat{A}_{\widehat{\Delta}_s}$. The third component $\widehat{A}_{\widehat{\Delta}_s}$ reads

$$\widehat{A}_{\widehat{\Delta}_s}(\eta) = \left(\widehat{A}_{\widehat{v_s}}-\widehat{A}_{\widehat{W}^+_{s,0}}\right)\left(1/\nu^{\frac{1}{2}}+\mathrm{i}\eta\right).$$

On one hand scaling (7.48),

$$\left|\widehat{A}_{\widehat{v_s}}\left(\nu^{-\frac{1}{2}}+\mathrm{i}\eta\right)\right| = \nu^2\left|A_{v_s}(\mathrm{i}\pi+\mathrm{i}\eta\nu+\nu^{\frac{1}{2}})\right| \le \nu^2\frac{2M_{\text{out}}\ \rho\ \nu^{\frac{3}{2}}}{\pi^3|\mathrm{i}\eta\nu+\nu^{\frac{1}{2}}|^3} \le \frac{2M_{\text{out}}\rho_s}{\pi^3}\ \nu^2.$$

On the other hand, recalling that $\delta \ge 1$ Proposition 7.4.34 gives

$$\left|\widehat{A}_{\widehat{W}^+_{s,0}}\left(1/\nu^{\frac{1}{2}}+\mathrm{i}\eta\right)\right| \le \frac{4M_0^+\rho}{\sqrt{\delta}}\frac{1}{\left|1/\nu^{\frac{1}{2}}+\mathrm{i}\eta\right|^4} \le 4M_0^+\rho_s\ \nu^2.$$

Grouping these two estimates, we conclude that there exists M_A such that for every $\nu \in]0, \nu_6]$ nd every $\eta \in]-1/\nu^{\frac{1}{4}}, -\delta[$,

$$|\widehat{A}_{\widehat{\Delta}_s}(\eta)| \leq M_A\, \nu^2. \tag{7.104}$$

Step 1.4. Estimates of $\widehat{\beta}_{\widehat{\Delta}_s}$. Similarly, we prove that there exists M_B such that for every $\nu \in]0, \nu_6]$ and every $\eta \in]-1/\nu^{\frac{1}{4}}, -\delta[$,

$$|\widehat{\beta}_{\widehat{\Delta}_s}(\eta)| \leq M_B\, \nu^2. \tag{7.105}$$

Step 2. Estimate of $\widehat{\mathcal{F}}^{\Delta}_{1,\alpha}$, $\widehat{\mathcal{F}}^{\Delta}_{1,\beta}$. Let us denote

$$(\widehat{\mathcal{F}}^{\Delta}_{1,\alpha}, \widehat{\mathcal{F}}^{\Delta}_{1,\beta}, 0, 0) = \langle \widehat{\mathbf{p}}^*(\tfrac{1}{\sqrt{\nu}}+\mathrm{i}\eta), \widehat{\Delta}_s(\eta)\rangle_* \, \widehat{\mathbf{p}}(\tau+\mathrm{i}\eta) + \langle \widehat{\mathbf{q}}^*(\tfrac{1}{\sqrt{\nu}}+\mathrm{i}\eta), \widehat{\Delta}_s(\eta)\rangle_* \, \widehat{\mathbf{q}}(\tau+\mathrm{i}\eta).$$

Using Lemma 7.4.39, 7.4.40 which give the explicit formula for $\widehat{\mathbf{p}}, \widehat{\mathbf{q}}, \widehat{\mathbf{p}}^*, \widehat{\mathbf{q}}^*$, we get

$$\begin{aligned}\widehat{\mathcal{F}}^{\Delta}_{1,\alpha}(\tau,\eta) = {} & \frac{1}{7}\Big[4(\tfrac{1}{\sqrt{\nu}}+\mathrm{i}\eta)^3 \widehat{\alpha}_{\widehat{\Delta}_s}(\eta) - (\tfrac{1}{\sqrt{\nu}}+\mathrm{i}\eta)^4 \widehat{\beta}_{\widehat{\Delta}_s}(\eta)\Big]\frac{1}{(\tau+\mathrm{i}\eta)^3} \\ & + \frac{1}{7}\Big[\frac{3}{(\frac{1}{\sqrt{\nu}}+\mathrm{i}\eta)^4}\widehat{\alpha}_{\widehat{\Delta}_s}(\eta) + \frac{1}{(\frac{1}{\sqrt{\nu}}+\mathrm{i}\eta)^3}\widehat{\beta}_{\widehat{\Delta}_s}(\eta)\Big](\tau+\mathrm{i}\eta)^4.\end{aligned}$$

and

$$\begin{aligned}\widehat{\mathcal{F}}^{\Delta}_{1,\beta}(\tau,\eta) = {} & \frac{-3}{7}\Big[4(\tfrac{1}{\sqrt{\nu}}+\mathrm{i}\eta)^3 \widehat{\alpha}_{\widehat{\Delta}_s}(\eta) - (\tfrac{1}{\sqrt{\nu}}+\mathrm{i}\eta)^4 \widehat{\beta}_{\widehat{\Delta}_s}(\eta)\Big]\frac{1}{(\tau+\mathrm{i}\eta)^4} \\ & + \frac{4}{7}\Big[\frac{3}{(\frac{1}{\sqrt{\nu}}+\mathrm{i}\eta)^4}\widehat{\alpha}_{\widehat{\Delta}_s}(\eta) + \frac{1}{(\frac{1}{\sqrt{\nu}}+\mathrm{i}\eta)^3}\widehat{\beta}_{\widehat{\Delta}_s}(\eta)\Big](\tau+\mathrm{i}\eta)^3.\end{aligned}$$

These explicit formula combined with the estimates of $\widehat{\alpha}_{\widehat{\Delta}_s}$ and $\widehat{\alpha}_{\widehat{\Delta}_s}$ gives

$$\sup_{(\tau,\eta)\in\widehat{\Sigma}^s_{\delta,\frac{1}{2},\nu}} \left(|\widehat{\mathcal{F}}^{\Delta}_{1,\alpha}||\tau+\mathrm{i}\eta|^3\right) \leq M'\nu^{\frac{1}{4}}, \qquad \sup_{(\tau,\eta)\in\widehat{\Sigma}^s_{\delta,\frac{1}{2},\nu}} \left(|\widehat{\mathcal{F}}^{\Delta}_{1,\beta}||\tau+\mathrm{i}\eta|^4\right) \leq M'\nu^{\frac{1}{4}}$$

since $|(\frac{1}{\sqrt{\nu}}+\mathrm{i}\eta)| \geq \frac{1}{\sqrt{\nu}}$ and $|\tau+\mathrm{i}\eta| \leq \frac{\sqrt{2}}{\sqrt{\nu}}$ for $(\tau,\eta) \in \widehat{\Sigma}^s_{\delta,\frac{1}{2},\nu}$.

Step 3. Estimate of $\widehat{\mathcal{F}}^{\Delta}_{1,A}$, $\widehat{\mathcal{F}}^{\Delta}_{1,B}$. Let us denote

$$\begin{aligned}&(\widehat{\mathcal{F}}^{\Delta}_{1,\alpha}, \widehat{\mathcal{F}}^{\Delta}_{1,\beta}, 0, 0) = \\ &\quad \langle \widehat{\mathbf{r}}^*_+(\tfrac{1}{\sqrt{\nu}}+\mathrm{i}\eta), \widehat{\Delta}_s(\eta)\rangle_* \, \widehat{\mathbf{r}}_+(\tau+\mathrm{i}\eta) + \langle \widehat{\mathbf{r}}^*_-(\tfrac{1}{\sqrt{\nu}}+\mathrm{i}\eta), \widehat{\Delta}_s(\eta)\rangle_* \, \widehat{\mathbf{r}}_-(\tau+\mathrm{i}\eta).\end{aligned}$$

Using Lemma 7.4.39, 7.4.40 which give the explicit formula for $\widehat{\mathbf{r}}_\pm, \widehat{\mathbf{r}}^*_\pm$, we get

$$\widehat{\mathcal{F}}^{\Delta}_{1,A} = \widehat{A}_{\widehat{\Delta}_s}\cos\big[\widehat{\psi}(\tau+\mathrm{i}\eta) - \widehat{\psi}(\tfrac{1}{\sqrt{\nu}}+\mathrm{i}\eta)\big] - \widehat{B}_{\widehat{\Delta}_s}\sin\big[\widehat{\psi}(\tau+\mathrm{i}\eta) - \widehat{\psi}(\tfrac{1}{\sqrt{\nu}}+\mathrm{i}\eta)\big]$$

$$\widehat{\mathcal{F}}^{\Delta}_{1,B} = \widehat{A}_{\widehat{\Delta}_s} \sin[\widehat{\psi}(\tau+\mathrm{i}\eta) - \widehat{\psi}(\tfrac{1}{\sqrt{\nu}}+\mathrm{i}\eta)] + \widehat{B}_{\widehat{\Delta}_s} \cos[\widehat{\psi}(\tau+\mathrm{i}\eta) - \widehat{\psi}(\tfrac{1}{\sqrt{\nu}}+\mathrm{i}\eta)]$$

Then, observing that

$$\widehat{\psi}(\tau+\mathrm{i}\eta) - \widehat{\psi}(\tfrac{1}{\sqrt{\nu}}+\mathrm{i}\eta) = \omega(\tau-\tfrac{1}{\sqrt{\nu}}) + \frac{b}{\tau+\mathrm{i}\eta} - \frac{b}{\frac{1}{\sqrt{\nu}}+\mathrm{i}\eta},$$

and

$$|\tau+\mathrm{i}\eta| \geq \delta \geq 1, \qquad |\mathrm{i}\eta + \tfrac{1}{\sqrt{\nu}}| \geq 1$$

we get

$$|\mathcal{I}m\left(\widehat{\psi}(\tau+\mathrm{i}\eta) - \widehat{\psi}(\tfrac{1}{\sqrt{\nu}}+\mathrm{i}\eta)\right)| \leq |b| \left(\frac{1}{|\tau+\mathrm{i}\eta|} + \frac{1}{|\frac{1}{\sqrt{\nu}}+\mathrm{i}\eta|}\right) \leq 2|b|$$

and

$$|\widehat{\mathcal{F}}^{\Delta}_{1,A}| \leq \mathrm{e}^{2|b|}(|\widehat{A}_{\widehat{\Delta}_s}| + |\widehat{B}_{\widehat{\Delta}_s}|), \qquad |\widehat{\mathcal{F}}^{\Delta}_{1,B}| \leq \mathrm{e}^{2|b|}(|\widehat{A}_{\widehat{\Delta}_s}| + |\widehat{B}_{\widehat{\Delta}_s}|). \quad (7.106)$$

Finally (7.104), (7.105), (7.106), give

$$\sup_{(\tau,\eta)\in\widehat{\Sigma}^s_{\delta,\frac{1}{2},\nu}} \left(|\widehat{\mathcal{F}}^{\Delta}_{1,A}||\tau+\mathrm{i}\eta|^3\right) \leq M''\nu^{\frac{1}{4}}, \qquad \sup_{(\tau,\eta)\in\widehat{\Sigma}^s_{\delta,\frac{1}{2},\nu}} \left(|\widehat{\mathcal{F}}^{\Delta}_{1,B}||\tau+\mathrm{i}\eta|^3\right) \leq M''\nu^{\frac{1}{4}}. \square$$

Lemma 7.I.2. *There exist $\delta_2 \geq 1$ such that for every $\delta \geq \delta_2$, $\rho \in [0, \rho_s[$, $(\tau,\eta) \in \widehat{\Sigma}^s_{\delta,\frac{1}{2},\nu}$ and every $\widehat{w} \in \mathsf{B}^s\widehat{\mathsf{C}}^0_{\delta,\frac{1}{2},\nu}[1]$, $\widehat{g}_1(\widehat{w},\tau,\eta,\rho)$ is well defined. Moreover there exists $\mathcal{M}_{13}$ such that for every $\delta \geq \delta_2$, $\rho \in [0,\rho_s[$, $(\tau,\eta) \in \widehat{\Sigma}^s_{\delta,\frac{1}{2},\nu}$ and every $\widehat{w}, \widehat{w}' \in \mathsf{B}^s\widehat{\mathsf{C}}^0_{\delta,\frac{1}{2},\nu}[1]$,*

(a) $$|\widehat{g}_1(\widehat{w},\tau,\eta,\rho)| \leq \mathcal{M}_{13}\left[\left(\frac{1}{\delta}|\widehat{w}|_{^s\widehat{\mathsf{C}}^0_{\delta,\frac{1}{2},\nu}} + |\widehat{w}|^2_{^s\widehat{\mathsf{C}}^0_{\delta,\frac{1}{2},\nu}}\right)\frac{1}{|\tau+\mathrm{i}\eta|^6} + \frac{\nu^2}{|\tau+\mathrm{i}\eta|^2}\right],$$

(b) $$|\widehat{g}_1(\widehat{w},\tau,\eta,\rho) - \widehat{g}_1(\widehat{w}',\tau,\eta,\rho)| \leq$$
$$\mathcal{M}_{13}\left[\left(|\widehat{w}|_{^s\widehat{\mathsf{C}}^0_{\delta,\frac{1}{2},\nu}} + |\widehat{w}'|_{^s\widehat{\mathsf{C}}^0_{\delta,\frac{1}{2},\nu}} + \frac{1}{\delta}\right)\frac{1}{|\tau+\mathrm{i}\eta|^6} + \frac{\nu^2}{|\tau+\mathrm{i}\eta|^3}\right]|\widehat{w}-\widehat{w}'|_{^s\widehat{\mathsf{C}}^0_{\delta,\frac{1}{2},\nu}}.$$

Proof. We split $\widehat{g}_1$ in four parts as follows

$$\begin{aligned}\widehat{g}_1(\widehat{w},\tau,\eta) = \quad & \widehat{Q}^F_1(\widehat{w},\tau,\eta) + \rho\widehat{Q}^R_1(\widehat{w},\tau,\eta) \\ & +\widehat{N}_1(\widehat{h}^+_0 + \widehat{W}^+_{s,0} + \widehat{w}) + \rho\widehat{R}_1(\widehat{h}^+_0 + \widehat{W}^+_{s,0} + \widehat{w})\end{aligned}$$

where

$$\begin{aligned}\widehat{Q}^F_1(\widehat{w},\tau,\eta) &= \widehat{N}_0(\widehat{h}^+_0 + \widehat{W}^+_{s,0} + \widehat{w}) - \widehat{N}_0(\widehat{h}^+_0 + \widehat{W}^+_{s,0}) - D\widehat{N}_0(\widehat{h}^+_0).\widehat{w}, \\ \widehat{Q}^R_1(\widehat{w},\tau,\eta) &= \widehat{R}_0(\widehat{h}^+_0 + \widehat{W}^+_{s,0} + \widehat{w}) - \widehat{R}_0(\widehat{h}^+_0 + \widehat{W}^+_{s,0}).\end{aligned}$$

We proceed in four steps: we compute the estimates of each part separately.

Step 1. Using (7.53) we compute explicitly $\widehat{Q}_1^F$ which reads

$$\widehat{Q}_1^F(\widehat{w},\tau,\eta) = \begin{pmatrix} 0 \\ -3\widehat{\alpha}_{\widehat{W}^+_{s,0}}\widehat{\alpha} - 2c\left(\widehat{A}_{\widehat{W}^+_{s,0}}\widehat{A}+\widehat{B}_{\widehat{W}^+_{s,0}}\widehat{B}\right) - \frac{3}{2}\widehat{\alpha}^2 - c\left(\widehat{A}^2+\widehat{B}^2\right) \\ -b\left(\widehat{B}_{\widehat{W}^+_{s,0}}\widehat{\alpha} + \widehat{\alpha}_{\widehat{W}^+_{s,0}}\widehat{B}\right) - b\widehat{\alpha}\widehat{B} \\ b\left(\widehat{A}_{\widehat{W}^+_{s,0}}\widehat{\alpha} + \widehat{\alpha}_{\widehat{W}^+_{s,0}}\widehat{A}\right) + b\widehat{\alpha}\widehat{A} \end{pmatrix}.$$

Recalling that $|\tau + i\eta| \geq |\eta| \geq \delta \geq 1$ and using Proposition 7.4.34 and Definition 7.4.44 we get

$$\left|\left[\widehat{Q}_1^F(\widehat{w},\tau,\eta)\right]_\beta\right| \leq M\left(\frac{|\widehat{w}|_{{}^s\widehat{C}^0_{\delta,\frac{1}{2},\nu}}}{\delta|\tau+i\eta|^6} + \frac{|\widehat{w}|^2_{{}^s\widehat{C}^0_{\delta,\frac{1}{2},\nu}}}{|\tau+i\eta|^6}\right),$$

$$\left|\left[\widehat{Q}_1^F(\widehat{w},\tau,\eta)\right]_{A\,\mathrm{or}\,B}\right| \leq M\left(\frac{|\widehat{w}|_{{}^s\widehat{C}^0_{\delta,\frac{1}{2},\nu}}}{\delta|\tau+i\eta|^6} + \frac{|\widehat{w}|^2_{{}^s\widehat{C}^0_{\delta,\frac{1}{2},\nu}}}{|\tau+i\eta|^6}\right),$$

$$\left|\left[\widehat{Q}_1^F(\widehat{w},\tau,\eta) - \widehat{Q}_1^F(\widehat{w}',\tau,\eta)\right]_\beta\right| \leq$$
$$M\left(\frac{1}{\delta|\tau+i\eta|^6} + \frac{|\widehat{w}|_{{}^s\widehat{C}^0_{\delta,\frac{1}{2},\nu}} + |\widehat{w}'|_{{}^s\widehat{C}^0_{\delta,\frac{1}{2},\nu}}}{|\tau+i\eta|^6}\right)|\widehat{w}-\widehat{w}'|_{{}^s\widehat{C}^0_{\delta,\frac{1}{2},\nu}},$$

$$\left|\left[\widehat{Q}_1^F(\widehat{w},\tau,\eta) - \widehat{Q}_1^F(\widehat{w}',\tau,\eta)\right]_{A\,\mathrm{or}\,B}\right| \leq$$
$$M\left(\frac{1}{\delta|\tau+i\eta|^6} + \frac{|\widehat{w}|_{{}^s\widehat{C}^0_{\delta,\frac{1}{2},\nu}} + |\widehat{w}'|_{{}^s\widehat{C}^0_{\delta,\frac{1}{2},\nu}}}{|\tau+i\eta|^6}\right)|\widehat{w}-\widehat{w}'|_{{}^s\widehat{C}^0_{\delta,\frac{1}{2},\nu}}.$$

Step 2 : Estimates of $\widehat{Q}_1^R$. We start with the computation of an estimate of

$$\widehat{R}_0(\widehat{h}_0^+ + \widehat{W}^+_{s,0} + \widehat{w}) - \widehat{R}_0(\widehat{h}_0^+ + \widehat{W}^+_{s,0} + \widehat{w}') = \sum_{|\overline{m}|\geq 3} a_{\overline{m},0}\,\Delta_{\overline{m},0}$$

where

$$\Delta_{\overline{m},0} := a_{\overline{m},0}\left[(\widehat{h}_0^+ + \widehat{W}^+_{s,0} + \widehat{w})^{\overline{m}} - (\widehat{h}_0^+ + \widehat{W}^+_{s,0} + \widehat{w}')^{\overline{m}}\right].$$

The difference $\left[(\widehat{h}_0^+ + \widehat{W}^+_{s,0} + \widehat{w})^{\overline{m}} - (\widehat{h}_0^+ + \widehat{W}^+_{s,0} + \widehat{w}')^{\overline{m}}\right]$ reads

$$
\begin{aligned}
&\left((\widehat{h}_0^+ + \widehat{W}_{s,0}^+ + \widehat{w})^{\overline{m}} - (\widehat{h}_0^+ + \widehat{W}_{s,0}^+ + \widehat{w}')^{\overline{m}}\right) = \\
&\Big[(\widehat{\alpha}_{\widehat{h}_0^+}+\widehat{\alpha}_{\widehat{W}_{s,0}^+}+\widehat{\alpha})^{m_\alpha} - (\widehat{\alpha}_{\widehat{h}_0^+}+\widehat{\alpha}_{\widehat{W}_{s,0}^+}+\widehat{\alpha}')^{m_\alpha}\Big](\widehat{\beta}_{\widehat{h}_0^+}+\widehat{\beta}_{\widehat{W}_{s,0}^+}+\widehat{\beta})^{m_\beta}(\widehat{A}_{\widehat{W}_{s,0}^+}+\widehat{A})^{m_A}(\widehat{B}_{\widehat{W}_{s,0}^+}+\widehat{B})^{m_B} \\
&+(\widehat{\alpha}_{\widehat{h}_0^+}+\widehat{\alpha}_{\widehat{W}_{s,0}^+}+\widehat{\alpha}')^{m_\alpha}\Big[(\widehat{\beta}_{\widehat{h}_0^+}+\widehat{\beta}_{\widehat{W}_{s,0}^+}+\widehat{\beta})^{m_\beta} - (\widehat{\beta}_{\widehat{h}_0^+}+\widehat{\beta}_{\widehat{W}_{s,0}^+}+\widehat{\beta}')^{m_\beta}\Big](\widehat{A}_{\widehat{W}_{s,0}^+}+\widehat{A})^{m_A}(\widehat{B}_{\widehat{W}_{s,0}^+}+\widehat{B})^{m_B} \\
&+(\widehat{\alpha}_{\widehat{h}_0^+}+\widehat{\alpha}_{\widehat{W}_{s,0}^+}+\widehat{\alpha}')^{m_\alpha}(\widehat{\beta}_{\widehat{h}_0^+}+\widehat{\beta}_{\widehat{W}_{s,0}^+}+\widehat{\beta}')^{m_\beta}\Big[(\widehat{A}_{\widehat{W}_{s,0}^+}+\widehat{A})^{m_A} - (\widehat{A}_{\widehat{W}_{s,0}^+}+\widehat{A}')^{m_A}\Big](\widehat{B}_{\widehat{W}_{s,0}^+}+\widehat{B})^{m_B} \\
&+(\widehat{\alpha}_{\widehat{h}_0^+}+\widehat{\alpha}_{\widehat{W}_{s,0}^+}+\widehat{\alpha}')^{m_\alpha}(\widehat{\beta}_{\widehat{h}_0^+}+\widehat{\beta}_{\widehat{W}_{s,0}^+}+\widehat{\beta}')^{m_\beta}(\widehat{A}_{\widehat{W}_{s,0}^+}+\widehat{A})^{m_A}\Big[(\widehat{B}_{\widehat{W}_{s,0}^+}+\widehat{B})^{m_B} - (\widehat{B}_{\widehat{W}_{s,0}^+}+\widehat{B}')^{m_B}\Big].
\end{aligned}
$$

Estimates (7.7) implies that

$$|a_{\overline{m},0}| \leq \frac{M_R}{r_\alpha^{m_\alpha} r_\beta^{m_\beta} r_A^{m_A} r_B^{m_B}}.$$

Let us denote

$$\Theta := \max\left(\frac{1+\frac{4M_0^+\rho_s}{\sqrt{\delta_1}}+4}{r_\alpha}, \frac{1+\frac{4M_0^+\rho_s}{\sqrt{\delta_1}}+8}{r_\beta}, \frac{1+\frac{4M_0^+\rho_s}{\sqrt{\delta_1}}}{r_A}, \frac{1+\frac{4M_0^+\rho_s}{\sqrt{\delta_1}}}{r_B}\right).$$

For every $\delta \geq \delta_1$, $\widehat{w} \in \mathsf{B}^s\widehat{\mathsf{C}}^0_{\delta,\frac{1}{2},\nu}[1]$ and every $(\tau,\eta) \in \widehat{\Sigma}^s_{\delta,\frac{1}{2},\nu}$, observing that $|\tau + \mathrm{i}\eta| \geq \delta \geq 1$, we get

$$
\begin{aligned}
\left|\frac{\widehat{\alpha}_{\widehat{h}_0^+}+\widehat{\alpha}_{\widehat{W}_{s,0}^+}+\widehat{\alpha}}{r_\alpha}\right|^{m_\alpha} &\leq \frac{1}{r_\alpha^{m_\alpha}}\left(\frac{4}{|\tau+\mathrm{i}\eta|^2}+\frac{4M_0^+\rho_s}{\sqrt{\delta}|\tau+\mathrm{i}\eta|^{3+\frac{1}{2}}}+\frac{|\widehat{w}|_{^s\widehat{\mathsf{C}}^0_{\delta,\frac{1}{2},\nu}}}{|\tau+\mathrm{i}\eta|^3}\right)^{m_\alpha} \\
&\leq \frac{\Theta^{m_\alpha}}{|\tau+\mathrm{i}\eta|^{2m_\alpha}}, \\
\left|\frac{\widehat{\beta}_{\widehat{h}_0^+}+\widehat{\beta}_{\widehat{W}_{s,0}^+}+\widehat{\beta}}{r_\beta}\right|^{m_\beta} &\leq \frac{\Theta^{m_\beta}}{|\tau+\mathrm{i}\eta|^{3m_\beta}}, \\
\left|\frac{\widehat{A}+\widehat{A}_{\widehat{W}_{s,0}^+}}{r_A}\right|^{m_A} &\leq \frac{1}{r_A^{m_a}}\left(\frac{|\widehat{w}|_{^s\widehat{\mathsf{C}}^0_{\delta,\frac{1}{2},\nu}}}{|\tau+\mathrm{i}\eta|^3}+\frac{4M_0^+\rho_s}{\sqrt{\delta}|\tau+\mathrm{i}\eta|^4}\right)^{m_A} \\
&\leq \frac{\Theta^{m_A}}{|\tau+\mathrm{i}\eta|^{3m_A}}, \\
\left|\frac{\widehat{B}+\widehat{B}_{\widehat{W}_{s,0}^+}}{r_B}\right|^{m_B} &\leq \frac{\Theta^{m_B}}{|\tau+\mathrm{i}\eta|^{3m_B}}.
\end{aligned}
\tag{7.107}
$$

Using (7.107) we get

$$\frac{\left|\left(\widehat{\alpha}_{\widehat{h}_0^+}+\widehat{\alpha}_{\widehat{W}_{s,0}^+}+\widehat{\alpha}\right)^{m_\alpha}-\left(\widehat{\alpha}_{\widehat{h}_0^+}+\widehat{\alpha}_{\widehat{W}_{s,0}^+}+\widehat{\alpha}'\right)^{m_\alpha}\right|}{r_\alpha^{m_\alpha}}$$
$$=\frac{|\widehat{\alpha}-\widehat{\alpha}'|}{r_\alpha^{m_\alpha}}\left|\sum_{k=0}^{m_\alpha-1}\left(\widehat{\alpha}_{\widehat{h}_0^+}+\widehat{\alpha}_{\widehat{W}_{s,0}^+}+\widehat{\alpha}\right)^{k}\left(\widehat{\alpha}_{\widehat{h}_0^+}+\widehat{\alpha}_{\widehat{W}_{s,0}^+}+\widehat{\alpha}'\right)^{m_\alpha-1-k}\right|$$
$$\le\frac{|\widehat{\alpha}-\widehat{\alpha}'|}{r_\alpha}\,m_\alpha\left(\frac{\Theta}{|\tau+\mathrm{i}\eta|^2}\right)^{m_\alpha-1},$$
$$\le\frac{m_\alpha\Theta^{m_\alpha-1}}{r_\alpha}\,\frac{|\widehat{w}-\widehat{w}'|_{{}_s\widehat{C}^0_{\delta,\frac12,\nu}}}{|\tau+\mathrm{i}\eta|^{2(m_\alpha-1)+3}}.$$

$$\frac{\left|\left(\widehat{\beta}_{\widehat{h}_0^+}+\widehat{\beta}_{\widehat{W}_{s,0}^+}+\widehat{\beta}\right)^{m_\beta}-\left(\widehat{\beta}_{\widehat{h}_0^+}+\widehat{\beta}_{\widehat{W}_{s,0}^+}+\widehat{\beta}'\right)^{m_\beta}\right|}{r_\beta^{m_\beta}}\le\frac{m_\alpha\Theta^{m_\alpha-1}}{r_\alpha}\,\frac{|\widehat{w}-\widehat{w}'|_{{}_s\widehat{C}^0_{\delta,\frac12,\nu}}}{|\tau+\mathrm{i}\eta|^{3(m_\alpha-1)+4}}$$

$$\frac{\left|(\widehat{A}_{\widehat{W}_{s,0}^+}+\widehat{A})^{m_A}-(\widehat{A}_{\widehat{W}_{s,0}^+}+\widehat{A}')^{m_A}\right|}{r_A^{m_A}}\le\frac{m_A\Theta^{m_A-1}}{r_A}\,\frac{|\widehat{w}-\widehat{w}'|_{{}_s\widehat{C}^0_{\delta,\frac12,\nu}}}{|\tau+\mathrm{i}\eta|^{3m_A}},$$

$$\frac{\left|(\widehat{B}_{\widehat{W}_{s,0}^+}+\widehat{B})^{m_B}-(\widehat{B}_{\widehat{W}_{s,0}^+}+\widehat{B}')^{m_B}\right|}{r_B^{m_B}}\le\frac{m_B\Theta^{m_B-1}}{r_B}\,\frac{|\widehat{w}-\widehat{w}'|_{{}_s\widehat{C}^0_{\delta,\frac12,\nu}}}{|\tau+\mathrm{i}\eta|^{3m_B}},$$

Denoting by $M=\max\left(r_\alpha^{-1},r_\beta^{-1},r_A^{-1},r_A^{-1}\right)$ we obtain

$$|\Delta_{\overline{m},0}|\le\frac{M_RM\Theta^{|\overline{m}|-1}|\widehat{w}-\widehat{w}'|_{{}_s\widehat{C}^0_{\delta,\frac12,\nu}}}{|\tau+\mathrm{i}\eta|^{2m_\alpha+3m_\beta+2(m_A+m_B)}}\left[\frac{m_\alpha+m_\beta}{|\tau+\mathrm{i}\eta|^1}+\frac{m_A}{|\tau+\mathrm{i}\eta|^{m_A}}+\frac{m_B}{|\tau+\mathrm{i}\eta|^{m_B}}\right]$$
$$\le M_RM\,|\overline{m}|\,\frac{\Theta^{|\overline{m}|-1}}{|\tau+\mathrm{i}\eta|^{2|\overline{m}|+1}}\,|\widehat{w}-\widehat{w}'|_{{}_s\widehat{C}^0_{\delta,\frac12,\nu}}$$

Choosing δ_2 such that $\delta_2:=\max\left(\sqrt{2\Theta},\delta_1\right)$, we get for $|\overline{m}|\ge 3$,

$$|\Delta_{\overline{m},0}|\le\frac{M_RM\Theta^2}{|\tau+\mathrm{i}\eta|^7}\,|\overline{m}|\,\left(\tfrac12\right)^{|\overline{m}|-3}\,|\widehat{w}-\widehat{w}'|_{{}_s\widehat{C}^0_{\delta,\frac12,\nu}}.$$

Summing these estimates, we obtain

$$
\begin{aligned}
|\widehat{R}_0(\widehat{h}_0^+ + \widehat{W}_{s,0}^+ + \widehat{w}) - \widehat{R}_0(\widehat{h}_0^+ + \widehat{W}_{s,0}^+ + \widehat{w}')| & \\
&\leq \frac{M_R M \Theta^2 |\widehat{w} - \widehat{w}'|_{{}^s\widehat{C}^0_{\delta,\frac{1}{2},\nu}}}{|\tau + i\eta|^7} \sum_{|\overline{m}|\geq 3} \left(\frac{1}{2}\right)^{|\overline{m}|-3} |\overline{m}| \\
&\leq 4 \frac{M_R M \Theta^2 |\widehat{w} - \widehat{w}'|_{{}^s\widehat{C}^0_{\delta,\frac{1}{2},\nu}}}{|\tau + i\eta|^7} \sum_{|\overline{m}|\geq 0} \left(\frac{1}{2}\right)^{|\overline{m}|-1} |\overline{m}| \\
&\leq 2^9 M_R M \Theta^2 \frac{|\widehat{w} - \widehat{w}'|_{{}^s\widehat{C}^0_{\delta,\frac{1}{2},\nu}}}{|\tau + i\eta|^7}.
\end{aligned}
\tag{7.108}
$$

Finally, recalling that $|\tau + i\eta| \geq \delta$ we get

$$
\begin{aligned}
|\widehat{Q}_1^R(\widehat{w}, \tau, \eta)| &\leq 2^9 M_R M \Theta^2 \frac{|\widehat{w}|_{{}^s\widehat{C}^0_{\delta,\frac{1}{2},\nu}}}{\delta |\tau + i\eta|^6} \\
|\widehat{Q}_1^R(\widehat{w}, \tau, \eta) - \widehat{Q}_1^R(\widehat{w}', \tau, \eta)| &\leq 2^9 M_R M \Theta^2 \frac{|\widehat{w} - \widehat{w}'|_{{}^s\widehat{C}^0_{\delta,\frac{1}{2},\nu}}}{\delta |\tau + i\eta|^6}.
\end{aligned}
\tag{7.109}
$$

Step 3: Estimates of $\widehat{N}_1$. Using the explicit expression of $\widehat{N}_1$ given by (7.53) and estimates (7.107) we get

$$
\begin{aligned}
\left| \left[\widehat{N}_1(\widehat{h}_0^+ + \widehat{W}_{s,0}^+ + \widehat{w}) \right]_\beta \right| &\leq M' \frac{\nu^2}{|\tau + i\eta|^2}, \\
\left| \left[\widehat{N}_1(\widehat{h}_0^+ + \widehat{W}_{s,0}^+ + \widehat{w}) \right]_{A \text{or} B} \right| &\leq M' \frac{\nu^2}{|\tau + i\eta|^3}, \\
\left| \left[\widehat{N}_1(\widehat{h}_0^+ + \widehat{W}_{s,0}^+ + \widehat{w}) - \widehat{N}_1(\widehat{h}_0^+ + \widehat{W}_{s,0}^+ + \widehat{w}') \right]_\beta \right| &\leq \frac{M' \nu^2}{|\tau + i\eta|^3} |\widehat{w} - \widehat{w}'|_{{}^s\widehat{C}^0_{\delta,\frac{1}{2},\nu}}, \\
\left| \left[\widehat{N}_1(\widehat{h}_0^+ + \widehat{W}_{s,0}^+ + \widehat{w}) - \widehat{N}_1(\widehat{h}_0^+ + \widehat{W}_{s,0}^+ + \widehat{w}') \right]_{A \text{or} B} \right| &\leq \frac{M' \nu^2}{|\tau + i\eta|^3} |\widehat{w} - \widehat{w}'|_{{}^s\widehat{C}^0_{\delta,\frac{1}{2},\nu}}.
\end{aligned}
$$

Step 4: Estimates of $\widehat{R}_1(\widehat{h}_0^+ + \widehat{W}_{s,0}^+ + \widehat{w}', \nu)$.

$$
\widehat{R}_1(\widehat{h}_0^+ + \widehat{W}_{s,0}^+ + \widehat{w}) = \sum_{\substack{|\overline{m}|\neq 0,\ k\neq 0 \\ |\overline{m}|+k\geq 3}} \Delta_{\overline{m},k}
$$

with

$$
\Delta_{\overline{m},k} = a_{\overline{m},k} (\widehat{h}_0^+ + \widehat{W}_{s,0}^+ + \widehat{w})^{\overline{m}} \nu^{2k}.
$$

Computations made in Step 2 ensure that

$$
|\Delta_{\overline{m},k}| \leq \frac{M_R \Theta^{|\overline{m}|}}{|\tau + i\eta|^{2m_\alpha + 3m_\beta + 3(m_A + m_B)}} \left(\frac{\nu}{\nu_0}\right)^{2k}.
$$

Denoting by $\sigma = (\nu_6/\nu_0)^2 < 1$ and recalling δ_2 has been chosen so that $\Theta/\delta_2^2 \le \frac{1}{2}$, we obtain that for $k \ge 1$ and $|\overline{m}| \ge 1$

$$\begin{aligned} |\Delta_{\overline{m},k}| &\le \frac{M_R\Theta}{\nu_0^2} \frac{\nu^2}{|\tau+\mathrm{i}\eta|^2} \left(\frac{\Theta}{|\tau+\mathrm{i}\eta|^2}\right)^{|\overline{m}|-1} \sigma^{k-1} \\ &\le \frac{2M_R\Theta}{\nu_0^2} \frac{\nu^2}{|\tau+\mathrm{i}\eta|^2} \left(\frac{1}{2}\right)^{|\overline{m}|} \sigma^{k-1}. \end{aligned}$$

Hence,

$$\left|\widehat{R}_1(\widehat{h}_0^+ + \widehat{W}_{\mathrm{s},0}^+ + \widehat{w})\right| \le \frac{2^5 M_R\Theta}{\nu_0^2(1-\sigma)} \frac{\nu^2}{|\tau+\mathrm{i}\eta|^2}. \tag{7.110}$$

Similarly, we prove that

$$\left|\widehat{R}_1(\widehat{h}_0^+ + \widehat{W}_{\mathrm{s},0}^+ + \widehat{w}) - \widehat{R}_1(\widehat{h}_0^+ + \widehat{W}_{\mathrm{s},0}^+ + \widehat{w}')\right| \le \frac{2^9 M_R M}{\nu_0^2(1-\sigma)} \frac{\nu^2|\widehat{w}-\widehat{w}'|_{{}^{\mathrm{s}}\widehat{\mathcal{C}}^0_{\delta,\frac{1}{2},\nu}}}{|\tau+\mathrm{i}\eta|^3}. \square$$

We need two technical lemmas, before starting the proof of Lemma 7.4.45.

Lemma 7.I.3. *For every $\delta \ge 1$, $(\tau,\eta) \in \widehat{\Sigma}^{\mathrm{s}}_{\delta,\frac{1}{2},\nu}$ and every $m \ge 2$*

$$\mathcal{K}_m(\tau,\eta) := \int_\tau^{1/\nu^{\frac{1}{2}}} \frac{ds}{|s+\mathrm{i}\eta|^m} \le \frac{\pi}{2}\left(\frac{1+(\sqrt{2})^{m-1}}{|\tau+\mathrm{i}\eta|^{m-1}}\right).$$

Proof.

$$|\mathcal{K}_m(\tau,\eta)| \le \int_\tau^{+\infty} \frac{ds}{|s+\mathrm{i}\eta|^m} \le \int_0^{+\infty} \frac{dt}{|t+(\tau+\mathrm{i}\eta)|^m}$$

Then, since $(\tau,\eta) \in \widehat{\Sigma}^{\mathrm{s}}_{\delta,\frac{1}{2},\nu} \subset \widehat{\Omega}^{\mathrm{s}}_\delta$, Lemma 7.H.2 gives the result. □

Lemma 7.I.4. *There exists $\mathcal{M}_{14}$ such that for every $\delta \ge 1$, $\nu \in]0,\nu_6]$ and every function*

$$\begin{aligned} \widehat{f}:\ \widehat{\Sigma}^{\mathrm{s}}_{\delta,\frac{1}{2},\nu} &\longrightarrow \mathbb{C}^4 \\ (\tau,\eta) &\longrightarrow (.,.,\widehat{f}_A(\tau,\eta),\widehat{f}_B(\tau,\eta)) \end{aligned}$$

satisfying $\sup_{\widehat{\Sigma}^{\mathrm{s}}_{\delta,\frac{1}{2},\nu}} |z^2\widehat{f}_A| < +\infty$, $\sup_{\widehat{\Sigma}^{\mathrm{s}}_{\delta,\frac{1}{2},\nu}} |z^2\widehat{f}_B| < +\infty$ *the inequality*

$$\left|\int_\tau^{1/\nu^{\frac12}} \left[\langle \widehat{\mathbf{r}}_+^*(s+i\eta), \widehat{f}(s,\eta)\rangle_* \widehat{\mathbf{r}}_+(\tau+i\eta) + \langle \widehat{\mathbf{r}}_-^*(s+i\eta), \widehat{f}(s,\eta)\rangle_* \widehat{\mathbf{r}}_-(\tau+i\eta)\right] ds\right|$$

$$\leq \mathcal{M}_{14}\int_\tau^{1/\nu^{\frac12}} (|\widehat{f}_A(s,\eta)| + |\widehat{f}_B(s,\eta)|)\, ds$$

holds for every $(\tau,\eta) \in \widehat{\Sigma}^{\mathrm{s}}_{\delta,\frac12,\nu}$.

The proof of this lemma is very similar to the one of Lemma 7.3.29. So the details are left to the reader.

Proof of lemma 7.4.45. Lemma 7.I.1 gives the estimate of $|\widehat{\mathcal{F}}_1^{\Delta}|_{{}_{\mathrm{s}}\widehat{\mathrm{C}}^0_{\delta,\frac12,\nu}}$. So must now bound $\widehat{\mathcal{F}}_{11}(\widehat{w})$. We proceed in several steps.

Step 1.1: Computation of the two first components. The two first components are given by

$$\begin{aligned}(\widehat{\mathcal{F}}_{11,\alpha}, \widehat{\mathcal{F}}_{11,\beta}, 0, 0) = & -\int_\tau^{1/\nu^{\frac12}} \langle \widehat{\mathbf{p}}^*(s), \widehat{g}_1(\widehat{w}, s, \eta)\rangle_* ds\; \widehat{\mathbf{p}}(\tau+i\eta) \\ & -\int_\tau^{1/\nu^{\frac12}} \langle \widehat{\mathbf{q}}^*(s), \widehat{g}_1(\widehat{w}, s, \eta)\rangle_* ds\; \widehat{\mathbf{q}}(\tau+i\eta).\end{aligned}$$

Using the explicit expressions of $\widehat{\mathbf{p}}$, $\widehat{\mathbf{q}}$, $\widehat{\mathbf{p}}^*$, $\widehat{\mathbf{q}}^*$ given by lemmas 7.4.39, 7.4.40 we get

$$\widehat{\mathcal{F}}_{11,\alpha} = J_\alpha - K_\alpha, \qquad \widehat{\mathcal{F}}_{11,\beta} = -\frac{3J_\alpha}{|\tau+i\eta|} - \frac{4K_\alpha}{|\tau+i\eta|} \tag{7.111}$$

where

$$J_\alpha(\tau,\eta) = \frac17\int_\tau^{1/\nu^{\frac12}} \frac{(s+i\eta)^4}{(\tau+i\eta)^3}\, \widehat{g}_{1,\beta}(\widehat{w}, s, \eta) ds,$$

$$K_\alpha(\tau,\eta) = \frac17\int_\tau^{1/\nu^{\frac12}} \frac{(\tau+i\eta)^4}{(s+i\eta)^3}\, \widehat{g}_{1,\beta}(\widehat{w}, s, \eta) ds.$$

Step 1.2: Estimates of J_α. Using lemma 7.I.2, we get

$$|J_\alpha| \leq \frac{\mathcal{M}_{13}}{7|\tau+i\eta|^3}\left(\left(\frac1\delta|\widehat{w}|_{{}_{\mathrm{s}}\widehat{\mathrm{C}}^0_{\delta,\frac12,\nu}} + |\widehat{w}|^2_{{}_{\mathrm{s}}\widehat{\mathrm{C}}^0_{\delta,\frac12,\nu}}\right) J'_\alpha + \nu^2 J''_\alpha\right)$$

where

$$J'_\alpha = \int_\tau^{1/\nu^{\frac12}} \frac{1}{|\tau+i\eta|^2} ds, \qquad J''_\alpha = \int_\tau^{1/\nu^{\frac12}} |\tau+i\eta|^2 ds.$$

Moreover, since for $(\tau,\eta)\in\widehat{\Sigma}^{s}_{\delta,\frac12,\nu}$, $|\tau|\le 1/\nu^{\frac12}$ and $1\le\delta\le|\eta|\le 1/\nu^{\frac12}$ holds,

$$J'_\alpha \le \int_{-1}^{+\infty}\frac{1}{(s^2+\eta^2)}ds \le \int_{-1}^{+\infty}\frac{1}{(s^2+1)}ds = \frac{3\pi}{4}$$

and

$$J''_\alpha \le \int_{\tau}^{1/\nu^{\frac12}}\frac{2}{\nu}ds \le \frac{4}{\nu^{\frac32}}$$

Hence, there exits M_J such that for every $\nu\in]0,\nu_6]$, $\delta\in[\delta_2,1/\nu^{\frac12}]$, and every $(\tau,\eta)\in\widehat{\Sigma}^{s}_{\delta,\frac12,\nu}$

$$|J_\alpha|\le\frac{M_J}{|\tau+i\eta|^3}\left(\frac{1}{\delta}|\widehat{w}|_{{}_s\widehat{C}^0_{\delta,\frac12,\nu}}+|\widehat{w}|^2_{{}_s\widehat{C}^0_{\delta,\frac12,\nu}}+\nu^{\frac12}\right)\tag{7.112}$$

Step 1.3: Estimates of K_α. Using lemmas 7.I.2 and 7.I.3, we obtain

$$\begin{aligned}|K_\alpha| &\le \frac{\mathcal{M}_{13}|\tau+i\eta|^4}{7}\left(\left(\frac{1}{\delta}|\widehat{w}|_{{}_s\widehat{C}^0_{\delta,\frac12,\nu}}+|\widehat{w}|^2_{{}_s\widehat{C}^0_{\delta,\frac12,\nu}}\right)\mathcal{K}_9+\nu^2\mathcal{K}_5\right)\\ &\le \frac{\mathcal{M}_{13}\pi(1+(\sqrt{2})^8)}{14|\tau+i\eta|^4}\left(\frac{1}{\delta}|\widehat{w}|_{{}_s\widehat{C}^0_{\delta,\frac12,\nu}}+|\widehat{w}|^2_{{}_s\widehat{C}^0_{\delta,\frac12,\nu}}\right)+\frac{\mathcal{M}_{13}\pi(1+(\sqrt{2})^4)}{14}\nu^2.\end{aligned}$$

Hence, there exits M_K such that for every $\nu\in]0,\nu_6]$, $\delta\in[\delta_2,1/\nu^{\frac12}]$, and every $(\tau,\eta)\in\widehat{\Sigma}^{s}_{\delta,\frac12,\nu}$

$$|K_\alpha|\le\frac{M_K}{|\tau+i\eta|^4}\left(\frac{1}{\delta}|\widehat{w}|_{{}_s\widehat{C}^0_{\delta,\frac12,\nu}}+|\widehat{w}|^2_{{}_s\widehat{C}^0_{\delta,\frac12,\nu}}\right)+M_K\nu^2\tag{7.113}$$

Recalling once again that $1\le|\tau+i\eta|\le\sqrt{2}\nu^{-\frac12}$ for $(\tau,\eta)\in\widehat{\Sigma}^{s}_{\delta,\frac12,\nu}$ and using (7.111), (7.112), (7.113) we obtain

$$|\widehat{\mathcal{F}}_{11,\alpha}|\le\frac{M_J+M_K}{|\tau+i\eta|^3}\left(\frac{1}{\delta}|\widehat{w}|_{{}_s\widehat{C}^0_{\delta,\frac12,\nu}}+|\widehat{w}|^2_{{}_s\widehat{C}^0_{\delta,\frac12,\nu}}\right)+\frac{M_J}{|\tau+i\eta|^3}\nu^{\frac12}+M_K\nu^2$$

and thus

$$\sup_{\widehat{\Sigma}^{s}_{\delta,\frac12,\nu}}|\widehat{\mathcal{F}}_{11,\alpha}||\tau+i\eta|^3\le(M_J+M_K)\left(\tfrac{1}{\delta}|\widehat{w}|_{{}_s\widehat{C}^0_{\delta,\frac12,\nu}}+|\widehat{w}|^2_{{}_s\widehat{C}^0_{\delta,\frac12,\nu}}\right)+\left(M_J+M_K(\sqrt{2})^3\right)\nu^{\frac12}.$$

Similarly,

$$\begin{aligned}\sup_{\widehat{\Sigma}^{s}_{\delta,\frac12,\nu}}|\widehat{\mathcal{F}}_{11,\beta}||\tau+i\eta|^4\le\ &(3M_J+4M_K)\left(\tfrac{1}{\delta}|\widehat{w}|_{{}_s\widehat{C}^0_{\delta,\frac12,\nu}}+|\widehat{w}|^2_{{}_s\widehat{C}^0_{\delta,\frac12,\nu}}\right)\\ &+(3M_J+4M_K(\sqrt{2})^3)\,\nu^{\frac12}.\end{aligned}$$

Step 2: Estimates of the two last components. The two first components are given by

$$(\widehat{\mathcal{F}}_{11,A},\widehat{\mathcal{F}}_{11,b},0,0) = -\int_{\tau}^{1/\nu^{\frac{1}{2}}} \langle \widehat{\mathbf{r}}_{+}^{*}(s),\widehat{g}_1(\widehat{w},s,\eta)\rangle_{*} ds\, \widehat{\mathbf{r}}_{+}(\tau+\mathrm{i}\eta)$$
$$-\int_{\tau}^{1/\nu^{\frac{1}{2}}} \langle \widehat{\mathbf{r}}_{-}^{*}(s),\widehat{g}_1(\widehat{w},s,\eta)\rangle_{*} ds\, \widehat{\mathbf{r}}_{-}(\tau+\mathrm{i}\eta).$$

Using lemmas 7.I.4, 7.I.3 we get

$$\begin{aligned}
|\widehat{\mathcal{F}}_{11,A\mathrm{or}B}| &\le \int_{\tau}^{1/\nu^{\frac{1}{2}}} (|\widehat{g}_{1,A}(\widehat{w},s,\eta)|+|\widehat{g}_{1,B}(\widehat{w},s,\eta)|)\,ds\\
&\le 2\mathcal{M}_{13}\mathcal{M}_{14}\left(\tfrac{1}{\delta}|\widehat{w}|_{{}_s\widehat{\mathrm{C}}^0_{\delta,\frac{1}{2},\nu}}+|\widehat{w}|^2_{{}_s\widehat{\mathrm{C}}^0_{\delta,\frac{1}{2},\nu}}\right)\mathcal{K}_6+2\mathcal{M}_{13}\mathcal{M}_{14}\nu^2\mathcal{K}_2\\
&\le \frac{\mathcal{M}_{13}\mathcal{M}_{14}\pi(1+(\sqrt{2})^5)}{|\tau+\mathrm{i}\eta|}\left(\tfrac{1}{\delta}|\widehat{w}|_{{}_s\widehat{\mathrm{C}}^0_{\delta,\frac{1}{2},\nu}}+|\widehat{w}|^2_{{}_s\widehat{\mathrm{C}}^0_{\delta,\frac{1}{2},\nu}}\right)\\
&\quad+\mathcal{M}_{13}\mathcal{M}_{14}\pi(1+\sqrt{2})\frac{\nu^2}{|\tau+\mathrm{i}\eta|}.
\end{aligned}$$

Thus,

$$\begin{aligned}
\sup_{\widehat{\Sigma}^s_{\delta,\frac{1}{2},\nu}} |\widehat{\mathcal{F}}_{11,A\mathrm{or}B}||\tau+\mathrm{i}\eta|^3 \le\ & \mathcal{M}_{13}\mathcal{M}_{14}\pi(1+(\sqrt{2})^5)\left(\tfrac{1}{\delta}|\widehat{w}|_{{}_s\widehat{\mathrm{C}}^0_{\delta,\frac{1}{2},\nu}}+|\widehat{w}|^2_{{}_s\widehat{\mathrm{C}}^0_{\delta,\frac{1}{2},\nu}}\right)\\
&+2\mathcal{M}_{13}\mathcal{M}_{14}\pi(1+(\sqrt{2})\nu.
\end{aligned}$$

Finally summing the estimates of each components of $\widehat{\mathcal{F}}_{11}$ previously obtained, and the estimates of $\widehat{\mathcal{F}}_1^{\Delta}$ given by lemma 7.I.1 we obtain that there exists M_{in} such that every $\nu\in]0,\nu_6]$, and every $\delta\in[\delta_2,1/\nu^{\frac{1}{2}}]$,

$$|\widehat{\mathcal{F}}_1(\widehat{w})|_{{}_s\widehat{\mathrm{C}}^0_{\delta,\frac{1}{2},\nu}} \le M_{\mathrm{in}}\left(\frac{|\widehat{w}|_{{}_s\widehat{\mathrm{C}}^0_{\delta,\frac{1}{2},\nu}}}{\delta}+|\widehat{w}|^2_{{}_s\widehat{\mathrm{C}}^0_{\delta,\frac{1}{2},\nu}}+\nu^{\frac{1}{4}}\right).$$

This ends the the proof of estimate (a) of lemma 7.4.45. The proof of estimate (b) is very similar. So the details are left to the reader. □

7.J Appendix. Proof of Lemmas 7.4.15, 7.4.16

Proof of Lemma 7.4.15. Lemma 7.4.48-(a) ensures that for every n, and every $\zeta\in]-\pi+\delta_s\nu,\pi+\delta_s\nu[$

$$SY_{\nu_n}(\mathrm{i}\zeta)=\overline{Y_{\nu_n}(\mathrm{i}\zeta)}.$$

Limiting ourselves to $\Omega^{s}_{\delta_s} \cap i\mathbb{R}$ and rewriting this symmetry condition in the inner system of coordinates recalling that on $\widehat{\Omega}^{s}_{\delta_s}$, $\widehat{Y}_s = \widehat{Y}^+_{s,0} + \widehat{w}_s$ we obtain

$$S\left(\widehat{Y}^+_{s,0}(-i\eta,\rho) + \widehat{w}_s(-i\eta,\rho,\nu_n)\right) = \overline{\widehat{Y}^+_{s,0}(-i\eta,\rho) + \widehat{w}_s(-i\eta,\rho,\nu_n)} \tag{7.114}$$

for $\eta \in]\delta_s, 1/\nu^{\frac{1}{2}}[$, $\rho \in [0,\rho_s[$, $n \in \mathbb{N}$. Proposition 7.4.43 ensures that

$$|\widehat{w}_s|_{{}^s\widehat{\mathsf{H}}_{\delta_s,\frac{1}{2},\nu}} \leq 4M_{\text{in}}\nu_n^{\frac{1}{4}}. \tag{7.115}$$

Since $\eta > \delta_s \geq 1$, (7.115) gives

$$|\widehat{w}_s(-i\eta,\rho,\nu_n)| \leq 4M_{\text{in}}\nu_n^{\frac{1}{4}} \qquad \text{for } \eta > \delta_s,\ \rho \in [0,\rho_s[,\ n \in \mathbb{N}. \tag{7.116}$$

Now pushing n to infinity in (7.114) and using (7.116) we get

$$S\left(\widehat{Y}^+_{s,0}(-i\eta,\rho)\right) = \overline{\widehat{Y}^+_{s,0}(-i\eta,\rho)} \qquad \text{for } \eta > \delta_s,\ \rho \in [0,\rho_s[. \ \square$$

Proof of Lemma 7.4.16. (a): (i) $\iff$ (ii) and (iii)$\Rightarrow$(iv) are clear since $\widehat{h}^+_0$ is symmetric about the imaginary axis. So we only prove (ii)$\Rightarrow$(iii) and (iv)$\Rightarrow$(i).

(ii)$\Rightarrow$(iii): $(\widehat{\mathbf{p}},\widehat{\mathbf{q}},\widehat{\mathbf{r}}_+,\widehat{\mathbf{r}}_-)$ is a basis of $\mathbb{C}^4$. Let us decompose $\widehat{W}^+_{s,0}$ in this basis,

$$\widehat{W}^+_{s,0}(z,\rho) = \lambda_p(z,\rho)\widehat{\mathbf{p}}(z) + \lambda_q(z,\rho)\widehat{\mathbf{q}}(z) + \lambda_+(z,\rho)\widehat{\mathbf{p}}(z) + \lambda_-(z,\rho)\widehat{\mathbf{p}}(z)$$

where

$$\lambda_p(z,\rho) = \langle\widehat{\mathbf{p}}^*(z),\widehat{W}^+_{s,0}(z,\rho)\rangle_*, \qquad \lambda_+(z,\rho) = \langle\widehat{\mathbf{r}}^*_+(z),\widehat{W}^+_{s,0}(z,\rho)\rangle_*,$$
$$\lambda_q(z,\rho) = \langle\widehat{\mathbf{q}}^*(z),\widehat{W}^+_{s,0}(z,\rho)\rangle_*, \qquad \lambda_-(z,\rho) = \langle\widehat{\mathbf{r}}^*_-(z),\widehat{W}^+_{s,0}(z,\rho)\rangle_*.$$

Lemma 7.4.14 ensures that (ii) is equivalent to

$$S\widehat{W}^+_{s,0}(-i\eta,\rho) = \overline{\widehat{W}^+_{s,0}(-i\eta,\rho)}, \qquad \text{for } \eta \geq \ell_3.$$

Using the decomposition of $\widehat{W}^+_{s,0}(z,\rho)$ in the basis $(\widehat{\mathbf{p}},\widehat{\mathbf{q}},\widehat{\mathbf{r}}_+,\widehat{\mathbf{r}}_-)$ and the symmetry properties of the basis given by lemma 7.4.39 we get

$$\mathcal{R}e\,(\lambda_p(-i\eta,\rho)) = \mathcal{I}m\,(\lambda_q(-i\eta,\rho)) = \mathcal{I}m\,(\lambda_+(-i\eta,\rho)) = \mathcal{R}e\,(\lambda_-(-i\eta,\rho)) = 0$$

The result follows observing that

$$\Lambda_p = \mathcal{R}e\,(\lambda_p), \qquad \Lambda_+ = \mathcal{I}m\,(\lambda_+) + \mathcal{R}e\,(\lambda_-),$$
$$\Lambda_q = \mathcal{I}m\,(\lambda_q), \qquad \Lambda_- = \mathcal{I}m\,(\lambda_+) - \mathcal{R}e\,(\lambda_-).$$

(iv)⇒(i). The computations previously made ensures that (iv) implies that $S\widehat{W}^+_{s,0}(i\eta_0) = \overline{\widehat{W}^+_{s,0}(i\eta_0)}$. Hence, since $\widehat{h}_0^+$ is symmetrical about the imaginary axis,

$$S\widehat{X}^+(i\eta_0) = \overline{\widehat{X}^+(i\eta_0)}.$$

Because of the symmetry properties of the inner system (7.55), the functions $\widehat{X}^+(z)$ and $\overline{S\widehat{X}^+(-\overline{z})}$ are two holomorphic solutions of the inner system respectively defined on $\widehat{\Omega}^{s}_{\delta_s}$ and $\widehat{\Omega}^{u}_{\delta_s} = -\overline{\widehat{\Omega}^{s}_{\delta_s}}$ which are equal for $z = i\eta_0$. Uniqueness of the solution of the Cauchy problem ensures that

$$\widehat{X}^+(z) = \overline{S\widehat{X}^+(-\overline{z})}.$$

(b): The analyticity of Λ_p, Λ_q, Λ_+, Λ_- with respect to ρ follows directly from the analyticity of $\widehat{W}^+_{s,0}$ (see Remark 7.4.36). The computation of the derivatives of Λ_p, Λ_q, Λ_+, Λ_- at the origin is very similar to the computation of Λ_1 in Lemma 7.4.52. Thus it is left to the reader.

(c): The proof of (c) follows directly from the proof of Lemma 7.4.51. □

8. The $0^{2+}i\omega$ resonance in infinite dimensions. Application to water waves

8.1 Introduction

Chapter 7 was devoted to the study of 4 dimensional, one parameter, reversible families vector fields admitting a $0^{2+}i\omega$ resonance at the origin. In this chapter, we study the $0^{2+}i\omega$ resonance in infinite dimensions, i.e. one parameter, reversible families of vector fields admitting the origin as a fixed point, and such that the *central spectrum* of the differential at the origin is $\{0, \pm i\omega\}$ where 0 is a double non semi simple eigenvalues and $\pm i\omega$ are simple eigenvalues. Our aim is to prove that the two persistence results obtain in Chapter 7 for periodic orbits and homoclinic connection to exponentially small periodic orbits still hold in infinite-dimensions provided that the vector field satisfies some assumptions given in subsection 8.1.1.

One example of a vector field satisfying such assumptions occurs when describing the irrotational flow of an inviscid fluid layer under the influence of gravity and small surface tension (Bond number $b < \frac{1}{3}$) for a Froude number F close to 1. In this context a homoclinic solution to a periodic orbit is called a generalized solitary wave. The study of this problem was our original motivation for this work. The general Theorem 8.1.13 of persistence of homoclinic connections to exponentially small periodic orbits ensures then that there exist solitary waves with oscillations at infinity of order less than $c(\ell)\exp\left(-\omega\ell\sqrt{\frac{\frac{1}{3}-b}{|F^{-2}-1|}}\right)$, ℓ being any number between 0 and π, and $\omega \neq 0$ satisfying $\omega\cosh\omega = (1 + b\omega^2)\sinh\omega$.

Section8.3 is devoted to the persistence of periodic orbits, Section 8.4 to the persistence of homoclinic connections to exponentially small periodic orbits and Section 8.5 deals with the water wave problem.

8.1.1 The $0^{2+}i\omega$ resonance in infinite dimensions

Let E be a real Hilbert space, with norm $\|.\|_E$. We study the evolution equation

$$\frac{dw}{dx} = \mathcal{A}(\mu)w + f(w, \mu) \tag{8.1}$$

in E,where μ is a real parameter lying in a compact neighborhood of 0 ($|\mu| \in [0, \mu_0]$, $\mu_0 > 0$).

$\mathcal{A}(\mu)$ and $f(w,\mu)$ are supposed to satisfy the following assumptions:

Assumption 8.1.1. *$\mathcal{A}(\mu)$ is a closed, densely defined, unbounded linear operator in E with domain $\mathcal{D}$ independent on μ.*

Assumption 8.1.2. *System (8.1) is reversible: there exists a reflection, i.e., a bounded linear operator in E satisfying $S^2 = \mathrm{Id}_E$, such that*

$$S\mathcal{A}(\mu)w = -\mathcal{A}(\mu)Sw, \qquad Sf(w,\mu) = -f(Sw,\mu)$$

for every $w \in \mathcal{D}$, and $\mu \in [-\mu_0, \mu_0]$.

We denote by $\mathcal{A}^o$ the operator $\mathcal{A}(0)$ and we norm $\mathcal{D}$ with the graph norm induced by $\mathcal{A}^o$:

$$\|w\|_{\mathcal{D}}^2 = \|w\|_{E}^2 + \|\mathcal{A}^o w\|_{E}^2.$$

Assumption 8.1.3.

(a) *$\mathcal{A}(\mu)$ has the form $\mathcal{A}(\mu) = \mathcal{A}^o + \mu\mathcal{A}^1(\mu)$ and there exists a constant $C_{\mathcal{A}}$ such that*

$$\|\mathcal{A}^1(\mu)w\|_{E} \le C_{\mathcal{A}} \|w\|_{\mathcal{D}}$$

for every $\mu \in [-\mu_0, \mu_0]$ and $w \in \mathcal{D}$.

(b) *The function $\mu \mapsto \mathcal{A}^1(\mu)$ from $[-\mu_0, \mu_0]$ to $\mathcal{L}(\mathcal{D}, E)$ is analytic, where $\mathcal{L}(X,Y)$ denotes the Banach space of bounded linear operators from Banach space X to Banach space Y.*

Assumption 8.1.4.

(a) *The central spectrum $\Sigma_0 = \Sigma\mathcal{A}^o \cap \mathrm{i}\mathbb{R}$ of $\mathcal{A}^o$ is $\{\pm\mathrm{i}\omega, 0\}$, $\omega > 0$, where 0 is a double non-semi-simple isolated eigenvalue, and $\pm\mathrm{i}\omega$ are two simple isolated eigenvalues.*

(b) *Moreover, there exist $\sigma_0 > 0$, $\gamma > 0$, such that for every σ real satisfying $|\sigma| \ge \sigma_0$, $\mathrm{i}\sigma$ lies in the resolvent set $\rho(\mathcal{A}^o)$ of $\mathcal{A}^o$, and $\|(\mathrm{i}\sigma - \mathcal{A}^o)^{-1}\|_{\mathcal{L}(E)} \le \gamma/|\sigma|$.*

For later use we denote by φ_0 an eigenvector belonging to 0 ($\mathcal{A}^o\varphi_0 = 0$), by φ_1 by a generalized eigenvector belonging to 0 ($\mathcal{A}^o\varphi_1 = \varphi_0$) and by $\varphi_\pm$ two eigenvectors belonging to $\pm\mathrm{i}\omega$ ($\mathcal{A}^o\varphi_\pm = \pm\mathrm{i}\omega\varphi_\pm$).

Since $\{0, \pm\mathrm{i}\omega\}$ are isolated eigenvalues of finite multiplicity, there exists $(\varphi_0^*, \varphi_1^*, \varphi_+^*, \varphi_-^*)$ such that

$$\pi_c(w) = (\varphi_0^*, w)_E\varphi_0 + (\varphi_1^*, w)_E\varphi_1 + (\varphi_+^*, w)_E\varphi_+ + (\varphi_-^*, w)_E\varphi_-$$

is a projection onto the finite-dimensional subspace E_c spanned by the vectors $(\varphi_0, \varphi_1, \varphi_+, \varphi_-)$ satisfying $\pi_c \in \mathcal{L}(E, \mathcal{D})$ and $\mathcal{A}^o \pi_c w = \pi_c \mathcal{A}^o w$ for $w \in \mathcal{D}$. Let $\pi_h = \mathrm{Id} - \pi_c$, $E_h = \pi_h E$, $\mathcal{D}_h = \pi_h \mathcal{D}$, $\mathcal{A}_c^o = \mathcal{A}^o|_{E_c}$, $\mathcal{A}_h^o = \mathcal{A}^o|_{E_h}$. We have $E = E_c \oplus E_h$ and $\mathcal{D} = E_c \oplus \mathcal{D}_h$.

Throughout this section we identify E with $\mathbb{R}^4 \times E_h$ and $\mathcal{D}$ with $\mathbb{R}^4 \times D_h$. Moreover we use two more tractable norms on E and $\mathcal{D}$ which are equivalent to the two previously defined ones

$$|w|_E^2 := |w_c|^2 + \|w_h\|_E^2, \text{ for } w \in E, \quad |w|_{\mathcal{D}}^2 := |w_c|^2 + \|w_h\|_D^2 \text{ for } w \in \mathcal{D}.$$

With this notation we rewrite (8.1) as

$$\frac{dw_c}{dt} = \mathcal{A}_c^o w_c + f_c(w, \mu), \tag{8.2}$$

$$\frac{dw_h}{dt} = \mathcal{A}_h^o w_h + f_h(w, \mu), \tag{8.3}$$

where $f_c(w, \mu) = \mu \pi_c \mathcal{A}^1(\mu) w + \pi_c f(w, \mu)$, and $f_h(w, \mu) = \mu \pi_h \mathcal{A}^1(\mu) w + \pi_h f(w, \mu)$.

Assumption 8.1.5.

(a) *f is an analytic function from $B(0,r) \times [-\mu_0, \mu_0]$ to E where $B(0,r)$ is some ball of radius r in $\mathcal{D}$.*

(b) *There exists a constant C_f such that*

$$\|f(w, \mu)\|_E \leq C_f \, \|w\|_{\mathcal{D}}^2$$

for every $w \in B(0,r)$, and $\mu \in [-\mu_0, \mu_0]$.

Assumption 8.1.5(b) is satisfied by semi and quasilinear systems.

Assumptions 8.1.4, 8.1.2 ensure that necessarily $S\varphi_0 = \pm\varphi_0$. The vector fields for which $S\varphi_0 = \varphi_0$ are said to admit a $0^{2+}i\omega$ resonance at the origin and the other ones are said to admit a $0^{2-}i\omega$ resonance. Here we assume

Assumption 8.1.6.

(a) $S\varphi_0 = \varphi_0$,

(b) $(\varphi_1^*, \mathcal{A}^1(0)\varphi_0)_E \neq 0$,

(c) $(\varphi_1^*, D_{ww}^2 f(0,0)[\varphi_0, \varphi_0])_E \neq 0$.

Assumptions 8.1.6 (a)–(c) corresponds respectively to hypothesis made for the 4-dimensional cases in chapter 7. In Part 8.5, we check that Assumptions 8.1.1–8.1.6 are fulfilled for the water wave problem.

As previously mentioned, our aim is to prove that the result of persistence of homoclinic connections to exponentially small periodic orbits obtained in Chapter 7 still hold for the vector fields satisfying Assumptions 8.1.1-8.1.6. Since our tool, Lemma 2.1.1 used to obtain exponential estimates requires analytic functions, we do not use the center-manifold reduction and consequently we stay in an infinite-dimensional frame. The scheme of the proof is exactly the same as that for the 4-dimensional system. But now, we must control the infinite-dimensional part of the system, which comes from the hyperbolic part of the spectrum of the differential. For that purpose, we give the spectral properties of $\mathcal{A}_h^o$ which can be deduced from those of $\mathcal{A}^o$.

Lemma 8.1.7. *There exist $\delta_h > 0$, $\kappa_h > 0$, $M_h > 0$ such that for all $\sigma \in \mathbb{C}$, if $|\mathcal{R}e\,(\sigma)| \leq \delta_h + \kappa_h |\mathcal{I}m\,(\sigma)|$, then $\sigma \in \rho(\mathcal{A}_h^o)$,*

$$\begin{aligned} \left\|(\sigma - \mathcal{A}_h^o)^{-1}\right\|_{\mathcal{L}(E_h)} &\leq \frac{M_h}{1+|\sigma|}, \\ \left\|(\sigma - \mathcal{A}_h^o)^{-1}\right\|_{\mathcal{L}(E_h,\mathcal{D}_h)} &\leq M_h. \end{aligned}$$

The proof of this lemma is left as an exercise (see [VI91]).

Remark 8.1.8. Observe that for all ν in $]0,1]$,

$$\delta_h(\frac{\mathcal{A}_h^o}{\nu}) = \frac{\delta_h(\mathcal{A}_h^o)}{\nu}, \quad \kappa_h(\frac{\mathcal{A}_h^o}{\nu}) = \kappa_h(\mathcal{A}_h^o), \quad M_h(\frac{\mathcal{A}_h^o}{\nu}) = M_h(\mathcal{A}_h^o).$$

Remark 8.1.9 (Optimal resolution of affine equations). For studying the nonlinear Equation (8.1), in particular to turn it into an integral equation for use in the Banach Fixed-Point Theorem, we need to solve affine equations of the form

$$\frac{dw}{dx} = \mathcal{A}_h^o w + g_h(x),$$

with optimal regularity to compensate for the loss of regularity in the nonlinear terms $f(w(x),\mu)$. Optimal regularity is not known in $C^k(\mathbb{R}, E)$ spaces when E is infinite-dimensional, but holds when E is finite-dimensional. This is one of the major differences between the study of the 4-dimensional vector field (7.1) and of the Equation (8.1). We need an optimal regularity result in an analytic function space to study (8.1) in the same way as we did for the 4-dimensional vector field in Chapter 7. The optimal regularity results are given in Section 8.2.

8.1.2 Direct normalization and scaling

8.1.2.1 Normalization. As already mentioned, we want to work with analytic functions. Since center-manifold reduction destroys the analyticity of functions, we do not use it. We start the study of (8.1) by a direct normalization. The aim is to transform (8.1) into a system of two coupled equations, one lying in $E_c \equiv \mathbb{R}^4$, similar to the 4-dimensional vector field studied in Chapter 7, and another one lying in E_h. This latter part can be very well controlled. The idea is to adapt the standard Theorem 3.2.10 on normal forms for finite-dimensional vector fields with non fully critical spectrum to obtain a similar one in infinite dimensions. In the original paper of Elphick & al. [ETBCI87], the infinite-dimensional case is mentioned but not studied in detail. A first use of direct normalization in an infinite-dimensional framework is given in [IL90]. Following the method developed in [IL90] and [ETBCI87] we show

Theorem 8.1.10. *Let $p \geq 1$. There exist polynomial functions of X_c of degree p whose coefficients are analytic functions of μ:*

$$\phi(X_c, \mu) = \sum_{k=1}^{p} \phi_\mu^k[X_c^{(k)}], \quad \phi_\mu^k \in \mathcal{L}_s(E_c^k, \mathcal{D}), \tag{8.4}$$

$$\mathcal{Q}(X_c, \mu) = \sum_{k=1}^{p} \mathcal{Q}_\mu^k[X_c^{(k)}], \quad \mathcal{Q}_\mu^k \in \mathcal{L}_s(E_c^k, E_c), \tag{8.5}$$

satisfying $\phi(0,0) = 0$, $\mathcal{Q}(0,0) = 0$, $D_{X_c}\phi(0,0) = \phi_0^1 = 0$, $D_{X_c}\mathcal{Q}(0,0) = \mathcal{Q}_0^1 = 0$, such that, for $w = X_c + X_h + \phi(X_c, \mu)$ with $X_c \in E_c$, $X_h \in E_h$ and for μ small enough, in a neighborhood of (0,0) in $E_c \times E_h$, (8.1) *reads*

$$\begin{aligned} \frac{dX_c}{dx} &= \mathcal{A}_c^o X_c + \mathcal{Q}(X_c, \mu) + \mathcal{R}_c(X_c, X_h, \mu), && (8.6) \\ \frac{dX_h}{dx} &= \mathcal{A}_h^o X_h + \mathcal{R}_h(X_c, X_h, \mu) && (8.7) \end{aligned}$$

with

$$\mathcal{Q}(e^{x(\mathcal{A}_c^o)^*} X_c, \mu) = e^{x(\mathcal{A}_c^o)^*} \mathcal{Q}(X_c, \mu), \ \textit{for } x \in \mathbb{R}, \tag{8.8}$$

with $\mathcal{R}_c$, $\mathcal{R}_h$, are analytic functions with respect to X_c, X_h, μ satisfying

$$|\mathcal{R}_c(X_c, X_h, \mu)|_{E_c} + |\mathcal{R}_h(X_c, X_h, \mu)|_{E_h} = \mathcal{O}(|X_h|_{\mathcal{D}} (|X_c| + |X_h|_{\mathcal{D}} + \mu) + |X_c|^{p+1}).$$

Moreover, reversibility is preserved by normalization. Indeed, for all $X_c \in E_c$,

$$\begin{aligned} S\phi(X_c, \mu) &= \phi(SX_c, \mu), \\ S\mathcal{Q}(X_c, \mu) &= -\mathcal{Q}(SX_c, \mu), \\ S\mathcal{R}_c(X_c, X_h, \mu) &= -\mathcal{R}_c(SX_c, SX_h, \mu), \\ S\mathcal{R}_h(X_c, X_h, \mu) &= -\mathcal{R}_h(SX_c, SX_h, \mu). \end{aligned}$$

$\mathcal{L}_s(E_c^k, \mathcal{X})$ is the space of k-linear symmetric bounded operator from E_c^k to $\mathcal{X}$. In (8.8) $(\mathcal{A}_c^o)^*$ is the adjoint matrix of $\mathcal{A}_c^o$. The proof of this theorem is given in Appendix 8.A.

Remark 8.1.11. The normal polynomial function $\mathcal{Q}$ is identical to the one obtained by the standard normal-form theorem for finite-dimensional vector field in Chapter 7. Thus for $p = 2$, setting $X_c = (\widetilde{\alpha}, \widetilde{\beta}, \widetilde{A}, \widetilde{B})$ we have

$$\mathcal{A}_c^o X_c + \mathcal{Q}(X_c, \mu) = \begin{pmatrix} \widetilde{\alpha} \\ \mu c_1(\mu)\widetilde{\alpha} + c_2(\mu)\widetilde{\alpha}^2 + d_2(\mu)(\widetilde{A}^2 + \widetilde{B}^2) \\ -\widetilde{B}(\omega + \mu\omega_1(\mu) + m(\mu)\widetilde{\alpha}) \\ \widetilde{A}(\omega + \mu\omega_1(\mu) + m(\mu)\widetilde{\alpha}) \end{pmatrix}$$

where c_1, c_2, d_2, ω_1, m are analytic function of μ. Moreover c_{11} and c_{20} are given by

$$c_{10} = (\varphi_1^*, \mathcal{A}^1(0)\varphi_0), \qquad c_{20} = (\varphi_1^*, D^2_{ww} f(0,0)[\varphi_0, \varphi_0])$$

Assumption 8.1.6 ensures that c_{10} and c_{20} are not zero.

8.1.2.2 Scaling. As in Chapter 7 we assume that $c_{10}\mu < 0$. Now we perform the scaling (the same as in Chapter 7)

$$\begin{aligned} \widetilde{\alpha} &= -\frac{3}{2c_{20}}\nu^2\alpha, & \widetilde{A} &= \nu^2 A & X_h &= \nu^3 Y_h, & t &= \nu x, \\ \widetilde{\beta} &= -\frac{3}{2c_{20}}\nu^3\beta, & \widetilde{B} &= \nu^2 B & \nu &= \sqrt{c_{10}\mu}. \end{aligned}$$

As in Chapter 7 incorporating in R_c the term of order $\mu|X_c|^2 + \mu^2|X_c|$, the rescaled system reads

$$\frac{dY_c}{dt} = N(Y_c, \nu) + R_c(Y_c, Y_h, \nu), \tag{8.9}$$

$$\frac{dY_h}{dt} = \frac{\mathcal{A}_h^o}{\nu} Y_h + R_h(Y_c, Y_h, \nu), \tag{8.10}$$

with $R_c = (R_\alpha, R_\beta, R_A, R_B)$ and

$$\begin{aligned} |R_\alpha(Y,\nu)| + |R_A(Y,\nu)| + |R_B(Y,\nu)| &= \mathcal{O}(\nu^3|Y_c| + \nu^2\,|Y_h|_{\mathcal{D}}), \\ |R_\beta(Y,\nu)| + |R_h(Y,\nu)|_{\mathcal{E}} &= \mathcal{O}(\nu^2|Y_c| + \nu\,|Y_h|_{\mathcal{D}}), \end{aligned}$$

where N has been defined in (7.5) and $Y_c = (\alpha, \beta, A, B)$, $Y = (Y_c, Y_h)$.

The symmetry properties now read

$$SN(Y_c,\nu) = -N(SY_c,\nu), SR_c(Y,\nu) = -R_c(SY,\nu), SR_h(Y,\nu) = -R_h(SY,\nu),$$

with $S|_{E_c} \equiv \{(\alpha, \beta, A, B) \longrightarrow (\alpha, -\beta, A, -B)\}$.

We check that R_0, R_A, R_B and all their derivatives are $\mathcal{O}(\nu^2)$ uniformly for Y in a fixed ball of $\mathcal{D}$, whereas R_1, R_h and all their derivatives are $\mathcal{O}(\nu)$ uniformly for Y in a fixed ball of $\mathcal{D}$.

Our aim is to study (8.9) as we studied (7.5), now taking (8.10) into account.

The proof of the existence of a one parameter family of periodic orbits can be done in the same way. The only difference is that we must work with appropriate Sobolev spaces instead of C^k spaces, because of the unboundness of $\mathcal{A}_h^o$. However the formalism is exactly the same.

As for the solutions homoclinic to the periodic orbits, the difference between the proofs is a little more significant, but the general scheme of the proof is unchanged. So we will not redo the whole proof in details. We follow the same scheme of proof, and at each step we point out the most significant changes induced by the coupling with the second equation in e_h and by the mixed terms in the remainder R_c, of the first equation. So we start with the truncated system.

8.1.3 Truncated system

Here, the truncated system reads

$$\frac{dY_c}{dt} = N(Y_c, \nu), \quad \frac{dY_h}{dt} = \frac{\mathcal{A}_h^o}{\nu} Y_h.$$

The spectrum of $\mathcal{A}_h^o$ is bounded away from the imaginary axis. Thus for ν small, the only bounded solutions of the second equation is 0. Thus the truncated system admits the same families of periodic orbits and of solutions homoclinic to periodic orbits with a component Y_h in E_h equal to 0. In particular the truncated system admits one solution $\mathfrak{h} = (h, 0)$ homoclinic to 0 where h is the homoclinic orbit to 0 of the truncated system induced by the 4-dimensional vector field (h is defined by (7.15)). Throughout this section we do not distinguish $\mathfrak{h}$ and h and we denote $\mathfrak{h}$ by h.

8.1.4 Statement of the results of persistence

As in Chapter 7, there arises the question of the persistence of the periodic orbits and of the homoclinic connections observed for the truncated system when it is perturbed by the higher order terms required to obtain the full system (8.9), (8.10).

In this chapter we prove that the two results of persistence obtained in the 4-dimensional case are still true in infinite dimensions: Section 8.3 is devoted to the proof of Theorem 8.1.12 below which ensures the persistence of the periodic orbits and Section 8.4 is devoted to the proof of Theorem 8.1.13 which gives the persistence of homoclinic connections to exponentially small periodic orbits.

Theorem 8.1.12. *There exist $k_0 > 0$, $\nu_1 > 0$ such that for any ν in $]0, \nu_1]$ the real full System* (8.1) *admits a one-parameter family of periodic solutions $(Y_{k,\nu}(t))_{k\in[0,k_0]}$ with*

$$Y_{k,\nu}(t) = \widehat{Y}(\underline{\omega}_{k,\nu} t, k, \nu)$$

and

(a) $$\begin{cases} \widehat{Y}(s,k,\nu) = \sum\limits_{n\geq 1} k^n\, \widehat{Y}_n(s,\nu), \\ \widehat{Y}_n \in \mathfrak{P}_{n,R},\ \ \widehat{Y}_{1,c} = (0,0,\cos(s),\sin(s)),\ \ \widehat{Y}_{1,h} = 0, \\ (k_0)^n \left\|\widehat{Y}_n\right\|_{\mathfrak{P}^2} \leq C_{\widehat{Y}}\ \ \textit{for every } \nu \in]0,\nu_0],\ \ n \geq 1, \end{cases}$$

where

$$\mathfrak{P}_{n,R} = \Big\{\widehat{Y}/\widehat{Y} = \sum_{p=0}^{n} a_p \cos(ps) + \sum_{p=1}^{n} b_p \sin(ps)\ \ (a_p, b_p) \in \mathcal{D}^2,$$
$$S(a_p) = a_p, S(b_p) = -b_p\Big\} \cap \Big\{\widehat{Y}/\int_0^{2\pi} \langle\widehat{Y}(s), \widehat{Y}_1(s)\rangle ds = 0\Big\}.$$

(b) $$\begin{cases} \underline{\omega}_{k,\nu} = \dfrac{\omega_{0,\nu}}{\nu} + \widehat{\omega}(k,\nu)\ \textit{where}\ \omega_{0,\nu} = \omega + a\nu^2 + \mathcal{O}(\nu^4), \\ \widehat{\omega}(k,\nu) = \sum\limits_{n\geq 1} \widehat{\omega}_n(\nu)\, k^n, \\ \widehat{\omega}(k,\nu) \leq C_\omega k\nu\ \textit{for every}\ \nu \in]0,\nu_0],\ \ k \in [0,k_0]. \end{cases}$$

Theorem 8.1.13. *For any $\ell \in]0,\pi[$ and any $\lambda \in]0,1[$ there exist M, ν_4 such that for any ν in $]0,\nu_4]$ the full System (8.9), (8.10) admits two reversible solutions analytic in $t \in \mathbb{R}$ taking values in $\mathcal{D}$ of the form*

$$Y(t) = h(t) + v(t) + Y_{k,\nu}(t + \varphi \tanh \tfrac{1}{2}t)$$

where

$$|h(t)|_{\mathcal{D}} + |v(t)|_{\mathcal{D}} = \underset{t\to\pm\infty}{\mathcal{O}}(e^{-\lambda|t|}),$$

$Y_{k,\nu}$ is a periodic solution of the full system satisfying

$$\forall t \in \mathbb{R},\ |Y_{k,\nu}(t)|_{\mathcal{D}} \leq M\nu^2 e^{-\ell\omega/\nu}.$$

Remark 8.1.14. For system (8.1), this last theorem proves the existence of two reversible homoclinic solutions (of size of order ν^2) decaying as $e^{-\lambda\nu|x|}$ to an exponentially small periodic orbit of size of order $\nu^4 e^{-\ell\omega/\nu}$ (for any fixed ℓ in $]0,\pi[$).

As in the finite dimensional case (see Theorem 7.1.18), we can deduce from the proof of this theorem a third one which ensures the persistence of a one parameter family of reversible homoclinic connections to periodic orbits the size of which are greater than $M\nu^2 e^{-\ell\omega/\nu}$.

Theorem 8.1.15. *There exist $K_\star$, $M_\star$ such that for every ℓ, $0 < \ell < \pi$ and every λ, $0 < \lambda < 1$ there exist $K(\ell)$, $\nu_5(\ell)$ such that for every ν in $]0, \nu_5(\ell)]$ and every $k \in [K(\ell)\nu^2 \mathrm{e}^{-\ell\omega/\nu}, K_\star[$ the full System* (8.9), (8.10) *admits two reversible solutions of the form*

$$Y(t) = h(t) + v(t) + Y_{k,\nu}(t + \varphi \tanh \tfrac{1}{2}t)$$

where

$$|h(t)|_{\mathcal{D}} + |v(t)|_{\mathcal{D}} \le M_\star \mathrm{e}^{-\lambda|t|} \qquad \text{for } t \in \mathbb{R},$$

and where $Y_{k,\nu}$ is a periodic solution of the full system satisfying

$$|Y_{k,\nu}(t)|_{\mathcal{D}} \le M_\star k \qquad \text{for } t \in \mathbb{R}.$$

The proof of this theorem is very similar to the one of Theorem 7.1.18. Thus it is left to the reader.

Remark 8.1.16. These three theorems are exactly the same as those in Chapter 7 with a simple change of norm. In Chapter 7, we prove a fourth theorem which ensures the generic non persistence of homoclinic connections to 0. Such a theorem could be obtained with an analogous proof in infinite dimensions.

8.2 Optimal resolution of affine equations

As already explained, to study the nonlinear Equation (8.1), in particular for turning it into an integral equation for the purpose of using the Banach Fixed-Point Theorem, we need to solve affine equations of the form

$$\frac{dw}{dx} = \mathcal{A}_h^\circ w + g_h(x), \tag{8.11}$$

with optimal regularity to compensate for the loss of regularity in the nonlinear terms $f(w(x), \mu)$. Optimal regularity is not known in $C^k(\mathbb{R}, E)$ spaces when E is infinite-dimensional, but holds when E is finite-dimensional. This is one of the major differences between the study of the 4-dimensional vector field (7.1) and of the Equation (8.1). We need an optimal regularity result in an analytic functions space to prove the exponential estimates in the same way as we did for the 4-dimensional vector field.

We start by recalling the classical solutions of equation (8.11). Then using the optimal regularity results given by Mielke in [Mi87] we prove an optimal regularity result in a space of analytic functions which decay exponentially at infinity.

8.2.1 Green's function and classical solutions

One classical way to solve (8.11) is to construct a Green's function using the spectral properties of $\mathcal{A}_h^o$. The construction of the Green's function is based on the construction of two "quasi"-semigroups. Their definition and properties are summed up in the following lemma:

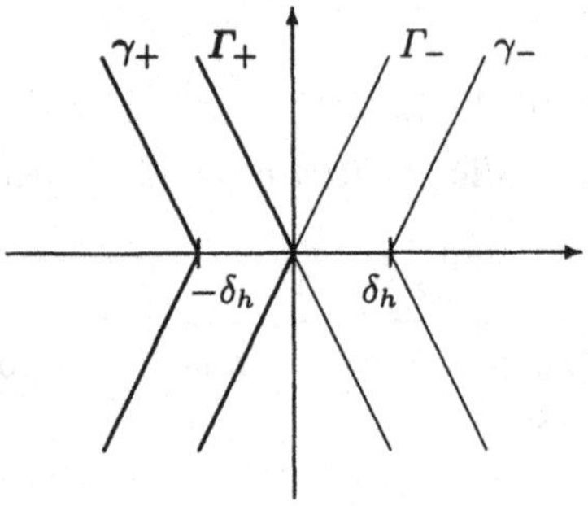

Fig. 8.1. Paths $\Gamma_\pm$ and $\gamma_\pm$.

Lemma 8.2.1. *Let*

$$\Gamma_\pm = \{\mp\kappa_h|q| \pm \mathrm{i}q,\ q \in \mathbb{R}\},\ \gamma_\pm = \{\mp\kappa_h|q| \pm \mathrm{i}q \mp \delta_h,\ q \in \mathbb{R}\}.$$

Let

$$\mathcal{S}_\pm(x) = \frac{1}{2\pi\mathrm{i}}\int_{\Gamma_\pm} \mathrm{e}^{\sigma x}(\sigma - \mathcal{A}_h^o)^{-1}\, d\sigma,\ \textit{for}\ \pm x > 0.$$

Then $\mathcal{S}_\pm$ satisfies

(a) $\mathcal{S}_\pm(x) \in \mathcal{L}(E_h, \mathcal{D}_h)$, *and is of class* C^∞,

(b) $\dfrac{d^n\mathcal{S}_\pm}{dx^n}(x) = (\mathcal{A}_h^o)^n\mathcal{S}_\pm(x)$, *for* $\pm x > 0$ *and* $n \in \mathbb{N}$,

(c) $\mathcal{A}_h^o\mathcal{S}_\pm(x)w_h = \mathcal{S}_\pm(x)\mathcal{A}_h^o w_h$, *for* $\pm x > 0$ *and* $w_h \in \mathcal{D}_h$,

(d) $\mathcal{S}_\pm(x+s) = \mathcal{S}_\pm(x)\mathcal{S}_\pm(s)$ *for* $\pm x > 0$ *and* $\pm s > 0$,

(e) *$\mathcal{S}_\pm$ has a logarithmic estimate for x near 0, more precisely, there exists M such that*

$$\|\mathcal{S}_\pm(x)\|_{\mathcal{L}(E_h)} \le M(1 - \ln|x|)\chi_{[0,1]}(|x|) + \frac{\mathrm{e}^{-\delta_h|x|}}{|x|}\chi_{[1,+\infty[}(|x|),$$

(f) *Let* $\Pi_\pm = \displaystyle\int_{\gamma_\pm} \sigma^{-1}(\sigma - \mathcal{A}_h^o)^{-1}\mathcal{A}_h^o\, d\sigma \in \mathcal{L}(\mathcal{D}_h, E_h)$. *Then*

$$\mathcal{S}_\pm(x)w_h \underset{x\longrightarrow\pm 0}{\longrightarrow} \Pi_\pm(w_h)\ \textit{and}\ (\Pi_+ + \Pi_-)(w_h) = w_h\ \textit{for every}\ w_h \in \mathcal{D}_h.$$

For any subset Ω of $\mathbb{R}$, χ_Ω is the function defined by $\chi_\Omega(x) = 1$ for $x \in \Omega$, $\chi_\Omega(x) = 0$ otherwise.

The logarithmic singularity (e) is due to the fact that the spectrum of $\mathcal{A}_h^o$ is unbounded on both sides of the imaginary axis. When w_h lies in $\mathcal{D}_h$ the singularity vanishes by (f).

The proof of this lemma can be easily deduced from the corresponding proofs in [VI91], §3.

Green's function is then defined by

$$\mathcal{K}(t) = \begin{cases} \mathcal{S}_+(x) & \text{for } x > 0, \\ -\mathcal{S}_-(x) & \text{for } x < 0, \end{cases}$$

and the solutions of (8.11) are given by

$$\mathfrak{F}(g_h)(x) := \int_{-\infty}^{+\infty} \mathcal{K}(s) f(x-s)\, ds,$$

when g_h lies in one of the two following Banach spaces:

$$\begin{aligned} \mathrm{C}_\alpha^k(\mathbb{R}, E_h) := \Bigg\{ & g_h : \mathbb{R} \longrightarrow E_h \text{ is of class } C^k, \\ & \|g_h\|_{C_\alpha^k(\mathbb{R},E_h)} := \sum_{i=0}^{k} \sup_{x\in\mathbb{R}} \left(\left| \frac{d^i g_h(x)}{dx^i} \right|_{E_h} e^{-\alpha|x|} \right) < +\infty \Bigg\}, \end{aligned} \tag{8.12}$$

$$\begin{aligned} \mathrm{C}_\alpha^{\omega,k}(\mathcal{B}_\ell, E_h) := \Bigg\{ & g_h : \mathcal{B}_\ell \longrightarrow E_h \text{ is holomorphic in } \mathcal{B}_\ell, \\ & \|g_h\|_{C_\alpha^{\omega,k}(\mathcal{B}_\ell,E_h)} := \sum_{i=0}^{k} \sup_{\xi\in\mathcal{B}_\ell} \left(\left| \frac{d^i g_h(\xi)}{d\xi^i} \right|_{E_h} e^{-\alpha|\mathcal{R}e(\xi)|} \right) < +\infty \Bigg\}, \end{aligned} \tag{8.13}$$

with $\alpha \in]-\delta_h, \delta_h[$. More precisely the following lemma holds:

Lemma 8.2.2. *Let* $\alpha \in]-\delta_h, \delta_h[$.

(a) $\mathfrak{F}$ *is a bounded operator in* $C_\alpha^k(\mathbb{R}, E_h)$ *for all* $k \geq 0$. *Moreover, for* $k \geq 1$, $g_h \in C_\alpha^k(\mathbb{R}, E_h)$, $\mathfrak{F}(g_h)$ *is the unique solution of* (8.11).

(b) $\mathfrak{F}$ *is a bounded operator in* $C_\alpha^{\omega,k}(\mathcal{B}_\ell, E_h)$ *for all* $k \geq 0$. *Moreover, for* $k \geq 0$, $g_h \in C_\alpha^{\omega,k}(\mathcal{B}_\ell, E_h)$, $\mathfrak{F}(g_h)$ *is the unique solution of* (8.11).

The proof of this lemma can be deduced from [HS85] by using Lemma 8.2.1. But this lemma does not give optimal regularity, because when g_h has finite C^k norm we expect $\mathfrak{F}(g_h)$ to have a finite C^{k+1} norm in the same space.

Remark 8.2.3. In the previous lemma, uniqueness comes from a more precise result which we shall need later (see [HS85]):

For all $\alpha \in]-\delta_h, \delta_h[$, *Equation* (8.11) *admits at most one solution in* $C_\alpha^0(\mathbb{R}, E_h) \cap C^1(\mathbb{R}, E_h)$.

Another important property of $\mathfrak{F}$ is that it preserves reversibility.

Lemma 8.2.4. *Let* $\mathfrak{S} \in \mathcal{L}(E_h)$, $\mathfrak{S}^2 = id$. *Assume that* $\mathfrak{S}\mathcal{A}_h^\circ w = -\mathcal{A}_h^\circ \mathfrak{S} w$, *for* $w \in \mathcal{D}_h$. *Then*

(a) $S_+(x)\mathfrak{S} = \mathfrak{S}S_-(-x)$ *for* $x > 0$.
(b) *For any* g_h *in* $C_\alpha^k(\mathbb{R}, E_h)$ *or in* $C_\alpha^{\omega,k}(\mathcal{B}_\ell, E_h)$ *satisfying* $g_h(x) = -\mathfrak{S}g_h(-x)$,
$$\mathfrak{F}(g_h)(x) = \mathfrak{S}\mathfrak{F}(g_h)(-x).$$

When we studied (7.5), we proved in Lemma 7.3.18 that the affine equation
$$\frac{dv}{dt} - DF(h)v = g,$$
induced by (7.26) can be solved in a space of reversible functions decaying exponentially at infinity, if and only if the function g satisfies the solvability condition
$$\int_0^{+\infty} \langle r_-^*(t), g(t)\rangle dt = 0.$$
Observe that the solvability condition is due to the antireversible solution $r_-(t)$, corresponding to the pair of pure imaginary eigenvalues $\pm i\omega$ of the differential of our initial vector field. The two solutions $p(t)$, $q(t)$ corresponding to the real eigenvalues $\pm\sqrt{-c_{10}\mu} + \mathcal{O}(\mu^{\frac{3}{2}})$ do not induce any solvability condition. Lemma 8.2.4 ensures that the same property holds for affine equations induced by $\mathcal{A}_h^\circ$ whose spectrum has no pure imaginary root. As in the finite-dimensional case, in the infinite-dimensional case there is only one solvability condition due to the pair of pure imaginary eigenvalues $\pm i\omega$. Thus we can study (8.1) with the same scheme of proof as for (7.5).

8.2.2 Optimal regularity

Our aim is now to prove an optimal regularity result for $\mathfrak{F}(g_h)$ when g_h lies in an appropriate space of analytic functions defined on $\mathcal{B}_\ell$, which decay exponentially at infinity. Moreover, this space should be an algebra if we want to use it later with nonlinear terms.

We do not give the right space at once. We construct it step by step. At each step we modify it, so as it satisfies one additional property.

- We start with L^2. Indeed, when g_h lies in L^2, (8.11) is solved with optimal regularity. The proof is elementary and based on the Fourier-Plancherel transform and on the resolvent estimates given in Lemma 8.1.7.
- To have an algebra we work with H^1 instead of L^2. The proof is exactly the same.

- Then to have an exponential decay at infinity, we add an exponential weight to obtain H^1_α ($\alpha < 0$). These three first optimal regularity results are proved in the paper of Mielke [Mi87].
- Finally, we complexify the space, to work with analytic functions defined on $\mathcal{B}_\ell$.

Let us give more details. For any given Hilbert space $\mathcal{X}$ of norm $|.|_{\mathcal{X}}$, we define

$$\begin{aligned} L^2(\mathcal{X}) &= \left\{ f : \mathbb{R} \longrightarrow \mathcal{X}, \|f\|^2_{L^2(\mathcal{X})} = \int_{-\infty}^{+\infty} |f(s)|^2_{\mathcal{X}}\, ds < +\infty \right\}, \\ H^p(\mathcal{X}) &= \left\{ f : \mathbb{R} \longrightarrow \mathcal{X}, \|f\|^2_{H^p(\mathcal{X})} = \sum_{i=0}^{p} \int_{-\infty}^{+\infty} \left| \frac{d^i f(s)}{ds^i} \right|^2_{\mathcal{X}} ds < +\infty \right\}. \end{aligned}$$

Lemma 8.2.5.

(a) *For any $g_h \in L^2(E_h)$, Equation (8.11) admits a unique solution w_{g_h} in $L^2(\mathcal{D}_h) \cap H^1(E_h)$. Moreover, there exists C such that for every $g_h \in L^2(E_h)$,*

$$\|w_{g_h}\|_{L^2(\mathcal{D}_h)} + \|w_{g_h}\|_{H^1(E_h)} \leq C \|g_h\|_{L^2(E_h)}.$$

(b) *If $g_h \in H^1(E_h)$, then $w_{g_h} \in H^1(\mathcal{D}_h) \cap H^2(E_h)$. Moreover, there exists C such that for every $g_h \in H^1(E_h)$,*

$$\|w_{g_h}\|_{H^1(\mathcal{D}_h)} + \|w_{g_h}\|_{H^2(E_h)} \leq C \|g_h\|_{H^1(E_h)}.$$

Proof. We give an elementary proof of this lemma to point out where the optimal regularity comes from. As mentioned earlier, the proof is based on the Fourier-Plancherel transform. We define

$$\hat{f}_h(\xi) = \frac{1}{\sqrt{2\pi}} \int_{-\infty}^{+\infty} f_h(t) e^{-i\xi t}\, dt.$$

Equation (8.11) is equivalent to $\hat{w}_h = (\mathrm{i}\xi - \mathcal{A}_h^\circ)^{-1} \hat{f}_h(\xi)$, which ensures the uniqueness of the solution. Then using the resolvent estimates given in Lemma 8.1.7 we have

$$\begin{aligned} \|w_h\|^2_{L^2(E_h)} + \|\mathcal{A}_h^\circ w_h\|^2_{L^2(E_h)} &= \|\hat{w}_h\|^2_{L^2(E_h)} + \|\mathcal{A}_h^\circ \hat{w}_h\|^2_{L^2(E_h)} \\ &= \left\|(\mathrm{i}\xi - \mathcal{A}_h^\circ)^{-1} \hat{f}_h\right\|^2_{L^2(E_h)} + \left\|\mathcal{A}_h^\circ (\mathrm{i}\xi - \mathcal{A}_h^\circ)^{-1} \hat{f}_h\right\|^2_{L^2(E_h)} \\ &\leq M_h^2 \left\|\hat{f}_h\right\|^2_{L^2(E_h)} = M_h^2 \|g_h\|^2_{L^2(E_h)}. \end{aligned}$$

Moreover $\left\| \dfrac{dw_h}{dt} \right\|_{L^2(E_h)} = \left\| \mathcal{A}_h^o w_h + g_h \right\|_{L^2(E_h)} \leq (M_h + 1) \left\| g_h \right\|_{L^2(E_h)}$.

(b) is proved in the same way. □

Now we add an exponential weight. For any Hilbert space $\mathcal{X}$, and any real number α we thus define

$$L^2_\alpha(\mathcal{X}) = \{f : \mathbb{R} \to \mathcal{X},\ f e^{-\alpha|x|} \in L^2(\mathcal{X})\},\quad \|f\|_{L^2_\alpha(\mathcal{X})} = \left\| f e^{-\alpha|x|} \right\|_{L^2(\mathcal{X})},$$

$$H^p_\alpha(\mathcal{X}) = \left\{ f : \mathbb{R} \to \mathcal{X}, \frac{d^i f}{dx^i} \in L^2_\alpha(\mathcal{X}),\ 0 \leq i \leq p \right\}, \|f\|^2_{H^p_\alpha(\mathcal{X})} = \sum_{i=1}^{p} \left\| \frac{d^i f}{dx^i} \right\|^2_{L^2_\alpha(\mathcal{X})}.$$

The following lemma ensures that $H^p_\alpha(\mathcal{X})$ is composed of functions which grow or decay exponentially at infinity, and that it is an "algebra" when α is negative.

Lemma 8.2.6.

(a) *For any $\alpha \in \mathbb{R}$ and any $p \geq 1$, $H^p_\alpha(\mathcal{X})$ is continuously embedded in $C^{p-1}_\alpha(\mathbb{R}, \mathcal{X})$.*

(b) *For any Hilbert space $\mathcal{X}'$, any number n and any n-linear bounded operator Q from $\mathcal{X}^n$ to $\mathcal{X}'$,*

$$Q(f_1, \cdots, f_p) \in H^p_{-\alpha}(\mathcal{X}')$$

for every $\alpha > 0$, $p \geq 1$, and $f_i \in H^p_{-\alpha}(\mathcal{X})$, $1 \leq i \leq p$.

Then we can solve (8.11) with optimal regularity in such spaces.

Lemma 8.2.7. *Let $\alpha \in]0, \delta_h[$.*

(a) *For any $g_h \in L^2_{-\alpha}(E_h)$, Equation (8.11) admits a unique solution w_{g_h} in $L^2_{-\alpha}(\mathcal{D}_h) \cap H^1_{-\alpha}(E_h)$. Moreover, there exists C such that for every $g_h \in L^2_{-\alpha}(E_h)$,*

$$\left\| w_{g_h} \right\|_{L^2_{-\alpha}(\mathcal{D}_h)} + \left\| w_{g_h} \right\|_{H^1_{-\alpha}(E_h)} \leq C \left\| g_h \right\|_{L^2_{-\alpha}(E_h)}.$$

(b) *If $g_h \in H^1_{-\alpha}(E_h)$, then $w_{g_h} \in H^1_{-\alpha}(\mathcal{D}_h) \cap H^2_{-\alpha}(E_h)$. Moreover, there exists C such that for every $g_h \in H^1_{-\alpha}(E_h)$,*

$$\left\| w_{g_h} \right\|_{H^1_{-\alpha}(\mathcal{D}_h)} + \left\| w_{g_h} \right\|_{H^2_{-\alpha}(E_h)} \leq C \left\| g_h \right\|_{H^1_{-\alpha}(E_h)}.$$

The proof of this lemma is given in [Mi87].

We now have enough material to construct a space of analytic functions defined on $\mathcal{B}_\ell$ which decay exponentially at infinity, in which (8.11) is solved with optimal regularity.

Let $\mathcal{X}$ be a complex Hilbert space. For any function $f : \mathcal{B}_\ell \longrightarrow \mathcal{X}$, and for any real number $\eta \in]-\ell,\ell[$ we denote by $f|_\eta : \mathbb{R} \longrightarrow \mathcal{X}$, $f|_\eta := f(\cdot + i\eta)$. With this notation, we can now define for any number $p \geq 1$ the Banach spaces

$$\mathcal{E}_\ell^{p,\alpha}(\mathcal{X}) := \{f : \mathcal{B}_\ell \to \mathcal{X},\ \text{holomorphic},\ f|_\eta \in H^p_{-\alpha}(\mathcal{X}),\ \eta \in]-\ell,\ell[,$$
$$\|f\|_{\mathcal{E}_\ell^{p,\alpha}(\mathcal{X})} = \sup_{\eta\in]-\ell,\ell[} \|f|_\eta\|_{H^p_{-\alpha}(\mathcal{X})} < +\infty\}. \tag{8.14}$$

Lemma 8.2.7 readily ensures that $\mathcal{E}_\ell^{p,\alpha}(\mathcal{X})$ is continuously embedded in $\mathrm{C}^{\omega,p-1}_{-\alpha}(\mathcal{B}_\ell, \mathcal{X})$ for $p \geq 1$ and that it is "an algebra" in the sense of Lemma 8.2.6. We can finally prove

Proposition 8.2.8. *Let $\alpha \in]0,\delta_h[$. For any $g_h \in \mathcal{E}_\ell^{1,\alpha}(E_h)$, (8.11) admits a unique solution w_{g_h} in $\mathcal{E}_\ell^{1,\alpha}(\mathcal{D}_h) \cap \mathcal{E}_\ell^{2,\alpha}(E_h)$. Moreover,*

$$w_{g_h}(\xi) = \mathfrak{F}(g_h)(\xi) = \int_{-\infty}^{+\infty} \mathcal{K}(s) g_h(\xi - s)\, ds,$$

and there exists C such that for every $g_h \in \mathcal{E}_\ell^{1,\alpha}(E_h)$,

$$\|w_{g_h}\|_{\mathcal{E}_\ell^{1,\alpha}(\mathcal{D}_h)} + \|w_{g_h}\|_{\mathcal{E}_\ell^{2,\alpha}(E_h)} \leq C \|g_h\|_{\mathcal{E}_\ell^{1,\alpha}(E_h)}.$$

Proof. $g_h \in \mathcal{E}_\ell^{1,\alpha}(E_h) \subset \mathrm{C}^{\omega,0}_{-\alpha}(\mathcal{B}_\ell, E_h)$. Lemma 8.2.2 ensures that there exists a unique solution of (8.11) in $\mathrm{C}^{\omega,0}_{-\alpha}(\mathcal{B}_\ell, E_h)$ given by

$$w_{g_h}(\xi) = \mathfrak{F}(g_h) = \int_{-\infty}^{+\infty} \mathcal{K}(s) g_h(\xi - s)\, ds.$$

We must now improve the regularity estimate of w_{g_h}. For any η in $]-\ell,\ell[$, $w_{g_h}|_\eta \in \mathrm{C}^0_{-\alpha}(\mathbb{R}, E_h) \cap C^1(\mathbb{R}, E_h)$ and satisfies

$$\frac{dw}{dx} - \mathcal{A}_h^o w = g_h|_\eta. \tag{8.15}$$

Since $g_h \in \mathcal{E}_\ell^{1,\alpha}(E_h)$, it follows that $g_h|_\eta \in H^1_{-\alpha}(E_h)$ and we denote by u the solution of (8.15) given by Lemma 8.2.7. This lemma ensures that $u \in H^1_{-\alpha}(\mathcal{D}_h) \cap H^2_{-\alpha}(E_h)$, and that there exists C which does not depend on g_h and η such that

$$\|u\|_{H^1_{-\alpha}(\mathcal{D}_h)} + \|u\|_{H^2_{-\alpha}(E_h)} \leq C \|g_h|_\eta\|_{H^1_{-\alpha}(E_h)}.$$

Thus, u, $w_{g_h}|_\eta$ are two solutions of (8.15) in $\mathrm{C}^0_{-\alpha}(\mathbb{R}, E_h) \cap C^1(\mathbb{R}, E_h)$. Then Remark 8.2.3 ensures that $u = w_{g_h}|_\eta$. Thus

$$\|w_{g_h}\|_{\mathcal{E}_\ell^{1,\alpha}(\mathcal{D}_h)} + \|w_{g_h}\|_{\mathcal{E}_\ell^{2,\alpha}(E_h)} \leq C \|g_h\|_{\mathcal{E}_\ell^{1,\alpha}(E_h)}. \quad \square$$

8.3 Persistence of periodic orbits, explicit form and complexification

8.3.1 Real periodic orbits : explicit form

This subsection is devoted to the proof of Theorem 8.1.12. As in Section 7.2, we look for a one parameter family of periodic orbits, expressed in terms of power series, which can be extended to the complex plane and whose terms we can control precisely. Theorem 8.1.12 is the same as Theorem 7.1.4 except that we have changed the norm with which we control the trigonometric polynomial functions $\widehat{Y}_n$. In Chapter 7, $\widehat{Y}_n$ is controlled with a C^1 norm, whereas here, $\widehat{Y}_n$ is controlled with a $H^1(\mathcal{D}) \cap H^2(E)$ norm. This is due to the unboundness of $\mathcal{A}_h^o$ and to the loss of regularity caused by the nonlinear terms f, in (8.1).

Theorem 7.1.4 is a direct consequence of the general Theorem 4.1.2 which ensures the existence of periodic orbits bifurcating from a pair of simple eigenvalues. The proof of Theorems 4.1.2 and 8.1.12 can be done with the same formalism once the problem is formulated as an implicit equation in appropriate Banach spaces : Equation (8.19) below is similar to equation (4.10). So, we redo only the beginning of the proof to formulate the problem of existence of periodic orbits as an implicit equation.

Using the analyticity of f in (8.1) from one space to a less regular one, we obtain the following lemma which is quite similar to Lemma 4.A.1.

Lemma 8.3.1. *System* (8.1) *is equivalent to*

$$\frac{dY}{dt} = \mathfrak{L}_\nu(Y) + \sum_{n\geq 2} \mathfrak{Q}_n(Y^{(n)}, \nu), \tag{8.16}$$

with

(a) $\mathfrak{L}_\nu = \mathfrak{L}_{0\nu} + \nu \mathfrak{L}_{1\nu}$ *where*

$$- \mathfrak{L}_{0\nu}(Y) = \left(L_{0\nu} Y_c, \frac{\mathcal{A}_h^o}{\nu}\, Y_h \right) \text{ with } L_{0\nu} = \begin{bmatrix} 0 & 1 & 0 & 0 \\ 1 & 0 & 0 & 0 \\ 0 & 0 & 0 & -\frac{\omega_{0,\nu}}{\nu} \\ 0 & 0 & \frac{\omega_{0,\nu}}{\nu} & 0 \end{bmatrix}$$

– $\mathfrak{L}_{1\nu}$ *has the form* $\mathfrak{L}_{1\nu}(Y) = (L_{1\nu} Y_c, \mathcal{B}_h^1(\nu)\, Y_h)$

with $L_{1\nu} = \begin{bmatrix} 0 & 0 & 0 & 0 \\ \mathcal{O}(\nu) & 0 & 0 & 0 \\ 0 & 0 & 0 & 0 \\ 0 & 0 & 0 & 0 \end{bmatrix}$ *and there exist* $\nu_0 < 1$, C_L, *such that* $\|\mathfrak{L}_{1\nu}\|_{\mathcal{L}(\mathcal{D},E)} \leq C_L$ *for every* $\nu \in]0, \nu_0]$.

(b) $\mathfrak{Q}_n(Y^{(n)},\nu) = \mathfrak{Q}_n(\underbrace{Y,\cdots,Y}_{\text{n times}},\nu)$ *is n linear symmetric from* $\mathcal{D}^n$ *to* E,

(c) *There exists* C_0, $r>0$ *such that*

$$|\mathfrak{Q}_n(Y_1,\cdots,Y_n,\nu)|_{_E} \leq C_0\frac{|Y_1|_{\mathcal{D}}}{r}\cdots\frac{|Y_n|_{\mathcal{D}}}{r},$$

for every $\nu\in[0,\nu_0]$, $n\geq 2$, *and* $Y_i\in\mathcal{D}$, $1\leq i\leq n$.

(d) *Denote* $\mathfrak{Q}_n = (\mathfrak{Q}_{n,c},\mathfrak{Q}_{n,h})$ *and* $\mathfrak{Q}_{n,c} = (\mathfrak{Q}_n^0,\mathfrak{Q}_n^1,\mathfrak{Q}_n^A,\mathfrak{Q}_n^B)$. *Then, there exits* C_1 *such that*

$$\begin{aligned}|\mathfrak{Q}_n^A(Y_1,\cdots,Y_n,\nu)| &\leq C_1\nu\,\frac{|Y_1|_{\mathcal{D}}}{r}\cdots\frac{|Y_n|_{\mathcal{D}}}{r},\\ |\mathfrak{Q}_n^B(Y_1,\cdots,Y_n,\nu)| &\leq C_1\nu\,\frac{|Y_1|_{\mathcal{D}}}{r}\cdots\frac{|Y_n|_{\mathcal{D}}}{r},\end{aligned}$$

for every $\nu\in[0,\nu_0]$, $n\geq 2$, *and* $Y_i\in\mathcal{D}$, $1\leq i\leq n$.

Observe the loss of regularity in (c), (d).

As in Chapter 4 we set $s = \underline{\omega}_{k,\nu}t$, $\underline{\omega}_{k,\nu} = \dfrac{\omega_{0,\nu}}{\nu}+\widehat{\omega}$, $Y(t)=\widehat{Y}(s) = \widehat{Y}(\underline{\omega}_{k,\nu}t)$. We define $\widehat{Y}_1 = (\widehat{Y}_{1,c},\widehat{Y}_{1,h})$ with $\widehat{Y}_{1,c}(s) = (0,0,\cos(s),\sin(s))$ and $\widehat{Y}_{1,h}=0$. We look for 2π-periodic solutions of the form

$$\widehat{Y}(s,k) = k\widehat{Y}_1(s) + k\widehat{Z}(s,k).$$

Under a convergence hypothesis, which will be specified later, system (8.16) is equivalent to

$$\mathcal{L}_\nu(\widehat{Z},\widehat{\omega}) = -\widehat{\omega}\frac{d\widehat{Z}}{ds} + \sum_{p\geq 1}k^p\sum_{m=0}^{p+1}C_{p+1}^m\mathfrak{Q}_{p+1}(\widehat{Y}_1^{(p+1-m)},\widehat{Z}^{(m)},\nu) \tag{8.17}$$

with

$$\mathcal{L}_\nu(\widehat{Z},\widehat{\omega}) = \frac{\omega_{0,\nu}}{\nu}\frac{d\widehat{Z}}{ds} - \mathfrak{L}_\nu\widehat{Z} + \widehat{\omega}\frac{d\widehat{Y}_1}{ds},$$

or equivalently $\mathcal{G}(k,\nu)=0$ where $\mathcal{G}$ was defined in Lemma 4.A.4, but now in the extended space.

In Chapter 4 this equation was solved by using Cauchy's method of undetermined coefficients with majorization of the power series involved. We looked for $u=(\widehat{Z},\widehat{\omega})$ in the form $u=\sum k^n u_n$ and the u_n were given by an induction of the form

$$\mathcal{L}_\nu(u_n) = g_n \tag{8.18}$$

where g_n depends on u_i, $1\leq i\leq n-1$, and $\mathfrak{Q}_p$. Because of the loss of regularity of $\mathfrak{Q}_p$, we must now solve (8.18) with optimal regularity for all the

u_n to have the same regularity. But because of the unboundedness of $\mathcal{A}_h^o$, optimal regularity cannot be obtained in $C^k(\mathbb{R}, E)$ whereas the Fourier series and the resolvent estimates given in Lemma 8.1.7 give it in $H^1(\mathcal{D}) \cap H^2(E)$.

So we define the following Banach spaces, which will play the same role as $\tilde{\mathcal{U}}_R^1$, $\mathcal{P}_{AR}^0$ defined in Chapter 4:

$$\begin{aligned}
T^1 &= \frac{\mathbb{R}}{2\pi\mathbb{Z}}, \\
\mathfrak{P}_R^2 &= \{\widehat{Y} / \widehat{Y} \in H^1(T^1, \mathcal{D}) \cap H^2(T^1, E), \widehat{Y} \text{ reversible}\}, \\
\mathfrak{P}_{AR}^1 &= \{\widehat{Y} / \widehat{Y} \in H^1(T^1, E), \widehat{Y} \text{ antireversible}\}, \\
\tilde{\mathfrak{P}}_R^2 &= \mathfrak{P}_R^2 \cap \{\widehat{Y} / \int_0^{2\pi} \langle \widehat{Y}(s), \widehat{Y}_1(s)\rangle ds = 0\}, \\
\tilde{\mathfrak{U}}_R^2 &= \{u = (\widehat{Y}, \widehat{\omega}) / \widehat{Y} \in \tilde{\mathfrak{P}}_R^2, \widehat{\omega} \in \mathbb{R}\}.
\end{aligned}$$

We norm these spaces with following norms:

$$\begin{aligned}
\left\|\widehat{Y}\right\|_{\mathfrak{P}^2}^2 &= \int_0^{2\pi} \left|\widehat{Y}\right|_E^2 + \left|\frac{d\widehat{Y}}{ds}\right|_E^2 + \left|\frac{d^2\widehat{Y}}{ds^2}\right|_E^2 + \left|\widehat{Y}\right|_{\mathcal{D}}^2 + \left|\frac{d\widehat{Y}}{ds}\right|_{\mathcal{D}}^2 ds \text{ for } \mathfrak{P}_R^2, \tilde{\mathfrak{P}}_R^2, \\
\left\|\widehat{Y}\right\|_{\mathfrak{P}^1}^2 &= \int_0^{2\pi} \left|\widehat{Y}\right|_E^2 + \left|\frac{d\widehat{Y}}{ds}\right|_E^2 ds \text{ for } \mathfrak{P}_{AR}^1, \\
\|u\|_{\mathfrak{U}^2}^2 &= \left\|\widehat{Y}\right\|_{\mathfrak{P}^2}^2 + |\widehat{\omega}|^2 \text{ for } \tilde{\mathfrak{U}}_R^2.
\end{aligned}$$

As in Chapter 4, we define now finite-dimensional subspaces of the previously defined ones. These are the 'trigonometric polynomial' subspaces:

$$\begin{aligned}
\mathfrak{P}_{n,R} &= \{\widehat{Y} / \widehat{Y} = \sum_{p=0}^{n} a_p \cos(ps) + \sum_{p=1}^{n} b_p \sin(ps), \\
&\qquad (a_p, b_p) \in \mathcal{D}^2,\ S(a_p) = a_p, S(b_p) = -b_p\}, \\
\mathfrak{P}_{n,AR} &= \{\widehat{Y} / \widehat{Y} = \sum_{p=0}^{n} a_p \cos(ps) + \sum_{p=1}^{n} b_p \sin(ps), \\
&\qquad (a_p, b_p) \in E^2,\ S(a_p) = -a_p, S(b_p) = b_p\}, \\
\tilde{\mathfrak{P}}_{n,R} &= \mathfrak{P}_{n,R} \cap \{\widehat{Y} / \int_0^{2\pi} \langle \widehat{Y}(s), \widehat{Y}_1(s)\rangle ds = 0\}, \\
\tilde{\mathfrak{U}}_{n,R} &= \tilde{\mathfrak{P}}_{n,R} \times \mathbb{R}.
\end{aligned}$$

Remark 8.3.2. $\mathfrak{P}_R^2$ is continuously embedded in $C^0(T^1, \mathcal{D})$. Thus, there exists C_{02} such that

$$\left\|\widehat{Y}\right\|_{C^0(T^1,\mathcal{D})} \leq C_{02} \left\|\widehat{Y}\right\|_{\mathfrak{P}^2}$$

for every $\widehat{Y} \in \mathfrak{P}_R^2$.

With this material we can now state: for every $\nu \in]0, \nu_0]$, k satisfying $|k| < \dfrac{r}{C_{02}(1 + \|\widehat{Y}_1\|_{\mathfrak{P}^2})}$ and $\widehat{Z} \in \tilde{\mathfrak{P}}^2_R$ satisfying $\|\widehat{Z}\|_{\mathfrak{P}^2} < 1$, (8.1) is equivalent to (8.17).

Finally the following lemma holds:

Lemma 8.3.3. *Let* $\rho \in \left]0, \dfrac{r}{C_{02}(1 + \|\widehat{Y}_1\|_{\mathfrak{P}^2})}\right[$ *be fixed. For every* $\nu \in]0, \nu_0]$, k *satisfying* $|k| < \rho$ *and* $u \in \tilde{\mathcal{U}}^1_R$ *satisfying* $\|u\|_{\mathfrak{U}^2} < 1$, *system* (8.1) *reads*

$$\mathcal{G}_\nu(k, u) = 0 \quad \text{with } u = (\widehat{Z}, \widehat{\omega}), \tag{8.19}$$

$$\mathcal{G}_\nu(k, u) = \sum_{p+q \geq 1} k^p \mathcal{G}_{pq}[u^{(q)}] = \mathcal{L}_\nu(u) + k\mathcal{G}_{10} + \sum_{p+q \geq 2} k^p \mathcal{G}_{pq}[u^{(q)}],$$

where

$$\mathcal{G}_{01}[u] = \mathcal{L}_\nu(u), \quad \mathcal{G}_{02}[u^{(2)}] = \widehat{\omega}\frac{d\widehat{Z}}{ds}, \quad \mathcal{G}_{0q}[u^{(q)}] = 0 \ \ \forall q \geq 3,$$

$$\forall p \geq 1, \ k^p \mathcal{G}_{pq}[u^{(q)}] = \begin{cases} -C^q_{p+1} k^p \mathfrak{Q}_{p+1}(\widehat{Y}_1^{(p+1-q)}, \widehat{Z}^{(q)}, \nu) & \text{if } q \leq p+1, \\ 0 & \text{if } q \geq p+2. \end{cases}$$

Moreover,

(a) *for every* (p, q), $\mathcal{G}_{pq}$ *is* q *linear from* $(\tilde{\mathfrak{U}}^2_R)^q$ *to* $\mathfrak{P}^1_{AR}$,
(b) *there exists* C_2, *such that*

$$\|k^p \mathcal{G}_{pq}[u_1, \cdots, u_q]\|_{\mathfrak{P}^1} \leq C_2 \|u_1\|_{\mathfrak{U}^2} \cdots \|u_q\|_{\mathfrak{U}^2} (k/\rho)^p.$$

for every $\nu \in]0, \nu_0]$, k *satisfying* $|k| < \rho$, $(p, q) \neq (0, 1)$, *and* $u_i \in \tilde{\mathfrak{U}}^2_R$ *satisfying* $\|u_i\|_{\mathfrak{U}^2} < 1$, $1 \leq i \leq q$.

Lemma 8.3.3 is the same as Lemma 4.A.5 with a change of norm. In Lemma 4.A.5, k had to satisfy $k \in]0, r/2[$ and now $k < \rho$ with $\rho = \dfrac{r}{C_{02}(1 + \|\widehat{Y}_1\|_{\mathfrak{P}^2})}$. This difference is only due to the change of norm. Indeed, in Lemma 4.A.5, $\|\widehat{Y}_1\| = 1$ and the embedding was trivial (see Remark 8.3.2). So, in Lemma 4.A.5 $C_{02} = 1$ and "$\rho = \dfrac{r}{1(1+1)} = \dfrac{r}{2}$".

The end of the proof of Theorem 8.1.12 is the same as the one of 4.1.2: as in Chapter 4 we proceed in three steps. The study of $\mathcal{L}_\nu$, and in particular the optimal regularity result is given in Appendix 8.B. The proof of the two last steps, (obtaining u_n by induction and majorization of the power series) are exactly the same as in Chapter 4. □

8.3.2 Complexification of the periodic orbits

As in Chapter 7, we want to extend $Y_{k,\nu}$ by analyticity in the complex field. Here again

$$Y_{k,\nu}(t) = \widehat{Y}(\underline{\omega}_{k,\nu}t, k, \nu), \quad \widehat{Y}(s, k, \nu) = \sum_{n\geq 1} k^n \widehat{Y}_n(s, \nu)$$

where $\widehat{Y}_n$ is a trigonometric polynomial which can be extended by analyticity to obtain an entire function. The following estimates give a lower bound for the radius of convergence of the complexified power series: there exists $C'_{\widehat{Y}}$ such that for every $k \in [0, k_0]$, $\nu \in]0, \nu_1]$, the following estimates holds

$$\forall \xi \in \mathbb{C}, \ \forall j \geq 1, \ k^n \left| \frac{d^j Y_n}{d\xi^j}(\xi, \nu) \right|_{\mathcal{D}} \leq C'_{\widehat{Y}} n^j \sqrt{2n+1} \left(\frac{k}{k_0} \right)^n e^{n|\mathcal{I}m(\xi)|}.$$

The proof of these estimates is the same as those of Proposition 7.2.1, using the embedding $\mathfrak{P}^2_R \subset C^0_b(T^1, \mathcal{D})$.

8.4 Persistence of homoclinic connections to exponentially small periodic orbits

Now we look for a reversible solution of (8.9), (8.10) homoclinic to the periodic orbit $Y_{k,\nu}$ in the same form as in Chapter 7:

$$Y(t) = h(t) + Y_{k,\nu}(t + \varphi \tanh \tfrac{1}{2} t) + v(t).$$

For the same reason as in Section 7.3 we complexify the problem and we make the same choice of parameters to have $h(\xi)$, $\tanh \frac{1}{2}\xi$, $Y_{k,\nu}(\xi + \varphi \tanh \frac{1}{2}\xi)$ well defined (see Subsection 7.3.1). The equation satisfied by v is now

$$\frac{dv_c}{d\xi} - DN(h, \nu) v_c = g_c(v(\xi), \varphi, K, \nu, \xi), \tag{8.20}$$

$$\frac{dv_h}{d\xi} - \frac{A^o_h}{\nu} v_h = g_h(v(\xi), \varphi, K, \nu, \xi), \tag{8.21}$$

with

$$\begin{aligned} g_c(v(\xi), \varphi, K, \nu, \xi) &= N'(v_c(\xi), \varphi, K, \nu, \xi) + R'_c(v(\xi), \varphi, K, \nu, \xi) \\ &\quad + \varphi \Phi_c(\xi, \varphi, K, \nu) \\ g_h(v(\xi), \varphi, K, \nu, \xi) &= R'_h(v(\xi), \varphi, K, \nu, \xi) + \varphi \Phi_h(\xi, \varphi, K, \nu) \end{aligned}$$

$$N'(v_c, \varphi, K, \nu, \xi) = N(h + Y_{k,\nu,c} + v_c, \nu) - N(h, \nu) - N(Y_{k,\nu,c}, \nu) - DN(h, \nu) v_c,$$

$$\begin{aligned}
R'_c(v,\varphi,K,\nu,\xi) &= R_c(h+Y_{k,\nu}+v) - R_c(Y_{k,\nu}),\\
R'_h(v,\varphi,K,\nu,\xi) &= R_h(h+Y_{k,\nu}+v) - R_h(Y_{k,\nu}),\\
\Phi(\xi,\varphi,K,\nu) &= \frac{-1}{2\cosh^2\frac{1}{2}\xi}\ \frac{dY_{k,\nu}}{d\xi}(\xi+\varphi\tanh\tfrac{1}{2}\xi),
\end{aligned}$$

where h, v are evaluated in ξ, and $Y_{k,\nu}$ in $\xi+\varphi\tanh\frac{1}{2}\xi$.

8.4.1 Choice of the space for v

We want to have the same scheme of proof as in Section 7.3. Thus v must lie in a space of reversible analytic functions defined on $\mathcal{B}_\ell$, which decay exponentially at infinity. As in Section 7.3, we want to get v by the Banach Fixed Point Theorem. Here v has two components, $v=(v_c,v_h)$, and $v_h(\xi)$ lies in an infinite-dimensional space. The new difficulty is that (8.21) is quasilinear, i.e., $\mathcal{A}_h^o$ is an unbounded operator and g_h is only bounded from $\mathcal{D}_h$ to E_h. To compensate for the loss of regularity due to g_h we must look for v_h in a space where the affine equations induced by $\mathcal{A}_h^o/\nu$ are solved with optimal regularity. Proposition 8.2.8 gives a family of appropriate spaces for v_h, $(\mathcal{E}_\ell^{1,\alpha}(\mathcal{D}_h)\cap\,\mathcal{E}_\ell^{2,\alpha}(E_h))_{\alpha>0}$.

The embedding $\mathcal{E}_\ell^{1,\alpha}(\mathcal{D}_h)\subset \mathrm{C}_{-\alpha}^{\omega,0}(\mathcal{B}_\ell,\mathcal{D}_h)$ ensures that $\mathcal{E}_\ell^{1,\alpha}(\mathcal{D}_h)\cap\mathcal{E}_\ell^{2,\alpha}(E_h)$ is composed of analytic functions defined on $\mathcal{B}_\ell$, that decay exponentially at infinity. For using Proposition 8.2.8, g_h should lie in $\mathcal{E}_\ell^{1,\alpha}(E_h)$. Thus v_c must lie in $\mathcal{E}_\ell^{1,\alpha}(\mathbb{C}^4)$, which is embedded in $\mathrm{C}_{-\alpha}^{\omega,0}(\mathcal{B}_\ell,\mathbb{C}^4)$. It remains to choose α.

Firstly, observe that the Banach space where v lies in Section 7.3 is $\mathrm{C}_{-\lambda}^{\omega,0}(\mathcal{B}_\ell,\mathbb{C}^4)$ with λ any number in $]0,1[$. Secondly, in Proposition 8.2.8 α must satisfy $\alpha<\delta_h(\mathcal{A}_h^o/\nu)$, where δ_h was defined in Lemma 8.1.7. Remark 8.1.8 ensures that $\delta_h(\mathcal{A}_h^o/\nu)=\delta_h(\mathcal{A}_h^o)/\nu$ for any ν in $]0,1[$. So we choose $\alpha=\lambda$ in $]0,1[$ an ν_2 such that for any ν in $]0,\nu_2]$, $\delta_h(\mathcal{A}_h^o)/\nu\geq 1$. Thus the restriction on α is fulfilled for any ν in $]0,\nu_2]$.

Let λ be fixed such that $0<\lambda<1$. We define the Banach space $\mathfrak{H}_\ell^\lambda=\mathfrak{H}_{\ell,c}^\lambda\times\mathfrak{H}_{\ell,h}^\lambda$ with norm $\|v\|_{\mathfrak{H}_\ell^\lambda}=\|v_c\|_{\mathfrak{H}_{\ell,c}^\lambda}+\|v_h\|_{\mathfrak{H}_{\ell,h}^\lambda}$ where

$$\begin{aligned}
\mathfrak{H}_{\ell,c}^\lambda &= \mathcal{E}_\ell^{1,\lambda}(\mathbb{C}^4)|_{\mathrm{R}}, & \|v_c\|_{\mathfrak{H}_{\ell,c}^\lambda} &= \|v_c\|_{\mathcal{E}_\ell^{1,\lambda}(\mathbb{C}^4)},\\
\mathfrak{H}_{\ell,h}^\lambda &= \mathcal{E}_\ell^{1,\lambda}(\mathcal{D}_h)|_{\mathrm{R}}\cap\mathcal{E}_\ell^{2,\lambda}(E_h)|_{\mathrm{R}}, & \|v_h\|_{\mathfrak{H}_{\ell,h}^\lambda} &= \|v_h\|_{\mathcal{E}_\ell^{2,\lambda}(E_h)}+\|v_h\|_{\mathcal{E}_\ell^{1,\lambda}(\mathcal{D}_h)},
\end{aligned}$$

and

$$\begin{aligned}
\mathcal{E}_\ell^{p,\alpha}(\mathcal{X})|_{\mathrm{R}} &= \mathcal{E}_\ell^{p,\alpha}(\mathcal{X})\cap\{f, f \text{ reversible}\},\\
\mathcal{E}_\ell^{p,\alpha}(\mathcal{X})|_{\mathrm{AR}} &= \mathcal{E}_\ell^{p,\alpha}(\mathcal{X})\cap\{f, f \text{ antireversible}\}.
\end{aligned}$$

Denote by $\mathfrak{B}\mathfrak{H}_\ell^\lambda(\delta\sqrt{\nu})$ the ball of radius $\delta\sqrt{\nu}$ of $\mathfrak{H}_\ell^\lambda$. We look for v in $\mathfrak{B}\mathfrak{H}_\ell^\lambda(\delta\sqrt{\nu})$ with $\delta\leq 1$.

8.4.2 Integral equation

In Section 7.3, the major difficulty of the study came from the fact that for an antireversible function g_c which decays exponentially at infinity, the affine equation

$$\frac{dv_c}{d\xi} - DN(h,\nu)v_c = g_c(\xi)$$

admits a reversible solution which decays exponentially at infinity if and only if g_c satisfies a solvability condition

$$\int_0^{+\infty} \langle r_-^*(s), g_c(s)\rangle ds = 0.$$

This solvability condition was studied in detail in subsection 7.3.5 and solved by an appropriate choice of the phase shift φ. Here, the second equation (8.21) does not add any solvability condition because Lemma 8.2.4 ensures that $\mathfrak{F}$ preserves reversibility. More precisely,

Proposition 8.4.1. *For every* $\nu \in]0,\nu_2]$, $\varphi \in [0, \frac{6\pi\nu}{\omega}]$, $\mathcal{K} \in [0, K_1]$ *and* $v \in \mathfrak{B}\mathfrak{H}_\ell^\lambda(\delta\sqrt{\nu})$, *if* $v = \mathcal{F}_{K,\varphi,\nu}(v)$, *with* $\mathcal{F}_{K,\varphi,\nu}(v) = (\mathcal{F}^c_{K,\varphi,\nu}(v), \mathcal{F}^h_{K,\varphi,\nu}(v))$ *where*

$$\begin{aligned}(\mathcal{F}^c_{K,\varphi,\nu}(v))(\xi) &= \int_{[0,\xi]} \langle p^*(s), g_c(v(s),\varphi,K,\nu,s)\rangle ds\ p(\xi)\\ &\quad - \int_{[\xi,+\infty)} \langle q^*(s), g_c(v(s),\varphi,K,\nu,s)\rangle ds\ q(\xi)\\ &\quad - \int_{[\xi,+\infty)} \langle r_+^*(s), g_c(v(s),\varphi,K,\nu,s)\rangle ds\ r_+(\xi)\\ &\quad - \int_{[\xi,+\infty)} \langle r_-^*(s), g_c(v(s),\varphi,K,\nu,s)\rangle ds\ r_-(\xi),\end{aligned}$$

and $\mathcal{F}^h_{K,\varphi,\nu}(v) = \mathfrak{F}(g_h(v,\varphi,K,\nu,s))$, *then* v *is a solution of Eqs.* (8.20), (8.21). *Moreover* $\mathcal{F}^c_{K,\varphi,\nu}(v)$ *is reversible if and only if* $J(v,\varphi,K,\nu) = 0$ *where*

$$J(v,\varphi,K,\nu) = \int_0^{+\infty} \langle r_-^*(t), g_c(v(t),\varphi,K,\nu,t)\rangle dt.$$

The functions $p, q, r_+, r_-, p^*, q^*, r_+^*, r_-^*$ are introduced in Subsection 7.3.3. The proof follows readily from Propositions 7.3.17, 8.2.8.

8.4.3 Choice of the phase shift

The study of the solvability condition can be done as in Section 7.3. The mixed terms in the remainder R_c do not pose any difficulty. Indeed, in Lemma 7.B.2, which is the cornerstone of the study we only need to bound the remainder in $C^{\omega,0}_{-\lambda}$. The embeddings of $\mathfrak{H}^{\lambda}_{\ell,c}$ in $\mathrm{C}^{\omega,0}_{-\lambda}(\mathcal{B}_\ell, \mathbb{C}^4)$ and of $\mathfrak{H}^{\lambda}_{\ell,h}$ in $\mathrm{C}^{\omega,0}_{-\lambda}(\mathcal{B}_\ell, \mathrm{d}_h)$ ensure that we can do this. Thus we can prove:

Theorem 8.4.2. *There exist M_1, $\nu_3 \leq \nu_2$ such that for every $\nu \in]0, \nu_3]$, $K \in [M_1\nu^2, K_1]$ and $v \in \mathfrak{B}\mathfrak{H}^{\lambda}_{\ell}(\delta\sqrt{\nu})$, there exists $\varphi(v, K, \nu)$ such that $\mathcal{F}_{K,\varphi(v,K,\nu),\nu}(v)$ is reversible. Moreover, there exists M_2 such that for every $\nu \in]0, \nu_3]$, $K \in [M_1\nu^2, K_1]$ and $v, v' \in \mathfrak{B}\mathfrak{H}^{\lambda}_{\ell}(\delta\sqrt{\nu})$,*

$$0 \leq \frac{\pi\nu}{8\omega} \leq \varphi(v, K, \nu) \leq \frac{4\pi\nu}{\omega} \leq \frac{6\pi\nu}{\omega},$$

$$|\varphi(v, K, \nu) - \varphi(v', K, \nu)| \leq M_2 \frac{\nu^2}{K}(\delta\sqrt{\nu} + K + \nu)\, \|v - v'\|_{\mathfrak{H}^{\lambda}_{\ell}}.$$

8.4.4 Fixed point

As in Section 7.3, we define $\widehat{\mathcal{F}}_{K,\nu}(v) = \mathcal{F}_{K,\varphi(v,K,\nu),\nu}(v)$ and we check that

Proposition 8.4.3. *For every $\nu \in]0, \nu_3]$, $K \in [M_1\nu^2, K_1]$ and $v \in \mathfrak{B}\mathfrak{H}^{\lambda}_{\ell}(\delta\sqrt{\nu})$, $\widehat{\mathcal{F}}_{K,\nu}(v)$ is holomorphic in $\mathcal{B}_\ell$, and is reversible. Moreover there exists M_3 such that for every $\nu \in]0, \nu_3]$, $K \in [M_1\nu^2, K_1]$ and $v, v' \in \mathfrak{B}\mathfrak{H}^{\lambda}_{\ell}(\delta\sqrt{\nu})$, $\widehat{\mathcal{F}}_{K,\nu}$ satisfies*

$$\left\|\widehat{\mathcal{F}}_{K,\nu}(v)\right\|_{\mathfrak{H}^{\lambda}_{\ell}} \leq M_3(\|v\|^2_{\mathfrak{H}^{\lambda}_{\ell}} + \nu + \frac{K}{\nu}),$$

$$\left\|\widehat{\mathcal{F}}_{K,\nu}(v) - \widehat{\mathcal{F}}_{K,\nu}(v')\right\|_{\mathfrak{H}^{\lambda}_{\ell}} \leq M_3[\,\|v\|_{\mathfrak{H}^{\lambda}_{\ell}} + \|v'\|_{\mathfrak{H}^{\lambda}_{\ell}} + \delta\sqrt{\nu} + \nu + K\,]\, \|v - v'\|_{\mathfrak{H}^{\lambda}_{\ell}}.$$

The details are left to the reader.

Proof of Theorem 8.1.13 We conclude this section by the proof of Theorem 8.1.13 which ensures the existence of reversible homoclinic connections to exponentially small periodic orbits.

Proposition 8.4.3 ensures that $\widehat{\mathcal{F}}_{K,\nu}$ is a contraction mapping from $\mathfrak{B}\mathfrak{H}^{\lambda}_{\ell}(\delta\sqrt{\nu})$ to $\mathfrak{B}\mathfrak{H}^{\lambda}_{\ell}(\delta\sqrt{\nu})$ if and only if

$$\nu \in]0, \nu_3], \quad K \in [M_1\nu^2, K_1],$$

$$M_3[\delta^2\nu+\nu+\frac{K}{\nu}]\leq\delta\sqrt{\nu},\quad M_3[\,2\delta+\delta\sqrt{\nu}+\nu+K]<1.$$

We can then choose $K=M_1\nu^2$, δ such that

$$\delta=2M_3\frac{[\nu+K/\nu]}{\sqrt{\nu}}=2M_3[M_1+1]\sqrt{\nu}$$

and $\nu_4\leq\nu_3$ such that for any ν in $]0,\nu_4]$,

$$M_3\delta\sqrt{\nu}\leq\frac{1}{2},\quad M_3[\,2\delta+\delta\sqrt{\nu}+\nu+K]<1.$$

This ensures that for any $\ell\in]0,\pi[$ and any $\lambda\in]0,1[$ there exist M, ν_4 such that for any ν in $]0,\nu_4]$ the full System (8.9), (8.10) admits two reversible solutions analytic in $t\in\mathbb{R}$ taking values in $\mathcal{D}$ of the form

$$Y(t)=h(t)+v(t)+Y_{k,\nu}(t+\varphi\tanh\tfrac{1}{2}t)$$

where

$$|h(t)|_{\mathcal{D}}+|v(t)|_{\mathcal{D}}=\underset{t\to\pm\infty}{\mathcal{O}}(\mathrm{e}^{-\lambda|t|}),$$

$Y_{k,\nu}$ is a periodic solution of the full system satisfying

$$\forall t\in\mathbb{R},\ \left|Y_{k,\nu}(t+\varphi\tanh\tfrac{1}{2}t)\right|_{\mathcal{D}}\leq M\nu^2\mathrm{e}^{-\ell\omega/\nu}.\ \square$$

8.5 Application to water waves

One example of a vector field satisfying Assumptions 8.1.1–8.1.6 occurs when describing the irrotational flow of an inviscid fluid layer under the influence of gravity and small surface tension (Bond number $b<\frac{1}{3}$) for a Froude number F close to 1. In this context a homoclinic solution to a periodic orbit is called a generalized solitary wave. Theorem 8.1.13 shows then that there exist solitary waves with oscillations at infinity of order less than $c(\ell)\exp\left(-\omega\ell\sqrt{\frac{\frac{1}{3}-b}{|F^{-2}-1|}}\right)$, ℓ being any number between 0 and π, and $\omega\neq0$ satisfying $\omega\cosh\omega=(1+b\omega^2)\sinh\omega$. Let us give more details. In [IK92], with the stream function as the transverse coordinate, Euler equations are rewritten as the following "evolution" equation:

$$\begin{aligned}\frac{d\beta}{dx}&=\frac{1}{b}(1+\beta^2)^{3/2}(W(\cdot,1)+(1+\mu)\,([\frac{1}{g}]-1))\quad\text{at } y=1,\\ \partial_x W&=K(W)\partial_y W\qquad\text{in }\mathbb{R}\times[0,1],\end{aligned}\tag{8.22}$$

where

$(x,y) \in \mathbb{R} \times [0,1]$, $W = (W_1, W_2)$, $\beta(x) = W_2(x,1)$,

$$g = \left(\frac{1+2W_1}{1+W_2^2}\right)^{1/2}, \quad K(W) = \begin{pmatrix} W_2 g & -g^3 \\ g^{-1} & W_2 g \end{pmatrix}, \quad [W_1] = \int_0^1 W_1 \, dy.$$

We must add the boundary condition $W_2 = 0$ at $y = 0$ which will be incorporated into the function-space setting.

In (8.22) b is the Bond Number, and $(1+\mu) = F^{-2}$ where F is the Froude number. Throughout this section we assume that b is fixed in $]0, \frac{1}{3}[$. We work with only one bifurcation parameter μ close to 0 (i.e., F close to 1). The System (8.22) is treated as a quasilinear evolution equation for

$$w = (\beta, W_1, W_2) \in \mathcal{D} = \mathbb{R} \times H^1(0,1) \times H^1(0,1) \cap \{W_2(0) = 0, W_2(1) = \beta\},$$

in the Hilbert space

$$E = \mathbb{R} \times L_2(0,1) \times L_2(0,1), \text{ with norm } \|w\|_E^2 = |\beta|^2 + |W_1|_{L^2}^2 + |W_2|_{L^2}^2.$$

We rewrite (8.22) in the abstract form

$$\frac{dw}{dx} = \mathcal{A}(\mu)w + f(w,\mu), \tag{8.23}$$

where $\mathcal{A}(\mu)$ is the linearization of the right-hand side of (8.22) in $w = 0$,

$$\mathcal{A}(\mu)w = \begin{pmatrix} \frac{1}{b}(W_1(1) - (1+\mu)[W_1]) \\ -\partial_y W_2 \\ \partial_y W_1 \end{pmatrix}. \tag{8.24}$$

We must now check that Assumptions 8.1.1-8.1.6 are fulfilled by $\mathcal{A}(\mu)$ and f. Firstly, $\mathcal{A}(\mu)$ is a closed, densely defined, unbounded operator on domain $\mathcal{D}$ which does not depend on μ. So Assumption 8.1.1 is satisfied. Secondly, observe that $\mathcal{A}(\mu) = \mathcal{A}^o + \mu\mathcal{A}^1$, where $\mathcal{A}^o = \mathcal{A}(0)$ and that $\mathcal{A}^1 \in \mathcal{L}(E)$, the Banach space of bounded operator in E ($\mathcal{A}^1$ does not depend on μ). Thus Assumption 8.1.3 is satisfied.

Now we are interested in the spectrum of $\mathcal{A}(\mu)$. For all μ, $\mathcal{A}(\mu)$ has a compact resolvent (see [IK92]). Thus its spectrum $\Sigma\mathcal{A}(\mu)$ consists of isolated eigenvalues σ of finite multiplicities. Their locations are determined by the equation

$$\sigma \cos\sigma = (1 + \mu - b\sigma^2)\sin\sigma, \quad \sigma \in \mathbb{C}. \tag{8.25}$$

For $\mu = 0$ the central spectrum Σ_0 of $\mathcal{A}^o$, $\Sigma_0 = \Sigma\mathcal{A}^o \cap i\mathbb{R}$ is $\{0, \pm i\omega\}$ where 0 is a double non-semi-simple eigenvalue, and $\pm i\omega$ are two simple eigenvalues. The corresponding eigenvectors and generalized eigenvectors are

$$\varphi_0 = \begin{pmatrix} 0 \\ 1 \\ 0 \end{pmatrix}, \quad \varphi_1 = \begin{pmatrix} -1 \\ 0 \\ -y \end{pmatrix}, \quad \varphi_+ = \begin{pmatrix} \mathrm{i}\sinh\omega \\ -\cosh\omega y \\ \mathrm{i}\sinh\omega y \end{pmatrix}, \quad \varphi_- = \overline{\varphi_+},$$

$$\mathcal{A}^o\varphi_0 = 0, \quad \mathcal{A}^o\varphi_1 = \varphi_0, \quad \mathcal{A}^o\varphi_+ = \mathrm{i}\omega\varphi_+, \quad \mathcal{A}^o\varphi_- = -\mathrm{i}\omega\varphi_-,$$

and $\omega > 0$ satisfies $\omega\cosh\omega = (1 + b\omega^2)\sinh\omega$. Hence, Assumption 8.1.4 (a) is satisfied. In addition, the adjoint $\mathcal{A}^{o*}$ of $\mathcal{A}^o$ is given by (see [IK92])

$$\mathcal{A}^{o*}w = \begin{pmatrix} -W_1(1) \\ -\partial_y W_2 + W_2(1) \\ \partial_y W_1 \end{pmatrix} \tag{8.26}$$

with $\mathcal{D}^* = \mathbb{R} \times H^1(0,1) \times H^1(0,1) \cap \{W_2(0) = 0, W_2(1) = -\frac{\beta}{b}\}$.

We have $\Sigma\mathcal{A}^{o*} = \Sigma\mathcal{A}^o$; the (generalized) eigenvectors φ_j^* belonging to the central part of $\Sigma\mathcal{A}^{o*}$ and satisfying $(\varphi_j^*, \varphi_k) = \delta_{j,k}$ read

$$\varphi_0^* = \frac{1}{b - \frac{1}{3}} \begin{pmatrix} 0 \\ b + \frac{1}{2}(y^2 - 1) \\ 0 \end{pmatrix}, \quad \varphi_1^* = \frac{1}{b - \frac{1}{3}} \begin{pmatrix} -b \\ 0 \\ y \end{pmatrix},$$

$$\varphi_+^* = \frac{1}{(b - 1/\omega^2)\sinh^2\omega + 1} \begin{pmatrix} \mathrm{i}b\sinh\omega \\ -\cosh\omega y + \dfrac{\sinh q}{q} \\ -\mathrm{i}\sinh\omega y \end{pmatrix}, \quad \varphi_-^* = \overline{\varphi_+^*}.$$

The resolvent estimates given in Assumption 8.1.4(b), are obtained for the Euler equation in [IK92]. The explicit form of f given by (8.22) and (8.24) ensures that it satisfies Assumption 8.1.5.

Now, observe that (8.22) is reversible, i.e., the right-hand side of (8.22) anticommutes with the reflection

$$S = \begin{pmatrix} -1 & 0 & 0 \\ 0 & 1 & 0 \\ 0 & 0 & -1 \end{pmatrix}.$$

This property results from the mirror symmetry $x \longrightarrow -x$ of the Euler equation even in the moving frame. So, Assumption 8.1.2 is fulfilled.

To conclude, we check that $S\varphi_0 = \varphi_0$, and that

$$(\varphi_1^*, \mathcal{A}^1(0)\varphi_0) = -\sigma, \quad (\varphi_1^*, D^2_{ww}f(0,0)[\varphi_0, \varphi_0]) = \tfrac{3}{2}\sigma,$$

where $\sigma = \left(\frac{1}{3} - b\right)^{-1} > 0$.

Thus Theorem 8.1.13 holds for the water wave problem. Returning to our original problem we compute the modulation of the free surface S using its expression given in [IK92]:

$$S = -1 + [\frac{1}{g}] = -[W_1] + \text{ higher order terms.}$$

Theorem 8.1.13 justifies the existence of a free surface S of the form

$$S = \left(-\alpha_0 + \frac{2}{\omega} \sinh(\omega)\ U \right)\ (1 + \mathcal{O}(\mu)), \tag{8.27}$$

where

$$\alpha_0(x) = \mu \cosh^{-2} \tfrac{\nu x}{2},\quad U(x) = M_1 \nu^4 \mathrm{e}^{-\ell\omega/\nu} \cos(\nu \underline{\omega}_{k,\nu} x + \varphi \tanh(\tfrac{\nu x}{2})),$$

with $\nu \underline{\omega}_{k,\nu} = \omega + \mathcal{O}(\nu^2)$ and $\varphi = 2\gamma\nu^2/\omega$. We recall that $\mu < 0$ and that $\nu = \sqrt{-(\frac{1}{3} - b)^{-1}\mu}$.

We observe in (8.27) that the shape of the free surface looks like a true solitary wave of elevation, with an exponentially small oscillating part superimposed and which is dominant at infinity.

Remark 8.5.1. The existence of generalized solitary wave with exponential ripples at infinity was established by Sun in [SS93] and by Lombardi in [Lo97]. Moreover in [Su99], Sun proved the non existence of true solitary wave for Froube number close to one and for Bond number less and close to $\frac{1}{3}$. The generic non persistence of homoclinic connections to 0 obtained in Chapter 7 suggests that there should not exist any true solitary wave for Froude number close to 1 and almost all Bond numbers between 0 and $\frac{1}{3}$.

8.A Appendix. Proof of Theorem 8.1.10

8.A.1 Substitution

We substitute the expression $w = X_c + X_h + \phi(X_c, \mu)$ in (8.1) using Equations (8.4 - 8.7), and we identify monomials in X_c. We obtain a hierarchy of equations:

$$(\mathrm{Id} + \phi_\mu^1)\mathcal{Q}_\mu^1 + \phi_\mu^1 \mathcal{A}_c^{\mathrm{o}} - \mathcal{A}(\mu)\phi_\mu^1 = (\mathcal{A}(\mu) - \mathcal{A}_c^{\mathrm{o}})\pi_c, \tag{8.28}$$

$$\forall X_c,\ (\mathrm{Id} + \phi_\mu^1)\mathcal{Q}_\mu^k[X_c^{(k)}] + D\phi_\mu^k[X_c^{(k)}](\phi_\mu^1 + \mathcal{A}_c^{\mathrm{o}})X_c - \mathcal{A}(\mu)\phi_\mu^k[X_c^{(k)}] = P_\mu^k[X_c^{(k)}], \tag{8.29}$$

where $P_\mu^k[X_c^{(k)}]$ only depends on f and ϕ_μ^i, $i \leq k-1$, given by the previous orders.

8.A.2 Study of equation (8.28)

This equation has the form $\mathcal{I}_1(\phi^1_\mu, \mathcal{Q}^1_\mu, \mu) = 0$ with

$$\begin{array}{rccc} \mathcal{I}_1 : & \mathcal{L}(E_c, \mathcal{D}) \times (\mathcal{L}(E_c) \cap \mathcal{N}) \times \mathbb{R} & \longrightarrow & \mathcal{L}(E_c, E) \\ & (\phi^1, \mathcal{Q}^1, \mu) & \mapsto & \mathcal{I}_1(\phi^1, \mathcal{Q}^1, \mu) \end{array}$$

where $\mathcal{I}_1(\phi^1, \mathcal{Q}^1, \mu) = (\mathrm{Id} + \phi^1)\mathcal{Q}^1 + \phi^1 \mathcal{A}^o_c - \mathcal{A}(\mu)\phi^1_\mu - (\mathcal{A}(\mu) - \mathcal{A}^o_c)\pi_c$, and $\mathcal{N}$ is the set of polynomial functions satisfying (8.8).

We observe that $\mathcal{I}_1(0,0,0) = 0$ and $\mathcal{I}_1$ is analytic in a neighborhood of 0, so we want to use the Analytic Implicit Function Theorem. Thus we must invert the differential

$$\begin{array}{rccc} D\mathcal{I}_1 : & \mathcal{L}(E_c, \mathcal{D}) \times (\mathcal{L}(E_c) \cap \mathcal{N}) & \longrightarrow & \mathcal{L}(E_c, E) \\ & (\phi^1, \mathcal{Q}^1) & \mapsto & \mathcal{Q}^1 + \phi^1 \mathcal{A}^o_c - \mathcal{A}^o \phi^1. \end{array}$$

So after the decomposition we must solve

$$\mathcal{Q}^1 + \pi_c \phi^1 \mathcal{A}^o_c - \mathcal{A}^o_c \pi_c \phi^1 = V_c, \tag{8.30}$$

$$\pi_h \phi^1 \mathcal{A}^o_c - \mathcal{A}^o_h \pi_h \phi^1 = V_h, \tag{8.31}$$

for any $V_c \in \mathcal{L}(E_c)$, $V_h \in \mathcal{L}(E_c, E_h)$. Equation (8.30) is the usual homological equation solved in [ETBCI87]. Equation (8.31) is equivalent to

$$\frac{d}{dx}(\pi_h \phi^1 \mathrm{e}^{x\mathcal{A}^o_c} X_c) - \mathcal{A}^o_h(\pi_h \phi^1 \mathrm{e}^{x\mathcal{A}^o_c} X_c) = V_h \mathrm{e}^{x\mathcal{A}^o_c} X_c \quad \text{for } x \in \mathbb{R}, X_c \in E_c.$$

Since all eigenvalues of $\mathcal{A}^o_c$ are purely imaginary, $x \mapsto V_h \mathrm{e}^{x\mathcal{A}^o_c} X_c$ has at most a polynomial growth. Thus it lies in $\mathrm{C}^1_\alpha(\mathbb{R}, E_h)$ for any α in $]0, \delta_h[$ (δ_h is defined in Lemma 8.1.7). Thus using Lemma 8.2.2 we get

$$\pi_h \phi^1 = \int_{-\infty}^{+\infty} \mathcal{K}(s) V_h \mathrm{e}^{-s\mathcal{A}^o_c} ds. \quad \pi_h \in \mathcal{L}(E_c, \mathcal{D}).$$

Thus $D\mathcal{I}_1$ is invertible and the Analytic Implicit Function Theorem ensures that the equation $\mathcal{I}_1(\phi^1, \mathcal{Q}^1, \mu) = 0$ has a unique solution $(\phi^1_\mu, \mathcal{Q}^1_\mu)$ analytic with respect to μ, for μ small enough.

8.A.3 Study of equation (8.29)

This equation has the form $L_{k,\mu}(\phi^k_\mu, \mathcal{Q}^k_\mu) = P^k_\mu$ where $L_{k,\mu}$ is the linear operator

$$\mathcal{L}_s(E^k_c, \mathcal{D}) \times (\mathcal{L}_s(E^k_c, E_c) \cap \mathcal{N}) \longrightarrow \mathcal{L}_s(E^k_c, E)$$

$$(\phi^k, \mathcal{Q}^k) \mapsto \Big\{ X_c \mapsto (\mathrm{Id} + \phi^1_\mu)\mathcal{Q}^k[X_c^{(k)}] + D\phi^k[X_c^{(k)}](\mathcal{A}^o_c + \phi^1_\mu)X_c - \mathcal{A}(\mu)\phi^k[X_c^{(k)}] \Big\}.$$

Observe that the functions $\mu \mapsto L_{k,\mu}$, $\mu \mapsto P_\mu^k$ are analytic. Thus if we prove that $L_{k,0}$ is invertible, then $L_{k,\mu}$ admits an inverse analytic with respect to μ, for μ small enough and (8.29) admits a unique solution given by $L_{k,\mu}^{-1} P_\mu^k$. Thus we study $L_{k,0}$ which is given by

$$L_{k,0} : \mathcal{L}_s(E_c^k, \mathcal{D}) \times (\mathcal{L}_s(E_c^k, E_c) \cap \mathcal{N}) \longrightarrow \mathcal{L}_s(E_c^k, E)$$
$$(\phi^k, \mathcal{Q}^k) \mapsto \{X_c \mapsto \mathcal{Q}^k[X_c^{(k)}] + D\phi^k[X_c^{(k)}]\mathcal{A}_c^o X_c - \mathcal{A}^o \phi^k[X_c^{(k)}]\}.$$

To invert $L_{k,0}$, we must solve

$$\mathcal{Q}^k[X_c^{(k)}] + D(\pi_c\phi^k)[X_c^{(k)}]\mathcal{A}_c^o X_c - \mathcal{A}_c^o \pi_c \phi^k[X_c^{(k)}] = W_c[X_c^{(k)}], \quad (8.32)$$
$$D(\pi_h\phi^k)[X_c^{(k)}]\mathcal{A}_c^o X_c - \mathcal{A}_h^o \pi_h \phi^k[X_c^{(k)}] = W_h[X_c^{(k)}], \quad (8.33)$$

for any $W_c \in \mathcal{L}_s(E_c^k, E_c)$, $W_h \in \mathcal{L}_s(E_c^k, E_h)$. Equation (8.32) is again the usual homologic equation. It admits a unique solution $(\pi_c\phi^k, \mathcal{Q}^k)$ because we look for $\mathcal{Q}^k$ in N (see [ETBCI87] or [IA92]).

We solve (8.33) as we solved (8.31). We get

$$\pi_h\phi^k[X_c^{(k)}] = \int_{-\infty}^{+\infty} \mathcal{K}(s) W_h[(e^{-s\mathcal{A}_c^o} X_c)^{(k)}] ds, \quad \pi_h\phi^k \in \mathcal{L}_s(E_c^k, \mathcal{D}).$$

We can then conclude that $L_{k,\mu}(\phi_\mu^k, \mathcal{Q}_\mu^k) = P_\mu^k$ admits a unique solution $(\phi_\mu^k, \mathcal{Q}_\mu^k)$ analytic with respect to μ, for μ small enough.

Finally the reversibility properties are proved by checking that the two functions $S\phi_\mu^k S$, $-S\mathcal{Q}_\mu^k S$ are solutions of the same affine equation as ϕ_μ^k, $\mathcal{Q}_\mu^k$. Uniqueness of the solution gives the result.

8.B Appendix. Study of $\mathcal{L}_\nu$

For studying $\mathcal{L}_\nu$ we split it into two parts $\mathcal{L}_\nu = \mathcal{L}_{0\nu} + \nu\mathcal{L}_{1\nu}$ where

$$\mathcal{L}_{0\nu}(u) = \frac{\omega_{0,\nu}}{\nu}\frac{d\widehat{Z}}{ds} - \mathfrak{L}_{0\nu}\widehat{Z} + \widehat{\omega}\frac{d\widehat{Y}_1}{ds}, \quad \mathcal{L}_{1\nu}(u) = -\mathfrak{L}_{1\nu}\widehat{Z}.$$

This appendix is devoted to the proof of the following lemma:

Lemma 8.B.1.

(a) *$\mathcal{L}_{0\nu}$ is an isomorphism from $\tilde{\mathfrak{U}}_R^2$ to $\mathfrak{P}_{AR}^1$.*
(b) *There exists C_4, $\nu_1 \le \nu_0$ such that $\left\|\mathcal{L}_{0\nu}^{-1}(f)\right\|_{\mathfrak{U}^2} \le C_4 \left\|f\right\|_{\mathfrak{P}^1}$ for $\nu \in]0, \nu_1]$, $f \in \mathfrak{P}_{AR}^1$.*
(c) *$\mathcal{L}_\nu$ is an isomorphism from $\tilde{\mathfrak{U}}_R^2$ to $\mathfrak{P}_{AR}^1$.*
(d) *For every $\nu \in]0, \nu_1]$, $f \in \mathfrak{P}_{AR}^1$, $\left\|\mathcal{L}_\nu^{-1}(f)\right\|_{\mathfrak{U}^2} \le 2C_4 \left\|f\right\|_{\mathfrak{P}^1}$.*

Proof. (a),(b): Let $f = (f_c, f_h) \in \mathfrak{P}^1_{AR}$. We look for $u = (\widehat{Z}, \widehat{\omega})$ such that $\mathcal{L}_{0\nu}(u) = f$. This equation reads

$$\frac{\omega_{0,\nu}}{\nu}\frac{d\widehat{Z}_c}{ds} - L_{0\nu}\widehat{Z}_c + \widehat{\omega}\frac{d\widehat{Y}_{1,c}}{ds} = f_c, \tag{8.34}$$

$$\frac{\omega_{0,\nu}}{\nu}\frac{d\widehat{Z}_h}{ds} - \frac{A_h^o}{\nu}\widehat{Z}_h = f_h. \tag{8.35}$$

As mentioned earlier we use Fourier series to solve this equation. We define the Fourier coefficients of $\widehat{Z}$, f:

$$\widehat{Z}_n = \frac{1}{2\pi}\int_0^{2\pi} \widehat{Z}(s)e^{-ins}\, ds, \qquad f_n = \frac{1}{2\pi}\int_0^{2\pi} f(s)e^{-ins}\, ds,$$

with $\widehat{Z}_n = (\widehat{Z}_{c,n}, \widehat{Z}_{h,n})$ and $f_n = (f_{c,n}, f_{h,n})$.

We start by the study of equation (8.35), which is equivalent to

$$in\frac{\omega_{0,\nu}}{\nu}\widehat{Z}_{h,n} - \frac{A_h^o}{\nu}\widehat{Z}_{h,n} = f_{h,n} \quad \text{for } n \in \mathbb{Z}.$$

Using the resolvent estimates given in Lemma 8.1.7 and taking ν_1 small enough to have $\omega_{0,\nu} > \omega/2$, we get

$$\left\|(in\frac{\omega_{0,\nu}}{\nu} - A_h^o/\nu)^{-1}\right\|_{\mathcal{L}(E_h)} \le \frac{4M_h\nu}{2+|n|\omega},$$

$$\left\|(in\frac{\omega_{0,\nu}}{\nu} - A_h^o/\nu)^{-1}\right\|_{\mathcal{L}(E_h,\mathcal{D}_h)} \le 2M_h\nu.$$

It follows that there exists C such that for all ν in $]0,\nu_1]$ (8.35) admits a unique solution in $H^2(T^1, E_h) \cap H^1(T^1, \mathcal{D}_h)$ and such that $\left\|\widehat{Z}_h\right\|_{\mathfrak{P}^2} \le C\nu \left\|f_h\right\|_{\mathfrak{P}^1}$ holds. The antireversibility of f_c induces automatically the reversibility of $\widehat{Z}_h$.

Now we look for $\widehat{Z}_c$ in $H^1(T^1, \mathbb{R}^4)$, reversible, satisfying $\int_0^{2\pi} \langle \widehat{Z}_c, \widehat{Y}_{1,c}\rangle = 0$, and for $\widehat{\omega}$ in $\mathbb{R}$ solving (8.34) which is equivalent to

$$(in\frac{\omega_{0,\nu}}{\nu} - L_{0\nu})\widehat{Z}_{c,n} = f_{c,n} \quad \text{for } n \neq \pm 1,$$

$$(i\frac{\omega_{0,\nu}}{\nu} - L_{0\nu}\widehat{Z}_{c,1} + \widehat{\omega}(0,0,\tfrac{1}{2}i,\tfrac{1}{2}) = f_{c,1} \quad \text{for } n = 1,$$

$$(-i\frac{\omega_{0,\nu}}{\nu} - L_{0\nu}\widehat{Z}_{c,-1} + \widehat{\omega}(0,0,-\tfrac{1}{2}i,\tfrac{1}{2}) = f_{c,-1} \quad \text{for } n = -1.$$

Moreover, we require that $\widehat{Z}_{c,n}$ is real reversible, i.e., $\widehat{Z}_{c,n} = \overline{\widehat{Z}}_{c,-n} = \widehat{Z}_{c,-n}$. We denote $\widehat{Z}_{c,n} = (\beta_{0,n}, \beta_{1,n}, A_n, B_n)$, $f_{c,n} = (f_{0,n}, f_{1,n}, f_{A,n}, f_{B,n})$. Since f is real antireversible, i.e., $f_{c,n} = \overline{f}_{c,-n} = -f_{c,-n}$, elementary computations lead to

$$\text{for } n \in \mathbb{N}, \quad \begin{pmatrix} \beta_{0,n} \\ \beta_{1,n} \end{pmatrix} = \begin{bmatrix} \dfrac{-in\nu\omega_{0,\nu}}{(n\omega_{0,\nu})^2+\nu^2} & \dfrac{\nu^2}{(n\omega_{0,\nu})^2+\nu^2} \\ \dfrac{\nu^2}{(n\omega_{0,\nu})^2+\nu^2} & \dfrac{-in\nu\omega_{0,\nu}}{(n\omega_{0,\nu})^2+\nu^2} \end{bmatrix} \begin{pmatrix} f_{0,n} \\ f_{1,n} \end{pmatrix},$$

$$\text{for } n \neq \pm 1, \quad \begin{pmatrix} A_n \\ B_n \end{pmatrix} = \begin{bmatrix} \dfrac{-in\nu}{\omega_{0,\nu}(1-n^2)} & \dfrac{\nu}{\omega_{0,\nu}(1-n^2)} \\ \dfrac{\nu}{\omega_{0,\nu}(1-n^2)} & \dfrac{-in\nu}{\omega_{0,\nu}(1-n^2)} \end{bmatrix} \begin{pmatrix} f_{A,n} \\ f_{B,n} \end{pmatrix},$$

$$\begin{aligned} \text{for } n = 1, \quad \begin{pmatrix} A_1 \\ B_1 \end{pmatrix} &= \begin{bmatrix} \dfrac{-\mathrm{i}\nu}{4\omega_{0,\nu}} & \dfrac{-\nu}{4\omega_{0,\nu}} \\ \dfrac{\nu}{4\omega_{0,\nu}} & \dfrac{-\mathrm{i}\nu}{4\omega_{0,\nu}} \end{bmatrix} \begin{pmatrix} f_{A,1} \\ f_{B,1} \end{pmatrix}, \\ \widehat{\omega} &= f_{B,1} - \mathrm{i} f_{A,1} = \frac{1}{2\pi}\int_0^{2\pi} \langle f, \frac{d\widehat{Y}_1}{ds}\rangle \, ds, \\ \text{for } n = -1, \quad \begin{pmatrix} A_{-1} \\ B_{-1} \end{pmatrix} &= \begin{pmatrix} A_1 \\ -B_1 \end{pmatrix}. \end{aligned}$$

We observe that there exists C' such that $|Z_{c,n}| \leq \dfrac{C'}{|n|+1}|f_{c,n}|$ for $\nu \in]0,\nu_1]$. It follows that there exists C'' such that for all ν in $]0,\nu_1]$ (8.34) admits a reversible unique solution $(\widehat{Z}_c, \widehat{\omega})$ in $H^2(T^1,\mathbb{R}^4) \times \mathbb{R}$ and that $\left\|\widehat{Z}_c\right\|_{\mathfrak{P}^2} \leq C'' \left\|f_c\right\|_{\mathfrak{P}^1}$.

(c), (d): Observe that $\mathcal{L}_{1\nu}$ is a bounded operator from $\tilde{\mathfrak{U}}^2_R$ to $\mathfrak{P}^1_{AR}$ and that $\left\|\mathcal{L}_{1\nu}(u)\right\|_{\mathfrak{P}^1} \leq C_L \left\|u\right\|_{\mathfrak{U}^2}$ holds for $\nu \in]0,\nu_0]$, $u \in \tilde{\mathfrak{U}}^2_R$.

We write $\mathcal{L}_\nu = \mathcal{L}_{0\nu}(\mathrm{Id} + \nu\mathcal{L}_{0\nu}^{-1}\mathcal{L}_{1\nu})$ and we choose ν_1 small enough to have $\nu_1 C_4 C_L \leq \frac{1}{2}$. This ensures that $\mathcal{L}_\nu$ is invertible and that the norm of its inverse is smaller than $2C_4$. □

Remark 8.B.2. The proof of Lemma 8.B.1 ensures that $\mathcal{L}_\nu$ is an isomorphism from $\tilde{\mathfrak{U}}_{n,R}$ to $\mathfrak{P}_{n,AR}$.

9. The $(\mathrm{i}\omega_0)^2\mathrm{i}\omega_1$ resonance

9.1 Introduction

This chapter is devoted to the $(\mathrm{i}\omega_0)^2\mathrm{i}\omega_1$ resonance in $\mathbb{R}^6$: we study *analytic* one parameter families of vector fields in $\mathbb{R}^6$,

$$\frac{du}{dx} = \mathcal{V}(u,\mu), \qquad u \in \mathbb{R}^6,\ \mu \in [-\mu_0, \mu_0],\ \mu_0 > 0 \tag{9.1}$$

near a fixed point placed at the origin , i.e.

$$\mathcal{V}(0,\mu) = 0 \qquad \text{for } \mu \in [-\mu_0, \mu_0]. \tag{H1}$$

In addition, the family is supposed to be *reversible*, i.e. there exists a reflection $S \in GL_6(\mathbb{R})$ such that for every u and μ,

$$\mathcal{V}(Su,\mu) = -S\mathcal{V}(u,\mu) \tag{H2}$$

holds. We assume that the origin is a $(\mathrm{i}\omega_0)^2\mathrm{i}\omega_1$ resonant fixed point, i.e. that the spectrum of the differential at the origin $D_u\mathcal{V}(0,0)$ is $\{\pm\mathrm{i}\omega_0, \pm\mathrm{i}\omega_1\}$ with $\omega_0, \omega_1 > 0$ and where $\pm\mathrm{i}\omega_0$ are double non semi-simple eigenvalues whereas $\pm\mathrm{i}\omega_1$ are simple eigenvalues. We denote by $(\varphi_0^+, \varphi_0^-, \psi_0^+, \psi_0^-, \varphi_1^+, \varphi_1^-)$ a basis of eigenvectors and generalized eigenvectors of $\mathcal{A} = D_u\mathcal{V}(0,0)$

$$\mathcal{A}\varphi_0^\pm = \pm\mathrm{i}\omega_0\varphi_0^\pm, \qquad \mathcal{A}\psi_0^\pm = \pm\mathrm{i}\omega\psi_0^\pm + \varphi_0^\pm, \qquad \mathcal{A}\varphi_1^\pm = \pm\mathrm{i}\omega_1\varphi_1^\pm, \tag{H3}$$

such that

$$\varphi_0^- = \overline{\varphi_0^+}, \qquad \psi_0^- = \overline{\psi_0^+}, \qquad \varphi_1^- = \overline{\varphi_1^+},$$

and by $(\varphi_0^{+,*}, \varphi_0^{-,*}, \psi_0^{+,*}, \psi_0^{-}, \varphi_1^{+,*}, \varphi_1^{-,*})$ the corresponding dual basis.

We also assume that the two frequencies are weakly non resonant, i.e.

$$\frac{\omega_1}{\omega_0} \neq \frac{p}{q} \quad \text{for } p+q \le 5 \quad \Leftrightarrow \quad \frac{\omega_1}{\omega_0} \neq \tfrac14, \tfrac13, \tfrac12, 1, 2, 3, 4, \tag{H4a}$$

$$\frac{\omega_1}{\omega_0} \notin \mathbb{N} \quad \Leftrightarrow \quad \omega_\star = \min_{p\in\mathbb{Z}} |\omega_1 - p\omega_0| > 0, \tag{H4b}$$

$$\frac{\omega_1}{\omega_0} \notin \frac{2}{\mathbb{N}} \quad \Leftrightarrow \quad \frac{\omega_1}{\omega_0} \neq \frac{2}{n} \quad \text{for } n \in \mathbb{N}. \tag{H4c}$$

Hypothesis (H4a) ensures that the Normal Form of $\mathcal{V}$ does not admit "exotic" resonant terms up to cubic terms; Hypothesis (H4b) is required to obtain exponentially small estimates of bi-oscillatory integrals (see Lemma 2.2.1); Hypothesis (H4c) enables us to use non resonant Floquet Theory when linearizing around the periodic orbits (see Theorem 5.1.1).

Two last generic hypotheses on the linear and the cubic parts of the vector field are made. The first one concerns the linear part of the vector field,

$$q_{10} = \langle \psi_0^{+,*}, D^2_{\mu,u}\mathcal{V}(0,0).\varphi_0^+ \rangle \neq 0. \tag{H5}$$

It ensures that the bifurcation described in Figure 9.1 really occurs: we do not study the hyper-degenerate case when the double eigenvalue 0 stays at 0 for $\mu \neq 0$.

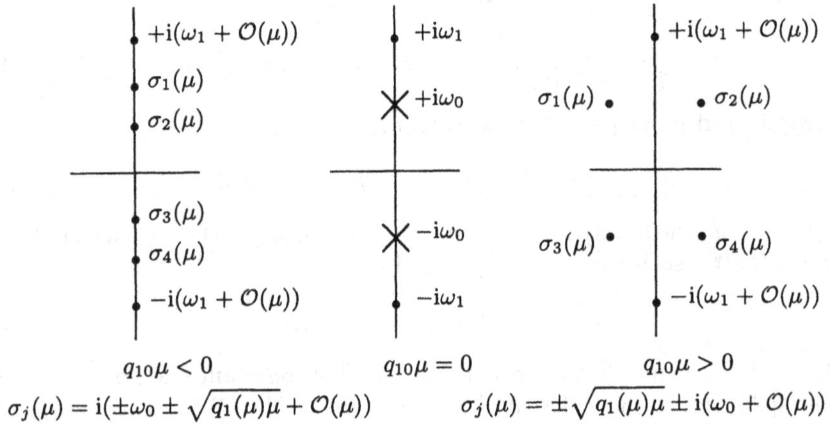

Fig. 9.1. Spectrum of $D_u\mathcal{V}(0,\mu)$ for different values of $q_{10}\mu$ ($q_1(\mu) = q_{10} + \mathcal{O}(\mu)$).

The last hypothesis,

$$q_{20} < 0 \tag{H6}$$

with

$$\begin{aligned} q_{20} = \langle \psi_0^{*,+}, 3D^3_{uuu}\mathcal{V}(0,0).[\varphi_0^+, \varphi_0^+, \varphi_0^-] \rangle \\ + \langle \psi_0^{*,+}, -4D^2_{uu}\mathcal{V}(0,0).[\varphi_0^+, \phi_{11}] + 2D^2_{uu}\mathcal{V}(0,0).[\varphi_0^-, \phi_{20}] \rangle, \\ \phi_{11} = \mathcal{A}^{-1}D^2_{uu}\mathcal{V}(0,0).[\varphi_0^+, \varphi_0^-],, \\ \phi_{20} = (2\mathrm{i}\omega_0 - \mathcal{A})^{-1}\, D^2_{uu}\mathcal{V}(0,0).[\varphi_0^+, \varphi_0^+], \end{aligned}$$

ensures that, in some sense, the cubic part of the vector field is not degenerated. In particular, it ensures that the normal form systems admit reversible homoclinic connections to 0. See (9.3), (9.4), (9.5) to identify what coefficient of the normal form q_{20} is.

Finally, since we are interested in the existence of homoclinic connections, we only study the "half bifurcation" corresponding to

$$q_{10}\mu > 0 \tag{H7}$$

for which the differential $D_u\mathcal{V}(0,\mu)$ admits eigenvalues with non zero real parts. This chapter is devoted to the proof of the following theorem:

Theorem 9.1.1. *Let* $\mathcal{V}(\cdot,\mu)$ *be an analytic, reversible one parameter family of vector fields in* $\mathbb{R}^6$ *admitting a* $(i\omega_0)^2 i\omega_1$ *resonance at the origin, i.e. satisfying Hypothesis (H1),$\cdots$,(H6).*

Then, there exist five constants $\sigma, \kappa_3, \kappa_2, \kappa_1 > 0$, $\kappa_0 \geq 0$ *such that for* $|\mu|$ *small enough with* $q_{10}\mu > 0$ *the vector field* $\mathcal{V}(\cdot,\mu)$ *admits near the origin*

(a) *a one parameter family of periodic orbits* $p_{\kappa,\mu}$ *of arbitrary small size* $\kappa \in [0, \kappa_3|\mu|^{\frac{7}{4}}]$;

(b) *for every* $\kappa \in [\kappa_1 e^{-\frac{\pi\omega_\star}{2\sqrt{q_{10}\mu}}}, \kappa_2|\mu|^{\frac{7}{4}}]$, *a pair of reversible homoclinic connections to* $p_{\kappa,\mu}$ *with one loop;*

(c) *for every* $\kappa \in [0, \kappa_0 e^{-\frac{\pi\omega_\star}{2\sqrt{q_{10}\mu}}}[$, *no reversible homoclinic connections to* $p_{\kappa,\mu}$ *with one loop; Generically (with respect to* $\mathcal{V}$), κ_0 *is positive, i.e.* $\kappa_0 > 0$;

where $\omega_\star = \min\limits_{p\in\mathbb{Z}} |\omega_1 - p\omega_0| > 0$.

More precise versions of this theorem are given below: see Theorem 9.1.4 for the periodic orbits, Theorems 9.1.8, Corollary 9.1.11 and Remark 9.1.12 for the existence of homoclinic connections to periodic orbits and Theorem 9.1.14 and Remark 9.1.20 for the non existence.

Periodic orbits $p_{\kappa,\mu}$ Reversible homoclinic connections to $p_{\kappa,\mu}$

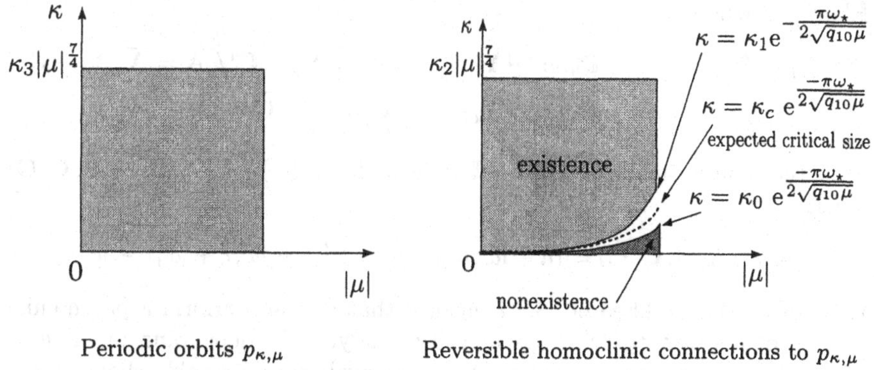

Fig. 9.2. Domain of existence of the reversible periodic orbits $p_{\kappa,\mu}$ and domains of existence and nonexistence of reversible homoclinic connections to $p_{\kappa,\mu}$

Figure 9.2 shows in the parameter plane $(\kappa, |\mu|)$ the domain of existence and non existence stated in Theorem 9.1.1. As for our first toy model 1.1 given in the introduction of the book, it appears a critical size $K_c(\mu)$ for the existence of reversible homoclinic connections to the periodic orbits. This critical size is exponentially small

$$\kappa_0 e^{-\frac{\pi\omega_*}{2\sqrt{q_{10}\mu}}} \le K_c(\mu) \le \kappa_1 e^{-\frac{\pi\omega_*}{2\sqrt{q_{10}\mu}}}.$$

We expect that generically, the critical size $K_c(\mu)$ is given by

$$K_c(\mu) = \kappa_c\, e^{-\frac{\pi\omega_*}{2\sqrt{q_{10}\mu}}}.$$

9.1.1 Full system, normal form and scaling

Step 1. Linear change of coordinates. The Theorem of classification of reversible matrices 3.1.10 ensures that up to a linear change of coordinates, we can assume that $L_0 = D_u\mathcal{V}(0,0)$ and the symmetry S read

$$L_0 = \begin{pmatrix} 0 & -\omega_0 & 1 & 0 & 0 & 0 \\ \omega_0 & 0 & 0 & 1 & 0 & 0 \\ 0 & 0 & 0 & -\omega_0 & 0 & 0 \\ 0 & 0 & \omega_0 & 0 & 0 & 0 \\ 0 & 0 & 0 & 0 & 0 & -\omega_1 \\ 0 & 0 & 0 & 0 & \omega_1 & 0 \end{pmatrix}, \qquad S = \mathrm{diag}(1,-1,-1,1,1,-1)$$

Step 2. Normal form and scaling. The previous linear change of coordinates leads to a simple linear part. We now wish to obtain a quadratic and a cubic parts as simple as possible. The theory of normal form was precisely designed for that purpose.

Step 2.1. Polynomial change of coordinates: normal form. Identifying $\mathbb{R}^6$ and $\mathbb{D}$ where

$$\mathbb{D} = \mathbb{D}_h \times \mathbb{D}_c \qquad \text{with } \begin{aligned} \mathbb{D}_h = &\ \{\mathbf{Y}_h = (\mathbf{A}, \mathbf{B}, \widetilde{\mathbf{A}}, \widetilde{\mathbf{B}}) \in \mathbb{C}^4 / \ \widetilde{\mathbf{A}} = \overline{\mathbf{A}}, \widetilde{\mathbf{B}} = \overline{\mathbf{B}}\}, \\ \mathbb{D}_c = &\ \{\mathbf{Y}_c = (\mathbf{C}, \widetilde{\mathbf{C}}) \in \mathbb{C}^2 / \ \widetilde{\mathbf{C}} = \overline{\mathbf{C}}\}, \end{aligned}$$

with the isomorphism $\mathbb{T} : \mathbb{R}^6 \to \mathbb{D} : (a, a', b, b', c, c') \mapsto (\mathbf{A}, \mathbf{B}, \widetilde{\mathbf{A}}, \widetilde{\mathbf{B}}, \mathbf{C}, \widetilde{\mathbf{C}})$ where

$$(\mathbf{A}, \mathbf{B}, \widetilde{\mathbf{A}}, \widetilde{\mathbf{B}}, \mathbf{C}, \widetilde{\mathbf{C}}) = (a + ia', b + ib, a - ia', b - ib', c + ic', c - ic'),$$

the Normal Form Theorem 3.2.1 ensures that we can perform a polynomial change of coordinates close to identity, analytically depending on μ, $u = \mathbf{Y} + \Phi(\mathbf{Y}, \mu)$ such that equation (9.1) is equivalent in a neighborhood of the origin to

$$\frac{d\mathbf{Y}}{dx} = \mathcal{N}(\mathbf{Y}, \mu) + \mathcal{R}(\mathbf{Y}, \mu), \qquad \mathbf{Y} \in \mathbb{D} \tag{9.2}$$

where $\mathcal{N}$ is the normal form of order 2 and $\mathcal{R}$ represents the new terms of order larger or equal to 4. The two vector fields $\mathcal{N}$ and $\mathcal{R}$ satisfy

$$\mathcal{N}(\mathbf{Y},\mu)=\begin{pmatrix} \mathrm{i}\underline{\omega}_0(\mu)\mathbf{A}+\mathbf{B}+\mathrm{i}\mathbf{A}\mathbf{P}(\mathbf{Y},\mu) \\ \mathrm{i}\underline{\omega}_0(\mu)\mathbf{B}+\mathrm{i}\mathbf{B}\mathbf{P}(\mathbf{Y},\mu)+\mathbf{A}\mathbf{Q}(\mathbf{Y},\mu) \\ -\mathrm{i}\underline{\omega}_0(\mu)\widetilde{\mathbf{A}}+\widetilde{\mathbf{B}}-\mathrm{i}\widetilde{\mathbf{A}}\mathbf{P}(\mathbf{Y},\mu) \\ -\mathrm{i}\underline{\omega}_0(\mu)\widetilde{\mathbf{B}}-\mathrm{i}\widetilde{\mathbf{B}}\mathbf{P}(\mathbf{Y},\mu)+\widetilde{\mathbf{A}}\mathbf{Q}(\mathbf{Y},\mu) \\ \mathrm{i}\underline{\omega}_1(\mu)\mathbf{C}+\mathrm{i}\mathbf{C}\mathbf{N}(\mathbf{Y},\mu) \\ -\mathrm{i}\underline{\omega}_1(\mu)\widetilde{\mathbf{C}}-\mathrm{i}\widetilde{\mathbf{C}}\mathbf{N}(\mathbf{Y},\mu) \end{pmatrix}, \tag{9.3}$$

with

$$\begin{aligned} \mathbf{P}(\mathbf{Y},\mu)=&\ \mathbf{p}_2(\mu)\mathbf{A}\widetilde{\mathbf{A}}+\mathbf{p}_3(\mu)\tfrac{\mathrm{i}}{2}(\mathbf{A}\widetilde{\mathbf{B}}-\widetilde{\mathbf{A}}\mathbf{B})+\mathbf{p}_4(\mu)\mathbf{C}\widetilde{\mathbf{C}} \\ \mathbf{Q}(\mathbf{Y},\mu)=&\ \mathbf{q}_1(\mu)\mu+\mathbf{q}_2(\mu)\mathbf{A}\widetilde{\mathbf{A}}+\mathbf{q}_3(\mu)\tfrac{\mathrm{i}}{2}(\mathbf{A}\widetilde{\mathbf{B}}-\widetilde{\mathbf{A}}\mathbf{B})+\mathbf{q}_4(\mu)\mathbf{C}\widetilde{\mathbf{C}} \\ \mathbf{N}(\mathbf{Y},\mu)=&\ \mathbf{n}_2(\mu)\mathbf{A}\widetilde{\mathbf{A}}+\mathbf{n}_3(\mu)\tfrac{\mathrm{i}}{2}(\mathbf{A}\widetilde{\mathbf{B}}-\widetilde{\mathbf{A}}\mathbf{B})+\mathbf{n}_4(\mu)\mathbf{C}\widetilde{\mathbf{C}} \end{aligned} \tag{9.4}$$

and

$$\mathcal{R}(\mathbf{Y},\mu)=\mathcal{O}(|\mathbf{Y}|^4),\ \mathcal{R}(\mathbb{D},\mu)\subset\mathbb{D}$$

where $\mathbf{Y}=(\mathbf{A},\mathbf{B},\widetilde{\mathbf{A}},\widetilde{\mathbf{B}},\mathbf{C},\widetilde{\mathbf{C}})$, and $\underline{\omega}_0,\underline{\omega}_1,\mathbf{p}_j,\mathbf{q}_j,\mathbf{n}_j$, are *real analytic function of* μ and

$$\underline{\omega}_0(\mu)=\omega_0+\mathcal{O}(\mu),\qquad \underline{\omega}_1(\mu)=\omega_1+\mathcal{O}(\mu).$$

Moreover hypothesis (H5) and (H6) ensure that

$$q_{10}=\mathbf{q}_1(0)\neq 0,\qquad q_{20}=\mathbf{q}_2(0)<0. \tag{9.5}$$

Reversibility is preserved by the two changes of coordinates. So system (9.2) has the following symmetry properties:

$$S\mathcal{N}(\mathbf{Y},\mu)=-\mathcal{N}(S\mathbf{Y},\mu),\qquad S\,\mathcal{R}(\mathbf{Y},\mu)=-\,\mathcal{R}(S\mathbf{Y},\mu)$$

with $S(\mathbf{A},\mathbf{B},\widetilde{\mathbf{A}},\widetilde{\mathbf{B}},\mathbf{C},\widetilde{\mathbf{C}})=(\widetilde{\mathbf{A}},-\widetilde{\mathbf{B}},\mathbf{A},-\mathbf{B},\widetilde{\mathbf{C}},\mathbf{C})$. Moreover, analyticity is preserved too.

Step 2.2. Scaling. The last technical transformation of the system before studying dynamics is a scaling in space and time. We are interested in the existence of homoclinic connections. For $q_{10}\mu>0$ the truncated system

$$\frac{d\mathbf{Y}}{dx}=\mathcal{N}(\mathbf{Y},\mu)$$

admits reversible homoclinic connections to 0 which depend on μ, whereas for $q_{10}\mu<0$ there is none. Observe that $q_{10}\mu>0$ corresponds to the half bifurcation where the differential $D_u\mathcal{V}(0,\mu)$ generates an "oscillatory" dynamics with a frequency of order 1 — due to two simple, opposite eigenvalues lying on the imaginary axis— and a "slow" hyperbolic dynamics — due to the set of eigenvalues $\pm\sqrt{q_1(\mu)\mu} \pm i(\omega_0 + \mathcal{O}(\mu))$. The homoclinic connections to 0 comes from the four hyperbolic eigenvalues and their two components along the oscillatory part of the truncated system are equal to 0. Since we want to use perturbation theory to study the persistence of the reversible homoclinic connections, we wish them not to depend on the bifurcation parameter μ. This can be achieved by performing the scaling

$$x = t/\nu, \qquad \begin{array}{l} \mathbf{A} = \sigma\nu A, \quad \mathbf{B} = \sigma\nu^2 B, \quad \mathbf{C} = \nu^{\frac{3}{2}} C, \\ \widetilde{\mathbf{A}} = \sigma\nu \widetilde{A}, \quad \widetilde{\mathbf{B}} = \sigma\nu^2 \widetilde{B}, \quad \widetilde{\mathbf{C}} = \nu^{\frac{3}{2}} \widetilde{C}, \end{array} \tag{9.6}$$

where

$$\nu^2 = q_1(\mu)\mu, \qquad \sigma = \sqrt{\frac{-2}{q_2(\mu)}}.$$

Since by hypothesis (H5), $q_{10} \neq 0$ holds, the identity

$$\nu^2 = q_1(\mu)\mu \tag{9.7}$$

can be inverted in a neighborhood of the origin using the Analytic Implicit Function Theorem. So, for sufficiently small ν, (9.7) is equivalent to

$$\mu = \theta(\nu^2)$$

where θ is a real analytic function in a neighborhood of the origin such that $\theta(0) = 0$, $\theta'(0) = (q_{10})^{-1}$. Hence by composition, all the analytic functions with respect to μ are analytic with respect to ν. Setting

$$Y = (Y_h, Y_c) \qquad \text{with } Y_h = (A, B, \widetilde{A}, \widetilde{B}),\ Y_c = (C, \widetilde{C})$$

we obtain for small $\nu \in]0, \nu_0]$, a rescaled full system equivalent to our initial equation (9.1) which reads

$$\frac{dY}{dt} = \mathcal{N}(Y,\nu) + \mathcal{R}(Y,\nu), \qquad Y \in \mathbb{D}, \tag{9.8}$$

with

$$\mathcal{N}(Y,\nu) = \begin{pmatrix} \frac{i\omega_0(\nu)}{\nu} A + B + iAP(Y,\nu) \\ \frac{i\omega_0(\nu)}{\nu} B + iBP(Y,\nu) + AQ(Y,\nu) \\ -\frac{i\omega_0(\nu)}{\nu} \widetilde{A} + \widetilde{B} - i\widetilde{A}P(Y,\nu) \\ -\frac{i\omega_0(\nu)}{\nu} \widetilde{B} - i\widetilde{B}P(Y,\nu) + \widetilde{A}Q(Y,\nu) \\ \frac{i\omega_1(\nu)}{\nu} C + iCN(Y,\nu) \\ -\frac{i\omega_1(\nu)}{\nu} \widetilde{C} - i\widetilde{C}N(Y,\nu) \end{pmatrix}, \mathcal{R}(Y,\nu) = \begin{pmatrix} \mathcal{R}_A(Y,\nu) \\ \mathcal{R}_B(Y,\nu) \\ \mathcal{R}_{\widetilde{A}}(Y,\nu) \\ \mathcal{R}_{\widetilde{B}}(Y,\nu) \\ \mathcal{R}_C(Y,\nu) \\ \mathcal{R}_{\widetilde{C}}(Y,\nu) \end{pmatrix}$$

with

$$\begin{aligned}
P(Y,\nu) &= p_2(\nu)\nu\, A\widetilde{A} + p_3(\nu)\nu^2\, \tfrac{\mathrm{i}}{2}(A\widetilde{B} - \widetilde{A}B) + p_4(\nu)\nu^2\, C\widetilde{C} \\
Q(Y,\nu) &= 1 - 2A\widetilde{A} + q_3(\nu)\nu\, \tfrac{\mathrm{i}}{2}(A\widetilde{B} - \widetilde{A}B) + q_4(\nu)\nu\, C\widetilde{C} \\
N(Y,\nu) &= n_2(\nu)\nu\, A\widetilde{A} + n_3(\nu)\nu^2\, \tfrac{\mathrm{i}}{2}(A\widetilde{B} - \widetilde{A}B) + n_4(\nu)\nu^2\, C\widetilde{C}
\end{aligned}$$

and

$$\begin{aligned}
\underline{\omega}_0(\nu) = \omega_0 + \nu^2\underline{\omega}_{01}(\nu), \qquad \underline{\omega}_1(\nu) = \omega_1 + \nu^2\underline{\omega}_{11}(\nu) \\
p_j(\nu) = p_{j0} + \mathcal{O}(\nu^2),\;\; q_j(\nu) = q_{j0} + \mathcal{O}(\nu^2),\;\; n_j(\nu) = n_{j0} + \mathcal{O}(\nu^2).
\end{aligned} \tag{9.9}$$

where $\underline{\omega}_{01}, \underline{\omega}_{11}, p_j, q_j, n_j$, are *real bounded analytic function of* $\nu \in [0, \nu_0]$.

The scaling preserves the stability of $\mathbb{D}$ by the rest, i.e.

$$\mathcal{R}(\mathbb{D}) \subset \mathbb{D}$$

and the reversibility. So system (9.8) has the following symmetry properties:

$$S\mathcal{N}(Y,\nu) = -\mathcal{N}(SY,\nu), \;\; S\mathcal{R}(Y,\nu) = -\mathcal{R}(SY,\nu), \tag{9.10}$$

with $S(A,B,\widetilde{A},\widetilde{B},C,\widetilde{C}) = (\widetilde{A},-\widetilde{B},A,-B,\widetilde{C},C)$. Moreover, analyticity is preserved too. Thus, near the origin, the components of $\mathcal{R}(Y,\nu)$ reads

$$\begin{aligned}
\mathcal{R}_{A\,\mathrm{or}\,\widetilde{A}}(Y,\nu) &= \sum_{(\overline{m},\ell)\in\mathcal{I}} a_{A\,\mathrm{or}\,\widetilde{A},\overline{m},\ell}\, Y^{\overline{m}}\, \nu^{|\overline{m}|+m_B+m_{\widetilde{B}}+\frac{1}{2}(m_C+m_{\widetilde{C}})+2\ell-2}, \\
\mathcal{R}_{B\,\mathrm{or}\,\widetilde{B}}(Y,\nu) &= \sum_{(\overline{m},\ell)\in\mathcal{I}} a_{B\,\mathrm{or}\,\widetilde{B},\overline{m},\ell}\, Y^{\overline{m}}\, \nu^{|\overline{m}|+m_B+m_{\widetilde{B}}+\frac{1}{2}(m_C+m_{\widetilde{C}})+2\ell-3}, \\
\mathcal{R}_{C\,\mathrm{or}\,\widetilde{C}}(Y,\nu) &= \sum_{(\overline{m},\ell)\in\mathcal{I}} a_{C\,\mathrm{or}\,\widetilde{C},\overline{m},\ell}\, Y^{\overline{m}}\, \nu^{|\overline{m}|+m_B+m_{\widetilde{B}}+\frac{1}{2}(m_C+m_{\widetilde{C}})+2\ell-\frac{5}{2}},
\end{aligned} \tag{9.11}$$

where

$$\begin{aligned}
\overline{m} &= (m_A, m_B, m_{\widetilde{A}}, m_{\widetilde{B}}, m_C, m_{\widetilde{C}}) \in \mathbb{N}^6, \\
|\overline{m}| &= m_A + m_B + m_{\widetilde{A}} + m_{\widetilde{B}} + m_C + m_{\widetilde{C}}, \\
\mathcal{I} &= \{(\overline{m},\ell) \in \mathbb{N}^4 \times \mathbb{N} /\ |\overline{m}| \geq 4\},
\end{aligned}$$

and

$$\begin{aligned}
a_{\overline{m},\ell} &= (a_{A,\overline{m},\ell}, a_{B,\overline{m},\ell}, a_{\widetilde{A},\overline{m},\ell}, a_{\widetilde{B},\overline{m},\ell}, a_{C,\overline{m},\ell}, a_{\widetilde{C},\overline{m},\ell}) \in \mathbb{D}, \\
Y^{\overline{m}} &= A^{m_A}\, B^{m_B}\, \widetilde{A}^{m_A}\, \widetilde{B}^{m_B} C^{m_B}\, \widetilde{C}^{m_A}.
\end{aligned}$$

We denote by (r_0, ν_0) (where $r_0 = (r_A, r_B, r_{\widetilde{A}}, r_{\widetilde{B}}, r_C, r_{\widetilde{C}}) \in (\mathbb{R}_+^*)^6$) the "polyradius" of convergence of this power series, i.e. there exists $M_{\mathcal{R}}$ such that

$$|a_{\overline{m},\ell}| \leq \frac{M_{\mathcal{R}}}{r_0^{\overline{m}}\nu_0^{\ell}} = \frac{M_{\mathcal{R}}}{r_A^{m_A}\, r_B^{m_B}\, r_{\widetilde{A}}^{m_{\widetilde{A}}}\, r_{\widetilde{B}}^{m_{\widetilde{B}}}\, r_C^{m_C}\, r_{\widetilde{C}}^{m_{\widetilde{C}}}\, \nu_0^{2\ell}} \quad \text{for } (\overline{m},\ell) \in \mathcal{I}. \tag{9.12}$$

Using the power expansion of R and the bound of $a_{\overline{m},\ell}$ given by (9.11) and (9.12) we check that

Lemma 9.1.2. *For every $d > 0$, there exist two positive constants $\nu_{\mathcal{R}}(d)$, $M(d)$ such that for every $Y \in \mathbb{C}^6$ satisfying $|Y| \le d$, and every $\nu \in]0, \nu_{\mathcal{R}}(d)]$, $R(Y,\nu)$ is well defined and satisfies*

$$\begin{aligned}
|\mathcal{R}_{A \text{ or } \widetilde{A}}(Y,\nu)| &\le M\nu^2|Y|^4, \\
|\mathcal{R}_{B \text{ or } \widetilde{B}}(Y,\nu)| &\le M\nu|Y|^4, \\
|\mathcal{R}_{C \text{ or } \widetilde{C}}(Y,\nu)| &\le M\nu^{\frac{3}{2}}|Y|^4.
\end{aligned}$$

Moreover, for every $(Y,Y') \in (\mathbb{C}^6)^2$ satisfying $|Y| \le d$, $|Y'| \le d$ and every $\nu \in]0, \nu_{\mathcal{R}}(d)]$, the following estimates hold:

$$\begin{aligned}
|\mathcal{R}_{A \text{ or } \widetilde{A}}(Y,\nu) - \mathcal{R}_{A \text{ or } \widetilde{A}}(Y',\nu)| &\le M\nu^2|Y - Y'|, \\
|\mathcal{R}_{B \text{ or } \widetilde{B}}(Y,\nu) - \mathcal{R}_{B \text{ or } \widetilde{B}}(Y',\nu)| &\le M\nu|Y - Y'|, \\
|\mathcal{R}_{C \text{ or } \widetilde{C}}(Y,\nu) - \mathcal{R}_{C \text{ or } \widetilde{C}}(Y',\nu)| &\le M\nu^{\frac{3}{2}}|Y - Y'|.
\end{aligned}$$

The study of the existence of homoclinic connections requires the exponential tools developed in Chapter 2. Thus a complexification of time is necessary. So, we shall study the following complex differential equation:

$$\frac{dY}{d\xi} = \mathcal{N}(Y,\nu) + \mathcal{R}(Y,\nu) \qquad \text{with } \xi = t + i\eta \in \mathbb{C}. \tag{9.13}$$

This equation makes sense because $\mathcal{N}(\cdot,\nu)$ is a holomorphic function in $\mathbb{C}^6$ and because $\mathcal{R}(\cdot,\nu)$ is holomorphic in some neighborhood of 0 in $\mathbb{C}^6$. Observe that the symmetry properties (9.10) are still true when Y lies in $\mathbb{C}^6$.

For studying the dynamics of (9.1) near the origin, we have performed a polynomial change of coordinates followed by a scaling to obtain an equivalent system (9.1) for which the "linear+quadratic+cubic" part $\mathcal{N}$ is as simple as possible. So, following strategy explained in Remark 3.2.7, the next step is the study of the truncated system (9.14). This study is done in the next subsection. The rest of the chapter is then devoted to *the problem of the persistence for the full system of the homoclinic connections obtained for the truncated system.*

9.1.2 Truncated system

The truncated system

$$\frac{dY}{dt} = \mathcal{N}(Y,\nu), \qquad Y \in \mathbb{D} \tag{9.14}$$

is not only reversible ($S\mathcal{N}(Y,\nu) = -\mathcal{N}(SY,\nu)$), but it is also equivariant with respect to two different rotations, i.e.

$$R^{\mathrm{h}}_{\omega}\mathcal{N}(Y,\nu)=\mathcal{N}(R^{\mathrm{h}}_{\omega}Y,\nu),\quad R^{\mathrm{c}}_{\omega}\mathcal{N}(Y,\nu)=\mathcal{N}(R^{\mathrm{c}}_{\omega}Y,\nu)\quad \text{for } Y\in\mathbb{C}^6,\ \omega\in\mathbb{C}^6.$$

where the two rotations R^{h}_{ω} and R^{c}_{ω} read

$$R^{\mathrm{h}}_{\omega}=\mathrm{diag}(\mathrm{e}^{\mathrm{i}\omega},\mathrm{e}^{\mathrm{i}\omega},\mathrm{e}^{-\mathrm{i}\omega},\mathrm{e}^{-\mathrm{i}\omega},1,1),\quad R^{\mathrm{c}}_{\omega}=\mathrm{diag}(1,1,1,1,\mathrm{e}^{\mathrm{i}\omega},\mathrm{e}^{-\mathrm{i}\omega}) \tag{9.15}$$

and satisfy for every $\omega\in\mathbb{C}$,

$$\begin{array}{lll} R^{\mathrm{h}}_{\omega}(\mathbb{D})\subset\mathbb{D}, & SR^{\mathrm{h}}_{\omega}=R^{\mathrm{h}}_{-\omega}S, & SJ^{\mathrm{h}}=-J^{\mathrm{h}}S,\\ R^{\mathrm{c}}_{\omega}\subset\mathbb{D}, & SR^{\mathrm{c}}_{\omega}=R^{\mathrm{c}}_{-\omega}S & SJ^{\mathrm{c}}=-J^{\mathrm{c}}S \end{array}$$

with

$$J^{\mathrm{h}}=\frac{dR^{\mathrm{h}}_{\omega}}{d\omega}(0)=\mathrm{diag}(\mathrm{i},\mathrm{i},-\mathrm{i},-\mathrm{i},0,0),\quad J^{\mathrm{c}}=\frac{dR^{\mathrm{c}}_{\omega}}{d\omega}(0)=\mathrm{diag}(0,0,0,0,\mathrm{i},-\mathrm{i}).$$

Moreover the truncated system happens to be integrable, which makes its study far more easy. This property of integrability, is not only true for the normal form of order 3, but it also holds for normal forms of any order. For the normal form system of order 3, three first integrals are given by

$$\begin{gathered} |C|=k,\qquad \tfrac{\mathrm{i}}{2}(A\overline{B}-B\overline{A})=K,\\ |B|^2=|A|^2(1+q_3(\nu)\nu K+q_4(\nu)\nu k^2)-|A|^4+2H. \end{gathered} \tag{9.16}$$

The truncated system admits near the origin a one parameter family of reversible periodic orbits $Y^*_{k,\nu}$ of size $k\geq 0$ explicitly given by

$$Y^*_{k,\nu}(t)=\left(0,0,0,0,k\mathrm{e}^{\mathrm{i}\omega^*_1(k,\nu)t},k\mathrm{e}^{-\mathrm{i}\omega^*_1(k,\nu)t}\right)$$

with $\omega^*_1(k,\nu)=\dfrac{\omega_1(\nu)}{\nu}+n_4(\nu)\nu^2k^2$. In other words, the truncated system admits, for each ν, periodic solutions of arbitrary small size. Moreover, each periodic orbit admits a two-parameters family of solutions homoclinic to itself

$$h^*_{k,\theta,\psi}(t)=R^{\mathrm{h}}_{\theta}h^*_k(t)+R^{\mathrm{c}}_{\psi}Y^*_{k,\nu}\big(t+\varphi^*(k,\nu)\tanh(\eta(k)t)\big),\quad k,\theta,\psi\in\mathbb{R} \tag{9.17}$$

with

$$\eta(k)=\sqrt{1+q_4(\nu)\nu k^2}=1+\mathcal{O}(\nu k^2),\qquad \varphi^*(k,\nu)=\frac{n_2(\nu)\nu}{\omega^*_1(k,\nu)},$$

and

$$\begin{aligned} h^*_k(t) &= (A_{h^*_k}(t),B_{h^*_k}(t),\widetilde{A}_{h^*_k}(t),\widetilde{B}_{h^*_k}(t),0,0)\\ A_{h^*_k}(t) &= \frac{\eta(k)}{\cosh(\eta(k)t)}\,\mathrm{e}^{\mathrm{i}\psi_k(t)},\qquad B_{h^*_k}(t)=\frac{-\eta^2(k)\tanh(\eta(k)t)}{\cosh(\eta(k)t)}\,\mathrm{e}^{\mathrm{i}\psi_k(t)},\\ \widetilde{A}_{h^*_k}(t) &= A_{h^*_k}(-t),\qquad \widetilde{B}_{h^*_k}(t)=-B_{h^*_k}(-t)\\ \psi_k(t) &= \left(\tfrac{\omega_0(\nu)}{\nu}+p_4(\nu)\nu^2k^2\right)t+p_2(\nu)\nu\eta(k)\tanh(\eta(k)t) \end{aligned}$$

Among these homoclinic solutions there are four one-parameter families of *reversible* solutions

$$h^*_{k,0,0},\ h^*_{k,\pi,0},\ h^*_{k,0,\frac{\pi}{\omega_1^*(k,\nu)}},\ h^*_{k,\pi,\frac{\pi}{\omega_1^*(k,\nu)}}. \tag{9.18}$$

Observe that

$$h^*_{k,0,0} \underset{\mathcal{R}e(t)\to+\infty}{\longrightarrow} Y^*_{k,\nu}\left(t+\varphi^*(k,\nu)\right).$$

Thus, the phase shift between $h^*_{k,0,0}$ and $Y^*_{k,\nu}$ at $+\infty$ is

$$\varphi = \varphi^*(k,\nu) + \frac{2n\pi}{\omega_1^*(k,\nu)},\quad n\in\mathbb{Z}. \tag{9.19}$$

Finally, the truncated system admits a one parameter family of homoclinic connections to 0

$$h^*_{0,\theta,0} = R^{\mathrm{h}}_\theta h, \qquad \theta\in\mathbb{R}$$

where h is explicitly given by

$$h(t) := \left(r_0^{\mathrm{h}}(t)\mathrm{e}^{\mathrm{i}\psi_0(t)}, r_1^{\mathrm{h}}(t)\mathrm{e}^{\mathrm{i}\psi_0(t)}, r_0^{\mathrm{h}}(t)\mathrm{e}^{-\mathrm{i}\psi_0(t)}, r_1^{\mathrm{h}}(t)\mathrm{e}^{-\mathrm{i}\psi_0(t)}, 0, 0\right) \tag{9.20}$$

where

$$r_0^{\mathrm{h}}(t) = \frac{1}{\cosh(t)},\quad r_1^{\mathrm{h}}(t) = -\frac{\tanh(t)}{\cosh(t)},\quad \psi_0(t) = \frac{\omega_0(\nu)}{\nu}\,t + p_2(\nu)\nu\tanh(t).$$

Among, this one parameter family of homoclinic connections to 0, there are only two solutions which are reversible

$$h, \qquad R^{\mathrm{h}}_\pi h = -h.$$

The orbits h and $R^{\mathrm{h}}_\pi h$ appear to be the limit case for $k\longrightarrow 0$ of the previously found families of homoclinic connections (9.18), since

$$Y^*_{0,\nu}=0,\qquad h^*_{k,0,0} = h^*_{0,0,\frac{\pi}{\omega_1^*(0,\nu)}} = h,\qquad h^*_{k,\pi,0} = h^*_{k,\pi,\frac{\pi}{\omega_1^*(k,\nu)}} = R^{\mathrm{h}}_\pi h.$$

Observe that the homoclinic connection h is induced by the set of hyperbolic eigenvalues of the linear part whereas the periodic orbits are induced by the pair of simple eigenvalues which stay on imaginary axis after bifurcation.

Remark 9.1.3. The truncated system also admits two other class of solutions : elliptic periodic orbits and quasi periodic orbits. However, this book is devoted to the problem of the persistence of homoclinic connections. So, we do not study the persistence of these two classes of solutions.

9.1.3 Statement of the results of persistence for periodic orbits

There remains the problem of the persistence for the full system (9.8) of the previously found periodic and homoclinic orbits. In other words, do the previously found orbits persist when the normal form system is perturbed by the higher order terms?

9.1.3.1 Persistence of periodic solutions. For the periodic solutions the answer is yes : the full system admits a one parameter family of reversible periodic orbits $Y_{k,\nu}$ of arbitrary small size $k \geq 0$ given by Theorem 9.1.4 below. This result follows directly from the general Theorem 4.1.2 given in chapter 4 which ensures for reversible systems the existence of a one parameter family of periodic orbits bifurcating from a pair of simple purely imaginary eigenvalues. This theorem gives a detailed description of the periodic orbits which is very useful when complexifying time for studying the persistence of homoclinic connections to exponentially small periodic orbits.

Theorem 9.1.4. *There exist $k_0 > 0$, $\nu_1 > 0$ such that for all ν in $]0, \nu_1]$ the real full system* (9.8) *admits a one parameter family of periodic solutions $\left(Y_{k,\nu}(t)\right)_{k\in[0,k_0]}$ with $Y_{k,\nu}(t) = \widehat{Y}(\underline{\omega}_{k,\nu} t, k, \nu)$ and*

$$
(a)\quad \begin{cases} \widehat{Y}(s,k,\nu) = \sum\limits_{n\geq 1} k^n\, \widehat{Y}_n(s,\nu), \\ \widehat{Y}_n \in \tilde{\mathcal{P}}_{n,R},\ \ \widehat{Y}_1(s) = (0,0,0,0,\mathrm{e}^{\mathrm{i}s},\mathrm{e}^{-\mathrm{i}s}),\ \ \widehat{Y}_2 = \widehat{Y}_3 = 0 \\ (k_0)^n \left\|\widehat{Y}_n\right\|_{\mathcal{P}1} \leq C_{\widehat{Y}}\nu^2 \ \ \textit{for every } \nu \in]0,\nu_1],\ n \geq 2. \end{cases}
$$

with $$\tilde{\mathcal{P}}_{n,R} = \Big\{ \widehat{Y} : \mathbb{R} \mapsto \mathbb{D},\ \widehat{Y}(s) = a_0 + \sum_{p=1}^{n} a_p \cos(ps) + b_p \sin(ps),$$

$$Sa_p = a_p,\ Sb_p = -b_p \Big\} \cap \left\{ \widehat{Y} / \int_0^{2\pi} \langle \widehat{Y}(s), \widehat{Y}_1(s)\rangle ds = 0 \right\}$$

and $$\left\|\widehat{Y}\right\|_{\mathcal{P}1} := \sup_{s\in[0,2\pi]} |\widehat{Y}(s)| + \sup_{s\in[0,2\pi]} \left|\frac{d\widehat{Y}}{ds}(s)\right|.$$

$$
(b)\quad \begin{cases} \underline{\omega}_{k,\nu} = \dfrac{\underline{\omega}_1(\nu)}{\nu} + \widehat{\omega}(k,\nu),\ \ \widehat{\omega}(k,\nu) = \sum\limits_{n\geq 2} \widehat{\omega}_n(\nu)\, k^n, \\ |\widehat{\omega}(k,\nu)| \leq C_\omega k^2 \nu \ \textit{for every } \nu \in]0,\nu_1],\ k \in [0,k_0]. \end{cases}
$$

(c) $\widehat{\omega}(-k,\nu) = \widehat{\omega}(k,\nu), \qquad \widehat{Y}(s+\pi,-k,\nu) = \widehat{Y}(s,k,\nu).$

The proof of this theorem is given in Section 9.2.

Remark 9.1.5. Observe that $C_{\widehat{Y}}$, C_ω are independent of k, ν. We have constructed a particular family of reversible periodic solutions $Y_{k,\nu}$ of (9.8).

Since (9.8) is autonomous, we can deduce from $Y_{k,\nu}$, other periodic solutions by an arbitrary shift in time. Among these periodic solutions, $Y_{k,\nu}(t), Y_{k,\nu}(t+\pi/\underline{\omega}_{k,\nu})$ are the only ones which are reversible.

9.1.4 Statement of the results of persistence for homoclinic connections

For the homoclinic connections the situation is far more intricate. One way to make up one's mind is to formulate the problem in terms of stable manifolds and to illustrate it geometrically. We denote by $W_s^*(k,\nu)$ the stable manifold of the periodic solution $Y_{k,\nu}^*$ of the truncated system. The manifold $W_s^*(k,\nu)$ looks like a three dimensional cylinder in $\mathbb{R}^6$ (one dimension for time and one dimension for the phase shift and a third one due to the invariance by the rotation R_θ^{h}, see (9.17)).The radius of this cylinder is essentially the size of the periodic orbit, i.e. k. The fundamental remark, is that *for a reversible system, a reversible periodic solution admits a reversible homoclinic connection to itself if and only if the intersection between its stable manifold and the symmetry space* $\mathcal{P}_+ = \{Y/SY = Y\}$ *is not empty* (see Lemma 3.1.8). In this case, $\mathcal{P}_+$ is a 3 dimensional vector space in $\mathbb{R}^6$. For $k > 0$, the intersection consists of four points which lead to the four reversible homoclinic connections to the periodic orbit $Y_{k,\nu}^*$. For $k = 0$, there are two points of intersection, which leads to the two reversible homoclinic connection to 0 of the truncated system.

Let us denote by $W_s(k,\nu)$ the stable manifold of the periodic solution $Y_{k,\nu}$ of the full system. The stable manifold $W_s(k,\nu)$ is obtained by perturbation of the stable manifold $W_s^*(k,\nu)$ of the periodic solution of the truncated system. On Figure 7.4 we can observe that for $k = 0$ the situation is not robust since

$$\dim W_s(0,\nu) + \dim \mathcal{P}_+ = 2 + 3 = 5$$

So, there might not exist any reversible homoclinic connection to 0 for the full system, whereas for k large enough, the four points of intersection between $W_s^*(k,\nu)$ and $\mathcal{P}_+$ should persist since

$$\dim W_s(k,\nu) + \dim \mathcal{P}_+ = 3 + 3 = 6.$$

Thus, there should exist four reversible homoclinic connections to $Y_{k,\nu}$ for k large enough. The natural question is then the following:

Question 9.1.6.

(a) *Determine, for a fixed value of the bifurcation parameter* ν, *the smallest size* $k_c(\nu)$ *of a periodic solution which admits a reversible homoclinic connection to itself : is it 0 or not? And if it is not 0 we would like to compute* k_c *with respect to* ν.

(b) *If* $k_c(\nu) > 0$, *what is the behavior in the past of stable manifold* $W_s(0,\nu)$ *of 0?*

Remark 9.1.7. We have to face the same kind of problem as for the $0^{2+}\mathrm{i}\omega$ resonance (see paragraph 7.1.4 and question 7.1.6).

A more analytical approach complete usefully the geometrical approach and points out that we have to face the same kind of difficulty as for our toy model (1.2) given in the introduction of the book: the truncated system can be seen as a system of two "uncoupled" two dimensional systems, which have very different behaviors. Using the three firsts integrals (9.16), the first four dimensional system can be written

$$\begin{aligned}\frac{dA}{dt} &= \frac{\mathrm{i}\omega_0(\nu)}{\nu}A + B + \mathrm{i}A\mathrm{P}(|A|^2, K, k^2, \nu),\\ \frac{dB}{dt} &= \frac{\mathrm{i}\omega_0(\nu)}{\nu}B + \mathrm{i}B\mathrm{P}(|A|^2, K, k^2, \nu) + A\mathrm{Q}(|A|^2, K, k^2, \nu)\end{aligned}$$

with

$$\begin{aligned}\mathrm{P}(|A|^2, K, k^2, \nu) &= \ p_2(\nu)\nu\ |A|^2 + p_3(\nu)\nu^2\ K + p_4(\nu)\nu^2\ k^2\\ \mathrm{Q}(|A|^2, K, k^2, \nu) &= \ 1 - 2|A|^2 + q_3(\nu)\nu\ K + q_4(\nu)\nu\ k^2\end{aligned}$$

and the second system reads

$$\frac{dC}{dt} = \frac{\mathrm{i}\omega_1(\nu)}{\nu}C + \mathrm{i}C\left(n_2(\nu)\nu\ |A|^2 + n_3(\nu)\nu^2\ K + n_4(\nu)\nu^2\ k^2\right).$$

The first system is "hyperbolic and slow". The differential at the origin admits four complex simple eigenvalues $\pm 1 \pm \mathrm{i}\frac{\omega_0(\nu)}{\nu}$. For $k = 0$, it admits two reversible connections to 0 given by

$$\pm\left(r_0^{\mathrm{h}}(t)\mathrm{e}^{\mathrm{i}\psi_0(t)}, r_1^{\mathrm{h}}(t)\mathrm{e}^{\mathrm{i}\psi_0(t)}, r_0^{\mathrm{h}}(t)\mathrm{e}^{-\mathrm{i}\psi_0(t)}, r_1^{\mathrm{h}}(t)\mathrm{e}^{-\mathrm{i}\psi_0(t)}\right)$$

(see (9.20)).

The second is "oscillatory with high frequency". The differential at the origin admits two purely imaginary eigenvalues $\pm\frac{\mathrm{i}\omega_1(\nu)}{\nu}$. The solutions of this system oscillate rapidly along the circle $|C|^2 = k^2$.

For the homoclinic orbit h, these two systems are really uncoupled because $k^2 = |C|^2 = 0$ holds for h. Now, when we add the rest "$\mathcal{R}$", i.e. higher order terms, we "couple" these two systems. The question is then to determine whether or not the homoclinic orbit h survives the oscillations induced by the eigenvalues $\pm\dfrac{\mathrm{i}\omega_1(\nu)}{\nu}$? So, as for our first toy model, we expect that the answer is no. Moreover, we expect the appearance of a *critical size* $k_c(\nu)$ *which is exponentially small* such that for $k \geq k_c(\nu)$ there exists reversible homoclinic connections to periodic orbits of size k, and for $k < k_c(\nu)$ there should not exists any reversible homoclinic connections to periodic orbits of size k and in particular there should not exists any homoclinic connections to 0.

For our toy model (1.2) the problem of the persistence was solved by explicit computations. Here, no explicit computation are possible for the full system. So we use the "exponential tools" developed in chapter 2 for studying such systems.

9.1.4.1 Persistence of homoclinic connections to exponentially small periodic orbits. Using the First Bi-frequency Exponential Lemma 2.2.1 we prove the persistence of homoclinic connections to exponentially small periodic orbits.

Theorem 9.1.8. *For all ℓ, $0 < \ell < \frac{\pi}{2}$ and there exist $M, K(\ell) > 0$ such that for all sufficiently small ν, the full System* (9.8) *admits four reversible solutions of the form*

$$Y(t) = X(t) + Y_{k,\nu}(t + \varphi \tanh(\lambda_{k,\nu} t))$$

where

$$|X(t)| \leq M e^{-\lambda_{k,\nu}|t|}, \qquad \textit{for } t \in \mathbb{R};$$

$Y_{k,\nu}$ is the periodic solution of the full system given by Theorem 9.1.4 *of size $k = K(\ell)\nu^{\frac{3}{2}} e^{-\frac{\ell\omega_\star}{\nu}}$ and where*

$$\lambda_{k,\nu} = 1 + \mathcal{O}(k^2\nu) = 1 + \mathcal{O}(\nu^4 e^{-\frac{2\ell\omega_\star}{\nu}}) \qquad \textit{with } \omega_\star = \min_{p\in\mathbb{Z}} |\omega_1 - p\omega_0| > 0.$$

The proof of this theorem is given in Section 9.3.

Remark 9.1.9. To come back to our original Equation (9.1) it suffices to perform the scaling (9.6) and the polynomial change of coordinates given by the Normal Form Theorem 3.2.1. The solution has the same form and the oscillations at infinity are less than $M\nu^6 e^{-\ell\omega_\star/\nu}$.

Remark 9.1.10. This theorem is very similar to the one obtained for the $0^{2+}i\omega$ resonance (see Theorem 7.1.7). However, observe that here, the frequency in the exponential is $\omega_\star$ and not ω_1 because of the interaction between the two frequencies ω_0 and ω_1. This is why we need the First Bi-frequency Exponential Lemma 2.2.1 instead of the First Mono-frequency Exponential Lemma 2.1.1 as for Theorem 7.1.7. Moreover the use of the First Bi-frequency Exponential Lemma 2.2.1 requires the determination of the exact decay rate $\lambda_{k,\nu}$ at infinity for the homoclinic connection whereas for Theorem 7.1.7 this decay rate is λ with $0 \leq \lambda \leq 1$. To determine this decay rate we have to use the constructive Floquet Theory given in Chapter 5.

As for the $0^{2+}i\omega$ resonance, we deduce from the proof of this theorem, a first corollary which ensures the existence of homoclinic connections to periodic orbits of size k lying in $[K(\ell)\nu^{\frac{3}{2}} e^{-\frac{\ell\omega_\star}{\nu}}, K_\star]$.

Corollary 9.1.11. *There exist* $M_\star, K_\star > 0$ *such that for every* $\ell \in]0, \frac{\pi}{2}[$, *there exist* $K(\ell), \nu_\star(\ell) > 0$ *such that* $\nu \in]0, \nu_\star(\ell)[$, *the full System* (9.8) *admits four reversible solutions of the form*

$$Y(t) = X(t) + Y_{k,\nu}(t + \varphi \tanh(\lambda_{k,\nu} t))$$

with

$$|X(t)| \leq M_\star \, \mathrm{e}^{-\lambda_{k,\nu}|t|} \quad \text{for } t \in \mathbb{R}, \qquad \lambda_{k,\nu} = 1 + \mathcal{O}(k^2\nu)$$

where $Y_{k,\nu}$ *is a periodic solution of the full system satisfying*

$$|Y_{k,\nu}(t)| \leq M_\star k \quad \text{for } t \in \mathbb{R}.$$

Remark 9.1.12. Improvement of the domain of existence. As for the $0^{2+}\mathrm{i}\omega$ resonance, working in a strip $\mathcal{B}_\ell = \{\xi \in \mathbb{C}, |\mathcal{I}m\,(\xi)| < \ell\}$ of width $\ell = \pi - \delta\nu$ instead of a strip of fixed width as for Theorem 9.1.8, it is possible to enlarge the domain of existence of homoclinic connections to periodic orbits: there exist $\sigma, K_1', \nu_0', M' > 0$ such that for every $\nu \in]0, \nu_0']$ and every $k \in [K_1' \mathrm{e}^{-\frac{\pi\omega_\star}{2\nu}}, K_\star[$, the full System (9.8) admits four solution of the form

$$Y(t) = X(t) + Y_{k,\nu}(t + \varphi \tanh \lambda_{k,\nu} t)$$

where $Y_{k,\nu}$ is a periodic solution of the full system satisfying $|Y_{k,\nu}(t)| \leq M_\star k$ for $t \in \mathbb{R}$. The domains of persistence of reversible homoclinic connection to $Y_{k,\nu}$ are drawn in figure 9.5 p. 378.

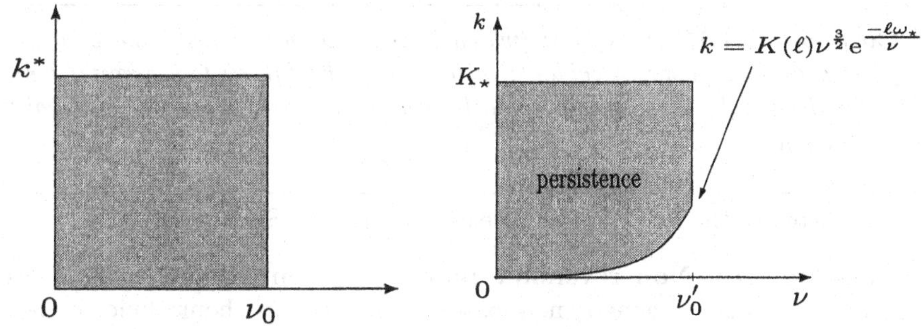

Truncated system — Full system (9.8)

Fig. 9.3. Domain of persistence of reversible homoclinic connections to a periodic orbit of size k for the full system (9.21)

Remark 9.1.13. Theorem 9.1.8, Corollary 9.1.11 and Remark 9.1.12 give a first partial answer to Question 9.1.6: they ensure that $k_c(\nu)$ is at least exponentially small.

9.1.4.2 Generic non persistence of reversible homoclinic connection to 0. A second partial answer which completes the previous one is given by the following theorem which ensures the generic non persistence of homoclinic connection to 0. To state the theorem we introduce in equation (9.8) an extra parameter $\rho \in [0,1]$

$$\frac{dY}{dt} = N(Y,\nu) + \rho R(Y,\nu). \tag{9.21}$$

For $\rho = 0$, we get the truncated system which admits two homoclinic connections h and $-h$ to 0. Now, we would like to understand what happens for $\rho \neq 0$ and in particular for $\rho = 1$ which corresponds to the full system (9.8).

Theorem 9.1.14. *There exists a real analytic function defined on [0,1]*

$$\Lambda(\rho) = \sum_{n \geq 1} \rho^n \Lambda_n$$

and $d_s > 0$ such that for any $\rho \in [0,1]$ with $\Lambda(\rho) \neq 0$, and for ν small enough, the full system does not admit any reversible homoclinic connection Y to 0, which principal part is $\pm h$, i.e. such that

$$\sup_{t \in \mathbb{R}} |Y(t) \pm h(t)| \leq d_s. \tag{9.22}$$

The first coefficient Λ_1 is explicitly known in terms of Bessel functions and in terms of the coefficients of the power expansion of the rest $\mathcal{R}$ (see Remark 9.1.19 for the explicit expression of Λ_1).

Corollary 9.1.15. *If $\Lambda_1 \neq 0$ (which is generically satisfied), there is at most a finite number of values of ρ for which the full system admits, for ν small enough, a reversible homoclinic connection to 0 whose principal part is $\pm h$.*

A sketch of the proof of this Theorem is given in Section 9.4.

Remark 9.1.16. Non reversible homoclinic connections to 0. This theorem ensures the generic non existence of *reversible* homoclinic connections to 0. The question of the existence of non reversible homoclinic connection to 0 remains open.

Remark 9.1.17. Equation (9.22) characterizes the homoclinic connections with one loop (see Figure 9.4). So, Theorem 9.1.14 ensures the generic non persistence of homoclinic connections to 0 with one loop whereas Theorem 9.1.8 ensures that there always exist homoclinic orbits with one loop connecting an exponentially small periodic orbits to itself. The question of the

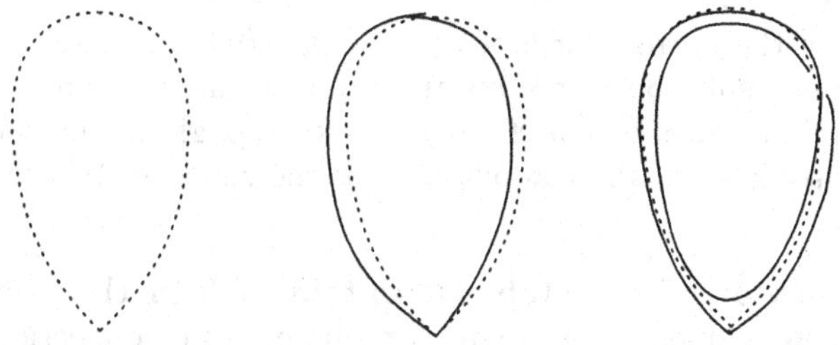

Fig. 9.4. Homoclinic connections with one or more loops

existence of homoclinic connections to 0 or to periodic orbits with more than one loop is not studied here.

Remark 9.1.18. For $\rho = 0$, $\Lambda(\rho) = 0$. Indeed, the truncated system does admit a reversible homoclinic connection to 0.

Remark 9.1.19. Among the coefficients Λ_n, Λ_1 is the only one which is known explicitly. Let us denote

$$\omega_\star = \min_{p\in\mathbb{Z}} |\omega_1 - p\omega_0|, \qquad \widetilde{\omega}_\star = \omega_1 - p_\star\omega_0$$

where $p_\star$ is the smallest integer such that the minimum is reached. The minimum is reached for a unique value of p when $\frac{\omega_1}{\omega_0} \notin \mathbb{N} + \frac{1}{2}$ and it is reached for two consecutive values of p for $\frac{\omega_1}{\omega_0} \in \mathbb{N} + \frac{1}{2}$. Then,

- for $\frac{\omega_1}{\omega_0} \notin \mathbb{N} + \frac{1}{2}$, $\omega_1 - p_\star\omega_0 > 0$, $\qquad \Lambda_1 = \Lambda^+(p_\star)$,
- for $\frac{\omega_1}{\omega_0} \notin \mathbb{N} + \frac{1}{2}$, $\omega_1 - p_\star\omega_0 < 0$, $\qquad \Lambda_1 = \Lambda^-(p_\star)$,
- for $\frac{\omega_1}{\omega_0} \in \mathbb{N} + \frac{1}{2}$, $\omega_1 = (p_\star + \frac{1}{2})\,\omega_0$, $\qquad \Lambda_1 = \Lambda^+(p_\star) + \Lambda^-(p_\star + 1)$

with

$$\Lambda^\pm(p_\star) = \sum_{\substack{m_A+m_B-m_{\widetilde{A}}-m_{\widetilde{B}}=p_\star \\ m_C+m_{\widetilde{C}}=0}} a_{C,\overline{m},0}\, \Lambda^\pm_{\overline{m}}(p_\star)$$

where

$$\Lambda^+_{\overline{m}}(p_\star) = -(\mathrm{i})^{m_B+m_{\widetilde{B}}+1} \sum_{n\geq 0} \frac{(p_{20} - p_\star n_{20})^n\, \omega_\star^{m_A+m_{\widetilde{A}}+2m_B+2m_{\widetilde{B}}+n-1}}{n!\,(m_A + m_{\widetilde{A}} + 2m_B + 2m_{\widetilde{B}} + n - 1)!},$$

$$\Lambda^-_{\overline{m}}(p_\star) = -(\mathrm{i})^{m_B+m_{\widetilde{B}}+1} \sum_{n\geq 0} \frac{(p_{20} - p_\star n_{20})^n\, \omega_\star^{m_A+m_{\widetilde{A}}+2m_B+2m_{\widetilde{B}}+n-1}}{n!\,(m_A + m_{\widetilde{A}} + 2m_B + 2m_{\widetilde{B}} + n - 1)!}.$$

where $a_{C,\overline{m},\ell}$ is the coefficient of of the power expansion of the fifth component of the rest $\mathcal{R}$, and where $p_{20} = p_2(0)$, $n_{20} = n_2(0)$ where p_2, n_2 are the coefficients of the normal form of order 3 (see (9.9)). So we have a generic non existence result, but for a given system we are not able to determine whether $\Lambda(\rho)$ vanishes or not. However, even though all the coefficients Λ_n were explicitly known, the computation of the zeros of $\Lambda(\rho)$ might be very difficult.

9.1.4.3 Behavior of the stable manifold of 0 in the past. Question 9.1.6-(b) remains open: when there is no homoclinic connection to 0, what is the behavior in the past of the stable manifold of 0? One can expect as for Toy Model (1.1), that the stable manifold develops exponentially small oscillations at $-\infty$. This is what happens with the toy model

$$\begin{aligned}
\frac{d\mathbf{a}}{dx} &= -\omega_0 \mathbf{a}' + \mathbf{b}, \\
\frac{d\mathbf{a}'}{dx} &= \omega_0 \mathbf{a} + \mathbf{b}', \\
\frac{d\mathbf{b}}{dx} &= \mu \mathbf{a} - \omega_0 \mathbf{b}' - 2\mathbf{a}(\mathbf{a}^2 + \mathbf{a}'^2) \\
\frac{d\mathbf{b}'}{dx} &= \mu \mathbf{a}' + \omega_0 \mathbf{b} - 2\mathbf{a}'(\mathbf{a}^2 + \mathbf{a}'^2), \\
\frac{d\mathbf{c}}{dx} &= -\omega_1 \mathbf{c}' - \mathbf{a}'^{p_*}(\mathbf{a}^2 + \mathbf{a}'^2)^2, \\
\frac{d\mathbf{c}'}{dx} &= +\omega_1 \mathbf{c} - \mathbf{a}^{p_*}(\mathbf{a}^2 + \mathbf{a}'^2)^2,
\end{aligned}$$

which is a particular example of reversible family vector fields admitting a $(i\omega_0)^2 i\omega_1$ resonance at the origin with $q_{10} = 1$ and $q_{20} = -2$. Applying the transformation $\mathbb{T}$ and the scaling (9.6), we get

$$\begin{aligned}
\frac{dA}{dt} &= \tfrac{i\omega_0}{\nu} A + B, \\
\frac{dB}{dt} &= \tfrac{i\omega_0}{\nu} B + A - 2A(A\widetilde{A}), \\
\frac{d\widetilde{A}}{dt} &= -\tfrac{i\omega_0}{\nu} \widetilde{A} + \widetilde{B}, \\
\frac{d\widetilde{B}}{dt} &= -\tfrac{i\omega_0}{\nu} \widetilde{B} + A - 2\widetilde{A}(A\widetilde{A}), \\
\frac{dC}{dt} &= \tfrac{i\omega_1}{\nu} C + i\rho\nu^{p_* + \frac{3}{2}} A^{p_*}(A\widetilde{A})^2, \\
\frac{d\widetilde{C}}{dt} &= -\tfrac{i\omega_1}{\nu} \widetilde{C} - i\rho\nu^{p_* + \frac{3}{2}} \widetilde{A}^{p_*}(A\widetilde{A})^2,
\end{aligned}$$

This systems corresponds to the rescaled normal form of order three with $P = N = 0$, $q_3 = q_4 = 0$, perturbed by a unique monomial

$$(0, 0, 0, 0, i\rho A^{p_*}(A\widetilde{A})^2, -i\rho\widetilde{A}^{p_*}(A\widetilde{A})^2).$$

The solutions of this system can be computed explicitly since the system is only partially coupled. We check that the two dimensional stable manifold of 0 is explicitly given by

$$Y_s(t,\theta) = (A_s(t,\theta), B_s(t,\theta), \overline{A_s(t,\theta)}, \overline{B_s(t,\theta)}, C_s(t,\theta), \overline{C_s(t,\theta)})$$

with

$$A_s(t,\theta) = e^{\frac{i\omega_0 t}{\nu}+i\theta}\cosh^{-1}(t), \qquad B_s(t,\theta) = -e^{\frac{i\omega_0 t}{\nu}+i\theta}\cosh^{-1}(t)\tanh(t),$$

$$C_s(t,\theta) = i\rho\nu^{p_\star+\frac{3}{2}}\, e^{\frac{i\omega_1 t}{\nu}+ip_\star\theta}\int_t^{+\infty} e^{-\frac{i(\omega_1-p_\star\omega_0)s}{\nu}}\frac{ds}{(\cosh s)^{p_\star+4}}.$$

and that Y_s develops exponentially small oscillations at $-\infty$

$$Y_s(t,\theta) \underset{t\to-\infty}{\sim} \left(0,0,0,0,iK(\nu)e^{\frac{i\omega_1 t}{\nu}+ip_\star\theta}, -iK(\nu)e^{-\frac{i\omega_1 t}{\nu}-ip_\star\theta}\right)$$

where

$$K(\nu) := \rho\nu^{p_\star+\frac{3}{2}}\int_{-\infty}^{+\infty} e^{\frac{i\omega_\star s}{\nu}}\frac{ds}{(\cosh s)^{p_\star+4}} \underset{\nu\to 0}{\sim} \rho\,\frac{2\pi\omega_\star^{p_\star+3}}{(p_\star+3)!}\,\frac{1}{\nu^{\frac{3}{2}}}\, e^{-\frac{\pi\omega_\star}{2\nu}}.$$

As for the $0^{2+}i\omega$ resonance, we also expect that for some vector fields admitting a $(i\omega_0)^2 i\omega_1$ resonance, the stable manifold of 0 might explode in the past. At the present time we do not know any general criteria to determine the behavior of the stable manifold of 0 in the past.

Remark 9.1.20. Sharper lower bound of the critical size. Moreover, for the previous particular example we can compute explicitly the critical size $k_c(\nu)$ of the periodic orbits which admits reversible homoclinic connections to themselves : we check that each periodic orbit of size $k > \frac{1}{2}K(\nu)$ admits four reversible homoclinic connection to itself, whereas none of the periodic orbits of size for $k < \frac{1}{2}K(\nu)$ admits reversible homoclinic connection to itself. Hence the critical size $k_c(\nu)$ is equal to $\frac{1}{2}K(\nu)$. However, in general we are not able to compute explicitly the critical size. Theorem 9.1.8, Corollary 9.1.11 and Remark 9.1.12 only give exponential upper bounds of the critical size and Theorem 9.1.14 ensures that generically there is no reversible homoclinic connection to 0. In fact as for the $0^{2+}i\omega$ resonance it is possible to obtain a sharper lower bound for the critical size (see Figure 9.5). Combining the proofs Theorems 9.1.8 and 9.1.14 we can check that there exists ν_0', d_s', $\alpha_0 \in]0,1[$ such that for every $\nu \in]0,\nu_0']$ and every $k \in [0, \alpha_0\Lambda(\rho)\, e^{\frac{-\pi\omega_\star}{2\nu}}[$ the full System (9.8) does not admit any reversible solutions of the form

$$Y(t) = X(t) + Y_{k,\nu}(t + \varphi\tanh\lambda_{k,\nu}t)$$

with

$$\sup_{t\in\mathbb{R}} |X(t) \pm h(\lambda_{k,\nu}t)| < d_s', \qquad \sup_{t\in\mathbb{R}} |X(t)|e^{\lambda_{k,\nu}|t|} < +\infty.$$

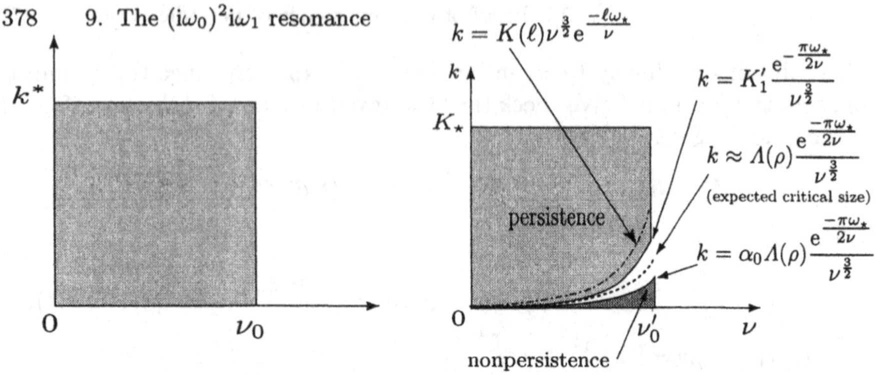

Truncated system ($\rho = 0$) Full system (9.21) ($\rho \neq 0$)

Fig. 9.5. Domain of persistence and non-persistence of reversible homoclinic connections to a periodic orbit of size k for the full system (9.21) ($\rho \neq 0$)

9.2 Proof of the persistence of periodic orbits: explicit form

This section is devoted to the persistence of periodic solutions. We prove Theorem 9.1.4 which follows directly from the general Theorem 4.1.2 given in chapter 4 which ensures for reversible systems the existence of a one parameter family of periodic orbits bifurcating from a pair of simple purely imaginary eigenvalues.

Proof of Theorem 9.1.4. The explicit form of $\mathcal{N}$, $\mathcal{R}$ given respectively by (9.8), (9.11) ensures that Equation (9.8) can be rewritten as

$$\frac{dY}{dt} = L_\nu Y + \mathcal{Q}(Y,\nu)$$

where

$$\text{(a)}\ L_\nu = \begin{bmatrix} \dfrac{L_{\mathrm{h}}(\nu)}{\nu} & 0 & 0 \\ 0 & 0 & -\dfrac{\underline{\omega}_1(\nu)}{\nu} \\ 0 & \dfrac{\underline{\omega}_1(\nu)}{\nu} & 0 \end{bmatrix} \text{ where}$$

$$L_{\mathrm{h}}(\nu) = \begin{bmatrix} L_{\mathrm{h}}^{\Delta}(\nu) & 0 \\ 0 & \overline{L_{\mathrm{h}}^{\Delta}(\nu)} \end{bmatrix} \text{ with } L_{\mathrm{h}}^{\Delta}(\nu) = \begin{bmatrix} \mathrm{i}\underline{\omega}_0(\nu) & \nu \\ \nu & \mathrm{i}\underline{\omega}_0(\nu) \end{bmatrix}.$$

Thus, (9.9) ensures that we can choose $\nu_1 \in]0,\nu_0]$ such that

$$\inf_{\nu\in]0,\nu_1]} \underline{\omega}_1(\nu) > 0, \quad \sup_{\nu\in]0,\nu_1]} |\nu\underline{\omega}_{11}(\nu)| < +\infty, \quad \sup_{\nu\in]0,\nu_1]} \|L_{\mathrm{h}}(\nu)\| < +\infty$$

Finally, since

$$\left(I_2 - \exp\left(\frac{2\pi L_{\mathrm{h}}^{\Delta}(\nu)}{\underline{\omega}_1(\nu)}\right)\right)^{-1} = \begin{bmatrix} \alpha(\nu) & \beta(\nu) \\ \beta(\nu) & \alpha(\nu) \end{bmatrix}$$

with

$$\alpha(\nu) = \frac{1}{2} - \frac{\mathrm{i}\sin\left(2\pi\frac{\underline{\omega}_0(\nu)}{\underline{\omega}_1(\nu)}\right)}{2\left(\cos\left(2\pi\frac{\underline{\omega}_0(\nu)}{\underline{\omega}_1(\nu)}\right) - \cosh\left(\frac{2\pi\nu}{\underline{\omega}_1(\nu)}\right)\right)},$$

$$\beta(\nu) = \frac{-\sinh\left(\frac{2\pi\nu}{\underline{\omega}_1(\nu)}\right)}{2\left(\cos\left(2\pi\frac{\underline{\omega}_0(\nu)}{\underline{\omega}_1(\nu)}\right) - \cosh\left(\frac{2\pi\nu}{\underline{\omega}_1(\nu)}\right)\right)},$$

hypothesis (H4c) and (9.9) ensure that ν_1 can be chosen sufficiently small so that there exists M such that for every $\nu \in]0, \nu_1]$

$$\left\|\left(I_4 - \exp\left(\frac{2\pi L_{\mathrm{h}}(\nu)}{\underline{\omega}_1(\nu)}\right)\right)^{-1}\right\| \leq M < +\infty.$$

(b) the nonlinear part $\mathcal{Q}$ satisfies $\mathcal{Q}(0,\nu) = D_Y\mathcal{Q}(0,\nu) = 0$; it is analytic in $D(0,r) = \{Y \in \mathbb{C}^6, |Y| < r\}$ with $0 < r < \min(r_A, r_B, r_{\widetilde{A}}, r_{\widetilde{B}}, r_C, r_{\widetilde{C}})$ and satisfies

$$\sup_{|Y|<r,\ \nu\in]0,\nu_1]} |\mathcal{Q}(Y,\nu)| < +\infty,$$

$$\sup_{|Y|<r,\ \nu\in]0,\nu_1]} |\nu^{-1}\mathcal{Q}_A(Y,\nu)| < +\infty, \qquad \sup_{|Y|<r,\ \nu\in]0,\nu_1]} |\nu^{-1}\mathcal{Q}_B(Y,\nu)| < +\infty.$$

Thus hypothesis (i) and (ii) of Theorem 4.1.2 are fulfilled. The result follows directly from this last theorem. The improvement of the estimates of $\left\|\widehat{Y}_n\right\|_{\mathcal{P}^1}$, i.e. $(k_0)^n \left\|\widehat{Y}_n\right\|_{\mathcal{P}^1} \leq C_{\widehat{Y}}\nu^2$ instead of $(k_0)^n \left\|\widehat{Y}_n\right\|_{\mathcal{P}^1} \leq C_{\widehat{Y}}$, for Theorem 4.1.2, has two origins

1. Here, $\left\|\left(I_4 - \exp\left(\frac{2\pi L_{\mathrm{h}}(\nu)}{\underline{\omega}_1(\nu)}\right)\right)^{-1}\right\|$ is bounded by M and not by $\frac{C_{\mathrm{h}}}{\nu}$ as for Theorem 4.1.2;
2. When looking for $\widehat{Y}$ in the form $\widehat{Y} = k\widehat{Y}_1 + k\widehat{Z}$ and when rewriting the equation for $u = (\widehat{Z}, \widehat{\omega})$ as an implicit equation $\mathcal{G}(u,k) = 0$ (see equation (4.10)), we check by explicit computation that

$$\left\|k^p\mathcal{G}_{pq}[u_1,\cdots,u_q]\right\|_{\mathcal{P}^0} \leq C_2\nu \left\|u_1\right\|_{\mathcal{U}^1} \cdots \left\|u_q\right\|_{\mathcal{U}^1} \left(\frac{2k}{r}\right)^p$$

Here again the estimates are better than the ones obtained in Lemma 4.A.5.

Finally, to check that $\widehat{Y}_2 = \widehat{Y}_3 = 0$, we simply put the expansion of $\widehat{Y}$ in powers of k in the equation satisfied by $\widehat{Y}$ and we identify the coefficients of the powers of k. □

9.3 Proof of the persistence of reversible homoclinic connections to exponentially small periodic solutions

A preliminary idea to find a reversible solution Y of the full system homoclinic to the periodic orbits $Y_{k,\nu}$ is to look for Y in the form

$$Y(t) = Y_{k,\nu}(t + \varphi \tanh(t)) + h(t) + v(t), \qquad t \in \mathbb{R},\ \varphi$$

where h is the homoclinic connection to 0 of the truncated system and where v is a perturbation term which is reversible and which tends to 0 at infinity. This choice of the form of Y is motivated by the form of the solutions of the truncated system homoclinic to the periodic orbit given by (9.17). We also used such a form for the $0^{2+}i\omega$ resonance (see Section 7.3). To build solution of this form, a natural idea is the following :

First rewrite the equation satisfied by v in the form

$$\frac{dv}{dt} - D\mathcal{N}(h, \nu).v = g(v, \varphi, k, \nu). \tag{9.23}$$

To transform this equation into an integral equation which can be solved with the Contraction Mapping Theorem, we must invert the linear part of (9.23). The crucial point of our analysis is the following : for an *antireversible* function f which *tends to 0 exponentially at infinity*, there exists a *reversible* function u which *tends to 0 exponentially at infinity* and which satisfies

$$\frac{du}{dt} - D\mathcal{N}(h, \nu).u = f$$

if and only if the function f satisfies the solvability condition

$$I(f, \nu) = \int_0^{+\infty} \langle r_-^*(t), f(t) \rangle_* = 0 \tag{9.24}$$

where the dual vector r_-^* reads

$$r_-^*(t) := (0, 0, 0, 0, -\tfrac{i}{2} e^{-i\psi_r(t)}, \tfrac{i}{2} e^{i\psi_r(t)}) \text{ with } \psi_r(t) = \tfrac{\omega_1(\nu)}{\nu} t + n_2(\nu)\nu \tanh(t).$$

This solvability condition is the typical one which occurs when the linear homogeneous system admits a pair of purely imaginary eigenvalues (see Section 6.4). Moreover, observe that the integral $I(g, \nu)$, is a *bi-oscillatory integral* since the dual vector rotates with a frequency of order ω_1/ν where ν is a small parameter and since g involves a second fast oscillation with a frequency of order ω_0/ν due to the terms coming from h. So, for studying this solvability condition, we have to use the Bi-Frequency Exponential Lemmas given in Section 2.2. For using these lemmas, we must find solutions v of (9.23) in the form $v(t) = \mathsf{v}(t, \frac{\omega_0 t}{\nu})$. For an antireversible function $\mathsf{f} : (\xi, s) \mapsto \mathsf{f}(\xi, s)$, which tends to 0 at infinity, Proposition 6.5.4 gives solutions u of the affine equation

$$\frac{du}{dt} - D\mathcal{N}(h,\nu).u = \mathsf{f}(t, \tfrac{\omega_0 t}{\nu})$$

of the form $u(t) = \mathsf{u}(t, \frac{\omega_0 t}{\nu})$ provided that f tends to 0 exponentially fast with an exponential decay rate larger than the one of h. To obtain such an exponential decay rate for $g(v, \varphi, k, \nu, t)$ we need it to be quadratic in v. This can be achieved by use of the constructive Floquet Theory given in Chapter 5. Before applying this Floquet Theory to the equation linearized around $Y_{k,\nu}$, it is more convenient to have a family of periodic orbits written as simply as possible. So finally the plan of the proof is as follows:

1. Using Theorem 4.3.1, we perform an analytic change of coordinates close to the identity, $Y = \Theta_\nu(Z)$ to obtain an equivalent equation

$$\frac{dZ}{dt} = \mathcal{N}(Z,\nu) + \mathcal{R}_z(Z,\nu) \tag{9.25}$$

   where $\mathcal{N}$ is the normal form of order 3 given by (9.8) and where $\mathcal{R}_z$ represents new higher order terms such that $Z_{k,\nu} = \Theta_\nu^{-1}(Y_{k,\nu})$ is a family of periodic solutions of (4.3.1) which are circles.

2. We look for an homoclinic connection Z to $Z_{k,\nu}$ of the form

$$\begin{aligned} Z(t) = {} & Z_{k,\nu}\big(t+\varphi\tanh(\lambda_{k,\nu}t)\big) \\ & +\big[\mathrm{Id}+Q_1\big(t+\varphi\tanh(\lambda_{k,\nu}t)\big)\big].\big(h'(\lambda_{k,\nu}t)+v'(\lambda_{k,\nu}t)\big) \end{aligned} \tag{9.26}$$

   where the decay rate $\lambda_{k,\nu} = 1 + \mathcal{O}(k^2\nu)$ and the Matrix Q_1 are obtained by the constructive Floquet Theory given in Chapter 5 applied to the equation linearized around $Z_{k,\nu}$ and where h' has the same form as h except that $\underline{\omega}_0(\nu)$ is replaced by $\underline{\omega}_0'(k,\nu) = \omega_0 + \mathcal{O}(\nu^2 + \nu k^2)$.
   So from now on, we look for the phase shift $\varphi \in \mathbb{R}$ and for a perturbation term v' which is reversible and which satisfies $\sup_{\tau\in\mathbb{R}}(|v(\tau)|\mathrm{e}^{|\tau|}) < +\infty$. The equation satisfied by v' reads

$$\frac{dv'}{d\tau} - D\mathcal{N}'(h'(\tau)).v' = g'(v', \varphi, k, \nu). \tag{9.27}$$

   This equation is very similar to (9.23) : $\mathcal{N}'$ and h' have the same form as $\mathcal{N}$ and h except that $\underline{\omega}_0(\nu)$ is replaced by $\underline{\omega}_0'(k,\nu)$. Moreover g' is quadratic with respect to v' whereas g contains linear terms in v. This is why we performed the linear time depending change of coordinates given by the Constructive Floquet Theory.

   As already explained the crucial point of the analysis is the study of the solvability condition that g' should satisfy to belong to the range of $\mathcal{L}$ with

$$\mathcal{L}(v') = \frac{dv'}{d\tau} - D\mathcal{N}'(h'(\tau)).v'.$$

This solvability condition is given by a bi-oscillatory integral, the analysis of which requires the Bi-Frequency Exponential Lemmas given in Section 2.2 and thus requires solution of (9.27) of the form $v'(\tau) = \mathsf{v}'(\tau, \frac{\omega_0\tau}{\nu})$. So we proceed as follows:

3. Using the inversion of affine equations with partially complexified time given in Section 6.5, we build an inverse of $\mathcal{L}$ for functions $\mathsf{f} : \mathcal{B}_\ell \times \mathbb{R} \to \mathbb{C}^6$ with $\mathcal{B}_\ell = \{\xi \in \mathbb{C},\ |\mathcal{I}m\,(\xi)\,| < \ell\}$. More precisely we prove that there exists a linear operator T such that for any antireversible function $\mathsf{f}' : \mathcal{B}_\ell \times \mathbb{R} \to \mathbb{C}^6$ satisfying $\sup_{\xi\in\mathcal{B}_\ell}(|\mathsf{f}'(\xi,s)|\mathrm{e}^{\lambda|\mathcal{R}e(\xi)|}) < +\infty$ for some $\lambda > 1$, the function $v'(\tau) = \mathsf{v}'(\tau, \frac{\omega_0\tau}{\nu})$ with $\mathsf{v}'(\xi, s) = \mathsf{T}(\mathsf{f}')(\xi, s)$ is a solution of
$$\mathcal{L}(v') = \mathsf{f}'(\tau, \tfrac{\omega_0\tau}{\nu})$$
provided that f satisfies the solvability condition
$$\int_0^{+\infty} \langle r'^*_-(\tau), \mathsf{f}'(\tau, \tfrac{\omega_0\tau}{\nu})\rangle_* d\tau = 0 \tag{9.28}$$
where r'^*_- has the same form as r^*_- except that $\underline{\omega}_1(\nu)$ is replaced by $\underline{\omega}'_1(k,\nu) = \omega_1 + \mathcal{O}(\nu^2 + \nu k^2)$.
4. Using the First Bi-Frequency Exponential Lemma 2.2.1, we prove that for each reversible function $\mathsf{v}' : \mathcal{B}_\ell \times \mathbb{R} \to \mathbb{C}^6$ which tends to 0 at infinity, there exists an appropriate phase shift $\varphi(\mathsf{v}', k, \nu)$ such that $g'(\mathsf{v}', \varphi(\mathsf{v}', k, \nu), k, \nu)$ satisfies the solvability condition (9.28) provided that the size of the periodic orbit k is larger than $K(\ell)\mathrm{e}^{-\frac{\omega_*\ell}{\nu}}$.
5. Finally, using the Contraction Mapping Theorem we build a reversible solution $\mathsf{v}' : \mathcal{B}_\ell \times \mathbb{R} \to \mathbb{C}^6$ of
$$\mathsf{u}' = \mathsf{T}\big(g'(\mathsf{u}', \varphi(\mathsf{u}', k, \nu), k, \nu)\big)$$
which tends to 0 at infinity. Then , $v'(\tau) = \mathsf{v}'(\tau, \frac{\omega_0\tau}{\nu})$ is a solution of (9.27).

For summarizing, for $k \geq K(\ell)\nu^{\frac{3}{2}}\mathrm{e}^{-\frac{\omega_*\ell}{\nu}}$, this proves the existence of a reversible solution of (9.25) given by (9.26) and homoclinic to the periodic orbit $Z_{k,\nu}$ of size k. Coming back to our original equation (9.8) with the analytical change of coordinates $Y = \Theta(Z)$ close to identity, this ensures the existence of reversible homoclinic connection to the periodic orbit $Y_{k,\nu}$ of size k provided that $k \geq K(\ell)\nu^{\frac{3}{2}}\mathrm{e}^{-\frac{\omega_*\ell}{\nu}}$.

9.3.1 Analytic conjugacy to circles

As previously explained, we first perform an analytic change of coordinates close to the identity to obtain an equivalent equation for which the periodic orbits are circles.

Proposition 9.3.1. *There exists $k_1 \in]0, k_0[$ and an analytic diffeomorphism Θ_ν close to identity such that setting $Y = \Theta_\nu(Z)$, the equation (9.8) is equivalent to*

$$\frac{dZ}{dt} = \mathcal{N}(Z,\nu) + \mathcal{R}_z(Z,\nu), \qquad Z = (Z_\mathrm{h}, Z_\mathrm{c}) \in \mathbb{D}, \tag{9.29}$$

with

(a) *For $k \in]0, k_1[$, $\nu \in]0, \nu_1]$, the periodic orbits $Y_{k,\nu}$ of (9.8) reads $Y_{k,\nu} = \Theta_\nu(Z_{k,\nu})$ where $Z_{k,\nu}(t) = (0,0,0,0,k\mathrm{e}^{\mathrm{i}\underline{\omega}_{k,\nu}t}, k\mathrm{e}^{-\mathrm{i}\underline{\omega}_{k,\nu}t})$ is a periodic solution of (9.29).*

(b) $S\Theta_\nu(Z) = \Theta_\nu(SZ), \qquad \Theta_\nu(\mathbb{D}) \subset \mathbb{D}.$

(c) *$|\Theta_\nu(Z_\mathrm{h}, Z_\mathrm{c})| = (Z_\mathrm{h}, Z_\mathrm{c}) + \Theta^1_\nu(Z_\mathrm{c})$, where $\Theta^1_\nu : (] - k_1, k_1[)^2 \mapsto \mathbb{D}$ satisfies*

$$|\Theta^1_\nu(Z_\mathrm{c})| \le M_\Theta \nu^2 |Z_\mathrm{c}|^4, \qquad |D\Theta^1_\nu(Z_\mathrm{c})| \le M_\Theta \nu^2 |Z_\mathrm{c}|^3,$$

for $\nu \in]0, \nu_1]$ and $Z_\mathrm{c} \in (D(0,k_1))^2$ where $D(0,k) = \{\xi \in \mathbb{C}, |\xi| < k\}$.

(d) *$\mathcal{N}$ is the normal form of order 3 given by (9.8) and $\mathcal{R}_z$ is a reversible analytic vector field which satisfies $\mathcal{R}(\mathbb{D}) \subset \mathbb{D}$ and*

$$\mathcal{R}_{z,A\,\mathrm{or}\,\widetilde{A}}(Z,\nu) = \mathcal{O}(\nu^2|Z_\mathrm{h}|\ |Z|^3),$$
$$\mathcal{R}_{z,B\,\mathrm{or}\,\widetilde{B}}(Z,\nu) = \mathcal{O}(\nu|Z_\mathrm{h}|\ |Z|^3),$$
$$\mathcal{R}_{z,C\,\mathrm{or}\,\widetilde{C}}(Z,\nu) = \mathcal{O}(\nu^{\frac{3}{2}}|Z|^4 + \nu|Z_\mathrm{c}|^4).$$

Proof. This proposition is a direct consequence of Theorem 4.3.1 since the periodic orbits $Y_{k,\nu}$ have been obtained by Theorem 9.1.4 which is a consequence of Theorem 4.1.2. Moreover, the proof of this Theorem gives an explicit formula for Θ_ν which reads in this particular case

$$\Theta_\nu(Z) = Z + \sum_{n \ge 4} \Theta^1_{\nu,n}(Z_\mathrm{c}) \tag{9.30}$$

The polynomials $\Theta^1_{\nu,n}(Z_\mathrm{c})$ are derived from the coefficients $\widehat{Y}_n$ of the power expansion of $Y_{k,\nu}(t) = \widehat{Y}(\underline{\omega}_{k,\nu}t, k, \nu)$ with $\widehat{Y}(s,k,\nu) = \widehat{Y}_1 + \sum_{n\ge 4} k^n\ \widehat{Y}_n(s,\nu)$ and

$$\widehat{Y}_n(s,\nu) = \sum_{p=-n}^{n} k^n(\widehat{a}_{n,p}\mathrm{e}^{\mathrm{i}ps}, \widehat{b}_{n,p}\mathrm{e}^{\mathrm{i}ps}, \widehat{a}_{n,p}\mathrm{e}^{-\mathrm{i}ps}, -\widehat{b}_{n,p}\mathrm{e}^{-\mathrm{i}ps}, \widehat{c}_{n,p}\mathrm{e}^{\mathrm{i}ps}, \widehat{c}_{n,p}\mathrm{e}^{-\mathrm{i}ps}).$$

where $(\widehat{a}_{n,p}, \widehat{b}_{n,p}, \widehat{c}_{n,p}), \mathbb{R} \times \mathrm{i}\mathbb{R} \times \mathbb{R}$. More precisely, we get

$$\Theta^1_{\nu,n}(Z_c) = \sum_{\substack{p=-n \\ p+n \text{ even}}}^{n} \Big(\widehat{a}_{n,p} C^{\frac{n+p}{2}} \widetilde{C}^{\frac{n-p}{2}}, \widehat{b}_{n,p} C^{\frac{n+p}{2}} \widetilde{C}^{\frac{n-p}{2}}, \widehat{a}_{n,p} C^{\frac{n-p}{2}} \widetilde{C}^{\frac{n+p}{2}}, -\widehat{b}_{n,p} C^{\frac{n-p}{2}} \widetilde{C}^{\frac{n+p}{2}}, \widehat{c}_{n,p} C^{\frac{n+p}{2}} \widetilde{C}^{\frac{n-p}{2}}, \widehat{c}_{n,p} C^{\frac{n-p}{2}} \widetilde{C}^{\frac{n+p}{2}} \Big).$$

Since $\widehat{Y}_2 = \widehat{Y}_3 = 0$, the power series in (9.30) starts with $n = 4$. Hence, when performing the changes of coordinates $Y = \Theta_\nu(Z)$ in (9.8), the terms of order less than 3, i.e. the normal form's terms, are kept unchanged. □.

9.3.2 Equation centered on periodic orbits

As already explained in the introduction of this section, using the constructive Floquet Theory given in Chapter 5 for computing the Floquet exponents $\lambda_{k,\nu}$ of $Z_{k,\nu}$, we can look for reversible homoclinic connections to $Z_{k,\nu}$ of the form

$$Z(t) = Z_{k,\nu}\big(t+\varphi \tanh(\lambda_{k,\nu} t)\big) + \big[\mathrm{Id} + \mathcal{Q}^1_\nu\big(t+\varphi \tanh(\lambda_{k,\nu} t)\big)\big].w'(\lambda_{k,\nu} t)$$

where $\mathcal{Q}^1_\nu$ is a $2\pi/\underline{\omega}_{k,\nu}$-periodic matrix-valued function given by Floquet Theory. We look for the phase shift φ and for $w' \in \mathbb{D}$ which is reversible and which satisfies $\sup_{\tau\in\mathbb{R}}(|w'(\tau)| e^{|\tau|}) < +\infty$. We perform the Floquet linear change of coordinates to remove all the w'-linear terms in the "hyperbolic component" $\mathcal{R}'_h$ of the rest $\mathcal{R}'$. More precisely, we obtain

Proposition 9.3.2. *There exist $\nu_2, k_2 > 0$, a numerical coefficient $\lambda_{k,\nu} = 1 + \mathcal{O}(\nu k^2)$ and an analytic matrix valued function $\check{\mathcal{Q}}^1_\nu$ reading*

$$\check{\mathcal{Q}}^1_\nu(C, \widetilde{C}) = \sum_{p+q\geq 2} \mathcal{Q}^1_{p,q}(\nu) C^p \widetilde{C}^q \quad \textit{with } \|\mathcal{Q}^1_{p,q}(\nu)\| \leq \frac{M_{\mathcal{Q}}\,\nu}{(k_2)^{p+q}} \textit{ for } \nu \in]0, \nu_2]$$

with $S\mathcal{Q}^1_{p,q}(\nu) = \mathcal{Q}^1_{q,p}(\nu) S$, such that for every $\nu \in]0, \nu_2]$ and $k \in [0, k_2]$, setting

$$\begin{aligned} Z(t) = {} & Z_{k,\nu}\big(t+\varphi \tanh \lambda_{k,\nu} t\big) \\ & + \Big[\mathrm{Id} + \check{\mathcal{Q}}^1_\nu\big(Z_{c,k,\nu}(t+\varphi \tanh \lambda_{k,\nu} t)\big)\Big].w'(\lambda_{k,\nu} t) \end{aligned} \tag{9.31}$$

with $Z_{k,\nu}(t) = (0, Z_{c,k,\nu}(t)) \in \mathbb{D}_h \times \mathbb{D}_c$, the equation (9.29) is equivalent to

$$\begin{aligned} \frac{dw'}{d\tau} = {} & \mathcal{N}'(w'(\tau), \nu) + \mathcal{R}'(w'(\tau), Z'_{c,k,\nu}(\tau+\varphi' \tanh \tau), \nu) \\ & + \varphi' \Phi'(\tau, k, \nu, \varphi') \end{aligned} \tag{9.32}$$

where

(a) $\mathcal{N}'$ *is derived from* $\mathcal{N}$ *by substituting* $\underline{\omega}'_i(k,\nu)$ *with* $\underline{\omega}_i(\nu)$ *where*

$$\underline{\omega}'_i(k,\nu) = \omega_i + \nu\underline{\omega}'_{i1}(k,\nu), \quad \underline{\omega}'_{i1}(k,\nu) = \mathcal{O}(\nu + k^2), \qquad \text{for } i = 0,1,$$

where $\underline{\omega}'_{i1}$ *are real analytic function of* $(k,\nu) \in [0,k_2] \times [0,\nu_2]$.

(b) $\tau{=}\lambda_{k,\nu}t$, $\varphi'{=}\lambda_{k,\nu}\varphi$, $Z'_{c,k,\nu}(\tau) = (k e^{i\underline{\omega}'_{k,\nu}\tau}, k e^{-i\underline{\omega}'_{k,\nu}\tau})$ *where*

$$\underline{\omega}'_{k,\nu} = \frac{\underline{\omega}'_1(k,\nu)}{\nu} + \widehat{\omega}'(k,\nu), \qquad \widehat{\omega}'(k,\nu) = \mathcal{O}(k^2\nu).$$

(c) *The rest* $\mathcal{R}' = (\mathcal{R}'_h, \mathcal{R}'_c) \in \mathbb{D}_h \times \mathbb{D}_c$ *is an analytic reversible vector field satisfying for* $Z'_c \in (D(0,k_2))^2$,

$$\mathcal{R}'_h(w', Z'_c){=}\mathcal{O}\Big(\nu|w'_h||w'_c||Z'_c|{+}\nu|w'_h||w'|(|w'|{+}|Z'_c|)^2{+}k^2\nu|w'_h||w'|^2\Big),$$

$$\mathcal{R}'_c(w', Z'_c) = \mathcal{O}\Big(\nu^{\frac{3}{2}}|w'|(|w'|{+}|Z'_c|)^3{+}\nu|w'_c|(|w'_c|{+}|Z'_c|)^3{+}\nu|w'_h|^2|Z'_c| + \nu^2|w'_c||Z'_c|(|w'_c|{+}|Z'_c|){+}\nu^2k^2|w'_c||w'|^2\Big).$$

where $D(0,k) = \{\xi \in \mathbb{C}, |\xi| < k\}$.

(d) $\Phi' = (\Phi'_h, \Phi'_c) \in \mathbb{D}_h \times \mathbb{D}_c$ *satisfies* $S\Phi'(-\tau,k,\nu,\varphi') = -\Phi'(\tau,k,\nu,\varphi')$ *and reads*

$$\Phi'_h(\tau,k,\nu,\varphi') = -\frac{1}{\cosh^2(\tau)}\Upsilon'_h\big(Z'_{c,k,\nu}(\tau{+}\varphi'\tanh\tau)\big)$$

$$\Phi'_c(\tau,k,\nu,\varphi') = -\frac{1}{\cosh^2(\tau)}\left(\frac{dZ'_{c,k,\nu}}{d\tau}\right)(\tau{+}\varphi'\tanh\tau)$$

where ϕ'_h *is an analytic function satisfying* $\Upsilon'_h(Z'_c) = \mathcal{O}(\nu|Z'_c|^2)$ *for* $Z'_c \in (D(0,k_2))^2$.

The proof of this proposition is given in Appendix 9.A. □

The truncated system

$$\frac{dw'}{d\tau} = \mathcal{N}'(w',\nu), \qquad w' \in \mathbb{D} \tag{9.33}$$

admits the same families of solutions as the truncated system (9.14) modulo the substitution of $\underline{\omega}'_i(k,\nu)$ with $\underline{\omega}_i(\nu)$ for $i = 0,1$. In particular, it admits two reversible homoclinic connections to 0

$$h', \qquad R^h_\pi h'.$$

where h' is explicitly given by

$$h'(\tau) := \left(r_0^{\mathrm{h}}(\tau)\mathrm{e}^{\mathrm{i}\psi_0'(\tau)}, r_1^{\mathrm{h}}(\tau)\mathrm{e}^{\mathrm{i}\psi_0'(\tau)}, r_0^{\mathrm{h}}(\tau)\mathrm{e}^{-\mathrm{i}\psi_0'(\tau)}, r_1^{\mathrm{h}}(\tau)\mathrm{e}^{-\mathrm{i}\psi_0'(\tau)}, 0, 0\right) \quad (9.34)$$

where

$$r_0^{\mathrm{h}}(\tau) = \frac{1}{\cosh(\tau)}, \quad r_1^{\mathrm{h}}(\tau) = -\frac{\tanh(\tau)}{\cosh(\tau)}, \quad \psi_0'(\tau) = \frac{\underline{\omega_0'}(k,\nu)}{\nu}\tau + p_2(\nu)\nu\tanh(\tau).$$

Observe that h' is holomorphic in $\mathbb{C} \setminus (\frac{\mathrm{i}\pi}{2} + \mathrm{i}\pi\mathbb{Z})$.

Then, we look for homoclinic connection to 0 of (9.32) in the form

$$w'(\tau) = h'(\tau) + v'(\tau).$$

the equation satisfied by v' reads

$$\frac{dv'}{d\tau} - D\mathcal{N}(h',\nu).v' = g'(v'(\tau), \varphi', k, \nu, \tau) \quad (9.35)$$

with

$$g'(v', \varphi', k, \nu, \tau) = \mathcal{N}_2'(v', \varphi', k, \nu, \tau) + \mathcal{R}'(h' + v', Z_{\mathrm{c},k,\nu}', \nu) + \varphi'\Phi'(\tau, \varphi', k, \nu)$$

and

$$\mathcal{N}_2'(v', \varphi, k, \nu, \tau) = \mathcal{N}'(h' + v', \nu) - \mathcal{N}'(h', \nu) - D\mathcal{N}'(h', \nu).v',$$

where h', v' are evaluated at τ, and $Z_{\mathrm{c},k,\nu}'$ at $\tau + \varphi'\tanh\tau$.

The classical way to find solutions of (9.35) is to transform (9.35) into an integral equation which can be solved using the Contraction Mapping Theorem. To perform such a transformation, we need to invert the corresponding linear part, i.e. to solve the corresponding affine equations. As we shall see in the following subsection, the inversion of such affine equations lead to a solvability condition given by a bi-oscillatory integral. The analysis of this solvability condition is the crucial point of the proof. As we shall see, it requires a partial complexification of the time.

9.3.3 Inversion of affine equation with partially complexified time

For solving affine equations, the first step is to compute a basis of solutions of the linear homogeneous equation. For describing the properties of such a basis we introduce the following spaces of functions:

Definition 9.3.3. *Let* $\lambda \in \mathbb{R}$, $\ell > 0$.

$$H_\ell^\lambda|_{\mathrm{R,D}} = H_\ell^\lambda(\mathbb{C}^6) \cap \{f : \mathbb{C} \to \mathbb{C}^6 /\ f \text{ is reversible and maps } \mathbb{R} \text{ in } \mathbb{D}\}$$
$$H_\ell^\lambda|_{\mathrm{AR,D}} = H_\ell^\lambda(\mathbb{C}^6) \cap \{f : \mathbb{C} \to \mathbb{C}^6 /\ f \text{ is antireversible and maps } \mathbb{R} \text{ in } \mathbb{D}\}$$

where $H_\ell^\lambda(\mathbb{C}^n)$ *is introduced in Definition 6.3.10.*

Lemma 9.3.4. *The homogeneous equation*

$$\frac{dv'}{d\tau} - D\mathcal{N}'(h',\nu).v' = 0$$

admits a basis of solutions $(p'_0, p'_1, q'_0, q'_1, r'_+, r'_-)$ *satisfying*

(a) $p'_j, q'_j \in \mathbb{C}^4 \times \{0_{\mathbb{C}^2}\}$ *and for every* $\ell \in]0, \frac{\pi}{2}[$,

$$\overset{\vee}{p}{}'_j = R^{\mathrm{h}}_{-\frac{\omega_0\tau}{\nu}} p'_j \in H^1_\ell|_{\mathrm{AR,D}}, \qquad \overset{\vee}{q}{}'_j = R^{\mathrm{h}}_{-\frac{\omega_0\tau}{\nu}} q'_j \in H^{-1}_\ell|_{\mathrm{R,D}},$$

where R^{h}_ω *is the rotation given by (9.15).*

(b) $r'_+ = (0,0,0,0,\mathrm{e}^{\mathrm{i}\psi'_r}, \mathrm{e}^{-\mathrm{i}\psi'_r})$, $r'_- = (0,0,0,0,\mathrm{i}\mathrm{e}^{\mathrm{i}\psi'_r}, -\mathrm{i}\mathrm{e}^{-\mathrm{i}\psi'_r})$ *where*

$$\psi'_r(\tau) = \frac{\omega'_1(k,\nu)}{\nu}\tau + n_2(\nu)\nu \tanh\tau$$

which ensures that r'_+ *is reversible whereas* r'_- *is antireversible.*

(c) *The dual basis* $(p'^*_0, p'^*_1, q'^*_0, q'^*_1, r'^*_+, r'^*_-)$ *(see Definition 6.3.5) satisfies*

$$\overset{\vee}{p}{}'^*_j = R^{\mathrm{h}}_{\frac{\omega_0\tau}{\nu}} p'^*_j \in H^{-1}_\ell|_{\mathrm{AR,D}}, \qquad \overset{\vee}{q}{}'_j = R^{\mathrm{h}}_{\frac{\omega_0\tau}{\nu}} q'^*_j \in H^1_\ell|_{\mathrm{R,D}},$$

and

$$r'^*_+ = (0,0,0,0,\tfrac{1}{2}\mathrm{e}^{-\mathrm{i}\psi'_r}, \tfrac{1}{2}\mathrm{e}^{\mathrm{i}\psi'_r}), \quad r'^*_- = (0,0,0,0,-\tfrac{\mathrm{i}}{2}\mathrm{e}^{-\mathrm{i}\psi'_r}, \tfrac{\mathrm{i}}{2}\mathrm{e}^{\mathrm{i}\psi'_r}).$$

Remark 9.3.5. Observe that the four first vectors of the basis are induced by the eigenvalues $\pm\frac{\mathrm{i}\omega_0(\nu)}{\nu} \pm 1$. So, p'_0 and p'_1 tends to 0 at infinity with an exponential decay rate equal to 1 and with a fast oscillation equal to $\frac{\omega_0}{\nu}$. The two next vectors (q'_0, q'_1) go to infinity at infinity with an exponential growth rate equal to 1 and a fast oscillation equal to $\frac{\omega_0}{\nu}$. Finally, the two last vectors $r'_\pm$ are induced by the pair of purely imaginary eigenvalues $\pm\frac{\omega'_1(k,\nu)}{\nu}$. So, these two vectors are purely oscillatory with a frequency of order $\frac{\omega_1}{\nu}$,

Proof. This lemma is a direct consequence of Lemma 6.3.11 since $D\mathcal{N}'(h'(\tau))$ as the form

$$D\mathcal{N}'(h'(\tau)) = \begin{pmatrix} D\mathcal{N}'_{\mathrm{h}}(h'(\tau)) & 0 & 0 \\ 0 & \mathrm{i}\Omega(\tau) & 0 \\ 0 & 0 & -\mathrm{i}\Omega(\tau) \end{pmatrix} \tag{9.36}$$

with

$$\Omega(\tau) = \tfrac{\omega'_1(\nu)}{\nu}\tau + n_2(\nu)\nu^2(r^{\mathrm{h}}_0(\tau))^2$$

and where $D\mathcal{N}'_{\mathrm{h}}(h')$ satisfies all the hypothesis of Lemma 6.3.11 modulo the identification of $\mathbb{R}^4$ and $\mathbb{D}_{\mathrm{h}}$. In particular the explicit formula (9.34) giving h' enables to check easily that

$$\overset{\vee}{h}{}'(\tau,\nu) = R^{\mathrm{h}}_{-\frac{\omega_0\tau}{\nu}} h'(\tau,\nu) \in H^1_\ell|_{\mathrm{R,D}}.$$

for every $\ell \in]0, \frac{\pi}{2}[$.

Observe that the symmetry of reversibility S is self adjoint, so the dual vectors are reversible or antireversible with respect to $S = {}^tS$.

In fact since the truncated system (9.33) is integrable, it is possible to compute explicitly the basis of solutions and its dual basis. □

Using the above basis of solutions and the variation of constant formula, we get

Lemma 9.3.6. *If v' is reversible and tends to 0 at infinity then the function $f' = \dfrac{dv'}{d\tau} - D\mathcal{N}'(h').v'$ is antireversible and satisfies*

$$\int_0^{+\infty} \langle r'_-{}^*(\tau), f'(\tau)\rangle_* d\tau = 0 \tag{9.37}$$

This solvability condition is the typical one which occurs when the linear homogeneous system admits a pair of simple purely imaginary eigenvalues (see Section 6.4). So if (9.35) admits a reversible solution v' which tends to 0 at infinity, then necessarily

$$\int_0^{+\infty} \langle r'_-{}^*(\tau), g'(v'(\tau),\varphi',k,\nu,\tau)\rangle_* d\tau = 0 \tag{9.38}$$

The analysis of this solvability condition, is the crucial point of the proof. Observe that the above integral is an *bi-oscillatory integral* since the dual vector rotates with a frequency of order ω_1/ν where ν is a small parameter and since g' involves a second fast oscillation with a frequency of order ω_0/ν due to the terms coming from h'. Hence, to study this solvability condition, we want to use the First Bi-Frequency Exponential Lemma 2.2.1. For using this lemma, we must find solutions v, of (9.35) in the form $v'(\tau) = \mathrm{v}'(\tau, \frac{\omega_0\tau}{\nu})$. So the first step is to solve the corresponding affine equations with partially complexified time. For that purpose we introduce the following Banach spaces of functions with partially complexified time :

Definition 9.3.7. *Let $\ell > 0$ and $\lambda, \lambda_{\mathrm{h}}, \lambda_{\mathrm{c}} \in \mathbb{R}$.*

$$\begin{aligned}{}^b\mathcal{H}^\lambda_\ell|_{\mathrm{R,D}} = {}^b\mathcal{H}^\lambda_\ell(\mathbb{C}^6) \cap \{\mathrm{f} : \mathcal{B}_\ell \times \mathbb{R} \to \mathbb{C}^6 / \\ S\mathrm{f}(-\xi,-s) = \mathrm{f}(\xi,s),\ \mathrm{f}(\tau, \tfrac{\omega_0\tau}{\nu}) \in \mathbb{D} \textit{ for } \tau \in \mathbb{R}\},\end{aligned}$$

....

$$^b\mathcal{H}_\ell^{\lambda_h,\lambda_c}|_{\mathrm{AR,D}} = {}^b\mathcal{H}_\ell^{\lambda_h}(\mathbb{C}^4) \times {}^b\mathcal{H}_\ell^{\lambda_c}(\mathbb{C}^2) \cap \{\mathsf{f} : \mathcal{B}_\ell \times \mathbb{R} \to \mathbb{C}^6 / \; S\mathsf{f}(-\xi,-s) = -\mathsf{f}(\xi,s),\ \mathsf{f}(\tau,\tfrac{\omega_0\tau}{\nu}) \in \mathbb{D} \text{ for } \tau \in \mathbb{R}\},$$

$$^b\mathcal{H}_\ell^{\lambda_h,\lambda_c}|^\perp_{\mathrm{AR,D}} = {}^b\mathcal{H}_\ell^{\lambda_h,\lambda_c}|_{\mathrm{AR,D}} \cap \{\mathsf{f} : \mathcal{B}_\ell \times \mathbb{R} \to \mathbb{C}^6 / \; \int_0^{+\infty} \langle \tau'^{\,*}_-(\tau), \mathsf{f}(\tau,\tfrac{\omega_0\tau}{\nu})\rangle_* d\tau = 0\}$$

where ${}^b\mathcal{H}_\ell^\lambda(\mathbb{C}^n)$ *is introduced in Definition 6.5.1.*

Proposition 9.3.8. *Let* $\ell \in]0,\frac{\pi}{2}[$. *For every* $\nu \in]0,\nu_2]$ *there exists a linear operator* T'_ν *which maps* ${}^b\mathcal{H}_\ell^{2,1}|^\perp_{\mathrm{AR,D}}$ *to* ${}^b\mathcal{H}_\ell^1|_{\mathrm{R,D}}$ *such that for every* $\mathsf{f}' \in {}^b\mathcal{H}_\ell^{2,1}|^\perp_{\mathrm{AR,D}}$ *the function*

$$v'_\mathsf{f}(\tau) = \mathsf{T}'_\nu(\mathsf{f}')(\tau,\tfrac{\omega_0\tau}{\nu})$$

is a solution of

$$\frac{dv'}{d\tau} - D\mathcal{N}'(h',\nu).v' = \mathsf{f}'(\tau,\tfrac{\omega_0\tau}{\nu}). \tag{9.39}$$

Moreover, the family of operators T'_ν *is uniformly bounded, i.e. there exists* $M_{\mathsf{T}'}(\ell)$ *such that*

$$|\mathsf{T}'_\nu(\mathsf{f}')|_{{}^b\mathcal{H}_\ell^1} \le M_{\mathsf{T}'}(\ell)\ |\mathsf{f}'|_{{}^b\mathcal{H}_\ell^{2,1}} \qquad \text{for } \mathsf{f}' \in {}^b\mathcal{H}_\ell^{2,1}|^\perp_{\mathrm{AR,D}},\ \nu \in]0,\nu_2].$$

Proof. Modulo the identifications of $\mathbb{R}^4$ with $\mathbb{D}_h$ and of $\mathbb{R}^2$ with $\mathbb{D}_c$, the proof of this proposition follows directly from Propositions 6.5.4, 6.5.7 and Lemma 6.3.11, observing here again that $D\mathcal{N}'(h')$ has the block diagonal form given by (9.36). □

Remark 9.3.9. This proposition ensures that if f is partially complexified and satisfies the solvability condition (9.37), then the affine equation (9.39) admits a solution with partially complexified time. Moreover, observe that this only works for functions $\mathsf{f} \in {}^b\mathcal{H}_\ell^{2,1}|_{\mathrm{AR,D}}$ and not for functions $\mathsf{f} \in {}^b\mathcal{H}_\ell^1|_{\mathrm{AR,D}}$. In other words. it only works for functions f with an exponential decay rate corresponding to the four first components of f equal to 2 and not to 1. In fact, it would work for any decay rate $\lambda > 1$. The constraint $\lambda > 1$ comes from Proposition 6.5.4. For obtaining a nonlinear term $\mathsf{g}' = (\mathsf{g}'_h, \mathsf{g}'_c)$ satisfying this constraint we had to perform the Floquet linear change of coordinates which removes in g'_h all the terms linear with respect to v'.

9.3.4 Partial complexification of time and choice of the parameters.

As already explain, the crucial point of the analysis is the study of the solvability condition (9.38) the study of which requires the Bi-frequency Exponential Lemmas given in Section 2.2. For that purpose, we look for solutions of (9.35) in the form $v'(\tau) = \mathsf{v}'(\tau, \frac{\omega_0 \tau}{\nu})$. So we introduce below the function $\mathsf{g}'(\mathsf{v}', \varphi', k, \nu)$ such that

$$\mathsf{g}'(\mathsf{v}', \varphi', k, \nu)(\tau, \tfrac{\omega_0 \tau}{\nu}) = g'(v'(\tau), \varphi', k, \nu, \tau) \tag{9.40}$$

where $v'(\tau) = \mathsf{v}'(\tau, \frac{\omega_0 \tau}{\nu})$. In this subsection we check that $\mathsf{g}'(\mathsf{v}', \varphi', k, \nu)$ belongs to ${}^b\mathcal{H}_\ell^{2,1}|_{\mathrm{AR,D}}$. Then, in Subsection 9.3.5, using the First Bi-frequency Exponential Lemma 2.2.1, we prove that for every $\mathsf{v}' \in {}^b\mathcal{H}_\ell^1|_{\mathrm{R,D}}$, every sufficiently small ν and every $k \in [K(\ell)\nu^{\frac{3}{2}} \mathrm{e}^{-\frac{\omega_* \ell}{\nu}}, K\mathrm{e}^{-\frac{\omega_* \ell}{\nu}}[$, there exists an appropriate phase shift φ' such that $\mathsf{g}'(\mathsf{v}', \varphi'(\mathsf{v}', k, \nu), k, \nu)$ satisfies the solvability condition

$$\int_0^{+\infty} \langle r_-'^*(\tau), \mathsf{g}'(\mathsf{v}', \varphi'(\mathsf{v}', k, \nu), k, \nu)(\tau, \tfrac{\omega_0 \tau}{\nu}) \rangle_* d\tau = 0. \tag{9.41}$$

Hence, $\mathsf{g}'(\mathsf{v}', \varphi'(\mathsf{v}', k, \nu), k, \nu)$ belongs to ${}^b\mathcal{H}_\ell^{2,1}|_{\mathrm{AR,D}}^{\perp}$. Then, Proposition 9.3.8 ensures that if v' satisfies

$$\mathsf{v}' = \mathsf{T}'\big(\mathsf{g}'(\mathsf{v}', \varphi'(\mathsf{v}', k, \nu), k, \nu)\big).$$

then $v'(\tau) = \mathsf{v}'(\tau, \frac{\omega_0 \tau}{\nu})$ is a solution of (9.35). Finally, in Subsection 9.3.6, we solve this equation using the Contraction Mapping Theorem.

So, from now on, we fix $\ell \in]0, \frac{\pi}{2}[$. In what follows, all the constants M_i depends on ℓ even if we write M_i instead of $M_i(\ell)$.

Moreover, we expect to find homoclinic connection to periodic orbits of size k of order $\mathrm{e}^{-\frac{\omega_* \ell}{\nu}}$.

So, we set $k = K\mathrm{e}^{-\frac{\omega_* \ell}{\nu}}$ *with* $K \in [0, k_2]$*; we look for the phase shift* φ' *in* $[0, \frac{6\pi\nu}{\omega_1}]$ *and we look for* v *in* $\mathcal{B}{}^b\mathcal{H}_\ell^1|_{\mathrm{R,D}}(\delta\sqrt{\nu})$ *with* $\delta \leq 1$ *and*

$$\mathcal{B}{}^b\mathcal{H}_\ell^1|_{\mathrm{R,D}}(\delta) = \{\mathsf{v}' \in {}^b\mathcal{H}_\ell^1|_{\mathrm{R,D}},\ |\mathsf{v}'|_{{}^b\mathcal{H}_\ell^1} \leq \delta\}.$$

As already explained, our aim in this subsection is to define $\mathsf{g}'(\mathsf{v}', \varphi', k, \nu)$ such that (9.40) holds. Observe that $g'(\mathsf{v}', \varphi, k, \nu)$ involves $\mathcal{R}'(h' + v', Z'_{\mathrm{c},k,\nu}, \nu)$ where $\mathcal{R}'(w', Z'_{\mathrm{c}}, \nu)$ is well defined for $Z'_{\mathrm{c}} \in (D(0, k_2))^2$. If we complexify "the full time" in $Z'_{\mathrm{c},k,\nu}(\tau + \varphi' \tanh \tau)$, then for every $\ell > 0$, $k = K\mathrm{e}^{\frac{-\ell\omega_*}{\nu}}$ and $\varphi' \in [0, \frac{6\pi\nu}{\omega_1}]$, we get

$$Z'_{c,k,\nu}(i\ell + \varphi' \tanh i\ell) \underset{\nu \to 0}{\sim} \sqrt{2} K e^{\frac{\ell(\omega_1 - \omega_\star)}{\nu}}$$

with $\omega_\star = \min_{p \in \mathbb{Z}} |\omega_1 - p\omega_0|$. Generically, $\omega_1 > \omega_\star$ and the norm of $Z'_{c,k,\nu}(\xi + \varphi' \tanh \xi)$ on $\mathcal{B}_\ell$ diverges when ν tends to 0. So to avoid this difficulty we perform a partial complexification of time for $Z'_{c,k,\nu}$.

Definition 9.3.10. *Denote*

$$\mathsf{Z}'_{c,K,\nu}(\xi, s, \varphi') = K e^{\frac{-\ell\omega_\star}{\nu}} (e^{i\psi'_c(\xi,s,\varphi')}, e^{-i\psi'_c(\xi,s,\varphi')})$$

with

$$\psi'_c(\xi, s, \varphi') = (\frac{\widetilde{\omega}_\star}{\nu} + \underline{\omega}'_{11}(k', \nu) + \widehat{\omega}'(k', \nu))\xi + \underline{\omega}'_{k,\nu} \varphi' \tanh \xi + p_\star s$$

where

$$\omega_\star = \min_{p \in \mathbb{Z}} |\omega_1 - p\omega_0|, \qquad \widetilde{\omega}_\star = \omega_1 - p_\star \omega_0$$

where $p_\star$ is the smallest integer such that the minimum is reached.

Remark 9.3.11. The minimum is reached for a unique value of p when $\frac{\omega_1}{\omega_0} \notin \mathbb{N} + \frac{1}{2}$ and it is reached for two consecutive values of p for $\frac{\omega_1}{\omega_0} \in \mathbb{N} + \frac{1}{2}$. Then, we readily check

Lemma 9.3.12.

(a) *There exits M_0 such that for every $\nu \in]0, \nu_2]$, $\varphi' \in [0, \frac{6\pi\nu}{\omega_1}]$, $K \in [0, k_2]$, $\xi \in \mathcal{B}_\ell$, $s \in \mathbb{R}$, $|\mathcal{I}m\,(\psi'_c(\xi, s, \varphi'))| \leq M_0 + \frac{\omega_\star \ell}{\nu}$ holds.*

(b) *Let K_1 be fixed in $]0, k_2 e^{-M_0}[$. Then for every $\nu \in]0, \nu_2]$, $\varphi' \in [0, \frac{6\pi\nu}{\omega_1}]$, $K \in [0, K_1]$, $\xi \in \mathcal{B}_\ell$, $s \in \mathbb{R}$, $\mathsf{Z}'_{c,K,\nu}(\xi, s, \varphi')$ lies in $(D(0, k_2))^2$.*

(c) *There exists M_1 such that for every $\nu \in]0, \nu_2]$, $\varphi' \in [0, \frac{6\pi\nu}{\omega_1}]$, $K \in [0, K_1]$, $\mathsf{Z}'_{c,K,\nu}(\cdot, \cdot, \varphi')$ belongs to ${}^b\mathcal{H}^0_\ell|_{\mathbb{R},\mathbb{D}}$ and*

$$|\mathsf{Z}'_{c,K,\nu}(\cdot, \cdot, \varphi')|_{{}^b\mathcal{H}^0_\ell|_{\mathbb{R},\mathbb{D}}} \leq M_1 K$$
$$|\mathsf{Z}'_{c,K,\nu}(\cdot, \cdot, \varphi') - \mathsf{Z}'_{c,K,\nu}(\cdot, \cdot, \varphi'_0)|_{{}^b\mathcal{H}^0_\ell|_{\mathbb{R},\mathbb{D}}} \leq M_1 \tfrac{K}{\nu} |\varphi' - \varphi'_0|.$$

(d) *For every $\nu \in]0, \nu_2]$, $\varphi' \in [0, \frac{6\pi\nu}{\omega_1}]$, $K \in [0, K_1]$ and every $\tau \in \mathbb{R}$,*

$$\mathsf{Z}'_{c,K,\nu}(\tau, \tfrac{\omega_0 \tau}{\nu}, \varphi') = Z'_{c,k,\nu}(\tau + \varphi' \tanh \tau)$$

where $k = K e^{\frac{-\ell\omega_\star}{\nu}}$.

We can now define

Definition 9.3.13. *For every* $\nu \in]0, \nu_2]$, $\varphi' \in [0, \frac{6\pi\nu}{\omega_1}], K \in [0, K_1]$, *let us define* $\mathsf{g}'(\mathsf{v}, \varphi', K, \nu)$ *by*

$$\mathsf{g}'(\mathsf{v}', \varphi', K, \nu) = \mathsf{N}_2'(\mathsf{v}, \nu) + \mathsf{R}'(\mathsf{v}, \varphi', K, \nu) + \varphi' \phi(\varphi', K, \nu)$$

where

$$\mathsf{N}_2'(\mathsf{v}, \nu) = \mathcal{N}'(R_s^{\mathrm{h}} \overset{\vee}{h}{}'(\xi) + \mathsf{v}'(\xi, s)) - \mathcal{N}'(R_s^{\mathrm{h}} \overset{\vee}{h}{}'(\xi)) - D\mathcal{N}'(R_s^{\mathrm{h}} \overset{\vee}{h}{}'(\xi)).\mathsf{v}',$$

$$\mathsf{R}'(\mathsf{v}, \varphi', K, \nu) = \mathcal{R}'(R_s^{\mathrm{h}} \overset{\vee}{h}{}'(\xi) + \mathsf{v}'(\xi, s), \mathsf{Z}_{\mathrm{c},K,\nu}(\xi, s, \varphi')),$$

and $\phi' = (\phi_{\mathrm{h}}', \phi_{\mathrm{c}}') \in \mathbb{C}^4 \times \mathbb{C}^2$ *with*

$$\Phi_{\mathrm{h}}'(K, \nu, \varphi')(\xi, s) = -\frac{1}{\cosh^2(\xi)} \Upsilon_{\mathrm{h}}'(\mathsf{Z}_{\mathrm{c},K,\nu}'(\xi, s, \varphi')),$$

$$\Phi_{\mathrm{c}}'(K, \nu, \varphi')(\xi, s) = -\frac{1}{\cosh^2(\xi)} K \underline{\omega}_{k,\nu}' \mathrm{e}^{\frac{-\ell\omega_*}{\nu}} (\mathrm{i}\mathrm{e}^{\mathrm{i}\psi_{\mathrm{c}}'(\xi,s,\varphi')}, -\mathrm{i}\mathrm{e}^{-\mathrm{i}\psi_{\mathrm{c}}'(\xi,s,\varphi')})$$

where $\overset{\vee}{h}{}'(\xi) = R^{\mathrm{h}}_{\frac{-\omega_0\xi}{\nu}} h'(\xi)$.

Remark 9.3.14. The explicit formula (9.34) giving h' ensures that $\overset{\vee}{h}{}'$ belongs to $H_\ell^1|_{\mathrm{R,D}}$ for $\ell \in]0, \frac{\pi}{2}[$, i.e. once the rotation is factorized out of h', $\overset{\vee}{h}{}'$ is holomorphic in the strip $\mathcal{B}_\ell$ and decays exponentially to 0 uniformly with respect to ν, when ξ tends to infinity. Thus, $(\xi, s) \mapsto R_s^{\mathrm{h}} \overset{\vee}{h}{}'(\xi)$ belongs to ${}^b\mathcal{H}_\ell^1|_{\mathrm{R,D}}$. Using Propositions 9.3.2 and Lemma 9.3.12, we obtain

Lemma 9.3.15.

(a) *For every* $\nu \in]0, \nu_2]$, $\varphi' \in [0, \frac{6\pi\nu}{\omega_1}], K \in [0, K_1]$ *and every* $\mathsf{v}' \in \mathcal{B}{}^b\mathcal{H}_\ell^1|_{\mathrm{R,D}}(\delta\sqrt{\nu})$, $\mathsf{g}'(\mathsf{v}', \varphi', K, \nu)$ *belongs to* ${}^b\mathcal{H}_\ell^{2,1}|_{\mathrm{AR,D}}$.

(b) *There exists* M_2 *such that for every* $\nu \in]0, \nu_2]$, $\varphi', \varphi_0', \in [0, \frac{6\pi\nu}{\omega_1}], K \in [0, K_1]$ *and every* $\mathsf{v}', \mathsf{v}_0' \in \mathcal{B}{}^b\mathcal{H}_\ell^1|_{\mathrm{R,D}}(\delta\sqrt{\nu})$,

$$|\mathsf{g}_{\mathrm{h}}'(\mathsf{v}', \varphi', K, \nu)|_{{}^b\mathcal{H}_\ell^{2,1}} \le M_2 \left(|\mathsf{v}'|^2_{{}^b\mathcal{H}_\ell^1} + \nu\right),$$

$$|\mathsf{g}_{\mathrm{c}}'(\mathsf{v}', \varphi', K, \nu)|_{{}^b\mathcal{H}_\ell^{2,1}} \le M_2 \left(\nu|\mathsf{v}'|^2_{{}^b\mathcal{H}_\ell^1} + \nu^{\frac{3}{2}} + K\right),$$

and

$$|\mathsf{g}'_{\mathrm{h}}(\mathsf{v}',\varphi',K,\nu)-\mathsf{g}'_{\mathrm{h}}(\mathsf{v}'_0,\varphi'_0,K,\nu)|_{{}^b\mathcal{H}^{2,1}_\ell} \leq M_2\left(\left(|\mathsf{v}'|_{{}^b\mathcal{H}^1_\ell}+|\mathsf{v}'_0|_{{}^b\mathcal{H}^1_\ell}+\nu\right)|\mathsf{v}'-\mathsf{v}'_0|_{{}^b\mathcal{H}^1_\ell}+K|\varphi'-\varphi'_0|\right),$$

$$|\mathsf{g}'_{\mathrm{c}}(\mathsf{v}',\varphi',K,\nu)-\mathsf{g}'_{\mathrm{c}}(\mathsf{v}'_0,\varphi'_0,K,\nu)|_{{}^b\mathcal{H}^{2,1}_\ell} \leq M_2\left(\left(\nu(|\mathsf{v}'|_{{}^b\mathcal{H}^1_\ell}+|\mathsf{v}'_0|_{{}^b\mathcal{H}^1_\ell})+\nu^{\frac{3}{2}}+\nu K\right)|\mathsf{v}'-\mathsf{v}'_0|_{{}^b\mathcal{H}^1_\ell}+\tfrac{K}{\nu}|\varphi'-\varphi'_0|\right).$$

(c) *For every* $\nu\in]0,\nu_2]$, $\varphi'\in[0,\frac{6\pi\nu}{\omega_1}]$, $K\in[0,K_1]$, $\mathsf{v}'\in\mathcal{B}{}^b\mathcal{H}^1_\ell|_{\mathrm{R},\mathbb{D}}(\delta\sqrt{\nu})$ *and every* $\tau\in\mathbb{R}$,

$$\mathsf{g}'(\mathsf{v}',\varphi',K,\nu)(\tau,\tfrac{\omega_0\tau}{\nu})=g'(v'(\tau),\varphi',k,\nu,\tau)$$

where $v'(\tau)=\mathsf{v}'(\tau,\frac{\omega_0\tau}{\nu})$ *and* $k=K\mathrm{e}^{\frac{-\ell\omega_*}{\nu}}$.

9.3.5 Choice and control of the phase shift

We now look for an appropriate phase shift $\varphi'(\mathsf{v}',K,\nu)$ such that the function $\mathsf{g}'(\mathsf{v}',\varphi'(\mathsf{v}',K,\nu),K,\nu)$ belongs to ${}^b\mathcal{H}^{2,1}_\ell|^{\perp}_{\mathrm{AR},\mathbb{D}}$, i.e. that g' satisfies the solvability condition (9.41), i.e.

$$J(\mathsf{v}',\varphi',K,\nu)=0$$

with

$$J(\mathsf{v}',\varphi',K,\nu)=\int_0^{+\infty}\langle {r'_-}^*(\tau),\mathsf{g}'(\mathsf{v}',\varphi',K,\nu)(\tau,\tfrac{\omega_0\tau}{\nu})\rangle_* d\tau.$$

This equation has exactly the same form as the one obtained for the $0^{2+}\mathrm{i}\omega$ resonance (see Lemma 7.3.18) except that here we have a bi-oscillatory integral instead of an oscillatory integral. So, as for the $0^{2+}\mathrm{i}\omega$ resonance (see Proposition 7.3.19) we can show

Lemma 9.3.16. *There exists* M_3 *such that for every* $\nu\in]0,\nu_2]$, $\varphi'\in[0,\frac{6\pi\nu}{\omega_1}]$, $K\in[0,K_1]$, $\mathsf{v}'\in\mathcal{B}{}^b\mathcal{H}^1_\ell|_{\mathrm{R},\mathbb{D}}(\delta\sqrt{\nu})$

$$J(\mathsf{v}',\varphi',K,\nu)=-\frac{\omega_1}{\nu}K\mathrm{e}^{-\frac{\omega_*\ell}{\nu}}\varphi'\frac{\sin(\underline{\omega}'_{k,\nu}\varphi'-n_2(\nu)\nu)}{\underline{\omega}'_{k,\nu}\varphi'-n_2(\nu)\nu}+J_2(\mathsf{v}',\varphi',K,\nu)$$

with $|J_2(\mathsf{v}',\varphi',K,\nu)|\leq M_3(\nu^{\frac{3}{2}}+\nu K)\mathrm{e}^{\frac{-\ell\omega_*}{\nu}}$.

Proof. The explicit leading term of J comes from ϕ'_c which is explicitly known and the perturbation term J_2 can be bounded using the First Bi-frequency Exponential Lemma 2.2.1. The proof is exactly the same as for the $0^{2+}i\omega$ resonance (see the proof of Proposition 7.3.19) except that in the later case we use the First Mono-frequency Exponential Lemma 2.1.1 instead of the Bi-frequency corresponding one. So the details are left to the reader. □

Then as for the $0^{2+}i\omega$ resonance (see Proposition 7.3.20) the above estimate of J ensures

Proposition 9.3.17. *There exists $\nu_3 \le \nu_2$ such that for every $\nu \in]0, \nu_3]$, $K \in [8M_4\nu^2, K_1]$ and $\mathrm{v}' \in \mathcal{B}^b\mathcal{H}^1_\ell|_{\mathrm{R,D}}(\delta\sqrt{\nu})$, there exists $\varphi'(\mathrm{v}', K, \nu)$ such that*

$$J(\mathrm{v}', \varphi'(\mathrm{v}', K, \nu), K, \nu) = 0.$$

Moreover, $0 \le \dfrac{\pi\nu}{8\omega_1} \le \varphi'(\mathrm{v}', K, \nu) \le \dfrac{4\pi\nu}{\omega_1} \le \dfrac{6\pi\nu}{\omega_1}$.

For using the contraction Mapping Theorem, we need an estimate of $|\varphi'(\mathrm{v}', K, \nu) - \varphi'(\mathrm{v}'_0, K, \nu)|$. Here again the computation of the estimates is very similar to the one obtained for the $0^{2+}i\omega$ resonance (see paragraph 7.3.5.2 and Proposition 7.3.27). So we skip the details and we can finally conclude

Proposition 9.3.18. *There exist M_5 and $\nu_4 > 0$ such that for every $\nu \in]0, \nu_4]$, $K \in [M_5\nu^{\frac{3}{2}}, K_1]$ and $\mathrm{v}' \in \mathcal{B}^b\mathcal{H}^1_\ell|_{\mathrm{R,D}}(\delta\sqrt{\nu})$, there exists $\varphi'(\mathrm{v}', K', \nu)$ such that $\mathrm{g}'(\mathrm{v}', \varphi'(\mathrm{v}', K, \nu), K, \nu)$ belongs to ${}^b\mathcal{H}^{2,1}_\ell|^{\perp}_{\mathrm{AR,D}}$, i.e.*

$$J(\mathrm{v}', \varphi'(\mathrm{v}', K, \nu), K, \nu) = 0.$$

Moreover, there exists M_6 such that for every $\nu \in]0, \nu_4]$, $K \in [M_5\nu^{\frac{3}{2}}, K_1]$ and $\mathrm{v}', \mathrm{v}'_0 \in \mathcal{B}^b\mathcal{H}^1_\ell|_{\mathrm{R,D}}(\delta\sqrt{\nu})$,

$$0 \le \frac{\pi\nu}{8\omega_1} \le \varphi'(\mathrm{v}', K, \nu) \le \frac{4\pi\nu}{\omega_1} \le \frac{6\pi\nu}{\omega_1}, \tag{9.42}$$

$$|\varphi'(\mathrm{v}', K, \nu) - \varphi'(\mathrm{v}'_0, K, \nu)| \le M_6\frac{\nu^2}{K}(\delta\sqrt{\nu} + K + \sqrt{\nu})|\mathrm{v}' - \mathrm{v}'_0|_{{}^b\mathcal{H}^1_\ell}. \tag{9.43}$$

9.3.6 fixed point

We recall here that we look for a solution of (9.35), i.e.

$$\frac{dv'}{d\tau} - D\mathcal{N}(h', \nu).v' = g'(v'(\tau), \varphi', k, \nu, \tau)$$

of the form $v'(\tau) = \mathsf{v}'(\tau, \frac{\omega_0 \tau}{\nu})$ with $\mathsf{v}' \in {}^b\mathcal{H}^1_\ell|_{\mathrm{R,D}}$. In subsection 9.3.4 we have introduced $\mathsf{g}'(\mathsf{v}', \varphi', K, \nu)$ such that

$$\mathsf{g}'(\mathsf{v}', \varphi', K, \nu)(\tau, \tfrac{\omega_0 \tau}{\nu}) = g'(v'(\tau), \varphi', k, \nu, \tau)$$

where $v'(\tau) = \mathsf{v}'(\tau, \frac{\omega_0 \tau}{\nu})$ and $k = K e^{\frac{-\ell\omega_*}{\nu}}$ (see Lemma 9.3.15).

In the previous subsection we have shown that for every $\nu \in]0, \nu_4]$, $K \in [M_5 \nu^{\frac{3}{2}}, K_1]$ and $\mathsf{v}' \in \mathcal{B}^b\mathcal{H}^1_\ell|_{\mathrm{R,D}}(\delta\sqrt{\nu})$, there exists $\varphi'(\mathsf{v}', K', \nu)$ such that $\mathsf{g}'(\mathsf{v}', \varphi'(\mathsf{v}', K, \nu), K, \nu)$ belongs to ${}^b\mathcal{H}^{2,1}_\ell|^{\perp}_{\mathrm{AR,D}}$. Then, Proposition 9.3.8 ensures that if v' satisfies

$$\mathsf{v}' = \mathcal{F}_{K,\nu}(\mathsf{v}') \qquad \text{with} \qquad \mathcal{F}_{K,\nu}(\mathsf{v}') = \mathsf{T}'\big(\mathsf{g}'(\mathsf{v}', \varphi'(\mathsf{v}', K, \nu), K, \nu)\big),$$

then $v'(\tau) = \mathsf{v}'(\tau, \frac{\omega_0 \tau}{\nu})$ is a solution of (9.35). So our aim is now to solve this later fixed point equation using the Contraction Mapping Theorem. Using Propositions 9.3.8, 9.3.15 and 9.3.18 we get

Lemma 9.3.19. *For every $\nu \in]0, \nu_4]$, $K \in [M_5 \nu^{\frac{3}{2}}, K_1]$ and $\mathsf{v}' \in \mathcal{B}^b\mathcal{H}^1_\ell|_{\mathrm{R,D}}(\delta\sqrt{\nu})$, $\mathcal{F}_{K,\nu}(\mathsf{v}')$ belongs to ${}^b\mathcal{H}^1_\ell|_{\mathrm{R,D}}$. Moreover, there exists M_7 such that for every $\nu \in]0, \nu_4]$, $K \in [M_5 \nu^{\frac{3}{2}}, K_1]$ and $\mathsf{v}', \mathsf{v}'_0 \in \mathcal{B}^b\mathcal{H}^1_\ell|_{\mathrm{R,D}}(\delta\sqrt{\nu})$, $\mathcal{F}_{K,\nu}$ satisfies*

$$|\mathcal{F}_{K,\nu}(\mathsf{v}')|_{{}^b\mathcal{H}^1_\ell} \leq M_7(|\mathsf{v}'|^2_{{}^b\mathcal{H}^1_\ell} + K + \nu),$$

$$|\mathcal{F}_{K,\nu}(\mathsf{v}') - \mathcal{F}_{K,\nu}(\mathsf{v}'_0)|_{{}^b\mathcal{H}^1_\ell} \leq M_7\big(|\mathsf{v}'|_{{}^b\mathcal{H}^1_\ell} + |\mathsf{v}'_0|_{{}^b\mathcal{H}^1_\ell} + \nu\big)\ |v - v'|_{{}^b\mathcal{H}^1_\ell}.$$

We can finally conclude

Proposition 9.3.20. *There exists $\nu_5 > 0$, M_8 such that for $\nu \in]0, \nu_5]$, the equation (9.32) centered on the periodic orbit $Z'_{k,\nu}$ admits for $k = M_5 \nu^{\frac{3}{2}} e^{\frac{-\omega_* \ell}{\nu}}$ an homoclinic connection to 0 of the form*

$$w'(\tau) = h'(\tau) + \mathsf{v}'(\tau, \tfrac{\omega_0 \tau}{\nu})$$

with $\mathsf{v}' \in {}^b\mathcal{H}^1_\ell|_{\mathrm{R,D}}$ and $|\mathsf{v}'|_{{}^b\mathcal{H}^1_\ell} \leq M_8 \nu$.

Remark 9.3.21. Coming back to our original equation (9.8), this proposition ensures the existence of an homoclinic connection to $Y_{k,\nu}$ of the form

$$Y(t) = Y_{k,\nu}\big(t + \varphi \tanh \lambda_{k,\nu} t\big) + X(t)$$

with $k = M_5 \nu^{\frac{3}{2}} e^{-\frac{\ell\omega_*}{\nu}}$, $\lambda_{k,\nu} = 1 + \mathcal{O}(\nu k^2)$ and

$$X(t) = \Theta_\nu \circ \Big(\mathrm{Id} + \check{\mathcal{Q}}^1_\nu\big(Z_{c,k,\nu}(t + \varphi \tanh \lambda_{k,\nu} t)\big)\Big).w'(\lambda_{k,\nu} t).$$

Hence,

$$|X(t)| \leq M(\ell)e^{-\lambda_{k,\nu}|t|}, \qquad \text{for } t \in \mathbb{R}.$$

Proof of Proposition 9.3.20. Lemma 9.3.19 ensures that $\mathcal{F}_{K,\nu}$ is a contraction mapping from $\mathcal{B}^b\mathcal{H}^1_\ell|_{\mathrm{R,D}}(\delta\sqrt{\nu})$ to $\mathcal{B}^b\mathcal{H}^1_\ell|_{\mathrm{R,D}}(\delta\sqrt{\nu})$ if

$$\nu \in]0, \nu_4], \ K \in [M_5\nu^{\frac{3}{2}}, K_1],$$
$$M_7(\delta^2\nu + K + \nu) \leq \delta\sqrt{\nu}, \qquad M_7(2\delta\sqrt{\nu} + \nu) < 1.$$

We then can choose $K = M_5\nu^{\frac{3}{2}}$, $\delta = 2M_7\sqrt{\nu}$, and $\nu_5 \in]0, \nu_4]$ such that for all ν in $]0, \nu_5]$,

$$4(M_7)^3\sqrt{\nu} + M_5\nu \leq M_7, \qquad (4(M_7)^2 + M_7)\nu < 1. \ \square$$

This completes the proof of Theorem 9.1.8.

9.4 Generic non persistence of reversible homoclinic connections to 0

This section is devoted to the question of the persistence of homoclinic connections to 0 for the full system

$$\frac{dY}{dt} = \mathcal{N}(Y, \nu) + \mathcal{R}(Y, \nu)$$

where $\mathcal{N}$, $\mathcal{R}$ are defined in (9.8), (9.11). For studying this question we introduce in the previous equation an additional parameter $\rho \in [0, 1]$

$$\frac{dY}{dt} = \mathcal{N}(Y, \nu) + \rho\mathcal{R}(Y, \nu) \tag{9.44}$$

For $\rho = 0$, we get the truncated system (9.14) which admits only two homoclinic connections h, $-h$ to 0 where h is given by (9.20). Now, we would like to understand what happens for $\rho \neq 0$ and in particular for $\rho = 1$ which corresponds to the full system (9.8). For studying the persistence of h, a classical way is to use an approach of Melnikov type, which can be summarized as follows:

If the system (9.44) admits a homoclinic connection Y to 0, then the perturbation term $v := Y - h$ satisfies the equation

$$\frac{dv}{dt} - D\mathcal{N}(h).v = g(v, t, \rho, \nu) \tag{9.45}$$

with

$$g(v, t, \nu) = Q(v) + \rho\mathcal{R}(h + v, \nu)$$

where

$$\mathcal{Q}(v,\nu) = \mathcal{N}(h+v,\nu) - \mathcal{N}(h,\nu) - D\mathcal{N}(h,\nu).v.$$

Here again, the crucial point of our analysis is the following: for an *antireversible* function f which *tends to 0 exponentially at infinity*, there exists a *reversible* function u which *tends to 0 exponentially at infinity* and which satisfies

$$\frac{du}{dt} - D\mathcal{N}(h,\nu).u = f$$

if and only if the function f satisfies the solvability condition

$$I(f,\nu) = \int_0^{+\infty} \langle r_-^*(t), f(t)\rangle_* dt = 0 \tag{9.46}$$

where the dual vector r_-^* reads

$$r_-^*(t) := (0,0,0,0,-\tfrac{\mathrm{i}}{2}\mathrm{e}^{-\mathrm{i}\psi_r(t)}, \tfrac{\mathrm{i}}{2}\mathrm{e}^{\mathrm{i}\psi_r(t)}) \text{ with } \psi_r(t) = \tfrac{\omega_1(\nu)}{\nu}t + n_2(\nu)\nu\tanh(t).$$

Thus, if the full system (9.44) admits a reversible homoclinic connection Y to 0, then Y necessarily satisfies

$$I(\rho,\nu) := \int_0^{+\infty} \langle r_-^*(t), \mathcal{Q}(v,\nu) + \rho\mathcal{R}(h(t)+v(t),\nu)\rangle dt = 0$$

where $v = Y - h$. The usual way to study such a solvability condition is to split the integral in two parts $I(\rho,\nu) = M_\mathrm{e}(\rho,\nu) + J(\rho,\nu)$ where $M_\mathrm{e}(\rho,\nu)$ which only depends on h is explicitly known

$$M_\mathrm{e}(\rho,\nu) = \rho\int_0^{+\infty} \langle r^*(t), \mathcal{R}(h(t),\nu)\rangle_* dt$$

whereas $J(\rho,\nu)$ involves v which is only characterized by the fact that it is a *reversible* solution of (9.44) which *tends to 0 at infinity*. The first part $M_\mathrm{e}(\rho,\nu)$ is called the Melnikov function. Observe that the *Melnikov function is a bi-oscillatory integral* since the dual vector r_-^* rotates with a high frequency of order $\frac{\omega_1}{\nu}$ and h rotates with a frequency of order $\frac{\omega_0}{\nu}$.

When performing such a splitting of the solvability condition, one hopes that the Melnikov function $M_\mathrm{e}(\rho,\nu)$ which comes from the leading part of Y on $\mathbb{R}$, is the leading part of the integral $I(\rho,\nu)$. So, we first compute explicitly $M_\mathrm{e}(\rho,\nu)$: expanding in power series the essential singularity in the oscillatory term and computing the residues, we get

$$M_\mathrm{e}(\rho,\nu) = \frac{\rho}{\nu^{\frac{3}{2}}}\mathrm{e}^{-\frac{\pi\omega_*}{2\nu}}(\Lambda_1 + \underset{\nu\to 0}{\mathcal{O}}(\nu)).$$

Then, classical perturbation theory enables us to bound J which involves the not explicitly known perturbation term v. This leads to

$$I(\rho,\nu) = \frac{\rho}{\nu^{\frac{3}{2}}}\mathrm{e}^{-\frac{\pi\omega_*}{2\nu}}(\Lambda_1 + \underset{\nu\to 0}{\mathcal{O}}(\nu)) + \underset{\nu\to 0}{\mathcal{O}}(\rho^2\nu^2).$$

However, such an estimate of $I(\rho,\nu)$ is not accurate enough to determine whether $I(\rho,\nu)$ vanishes or not, since the upper bound of the not explicitly known part J is far bigger than the expected leading part $M_e(\rho,\nu)$ which is exponentially small. This exponential smallness comes from the fact that the solvability condition is given by a bi-oscillatory integral.

The Exponential Tools given in Section 2.2 were precisely developed to study such solvability conditions given by bi-oscillatory integrals. The Third Bi-frequency Exponential Lemma 2.2.9, coupled with the strategy proposed in Subsection 2.1.4 enables us to obtain equivalents of oscillatory integrals involving solutions of nonlinear differential equations. Indeed, combining the strategy of proof used for the $0^{2+}\mathrm{i}\omega$ resonance (see section 7.4) and the partial complexification of time used for the proof of Theorem 9.1.8, we get that if Y is a homoclinic connection to 0 of (9.44), then it admits a holomorphic continuation in a space of type $\prod_{i=1}^{6} {}^{b}E^{1,\gamma_i}_{\frac{\pi}{2},\delta}$ where ${}^{b}E^{\gamma,\lambda}_{\sigma,\delta}$ is introduced in Definition 2.2.6.

So using the Third Bi-frequency Exponential Lemma 2.2.9, we get that if the full system (9.44) admits a reversible homoclinic connection Y to 0, then

$$I(\rho,\nu) = \frac{1}{\nu^{\frac{3}{2}}}\mathrm{e}^{-\frac{\pi\omega_*}{2\nu}}\left(\Lambda(\rho) + \underset{\nu\to 0}{O}(\nu^{\frac{1}{4}})\right)$$

where $\rho \mapsto \Lambda(\rho)$ is a real analytic function on $[0,1]$ such that $\Lambda(\rho) = \Lambda_1\rho + \underset{\rho\to 0}{O}(\rho^2)$. Such an estimate of $I(\rho,\nu)$ ensures that if $\Lambda(\rho)\neq 0$ (which is generically the case), then for ν small enough, $I(\rho,\nu)$ is exponentially small but does not vanish and thus there is no reversible homoclinic connection to 0.

Observe that for a fixed $\rho>0$, the Melnikov function $M_e(\rho,\nu)$ is not the leading part of the integral $I(\rho,\nu)$ for ν small. The homoclinic connection of the truncated system h and the perturbation term v contributes to the size of $I(\rho,\nu)$ at the same order in ν.

9.A Appendix. Floquet linear change of coordinates. Proof of Proposition 9.3.2

In this section, it is more convenient to rewrite the system (9.29) under the following equivalent form for $Z=(Z_h,Z_c)\in\mathbb{D}=\mathbb{D}_h\times\mathbb{D}_c$,

$$\begin{aligned}\frac{dZ_h}{dt} &= L_{h,\nu}Z_h + \mathcal{T}_h(Z_h,Z_c,\nu) + \mathcal{R}_{z,h}(Z_h,Z_c,\nu),\\ \frac{dZ_h}{dt} &= L_{c,\nu}Z_c + \mathcal{T}_c(Z_h,Z_c,\nu) + \mathcal{R}_{z,c}(Z_h,Z_c,\nu).\end{aligned} \tag{9.47}$$

where the higher order terms reads $\mathcal{R}_z = (\mathcal{R}_{z,h},\mathcal{R}_{z,c})$; the linear part are given by

$$L_{\mathrm{h},\nu} = \begin{pmatrix} \mathrm{i}\frac{\omega_0(\nu)}{\nu} & 1 & 0 & 0 \\ 1 & \mathrm{i}\frac{\omega_0(\nu)}{\nu} & 0 & 0 \\ 0 & 0 & -\mathrm{i}\frac{\omega_0(\nu)}{\nu} & 1 \\ 0 & 0 & 1 & -\mathrm{i}\frac{\omega_0(\nu)}{\nu} \end{pmatrix}, \quad L_{\mathrm{c},\nu} = \begin{pmatrix} \mathrm{i}\frac{\omega_1(\nu)}{\nu} & 0 \\ 0 & -\mathrm{i}\frac{\omega_1(\nu)}{\nu} \end{pmatrix},$$

and the cubic terms by

$$\mathcal{T}_{\mathrm{h}}(Z,\nu) = \begin{pmatrix} \mathrm{i}AP(Z,\nu) \\ \mathrm{i}BP(Z,\nu) + A(Q(Z,\nu)-1) \\ -\mathrm{i}\widetilde{A}P(Z,\nu) \\ -\mathrm{i}\widetilde{B}P(Z,\nu) + \widetilde{A}(Q(Z,\nu)-1) \end{pmatrix}, \quad \mathcal{T}_{\mathrm{c}}(Z,\nu) = \begin{pmatrix} \mathrm{i}CN(Z,\nu) \\ -\mathrm{i}\widetilde{C}N(Z,\nu) \end{pmatrix},$$

where $Z = (Z_{\mathrm{h}}, Z_{\mathrm{c}}) = (A, B, \widetilde{A}, \widetilde{B}, C, \widetilde{C})$ and P, Q, N are given by (9.8). Similarly, we rewrite S as $S = (S_{\mathrm{h}}, S_{\mathrm{c}})$ with $S_{\mathrm{h}}(A, B, \widetilde{A}, \widetilde{B}) = (\widetilde{A}, -\widetilde{B}, A, -B)$ and $S_{\mathrm{c}}(C, \widetilde{C}) = (\widetilde{C}, C)$.

For determining the existence of homoclinic connections to the periodic orbits $Z_{k,\nu}$ we first study equation centered at these periodic orbits. It is explicitly given by

Lemma 9.A.1. *Setting* $Z(t) = Z_{k,\nu}(t) + X(t)$, *i.e.* $X_{\mathrm{h}} = Z_{\mathrm{h}}$, $X_{\mathrm{c}} = Z_{\mathrm{c},k,\nu} + X_{\mathrm{c}}$, *with* $Z_{\mathrm{c},k,\nu}(t) = (k\mathrm{e}^{\mathrm{i}\frac{\omega_{k,\nu}}{\nu}t}, k\mathrm{e}^{-\mathrm{i}\frac{\omega_{k,\nu}}{\nu}t}) \in \mathbb{D}_{\mathrm{c}}$, *system (9.29) is equivalent to*

$$\begin{aligned} \frac{dX_{\mathrm{h}}}{dt} &= L_{\mathrm{h},\nu}X_{\mathrm{h}} + \mathcal{M}_{\mathrm{h},\nu}(Z_{\mathrm{c},k,\nu}).X_{\mathrm{h}} + \mathcal{T}_{\mathrm{h}}(X_{\mathrm{h}}, X_{\mathrm{c}}, \nu) + \mathcal{R}_{X,\mathrm{h}}(X_{\mathrm{h}}, X_{\mathrm{c}}, Z_{\mathrm{c},k,\nu}, \nu), \\ \frac{dX_{\mathrm{c}}}{dt} &= L_{\mathrm{c},\nu}X_{\mathrm{c}} + \mathcal{T}_{\mathrm{c}}(X_{\mathrm{h}}, X_{\mathrm{c}}, \nu) + \mathcal{R}_{X,\mathrm{c}}(X_{\mathrm{h}}, X_{\mathrm{c}}, Z_{\mathrm{c},k,\nu}, \nu), \end{aligned} \tag{9.48}$$

where

(a) $\mathcal{M}_{\mathrm{h},\nu}(Z_{\mathrm{c}})$ *is a* 4×4 *complex matrix which reads*

$$\mathcal{M}_{\mathrm{h},\nu}(Z_{\mathrm{c}}) = \sum_{p+q\geq 2} \mathcal{M}_{p,q}(\nu)C^p\widetilde{C}^q \quad \text{with } |\mathcal{M}_{p,q}(\nu)| \leq \frac{M_{\mathcal{M}}\,\nu}{(k_1)^{p+q}} \text{ for } \nu \in]0, \nu_1].$$

and satisfies for every $Z_{\mathrm{c}} = (C, \widetilde{C}) \in \mathbb{D}_{\mathrm{c}}$,

$$\mathcal{M}_{\mathrm{h},\nu}(Z_{\mathrm{c}}).\mathbb{D}_{\mathrm{c}} \subset \mathbb{D}_{\mathrm{c}}, \qquad S_{\mathrm{h}}\mathcal{M}_{\mathrm{h},\nu}(Z_{\mathrm{c}}) = -\mathcal{M}_{\mathrm{h},\nu}(S_{\mathrm{c}}Z_{\mathrm{c}}).S_{\mathrm{h}}.$$

(b) $\mathcal{R}_{X,\mathrm{h}}$ *and* $\mathcal{R}_{X,\mathrm{c}}$ *are analytic functions which satisfies*

$$\begin{aligned} &\mathcal{R}_{X,\mathrm{h}}(\mathbb{D}, \mathbb{D}_{\mathrm{c}}, \nu) \subset \mathbb{D}_{\mathrm{h}}, \qquad S_{\mathrm{h}}\mathcal{R}_{X,\mathrm{h}}(X, Z_{\mathrm{c}}, \nu) = -\mathcal{R}_{X,\mathrm{h}}(SX, S_{\mathrm{c}}Z_{\mathrm{c}}, \nu), \\ &\mathcal{R}_{X,\mathrm{c}}(\mathbb{D}, \mathbb{D}_{\mathrm{c}}, \nu) \subset \mathbb{D}_{\mathrm{c}}, \qquad S_{\mathrm{c}}\mathcal{R}_{X,\mathrm{c}}(X, Z_{\mathrm{c}}, \nu) = -\mathcal{R}_{X,\mathrm{c}}(SX, S_{\mathrm{c}}Z_{\mathrm{c}}, \nu), \end{aligned}$$

and

$$\mathcal{R}_{X,\mathrm{h}}(X,Z_\mathrm{c}) = \mathcal{O}\Big(\nu|X_\mathrm{h}||X_\mathrm{c}||Z_\mathrm{c}|+\nu|X_\mathrm{h}||X|(|X|+|Z_\mathrm{c}|)^2\Big)$$

$$\mathcal{R}_{X,\mathrm{c}}(X,Z_\mathrm{c}) = \mathcal{O}\Big(\nu^{\frac{3}{2}}|X|(|X|+|Z_\mathrm{c}|)^3+\nu|X_\mathrm{c}|(|X_\mathrm{c}|+|Z_\mathrm{c}|)^3 +\nu|X_\mathrm{h}|^2|Z_\mathrm{c}|+\nu^2|X_\mathrm{c}||Z_\mathrm{c}|(|X_\mathrm{c}|+|Z_\mathrm{c}|)\Big)$$

Using the constructive Floquet Theory given in Chapter 5, we prove

Proposition 9.A.2. *There exists $\nu_2, k_2 > 0$ and a linear change of coordinates $X_\mathrm{h} = \big[\mathrm{Id} + \mathcal{Q}^1_{\mathrm{h},\nu}(t,k)\big].w_\mathrm{h}$ such that for every $\nu \in]0,\nu_2]$ and $k \in [0,k_2]$, the equation*

$$\frac{dX_\mathrm{h}}{dt} = L_{\mathrm{h},\nu}X_\mathrm{h} + \mathcal{M}_{\mathrm{h},\nu}(Z_{\mathrm{c},k,\nu}).X_\mathrm{h} \tag{9.49}$$

is equivalent to

$$\frac{dw_\mathrm{h}}{dt} = \big(L_{\mathrm{h},\nu} + L^1_{\mathrm{h},\nu}(k)\big)w_\mathrm{h}$$

where

(a) *$\mathcal{Q}^1_{\mathrm{h},\nu}$ satisfies $\mathcal{Q}^1_{\mathrm{h},\nu}(t,k).\mathbb{D}_\mathrm{h} \subset \mathbb{D}_\mathrm{h}$, $S_\mathrm{h}\mathcal{Q}^1_{\mathrm{h},\nu}(t,k) = \mathcal{Q}^1_{\mathrm{h},\nu}(-t,k)S_\mathrm{h}$ and reads $\mathcal{Q}^1_{\mathrm{h},\nu}(t,k,\nu) = \check{\mathcal{Q}}^1_{\mathrm{h},\nu}(Z_{\mathrm{c},k,\nu}(t))$ where*

$$\check{\mathcal{Q}}^1_{\mathrm{h},\nu}(Z_\mathrm{c}) = \sum_{p+q\geq 2} \mathcal{Q}^1_{p,q}(\nu)C^p\widetilde{C}^q \quad \textit{with } |\mathcal{Q}^1_{p,q}(\nu)| \leq \frac{M_\mathcal{Q}\,\nu^2}{(k_2)^{p+q}} \textit{ for } \nu \in]0,\nu_2].$$

(b) *$L^1_{\mathrm{h},\nu}(k)$ commutes with $L_{\mathrm{h},\nu}$, satisfies $S_\mathrm{h}L^1_{\mathrm{h},\nu}(k) = -L^1_{\mathrm{h},\nu}(k)S_\mathrm{h}$, $L^1_{\mathrm{h},\nu}\mathbb{D}_\mathrm{h} \subset \mathbb{D}_\mathrm{h}$ and reads*

$$L^1_{\mathrm{h},\nu}(k) = \sum_{n\geq 1} k^{2n}L^1_{2n}(\nu) \qquad \textit{with } |L^1_{2n}(\nu)| \leq \frac{M_L\,\nu}{(k_2)^{2n}} \textit{ for } \nu \in]0,\nu_2],$$

(c) *Moreover, for every $n \geq 1$ and $\nu \in]0,\nu_2]$, the matrix $L^1_{2n}(\nu)$ reads*

$$L^1_{2n}(\nu) = \begin{pmatrix} \mathrm{i}\omega^1_{2n}(\nu) & \alpha^1_{2n}(\nu) & 0 & 0 \\ \alpha^1_{2n}(\nu) & \mathrm{i}\omega^1_{2n}(\nu) & 0 & 0 \\ 0 & 0 & -\mathrm{i}\omega^1_{2n}(\nu) & \alpha^1_{2n}(\nu) \\ 0 & 0 & \alpha^1_{2n}(\nu) & -\mathrm{i}\omega^1_{2n}(\nu) \end{pmatrix},$$

with $\alpha^1_{2n}(\nu), \omega^1_{2n}(\nu) \in \mathbb{R}$.

Thus, $L^1_{\mathrm{h},\nu}(k)$ has the same form and we denote by

$$\alpha^1_{\mathrm{h},\nu}(k) = \sum_{n\geq 1} k^{2n}\alpha^1_{2n}(\nu) = \mathcal{O}(k^2\nu), \quad \omega^1_{\mathrm{h},\nu}(k) = \sum_{n\geq 1} k^{2n}\omega^1_{2n}(\nu) = \mathcal{O}(k^2\nu).$$

Observe that $\pm(1+\alpha^1_{\mathrm{h},\nu}(k)) \pm \mathrm{i}(\frac{\underline{\omega}_0(\nu)}{\nu} + \omega^1_{\mathrm{h},\nu}(k))$ are four Floquet Exponents of the matrix $L_{\mathrm{h},\nu} + \mathcal{M}_{\mathrm{h},\nu}(Z_{\mathrm{c},k,\nu})$.

Proof. (a),(b) : For working with functions with a fixed period, we perform the scaling of time

$$s = \underline{\omega}_{k,\nu} t, \qquad \text{with } X_{\mathrm{h}}(t) = \widehat{X}_{\mathrm{h}}(\underline{\omega}_{k,\nu} t). \tag{9.50}$$

Then, (9.49) is equivalent to

$$\frac{d\widehat{X}_{\mathrm{h}}}{ds} = \frac{L_{\mathrm{h},\nu}}{\underline{\omega}_{k,\nu}}\widehat{X}_{\mathrm{h}} + \frac{\mathcal{M}_{\mathrm{h},\nu}(\widehat{Z}_{\mathrm{c},k,\nu}(s))}{\underline{\omega}_{k,\nu}}\widehat{X}_{\mathrm{h}} \tag{9.51}$$

where $\widehat{Z}_{\mathrm{c},k,\nu}(s) = (k\mathrm{e}^{\mathrm{i}s}, k\mathrm{e}^{-\mathrm{i}s})$. From Theorem 9.1.4, we get that

$$\underline{\omega}_{k,\nu} = \frac{\underline{\omega}_1(\nu)}{\nu} + \widehat{\omega}(k;\nu) = \frac{\underline{\omega}_1(\nu)}{\nu} + \sum_{n\geq 1}\widehat{\omega}_{2n}(\nu)k^{2n}$$

where $|\widehat{\omega}_{2n}(\nu)| \leq \dfrac{C_\omega \nu}{(k_0)^{2n}}$. Thus we can rewrite (9.51) in the form

$$\frac{d\widehat{X}_{\mathrm{h}}}{ds} = \widehat{L}_{\mathrm{h},\nu}\widehat{X}_{\mathrm{h}} + \widehat{\mathcal{M}}_{\mathrm{h},\nu}(k\mathrm{e}^{\mathrm{i}s}, k\mathrm{e}^{-\mathrm{i}s})\widehat{X}_{\mathrm{h}} \tag{9.52}$$

where

$$\widehat{L}_{\mathrm{h},\nu} = \frac{\nu L_{\mathrm{h},\nu}}{\underline{\omega}_1(\nu)} = \begin{pmatrix} \mathrm{i}\frac{\underline{\omega}_0(\nu)}{\underline{\omega}_1(\nu)} & \frac{\nu}{\underline{\omega}_1(\nu)} & 0 & 0 \\ \frac{\nu}{\underline{\omega}_1(\nu)} & \mathrm{i}\frac{\underline{\omega}_0(\nu)}{\underline{\omega}_1(\nu)} & 0 & 0 \\ 0 & 0 & -\mathrm{i}\frac{\underline{\omega}_0(\nu)}{\underline{\omega}_1(\nu)} & \frac{\nu}{\underline{\omega}_1(\nu)} \\ 0 & 0 & \frac{\nu}{\underline{\omega}_1(\nu)} & -\mathrm{i}\frac{\underline{\omega}_0(\nu)}{\underline{\omega}_1(\nu)} \end{pmatrix},$$

and

$$\widehat{\mathcal{M}}_{\mathrm{h},\nu}(C,\widetilde{C}) = \left(\frac{\nu}{\underline{\omega}_1(\nu) + \sum_{n\geq 1}\nu\widehat{\omega}_{2n}(\nu)C^n\widetilde{C}^n} - \frac{\nu}{\underline{\omega}_1(\nu)}\right)\widehat{L}_{\mathrm{h},\nu}$$
$$+\frac{\nu\mathcal{M}_{\mathrm{h},\nu}(C,\widetilde{C})}{\underline{\omega}_1(\nu) + \sum_{n\geq 1}\nu\widehat{\omega}_{2n}(\nu)C^n\widetilde{C}^n}$$

Hence,

$$\widehat{\mathcal{M}}_{h,\nu}(C,\widetilde{C}) = \sum_{p+q\geq 2} \widehat{\mathcal{M}}_{p,q}(\nu)C^p\widetilde{C}^q \text{ with } |\mathcal{M}_{p,q}(\nu)| \leq \frac{M_{\widehat{\mathcal{M}}}\,\nu^2}{k_1^{p+q}} \text{ for } \nu \in]0,\nu_1].$$

Now we want to apply Theorem 5.1.1 and Lemma 5.1.2 to (9.52). For that purpose, we must check that the eigenvalues $(\sigma_j)_{1\leq j\leq 4}$ of $\widehat{\mathcal{L}}_{h,\nu}$ satisfies the non resonance criteria

$$\sigma_j - \sigma_k \notin i\mathbb{Z}\setminus\{0\}. \tag{9.53}$$

The eigenvalues of $\widehat{\mathcal{L}}_{h,\nu}$ are $\pm i\frac{\omega_0(\nu)}{\underline{\omega}_1(\nu)} \pm \frac{\nu}{\underline{\omega}_1(\nu)}$. Thus , the possible differences of eigenvalues are

$$0, \qquad \pm\frac{2\nu}{\underline{\omega}_1(\nu)}, \qquad \pm 2\left(i\frac{\underline{\omega}_0(\nu)}{\underline{\omega}_1(\nu)} + \frac{\nu}{\underline{\omega}_1(\nu)}\right), \qquad \pm i\frac{\underline{\omega}_0(\nu)}{\underline{\omega}_1(\nu)}.$$

Then , the hypothesis of non resonance (H4c) ensures that the non resonance criteria (9.53) is satisfied for every $\nu \in [0,\nu_2]$ for some sufficiently small $\nu_2 > 0$. Moreover, S_h is an orthogonal symmetry and

$$S_h\widehat{\mathcal{L}}_{h,\nu} = -\widehat{\mathcal{L}}_{h,\nu}S_h, \qquad S_h\widehat{\mathcal{M}}_{h,\nu}(ke^{-is},ke^{is}) = -\widehat{\mathcal{M}}_{h,\nu}(ke^{is},ke^{-is})S_h.$$

Thus Theorem 5.1.1 and Lemma 5.1.2 ensure that there exit $k_2 \in]0,k_1]$ and a pair of matrices $\widehat{\mathcal{L}}^1_{h,\nu}$, $\widehat{\mathcal{Q}}^1_{h,\nu}(s,k)$ 2π-periodic such that for every $\nu \in]0,\nu_2]$ and every $k \in [0,k_2]$, setting

$$\widehat{X}_h(s) = (\mathrm{Id} + \widehat{\mathcal{Q}}^1_{h,\nu}(s,k))\widehat{w}_h(s),$$

the equation (9.52) is equivalent to

$$\frac{d\widehat{w}_h}{dt} = (\widehat{\mathcal{L}}_{h,\nu} + \widehat{\mathcal{L}}^1_{h,\nu}(k))\widehat{w}_h$$

where

(a) $\widehat{\mathcal{Q}}^1_{h,\nu}$ satisfies $\widehat{\mathcal{Q}}^1_{h,\nu}(s,k).\mathbb{D}_h \subset \mathbb{D}_h$, $S_h\widehat{\mathcal{Q}}^1_{h,\nu}(s,k) = \mathcal{Q}^1_{h,\nu}(-s,k)S_h$ and reads $\widehat{\mathcal{Q}}^1_{h,\nu}(t,k,\nu) = \check{\mathcal{Q}}^1_{h,\nu}(ke^{is},ke^{-is})$ where

$$\check{\mathcal{Q}}^1_{h,\nu}(C,\widetilde{C}) = \sum_{p+q\geq 2} \mathcal{Q}^1_{p,q}(\nu)C^p\widetilde{C}^q \text{ with } |\mathcal{Q}^1_{p,q}(\nu)| \leq \frac{m_2\,\nu^2}{(k_2)^{p+q}} \text{ for } \nu \in]0,\nu_2].$$

(b) $\widehat{\mathcal{L}}^1_{h,\nu}(k)$ satisfies $S_h\widehat{\mathcal{L}}^1_{h,\nu}(k) = -\widehat{\mathcal{L}}^1_{h,\nu}(k)S_h$, $\widehat{\mathcal{L}}^1_{h,\nu}\mathbb{D}_h \subset \mathbb{D}_h$ and reads

$$\widehat{\mathcal{L}}^1_{h,\nu}(k) = \sum_{n\geq 1} k^{2n}\widehat{\mathcal{L}}^1_{2n}(\nu) \qquad \text{with } |\widehat{\mathcal{L}}^1_{2n}(\nu)| \leq \frac{m_2\,\nu^2}{(k_2)^{2n}} \text{ for } \nu \in]0,\nu_2].$$

and for every $n \geq 1$, $\widehat{\mathcal{L}}^1_{2n}(\nu)$ commutes with $\widehat{\mathcal{L}}^*_{h,\nu}$.

Since $\widehat{L}_{\mathrm{h},\nu}$ and $\widehat{\mathcal{M}}_{\mathrm{h},\nu}$ are smooth with respect to ν, we check that all the estimates made in the proof of Theorem 5.1.1 and Lemma 5.1.2 are uniform with respect to $\nu \in]0, \nu_2]$ and thus we get that the constant m_2 does not depend on ν.

Observe that $\widehat{L}_{\mathrm{h},\nu}$ commutes with its adjoint $\widehat{L}^*_{\mathrm{h},\nu}$ thus it is diagonalizable in an orthonormal basis. Moreover, since it has only simple eigenvalues, we check that commuting with $\widehat{L}_{\mathrm{h},\nu}$ is equivalent to commuting with $\widehat{L}^*_{\mathrm{h},\nu}$. Hence,

$$\widehat{L}^1_{2n}(\nu)\widehat{L}_{\mathrm{h},\nu} - \widehat{L}_{\mathrm{h},\nu}\widehat{L}^1_{2n}(\nu) = 0, \qquad \text{for } n \geq 1.$$

Finally, to come back to our original equation, we perform the scaling (9.50): setting

$$X_{\mathrm{h}}(t) = (\mathrm{Id} + \widehat{\mathcal{Q}}^1_{\mathrm{h},\nu}(\underline{\omega}_{k,\nu}t, k))w_{\mathrm{h}}(t) = (\mathrm{Id} + \check{\mathcal{Q}}^1_{\mathrm{h},\nu}(Z_{\mathrm{c},k,\nu}(t))w_{\mathrm{h}}(t),$$

equation (9.49) is equivalent to

$$\frac{dw_{\mathrm{h}}}{dt} = \left(L_{\mathrm{h},\nu} + L^1_{\mathrm{h},\nu}(k)\right)w_{\mathrm{h}}$$

with

$$L^1_{\mathrm{h},\nu}(k) = \frac{\nu\widehat{\omega}(k,\nu)}{\underline{\omega}_1(\nu)}L_{\mathrm{h},\nu} + \left(\frac{\underline{\omega}_1(\nu)}{\nu} + \widehat{\omega}(k,\nu)\right)\widehat{L}^1_{\mathrm{h},\nu}(k)$$

and where $L^1_{\mathrm{h},\nu}(k)$ and $\check{\mathcal{Q}}^1_{\mathrm{h},\nu}$ satisfies the statements (a) and (b) of Proposition 9.A.2.

(c): The above proof ensures that

$$L^1_{\mathrm{h},\nu}(k) = \sum_{n\geq 1} k^{2n} L^1_{2n}(\nu)$$

where $L^1_{2n}(\nu)$ commutes with $L_{\mathrm{h},\nu}$ which is diagonalizable with simple eigenvalues. Hence

$$L^1_{2n}(\nu) = \sum_{j=0}^{3} a_j(\nu)(L_{\mathrm{h},\nu})^j.$$

Moreover, $L^1_{2n}(\nu)$ anticommutes with S_{h} and satisfies $L^1_{2n}(\nu)\mathbb{D}_{\mathrm{h}} \subset \mathbb{D}_{\mathrm{h}}$. Thus,

$$L^1_{2n}(\nu) = a_1(\nu)L_{\mathrm{h},\nu} + a_3(\nu)(L_{\mathrm{h},\nu})^3$$

with $a_1(\nu), a_3(\nu) \in \mathbb{R}$. Computing explicitly $(L_{\mathrm{h},\nu})^3$ we check that $L^1_{2n}(\nu)$ has the form

$$L^1_{2n}(\nu) = \begin{pmatrix} \mathrm{i}\omega^1_{2n}(\nu) & \alpha^1_{2n}(\nu) & 0 & 0 \\ \alpha^1_{2n}(\nu) & \mathrm{i}\omega^1_{2n}(\nu) & 0 & 0 \\ 0 & 0 & -\mathrm{i}\omega^1_{2n}(\nu) & \alpha^1_{2n}(\nu) \\ 0 & 0 & \alpha^1_{2n}(\nu) & -\mathrm{i}\omega^1_{2n}(\nu) \end{pmatrix},$$

with $\alpha^1_{2n}(\nu), \omega^1_{2n}(\nu) \in \mathbb{R}$. This completes the proof of Proposition 9.A.2. □

References

[AK89] Amick C.J., Kirchgässner K. (1989) A theory of solitary-waves in the presence of surface tension. *Arch. Rat. Mech. and Anal.* **105**, p. 1–49.

[AL90] Amick C.J., McLeod J.B. (1990) A singular perturbation problem in needle crystals *Arch. Rational Mech. Anal.* **109**, no 2, p. 139-171.

[AL92] Amick C.J., McLeod J.B. (1992) A singular perturbation problem in water waves *Stab. Appl. Anal. of Cont. Media*, p. 127-148.

[AT92] Amick C. J., Toland J. F. (1992) Solitary waves with surface tension. I. Trajectories homoclinic to periodic orbits in four dimensions. *Arch. Rational Mech. Anal.* **118**, no. 1, 37–69.

[Ar64] Arnold V.I. (1964) Instability of dynamical systems with several degress of freedom. Doklady Akad Nauk SSSR **156**, p. 581–585.

[Ar83] Arnold V.I. (1983) *Geometrical methods in the theory of ordinary differential equations.* Springer-Verlag, New York-Berlin.

[Au97] Aubry S. (1997) Breathers in nonlinear lattices: existence, linear stability and quantization *Physica D* **103**, p. 201–250.

[Be91] Beale J.T. (1991) Exact solitary water waves with capillary ripples at infinity. *Comm. Pure Appl. Math.* **44**, p. 211-257.

[Ch2000] Champneys A.R. (2000) Codimension-one persistence beyond all orders of homoclinic orbits to singular saddle centres in reversible systems. *Submitted to Nonlinearity.*

[CL55] Coddington E.A., Levinson N. (1955) *Theory of Ordinary Differential Equations.* Robert E. Krieger Publishing company, Florida.

[CHDPP88] Combescot R., Hakim V., Dombre T., Pomeau Y. Pumir A. (1988) Analytic theory of the Saffman-Taylor fingers. *Phys. Rev. A* (3) **37**, no. 4, p. 1270–1283.

[Co78] Coppel W.A. (1978) *Dichotomies in Stability Theory.* Springer Lecture Notes in Mathematics **vol. 629** Springer-Verlag: New York, Berlin, Heidelberg.

[DKLS86] Dashen R.F., Kessler H., Levine H., Savit R. (1986) *Phys. D* **21**, p. 371–380.

[DGJS97] Delsham A., Gelfreich V., Jorba A. and Seara T. (1997) Exponentially small splitting of separatrices under fast quasi periodic forcing. *Commun. Math Phys.*, **189**, p. 35–71.

[DS92] Delshams A., Seara T.M. (1992) An asymptotic expression for the splitting of separatrices of rapidly forced pendulum. *Com. Math. Phys.* **150**, p. 433–463.

[De76] Devaney R.L. (1976) Reversible diffeomorphisms and flows. *Trans. Am. Math. Soc.*, **218**, p. 89-113.

[Du96] Duistermaat, J. J. (1996) *Fourier integral operators.* Progress in Mathematics, 130. Birkhäuser Boston, Inc., Boston, MA.

[DH72] Duistermaat J. J., Hörmander L. (1972) Fourier integral operators. II. *Acta Math.* **128** , no. 3-4, p. 183–269.

[Ea84] Easton, R. W. (1984) Computing the dependence on a parameter of a family of unstable manifolds: generalized Melnikov formulas. *Nonlinear Anal.* **8**, no. 1, p. 1–4.

[Ec92] Eckhaus W. (1992) Singular perturbations of homoclinic orbits in $\mathbb{R}^4$. *SIAM J. Math. Anal.* vol 23 No. 5, p. 1269-1290.

[ETBCI87] Elphick C., Tirapegui E., Brachet M.E., Coullet P., Iooss G. (1987) A simple global characterization for normal forms of singular vector fields. *Physica D* **29** p. 95–127.

[Er56] Erdélyi A. (1956) *Asymptotic expansions.* Dover Publications, Inc., New York.

[Fo93] Fontich E. (1993) Exponentially small upper bounds for the splitting of separatrices for high frequency periodic perturbations. Nonlinear Anal. 20, no. 6, p. 733–744.

[Fo95] Fontich (1995), E. Rapidly forced planar vector fields and splitting of separatrices. *J. Differential Equations* **119**, no. 2, p. 310–335

[FS90] Fontich E., Simò C. (1990) Invariant manifolds for near identity differentiable maps and splitting of separatrices. *Ergodic Theory Dynamical Systems***10** , no. 2, p. 319–346.

[Ga83] Gambaudo J.M. (1983) Perturbation de "l'application temps τ" d'un champ de vecteurs intégrable de R^2. *C. R. Acad. Sci. Paris Sér. I Math.* **297**, no. 4, p. 245–248.

[Ga85] Gambaudo J.-M. (1985) Perturbation of a Hopf bifurcation by an external time-periodic forcing. *J. Differential Equations* **57**, no. 2, p. 172–199.

[GLS94] Gelfreich V. G., Lazutkin V. F., Svanidze N. V. (1994) A refined formula for the separatrix splitting for the standard map. *Phys. D* **71**, no. 1-2, 82–101.

[Ge97] Gelfreich V.G. (1997) Reference system for splitting of separatrix. *Nonlinearity* **10**, p. 175–193
Gelfreich V.G. (1996) Separatrix splitting for a high-frequency perturbation of the pendulum (unpublished).

[Ge99] Gelfreich V.G. (1999) A proof of the exponentially small transversality of the separatrices for the standard map. *Comm. Math. Phys.* **201** , no. 1, p. 155–216.

[Gr92] Grimshaw R. (1992) The use of Borel-summation in the establishment of nonexistence of certain travelling-wave solutions of the Kuramoto-Sivashinsky equation. *Wave Motion* **15**, no. 4, p. 393–395.

[GJ95] Grimshaw R., Joshi (1995) N. Weakly nonlocal solitary waves in a singularly perturbed Korteweg-de Vries equation. *SIAM J. Appl. Math.* **55**, no. 1, p. 124–135.

[GM99] Groves M.D. and Mielke A. (1999) A spatial dynamics approach to the three-dimensional gravity-capillary steady water waves. Preprint Universität Stuttgart 99-7.

[GH83] Guckenheimer J., Holmes P. (1983) *Nonlinear oscillations, dynamical systems and bifurcations of vector fields.* Applied Mathematical Sciences vol.**42** Berlin, Heidelberg, New York: Springer-Verlag

[Hk91] Hakim V. (1991) Computation of transcendental effects in growth problems: linear solvability conditions and nonlinear methods—the example of the geometric model. *Asymptotics beyond all orders (La Jolla, CA, 1991)*, 15–28, NATO Adv. Sci. Inst. Ser. B Phys., 284, Plenum, New York.

[HM93] Hakim V., Mallick K. (1993) Exponentially small splitting of separatrices, matching in the complex plane and Borel summation. *Nonlinearity* **6** , no. 1, p. 57–70.

[Ha78] Hale J.K. (1978) Introduction to dynamic bifurcation. *Springer Lecture Notes in Math.* Vol 1057 Springer-Verlag New York/Berlin.

[HS85] Hale J.K.,Scheurle J. (1985) Smoothness of bounded Solutions of nonlinear evolution equations. *J. Differential Equations* **56**, p. 142–163.

[HM89a] Hammersley J.M., Mazzarino G. (1989) Computational aspects of some autonomous differential equations. *Proc. Roy. Soc. London Ser. A* **424**

[HM89b] Hammersley J.M., Mazzarino G. (1989) A differential equation connected with the dendritic growth of crystals. *IMA J. Appl. Math.* **42**, p. 43–75.

[HI98] Haragus-Courcelle M., Illichev A. (1998) Three dimensional solitary waves in the presence of additional surface effects. *Eur. J. Mech. B. Fluids* **17**, p. 739–768.

[HPS77] Hirsch M.W., Pugh C.C. Shub M. (1997) *Invariant Manifolds.* Springer Lecture Notes in Mathematics **vol. 583** Springer-Verlag: New York, Berlin, Heidelberg.

[HMS88] Holmes P., Marsden J., Scheurle J. (1988) Exponentially small splittings of separatrices with applications to KAM theory and degenerate bifurcations. *Contemporary Mathematics* Vol. 81, p. 213–244.

[Ho71] Hörmander L. (1971) Fourier integral operators. I. *Acta Math.* **127**, no. 1-2, p. 79–183.

[Ho9094] Hörmander L. (1990–1994) The analysis of linear partial differential operators. I–IV Springer-Verlag, Berlin.

[HS88] Hunter J.K., Scheurle J. (1988) Existence of perturbed solitary wave solutions to a model equation for water waves. *Phys. D* **32** , no. 2, p. 253–268.

[IA92] Iooss G., Adelmeyer M. (1992) *Topics in bifurcation theory and applications.* Advanced Series in Non Linear Dynamics **3** World Scientific.

[IK90] Iooss G., Kirchgässner K. (1990) Bifurcation d'ondes solitaires en présence d'une faible tension superficielle. *C.R. Acad. Sci. Paris* **311** I, p. 265–268.

[IK92] Iooss G., Kirchgässner K. (1992) Water waves for small surface tension: An approach via normal form. *Proceedings of the Royal Society of Edinburgh* **122A**, p. 267-299.

[IK99] Iooss G., Kirchgässner K. (1999) Travelling waves in a chain of coupled nonlinear oscillators. *To appear in Commun. Math Phys.*

[IL90] Iooss G., Los J. (1990) Bifurcation of spatially quasi-periodic solutions in hydrodynamic stability problems. *Nonlinearity* **3**, p. 851–871.

[IMD89] Iooss G., Mielke A., Demay Y. Theory of steady Ginzburg-Landau equation in Hydrodynamic stability problems. *Eur. J. B/Fluids* **8**, p. 229–268.

[IP93] Iooss G., Pérouème M.C. (1993) Perturbed homoclinic solutions in 1:1 resonance vector fields. *Journal of Differential Equations.* Vol 102 No.1

[Ke67] Kelley A. (1967) The stable, center stable, center, center unstable, unstable manifolds. *J. Diff. Equ.* **3**, p. 546–570.

[Ki82] Kirchgässner K. (1982) Wave solutions of reversible systems and applications. *J.Diff. Equ.* **45**, p. 113–127.

[KS91] Kruskal M.D., Segur H. (1991) Asymptotics beyond all orders in a model of crystal growth. *Studies in Applied Mathematics* **85**, p. 129–181.

[La84] Lazutkin V.F. (1984) Splitting of separatrices for the Chirikov's standard map. VINITI no 6372/84, (russian).

[La93] Lazutkin V.F. (1993) Resurgent approach to the separatrices splitting. International Conference on Differential Equations, Vol. 1, 2 (Barcelona, 1991), 163–176, World Sci. Publishing, River Edge, NJ.

[Lo96a] Lombardi E. (1996) Homoclinic orbits to small periodic orbits for a class of reversible systems *Proc. Roy. Soc. Edinburgh* **126A**, p. 1035–1054.

[Lo97] Lombardi E. (1997) Orbits homoclinic to exponentially small periodic orbits for a class of reversible systems. Application to water waves. *Arch. Rational Mech. Anal.* **137**, p. 227-304.

[Lo99] Lombardi E. (1999) Non-persistence of homoclinic connections for perturbed integrable reversible systems *Journal of Dynamics and Differential Equations* **11** , p. 124–208.

[MKA94] Mackay R.S., Aubry S. Proof of the existence of breathers for time-reversible or hamiltonian networks of weakly coupled oscillators, *Nonlinerarity* **7**, p. 1623–1643.

[Me63] Melnikov V.K. (1963) On the stability of the center for time periodic perturbations. *Trans. Moscow Math. soc.* **12**, p. 1–57.

[Mi87] Mielke A. (1987) Über maximale L^P-Regularität für Differentialgleichungen in Banach- und Hilberträumen. *Math. Ann.***227**, p. 121–133.

[Mi88] Mielke A. (1988) Reduction of quasilinear elliptic equation in cylindrical domains with applications. *Math. Meth. Appl. Sci.* **10**, p. 51–66.

[Ne84] Neishtadt A. I. (1984) The separation of motions in systems with rapidly rotating phase. *J. Appl. Math. Mech.* **48** , no. 2, p. 133–139 (1985).; translated from *Prikl. Mat. Mekh.* **48** (1984), no. 2, p. 197–204 (Russian).

[Po1893] Poincaré H. (1893) *Les méthodes nouvelles de la mécanique céleste* (vol. 2) . Paris: Gauthier-Villars.

[PYG88] Pomeau Y., Ramani A., Grammaticos B. (1988) Structural stability of the Korteweg-De Vries solitons under a singular perturbation. *Physica D* **31**, p. 127–134.

[Sa82] Sanders J. (1982) Melnikov's method and averaging. *Cel Mech.* **28**, p. 171–181.

[Sz95] Sauzin D. (1995) Résurgence paramétrique et exponentielle petitesse de l'écart des séparatrices du pendule rapidement forcé. *Ann. Inst. Fourier (Grenoble)*, **45**, no. 2, p. 453–511.

[Sc89] Scheurle J.(1989) Chaos in a rapidly forced pendulum equation. *Dynamics and control of multibody systems (Brunswick, ME, 1988), 411–419, Contemp. Math.*, **97**, Amer. Math. Soc., Providence, RI.

[Sz99] Sauzin D. (1999) A new method for measuring the splitting of invariant manifolds. *Preprint 066 Bureau des Longitudes.*

[STL91] Segur H., Tanveer S., Levine H., (1991) *Asymptotics beyond all orders* NATO ASI Series B Vol. 284

[Se86] Sevryuk M.B. (1986) *Reversible Systems.* Lecture Notes in Mathematics **vol. 1211** Springer-Verlag: New York, Berlin, Heidelberg.

[SM71] Siegel C.L. and Moser J.K. (1971) *Lectures on celestial mechanics.* Springer.

[Si94] Simó C. (1994) Averaging under fast quasiperiodic forcing. Hamiltonian mechanics (Toruń, 1993), p.13–34, NATO Adv. Sci. Inst. Ser. B Phys., 331, Plenum, New York.

[Su98] Sun S.M. (1998) On the oscillatory tails with arbitrary phase shift for solutions of the perturbed KdV Equation. *SIAM J. Appl, Math.* **58**, no. 4, p. 1163–1177.

[Su99] Sun S.M. (1999) Non-existence of truly solitary waves in water with small surface tension. *Proc. Roy. London A* **455**, p. 2191-2228.

[SS93] Sun S.M., Shen M.C. (1993) Exponentially small estimate for the amplitude of capillary ripples of generalized solitary wave *J. Math. Anal. Appl.* **172** p. 533–566.

[To94] Tovbis A. (1994) Asymptotics beyond all orders and analytic properties of inverse Laplace transforms of solutions. *Comm. Math. Phys.* **163** , no. 2, p. 245–255.

[TTJ98] Tovbis A. Tsuchiya M., Jaffé C. (1998) Exponential asymptotic expansions and approximations of the unstable and stable manifolds of singularly perturbed systems with the Hénon map as an example. *Chaos* **8**, no. 3, p. 665–681.

[Va92] Vanderbauwhede A. (1992) Center manifolds, normal forms and elementary bifurcations. *Dynamics Reported* **Vol. 2** p. 89-169.

[VI91] Vanderbauwhede A. and Iooss G. (1991) Center manifold theory in infinite dimensions. *Dynamics Reported* **Vol. 1**, p. 124–163.

[Wi90] Wiggins S. (1990) *Introduction to applied nonlinear dynamical systems and chaos. Texts in Applied Mathematics* **vol. 2** Springer-Verlag: New York, Berlin, Heidelberg.

[YA96] Yang T.S. and Akylas T.R. (1996) Weakly nonlocal gravity-capillary solitary waves. *Phys. Fluids.* **8**(6), p. 1506–14.

Index

List of symbols

Vol. 1654: R. W. Ghrist, P. J. Holmes, M. C. Sullivan, Knots and Links in Three-Dimensional Flows. X, 208 pages. 1997.

Vol. 1655: J. Azéma, M. Emery, M. Yor (Eds.), Séminaire de Probabilités XXXI. VIII, 329 pages. 1997.

Vol. 1656: B. Biais, T. Björk, J. Cvitanic, N. El Karoui, E. Jouini, J. C. Rochet, Financial Mathematics. Bressanone, 1996. Editor: W. J. Runggaldier. VII, 316 pages. 1997.

Vol. 1657: H. Reimann, The semi-simple zeta function of quaternionic Shimura varieties. IX, 143 pages. 1997.

Vol. 1658: A. Pumarino, J. A. Rodrıguez, Coexistence and Persistence of Strange Attractors. VIII, 195 pages. 1997.

Vol. 1659: V. Kozlov, V. Maz'ya, Theory of a Higher-Order Sturm-Liouville Equation. XI, 140 pages. 1997.

Vol. 1660: M. Bardi, M. G. Crandall, L. C. Evans, H. M. Soner, P. E. Souganidis, Viscosity Solutions and Applications. Montecatini Terme, 1995. Editors: I. Capuzzo Dolcetta, P. L. Lions. IX, 259 pages. 1997.

Vol. 1661: A. Tralle, J. Oprea, Symplectic Manifolds with no Kähler Structure. VIII, 207 pages. 1997.

Vol. 1662: J. W. Rutter, Spaces of Homotopy Self-Equivalences – A Survey. IX, 170 pages. 1997.

Vol. 1663: Y. E. Karpeshina; Perturbation Theory for the Schrödinger Operator with a Periodic Potential. VII, 352 pages. 1997.

Vol. 1664: M. Väth, Ideal Spaces. V, 146 pages. 1997.

Vol. 1665: E. Giné, G. R. Grimmett, L. Saloff-Coste, Lectures on Probability Theory and Statistics 1996. Editor: P. Bernard. X, 424 pages, 1997.

Vol. 1666: M. van der Put, M. F. Singer, Galois Theory of Difference Equations. VII, 179 pages. 1997.

Vol. 1667: J. M. F. Castillo, M. González, Three-space Problems in Banach Space Theory. XII, 267 pages. 1997.

Vol. 1668: D. B. Dix, Large-Time Behavior of Solutions of Linear Dispersive Equations. XIV, 203 pages. 1997.

Vol. 1669: U. Kaiser, Link Theory in Manifolds. XIV, 167 pages. 1997.

Vol. 1670: J. W. Neuberger, Sobolev Gradients and Differential Equations. VIII, 150 pages. 1997.

Vol. 1671: S. Bouc, Green Functors and *G*-sets. VII, 342 pages. 1997.

Vol. 1672: S. Mandal, Projective Modules and Complete Intersections. VIII, 114 pages. 1997.

Vol. 1673: F. D. Grosshans, Algebraic Homogeneous Spaces and Invariant Theory. VI, 148 pages. 1997.

Vol. 1674: G. Klaas, C. R. Leedham-Green, W. Plesken, Linear Pro-*p*-Groups of Finite Width. VIII, 115 pages. 1997.

Vol. 1675: J. E. Yukich, Probability Theory of Classical Euclidean Optimization Problems. X, 152 pages. 1998.

Vol. 1676: P. Cembranos, J. Mendoza, Banach Spaces of Vector-Valued Functions. VIII, 118 pages. 1997.

Vol. 1677: N. Proskurin, Cubic Metaplectic Forms and Theta Functions. VIII, 196 pages. 1998.

Vol. 1678: O. Krupková, The Geometry of Ordinary Variational Equations. X, 251 pages. 1997.

Vol. 1679: K.-G. Grosse-Erdmann, The Blocking Technique. Weighted Mean Operators and Hardy's Inequality. IX, 114 pages. 1998.

Vol. 1680: K.-Z. Li, F. Oort, Moduli of Supersingular Abelian Varieties. V, 116 pages. 1998.

Vol. 1681: G. J. Wirsching, The Dynamical System Generated by the 3n+1 Function. VII, 158 pages. 1998.

Vol. 1682: H.-D. Alber, Materials with Memory. X, 166 pages. 1998.

Vol. 1683: A. Pomp, The Boundary-Domain Integral Method for Elliptic Systems. XVI, 163 pages. 1998.

Vol. 1684: C. A. Berenstein, P. F. Ebenfelt, S. G. Gindikin, S. Helgason, A. E. Tumanov, Integral Geometry, Radon Transforms and Complex Analysis. Firenze, 1996. Editors: E. Casadio Tarabusi, M. A. Picardello, G. Zampieri. VII, 160 pages. 1998

Vol. 1685: S. König, A. Zimmermann, Derived Equivalences for Group Rings. X, 146 pages. 1998.

Vol. 1686: J. Azéma, M. Émery, M. Ledoux, M. Yor (Eds.), Séminaire de Probabilités XXXII. VI, 440 pages. 1998.

Vol. 1687: F. Bornemann, Homogenization in Time of Singularly Perturbed Mechanical Systems. XII, 156 pages. 1998.

Vol. 1688: S. Assing, W. Schmidt, Continuous Strong Markov Processes in Dimension One. XII, 137 page. 1998.

Vol. 1689: W. Fulton, P. Pragacz, Schubert Varieties and Degeneracy Loci. XI, 148 pages. 1998.

Vol. 1690: M. T. Barlow, D. Nualart, Lectures on Probability Theory and Statistics. Editor: P. Bernard. VIII, 237 pages. 1998.

Vol. 1691: R. Bezrukavnikov, M. Finkelberg, V. Schechtman, Factorizable Sheaves and Quantum Groups. X, 282 pages. 1998.

Vol. 1692: T. M. W. Eyre, Quantum Stochastic Calculus and Representations of Lie Superalgebras. IX, 138 pages. 1998.

Vol. 1694: A. Braides, Approximation of Free-Discontinuity Problems. XI, 149 pages. 1998.

Vol. 1695: D. J. Hartfiel, Markov Set-Chains. VIII, 131 pages. 1998.

Vol. 1696: E. Bouscaren (Ed.): Model Theory and Algebraic Geometry. XV, 211 pages. 1998.

Vol. 1697: B. Cockburn, C. Johnson, C.-W. Shu, E. Tadmor, Advanced Numerical Approximation of Nonlinear Hyperbolic Equations. Cetraro, Italy, 1997. Editor: A. Quarteroni. VII, 390 pages. 1998.

Vol. 1698: M. Bhattacharjee, D. Macpherson, R. G. Möller, P. Neumann, Notes on Infinite Permutation Groups. XI, 202 pages. 1998.

Vol. 1699: A. Inoue,Tomita-Takesaki Theory in Algebras of Unbounded Operators. VIII, 241 pages. 1998.

Vol. 1700: W. A. Woyczyński, Burgers-KPZ Turbulence,XI, 318 pages. 1998.

Vol. 1701: Ti-Jun Xiao, J. Liang, The Cauchy Problem of Higher Order Abstract Differential Equations, XII, 302 pages. 1998.

Vol. 1702: J. Ma, J. Yong, Forward-Backward Stochastic Differential Equations and Their Applications. XIII, 270 pages. 1999.

Vol. 1703: R. M. Dudley, R. Norvaiša, Differentiability of Six Operators on Nonsmooth Functions and p-Variation. VIII, 272 pages. 1999.

Vol. 1704: H. Tamanoi, Elliptic Genera and Vertex Operator Super-Algebras. VI, 390 pages. 1999.

Vol. 1705: I. Nikolaev, E. Zhuzhoma, Flows in 2-dimensional Manifolds. XIX, 294 pages. 1999.

Vol. 1706: S. Yu. Pilyugin, Shadowing in Dynamical Systems. XVII, 271 pages. 1999.

Vol. 1707: R. Pytlak, Numerical Methods for Optimal Control Problems with State Constraints. XV, 215 pages. 1999.

Vol. 1708: K. Zuo, Representations of Fundamental Groups of Algebraic Varieties. VII, 139 pages. 1999.

Vol. 1709: J. Azéma, M. Émery, M. Ledoux, M. Yor (Eds), Séminaire de Probabilités XXXIII. VIII, 418 pages. 1999.

Vol. 1710: M. Koecher, The Minnesota Notes on Jordan Algebras and Their Applications. IX, 173 pages. 1999.

Vol. 1711: W. Ricker, Operator Algebras Generated by Commuting Projections: A Vector Measure Approach. XVII, 159 pages. 1999.

Vol. 1712: N. Schwartz, J. J. Madden, Semi-algebraic Function Rings and Reflectors of Partially Ordered Rings. XI, 279 pages. 1999.

Vol. 1713: F. Bethuel, G. Huisken, S. Müller, K. Steffen, Calculus of Variations and Geometric Evolution Problems. Cetraro, 1996. Editors: S. Hildebrandt, M. Struwe. VII, 293 pages. 1999.

Vol. 1714: O. Diekmann, R. Durrett, K. P. Hadeler, P. K. Maini, H. L. Smith, Mathematics Inspired by Biology. Martina Franca, 1997. Editors: V. Capasso, O. Diekmann. VII, 268 pages. 1999.

Vol. 1715: N. V. Krylov, M. Röckner, J. Zabczyk, Stochastic PDE's and Kolmogorov Equations in Infinite Dimensions. Cetraro, 1998. Editor: G. Da Prato. VIII, 239 pages. 1999.

Vol. 1716: J. Coates, R. Greenberg, K. A. Ribet, K. Rubin, Arithmetic Theory of Elliptic Curves. Cetraro, 1997. Editor: C. Viola. VIII, 260 pages. 1999.

Vol. 1717: J. Bertoin, F. Martinelli, Y. Peres, Lectures on Probability Theory and Statistics. Saint-Flour, 1997. Editor: P. Bernard. IX, 291 pages. 1999.

Vol. 1718: A. Eberle, Uniqueness and Non-Uniqueness of Semigroups Generated by Singular Diffusion Operators. VIII, 262 pages. 1999.

Vol. 1719: K. R. Meyer, Periodic Solutions of the N-Body Problem. IX, 144 pages. 1999.

Vol. 1720: D. Elworthy, Y. Le Jan, X-M. Li, On the Geometry of Diffusion Operators and Stochastic Flows. IV, 118 pages. 1999.

Vol. 1721: A. Iarrobino, V. Kanev, Power Sums, Gorenstein Algebras, and Determinantal Loci. XXVII, 345 pages. 1999.

Vol. 1722: R. McCutcheon, Elemental Methods in Ergodic Ramsey Theory. VI, 160 pages. 1999.

Vol. 1723: J. P. Croisille, C. Lebeau, Diffraction by an Immersed Elastic Wedge. VI, 134 pages. 1999.

Vol. 1724: V. N. Kolokoltsov, Semiclassical Analysis for Diffusions and Stochastic Processes. VIII, 347 pages. 2000.

Vol. 1725: D. A. Wolf-Gladrow, Lattice-Gas Cellular Automata and Lattice Boltzmann Models. IX, 308 pages. 2000.

Vol. 1726: V. Marić, Regular Variation and Differential Equations. X, 127 pages. 2000.

Vol. 1727: P. Kravanja, M. Van Barel, Computing the Zeros of Analytic Functions. VII, 111 pages. 2000.

Vol. 1728: K. Gatermann, Computer Algebra Methods for Equivariant Dynamical Systems. XV, 153 pages. 2000.

Vol. 1729: J. Azéma, M. Émery, M. Ledoux, M. Yor, Séminaire de Probabilités XXXIV. VI, 431 pages. 2000.

Vol. 1730: S. Graf, H. Luschgy, Foundations of Quantization for Probability Distributions. X, 230 pages. 2000.

Vol. 1731: T. Hsu, Quilts: Central Extensions, Braid Actions, and Finite Groups,. XII, 185 pages. 2000.

Vol. 1732: K. Keller, Invariant Factors, Julia Equivalences and the (Abstract) Mandelbrot Set. X, 206 pages. 2000.

Vol. 1733: K. Ritter, Average-Case Analysis of NumericalProblems. IX, 254 pages. 2000.

Vol. 1734: M. Espedal, A. Fasano, A. Mikelić, Filtration in Porous Media and Industrial Applications. Cetraro 1998. Editor: A. Fasano. 2000.

Vol. 1735: D. Yafaev, Scattering Theory: Some Old and New Problems. XVI, 169 pages. 2000.

Vol. 1736: B. O. Turesson, Nonlinear Potential Theory and Weighted Sobolev Spaces. XIV, 173 pages. 2000.

Vol. 1737: S. Wakabayashi, Classical Microlocal Analysis in the Space of Hyperfunctions. VIII, 367 pages. 2000.

Vol. 1738: M. Emery, A. Nemirovski, D. Voiculescu, Lectures on Probability Theory and Statistics. XI, 356 pages. 2000.

Vol. 1739: R. Burkard, P. Deuflhard, A. Jameson, J.-L. Lions, G. Strang, Computational Mathematics Driven by Industrial Problems. Martina Franca, 1999. Editors: V. Capasso, H. Engl, J. Periaux. VI, 414 pages. 2000.

Vol. 1740: B. Kawohl, O. Pironneau, L. Tartar, J.-P. Zolesio, Optimal Shape Design. Troia 1999. Editors: A. Cellina, A. Ornelas. IX, 393 pages. 2000.

Vol. 1741: E. Lombardi, Oscillatory Integrals and Phenomena Beyond all Algebraic Orders. XV, 413 pages. 2000.

4. Lecture Notes are printed by photo-offset from the master-copy delivered in camera-ready form by the authors. Springer-Verlag provides technical instructions for the preparation of manuscripts. Macro packages in TEX, LATEX2e, LATEX2.09 are available from Springer's web-pages at

http://www.springer.de/math/authors/b-tex.html.

Careful preparation of the manuscripts will help keep production time short and ensure satisfactory appearance of the finished book.

The actual production of a Lecture Notes volume takes approximately 12 weeks.

5. Authors receive a total of 50 free copies of their volume, but no royalties. They are entitled to a discount of 33.3 % on the price of Springer books purchase for their personal use, if ordering directly from Springer-Verlag.

Commitment to publish is made by letter of intent rather than by signing a formal contract. Springer-Verlag secures the copyright for each volume. Authors are free to reuse material contained in their LNM volumes in later publications: A brief written (or e-mail) request for formal permission is sufficient.

Addresses:

Professor F. Takens, Mathematisch Instituut,
Rijksuniversiteit Groningen, Postbus 800,
9700 AV Groningen, The Netherlands
E-mail: F.Takens@math.rug.nl

Professor B. Teissier
Université Paris 7
UFR de Mathématiques
Equipe Géométrie et Dynamique
Case 7012
2 place Jussieu
75251 Paris Cedex 05
E-mail: Teissier@math.jussieu.fr

Springer-Verlag, Mathematics Editorial, Tiergartenstr. 17,
D-69121 Heidelberg, Germany,
Tel.: *49 (6221) 487-701
Fax: *49 (6221) 487-355
E-mail: lnm@Springer.de